Sustainable Remediation for Pollution and Climate Resilience

Arafat Abdel Hamed Abdel Latef
Ehab M. Zayed · Ahmad Alsayed Omar
Editors

Sustainable Remediation for Pollution and Climate Resilience

Springer

Editors
Arafat Abdel Hamed Abdel Latef
Department of Botany and Microbiology, Faculty of Science
South Valley University
Qena, Egypt

Ehab M. Zayed
Department of Cell Study Research
Field Crops Research Institute
Agricultural Research Center
Giza, Egypt

Ahmad Alsayed Omar
Biochemistry Department, Faculty of Agriculture
Zagazig University
Zagazig, Egypt

Citrus Research and Education Center, Department of Horticultural Sciences, Institute of Food and Agricultural Sciences
University of Florida
Lake Alfred, FL, USA

ISBN 978-981-96-5673-8 ISBN 978-981-96-5674-5 (eBook)
https://doi.org/10.1007/978-981-96-5674-5

This Springer imprint is published by the registered company Springer Nature Singapore Pte Ltd.
The registered company address is: 152 Beach Road, #21-01/04 Gateway East, Singapore 189721, Singapore

Contents

The Comparison of Bioremediation and Nanoremediation and Their History Background

V. Edwin Hillary, T. P. Ajeesh Krishna, T. Maharajan, and S. Antony Ceasar

Abstract The proliferation of population and industrialization has resulted in the emergence and survival of many forms of organic and inorganic contaminants, which substantially threaten the environment and human well-being. Hence, there is an urgent need to look for and apply new technology to address the problem of environmental pollution. Scientists have devised the method of bioremediation to effectively minimize environmental risks by using biodegradation or bioreduction to eliminate pollutants in water, soil, and air. Furthermore, nanoparticles have attracted considerable interest for their ability to tackle environmental cleanup alongside bioremediation. This chapter provides an early overview of the importance of bioremediation and nanoremediation in the context of environmental pollution removal. Next, we have discussed the benefits and limitations of bioremediation and nanoremediation as methods for mitigating environmental contamination. This chapter might help the researchers gain knowledge on bioremediation and nano remediation to reduce environmental risks in the future.

Keywords Air · Bioremediation · Environmental risks · Pollution · Nanoremediation

1 Introduction

The global population is increasing at an astounding rate. Projections indicate that nearly nine billion individuals will inhabit the Earth by 2050 (Adeleye et al. 2016; Alori and Fawole 2017). The exponential increase in population necessitates highly intensive agricultural and industrial systems, ultimately resulting in increased environmental risks, including air, water, and soil pollution. Estimates indicate that pollution is responsible for 602 million deaths per year, accounting for 40% of all

V. Edwin Hillary · T. P. Ajeesh Krishna · T. Maharajan · S. Antony Ceasar (✉)
Department of Biosciences, Rajagiri College of Social Sciences, Cochin, Kerala, India

A. A. H. Abdel Latef et al. (eds.), *Sustainable Remediation for Pollution and Climate Resilience*, https://doi.org/10.1007/978-981-96-5674-5_1

global mortality. However, the World Health Organization (WHO) has claimed that air pollution causes around seven million deaths yearly (Alori et al. 2022).

Apart from this, ever-increasing releases of greenhouse gases (GHGs) have contributed to climate change, which has degraded ecosystems through increased temperatures, droughts, floods, etc. In recent years, extremely high or very low rainfalls worldwide have often caused flash floods and droughts (Arora 2018). For example, floods in Kerala (2018) and Tamil Nadu (2023) in India have destroyed agricultural fields and crops due to silt accumulation and washing out of the topsoil. Similarly, wildfires in California have severely damaged the soil and the area by destroying a considerable amount of the forest, environment, and human life. In addition, heat waves are far more prevalent nowadays in Europe and other world regions (Arora 2018). Overall, due to climate change, land has become salinized and decertified, resulting in decreased productivity and biological resource availability. In order to tackle this, it is necessary to eliminate toxic pollutants from ecosystems, clean up the debris, reclaim marginal lands and refuse saline soils, revitalize freshwater bodies, and decontaminate oceans (Arora 2018).

Over the past few years, several traditional techniques, including physical, chemical, and thermal treatments, have been employed to remove pollutants and restore ecosystems (Alori et al. 2022). However, these techniques have encountered many significant problems, including generating hazardous intermediates, transporting polluted soil or water for treatment, the high expenses involved in treatment, and the ineffective restoration of natural flora and fauna. To address these environmental risks, scientists have initiated a search for an alternative approach and discovered that bioremediation can effectively reduce pollutants through the processes of biodegradation or bioreduction (Azubuike et al. 2016; Divya et al. 2015). This involves breaking down or removing complex toxic compounds into simpler, less harmful compounds with the assistance of microorganisms, plants, and animals (Doran et al. 2002). However, a limited understanding of the genetics and genome-level traits of the employed organisms impedes progress in bioremediation research and application (Azubuike et al. 2016; Divya et al. 2015). As a result, there is a lack of capability to accurately simulate and forecast the behavior of these processes, which poses challenges to the advancement of practical bioremediation techniques in field applications (Alori et al. 2022). In addition, bioremediation, which employs microorganisms (bacteria, fungi, yeasts, and algae) to detoxify and break down environmental pollutants, has gained significant attention in recent years for restoring a contaminated environment (Alori et al. 2022).

In addition to bioremediation, nanomaterials have also received significant attention for their potential to address environmental clean-up (Marlon et al. 2019; Khan 2020; Kaur and Roy 2021). Nanomaterials outperformed conventional methods in environmental remediation due to their extensive surface area and heightened reactivity (Marlon et al. 2019; Khan 2020; Kaur and Roy 2021).

This chapter primarily concentrated on the significance of bioremediation and nanoremediation, together with their historical background, in effectively eliminating environmental pollutants. In the end, we executed a comparative analysis of the benefits and drawbacks of bioremediation and nanoremediation as methods for mitigating environmental contamination. This chapter aims to provide academics with valuable insights on bioremediation and nano remediation, enabling them to effectively mitigate environmental risks in the future.

2 Bioremediation

Humanity is increasingly seeking remedies from nature amidst environmental deterioration and contamination. In recent years, bioremediation has become increasingly popular as a remedy. This novel strategy harnesses the capabilities of living creatures to purify polluted surroundings, providing a sustainable and economically efficient substitute for conventional remediation techniques (Bala et al. 2022). Therefore, scientists have widely adopted bioremediation to remove pollution from the environment, which is discussed in detail below.

2.1 Concept of Bioremediation

Bioremediation is the utilization of living organisms (particularly microorganisms) to eliminate or counteract pollutants from soil, water, or air. In contrast to conventional cleanup procedures, bioremediation does not require substantial excavation or chemical treatment. Instead, it harnesses the power of nature to address the issue. Microorganisms, including bacteria, fungi, and plants, are utilized to decompose contaminants into innocuous substances via natural biological processes (Wang et al. 2023). The microorganisms can break down various organic and inorganic molecules, transforming them into less complex and harmless substances (Ławniczak et al. 2020). To facilitate the growth and functioning of microorganisms, bioremediation frequently entails the provision of vital nutrients like nitrogen, phosphorus, and oxygen to polluted areas. These nutrients act as catalysts for microbial metabolism, speeding up the breakdown of contaminants (Adewoyin and Arimoro 2023). Environmental conditions such as temperature, pH, moisture levels, and oxygen availability are significant elements that influence the efficacy of bioremediation. Therefore, it is essential to maintain optimal conditions for the survival of microbial communities and facilitate the effective removal of pollutants in the environment (Mishra et al. 2022). Two types of bioremediation methods (in situ and ex situ) address specific types of toxins and environmental circumstances in the environment.

2.1.1 In Situ Bioremediation Method

This method entails treating pollutants directly at the source of contamination, eliminating the need for excavation or transportation. To clean up polluted areas, in situ bioremediation methods include bioaugmentation, which adds special microbes to help pollutants break down, and natural attenuation, which relies on the existing microbial populations to fix the problem (Sharma 2020; Urionabarrenetxea et al. 2021).

2.1.2 Ex Situ Bioremediation Method

This method utilizes various bioreactors to remove soil, water, or sediment pollution in the environment (Hussain et al. 2022).

2.2 Historical Background of Bioremediation and Recent Advancement

Bioremediation has a long and rich history (Koenigsberg et al. 2005). For instance, urban and industrialized areas experienced extraordinary amounts of pollution due to fast industrialization during the nineteenth and twentieth centuries. Due to increasing worries about the environmental and public health consequences of industrial waste, scientists initiated research into biological approaches for pollution management. These initial investigations into sewage treatment, microbial fermentation, and biological waste management laid the foundation for advancing bioremediation as an effective approach for cleaning up contaminated environments (Mishra et al. 2022). Similarly, in the second century, the Exxon Valdez oil spill (1989) and the Deepwater Horizon oil leak (2010) emphasized the urgent requirement for efficient cleanup technology. During these searches for efficient cleanup technology, scientists generated specialized methods such as bioaugmentation, biostimulation, and bioventing to treat soil and water contaminated with oil (Barron et al. 2020). Following that, researchers observed microbial species' ability to purify polluted surroundings (Gentry et al. 2021).

Following this research, government bodies, including the Environmental Protection Agency (EPA) in the United States, started supporting microbial bioremediation as the preferred method for cleaning up polluted sites. This endorsement has led to an expansion of the bioremediation sector (Lin et al. 1996). Nowadays, genetic engineering, nanotechnology, and microbial consortia design are emerging concepts that have the potential to improve the effectiveness and adaptability of bioremediation procedures.

The historical progression of bioremediation demonstrates a remarkable process of exploration, ingenuity, and adaptability in response to environmental challenges. The history of bioremediation, spanning from traditional farming methods to modern biotechnologies, illustrates humanity's resourcefulness and dedication to safeguarding the environment. In order to fully harness the promise of bioremediation as a sustainable solution to environmental contamination, it is crucial to maintain ongoing investment in research.

2.3 *Mechanism of Bioremediation*

Gaining a comprehensive understanding of the complex mechanisms that drive bioremediation is crucial for fully harnessing its capacity to alleviate environmental deterioration. Microbial metabolism is the central process in bioremediation, where microorganisms convert pollutants into less toxic compounds. Microbial genomes contain a variety of enzymes that enable them to use contaminants as sources of energy and nutrients, therefore driving their metabolic activity (Farias et al. 2023). Important microbial metabolism pathways that play a role in bioremediation include the following mechanisms.

2.3.1 Aerobic Degradation

Aerobic bacteria thrive in oxygen-rich environments and use molecular oxygen to break down organic contaminants like hydrocarbons and pesticides. Oxygenases are enzymes that aid in the oxidation of pollutants, causing them to decompose into simpler molecules such as carbon dioxide and water (Roy et al. 2023).

2.3.2 Anaerobic Respiration

When there is a lack of oxygen, anaerobic microbes use different substances like nitrates, sulfates, or carbon dioxide to respire and break down pollutants. Anaerobic respiration, a highly efficient process, effectively treats contaminants like chlorinated solvents and heavy metals (Pradhan and Punji 2024). In co-metabolism, certain bacteria can break down contaminants as secondary substrates while utilizing more advantageous substances as primary carbon and energy sources. Co-metabolism is a phenomenon that improves the breakdown of stubborn pollutants by increasing the metabolic flexibility of microbial communities (Shukla et al. 2023). The plant-microbe interaction is another bioremediation mechanism that utilizes microbial mechanisms and the symbiotic link between plants and bacteria to restore polluted environments (Mishra et al. 2022).

2.3.3 Phytoremediation

Phytoremediation, a branch of bioremediation, utilizes the inherent capabilities of plants & algae to absorb, collect, and detoxify contaminants through diverse mechanisms (Pilon-Smits 2005).

2.3.4 Phytoextraction

Phytoextraction is a process where hyperaccumulating plants absorb pollutants from soil or water through their roots and transport them to aboveground tissues for storage or release as gas. This technique efficiently addresses heavy metal pollution in mining sites and industrial areas (Muthusaravanan et al. 2018).

2.3.5 Rhizodegradation

Rhizodegradation is another process where plant roots produce a variety of substances, such as organic acids, enzymes, and sugars, that enhance the growth of microorganisms in the rhizosphere, which is the area of soil around the roots. This rhizodegradation is an advanced process that boosts the microbial decomposition of organic pollutants near plant roots, leading to their faster disintegration (Moklyachuk et al. 2012).

2.3.6 Phytostabilization

Phytostabilization is a process where specific plant species trap pollutants in the soil through adsorption, precipitation, or root absorption. This reduces the pollutants' ability to enter the groundwater, making them less available for biological processes (Awa and Hadibarata 2020).

Overall, bioremediation processes involve a wide range of complex biochemical pathways and ecological interactions, all coordinated by the astonishing ability of microbial and plant populations to adapt. Through the process of unraveling these mechanisms, scientists and environmental engineers can create customized bioremediation systems that may effectively target a variety of pollutants and environmental circumstances. To fully harness the potential of bioremediation as a sustainable cleanup method, it is crucial to promote interdisciplinary research and innovation. This will play a vital role in achieving its revolutionary effects on environmental conservation and human well-being (Pinto et al. 2015).

2.4 Applications of Bioremediation in Various Fields

Bioremediation has shown its adaptability in dealing with various types of environmental pollutants.

2.4.1 Hydrocarbons

Petroleum spills and industrial activity frequently lead to the discharge of hydrocarbon contaminants into soil and water. Bioremediation provides an efficient method for breaking down these substances, as specific bacterial strains can metabolize hydrocarbons as their main source of energy (Margesin and Schinner 2001). Toxic heavy metals, including lead, mercury, and cadmium, substantially harm ecosystems and human well-being. Bioremediation techniques, like phytoremediation, employ plants to absorb and isolate metals, providing a sustainable solution for cleaning up polluted environments (Wu et al. 2010). Chlorinated solvents, which are frequently utilized in industrial operations, can last in the environment for prolonged periods, presenting enduring risks to groundwater quality. Bioremediation methods, like reductive dechlorination, use microbes to break down these chemicals into molecules that are not harmful (Ferguson and Pietari 2000). The various approaches to bioremediation can be applied to remediate soil, water, and air pollutants. Below, we describe the significant applications of bioremediation techniques.

2.4.2 Soil Remediation

Contaminated soil presents substantial hazards to ecosystems and human well-being, requiring efficient methods for remediation. Bioremediation provides a cost-efficient and ecologically sound method for addressing soil contamination (Margesin et al. 2000; Raffa and Chiampo 2021).

2.4.3 Petroleum Hydrocarbons

During industrial activity, oil spills and leaks can contaminate the soil with hydrocarbon pollutants. Therefore, utilizing bioremediation methods like land farming, bioventing, and biopiles can artificially improve the conversion of hydrocarbons into harmless substances like carbon dioxide and water (Stepanova et al. 2022).

2.4.4 Pesticides and Agrochemicals

Agricultural activities frequently employ pesticides and fertilizers that have the potential to endure in soil and seep into groundwater. Some microbes and plants are used in bioremediation methods like bioaugmentation and phytoremediation to break down or separate pesticide residues and improve the soil's overall quality and productivity (Raffa and Chiampo 2021).

2.4.5 Water Contamination

Purifying water is critical for human consumption, crop growth, and aquatic ecosystem preservation. Therefore, researchers used bioremediation processes like dechlorinating bacteria and biofilm reactors, which effectively break down and get rid of pollutants and slowly improve groundwater quality (Martino et al. 2019).

2.4.6 Enhancing Air Quality

The negative impact of air pollution on human health and the environment underscores the significance of implementing effective measures to address this issue. Bioremediation provides novel methods for enhancing air quality by breaking down volatile organic compounds (VOCs) into simpler and less harmful substances: Industrial emissions and urban activities release VOCs such as benzene, toluene, and xylene into the air. These compounds contribute to smog production and can cause respiratory problems. Bioremediation systems, including biofilters, bioscrubbers, and biotrickling filters, can be employed to break down volatile organic compounds (VOCs), leading to a significant decrease in air pollution and an enhancement in air quality in both indoor and outdoor settings (Sheoran et al. 2022).

3 Nanoremediation

Human activity, including industrial, agricultural, and other anthropogenic activities, has primarily been responsible for environmental pollutants in recent years. In addition, they generate high amounts of hazardous substances, such as heavy metals, organic pollutants (including pesticide and herbicide chemicals), and wastewater discharges from both private and public sources. Therefore, in order to expedite the cost-effective removal of hazardous pollutants, scientists started to integrate nanotechnology with bioremediation (Marlon et al. 2019; Khan 2020; Kaur and Roy 2021). These technologies help strengthen environmental sustainability and have the ability to effectively mitigate harmful substances, surpassing the capabilities of traditional remediation approaches.

3.1 Historical Background of Nanoremediation and Recent Advancement

In the early twenty-first century, researchers began studying nanoparticles for environmental cleanup. These studies successfully used nanoparticles for environmental treatment, which they called "nanoremediation" (Marlon et al. 2019; Khan 2020; Kaur and Roy 2021). Nanoparticle synthesis and analysis for remedial applications are the main focus of nanoremediation research. Researchers next examined the nanoparticles (zero-valent iron (nZVI), carbon nanotubes, and nanoscale titanium dioxide) ability to breakdown or immobilize pollutants in soil, groundwater, and sewage in 2003 (Zhang 2003). Next, in 2007, a polluted site was successfully nanoremediated in the United States. This implementation shows if nanoparticles may be used for environmental restoration in real life (Zhang 2007). Researchers then improved nanoparticles, delivery systems, and nanoremediation technologies' long-term effects on ecology (Belal and El-Ramady 2016).

The National Science Foundation (NSF) from the United States launched the Nanoscale Science and Engineering for Agriculture and Food Systems program in 2013 (Scott and Chen 2013). This initiative involves research on nanoremediation as one of its components. This illustrates the rising acknowledgment of nanoremediation as a potentially useful option for tackling the issue of environmental pollution. In 2015, the Environmental Protection Agency (EPA) issued rules for the utilization of nanotechnology in environmental remediation, offering regulatory direction and suggestions for the secure and accountable implementation of nanoremediation methods. Then in 2019, the world's top universities held several International Conferences on Nanotechnology for Environmental Remediation to discuss the latest advancements and challenges in the field of nanoremediation (Alori et al. 2022). Researchers are now focusing on enhancing the efficiency, environmental friendliness, and growth potential of nanoremediation technologies in order to boost their global use to combat ecological degradation.

3.2 *Nanomaterials*

Polluted water, soil, and air pose a significant global issue, presenting severe environmental and human health hazards (Table 1). Nanoremediation after bioremediation has received significant attention after bioremediation for its potential to address environmental cleanup. Nanoremediation employs engineered nanoparticles to remediate contaminated environments, offering a cost-effective and highly efficient alternative to conventional approaches. Besides being affordable, nanomaterials such as metal oxides, carbon nanotubes, quantum dots, and biopolymers are also highly efficient for environmental cleanup, depending on the properties of the nanostructure. Due to their small size, they have been efficiently applied to remove pollutants from soil, water, and air.

3.2.1 Water Nanoremediation

In recent years, there has been a significant increase in the research and development of nanomaterials for use in water and wastewater treatment (Table 2). Contaminated groundwater is a major concern for environmental researchers due to its significant threats to many ecosystems. Clean water is essential for the survival

Table 1 Advantages of nanomaterials employed in soil, water, and air remediation. The details on the name of the nanomaterials, remediation mechanism, media, and advantages with their risks

Type of nanomaterial	Remediation mechanism	Media	Advantages	Risks
Carbon-based	Adsorption	Soil, water, and air	High surface area, eco-friendly nature, microporosity, sorption properties, and compatibility with other treatments	Toxicity and costly
Metal-based	Adsorption and oxidation	Soil and water	High specific surface area, removal of diverse pollutants, and compatibility with other treatments	Effect human and environmental health
Silica-based	Catalysis and adsorption	Air and water	Versatility in surface modification, adaptable pore size, compatibility with other treatments	Distribution scattered
Polymer-based	Nanofiltration	Water	Employment of polymer derived from waste materials and compatibility with other treatments	Denaturation by extreme temperature and pH dependability

of living species, making the issue of contaminated groundwater particularly worrisome (Schweitzer and Noblet 2018). Water supplies are at risk of contamination from multiple chemicals, including ions, heavy metals, petroleum hydrocarbons, pesticides, radioactive materials, and new pollutants like pharmaceuticals and personal care items. In the current situation, it is crucial to conduct research and development to create effective techniques for water cleanup. Therefore, scientists employed nanomaterials such as metallic nanoparticles, biopolymeric membranes, and carbon-derived materials for efficient water remediation (Jadhav et al. 2015).

3.2.2 Soil Nanoremediation

The first use of nanoparticles (NPs) in soil and groundwater remediation on a large scale was documented two decades ago (Table 2). The findings indicated that NPs could maintain their effectiveness in the injected soil for a maximum of 56 days and could travel with groundwater for distances exceeding 20 m (Zhang 2003). Zhang (2003) found that over 99% of trichloroethene (TCE) could be eliminated from polluted areas in just a few days (Zhang 2003). These studies proved that utilizing nanomaterials for soil remediation would remove all the toxic elements from the soil and would protect the world from environmental risks in the future.

Table 2 Nanoremediation methods used for soil and groundwater remediation. The details on the name of the media, nanomaterial, contamination, experimental condition, and their results with references are added

Media	Nanomaterial	Contamination	Experimental condition	Results	Reference
Soil	PVP-nZVI	Trichloroethylene	TCE initial concentration 1 mg kg^{-1} PVP-nZVI dosage 0.01 g g^{-1}	TCE removal efficiency was around 84.73%.	Tian et al. (2020)
	nZVI	Cr(VI)	Five ratios were tested (1000/140; 1000/70; 1000/35; 1000/23, 1000/11 mg mg^{-1}) nZVI injection pressure (10, 30, 100) kPa Cr(VI) initial concentration 800 mg kg^{-1}	Efficiency at (1000/11) mg mg^{-1} ratio was 98%.	Reginatto et al. (2020)
	nZVI	Ni	Initial nitrate concentration (45 mg L^{-1}) NO3-N nZVI and Cu modified (10 g per layer); Flow rate (5 mL min^{-1}); Residence time (99 min)	Two layer Cu-modified nZVI achieved complete contaminant removal	Xue et al. (2018)
	nZVI	Cd(II), Pb(II)	Sediment weight 0.5, 2 g; R-nZVI concentration 2.5, 10 mL with 0.01, 0.03, 0.05, 0.1, 0.2 wt%; Reaction time 0, 7, 14, 21, 28, 35, 42 days	R-nZVI was effective for heavy metal immobilization on river sediment.	Blundell and Owens (2021)
	nZVI	DDT	nZVI dosage (5% w/w); Reaction time (60 min)	DDT removal from contaminated soil	Apul et al. (2016)
	MWCNTs	TPH	TPH initial concentration 11,000 mg kg^{-1}; Microwave treatment 15, 30, 60 s; CTNs concentration 1, 2.5, 5 wt%	Removal of TPH	Zhang et al. (2018b)
	MWCNTs	Cr(VI)	Cr(VI) initial concentration 5–60 mg L^{-1}; Citric acid concentration 25–250 mg L^{-1}; MWCNTs concentration 1.25 mg L^{-1}	Catalysis function of CNTs on the reduction of Cr(VI) was decreased by increasing pH	Cheng et al. (2019)
	MCBN	Heavy metal and petroleum	Cd initial concentration 10 mg kg^{-1}; Ni initial concentration (100 mg kg^{-1}); Petroleum initial concentration (2000 mg kg^{-1}); Remediation time 60 days; MCBN dosage (1% w/w)	Heavy metal contaminants decrease in soil and enhance the growth of the plant	Zhang et al. (2017)
	SWCNTs and MWCNTs	DDT and HCH	SWCNTs and MWCNTs dosage 0.058, 0.145, 0.29 wt%; DDT initial concentration 3 mg kg^{-1}; HCH initial concentration 1 mg kg^{-1}	The efficiency of CNTs remediation was highly dependent on dose	Abbasian et al. (2016)

(continued)

Table 2 (continued)

Media	Nanomaterial	Contamination	Experimental condition	Results	Reference
	MWCNTs	Crude oil	MWNTs concentration 0.1, 0.5, 1 wt%; Crude oil concentration 0.5, 1, 2.5, 5 wt%; Remediation time 30 days	Enhanced the degradation of hydrocarbons by increasing the total microbial population	Hua et al. (2017)
	MWCNTs	DDT, HCH	DDT initial concentration 14 ng g^{-1}; HCH initial concentration (5.5 ng g^{-1})	AC was more effective than MWCNTS due to its great specific surface area.	Naderi Peikam and Jalali (2019)
	Al2O3, SiO2, TiO2	Zn, Ni, and Cd	Metal NPs (Al2O3, SiO2, TiO2) dosage 1, 3 wt%; Remediation time 30 days	SiO_2 NPs were feasible sorbent for reduction of the metals	Qiao et al. (2017)
	BC	Cd	Remediation time 28 days; Cd initial concentration (4.25–132.23) mg kg^{-1}	81.3% of Cd was reduced after 25 days treatment	Baragaño et al. (2020)
	nGoethite and nZVI	As	nZVI dosage (0.5, 2, 5,10) wt%; nGoethite dosage 0.2, 1, 2.5 wt%; As initial concentration 85 mg kg^{-1}	Both nGoethite at small dosage and nZVI are promising nanomaterials for immobilization	Fan et al. (2016)
	MNPs	Cd, Pb	Cd initial concentration 10.91 mg kg^{-1}; Pb initial concentration 190 mg kg^{-1}	MNPs did not change the chemical composition of the soil	Ma et al. (2020)
	Fe3O4@C--COOH MNPs	Pb	Pb initial concentration (737.34) mg kg^{-1}; Fe3O4@C-COOH MNPs dosage (0.6, 1.3, 2.0, 2.6, 3.3, 4.0) wt%	Fe_3O_4@C-COOH MNPs was a promising remediations for lead contaminant soil	Ma et al. (2020)
	MNPs	As, PAH, TPH	MNPs dosage 0.2, 1, 2, 5 wt%; As initial concentration 1305 mg kg^{-1}; PAHs initial concentration 6777 μg kg^{-1}; TPH initial concentration 384 mg kg^{-1}	Immobilized 92.3% at 5% dosage	Lin et al. (2018)
Groundwater	PEI-nZVI	PCE, TCE, and 1,2-DCE	TCE initial concentration 94,000 μg L^{-1}; 1,2-DCE initial concentration 2800 μg L^{-1}	Total removal of DNAPLs after 1 day from the (PEI-nZVI) injection	Chen et al. (2020)

	S-nZVI	TCE	TCE initial concentration 20 mg L^{-1}; S-nZVI to BC mass ratio 1:1, 1:3, 3:3	S-nZVI to BC and pyrolysis temperatures could satisfy the amount of degradation and adsorption of TCE	Dong et al. (2018b)
	S-nZVI	TCE	Initial pH 5.57, 7.10, 8.02; TCE initial concentration 30 mg L^{-1}	Higher TCE removal	Zhu et al. (2018)
	nZVI/Cu	Cr(VI)	Cr(VI) initial concentration 10, 10, 15 mg L^{-1}; nZVI/Cu concentration 0.4 g L^{-1}	At pH = 5, temperature 303 K, the Cr(VI) removal efficiency was 94.7%	Gil-Díaz et al. (2020)
	nZVI	Cu(II), Ni(II)	nZVI dosage 0, 1, 5 wt%; Cu initial concentration 1000 mg kg^{-1}; Ni initial concentration 2000 mg L^{-1}	5% complete removal of Cu and Ni removal from water samples is 54% and 21%	Mpouras et al. (2021)
	MWCNTs	Cr(VI)	MWCNTs dosage 10–50 g L^{-1}; Cr(VI) initial concentration 11 mg L^{-1} and 250 μg L^{-1}	Maximum adsorption percentage at pH 8, with MWCNTs concentration at 50 g L^{-1}	Lico et al. (2019)
	MWCNTs	Unleaded gasoline	MWCNTs dosage 0.2–0.8 g; Reaction time 5–120 min	MWCNTs (0.7 g) within 5 min could remove a high percentage of unleaded gasoline	Liang et al. (2015)
	MWCNTs	Cd(II), TCE	Al2O3/MWCNTs dosage 1 g L^{-1}; Cd(II)/TCE initial concentration 1 mg L^{-1}; pH (4–10)	Cd(II)- and TCE-contaminated groundwater remediations	Ou et al. (2020)
	Fe/Al BNPs	Cr(VI)	Cr(VI) initial concentration 4–200 mg L^{-1}; (Fe/Al) BNPs and (Al/Fe) BNPs dosage (2.5) g L^{-1}	Removal efficiency was 1.47 g g^{-1}	Wang et al. (2019)
	FeS NPs	Cr(VI)	Cr(VI) initial concentration 204.84, 3464 mg kg^{-1}; NPs dosage (28.2, 42.3, 67.7, 84.6 m 117.0) mg L^{-1}	High removal efficiency (1046.1 mg Cr (VI) per gram FeS NPs)	Xie et al. (2015)
	Fe-Me	Se(IV)	Se(IV) initial concentration 0–10 mg L^{-1}; Fe-Mn NPs dosage 0.05 g L^{-1}	High Se (IV) uptake was noticed at a pH range of 5–8	Dong et al. (2018a)

(continued)

Table 2 (continued)

Media	Nanomaterial	Contamination	Experimental condition	Results	Reference
	Fe/Ni BNPs	Tetracycline	Ageing time 5–90 days; Fe/Ni BNPs dosage 1 g L^{-1}; Ni content (1, 3, 5) wt%; TC initial concentration 100 mg L^{-1}	Removal efficacy of TC by 90 days was around 30%	Gong et al. (2017)
	Fe/FeS	Cr(VI)	S/Fe molar ratio 0, 0.070, 0.138, 0.207; Cr(VI) initial concentration 10, 15, 25, 35, 50, 80 mg L^{-1}	Increases the S/Fe ratio	Huang and Keller (2015)
	MLN	Cd(II), Pb(II)	Cd, Pb initial concentration 1–100 mg L^{-1}; pH 4–10; Mag-Ligand dosage 0.2 g L^{-1}	Removal of Cd and Pb from the contaminated aquatic system	Li et al. (2016)

Abbreviations: *CNTs-nZVI* carbon nanotubes-nanoscale zero-valent iron, *CMC* carboxymethyl cellulose, *BC* biochar, *TPH* total petroleum hydrocarbons, *ACF* activated carbon fiber, *TCE* trichloroethylene, *PS* per sulfate, *MCBN* modified carbon black nanoparticle, *S-Nzvi* sulfide-modified nanoscale zero-valent iron, *MWCNTs* multi-walled carbon nanotubes, *PCE* perchloroethene, *DCE* dichloroethene, *PVP* polyvinylpyrrolidone, *DDT* dichlorodiphenyltrichloroethane, *HCH* hexachlorocyclohexane, *SWCNTs* single-walled carbon nanotubes, *MCBN* modified carbon black NPs, *Cd* cadmium, *Pb* lead, *Cr* chromium, *Fe* iron, *FeS* ferrous sulfate, *Zn* zinc, *Ni* nickel

3.2.3 Air Nanoremediation

Air pollution is another major global issue in the twenty-first century, due to its adverse effects on climate change and human health. The six primary and harmful outdoor air pollutants are nitrogen oxides (NOx), sulfur dioxide (SO_2), carbon monoxide (CO), carbon dioxide (CO_2), lead, and ground-level ozone (produced through chemical reactions between NOx and volatile organic compounds) (Manisalidis et al. 2020). CO_2 is not considered a pollutant; however, it is the primary greenhouse gas produced as an outcome of human action. To address this issue, researchers have explored various alternatives, such as employing graphene oxides (GOs), graphite oxides, and carbon nanotubes (CNTs) with highly reactive surface sites, as well as utilizing mesoporous silica materials to remove pollution from the air effectively, which ultimately protects the world from environmental risks (Guerra et al. 2018a).

3.3 *Nano Materials-Based Remediation*

3.3.1 Metal-Based Nanomaterials

Several types of metal oxide nanoparticles, including iron oxide (Fe_2O_3/Fe_3O_4), zinc oxide (ZnO), and titanium dioxide (TiO_2), are used for water purification due to their high reactivity and photolytic properties. For example, iron nanoparticles have been used to clean up colors in wastewater from the paint, textile, and paper industries because they stay stable in liquids and can absorb a lot of dye (Yang et al. 2017; Zhou et al. 2018). Recently, these NPs are very effective at adsorbing dyes like methylene blue and methyl orange, two of the most widely used dyes in industry and those with the worst impacts on human health and the environment (Yang et al. 2017; Zhou et al. 2018). In this context, the removal efficiency of magnetic iron oxide nanoparticles in combination with carbon has been investigated. The results of this investigation have shown that the nanocomposites exhibit stronger interactions with the dye, with the concentration of carbon being a determining factor in the NPs' adsorbent behavior. In addition to colors, heavy metals such as chromium (VI) are another important category of contaminants that are present in water (Yang et al. 2017; Zhou et al. 2018). Recent studies have shown that the inclusion of iron oxide or zero-valent iron nanoparticles and organic acids (like citric acid) could reduce the environmental risk posed by chromium (VI) (Yang et al. 2017; Zhou et al. 2018).

3.3.2 Carbon-Based Nanomaterials

Nanotechnology offers advantages in the field of water, soil, and air pollution management because of its capabilities in cleanup and treatment, pollution prevention, and detection and sensing. Recently, carbon-based nanomaterials have shown great potential to avoid environmental risks. To increase their efficiency, researchers combined carbon-based nanomaterials with other nanomaterials to create nanocomposites that show significant potential for pollution remediation. For instance, researchers have successfully used zirconium hydroxide/graphene composites to adsorb SO_2 from the air. Moreover, UV light successfully reduces graphene oxide (GO) through photoreduction. Usually, carbon nanotubes require modification or coating with reactive compounds that possess suitable functional groups or charges (Guerra et al. 2018b). So, Zhang et al. (2012) used modified carboxylated multiwalled carbon nanotubes (MWCNTs) or single-walled carbon nanotubes (SWCNTs) to find H_2S and SO_2 (Zhang et al. 2012), Dong et al. (2013) found CO and NH_3 (Lv et al. 2016), Kim et al. (2016) found NO_2, and Amade et al. (2014) found VOCs (Amade et al. 2014).

3.3.3 Silica-Based Nanomaterial

Silica-based nanoparticles are highly versatile due to their changeable pore size and easily customizable coverings (Shukla et al. 2020). Additionally, these nanomaterials' ability to speed up chemical reactions and attract molecules has recently drawn more attention to their use in cleaning dirty air and removing impurities in the gaseous state (Guerra et al. 2018b). The surface modification of silica nanoparticles also significantly improves their physicochemical properties. For instance, adding hydroxyl groups to the surface of silica nanoparticles enhanced gas adsorption and wetting. This method efficiently generates catalysts and adsorbents, as well as improving gas adsorption and wetting. Another study conducted by Huang et al. (2003) examined the adsorption capacity of modified mesoporous silica. The study found that the presence of amine groups on the surface of the mesoporous silica enhances its ability to effectively collect H_2S and CO_2 from natural gas. Furthermore, the amine-based silica nanomaterial exhibited efficient removal of several organic pollutants, including aldehydes and ketones, in addition to CO_2 (Nomura and Jones 2013, 2014). On the other hand, the presence of lead (Pb) in the atmosphere is becoming a significant global environmental and health concern. In order to overcome these issues, Yang and Yu (2016) designed silica nanoparticles as a solution to address this environmental issue (Yang and Yu 2016). The findings effectively eliminated ambient Pb in contaminated air. Consequently, silica-based nanoparticles could serve as effective environmental agents in the future, combating industrial pollution resulting from Pb and other heavy metals.

3.3.4 Polymer-Based Nanomaterials

Polymer nanotechnology offers several options for pollution removal from the environment (Abdelbasir and Shalan 2019). In particular, polymeric nanomembranes are specifically used to remove undesirable nanoparticles from water by redirecting them via the pores of the membrane and through chemical interactions with the membrane, resulting in the immobilization of the contaminants. Chitosan is now one of the most commonly used polymers for producing nanofiltration (NF) membranes to remove contamination from water. These NF membranes serve as a method for purifying textile effluent (Long et al. 2020); however, the dyes' physical dimensions lead to low NF efficacy (Long et al. 2020). Therefore, researchers have found them with various components, such as triethanolamine, metal oxide nanoparticles, and carbon nanotubes, to enhance their stability and efficacy (Lakhotia et al. 2018). MWCNTs in polyamide membranes have been highly beneficial for removing industrial salts from textile effluents (Al-Hobaib et al. 2017). Similarly, the utilization of polyethersulfone membranes that have been modified with MWCNTs, graphene, or other polymers exhibited remarkable efficiency in removing heavy metals and dyes from water-based solutions (Peydayesh et al. 2020).

3.4 Combined Nanoremediation with Other Remediation

Several studies employed a combination of various nanoremediation with other remediation to control environmental risks (Table 3). As an example, Vilardi et al. (2018) did a batch experiment to see how well nZVI and CNTs worked together to get rid of Cr (VI), Se (Se), and Co (Co) from water solutions (Vilardi et al. 2018). The findings revealed that the primary process for removing Cr (VI) was reduction, while adsorption was the dominant mechanism for other metals. The findings indicated that the elimination rate of Cr (VI) was 100% when solely employing nZVI without altering the pH. However, the efficiency was reduced to approximately 90% when utilizing the CNTs-nZVI nanocomposite. In addition, the CNTs-nZVI demonstrated a significant capacity for removing Se and Co, achieving removal efficiencies of 90% and 80%, respectively. The findings indicate that the nanocomposite of carbon nanotubes and nano zero-valent iron (CNTs-nZVI) has a notable capacity for adsorbing heavy metals in water contaminated with such pollutants. Next, Zhang et al. (2018a) conducted a study to evaluate the effectiveness of carboxymethyl cellulose (CMC)-stabilized nZVI combined with biochar (BC) (referred to as CMC-nZVI/BC) for the remediation of soil contaminated with Cr (VI) (Zhang et al. 2018a). The study looked at how well Cr (VI) was immobilized after 21 days. The results showed that the doses of CMC-nZVI/BC that worked best were 11, 27.5, and 55 g kg^{-1}. The findings indicate that including BC in CMC-nZVI may lead to a minor reduction in the transformation of Cr (VI) since a fraction of CMC-nZVI has the potential to be adsorbed onto charcoal. In a recent study, Qian et al. (2020) conducted the first investigation into the efficacy of biochar-nZVI for field remediation of chlorinated hydrocarbons. The field study findings revealed a significant decrease in chlorinated solvents within 24 h, successfully following the initial injection of nZVI. In addition, the utilization of biochar-nZVI significantly enhanced the extraction of the chlorinated solvent from groundwater over 42 days (Qian et al. 2020). The findings indicate that the use of biochar-nZVI has great potential as a combination method for treating groundwater contaminated with chlorinated solvents.

Table 3 Combination of nanoremediation technologies with other remediation methods. The details on name of the nanomaterial, contamination, experimental conditions, results with their references are added

Nanomaterial	Contamination	Experimental conditions	Important results	Reference
CNTs-nZVI	Cr(VI), Se, Co	Dosage 3 g L^{-1}; Initial concentration 1–10 mg L^{-1}; pH 6–8	Reduction mechanism for Cr(VI), adsorption mechanism for the metals. The Cr(VI) removal efficiency was 100% when nZVI was used alone and around 90% when CNTs-nZVI nanocomposite was used. CNTs-nZVI showed high removal efficiency for Se and Co at 90% and 80%, respectively.	Vilardi et al. (2018)
CMC-nZVI/BC	Cr(VI)	CMC-nZVI/BC dosage 11, 27.5, 55 kg g^{-1}; Cr(VI) initial concentration 800 mg kg^{-1}	The maximum immobilization efficiency of 100% at 55 g kg^{-1} dosage after 21 days.	Zhang et al. (2018b)
BC-nZVI	Chlorinated solvents	nZVI dosage (30) g L^{-1}	Implementation of biochar-nZVI highly improved the removal of the chlorinated solvent from groundwater for 42 days	Qian et al. (2020)
nZVI	TPH, PAHs, and heavy metals	TPH initial concentration (104.3) mg kg^{-1}; PAHs initial concentration (2.25) mg kg^{-1}	Effectively decreased the aliphatic hydrocarbons concentration by up to 60%.	Galdames et al. (2017)
	Crude oil	Oil initial concentration (375) mg L^{-1}; NPs dosage (18) mg L^{-1}	The combination of NPs and oil-degrading bacteria could completely remove the oil within 48 h	Alabresm et al. (2018)
nZVI-EK	Chlorinated ethenes (CEs)	nZVI dosage 3 g L^{-1}; DC voltage 24 V	nZVI-EK remediation technology is a promising method for CE remediation and also enhanced the bacteria availability in soil and groundwater	Czinnerová et al. (2020)
Soil washing assessed nZVI	As, Cu, Hg, Pb, Sb	nZVI dosage (16) wt%	Soil washing efficiency was enhanced by adding nZVI, especially for a larger fraction.	Boente et al. (2018)
ACF-nZVI	Cr(VI)	Cr(VI) initial concentration (5, 10) mg L^{-1}	In an acidic environment, full removal (100%) of Cr(VI) was observed in 1 h reaction time.	Qu et al. (2017)
PS-Z/nZVI	TCE	TCE initial concentration (0.15) mM; Z/nZVI dosage (84) mg L^{-1}; PS concentration (1.5) mM	The PS-Z/nZVI system showed high efficiency in terms of TCE for a wide range of pH (4–7).	Huang et al. (2019)

Abbreviations: *CNTs-nZVI* carbon nanotubes-nanoscale zero-valent iron, *CMC* carboxymethyl cellulose, *BC* biochar, *TPH* total petroleum hydrocarbons, *ACF* activated carbon fiber, *TCE* trichloroethylene, *PS* per sulfate

Czinnerová et al. (2020) conducted a long-term field investigation to examine the degradation of chlorinated ethenes (CEs) utilizing nZVI supported by electrokinetic (EK) treatment (nZVI-EK). These findings demonstrated a significant and swift decline of around 70% in cis-1,2-dichloroethene (cDCE). The findings indicate that the nZVI-EK remediation technique has the potential to remediate CE from soil and groundwater and increase the availability of bacteria in soil and groundwater (Czinnerová et al. 2020). In another investigation, Boente et al. (2018) examined the effectiveness of a combination of soil washing and nZVI for the extraction and recovery of hazardous elements (arsenic, copper, mercury, lead, and antimony) from contaminated soil (Boente et al. 2018). The results showed that the magnetically separated fraction significantly recovered Pb, Cu, and Sb, while a nonmagnetic fraction concentrated Hg. Overall, the addition of nZVI significantly improved the efficiency of soil cleaning, particularly for a larger fraction. The results indicate that the examined methodology provides an opportunity for the utilization of nanoparticles in soil-washing remediation. Huang et al. (2019) conducted a study to investigate the activation of persulfate (PS) utilizing a system consisting of zeolite-supported nZVI composites (PS-Z/nZVI) (Huang et al. 2019). The study aimed to assess the effectiveness of this system in degrading TCE. The findings demonstrated that Z/nZVI had a significant capacity for PS activation at a concentration of 1.5 mM. Moreover, it displayed a remarkable elimination efficiency of 98.8% for TCE within a timeframe of 2 h at a pH level of 7. Furthermore, the PS-Z/nZVI system exhibited exceptional effectiveness in removing TCE across a broad pH range of 4–7.

4 Limitations of Bioremediation and Nanoremediation

Nanoremediation and bioremediation strategies are commonly used to remediate polluted environments, but they possess distinct constraints, which are explained in detail below.

4.1 Bioremediation

1. Bioremediation may be highly targeted at specific pollutants and ecosystems. Certain bacteria, for example, may be particularly successful at degrading one type of pollution but ineffective at another. However, nano remediation, while adaptable, may not always be as selective and could accidentally damage nontarget compounds (Qian et al. 2020).
2. Bioremediation procedures require more time to produce the intended outcomes in comparison to nanoremediation. The repair process might be slowed by factors such as temperature, pH, and nutrition availability, which can affect micro-

bial activity. However, nano remediation has the potential to be more speedy and effective in eliminating pollutants from the environment (Sheoran et al. 2022).

3. Besides bioremediation, an important issue is cost. While bioremediation is generally considered a cost-effective strategy, nano remediation requires less money for manufacturing and employing designed nanoparticles (Shukla et al. 2023).

Both nanoremediation and bioremediation have limitations in effectively cleaning up polluted settings, necessitating a demand for adequate remediation methods in the future.

4.2 Nanoremediation

Nanoremediation, the use of nanoparticles for ecological cleaning, particularly in bioprocessing, offers numerous applications such as environmental sensing and monitoring, nanofiltration and membrane technologies, soil stabilization and erosion control, and nano-enabled waste management to mitigate environmental risks; however, there are numerous limitations, which are listed below.

1. Nanoremediation frequently uses engineered nanoparticles, which, if not properly monitored, could potentially endanger nontarget organisms and the environment. Bioremediation, in contrast, commonly utilizes indigenous microbes or plants, which have less ability to create ecological damage than nanoremediation (Huang et al. 2019).
2. While nanoparticles are successful in removing pollutants, there are questions about their possible environmental impact. For example, nanoparticle toxicity to nontarget organisms remains little known, which requires further research (Qian et al. 2020).
3. Nanoremediation is also not applicable to all categories of pollutants or environmental circumstances. For example, various variables, such as pH, temperature, or the presence of other compounds in the environment, may prevent nanoparticles or the remediation process from efficiently targeting certain pollutants (Peydayesh et al. 2020).
4. The use of nanoparticles for environmental restoration and the regulatory frameworks governing their use are not yet fully developed (Peydayesh et al. 2020). Therefore, getting regulatory permission for the utilization of nanoremediation technology in bioprocessing activities is a difficult process (Alori et al. 2022).
5. If nanoparticles are unintentionally released into the environment during or after the nanoremediation process, it will result in unexpected outcomes, such as ecological disturbance or risks to human health, which require further research (Martínez et al. 2020; Devasena et al. 2022).

However, addressing these barriers may pave the way for exciting nanoprocessing applications in the future. Further advances in nanomaterial synthesis, characterization, and engineering will make it easier to create novel nanoproducts and nanodevices with better functionality and performance. Furthermore, integrating nanotechnology with other emerging technologies such as artificial intelligence, biotechnology, and materials science might help to improve and widen the scope and effect of nanoprocessing in a variety of sectors.

5 Conclusion

Several studies have shown the potential application of bioremediation and nanoremediation for environmental cleanup. However, further research is required to develop an efficient method to control the environmental risks associated with pollution. Another major hurdle with bioremediation and nanoremediation is translating the laboratory research results into real-world settings. Therefore, additional research is needed to establish standardized procedures and dosages for the future field application of bioremediation and nanoremediation for environmental cleanup.

Acknowledgements Authors thank the Rajagiri College of Social Sciences for the research support.

Ethics Approval and Consent to Participate Not applicable.

Consent for Publication Not applicable.

Availability of Data and Materials Not applicable.

Declaration of Interest None.

Funding This work was financially supported by Rajagiri College of Social Sciences (Autonomous), Kerala, India, under Seed Money for Faculty Minor Research.

Authors' Contributions V. Edwin Hillary wrote the article. S Antony Ceasar edited and revised the chapter. All authors listed have significantly contributed to the investigation, the development and the writing of this chapter.

References

Abbasian F, Lockington R, Palanisami T et al (2016) Multiwall carbon nanotubes increase the microbial community in crude oil contaminated fresh water sediments. Sci Total Environ 539:370–380

Abdelbasir SM, Shalan AE (2019) An overview of nanomaterials for industrial wastewater treatment. Korean J Chem Eng 36:1209–1225

Adeleye AS, Conway JR, Garner K et al (2016) Engineered nanomaterials for water treatment and remediation: costs, benefits, and applicability. Chem Eng J 286:640–662

Adewoyin JA, Arimoro FO (2023) Animal manure as a biostimulant in bioremediation of oil-contaminated soil: the role of earthworms. Environ Monit Assess 195:293

Alabresm A, Chen YP, Decho AW, Lead J (2018) A novel method for the synergistic remediation of oil-water mixtures using nanoparticles and oil-degrading bacteria. Sci Total Environ 630:1292–1297. https://doi.org/10.1016/j.scitotenv.2018.02.277

Al-Hobaib AS, Al-Sheetan KM, Shaik MR, Al-Suhybani MS (2017) Modification of thin-film polyamide membrane with multi-walled carbon nanotubes by interfacial polymerization. Appl Water Sci 7:4341–4350

Alori ET, Fawole OB (2017) Microbial inoculants-assisted phytoremediation for sustainable soil management. In: Phytoremediation: management of environmental contaminants, vol 5. Springer, Cham, pp 3–17

Alori ET, Gabasawa AI, Elenwo CE, Agbeyegbe OO (2022) Bioremediation techniques as affected by limiting factors in soil environment. Front Soil Sci 2:937186

Amade R, Hussain S, Ocaña IR, Bertran E (2014) Growth and functionalization of carbon nanotubes on quartz filter for environmental applications. J Environ Eng Ecol Sci 3:1–7

Apul OG, Delgado AG, Kidd J et al (2016) Carbonaceous nano-additives augment microwave-enabled thermal remediation of soils containing petroleum hydrocarbons. Environ Sci Nano 3:997–1002

Arora NK (2018) Bioremediation: a green approach for restoration of polluted ecosystems. Environ Sustain 1:305–307

Awa SH, Hadibarata T (2020) Removal of heavy metals in contaminated soil by phytoremediation mechanism: a review. Water Air Soil Pollut 231:47

Azubuike CC, Chikere CB, Okpokwasili GC (2016) Bioremediation techniques–classification based on site of application: principles, advantages, limitations and prospects. World J Microbiol Biotechnol 32:180

Bala S, Garg D, Thirumalesh BV et al (2022) Recent strategies for bioremediation of emerging pollutants: a review for a green and sustainable environment. Toxics 10:484

Barагаño D, Alonso J, Gallego JR et al (2020) Zero valent iron and goethite nanoparticles as new promising remediation techniques for As-polluted soils. Chemosphere 238:124624

Barron MG, Vivian DN, Heintz RA, Yim UH (2020) Long-term ecological impacts from oil spills: comparison of Exxon Valdez, Hebei Spirit, and Deepwater Horizon. Env Sci & Tech 54:6456–6467

Belal E-S, El-Ramady H (2016) Nanoparticles in water, soils and agriculture. Nanosci Food Agric 2:311–358

Blundell SP, Owens G (2021) Evaluation of enhancement techniques for the dechlorination of DDT by nanoscale zero-valent iron. Chemosphere 264:128324

Boente C, Sierra C, Martínez-Blanco D et al (2018) Nanoscale zero-valent iron-assisted soil washing for the removal of potentially toxic elements. J Hazard Mater 350:55–65

Chen J, Dong H, Tian R et al (2020) Remediation of trichloroethylene-contaminated groundwater by sulfide-modified nanoscale zero-valent iron supported on biochar: investigation of critical factors. Water Air Soil Pollut 231:1–13

Cheng J, Sun Z, Yu Y et al (2019) Effects of modified carbon black nanoparticles on plant-microbe remediation of petroleum and heavy metal co-contaminated soils. Int J Phytoremediation 21:634–642

Czinnerová M, Vološčuková O, Marková K et al (2020) Combining nanoscale zero-valent iron with electrokinetic treatment for remediation of chlorinated ethenes and promoting biodegradation: a long-term field study. Water Res 175:115692

Devasena T, Iffath B, Renjith Kumar R et al (2022) Insights on the dynamics and toxicity of nanoparticles in environmental matrices. Bioinorg Chem Appl 2022:4348149. https://doi.org/10.1155/2022/4348149

Dong KY, Choi J, Lee YD, Kang BH, Yu YY, Choi HH, Ju BK (2013) Detection of a CO and NH 3 gas mixture using carboxylic acid-functionalized single-walled carbon nanotubes. Nanoscale res lett 8:1–6

Divya M, Aanand S, Srinivasan A, Ahilan B (2015) Bioremediation–an eco-friendly tool for effluent treatment: a review. Int J Appl Res 1:530–537

Dong H, Jiang Z, Deng J et al (2018a) Physicochemical transformation of Fe/Ni bimetallic nanoparticles during aging in simulated groundwater and the consequent effect on contaminant removal. Water Res 129:51–57

Dong H, Zhang C, Deng J et al (2018b) Factors influencing degradation of trichloroethylene by sulfide-modified nanoscale zero-valent iron in aqueous solution. Water Res 135:1–10
Doran JW, Stamatiadis SI, Haberern J (2002) Soil health as an indicator of sustainable management. Agric Ecosyst Environ 88:107–110
Fan L, Song J, Bai W et al (2016) Chelating capture and magnetic removal of non-magnetic heavy metal substances from soil. Sci Rep 6:21027
Farias JP, Okeke BC, De Ávila FD et al (2023) Biotechnology process for microbial lipid synthesis from enzymatic hydrolysate of pre-treated sugarcane bagasse for potential bio-oil production. Renew Energy 205:174–184
Ferguson JF, Pietari JMH (2000) Anaerobic transformations and bioremediation of chlorinated solvents. Environ Pollut 107:209–215
Galdames A, Mendoza A, Orueta M et al (2017) Development of new remediation technologies for contaminated soils based on the application of zero-valent iron nanoparticles and bioremediation with compost. Resour Effic Technol 3:166–176. https://doi.org/10.1016/j.reffit.2017.03.008
Gentry T, Fuhrmann JJ, Zuberer DA (2021) Principles and applications of soil microbiology. Elsevier, Amsterdam
Gil-Díaz M, Alvarez MA, Alonso J, Lobo MC (2020) Effectiveness of nanoscale zero-valent iron for the immobilization of Cu and/or Ni in water and soil samples. Sci Rep 10:15927
Gong Y, Gai L, Tang J et al (2017) Reduction of Cr (VI) in simulated groundwater by FeS-coated iron magnetic nanoparticles. Sci Total Environ 595:743–751
Guerra AM, Aimedieu P, Bornert M et al (2018a) Analysis of the structural changes of a pellet/powder bentonite mixture upon wetting by X-ray computed microtomography. Appl Clay Sci 165:164–169
Guerra FD, Attia MF, Whitehead DC, Alexis F (2018b) Nanotechnology for environmental remediation: materials and applications. Molecules 23:1760
Hua S, Gong J-L, Zeng G-M et al (2017) Remediation of organochlorine pesticides contaminated lake sediment using activated carbon and carbon nanotubes. Chemosphere 177:65–76
Huang HY, Yang RT, Chinn D, Munson CL (2003) Amine-grafted MCM-48 and silica xerogel as superior sorbents for acidic gas removal from natural gas. Ind Eng Chem Res 42:2427–33
Huang Y, Keller AA (2015) EDTA functionalized magnetic nanoparticle sorbents for cadmium and lead contaminated water treatment. Water Res 80:159–168
Huang J, Yi S, Zheng C, Lo IMC (2019) Persulfate activation by natural zeolite supported nanoscale zero-valent iron for trichloroethylene degradation in groundwater. Sci Total Environ 684:351–359
Hussain N, Bilal M, Iqbal HMN (2022) Carbon-based nanomaterials with multipurpose attributes for water treatment: greening the 21st-century nanostructure materials deployment. Biomater Polym Horiz 1:48–58
Jadhav SV, Bringas E, Yadav GD et al (2015) Arsenic and fluoride contaminated groundwaters: a review of current technologies for contaminants removal. J Environ Manag 162:306–325
Kaur S, Roy A (2021) Bioremediation of heavy metals from wastewater using nanomaterials. Environ Dev Sustain 23:9617–9640
Khan SH (2020) Green nanotechnology for the environment and sustainable development. In: Green materials for wastewater treatment. Springer International Publishing, Cham, pp 13–46
Kim DS, Jeong J, Ahn J (2016) Characteristics in atmospheric chemistry between NO, NO 2 and O 3 at an urban site during MAPS (Megacity air pollution study)-Seoul, Korea. J Korean Soc Atmos Env 32:422–434
Koenigsberg SS, Hazen TC, Peacock AD (2005) Environmental biotechnology: a bioremediation perspective. J Env Cleanup Costs, Tech & Tech 15:5–25
Lakhotia SR, Mukhopadhyay M, Kumari P (2018) Cerium oxide nanoparticles embedded thin-film nanocomposite nanofiltration membrane for water treatment. Sci Rep 8:4976
Ławniczak Ł, Woźniak-Karczewska M, Loibner AP et al (2020) Microbial degradation of hydrocarbons—basic principles for bioremediation: a review. Molecules 25:856

Li D, Egodawatte S, Kaplan DI et al (2016) Functionalized magnetic mesoporous silica nanoparticles for U removal from low and high pH groundwater. J Hazard Mater 317:494–502
Liang J, Liu J, Yuan X et al (2015) Facile synthesis of alumina-decorated multi-walled carbon nanotubes for simultaneous adsorption of cadmium ion and trichloroethylene. Chem Eng J 273:101–110
Lico D, Vuono D, Siciliano C et al (2019) Removal of unleaded gasoline from water by multiwalled carbon nanotubes. J Environ Manag 237:636–643
Lin G-H, Sauer NE, Cutright TJ (1996) Environmental regulations: a brief overview of their applications to bioremediation. Int Biodeterior Biodegradation 38:1–8
Lin K-S, Mdlovu NV, Chen C-Y et al (2018) Degradation of TCE, PCE, and 1, 2–DCE DNAPLs in contaminated groundwater using polyethylenimine-modified zero-valent iron nanoparticles. J Clean Prod 175:456–466
Long Q-X, Liu B-Z, Deng H-J et al (2020) Antibody responses to SARS-CoV-2 in patients with COVID-19. Nat Med 26:845–848
Lv D, Zhu T, Liu R et al (2016) Effects of co-processing sewage sludge in cement kiln on NOx, NH3 and PAHs emissions. Chemosphere 159:595–601
Ma C, Liu F-Y, Wei M-B et al (2020) Synthesis of novel core-shell magnetic Fe3O4@C nanoparticles with carboxyl function for use as an immobilisation agent to remediate lead-contaminated soils. Pol J Environ Stud 29:2273–2283
Manisalidis I, Stavropoulou E, Stavropoulos A, Bezirtzoglou E (2020) Environmental and health impacts of air pollution: a review. Front Public Health 8:505570
Margesin R, Schinner F (2001) Biodegradation and bioremediation of hydrocarbons in extreme environments. Appl Microbiol Biotechnol 56:650–663
Margesin R, Zimmerbauer A, Schinner F (2000) Monitoring of bioremediation by soil biological activities. Chemosphere 40:339–346
Marlon JR, Bloodhart B, Ballew MT et al (2019) How hope and doubt affect climate change mobilization. Front Commun 4:20
Martínez G, Merinero M, Pérez-Aranda M et al (2020) Environmental impact of nanoparticles' application as an emerging technology: a review. Materials (Basel) 14:166
Martino C, Morton JT, Marotz CA et al (2019) A novel sparse compositional technique reveals microbial perturbations. mSystems 4:10–1128
Mishra S, Huang Y, Li J et al (2022) Biofilm-mediated bioremediation is a powerful tool for the removal of environmental pollutants. Chemosphere 294:133609
Moklyachuk L, Petryshyna V, Slobodenyuk O, Zatsarinna Y (2012) Sustainable strategies of phytoremediation of the sites polluted with obsolete pesticides. Springer, Dordrecht
Mpouras T, Polydera A, Dermatas D et al (2021) Multi wall carbon nanotubes application for treatment of Cr (VI)-contaminated groundwater; modeling of batch & column experiments. Chemosphere 269:128749
Muthusaravanan S, Sivarajasekar N, Vivek JS et al (2018) Phytoremediation of heavy metals: mechanisms, methods and enhancements. Environ Chem Lett 16:1339–1359
Naderi Peikam E, Jalali M (2019) Application of three nanoparticles (Al2O3, SiO2 and TiO2) for metal-contaminated soil remediation (measuring and modeling). Int J Environ Sci Technol 16:7207–7220
Nomura A, Jones CW (2013) Amine-functionalized porous silicas as adsorbents for aldehyde abatement. ACS appl mater interfaces 5:5569–5577
Nomura A, Jones CW (2014) Enhanced formaldehyde-vapor adsorption capacity of polymeric amine-incorporated aminosilicas Chem Euro J 20:6381–90
Ou J-H, Sheu Y-T, Tsang DCW et al (2020) Application of iron/aluminum bimetallic nanoparticle system for chromium-contaminated groundwater remediation. Chemosphere 256:127158
Peydayesh M, Mohammadi T, Nikouzad SK (2020) A positively charged composite loose nanofiltration membrane for water purification from heavy metals. J Membr Sci 611:118205
Pilon-Smits E (2005) Phytoremediation. Annu Rev Plant Biol 56:15–39

Pinto LM, Alghamdi M, Benedetti A et al (2015) Derivation and validation of clinical phenotypes for COPD: a systematic review. Respir Res 16:1–13

Pradhan C, Punji B (2024) Advancement in the C–H bond alkylation of (hetero) arenes catalyzed by the most abundant transition metal–iron. Org Chem Front 11:2397-417

Qian L, Chen Y, Ouyang D et al (2020) Field demonstration of enhanced removal of chlorinated solvents in groundwater using biochar-supported nanoscale zero-valent iron. Sci Total Environ 698:134215

Qiao Y, Wu J, Xu Y et al (2017) Remediation of cadmium in soil by biochar-supported iron phosphate nanoparticles. Ecol Eng 106:515–522

Qu G, Kou L, Wang T et al (2017) Evaluation of activated carbon fiber supported nanoscale zero-valent iron for chromium (VI) removal from groundwater in a permeable reactive column. J Environ Manag 201:378–387. https://doi.org/10.1016/j.jenvman.2017.07.010

Raffa CM, Chiampo F (2021) Bioremediation of agricultural soils polluted with pesticides: a review. Bioengineering 8:92

Reginatto C, Cecchin I, Heineck KS et al (2020) Use of nanoscale zero-valent iron for remediation of clayey soil contaminated with hexavalent chromium: batch and column tests. Int J Environ Res Public Health 17:1001

Roy A, Vajpayee P, Srivastava S, Srivastava PK (2023) Revelation of bioremediation approaches for hexachlorocyclohexane degradation in soil. World J Microbiol Biotechnol 39:243

Schweitzer L, Noblet J (2018) Water contamination and pollution. In: Green chemistry. Elsevier, Amsterdam, pp 261–290

Scott N, Chen H (2013) Nanoscale science and engineering for agriculture and food systems. Ind Biotechnol 9:17–18

Sharma I (2020) Bioremediation techniques for polluted environment: concept, advantages, limitations, and prospects. In: Trace metals in the environment-new approaches and recent advances. IntechOpen, London

Sheoran K, Siwal SS, Kapoor D et al (2022) Air pollutants removal using biofiltration technique: a challenge at the frontiers of sustainable environment. ACS Eng Au 2:378–396

Shukla N, Singh B, Kim H et al (2020) Combinational chemotherapy and photothermal therapy using a gold nanorod platform for cancer treatment. Part Part Syst Charact 37:2000099

Shukla P, Srivastava P, Mishra A (2023) New developments in the production and recovery of amino acids, vitamins, and metabolites from microbial sources. In: Industrial microbiology and biotechnology: emerging concepts in microbial technology. Springer, Singapore, pp 175–241

Stepanova AY, Gladkov EA, Osipova ES et al (2022) Bioremediation of soil from petroleum contamination. Processes 10:1224

Tian H, Liang Y, Yang D, Sun Y (2020) Characteristics of PVP–stabilised NZVI and application to dechlorination of soil–sorbed TCE with ionic surfactant. Chemosphere 239:124807

Urionabarrenetxea E, Garcia-Velasco N, Anza M et al (2021) Application of in situ bioremediation strategies in soils amended with sewage sludges. Sci Total Environ 766:144099

Vilardi G, Mpouras T, Dermatas D et al (2018) Nanomaterials application for heavy metals recovery from polluted water: the combination of nano zero-valent iron and carbon nanotubes. Competitive adsorption non-linear modeling. Chemosphere 201:716–729

Wang T, Liu Y, Wang J et al (2019) In-situ remediation of hexavalent chromium contaminated groundwater and saturated soil using stabilized iron sulfide nanoparticles. J Environ Manag 231:679–686

Wang Z, Wang H, Nie Q et al (2023) Pb (II) bioremediation using fresh algal-bacterial aerobic granular sludge and its underlying mechanisms highlighting the role of extracellular polymeric substances. J Hazard Mater 444:130452

Wu G, Kang H, Zhang X et al (2010) A critical review on the bio-removal of hazardous heavy metals from contaminated soils: issues, progress, eco-environmental concerns and opportunities. J Hazard Mater 174:1–8

Xie W, Liang Q, Qian T, Zhao D (2015) Immobilization of selenite in soil and groundwater using stabilized Fe–Mn binary oxide nanoparticles. Water Res 70:485–494

Xue W, Huang D, Zeng G et al (2018) Nanoscale zero-valent iron coated with rhamnolipid as an effective stabilizer for immobilization of Cd and Pb in river sediments. J Hazard Mater 341:381–389

Yang Y, Yu C (2016) Advances in silica based nanoparticles for targeted cancer therapy. Nanomedicine 12:317–332

Yang B, Fu X, Sidiropoulos ND, Hong M (2017) Towards k-means-friendly spaces: simultaneous deep learning and clustering. In: International conference on machine learning. PMLR, Sydney, pp 3861–3870

Zhang W (2003) Nanoscale iron particles for environmental remediation: an overview. J Nanopart Res 5:323–332

Zhang G (2007) Soil nanoparticles and their influence on engineering properties of soils. In: Advances in measurement and modeling of soil behavior. American Society of Civil Engineers, Reston, pp 1–13

Zhang T, Tang M, Kong L et al (2012) Comparison of cytotoxic and inflammatory responses of pristine and functionalized multi-walled carbon nanotubes in RAW 264.7 mouse macrophages. J Hazard Mater 219:203–212

Zhang J, Gong J-L, Zeng G-M et al (2017) Carbon nanotube amendment for treating dichlorodiphenyltrichloroethane and hexachlorocyclohexane remaining in Dong-ting Lake sediment—an implication for in-situ remediation. Sci Total Environ 579:283–291

Zhang R, Zhang N, Fang Z (2018a) In situ remediation of hexavalent chromium contaminated soil by CMC-stabilized nanoscale zero-valent iron composited with biochar. Water Sci Technol 77:1622–1631

Zhang Y, Yang J, Zhong L, Liu L (2018b) Effect of multi-wall carbon nanotubes on Cr (VI) reduction by citric acid: implications for their use in soil remediation. Environ Sci Pollut Res 25:23791–23798

Zhou X, Li N, Luo Y et al (2018) Emergence of African swine fever in China, 2018. Transbound Emerg Dis 65:1482–1484

Zhu F, Ma S, Liu T, Deng X (2018) Green synthesis of nano zero-valent iron/Cu by green tea to remove hexavalent chromium from groundwater. J Clean Prod 174:184–190

Phytoremediation of Different Types of Domestic Wastewater Using Constructed Wetlands: Case Studies in Egypt

Mohamed A. El-Khateeb, Sozan El-Abeid, Maha S. A. Haridy, Hussein M. Ahmed, and H. I. Abdel-Shafy

Abstract The development of unconventional water resources is necessary to bridge the gap between supply and demand. Wastewater from homes is one of these resources. Therefore, treated wastewater is an excellent non-traditional resource for alleviating water scarcity, particularly in arid regions. Due to agricultural growth and rising drinking water demand, water supplies are becoming more and more limited in many arid and semi-arid locations. Although sewage treatment by conventional means is very effective, it requires high cost of energy, skilled personnel, huge budget for engineering work such as capital cost, the necessary maintenance and operation. These represent obstacles towards implementing such methods to be applied in rural and urban areas. Simple, effective, and reasonably priced on-site wastewater treatment methods are essential. Using artificial wetlands as a phytoremediation technology is a good choice for this. At a fair price, the implementation of such a system will address several environmental issues. Random releases of untreated or inadequately treated wastewater are one of these issues, which can harm the ecosystem, worsen surface water channel eutrophication, and seriously harm human health. Egypt receives 55.5 billion cubic meters of water annually from the Nile River, compared to its annual water demand of over 114 billion cubic

M. A. El-Khateeb (✉) · H. I. Abdel-Shafy
Water Pollution Research Department, National Research Centre, Dokki, Cairo, Egypt

S. El-Abeid
Nanotechnology & Advanced Nano-Materials Laboratory (NANML), Plant Pathology Research Institute, Agricultural Research Center, Giza, Egypt

Mycology and Disease Survey Research Department, Plant Pathology Research Institute, Agricultural Research Center, Giza, Egypt
e-mail: sozanalabeid@arc.sci.eg

M. S. A. Haridy
Central Lab of Organic Agriculture, Agricultural Research Center (ARC), Giza, Egypt

H. M. Ahmed
Housing and Building Research Center (HBRC), Sanitary and Environmental Institute, Dokki, Giza, Egypt

A. A. H. Abdel Latef et al. (eds.), *Sustainable Remediation for Pollution and Climate Resilience*, https://doi.org/10.1007/978-981-96-5674-5_2

meters. About 20 billion cubic meters of treated wastewater are produced annually through reuse, which is a respectable percentage if done properly.

Keywords Plant · Constructed wetland · Water resources · Treatment · Reuse

1 Introduction

In many parts of the world, reclaimed wastewater is a tested and trustworthy supply of water. The methods that are now available to purify household wastewater to levels that allow for the unrestricted reuse of recovered water are costly to build and run. Over the past few decades, the usage of macrophyte (wetland) systems for wastewater treatment has grown as a low-cost method (El-Khateeb and El-Gohary 2003). Reclaimed wastewater is a dependable and tested supply of water in many parts of the world. Construction and operation of the systems available for treating household wastewater to levels adequate for unrestricted reuse of recovered water are costly. As a low-cost technique, the usage of macrophyte (wetland) systems for wastewater treatment has grown during the past few decades (El-Khateeb et al. 2009).

2 Constructed Wetlands (CWs)

As a low-cost technique, the usage of macrophyte (wetland) systems for wastewater treatment has grown during the past few decades (Perkins and Hunter 2000). Artificial wastewater treatment systems, called constructed wetlands, are made up of shallow ponds or canals that are typically less than 1 m deep and have been seeded with aquatic vegetation. To regulate the flow direction, liquid detention duration, and water level, CWs usually feature structural structures and impervious clay or synthetic liners. Rock, gravel, or sand are examples of 'inert porous media' that may or may not be included in CWs, depending on the type of system (El-Khateeb and El-Gohary 2003). Unlike enhancement systems or polishing wetlands, which accept secondary effluent and further treat it before releasing it into the environment, these treatment systems receive initial wastewater and treat it to secondary effluent standards and better (EPA 1999).

Since the 1970s, CWs have only been utilised for wastewater treatment, making them a relatively new technology. However, there is now a lot of interest in using them. The most often suggested topic for future articles, for instance, was constructed wetland technologies. Wetland systems are also a hot topic among homeowners, health officials, and community leaders (EPA 1999; Wu et al. 2016).

3 CW Components

CWs have four principal components:

- Plants acclimated to anaerobic/aerobic water environment
- Substrates with different follow rates
- Miscellaneous microbial communities
- The water or wastewater stream

3.1 Plants in CWs

In contrast to what is commonly believed, plants in wetland systems are an integral part of the process, although they do not significantly contribute to the uptake of contaminants by plants. The plants' physical presence is crucial because they serve as host surfaces for the attached growing organisms, and in the case of free water surface (FWS), the leaf canopy shadows the surface and inhibits the growth of harmful algae. The subsurface flow (SSF) example likewise views the plant roots as a partial source of oxygen (Shelef et al. 2013; Vymazal 2013).

When collected for compost, animal feed, papermaking fibre, and biogas, the increase in vegetation biomass in artificial wetlands can yield financial benefits for local communities (Vasudevan et al. 2011; El-Khateeb and El-Bahrawy 2013). In nations with limited resources, where more incentives are needed to motivate communities to maintain treatment wetlands, the economic benefits of constructed wetlands are significant. Currently, bulrush (*Scirpus* sp.), cattail (*Typha* sp.), and reeds (*Phragmites* sp.) are the most often utilized aquatic plants in subsurface wetlands. Other kinds of moisture-tolerant plants could also be used in artificial wetlands, though. In addition to serving as a vital component of a wastewater treatment system, ornamental plants could boost the local economy. If ornamental plants in a wetland were contaminated with harmful substances, they would not pose a health danger because they are not eaten (Zurita et al. 2009; Sandoval et al. 2019).

Because of its inexpensive operating costs and straightforward technology, CWs are efficient treatment devices that can be especially helpful in developing nations. The majority of the time, local resources may be used to build the wetlands, which drastically reduces building costs. These treatment systems are also effective in eliminating organic contaminants, hazardous metals, and harmful nutrients (Belmont et al. 2018). Despite the tremendous potential and the urgent need for low-cost treatment methods, using artificial wetlands in developing nations is undoubtedly lower than in the US or Europe. The use of unconventional species, like decorative plants with economic value, as emergent plants in artificial wetlands is made possible by the warmer climate and greater biodiversity found in most developing nations. In addition to enhancing the environment of the treatment system, these lovely plants can assist the community economically by producing flowers (Metcalfe et al. 2018; Zurita and Vymazal 2023).

The plants in the wetlands play the most significant role. In temperate climates, the plant litter acts as an insulating layer against frost during the winter, while the roots offer a large surface area for associated microbial development. By releasing oxygen into the rhizosphere, plants can also promote aerobic degradation; however, the overall impact on pollutant removal is likely to vary, and oxygen release rates are hard to measure (Toscano et al. 2009; Malyan et al. 2021). Numerous studies conducted in temperate regions have demonstrated that the quantity of nitrogen (N) and phosphorus (P) that may be extracted through harvesting is typically negligible (Tanner and Sukias 2011). However, as other studies have shown, fertilizer intake can likely lead to much higher nutrient losses in tropical regions where plants grow more quickly and all year round. Nevertheless, throughout the biomass breakdown, the absorbed nutrients will be released again if the plants are not harvested (Greenway et al. 2022).

Heliconia and canna are examples of ornamental plants that enhance the wastewater treatment wetland's aesthetics. This is another purpose of the plants that has nothing to do with treatment effectiveness. Some recently constructed tropical CWs around the world have a strong emphasis on this purpose. They are intended to raise locals' understanding of wastewater treatment by creating park-like spaces in the villages (Zurita et al. 2009). The long-term functioning of the systems is expected to gain from increased interest in the upkeep and operation of the attractive systems. Another advantage is that flowers like heliconias may be sold in markets, which gives them economic potential. Studies are necessary to determine the suitability of these tropical ornamental plants for use in CWs, though.

There seem to be two significant indirect roles for wetland plants:

1. Stems and leaves expand the surface area of the water column where microbial populations can attach.
2. To survive in an anaerobic environment, wetland plants have the ability to transfer air gases, such as oxygen, into their roots.

Each root hair is surrounded by a thin layer called the rhizosphere (in the aerobic area). Around the rhizosphere, there is some chemical oxidation, but more significantly, it is home to enormous microbial populations. As such, anaerobic zones encircle the aerobic region. in order for nitrification-denitrification and other transformations of many good pollutants to occur (Gersberg et al. 1986, 2020; EPA 1999).

The following criteria should be met by macrophytes chosen for growth in artificial wetlands:

1. Consistently has a high biomass and
2. Easily accessible in the neighborhood.

3.2 Substrate

Plants are physically supported by substrates such as different types of soil, sand, or gravel; they also serve as attachment surfaces for microbial populations and a reactive area for complexing ions, anions, and other chemicals (Wang et al. 2018; Bastian and Hammer 2020). The properties of the media utilized in the CW systems are displayed in Table 1.

Table 1 Media characteristics for wetland system

Type	n Porosity %	k Hydraulic conductivity ($m^3/m^2/d$)
Rough sand	32	1000
Gravelly sand	35	5000
Fine gravel	38	7500
Medium gravel	40	10,000
Coarse rock	45	100,000

4 Benefits of Treated Wastewater Reuse

Reusing wastewater that has been treated can have many advantages. The following are the advantages of treated wastewater irrigation, according Ofori et al. (2021) and Scott et al. (2000):

1. In a basin setting, irrigation water is an example of recycled urban wastewater. According to related research, irrigation produces about US$1800 per hectare and US$0.16 (1994 USD) per cubic metre of water (Kloezen et al. 1998). As a result, society and water users both profit greatly from the water value of wastewater utilized for agriculture.
2. The waste flow contains nutrients that lower the cost of agricultural production. In Santa Catarina, the estimated yearly gross values of the wastewater and waste load to farmers in San José de Cervera were US$18,900 and US$252,000, respectively.
3. Crop yield is greatly increased by the high cost of synthetic fertilizers and the realization that wastewater contains valuable nutrients.
4. Wastewater is utilized to harvest nutrients for crops (such as N, P, and K) and to provide trace elements (such as Na) that are absent from surface water, which:
    - Increases agricultural output.
    - Reduces or stops fertilizers and organic products use.
    - Decreases or eliminates the need for fertilizers and organic products.
5. Farmers can now get money from wastewater.
6. Reusing wastewater following treatment reduces negative effects on the environment, such as water contamination.

5 Types of Wastewater

The three primary categories of wastewater are industrial, agricultural, and residential. Toilets, baths, and kitchens are the main sources of domestic wastewater, which is generated by residential settlements and services, like homes (Abdel-Shafy et al. 2014).

5.1 Domestic Wastewater Classification

Separating municipal wastewater into black and greywater has been shown to be an effective way to lower the amount of fecal-contaminated wastewater, prevent greywater pollution, and lower treatment costs (Regelsberger et al. 2007; Abdel-Shafy et al. 2009). Greywater reuse on-site has the potential to be quite important. In affluent nations, the average indoor residential water demand (not including garden irrigation and other exterior uses) is between 100 and 180 L/d PE, or 36 and 66 m^3/y PE. This amounts to 30–70% of the overall urban water demand (Birks et al. 2003; Abdel-Shafy et al. 2009).

There are four main types of wastewater in our homes, and they are all denoted by a color—black, grey, and yellow (Fig. 1).

However, it has been demonstrated that dividing municipal wastewater into black, yellow, and greywater is an effective way to reduce the volume of wastewater contaminated by faeces, prevent greywater pollution, and save treatment costs (Abdel-Shafy et al. 2009; Wulan et al. 2022). Reusing greywater on-site could be quite significant. Except for garden irrigation and other external uses, the amount of water needed for indoor household use in industrialized nations usually ranges from 100 to 180 L/d PE, or 36 to 66 m^3/y PE (Birks et al. 2003; Muthukumaran et al. 2011). Between 30% and 70% of the total water demand in urban areas is accounted for by this amount. In urban settings, flushing toilets is the most feasible approach to reuse greywater, reducing the net water use of each home by 40–60 L/d PE. This approach can lower urban water demand by 10–25% if it is extensively used. With a 30% penetration ratio (i.e., 30% of homes having greywater reuse units installed), the use of greywater reuse for toilet flushing in the home sector might save about 50 MCM/y in Mediterranean countries by 2023 (Maksimovic et al. 2003). This accounts for about 5% of the projected national urban water demand and is equal to the capacity of a medium-sized saltwater desalination plant. The authors also shown that it is possible to reach a 30% penetration rate in 20 years with government encouragement and support (Maksimovic et al. 2003). In fact, on-site greywater reuse has been extensively studied over the last decade, especially in the United

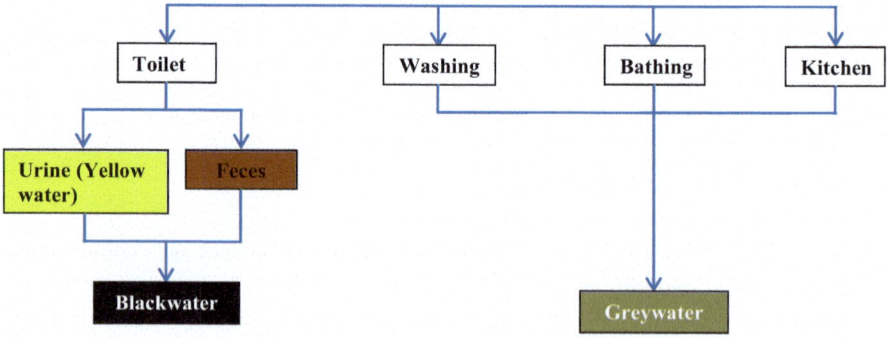

Fig. 1 Domestic wastewater components

States, Japan, Australia, and the European Union. However, fully operational commercial solutions are not being adopted by many (Regelsberger et al. 2007).

6 Phytoremediation of Blackwater

All wastewater sources are combined in conventional centralized sanitation systems, resulting in huge wastewater volumes with diluted organic and nutrient concentrations that are not conducive to the recovery of nutrients and energy. Wastewater source diversion is an alternative method for enhancing energy and nutrient recovery from domestic wastewater. When blackwater is collected separately and combined with water-conserving toilets (such as Hoover toilets), the result is a concentrated stream that contains the majority of contaminants (i.e., 80–95% of nutrients and more than 50% of organic content can be recovered) and accounts for less than 30% of the total household wastewater consumption (Kujawa-Roeleveld and Zeeman 2006). Greywater, the residual home water, is a rather clean stream that may be easily treated using various sustainable techniques that are appropriate for its intended use (Gao et al. 2019b). Treating blackwater separately increases the efficiency of energy and nutrient recovery since it contains the bulk of organics, nutrients, and pathogens (Gallagher and Sharvelle 2010; Gao et al. 2019a).

Abdel-Shafy et al. (2009) investigated the effectiveness of treating blackwater using an integrated system that included a contructed wetland (CW) after an upflow anaerobic sludge blanket (UASB). The UASB reactor had a hydraulic residence time (HRT) of 24 hours and an effective volume of 0.25 m^3. CW received the UASB reactor's wastewater. Subsurface flow type (SSF-CW) was the CW that was utilized. With a comparable organic loading rate (OLR) of 70.15 kg chemical oxygen demand (COD)/ha/day, the HRT in the SSF-CW was 10 days. The configuration of the treatment system is depicted in Fig. 2. The substrate in the SSF-CW was sand (0.5–1.0 mm), and the plant employed was reed.

The entire treatment system (UASB followed by SSF-h CW) for treating greywater is incredibly effective, as seen by the data's reduction in COD, BOD, TSS, and sulfides. In total, the removal rates were 89.5%, 94.0%, 77.8%, and 87.7%. Improvements were also made to the final effluent's turbidity, which decreased from 126.7 to 8.4 NTU. Meanwhile, total Kjeldahl nitrogen (TKN) and total phosphorus (TP) decreased by 52.7% and 44%, respectively. The values were 4.6 and 1.7 mg/L, respectively. There were less than 10^3 MPN of fecal coliform (FC) in 100 mL. As a result, it was determined that the final treated effluent was appropriate for irrigating various crops and trees.

It was examined by Abdel-Shafy et al. (2017) how well hybrid-constructed wetlands (CWs) treated blackwater. A pilot investigation was carried out using actual blackwater samples. In order to treat blackwater (strong wastewater), the study developed an integrated system that includes a hybrid horizontal-vertical flow-constructed wetland after the planned sedimentation technique (primary treatment) (Fig. 3).

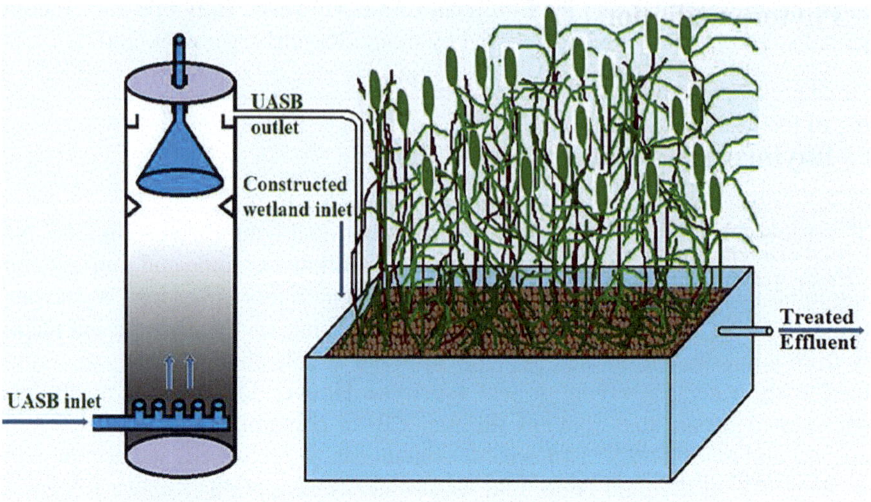

Fig. 2 The SSF-CW and UASB reactor combined for wastewater treatment

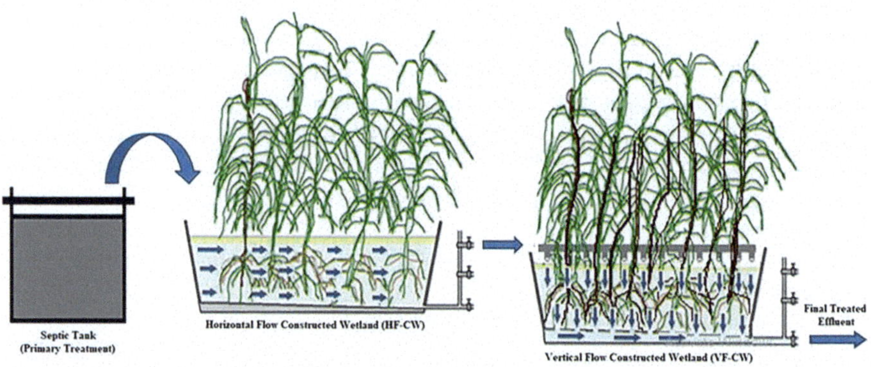

Fig. 3 The integrated septic tank and the hybrid CW system

Total suspended solids (TSS), biological oxygen demand (BOD), and COD were found to be eliminated by the sedimentation tank at rates of roughly 56.8%, 64.8%, and 58.0%, respectively. The removal efficiency of TSS, BOD, and COD rose to 82.9%, 88.0%, and 87.1%, respectively, when the sedimentation tank's effluent was further treated by the subsurface horizontal built wetland. The treated wastewater was additionally exposed to a vertical wetland to enhance it. For TSS, BOD, and COD, the combined system's total elimination of the pollutant parameters was 97.4%, 98.0%, and 98.5%, respectively. For unlimited water reuse, the resulting effluent met the National Regulatory Standards. The study came to the conclusion that, given the climates of arid and semiarid regions, the Middle East, and Africa, hybrid CWs provide an affordable wastewater treatment option. If there is land available, a hybrid system like this may be put into place with ease.

7 Phytoremediation of Greywater

To treat greywater, Abdel-Shafy et al. (2009) investigated the integration of the coupling of the UASB reactor and CW. The UASB reactor's HRT was 6 hours, and its OLR was 1.16 kg/m^3/day. The National Regulatory Standards were not being met by the UASB reactor's effluent (Abdella Ahmed et al. 2022). One well-known technique that is secondary handled with CW is the combination of UASB and CW. In the final effluent, the average removal of the major elements, which are represented by COD, BOD, and total suspended solids (TSS), was 87.7%, 89.5%, and 94%. COD was separated into three different forms by the authors Abdel-Shafy et al. (2009) dissolved, suspended, and colloidal. Using UASB and CW, the authors investigated the fate of several COD elements during the treatment process. The settleable portion of the raw sewage is the largest percentage, according to an assessment of the COD produced by the various COD constituents. The soluble and colloidal fractions in the UASB reactor's effluent rose as a result of the hydrolysis reactions that occur there. The soluble component in the CW effluent has been observed to grow further. This demonstrated the development of metabolic reactions taking place in anaerobic, anoxic, and/or aerobic environments. El-Khateeb and El-Gohary (2003) endorsed these findings. For wastewater reclamation and reuse in arid and semi-arid regions, the authors came to the conclusion that employing CW as a post-treatment step following a UASB reactor is a promising approach.

7.1 Phytoremediation of Domestic Wastewater

A common polluted effluent, residential wastewater, was treated using phytoremediation. El-Khateeb and El-Gohary (2003) investigated the use of CW systems to improve the quality of domestic wastewater. The UASB reactor and either free surface flow (FSF) or subsurface flow (SSF) built wetlands were the treatment techniques. *Typha latifolia*, or cattail, was utilized as the macrophyte and was planted at a density of three rhizomes/m^2. An unplanted gravel bed that was the same as the SSF and served as a control was used to assess the contribution of plants to the treatment process. The UASB effluent was supplied to the CWs at an organic loading rate that varied between 17.3 and 46.8 kg BOD5/ha·d (55.1 and 134.6 kg COD/ha·d). The effectiveness of the system in eliminating COD, BOD, TSS, nutrients, and FC was examined. The findings show that the SSF's final effluent had lower levels of TSS and COD than the FWS. SSF-CW outperformed the unplanted control in terms of total efficiency. Four log units of FC reduction were attained. The application of CWs for the elimination of biological and physico-chemical pollutants was assessed in 2009 by El-Khateeb et al. (2009). The OLR varied from 41.4 to 74.5 kg BOD/ha.day to 84.5 to 152 kg COD/ha.day. COD, BOD, TSS, and nutrients were successfully eliminated by the CWs. Additionally, *Pseudomonas aeruginosa*, *Salmonellae*, total *Staphylococci*, *Listeria monocytogenes*, *fecal streptococci*, *total coliform*, and *fecal coliform* were significantly reduced.

Kamel et al. (2010) investigated the performance of CW with various intake designs in an effort to enhance CW performance. The treatment plan included a filter unit, a CW unit, and either subsurface flow (SSF) or free water surface (FWS). We looked at two distinct entrance shapes: triangles and rectangles. In terms of COD, BOD, TSS, fecal coliform (FC), fecal streptococci (FS), Pseudomonas aeruginosa (PS), and Salmonellae (Sal.), the findings showed that the triangle-shaped entry improves CW's performance. The FWS with triangular entrance performed better than 73%, 83%, and 81% for the elimination of COD, BOD, and TSS, respectively. The removal of Salmonellae was total, but the removal of FC, FS, and PS was 10^4, 10^3, and 10^2 MPN/100 mL, respectively.

Abdel-Shafy and El-Khateeb (2013) built a pilot scale CW for the treatment of domestic wastewater in Sharquiya Governorate, Egypt, based on the findings of Kamel et al. (2010) and Masi and Martinuzzi (2007). A septic tank was used as the primary treatment. Following the construction of a 56 m^3 three-chamber septic tank, a 200 m^2 SSF-CW was installed. For the septic tank and CW, the estimated HRT was roughly 5.6 and 7 days, respectively. The physico-chemical and bacteriological properties of CW effluents, septic tank effluents, and household wastewater were examined. The findings showed that the faecal coliform count decreased by roughly 5 log units, while the levels of COD and BOD decreased by 87% and 89%, respectively. The quality of the treated wastewater was found to be within Egypt's permitted irrigation standards. There are no scent or bug problems.

8 Phytoremediation of Agricultural Drain Water

The agricultural drain wastewater could be reused for irrigation purposes after proper treatment. The quantity of drain water is estimated about 17 billion cubic metres (BCM) could be non-conventional water resource (Assar et al. 2019). In Egypt, constructed wetland systems have been used to improve drain water quality. Galal et al. (2014) investigated the effectiveness of a hybrid CW system that included floating aquatic plants (FAP) after free water surface flow (FWS). The plants employed in the study were water hyacinths (FAP) and Typha latofolia (cattail), respectively. The effectiveness of the system in eliminating COD, BOD, TSS, ammonia, and phosphorus has been thoroughly studied. They decreased from 115.2, 71.4, 79.4, 4.7, and 1.4 mg/L to 41.8, 13.9, 13.5, 1.8, and 0.5 mg/L for COD, BOD, TSS, ammonia, and phosphorus, respectively. The final effluent complied with national regulations pertaining to the reuse of treated wastewater. Consequently, CW systems could be used for valorize the drain water to be applied for irrigation purposes in arid and semi-aired countries.

9 Conclusions

Constructed wetland systems are efficient techniques for the treatment of wastewater. These devices act as gravel filters, capturing and adsorbing suspended particles onto the biomass film that is affixed to the gravel and root system.

High-quality effluent must be produced by combining CW with the appropriate first treatment step.

Using the subsurface CW did not result in any odor or pest issues. Wastewater treatment using a combination of horizontal CW (HF-CW) and vertical CW (VF-CW) has shown promise as a technology.

The high hydraulic and biological load of wastewater can be managed by the hybrid wetland system.

The technology of CW can be easily constructed, maintained, and used at a low cost. Furthermore, the systems can be deployed at any scale because to their lower energy consumption, which enables a decentralized (on-site) method of wastewater treatment.

Reusing the treated effluent for irrigation is an option because the final effluent contains essential nutrients like phosphates and nitrates that have a great potential for crop growth.

References

Abdella Ahmed AK, Shalaby M, Negim O, Abdel-Wahed T (2022) Egyptian reuse standards of treated wastewater for irrigation. MEJ Mansoura Eng J 47(2):43–59

Abdel-Shafy HIA, El-Khateeb MA (2013) Integration of septic tank and constructed wetland for the treatment of wastewater in Egypt. Desalin Water Treat 51(16–18):3539–3546

Abdel-Shafy HI, El-Khateeb M, Regelsberger M, El-Sheikh R, Shehata M (2009) Integrated system for the treatment of blackwater and greywater via UASB and constructed wetland in Egypt. Desalin Water Treat 8(1–3):272–278

Abdel-Shafy H, El-Khateeb M, Shehata M (2014) Greywater treatment using different designs of sand filters. Desalin Water Treat 52(28–30):5237–5242

Abdel-Shafy HI, El-Khateeb MA, Shehata M (2017) Blackwater treatment via combination of sedimentation tank and hybrid wetlands for unrestricted reuse in Egypt. Desalin Water Treat 71:145–151

Assar W, Ibrahim MG, Mahmod W, Fujii M (2019) Assessing the agricultural drainage water with water quality indices in the El-Salam Canal mega project, Egypt. Water 11(5):1013

Bastian R, Hammer D (2020) The use of constructed wetlands for wastewater treatment and recycling. Constructed wetlands for water quality improvement. CRC Press, pp 59–68

Belmont MA, Cantellano E, Ramirez-Mendoza N (2018) Ornamental flowers and fish in indigenous communities in Mexico: an incentive model for pollution abatement using constructed wetlands. In: Multifunctional wetlands: pollution abatement and other ecological services from natural and constructed wetlands. Springer, pp 241–252

Birks R, Hills S, Diaper C, Jeffrey P (2003) Assessment of water savings from single house domestic greywater recycling systems. In: 2nd international conference on efficient use and management of urban water supply. IWA, AWWA and AEAS, Tenerife, Canary Islands, Spain. www.iwatenerife2003.org

El-Khateeb M, El-Bahrawy A (2013) Extensive post treatment using constructed wetland. Life Sci J 10(2):560–568
El-Khateeb M, El-Gohary F (2003) Combining UASB technology and constructed wetland for domestic wastewater reclamation and reuse. Water Sci Technol Water Supply 3(4):201–208
El-Khateeb M, Al-Herrawy A, Kamel M, El-Gohary F (2009) Use of wetlands as post-treatment of anaerobically treated effluent. Desalination 245(1–3):50–59
EPA (1999) Constructed wetlands treatments of municipal wastewaters. EPA/625/r-99/010. USEPA, Cincinnati
Galal AA-AE, El-Gohary FA, Medhat MA, Saleh, El-Khateeb MA, Mahmoud UF, Moawad A (2014) Treatment of drain wastewater using constructed wetlands in Egypt. Pollut Res 33(3):865–870
Gallagher NT, Sharvelle SE (2010) Decentralized anaerobic blackwater management: a sustainable development technology concept for urban water management. In: World Environmental and Water Resources Congress 2010: challenges of change
Gao M, Guo B, Zhang L, Zhang Y, Liu Y (2019a) Microbial community dynamics in anaerobic digesters treating conventional and vacuum toilet flushed Blackwater. Water Res 160:249–258
Gao M, Zhang L, Florentino AP, Liu Y (2019b) Performance of anaerobic treatment of Blackwater collected from different toilet flushing systems: can we achieve both energy recovery and water conservation? J Hazard Mater 365:44–52
Gersberg RM, Elkins B, Lyon SR, Goldman CR (1986) Role of aquatic plants in wastewater treatment by artificial wetlands. Water Res 20(3):363–368
Gersberg RM, Gearheart R, Ives M (2020) Pathogen removal in constructed wetlands. Constructed wetlands for wastewater treatment. CRC Press, pp 431–445
Greenway M, de Rozari P, El Hanandeh A (2022) Plant growth and nutrient accumulation in Melaleuca quinquenervia and Cymbopogon citratus treating high strength sewage effluent in constructed wetland systems with biochar media. Ecol Eng 180:106667
Kamel M, El-Khateeb M, Megahed R, Abdel-Shafy E (2010) Effect of entrance shape on the performance of constructed wetland. J Am Sci 6(9):787–795
Kloezen WH, Kloezen WH, Garces-Restrepo C (1998) Assessing irrigation performance with comparative indicators: the case of the alto Rio Lerma Irrigation District, Mexico. IWMI
Kujawa-Roeleveld K, Zeeman G (2006) Anaerobic treatment in decentralised and source-separation-based sanitation concepts. Rev Environ Sci Biotechnol 5:115–139
Maksimovic C, Butler D, Memon FA (2003) Advances in water supply management: proceedings of the CCWI'03 conference, London, 15–17 September 2003. CRC Press
Malyan SK, Yadav S, Sonkar V, Goyal V, Singh O, Singh R (2021) Mechanistic understanding of the pollutant removal and transformation processes in the constructed wetland system. Water Environ Res 93(10):1882–1909
Masi F, Martinuzzi N (2007) Constructed wetlands for the Mediterranean countries: hybrid systems for water reuse and sustainable sanitation. Desalination 215(1–3):44–55
Metcalfe CD, Nagabhatla N, Fitzgerald SK (2018) Multifunctional wetlands: pollution abatement by natural and constructed wetlands. In: Multifunctional wetlands: pollution abatement and other ecological services from natural and constructed wetlands. Springer, pp 1–14
Muthukumaran S, Baskaran K, Sexton N (2011) Quantification of potable water savings by residential water conservation and reuse–a case study. Resour Conserv Recycl 55(11):945–952
Ofori S, Puškáčová A, Růžičková I, Wanner J (2021) Treated wastewater reuse for irrigation: pros and cons. Sci Total Environ 760:144026
Perkins J, Hunter C (2000) Removal of enteric bacteria in a surface flow constructed wetland in Yorkshire, England. Water Res 34(6):1941–1947
Regelsberger M, Baban A, Bouselmi L, Shafy HA, El Hamouri B (2007) Zer0-M, sustainable concepts towards a zero outflow municipality. Desalination 215(1–3):64–72
Sandoval L, Zamora-Castro SA, Vidal-Álvarez M, Marín-Muñiz JL (2019) Role of wetland plants and use of ornamental flowering plants in constructed wetlands for wastewater treatment: a review. Appl Sci 9(4):685

Scott CA, Zarazúa JA, Levine G (2000) Urban-wastewater reuse for crop production in the water-short Guanajuato river basin, Mexico. IWMI

Shelef O, Gross A, Rachmilevitch S (2013) Role of plants in a constructed wetland: current and new perspectives. Water 5(2):405–419

Tanner CC, Sukias JP (2011) Multiyear nutrient removal performance of three constructed wetlands intercepting tile drain flows from grazed pastures. J Environ Qual 40(2):620–633

Toscano A, Langergraber G, Consoli S, Cirelli GL (2009) Modelling pollutant removal in a pilot-scale two-stage subsurface flow constructed wetlands. Ecol Eng 35(2):281–289

Vasudevan P, Griffin P, Warren A, Thapliyal A, Tandon M (2011) Localized domestic wastewater treatment: part I-constructed wetlands (an overview)

Vymazal J (2013) Emergent plants used in free water surface constructed wetlands: a review. Ecol Eng 61:582–592

Wang HX, Xu JL, Sheng LX, Liu XJ (2018) A review of research on substrate materials for constructed wetlands. Mater Sci Forum. Trans Tech Publ 913:917

Wu S, Carvalho PN, Müller JA, Manoj VR, Dong R (2016) Sanitation in constructed wetlands: a review on the removal of human pathogens and fecal indicators. Sci Total Environ 541:8–22

Wulan DR, Hamidah U, Komarulzaman A, Rosmalina RT, Sintawardani N (2022) Domestic wastewater in Indonesia: generation, characteristics and treatment. Environ Sci Pollut Res Int 29(22):32397

Zurita F, Vymazal J (2023) Oportunities and challenges of using constructed wetlands for the treatment of high-strength distillery effluents: a review. Ecol Eng 196:107097

Zurita F, De Anda J, Belmont M (2009) Treatment of domestic wastewater and production of commercial flowers in vertical and horizontal subsurface-flow constructed wetlands. Ecol Eng 35(5):861–869

Soil-Poor Challenges and Its Measures for Bioremediation

Raja Husain, Nitin Vikram, Sonika Pandey, Kasinam Doruk, Sonbeer Chack, Atik Ahamad, Lipi Rina, Varun Dhiman, Kunvar Gyanendra Kumar, Deepak Kumar, Anurag Mishra, and Touseef Hussain

Abstract Bioremediation is the process of reducing pollutant complexity by biodegradation. Biodegradation is a general word that includes almost any biologically driven substrate change. Microorganisms are vital for the process of bioremediation, which occurs when microbes interact in their natural habitat under ideal parameters. The usage of biofilms, biostimulation, bioaugmentation, bioaccumulation, and biosorption are some of the latest technological developments in microbial bioremediation. Many microorganisms such as fungi, bacteria, yeast, and microalgae, biotransform, break down, or degrade various pollutants. However, controlling and optimizing bioremediation is a difficult process that depends on biotic and abiotic variables. An overview of the many forms of bioremediation, the kinds of bac-

R. Husain · K. Doruk · S. Chack · L. Rina
Department of Agriculture, Himalayan University, Itanagar, Arunachal Pradesh, India

N. Vikram
Zila Parishad Krishi Mahavidyalaya, Banda, Uttar Pradesh, India

S. Pandey
ICAR, Indian Institute of Pulses Research, Kanpur, Uttar Pradesh, India

A. Ahamad
KVK Bharari Jhansi, Banda University of Agriculture and Technology, Banda, Uttar Pradesh, India

V. Dhiman
Department of Environmental Science, Government College, Una, Himachal Pradesh, India

K. G. Kumar
Faculty of Agriculture, Bhagwant University, Ajmer, Rajasthan, India

D. Kumar
R & D, Nextnode Bioscience Pvt Ltd., Kadi, Gujarat, India

A. Mishra
Department of Agricultural Biotechnology and Molecular Biology, Dr. Rajendra Prasad Central Agricultural University, Pusa, Samastipur, Bihar, India

T. Hussain (✉)
Sevama Agri Clinic and Laboratory, Mati Mate AgroMart Pvt. Ltd., Bhavnagar, Gujarat, India

A. A. H. Abdel Latef et al. (eds.), *Sustainable Remediation for Pollution and Climate Resilience*, https://doi.org/10.1007/978-981-96-5674-5_3

teria connected to bioremediation, and the variables impacting the remediation process are given in this chapter. The employment of microbes for bioremediation of contaminated soil, pesticides, heavy metals, hydrocarbons, antibiotics, dyes, radioactive wastes, etc., has been explained in detail from the perspective of sustainable environmental management.

Keywords Bioremediation · Biosurfactants · Fungicides · Pollution · Pesticides · Soil contamination

1 Introduction

Growing industrialization and urbanization have exposed the environment to a wide range of toxins that are harmful to all living creatures. Significant contamination caused to aquatic and soil environments includes pollutants originating from various industrial processes. Heavy metals are emitted in various forms and concentrations both during and after industrial production when they are used as effluents. For instance, antimony, chromium, and mercury have been linked to the effluent of dye manufacturing enterprises (Kapahi and Sachdeva, 2019; Methneni et al. 2021). Through the application of fertilizers, insecticides, and herbicides, the agricultural sector releases aluminum, copper, zinc, nickel, lead, and arsenic into the environment (Tsadilas et al., 2018; Ayilara et al. 2020). Bioremediation is a relatively accessible approach that can be a great tool for mitigating the negative effects of pollution and reducing the pollution of contaminated soil that does not contain any toxic or refractory materials. Biological treatments are employed in the metabolism of fungi, bacteria, algae, or plants to reduce or oxidize inorganic and organic compounds.

2 Sources of Soil Pollutant

The term "soil pollution" describes the occurrence of abnormally high amounts of harmful compounds in soil. Controlling disease-carrying vectors in humans and animals is a major environmental concern, and its extensive usage worldwide has the potential to affect plants, animals, soil organisms, and human health (Mishra et al., 2016; FAO and ITPS 2017).

2.1 Man-Made Pollutants

1. *Excessive industrial activity*

Different industrial wastes contain varying amounts of harmful chemicals and poisonous substances, which, when deposited in soil, alter the strength of the top soil layer, lowering soil fertility and biological activity and causing problems for crop development. Environmental pollution tends to increase in areas where more polluting industries, such as cement manufacturing and petrochemicals, are increasing their production share (Bekabil, 2020). This puts extreme strain on the ecosystem as a whole as well as on the elements of the natural system, such as soil, water, air, and biodiversity (Bakhshipour et al. 2016). A considerable portion of soil contamination is caused by small-scale industries, such as lead battery recycling, dry cleaning, auto-mechanical workshops, and cottage industries (FAO and UNEP 2021). Paints, solvents, paint primers, outdated hydraulic fluid, grease, and lubricants are among the possible contaminants that have been discharged (Ekeocha et al. 2017).

2. *Acid rain*

Since the Industrial Revolution, acid rain has occurred more frequently and intensely. Sulfur dioxide and nitrogen oxides are released into the environment during the burning of fossil fuels in power plants and automobiles. Rainwater gets unduly acidic from solutions containing sulfuric and nitric acids. According to Mueller (2018), acid rain is a combination of hydrochloric acid (HCl_2), nitric acid ($4HNO_3$), and sulfuric acid (H_2SO_4).

Acid rain causes the soil to become more acidic, which increases the interchange of hydrogen ions with nutritional cations in the soil, such as potassium (K), calcium (Ca), and magnesium (Mg) (Fig. 1). These cations are released into the soil and, coupled with sulfate from acid input, can be quickly leached out in the soil solution (Gratchev and Towhata 2011).

Europe, China, and Eastern North America have all experienced soil acidification (Watmough and Dillon 2003). It is challenging to pinpoint the exact amount of acidic deposition that negatively affects soil qualities because a variety of causes, such as vegetation changes, can contribute to soil acidification (Singh and Agrawal 2008). Additionally demonstrated a maximum dry intensity,

Fig. 1 Impact of industrial pollution on the environment

and decrease in soil strength along with low pH and significant acid rain fluxes (Bakhshipour et al. 2016).

3. *Agricultural waste*

The contamination we discharge into the environment as a result of cultivating and rearing livestock, food crops, animal feed, and biofuel crops is known as agricultural pollution. The primary sources of soil and sediment contamination in agricultural environments are organic wastes, fertilizers, pesticides and natural pollutants.

(a) *Organic waste*

Animal dung, compost, septic and sewage sludge, municipal biosolids, and food processing wastes are examples of organic fertilizers that are frequently utilized in agriculture (Khan et al. 2018). Feeds used in intensive livestock production may include nonessential elements like lead, chromium, and other potentially hazardous metals, together with essential elements like copper and zinc. Higher amounts of copper and zinc in manure are caused by the use of copper and zinc in animal feed, which increases the high risk to agricultural soils (Xiong et al. 2010). Fermenting and composting the biological wastes from homes, businesses, and municipalities—such as green waste and food waste—can be turned into organic fertilizers, such fertilizers are a source of difficult-to-remove microplastic particles (Weithmann et al. 2018).

(b) *Pesticides*

When the quantities of pesticides or any other chemicals and components in the soil exceed their natural levels, they become pollutants (Usman et al. 2017). Several methods exist for pesticides to get into the soil through a spray drill used for treating foliage, through wash-off from treated foliage, or through the release of treated seeds and granulates from the soil (Burauel and Bassmann 2005). In order to control soil-borne pests, pesticides can also be directly applied to the soil, known as fumigant treatment. For instance, persistent organic pollutants like heptachlor, endosulfan, DDT, endrin, dieldrin, and chlordane were later found to be resistant but had a noticeably slow degradation rate, leading to risk to the health of animals, plants, and humans when contacted through contamination of soil (Mahugija 2013).

(c) *Fertilizers*

Fertilizers of mineral, synthetic, inorganic, and organic origin have all proven crucial in providing enough nutrients to boost crop yields. Environmental issues like soil nutrient saturation and fertilizer loss through leaching into groundwater and runoff into surface water, which contaminates drinking water and eutrophicates freshwater rivers, lakes, and oceans, are the result of excessive fertilizer use in some areas and nations. Enhancing fertilizer use efficiency is critical to addressing the world's growing food needs, addressing the depletion of critical nutrients like P, and reducing the environmental concerns associated with the overuse of fertilizers (e.g., eutrophication) (Rahman and Zhang 2018).

When nitrogen fertilizer is used excessively, the soil loses nitrogen due to erosion, volatilization, denitrification, leaching into groundwater, and surface runoff. According to Shen et al. (2018), soil acidification is an environmental condition that can decrease the number of nutrients available to plants for growth, increase the risk of soil-borne plant diseases, produce direct toxic effects from soil elements like aluminum, and reduce the plant's ability to fight off pathogen attacks.

Crop uptake and erosion cause soil phosphorus to be depleted, while air deposition and rock weathering take a while to restore it. Phosphorus fertilizers used excessively may cause croplands to lose phosphorus due to runoff or erosion (Smith et al. 2016). Phosphorus fertilizers used excessively may cause croplands to lose phosphorus due to runoff or erosion (Smith et al. 2016). Wet acid extraction is the process used to extract phosphate to create phosphate fertilizers, yielding phosphogypsum as a by-product (Saadaoui et al. 2017).

(d) *Natural pollutants*

Soil naturally contains a variety of contaminants, including asbestos, radionuclides, and trace elements. Natural forest fires also release organic pollutants into the soil, such as polycyclic aromatic hydrocarbons, or PAHs. In addition to having a detrimental effect on microbes, PAHs change the porosity, grain size, and soil's ability to store water. After sustained exposure, this pollutant may be a substantial generator of PAHs, which can cause cancer (Kim et al. 2011).

Phosphoric fertilizers contain trace metals (like cadmium) that come from phosphate rocks that are used to make P fertilizers. The trace elements cadmium (Cd), arsenic (As), lead (Pb), uranium (U), and radium (Ra) are frequently found in phosphate rock ores and, as a result, in phosphatic fertilizers (Zhang et al. 2017). Depending on where they came from, naturally occurring phosphate rocks may or may not contain harmful trace elements (Faridullah et al. 2017). The trace element concentrations increase to hazardous levels when the resulting fertilizers are consistently applied to soils for crop production, posing a risk to the health of people, animals, and plants (Gupta et al. 2014).

Many radioactive elements are found in nature and pose risks to human health and the environment (Cygan et al. 2007). Radon (Rn) is the element that exposes humans to the greatest amount of radiation naturally (Appleton 2007). Radon is a radioactive, colorless, odorless, tasteless noble gas that is created when uranium (U), thorium (Th), and radium (Ra) naturally decay. It travels through soil pores from bedrock to soil and can build up within structures, where it can seriously harm people's health (WHO 2009). Fibrous silicate minerals, including asbestos, are found in soils that were derived from ultramafic rocks like amphibole and serpentine (Rodríguez Eugenio et al. 2018).

Asbestos is a known carcinogen that can cause Mesothelioma; industrial and mining use of the material is controlled (Lee et al. 2008). Particulate matter or ash-bound materials may be released into the atmosphere as a result of wildfires. These pollutants can travel great distances in the atmosphere or become mobilized in the soil profile (Campos et al. 2016). This is particularly problematic for peatlands and other soils rich in organic carbon since they are more likely to experience severe and frequent fires as a result of human management and climate change (Evangeliou et al. 2014).

3 Effect of Different Types of Soil Pollutants on Plants and Animals

Soils act as an amazing life ecosystem. Those substances that, when present in the soil, cause harm to the environment, living organisms, and human health are referred to as soil pollutants. What is a soil pollutant that can be branded as treated organic fertilizers, overuse of mineral fertilizers, use of polluted water, urban waste, releases from industries, smelting and waste incineration, surface runoff, water, and wind erosion, etc. (FAO and ITPS 2015). Some common soil pollutants include (Syakila and Kroeze 2011):

Heavy Metals: Mercury, lead, cadmium, arsenic, and chromium can accumulate in soil as a result of mining, industrial operations, and improper disposal of electronic trash. They are toxic to people, animals, and plants, even in low quantities.

Pesticides and Herbicides: Chemicals popularly used in agriculture limiting weeds and pest populations, birds, and mammals.

Petroleum Products: Oil spills and improper disposal of oil-based products can contaminate soil, leading to long-term damage to ecosystems and affecting the growth of plants.

Industrial Waste: Chemicals and by-products from industries, such as solvents, acids and other toxic substances, if not disposed of properly, can seep into the soil, contaminating and affecting its ability to support plant life.

Nuclear Waste: Radioactive substances from nuclear accidents or improper disposal of radioactive waste can contaminate soil, posing serious health risks to living organisms.

Plastics and Microplastics: Inadequate disposal of plastic waste can cause nonbiodegradable materials to accumulate in the soil, which can alter the structure of the soil, water retention, and eventually plant development.

4 Effects of Soil Pollutants on Plants and Animals

Pollutants in the soil have a major effect on the health, growth, and general well-being of both plant and animal ecosystems. These vital environmental components can be negatively impacted by a variety of soil contaminants, disrupting ecological balance and threatening biodiversity.

Impact on Plants:

Decreased Growth and Productivity: Heavy metals, herbicides, and petroleum products are examples of soil contaminants that can hinder plant growth by interfering with nutrient intake, influencing photosynthesis, and producing physiological stress.

Toxicity and Damage: High concentrations of pollutants like heavy metals can be toxic to plants, leading to cellular damage, chlorosis, wilting, and, in severe cases, plant death.

Altered Soil Fertility: The pH, nutrient availability, and microbial activity of soil can all be changed by contaminants, which lower the soil's fertility and ability to support strong plant development.

5 Bioaccumulation and Biomagnification

Certain contaminants can build up in plant tissues, including heavy metals and persistent organic compounds. These pollutants can endanger both humans and animals by bioaccumulating and biomagnifying the food chain.

6 Impact on Animals

1. *Direct Toxicity*: Soil pollutants can directly harm animals by ingesting contaminated soil or plants, leading to toxicity, organ damage, reproductive issues, and even death.
2. *Disruption of Ecosystems*: Contaminants in soil can disrupt food chains and ecosystems by affecting the populations of soil organisms, insects, birds, and mammals, leading to imbalances and ecosystem collapse.
3. *Bioaccumulation and Biomagnification*: Higher-level predators may be at risk as a result of the accumulation of pollutants in the tissues of animals that consume polluted plants or other animals. As these species go up the food chain, their levels of toxins may rise.
4. *Habitat Degradation*: Soil pollutants can alter habitats, reducing biodiversity and impacting the availability of suitable habitats for various animal species.

7 Remediating of Contaminated Soil

Involves various techniques to reduce or eliminate pollutants to restore the soil's health and usability.

Some common remediation methods include:

1. *Bioremediation*: This technique breaks down or metabolizes pollutants into less dangerous forms by using microorganisms such as bacteria, fungi, and plants. Microorganisms can degrade organic pollutants, while plants (phytoremediation) can absorb, stabilize, or accumulate contaminants in their tissues, reducing their presence in the soil.
2. *Soil Vapor Extraction*: By using a vacuum on the soil, this method removes volatile pollutants from the soil. In order to gather them, it vaporizes several molecules, including volatile organic compounds (VOCs).
3. *Soil Washing*: Soil washing physically separates contaminants from soil particles. It involves the use of water or solutions to flush out contaminants, which are then treated separately, leaving behind cleaner soil.
4. *Chemical Treatment*: Techniques like oxidation, chemical immobilization, or using additives can help chemically degrade, immobilize, or neutralize contaminants in the soil.
5. *Thermal Treatment*: High-temperature methods like incineration or thermal desorption can be used to volatilize or destroy contaminants in the soil. However, this method can be energy-intensive and unsuitable for all contaminants.
6. *Landfilling or Containment*: In cases where contaminants cannot be easily removed, the soil might be contained or isolated by capping the area or creating barriers to prevent the further spreading of pollutants.
7. *Electrokinetic Remediation*: This method uses an electric current to move charged contaminants through the soil towards electrodes, where they can be collected, treated, or removed. These contaminants have the ability to linger in the soil for long periods of time, upset ecosystems, contaminate groundwater, influence plant growth, and endanger the health of humans and animals by way of the food chain. Various methods, such as soil washing, phytoremediation, and bioremediation, are frequently used in attempts to clean up contaminated soil in order to lessen the negative impacts of pollutants. Numerous factors, including the kind and degree of contamination, the type of soil, the location, the cost, and the environmental impact, influence the remediation method selection. A combination of techniques may often be used in a remediation project to address multiple contaminants or complex soil conditions. It's crucial to remember that soil remediation may be a difficult and drawn-out process that calls for careful planning, close observation, and occasionally continuing upkeep to guarantee the efficacy of the selected remediation approach.

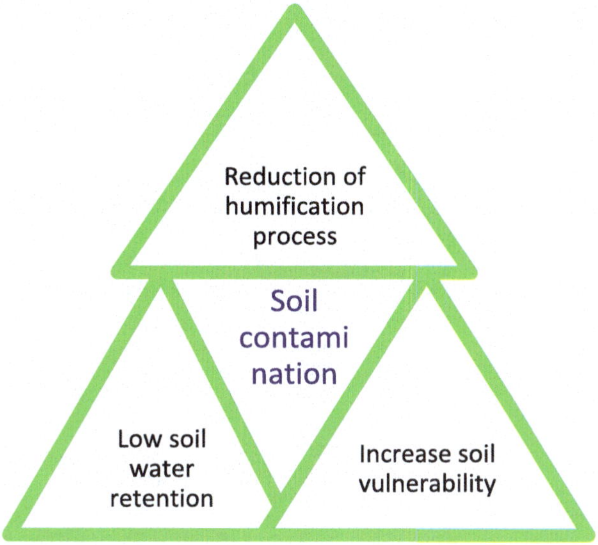

Fig. 2 Impact of soil contamination on soil property

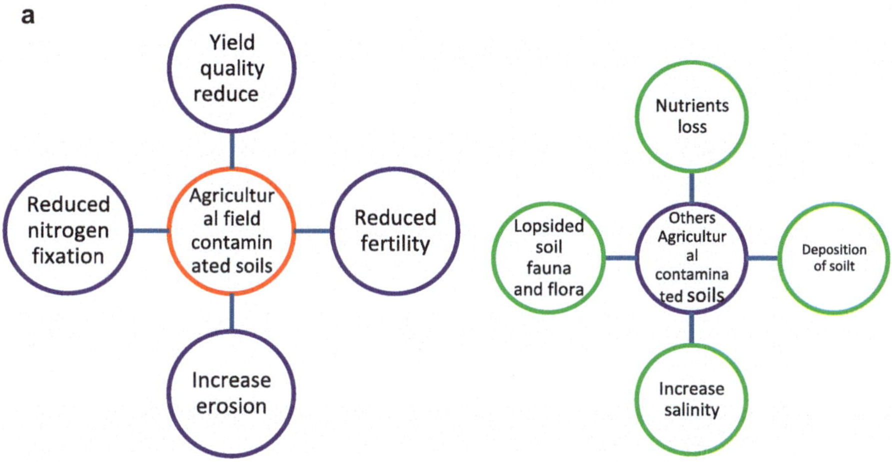

Fig. 3 (**a**) In the industrial region, the following challenges were associated. (**b**) Under environmental challenges

8 Challenges for Contaminated Soils

Soil contamination is an old and new environmental problem. It is a worldwide issue; estimations can be done based on local contamination, but the extent and content of diffuse contamination are largely unknown (Fig. 2). Although soil is able to store, filter, and reduce contamination. It can be called a negative legacy to industrial activities, and it changes according to periods and industrial structures, as

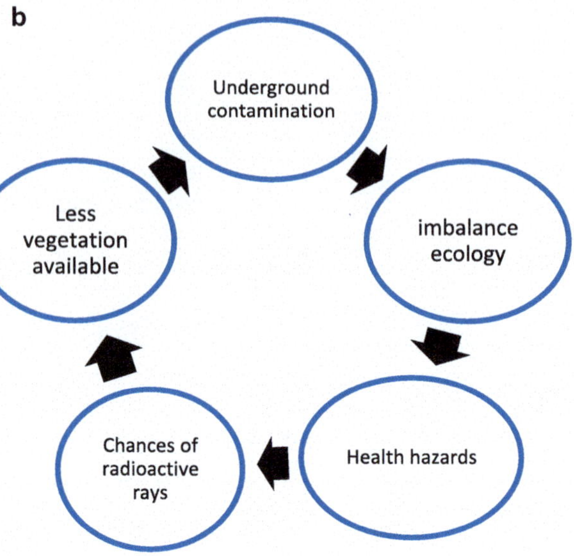

Fig. 3.3 (continued)

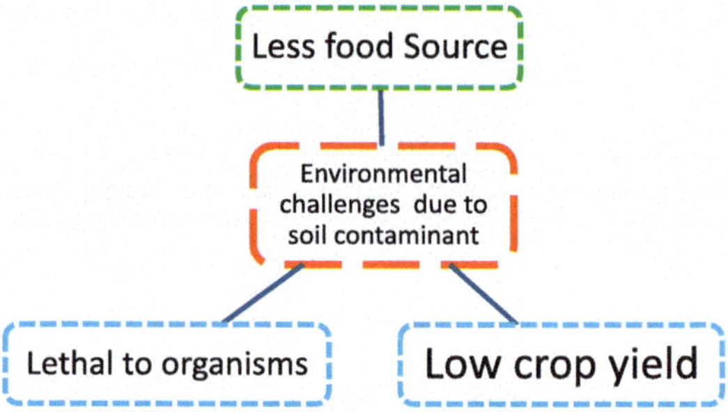

Fig. 4 Challenges of environment on soils

displayed in Fig. 3a, b. In people, reproductive, developmental-behavioral, neurologic, endocrine, and immunologic adverse health effects have been linked to persistent organic pollutants. People are mainly exposed to persistent organic pollutants through contaminated food ingestion coming from contaminated soil and water (Carre et al. 2018).

Soil contamination is caused by fertilizers, pesticides, chemicals, organic manure, radioactive wastes, discarded food, plastics, carcasses, and leathers. Industrial soils contain contaminants like copper, iron, zinc, lead, mercury, cyanides, cadmium, aluminum, acids, and alkalis, which reach the soil indirectly through the air and directly with water. The basic composition of soils is getting altered and becoming toxic for plant growth due to continuous and improper use of herbicides, pesticides, and fungicides to protect the crops from pests and diseases, as displayed in Fig. 4. Soil testing can help provide more information about the levels of a particular contaminant (Shayler et al. 2009).

Key Challenges

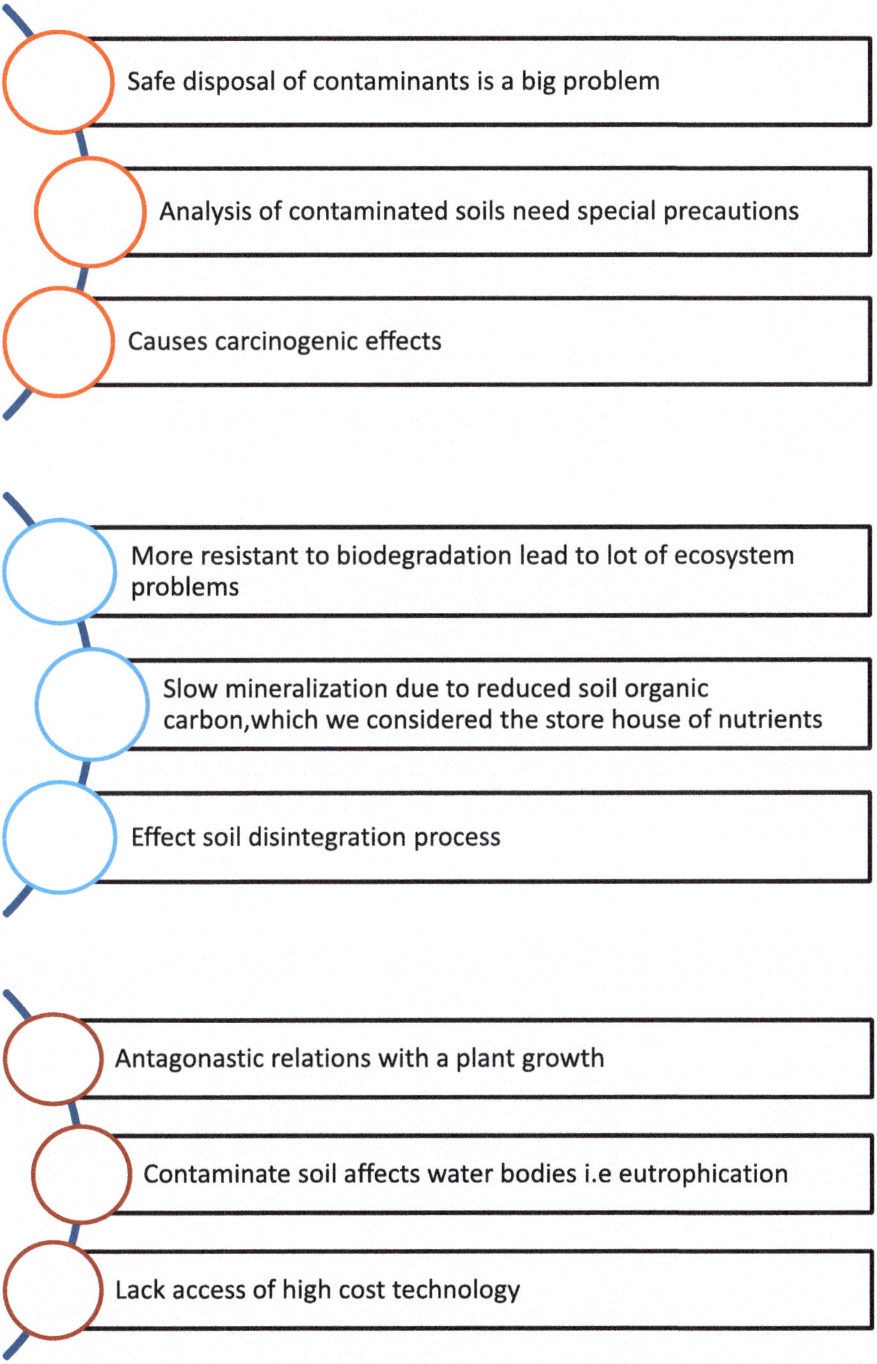

- In agricultural fields, contaminated soils lead to the following challenges

9 Bioremediation and Its Types

This biotechnical procedure for decreasing or removing contamination is referred to as bioremediation. This kind of waste management approach uses organisms to either absorb or remove pollutants from a polluted environment. In simple terms, Any procedure that uses a biological system—living or dead—to remove contaminants from the air, water, soil, flue gases, industrial effluents, etc., in either natural or man-made environments is known as bioremediation. Microalgae, fungi, bacteria, and plants are commonly used in mycoremediation and phytoremediation, respectively (Yuvraj, 2022). The natural ability of organisms to absorb, collect, and degrade both new and conventional contaminants has led to the use of biological resources in the treatment of contaminated conditions. Because bioremediation strives to be inexpensive, scalable, environmentally benign, and sustainable, it may offer significant advantages over traditional physicochemical methods for remediation. Oxidations are an essential component of bioremediations (Hasanuzzaman and Prasad, 2020). Organic molecules that have undergone oxidation are more soluble in water and more vulnerable to subsequent hydrolysis and oxidation-induced breakdown. Hydrocarbons ultimately decompose into carbon dioxide and water through biodegradation. Bioremediation techniques fall into one of the following categories: (i) In situ methods, which address contaminated sites directly, and (ii) *Ex situ methods*, which work with materials that have been excavated.

Both of these approaches provide additional nutrients, vitamins, minerals, and pH buffers to enhance the bacterial growth and metabolism. In some cases, specialized microbial cultures are added (biostimulation). Some techniques related to bioremediation include composting (biopiles and windrows), landfarming, bioventing, bioattenuation, biosparging, and phytoremediation. Thermal desorption, vitrification, air stripping, bioleaching, rhizofiltration, and soil washing are other remedial techniques (Garcia Frutos et al., 2010).

9.1 Types of Bioremediation

1. *Biostimulation*

The bacteria are triggered to start the process, as the name implies. First, the polluted soil combines unique nutrients and other essential elements in liquid or gaseous form. It enables microorganisms to proliferate, which allows them and other bacteria to swiftly and efficiently eliminate environmental contaminants.

2. *Bioaugmentation*

Sometimes, in order to remove the pollutants, specific locations call for the use of microorganisms. Take municipal wastewater as an example. The bioaugmentation method is employed in these unique circumstances. There is just one significant flaw in this procedure. Controlling the proliferation of microorganisms by eliminating a specific contamination becomes nearly difficult.

3. *Intrinsic Bioremediation*

Since soil and water are the two biomes most likely to contain pollutants and harmful substances, these are the environments where intrinsic bioremediation works best. Intrinsic bioremediation is mostly utilized in subterranean environments, such as underground fuel storage facilities. Locating leaks in such areas can be challenging, and pollutants and other impurities may seep in and taint the gasoline. Solely microorganisms can cleanse the tanks and eliminate the poisons.

10 Other Methods of Waste Management

1. *Incineration*

During this process, waste and other undesirable things are burned. Ash, flue gas, and heat are produced as the organic waste burns. The waste's inorganic components are still present as ash. Another name for it is thermal therapy.

2. *Phytoremediation*

In this case, pollutants in the soil are directly cleaned up or contained by plants. Without having to remove the contaminated material from the site and dispose of it elsewhere, the environmental problem can be reduced by using this bioremediation process. Remedium (or re-establishing equilibrium) plus phyto (plant) equals phytoremediation.

11 Use of Biosurfactants for Bioremediation

Potential applications include formulations for uranium ore processing, paper and pulp, coal, textiles, food and ceramic industries, insecticides and herbicides, detergents, and mechanical dewatering of peat. Sometimes, surfactants that have a microbiological origin are called biosurfactants (Ron and Rosenberg, 2022). It is known that a large number of bacteria produce biosurfactants, most of which are produced extracellularly (Mulligan 2005; Desai and Banat 1997; Shulga et al. 1999).

1. *Common Biosurfactants Include*

Combinations of micelle-forming substances known as "bile salts" encapsulate food to facilitate absorption via the small intestine. Lecithin is a common food ingredient that can be derived from soybean or egg yolk. Certain *Pseudomonas* and *Bacillus* species, such as *Pseudomonas aeruginosa* and *Bacillus subtilis*, can create rhamnolipids and lipopetides. Several nonpathogenic yeasts synthesize sophorolipids. *Acinetobacter calcoaceticus* produces emulsan.

Biosurfactants, which are naturally occurring surface-active compounds produced by microbial metabolism, have been more prevalent recently. Recent years have seen a steady rise in the number of research publications devoted to the isolation, characterization, and optimization of bacteria that create biosurfactants (Das et al. 2018; Rawat et al. 2020). The term "glycobiotechnology" describes the study of the structure, function, and interactions of these "glycoconjugates (combination of glycoprotein and glycolipids)" with living organisms (Messner et al. 2013; Enaime et al. 2019). Microbes create a variety of extracellular biosurfactants, such as exopeptidases, rhamnolipids, glycol-lipopeptides, and sophorolipids. In the sectors of biofuel production and xenobiotic cleanup, a variety of anionic and non-ionic synthetic surfactants, including sodium dodecyl sulfate, tween-80, tergitol NP10, brij35, and Triton X-100, have long been employed to accelerate microbial activity (Da Rocha et al. 2010).

The majority of biosurfactants are either neutral or anionic. However, cationic biosurfactants contain amine groups. Long-chain fatty acids, which can also be any amino acid, glycogen, cyclic peptide, alcohol, or phosphate carboxyl acid, make up the hydrophilic component of the biosurfactant. Proteins, polysaccharides, peptides, or monosaccharides make up the hydrophobic section. Biosurfactants typically have a molar mass of 500–1500 Dalton (Bhati et al. 2019).

Chemicals referred to as biosurfactants are produced by microorganisms and help break down and absorb hydrophobic substances. While a lot of study has focused on bacterial strains for biosurfactant generation, new studies have also found yeast synthesis. As generally recognized as safe (GRAS) microorganisms, yeasts are used extensively in the food and pharmaceutical industries as well as in bioremediation.

Once in the environment, waste or used lubricating oils can form lasting and toxic oil components by bonding to organic materials, mineral particles, and living things. There is now a serious environmental risk as a result. According to research on the interaction between microbes and hydrocarbons, petroleum and its derivatives are degraded by microbes. The presence of emulsifying agents in the environment increases the accessibility of hydrocarbons for degradation. Mercade et al. (1996) found that the bigger the hydrocarbon-water interface, the faster the microbial population will break down hydrocarbons (Muhlbachova et al., 2015).

In order to speed up the natural biodegradative processes in contaminated environments, bioremediation involves boosting the availability of materials (like oxygen and nutrients), conditions (like pH and moisture content), and common microorganisms. Biosurfactants may improve the effectiveness of bioremediation in the two ways listed below. The first approach entails making the substrate more bioavailable to microorganisms; for bacteria that feed on hydrocarbons, the surface area between water and oil may restrict the rate of growth. Biomass multiplies arithmetically rather than exponentially when microorganisms with hydrophilic solvents, like water, have a small surface area. The second process involves interaction with the cell surface, which increases the hydrophobicity of the microbial cell wall and makes it simpler for bacteria to associate with hydrophobic substrates (Pacwa-Plociniczak et al. 2012).

12 Role of Microbes in Bioremediation

In the process of bioremediation, microbes play a crucial role. They remove or neutralize pollutants from contaminated environmental spheres (Indu 2020). Microbes in bioremediation serve various purposes and employ a wide range of processes. Some bacteria can break down a wide range of pollutants. For example, bacteria and fungi can absorb oil, hydrocarbons, pesticides, and other organic substances to get energy and nutrition (Xu et al. 2018). Mutualistic connections between some microorganisms and plants (such as arbuscular mycorrhizal fungi) improve the plant's capacity to absorb nutrients and fend off contaminants (Li et al. 2023a, b). The process of phytoremediation involves the cleaning of polluted environments through a group of microbial and plant genera. Some microbes have developed adaptations to withstand extreme conditions, like high salinity, low pH, or high temperatures. These extremophiles can be employed for bioremediation in circumstances where conventional methods could be less effective. These microorganisms play a pivotal role in bioremediation due to their significant level of enzymatic activity, metabolic diversity, and potentiality to adapt to changing environments. By making use of the innate abilities of microbes, it is possible to reduce the pollution's detrimental impacts on the ecosystem can be reduced in a sustainable and environment-responsible manner. The following subsections describe how the microbial community functions in various environmental conditions:

12.1 Role in Removing Toxic Metals from the Soil

Through various processes, including phytoremediation, metal reduction, bioaccumulation, biomineralization, and chelation, the microbial community undoubtedly contributes to the efficient elimination of toxic heavy metals from the terrestrial environment (Dhiman et al. 2022). Within the microbial cells, metals can concentrate or accumulate. The process known as "bioaccumulation" helps to sequester these harmful metals and reduces their availability in the soil. Furthermore, the biomineralization process involves the precipitation of the metal ions of environmental concern, which ultimately lowers their movability and availability within the soil (Dhiman 2023a). *Pseudomonas*, *Arthrobacter*, *Citrobacter*, *Vibrio*, and *Clostridium* species are among the bacteria that can convert hazardous metal ions into less toxic or insoluble forms through metal reduction processes. As a result, the metals have less influence on the environment and are less available for plant uptake. Emerging and underdeveloped nations face serious challenges of toxic metal pollution due to their socioeconomic and developmental difficulties. The integrations of genomics, transcriptomics, signaling pathways, and proteomics pave the way for more advanced biotechnology. Microorganisms, through their extracellular barriers, intra- and extracellular sequestration, enzymatic detoxification, and metal ions active transport, build up the ability to withstand heavy metal stress (Pande et al.

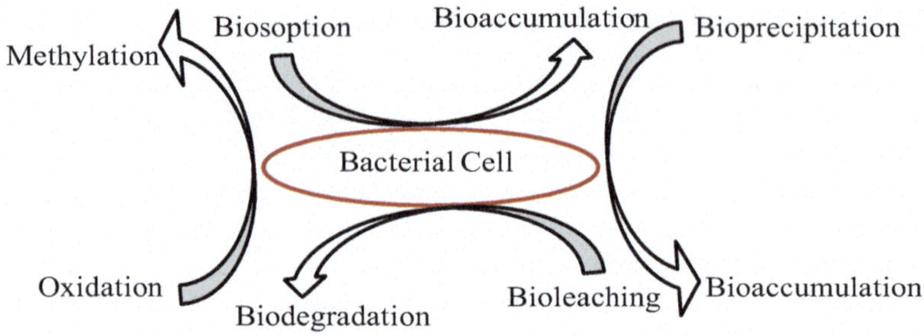

Fig. 5 Different microbial practices for toxic metals removal from soil

2022), as displayed in Fig. 5. Certain bacterial and fungal isolates can adsorb heavy metals such as chromium, copper, and lead from polluted soil (Dhiman and Pant 2022). *Pseudomonas aeruginosa* produces biofilms that resist lead, copper, and zinc. It is well known that the cell membranes of bacteria include a vast array of proteins and metabolic products, which aid in their ability to carry out intricate chelation processes with various metal ions.

12.2 Role in Removing Plastics

Recent studies confirm the role of numerous bacterial consortiums; fungal communities may prove beneficial in removing plastic polymers (Srikanth et al. 2022). Plastic polymers can be degraded based on polymer type and molecular weight. The scientific community will now be able to develop some engineered microbes that efficiently remove plastics from the environment. Microbes biodeteriorate plastic surfaces through chemical and structural changes, altering the polymers' characteristics (Tareen et al. 2022). After this process, the oxygenases, lipases, and esterases secreted by microbes perform plastic fragmentation as displayed in (Fig. 6).

12.3 Degradation of Organic Matter Content

Since microorganisms have metabolic machinery and can adapt to harsh environments, they can break down a wide variety of organic contaminants. Although several different types of bacteria are capable of decomposing organic pollutants, none of them possess the enzymatic ability to degrade all or even the majority of the organic compounds present in polluted soil. Because it takes the genetic makeup of multiple microorganisms to disintegrate the intricate mixes of natural chemicals present in highly polluted places, mixed microbial communities have the greatest potential for biodegradation. The biodegradation of harmful organic compounds in polluted soil has been demonstrated to be facilitated by plant-associated bacteria, which may enhance phytoremediation (Dotaniya et al., 2020). These include

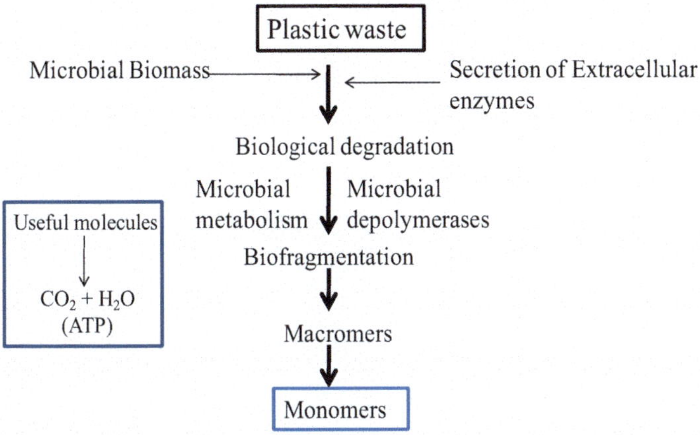

Fig. 6 Flow chart showing plastic removal with the help of the microbial community

endophytic bacteria, which are naturally occurring, nonpathogenic bacteria found in plants, and rhizospheric bacteria, which reside on and around plant roots. Nevertheless, bacterial degradation activity is greatly impacted by the suppression of enzyme activity and the processes of the proliferation of degrading microorganisms.

12.4 Degradation of Soil Pollutants

Several soil contaminants were found to be effectively degraded by distinct bacterial and fungal consortiums (Dhiman 2023b). Numerous complicated interactions in the terrestrial environment have formed various intermediate hazardous pollutants, including benzoquinone, phenols, and catechol. The bacterial community's many metabolic pathways aid in the breakdown of contaminants. *Acinetobacter* and *Saccharomyces cerevisiae* are two examples of exogenous microbes that contribute to the soil's elevated pH and DHA stimulation (Li et al. 2023a, 2023b). *Saccharomyces cerevisiae* is useful in efficiently removing pollutants such as hydrocarbons (Asemoloye and Marchisio 2022). Through both deleterious and advantageous interactions with plants, some rhizospheric bacteria (e.g. *Betaproteobacteria*) found in plant roots remove soil contaminants (Nwachukwu et al. 2021).

13 Management of Contaminated Soil

In the current situation, managing polluted soil is a major challenge. Soil pollution has resulted from mining operations, growing industrialization, agricultural methods, etc. Laws pertaining to soil monitoring have been created with this in mind in order to control possible soil hazards. In addition to endangering the ecosystem, soil

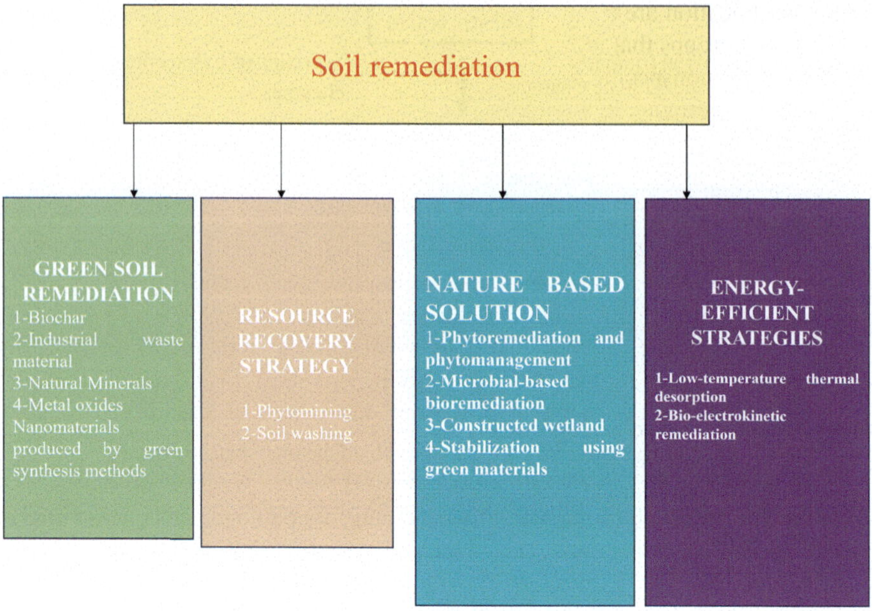

Fig. 7 Various strategies for soil remediation

pollution directly endangers people and animals through the use of explosives, pesticides, halogenated hydrocarbons, flame retardants, and surfactants. The primary contaminants that pollute soil are polyaromatic hydrocarbons found in pharmaceutical and medical waste. Management of contaminated soil starts with identifying the extent of contaminants, their concentration, and stability, as displayed in Fig. 7.

13.1 Green Soil Remediation

1. *Biochar*

Soil remediation frequently uses biochar, a carbon-rich solid made by pyrolyzing biomass during hydrothermal carbonization. This is because of its dynamic surface functionality, well-developed porous structure, and affordable price when compared to activated carbon. A variety of biological wastes, including wood, grass, sewage sludge, crop leftovers, and animal manure, can be converted into biochar. It interacts with heavy metals by a variety of processes, such as ion exchange, surface precipitation, electrostatic interactions, inner-sphere complexation, and outer-sphere complexation (Alloway, 2013). To improve the metal-binding properties of biochar, new trends in pyrolysis and post-pyrolysis modification techniques have surfaced. NH3-ambiance pyrolysis, co-pyrolysis, alkaline, acid, and oxidant activation, steam activation-microwave-assisted pyrolysis, clay mineral modification, and acid/

alkaline modification are a few of these. These techniques improve pore structure, add functional groups that include nitrogen or oxygen to the surface of the biochar, and strengthen nonspecific interactions, including ion exchange and physical adsorption. Furthermore, by promoting the activity of rhizosphere microorganisms, biochar can aid in phytoremediation processes (El-Naggar et al. 2018).

13.2 Industrial Waste Material

Soil remediation also makes use of industrial waste products. Fly ash, a typical by-product of the coal industry with an estimated annual production of 780 million tons, is one example. Fly ash's chemical makeup can change based on the coal's composition and source, but it always contains significant levels of SiO_2, Al_2O_3, Fe_2O_3, and CaO. These components are present in fly ash, according to several research (Leelarungroj et al. 2018; Palansooriya et al. 2019; Robl et al. 2017; Zha et al. 2019). Furthermore, fly ash and oxides both use surface complexation, liming, and precipitation as methods of metal immobilization (Sect. 4). Another example is red mud, which is a by-product of the Bayer process used in the alumina industry. Red mud is a mixture of oxides, such as Fe_2O_3, Al_2O_3, SiO_2, Na_2O, CaO, and TiO_2, just like fly ash (Zia-ur-Rehman et al. 2019). Because of its similar remedial mechanisms, red mud can also be used as an immobilizing agent for heavy metals (Huang and Hao 2012; Taneez and Hurel 2019; Wang et al. 2018c). Additionally, steel slag and coal gangue, two major industrial wastes that contain oxides, have been used for sustainable heavy metal cleanup (Leon-Romero et al. 2018; Ning et al. 2016; Chu et al. 2020). It's crucial to remember that industrial wastes might include high concentrations of organic pollutants and hazardous metals. Following their application to the soil, these toxins have the possibility of escaping and migrating over time, which could present environmental risks.

13.3 Natural Minerals

Because of their affordability and eco-friendliness, minerals show a lot of promise as possibilities for GSR (geochemical stabilization and remediation). The capacity of clay minerals to stabilize metals has attracted a lot of interest among these minerals. Their abundance of surface hydroxyl groups, high specific area, pH-increasing effect, and superior ion exchange capacity are the reasons for this (Chen et al. 2019; Doni et al. 2020; Xu et al. 2017). Liming, surface complexation, and precipitation are the main stabilizing mechanisms in clay-based S/S (solidification/stabilization) techniques (Wang et al. 2019a, 2020d; Xu et al. 2017; Yang et al. 2017). However, inadequate adsorption capacity, poor selectivity towards various metal types, and a gradual deterioration in stabilizing efficiency are some of the drawbacks of unaltered clay minerals (Wu et al. 2011; Xu et al. 2017). Minerals have a lot of potential as GSR (geochemical stabilization and remediation) options due to their low cost and environmental friendliness. There is a lot of interest in clay minerals because of their ability to stabilize metals. Their high specific area, pH-increasing effect, superior ion exchange capacity, and abundance of surface hydroxyl groups are the causes of this (Chen et al. 2019; Doni et al. 2020; Xu et al. 2017). In clay-based S/S (solidification/stabilization) approaches, the primary stabilizing mechanisms are precipitation, surface complexation, and limen (Wang et al. 2019a, 2020d; Xu et al. 2017; Yang et al. 2017). However, the disadvantages of unaltered clay minerals include a slow decline in stabilizing efficiency, poor selectivity towards different metal kinds, and insufficient adsorption capacity. Liang et al. (2019) immobilized soil Cd using a new mercapto-modified attapulgite. Because the modified attapulgite improved the sorption of bioavailable Cd onto the modified clay mineral, it showed a 75% reduction in plant accumulation of Cd. It's interesting to note that Wang et al. (2019a) found that sulfur-modified organoclay mobilized soil mercury, increasing the water and acid-soluble portions of soil mercury by 700%. Consequently, this enhanced the effectiveness of phytoextraction. In addition to clay, a variety of other natural mineral types can aid in heavy metal remediation. An alkaline aluminosilicate mineral with a well-developed microporous structure is zeolite, for instance. Through the liming action and nonspecific adsorption, it has been supported by studies conducted by Lee et al. (2019), Shi et al. (2009), and Tahervand and Jalali (2017). On the other hand, diatomite, a naturally occurring siliceous mineral, stabilizes metals through surface complexation and nonspecific adsorption. Additionally, it promotes plant growth due to its high Si content. This has been demonstrated in research by Radziemska et al. (2020) and Ye et al. (2015). Compared to amendments derived from metal-rich industrial waste, minerals have minimal adverse effects on soil. Once the challenge of long-term stability is addressed through appropriate modification strategies, mineral-based materials hold significant potential for the sustainable remediation of soil metals.

13.4 Metal Oxides

Because of their high absorption affinities, oxides—including hydroxides, oxyhydroxides, and hydrous oxides—have been thoroughly investigated as heavy metal immobilizing agents. Iron oxides (such as α-FeOOH, γ-FeOOH, and α-Fe_2O_3), manganese oxides (like γ-MnOOH and β-MnO_2), aluminum oxides (like γ-AlOOH, α-AlOOH, and γ-$Al(OH)_3$), and magnesium oxide (MgO) are the main metal oxide types employed in heavy metal stabilization (Komárek et al. 2013; Shen et al. 2019a). Inner-sphere complexation is responsible for the long-term stability of metal oxides, making them promising immobilization agents. When compared to nonspecific adsorption processes like Van der Waals force and electrostatic interactions, complexes offer higher long-term stability by stabilizing both cationic and anionic metal(loid)s (Hou 2020). For oxide-based metal stabilization methods, precipitation is just as important as complexation. For instance, MgO-based binders can efficiently stabilize Cd and Pb by precipitation because of the alkaline environment produced during the hydration process of magnesia. However, depending on the type of amendment, ferrous sulfates and other pH-increasing (liming) additions may or may not be successful. Acidification of the soil results from the oxidation of ferrous sulfates, which releases protons. Alkaline amendments such as lime and charcoal should be provided concurrently to stop the remobilization of cationic metals such as Cd, Hg, Cu, and Zn (Di Palma et al. 2015; Komárek et al. 2013).

13.5 Nanomaterials Produced by Green Synthesis Methods

There has been criticism of nano-sized cleanup agents produced with harmful chemicals (e.g., NaBH4) in conventional methods (US EPA 2017a; Zhang et al. 2020). Green synthesis is a concept that seeks to prevent pollution at the molecular level and reduce the environmental effect (US EPA 2017a). Either (a) using environmentally friendly materials as precursors, such as plant extracts and other chemicals originating from nature, or (b) carrying out chemical reactions in mild settings can accomplish this (ACS 2020; US EPA 2017a). For green remediation, green-synthesized nanomaterials have a number of advantages. Nanoparticles based on iron can act as immobilizing agents directly. For instance, Cr(VI) in the soil can be decreased by using nano zerovalent iron (nZVI) made from waste tea (Chrysochoou et al. 2012). Through co-precipitation, iron oxide nanoparticles made from leaf extract can stabilize As and Cd in soil (Lin et al. 2019; Su et al. 2020). Additionally, green nanoparticles can help remediate soil in an indirect way. Ag nanoparticles mediated by plant extracts can improve plant development by raising the pH, water-holding capacity, and nutritional bioavailability of the soil (Das et al. 2018). Feldspar and lime can be combined in a moderate hydrothermal process to create environmentally acceptable nano-sized mineral-based soil conditioners (Liu et al. 2017). However, because of their possible harmful impacts on humans, plants, and soil microbes, nanomaterials are frequently criticized. Furthermore, the practical applications of nanomaterials have been impeded by their relatively high cost.

14 Resource Recovery Strategies

14.1 Phytomining

When exposed to heavy metals in soil, plants show a variety of reaction patterns. Some plant species have evolved tolerance and resistance mechanisms and can even accumulate these metals in their tissues; however, the majority of plant species are vulnerable to high metal concentrations (Jia et al. 2019; O'Connor et al. 2019; Wang et al. 2020c). Hyperaccumulation is the process by which plants build up abnormally high concentrations of harmful metals in their tissues. According to Lopez et al. (2019), Rosenkranz et al. (2017), and Tognacchini et al. (2020), hyperaccumulators are plants that have the ability to collect metals to concentrations more than 100 times higher than those of regular plants in the same environment. The technique known as "phytomining" uses hyperaccumulators to recover valuable metals from metal-rich soils, including metalliferous soils, low-grade ores, and mine tailings. The metals are taken up by these plants from the soil and moved to their aboveground tissues. The plants are turned into ash after being harvested and dried, and the metals in the plant tissues can be recovered by sintering or smelting the ash (Anjum et al. 2012; Novo et al. 2017; Zhang et al. 2014). A large biomass production and metal concentration in the crop tissues is necessary for phytomining to be successful.

A number of parameters must be taken into account in order to guarantee a successful phytomining operation. The relationship between higher metal deposition in plant tissues and increased metal bioavailability is well-established. Thus, phytomining efficiency can be increased by encouraging the presence of labile forms of metals, which can be accomplished by altering the pH of the soil (e.g., raising pH to mobilize oxyanions). A drop in pH (Huang and Chen 2003; Zheng et al. 2019) or the application of chelating chemicals such as ethylenediaminetetraacetic acid (EDTA) (Nedelkoska and Doran 2000; Robinson et al. 1999) can mobilize metallic cations.

Adding organic amendments like compost, charcoal, and animal waste, as well as inorganic fertilizers, can also increase biomass. Choosing the right planting density and using efficient planting and harvesting techniques, such as double cropping and double harvesting, can also result in higher biomass yields. There has been much discussion of the processes underlying these techniques by Wang et al. (2020c). In contrast to open-pit mining operations, phytomining is a low-impact metal mining technique that is environmentally benign. Additionally, it can be used as a remediation method for soils that are extremely polluted. Phytomining provides a comparatively quick method of repairing degraded lands, cutting the restoration period to a few years, whereas natural revegetation of mining-affected areas may take centuries (Sheoran et al. 2009).

The rhizosphere stops harmful substances from migrating to groundwater, and the plant cover reduces wind erosion and runoff. Consequently, the public finds phytomining to be a visually beautiful, nondestructive, and environmentally friendly approach. However, numerous studies have frequently ignored the economic aspects of this "mining" process. The most important determinant of the operation's economics is the price of the metal being phytomined. The total value that can be obtained in a phytomining process is also influenced by biomass output and metal concentrations. In plant tissues, hyperaccumulators can only accumulate a certain amount of heavy metals (Alford et al. 2012; Burge and Barker 2010; Clemens 2017; Deinlein et al. 2012; Sheoran et al. 2009). Nonetheless, the most appropriate metals for phytomining techniques are precious metals such as Pb, Au, Pt, and Ag. The value of hyperaccumulators for other metals in environmental remediation is unaffected by this, though.

14.2 Soil Washing

By lowering pollutant concentrations considerably, soil washing can lower the soil's level below the legal environmental threshold. Screening the soil to get rid of the less polluted coarse particles is usually the first step. The dangerous metals are subsequently extracted by mixing the residual fine material with a washing solution. It is crucial to remember that this procedure varies depending on the metal, and the best washing solution should increase metal mobility while maintaining the physicochemical characteristics of the soil for possible reuse. Thus, by reusing the soil, soil washing can be viewed as a resource recovery tactic.

For soil cleaning to be successful, the right extractant must be chosen. Water, acids, bases, solvents, chelating agents, and surfactants are just a few of the materials that might be used in the washing process. It is important to note, nevertheless, that traditional cleaning methods could have an adverse effect on the environment. For instance, it can be detrimental to soil organisms if there is residual EDTA in the soil following chelant-assisted washing since it is difficult to break down. Furthermore, as EDTA is regarded as a significant contaminant in water bodies, it may move from the remediated soil to the surface or groundwater, causing water pollution. Using powerful acids, such as HCl or H_2SO_4, can cause soil to become acidic, making it unfit for use in agriculture.

Surfactants may have negative impacts on soil microorganisms, even though they can help remediate soils contaminated with both metals and organic contaminants. Many attempts have been made to reduce the impact on soil qualities by finding more ecologically friendly washing solutions. Different alternatives have emerged as innovative extractants in the soil washing process, including plant-derived washing agents, biosurfactants, biodegradable chelants like imminodisuccinic acid, methylglycinediacetic acid, citric acid, and poly-γ-glutamic acid, and natural organic substances like humic substances (Kulikowska et al. 2015; Meng et al. 2017; Begum et al. 2012; Guo et al. 2018; Cao et al. 2017a, b; Singh and Cameotra 2013). But thus far, these new extractants have only been examined in lab conditions. It should be noted that the ecosystem, including soil, groundwater, plants, and organisms, may suffer if these solutions are applied in situ. To assess the primary, secondary, and tertiary consequences of these novel techniques, pilot-scale research and the use of sustainability assessment instruments are therefore essential (Jorgensen 2007).

15 Nature-Based Solutions

15.1 Phytoremediation and Phytomanagement

Utilizing plants to carry out a variety of tasks associated with hazardous metals is known as phytoremediation. Metals can be extracted from the soil (phytoextraction), immobilized (phytostabilization), or released into the atmosphere (phytovolatilization) as part of these processes. The main goal of phytoremediation is to eliminate or immobilize pollutants using plants, as opposed to phytomining, which concentrates on using plants to recover valuable components. Metallophytes, or metal-tolerant plant species, are thought to be the best options for phytoremediation. Based on how they react to elevated soil metal concentrations, metallophytes can be divided into three groups: accumulators, indicators, and excluders. In their aboveground portions, accumulators have the ability to absorb and move harmful metals; indicators control the bioaccumulation of metals to mirror concentrations outside the ground; and excluders limit the uptake and movement of harmful metals, producing a stabilizing effect.

In metal phytoremediation, three primary approaches are utilized. (1) The most popular method, phytoextraction, uses accumulators' capacity to absorb metals to move them from the soil into plant tissues. (2) The release of highly volatile metals, such as mercury (Hg), into the environment from plant tissues is known as phytovolatilization. (3) Another popular tactic to lower metal mobility in the rhizosphere—the area of soil affected by plant roots—is phytostabilization.

Several techniques, such as complexation, metal precipitation, or physical adsorption brought on by the plant itself, can be used to achieve this goal. This process may also be aided by the immobilization effects of stabilizing substances including minerals, charcoal, and industrial waste. In the latter case, plants help the soil immobilization agents in a secondary capacity. We suggest consulting the works of Anjum et al. (2012), Ashraf et al. (2019), and Wang et al. (2020c) for a more thorough understanding of the mechanisms involved in remediation, techniques for plant selection, and factors that influence the process.

Although phytoremediation has been thoroughly investigated in field, greenhouse, and pot research, the commercialization of this natural remediation method has not yet accelerated. Indeed, of the 188 Superfund remediation operations in the United States from 2012 to 2014, only one used phytoremediation (US EPA 2017b). The fact that it frequently takes hundreds or even thousands of years for plants to absorb harmful metals to reach regulatory levels at heavily contaminated sites is one of the difficulties (Burges et al. 2020; Cheng et al. 2015).

Additionally, a major difficulty is how to properly handle contaminated biomass after harvest (Michelon et al. 2019; Vocciante et al. 2019). Since harmful metals are not eliminated from phytostabilized sites, long-term monitoring is required to evaluate the dangers related to metal leaching (Epelde et al. 2014; Houben et al. 2013; Labidi et al. 2017). "Phytomanagement" or "phytoattenuation" is a novel idea that aims to provide economic, social, and environmental benefits all at the same time. With metal remediation as a secondary goal, this idea places a higher priority on sustainable risk-based land use.

Profitable crops, which can be used for a variety of reasons, such as bio-energy production, animal feed, and soil fertility enhancement after harvest, are grown on polluted land instead of hyperaccumulators. Numerous site benefits are provided by phytomanagement, such as greenhouse gas reduction, water flow control, landscape preservation, and the production of renewable energy. Phytomanagement offers a chance to accomplish risk reduction, land restoration, and financial benefit all at once by fusing delicate remediation techniques with lucrative site usage.

15.2 Microbial-Based Bioremediation

Organic pollutants can be broken down, but heavy metals cannot. Stabilizing, bioaccumulating, or detoxifying these harmful substances is essential to microbial-based bioremediation. The production of oxides or carbonates for surface complexation or co-precipitation is one technique of microbially induced stability. For instance, biogenic manganese oxides (BioMnOx), which are potent stabilizing agents with a great affinity for hazardous metals, can be produced by specific bacteria, such as *Providencia* sp. (Li et al. 2020). According to Kang et al. (2016), ureolytic bacteria have the ability to cause calcite precipitation, which makes it easier for metals like $PbCO_3$, $CdCO_3$, and $CuCO_3$ to co-precipitate with $CaCO_3$. Toxic metals can be directly absorbed by organisms, causing bioaccumulation. Earthworms and other soil creatures are ideal for this procedure (Boughattas et al. 2019). However, more research is required to determine the long-term durability of microbial immobilization techniques. The stabilizing agents' efficacy can be greatly impacted by environmental conditions because they are produced during microbial metabolism, suggesting a limited capacity for adaptation. It is still difficult to handle organisms with high metal concentrations properly because improper treatment of contaminated biomass might exacerbate the secondary effect. A more practical method for the ecologically acceptable cleanup of Hg, As, and Cr seems to be microbial-based detoxification. By oxidizing or decreasing harmful metal(loid)s, this technique lessens their toxicity. Hg(II) can be microbially reduced to Hg(0) by bacteria possessing the mer operon (Mahbub et al. 2016; Wang et al. 2020b). During this process, transporter proteins that are encoded by the merP and merT genes help Hg(II) enter the cytoplasm. Hg(II) is then changed into the less hazardous and extremely volatile Hg(0) by the mercuric reductase enzyme, which is the expression product of the merA gene (Naguib et al. 2018; Petrus et al. 2015; Wang et al. 2020b).

Because elemental mercury (Hg) is extremely volatile, it volatilizes from soil into the air and passively diffuses out of cells. Nonetheless, mercury in the atmosphere might travel throughout the world before returning to the soil through dry or wet deposition. Microbial oxidation of As (III) to less hazardous As(V) is one way to detoxify arsenic (As). Anaerobic oxidation employing nitrate as an electron acceptor can also do this in oxygen-depleted flooded paddy soils. As (III) oxidase gene abundance is increased throughout this process, the relative abundances of the Acidovorax and Azoarcus genera rise. More than 100 times more poisonous and carcinogenic than Cr(III), chromium (Cr(VI)) is a carcinogen, teratogen, and mutagen. It may be possible to clean up Cr(VI)-contaminated soil using a variety of Cr-reducing bacteria, including *Bacillus*, *Aeromonas*, *Pseudomonas*, and *Enterobacter*. When applied to heavy metals, microbial-based bioremediation may not be ecologically benign, even while it offers a cost-effective balance and preserves site functionality for organic toxins. Metals cannot be entirely eliminated from the polluted soil during this procedure, thus ongoing maintenance and careful observation are required. Microbial-induced stabilization is more susceptible to fluctuations than solidification/stabilization based on green materials. Furthermore, there is ambiguity in microbial detoxification since the reversible "detoxification" response could turn into "toxification" if the environmental conditions alter.

For instance, the Hg(II) reduction process may be suppressed by an increase in Eh (redox potential), which would prevent Hg from being volatilized by microbes. Microorganisms have the potential to be very important in the phytoremediation process. Instead of operating independently, a mix of plants and microbes can produce encouraging remediation outcomes. By controlling phytohormone levels and boosting food availability, plant growth-promoting bacteria (PGPB) can either directly or indirectly promote plant development while acting as a biocontrol agent against plant diseases. By boosting biomass, enhancing metal tolerance, and affecting metal buildup in plants, PGPB can also aid in phytoremediation. As a result, metal bioaccumulation may be impacted by changes in metal bioavailability. PGPB can help in the processes of phytostabilization and phytoextraction. The effectiveness of metal extraction or stabilization can be greatly increased with the aid of microorganisms. For instance, *Brassica juncea's* phytoextraction efficiency of Zn, Cd, and Pb rose by 37%, 21%, and 10%, respectively, after being inoculated with the PGPB Burkholderiaphytofirmans. Similarly, *Bacillus proteolyticus*-inoculated sunflower demonstrated a 20% decrease in Pb accumulation and a 40% decrease in Cd accumulation in aboveground tissues, suggesting improved phytostabilization made possible by PGPB.

15.3 Constructed Wetland

It is suggested that this artificial system can be used for land reclamation even though a created wetland is usually used for wastewater treatment. Absorption, bioaccumulation, precipitation, biotransformation, and sedimentation are some of the processes that contribute to metal retention or removal in a built wetland (Marchand et al. 2010). The main process that affects how metals behave in the ecosystem in a man-made wetland is absorption. Both short-term retention and long-term immobilization can be accomplished by a variety of absorption methods, such as physisorption and chemisorption (Hu et al. 2020; Tran et al. 2017; Walaszek et al. 2018b). Although only a small number of aquatic plants, or macrophytes, are hyperaccumulators (such as the Ni hyperaccumulator Limnobiumlaevigatum and the Cd hyperaccumulator Potamogetonpectinatus), they are essential to metal bioaccumulation (Arán et al. 2017; Lu et al. 2018). This implies that alternate methods of metal retention or removal may take precedence over metal uptake. Other metals, such as Cu, Zn, Ni, As, Cd, and Pb, can co-precipitate with iron oxides and carbonates in wetland environments (Du Laing et al. 2009; Marchand et al. 2010; Sochacki et al. 2014). Wetlands' biotransformation and metal detoxification processes share many of the same mechanisms as microbial-based bioremediation. The biogeochemical processes of heavy metals in artificial wetlands are greatly influenced by oxidation or reduction mediated by microorganisms. When macrophytes decrease water flow, sedimentation takes place. Accordingly, a wetland under study resembled a stagnant pond, where high suspended matter concentrations encourage the production of flocs, which lowers metal mobility through sedimentation and coagulation (Tao and Haynes 2016; Walaszek et al. 2018a). Flood control, biodiversity preservation, a low life-cycle carbon footprint, and scenic enhancement are just a few of the many environmental, social, and economic advantages that constructed wetlands provide. However, there has been no report of any practical applications so far, and the use of artificial wetlands as a water treatment technology for soil remediation is rare. Furthermore, removing metal from a built wetland might take years (Marchand et al. 2010; Song et al. 2017), and uncertainties like the possible death of macrophytes (García-Mercadoa et al. 2017).

15.4 Stabilization Using Green Materials

An established remediation technique that dates back to the late 1950s, solidification/stabilization (S/S) is still in use today (Shen et al. 2019b). Nearly half of remediation projects in China (48.5%) use S/S, demonstrating the method's widespread adoption. In 2018, the remediation market in China was valued at 2.9 billion US dollars (Shen et al. 2019b). S/S is mostly caused by the conversion of metal geochemical fractions from labile to stable forms, such as binding them to residual fractions or organic materials.

Numerous techniques, such as surface complexation, precipitation, physical adsorption, ion exchange, or electrostatic interactions, can accomplish this transformation (Hou 2020). For metal immobilization, Portland cement (PC) is frequently utilized as the S/S agent. Nonetheless, PC manufacturing accounts for 8% of worldwide CO_2 emissions, making it a major contributor (Andrew 2019). It is critical to create alternative S/S agents with a lower environmental impact in view of the growing concern about climate change. Pulverized fly ash (Wang et al. 2019b), silica fumes (Goodarzi and Zandi 2016), ground granulated blast-furnace slag (Wang et al. 2018a), incinerated sludge ash (Su and Shih 2017), calcium carbide residue (Darikandeh 2018), and reactive magnesium oxide cement (Wang et al. 2018b) are some of the low-carbon supplementary cementitious materials that have been investigated.

However, because the enhanced mechanical strength brought on by S/S treatment may impair plant development and yields, the use of cement-like stabilizing agents may make it more difficult to remediate agricultural soils sustainably. Additionally, as the US EPA noted in 2015, the loss of soil biodiversity may have detrimental effects on the agroecosystem and soil health. Stabilizing agents obtained from biological waste may have a larger "net environmental benefit" when it comes to nature-based solutions (NBS). Using materials made from biological waste, such as compost and biochar, can improve soil fertility and encourage plant development while also reducing the negative environmental effects of cement-like substances.

It's crucial to remember that employing green immobilization agents does not always mean that a practice is sustainable and kind to the environment. A number of sustainability issues must be addressed in order to implement the concepts of green and sustainable remediation (GSR). For instance, employing specific agents, such as compost and sewage sludge that is rich in metals, may raise the possibility of metal leaching and provide environmental risks. Furthermore, long-term stability beyond the immediate future should be taken into account in a sustainable remediation method. Evaluating the long-term durability of nature-based metal immobilization methods is essential. The long-term efficacy of stabilizing strategies can be impacted by a number of environmental conditions, such as plant growth, UV irradiation, microbial metabolism, freeze-thaw cycles, and wetting-drying cycles (Shen et al. 2019a; Wang et al. 2020d, e). Metal migration and release may result from a decline in immobilization reliability. Therefore, in order to fully assess the effectiveness of "green" material-based solidification/stabilization (S/S) techniques, longer-term field investigations are required.

16 Energy-Efficient Strategies

16.1 Low-Temperature Thermal Desorption

Thermal desorption has been shown to be successful in cleaning up mercury-contaminated soil, despite its considerable volatility. Nevertheless, the procedure necessitates temperatures over 600 °C, which leads to substantial expenses (480 USD t^{-1}) and a substantial carbon footprint. Furthermore, the physicochemical characteristics of the treated soil are significantly altered, rendering it unfit for use in agriculture. Low-temperature thermal desorption techniques are required to optimize the total benefit and reduce any negative effects on the environment. It has shown promise to lower the heating temperature to 400 °C while retaining the effectiveness of mercury removal by using chemical agents such as citric acid and $FeCl_3$. Another efficient method is microwave-induced thermal desorption, which uses granule-activated carbon as the receptor and microwave irradiation to remove mercury with a high removal efficiency (87%). In contrast to traditional desorption methods, this approach makes it easier for off-gas to migrate from the soil. However, low-temperature thermal desorption techniques take a lot longer to achieve the same Hg removal rate. The sustainability of the remediation process as a whole may potentially be threatened by the use of chemical agents or GAC. As a result, it is essential to fully examine the entire effects of low-temperature thermal desorption processes. A life cycle assessment of a thermal desorption treatment on agricultural soil by Hou et al. (2016) found that low-temperature thermal desorption facilitated by citric acid successfully decreased greenhouse gas emissions from 357 kg CO_2-eq to 264 kg CO_2-eq as a result of lower electricity consumption. Additionally, the soil treated by low-temperature thermal desorption might be used for farming right away on the spot. However, only at the bench scale has the viability of this approach been verified. To obtain more profound insights and better knowledge, further field-scale research should be carried out.

16.2 Bio-electrokinetic Remediation

Using an external electric field to mobilize soil pollutants is a very successful technique known as electrokinetic remediation. Toxic metals can be extracted from the soil using this method (López Vizcaíno et al. 2018; US EPA 2017b). However, this technique's relatively high energy consumption has made it difficult to use in practice. Using renewable energy sources to create the necessary electric field is essential to overcoming this obstacle. Microbial fuel cells (MFCs) are cutting-edge bioelectrochemical systems that provide long-term solutions for soil remediation and wastewater treatment (Aarthy et al. 2019; Fang and Achal 2019). The microbial digestion of organic matter in an anaerobic environment produces energy within an MFC. As a result of this process, electrons are released at the anode and move to the cathode via an external circuit. With the help of an electron acceptor, such as oxygen, the combining of protons and electrons at the cathode creates water.

The anode's microbial oxidation process is essential to this system's ability to generate power. The composition of the anode's microbial community has been the subject of numerous investigations. This technology offers a quicker and more comprehensive way to remove metals from soil than S/S and phytoremediation, which need longer periods for remediation or monitoring. Proteobacteria, actinobacteria, and bacteroidetes are just a few of the phyla that can participate in this process. Inorganic nutrients, including Ca^{2+}, Mg^{2+}, K^{+}, PO^{4-3}, and NH^{4+}, can be introduced to the anode chamber to increase microbial activity in the anode (Chen et al. 2015; Habibul et al. 2016; Wang et al. 2016).

Furthermore, it is possible to incorporate exogenous organic carbon sources in various forms, such as leaves (Zhang et al. 2019) and straw (Song et al. 2018). Despite being in its infancy, MFC has developed into a variety of bioelectrochemical systems.

It is possible to use a standard three-chambered MFC in a horizontal or vertical orientation. In this configuration, negatively charged metalloids like Cr(VI) and As(III) will migrate toward the anode, whereas positively charged metals like Cu, Zn, Cd, and Pb will migrate toward the cathode. In a plant-based MFC (P-MFC), rhizodeposits, root exudates, and other organic materials created during photosynthesis can serve as carbon sources for the anodic processes. In this situation, plants can improve the production of power, but they do not have complete control over the process. The adsorbent close to the electrodes efficiently adsorbs metal cations or anions in this system once the impurities have moved in the direction of the cathode or anode. For example, Zhang et al. (2019) created an A-MFC to eliminate Cr(VI) from soil.

For the anode, they used fallen leaves as an organic carbon source and a natural biosorbent. With an adsorption capacity of 14 mg g^{-1}, absorption studies verified the biosorbent's high affinity for Cr(VI), leading to a 40% removal efficiency of Cr(VI) in soil. Heavy metal migration is influenced by their geochemical components. In contrast to less labile forms, such as residual and organic matter-bound metals, labile forms—such as exchangeable and acid-soluble metals—are more likely to migrate when exposed to an external electric field (Chen et al. 2015; Wang et al. 2020a). Therefore, remediation may be more successful if there is a larger percentage of labile fractions. This can be accomplished by lowering the pH of the soil with auxiliary reagents such as hydrochloric acid, acetic acid, or citric acid, which will mobilize the metals (Zhang et al. 2020). Microbial fuel cells (MFCs) are a cutting-edge technology used in soil remediation. However, because of soil variability and the intricate relationships between many environmental conditions, its field use has not been as successful thus far. Therefore, more work is needed to evaluate this technique's viability through field research. It will take time for the public to accept bioelectrokinetic remediation because it is a new technology. Present research primarily consists of proof-of-concept studies for this technology. Field demonstrations should be used to assess this technology's practicality.

17 Future Prospectus

Artificial intelligence, unmanned aerial vehicles, quantum computing, and other cutting-edge methodologies are examples of how quickly science is developing. In-depth research should be done on the following topics: the ability to detect contaminants, the identification of the routes from which contaminants originate to the final disposal, the interaction with all soil nutrient elements and particles, the effects on long-term human inheritance, the actual timeline of degraded contaminants, and the health risks associated with contaminate crop plant uptake. To find changes in contaminated soils over time, temporal and spatial investigations are employed. Researchers should conduct a thorough study and expand the understanding of soil capital by determining the critical limit of polluted soil and plants in comparison to normal crop development.

Implementing solutions with shorter time frames is frequently selected to provide a rapid return on investment and assure economic sustainability, even though using plant-based and microbial-based remediation technologies appears to increase environmental sustainability. Therefore, while thinking about remediation techniques, it's critical to balance the economic and environmental benefits. To develop more successful remediation solutions, future research should concentrate on combining several "green" approaches, as depending exclusively on one sort of green remediation strategy may have drawbacks. For instance, by improving metal immobilization and encouraging root growth, green stabilization ingredients like biochar and minerals, either alone or in combination, can act as both soil nutrients and immobilizing agents, supporting phytostabilization. Furthermore, because they improve metal mobility, new plant-based chelating agents can be used in phytoextraction, low-temperature thermal desorption, soil washing and phytomining, and bioelectrokinetic remediation. Additionally, in microbial-based and bioelectrokinetic remediation techniques, carbon-rich remediation materials like biochar can function as a carbon source, boosting the activities of microorganisms. A synergistic impact can be produced by integrating various green strategies, which will ultimately increase remediation efficiency.

18 Conclusion

Degradation of contaminants through biological means is known as bioremediation. The environment is becoming alarmingly contaminated with several toxins due to rapid agricultural and industrial expansion. Microbes are essential in the eradication of contaminants in order to overcome the catastrophic condition. Waste breakdown results from the release of toxins from pollutants, which increases the metabolic activity of a few hardy microorganisms. A diverse population of bacteria and fungi is essential to this process. The combined action of specialized enzyme production and genetic control guides the metabolic elimination of toxicants. To solve the problem of contaminated soils that are becoming more diverse and complex, it is necessary to systematically develop technologies for appropriately investigating the types and forms of soil contaminants. Countermeasures, low-cost technologies, low environmental damages and their consequences, and technologies may risk for communication based on risk assessments. Furthermore, science and technology research must be passed to the last person in the queue. It is recommended to fight soil pollution using both cutting-edge methods and long-standing customs. In order to determine the toxicity before disposal, many agencies, scientists, academicians, researchers, and civil society organizations should launch awareness campaigns among the general public, especially schoolchildren. Healthy agricultural production and improved eco-services for human well-being are the results of healthy soil.

Acknowledgments The Himalayan University at Jullang, Papumpare, Itanagar, Arunachal Pradesh, provided support for this article, and the authors are grateful. They also like to express their gratitude to the article's major authors, who made significant contributions to a number of the study's features that are addressed here. The co-authors all express their gratitude to their senior colleagues worldwide in the domains of microbiology, plant pathology, mycology, environment, and biochemistry for their inspiration.

References

Aarthy M, Rajesh T, Thirunavoukkarasu M (2019) Critical review on microbial fuel cells for concomitant reduction of hexavalent chromium and bioelectricity generation. J Chem Technol Biotechnol 95:1298. https://doi.org/10.1002/jctb.6257

ACS (2020) Green chemistry. Retrieved from https://www.acs.org/content/acs/en/greenchemistry.html

Alford ÉR, Pilon-Smits EAH, Fakra SC, Paschke MW (2012) Selenium hyperaccumulation by Astragalus (Fabaceae) does not inhibit root nodule symbiosis. Am J Bot 99:1930–1941

Alloway BJ (2013) Heavy metals in soils: trace metals and metalloids in soils and their bioavailability, 3rd edn. Springer

Andrew RM (2019) Global CO2 emissions from cement production, 1928–2018. Earth Syst Sci Data 11:1675–1710. https://doi.org/10.5194/essd-11-1675-2019

Anjum NA, Pereira ME, Ahmad I, Duarte AC, Umar S, Khan NA (2012) Phytotechnologies: remediation of environmental contaminants. CRC Press

Appleton JD (2007) Radon: sources, health risks, and hazard mapping. Ambio 36(1):85–89

Arán DS, Harguinteguy CA, Fernandez-Cirelli A, Pignata ML (2017) Phytoextraction of Pb, Cr, Ni, and Zn using the aquatic plant Limnobium laevigatum and its potential use in the treatment of wastewater. Environ Sci Pollut Res 24:18295–18308

Asemoloye MD, Marchisio MA (2022) Synthetic Saccharomyces cerevisiae tolerate and degrade highly pollutant complex hydrocarbon mixture. Ecotoxicol Environ Saf 241:113768

Ashraf S, Ali Q, Zahir ZA, Ashraf S, Asghar HN (2019) Phytoremediation: environmentally sustainable way for reclamation of heavy metal polluted soils. Ecotoxicol Environ Saf 174:714

Ayilara MS, Olanrewaju OS, Babalola OO, Odeyemi O (2020) Waste management through composting: challenges and potentials. Sustain For 12(11):4456. https://doi.org/10.3390/su12114456

Bakhshipour Z, Asadi A, Huat BBK, Sridharan A, Kawasaki S (2016) Effect of acid rain on geotechnical properties of residual soils. Soils Found 56(6):1008–1020

Begum ZA, Rahman IMM, Tate Y, Sawai H, Maki T, Hasegawa H (2012) Remediation of toxic metal contaminated soil by washing with biodegradable aminopolycarboxylatechelants. Chemosphere 87:1161–1170

Bekabil UT (2020) Industrialization and environmental pollution in Africa: an empirical review. J Resour Dev Manag 69:18–21

Bhati S, Kumar V, Singh S, Singh J (2019) Synthesis, biological activities and docking studies of piperazine incorporated 1, 3, 4-oxadiazole derivatives. J Mol Struct 1191:197–205. https://doi.org/10.1016/j.molstruc.2019.04.106

Boughattas I, Hattab S, Alphonse V, Livet A, Giusti-Miller S, Boussetta H, Banni M, Bousserrhine N (2019) Use of earthworms Eisenia andrei on the bioremediation of contaminated area in north of Tunisia and microbial soil enzymes as bioindicator of change on heavy metals speciation. J Soils Sediments 19:296–309

Burauel P, Bassmann F (2005) Soils as filter and buffer for pesticides: experimental concepts to understand soil functions. Environ Pollut 133:11–16. https://doi.org/10.1016/j.envpol.2004.04.011

Burge DO, Barker WR (2010) Evolution of nickel hyperaccumulation by Stackhousia tryonii (Celastraceae), a serpentinite-endemic plant from Queensland, Australia. Aust Syst Bot 23:415–430

Burges A, Fievet V, Oustriere N, Epelde L, Garbisu C, Becerril JM, Mench M (2020) Long-term phytomanagement with compost and a sunflower – tobacco rotation influences the structural microbial diversity of a Cu-contaminated soil. Sci Total Environ 700:134

Campos I, Abrantes N, Keizer JJ, Vale C, Pereira P (2016) Major and trace elements in soils and ashes of eucalypt and pine forest plantations in Portugal following a wildfire. Sci Total Environ 572:1363–1376

Cao Y, Zhang S, Wang G, Huang Q, Li T, Xu X (2017a) Removal of Pb, Zn, and Cd from contaminated soil by new washing agent from plant material. Environ Sci Pollut Res 24:8525–8533. https://doi.org/10.1007/s11356-017-8542-3

Cao Y, Zhang S, Wang G, Li T, Xu X, Deng O, Zhang Y, Pu Y (2017b) Enhancing the soil heavy metals removal efficiency by adding HPMA and PBTCA along with plant washing agents. J Hazard Mater 339:33–42

Carre F, Caudeville J, Bonnard R, Bert V, Boucard P, Ramel M (2018) Soil contamination and human health: a major challenge for global soil security. In: Global soil security. Progress in soil science. Springer, Cham, pp 275–295

Chen Z, Zhu BK, Jia WF, Liang JH, Sun GX (2015) Can electrokinetic removal of metals from contaminated paddy soils be powered by microbial fuel cells? Environ Technol Innov 3:63–67. https://doi.org/10.1016/j.eti.2015.02.003

Chen G, Shah KJ, Shi L, Chiang PC, You Z (2019) Red soil amelioration and heavy metal immobilization by a multi-element mineral amendment: performance and mechanisms. Environ Pollut 254:112964

Cheng SF, Huang CY, Lin SC, Chen KL, Lin YC (2015) Feasibility of using peanut (Arachis hypogaea L.) for phytoattenuation on lead-contaminated agricultural land-an in situ study. Agric Ecosyst Environ 202:25–30

Chrysochoou M, Johnston CP, Dahal G (2012) A comparative evaluation of hexavalent chromium treatment in contaminated soil by calcium polysulfide and green-tea nanoscale zero-valent iron. J Hazard Mater 201–202:33–42

Chu Z, Wang X, Wang Y, Zha F, Dong Z, Fan T, Xu X (2020) Influence of coal gangue aided phytostabilization on metal availability and mobility in copper mine tailings. Environ Earth Sci 79:68

Clemens S (2017) How metal hyperaccumulating plants can advance Zn biofortification. Plant Soil 411:111–120

Cygan RT, Stevens CT, Puls RW, Yabusaki SB, Wauchope RD, McGrath CJ, Curtis GP, Siegel MD, Veblen LA, Turner DR (2007) Research activities at U.S. Government agencies in subsurface reactive transport modeling. Vadose Zone J 6(4):805

Da Rocha UN, Andreote FD, Azevedo JL, van Elsas JD, van Overbeek LSm (2010) Cultivation of hitherto-uncultured bacteria belonging to the Verrucomicrobia subdivision 1 from the potato (Solanum tuberosum L.) rhizosphere Journal of Soils and Sediments 10(2):326–339. https://doi.org/10.1007/s11368-009-0160-3

Darikandeh F (2018) Expansive soil stabilised by calcium carbide residue-fly ash columns. Proc Inst Civ Eng Ground Improv 171:49–58. https://doi.org/10.1680/jgrim.17.00033

Das P, Barua S, Sarkar S, Karak N, Bhattacharyya P, Raza N, Kim KH, Bhattacharya SS (2018) Plant extract–mediated green silver nanoparticles: efficacy as soil conditioner and plant growth promoter. J Hazard Mater 346:62–72

Deinlein U, Weber M, Schmidt H, Rensch S, Trampczynska A, Hansen TH, Husted S, Schjoerring JK, Talke IN, Krämer U, Clemens S (2012) Elevated nicotianamine levels in Arabidopsis halleri roots play a key role in zinc hyperaccumulation. Plant Cell 24:708–723

Desai JD, Banat IM (1997) Microbial production of surfactants and their commercial potential. Microbiol Mol Biol Rev 61(1):47–64. https://doi.org/10.1128/mmbr.61.1.47-64.1997

Dhiman V (2023a) Remediation strategies of environmental mercury: an overview of its environmental persistence, associated threats, and health impacts. In: Kumar N (ed) Mercury toxicity: challenges and solutions. Springer Nature Singapore, Singapore, pp 235–247

Dhiman V (2023b) Metal bioleaching from E-waste using fungal communities. In: Debbarma P, Kumar S, Suyal DC, Soni R (eds) Microbial technology for sustainable E-waste management. Springer International Publishing, Cham, pp 201–211

Dhiman V, Pant D (2022) Bioindication and biomarker responses of earthworms: a tool for soil pollution assessment. In: Malik JA (ed) Advances in bioremediation and phytoremediation for sustainable soil management: principles, monitoring and remediation. Springer International Publishing, Cham, pp 365–378

Dhiman V, Giri A, Pant D (2022) Pollutants bioremediation using biosurfactants: a novel approach for improving soil health. In: Shit PK, Adhikary PP, Bhunia GS, Sengupta D (eds) Soil health and environmental sustainability: application of geospatial technology. Springer International Publishing, Cham, pp 489–502

Di Palma L, Gueye MT, Petrucci E (2015) Hexavalent chromium reduction in contaminated soil: a comparison between ferrous sulphate and nanoscale zero-valent iron. J Hazard Mater 281:70–76

Doni S, Gispert M, Peruzzi E, Macci C, Mattii GB, Manzi D, Masini CM, Masciandaro G (2020) Impact of natural zeolite on chemical and biochemical properties of vineyard soils. Soil Use Manag 37:832–842

Dotaniya ML, Dotaniya CK, Kumar K, Doutaniya RK, Meena AM, Shirale AO, Meena MD, Meena VD, Kumar R, Meena BP, Kumawat N, Lal R, Lata M, Singh M, Singh U, Meena AL, Kuri BR, Rai PK (2020) Type of soil pollutant and their degradation: methods and challenges. In: Handbook of nanomaterials and nanocomposites for energy and environmental applications. Springer. https://doi.org/10.1007/978-3-030-11155-7_65-1

Du Laing G, Rinklebe J, Vandecasteele B, Meers E, Tack FMG (2009) Trace metal behaviour in estuarine and riverine floodplain soils and sediments: a review. Sci Total Environ 407:3972–3985. https://doi.org/10.1016/j.scitotenv.2008.07.025

Ekeocha C, Nwoko IC, Onyeke L (2017) Impact of automobile repair activities on physicochemical and microbial properties of soils in selected automobile repair sites in Abuja, central Nigeria. Chem Sci Int J 20(2):1–15. https://doi.org/10.9734/CSJI/2017/36065

El-Naggar A, Lee SS, Awad YM, Yang X, Ryu C, Rizwan M, Rinklebe J, Tsang DCW, Ok YS (2018) Influence of soil properties and feedstocks on biochar potential for carbon mineralization and improvement of infertile soils. Geoderma 332:100–108. https://doi.org/10.1016/j.geoderma.2018.06.017

Enaime G, Nettmann E, Berzio S, Baçaoui A, Yaacoubi A, Wichern M, et al. (2019). Performance and microbial analysis during long-term anaerobic digestion of olive mill wastewater in a packed-bed biofilm reactor. J. Chem. Technol. Biotechnol. 95:850–861

Epelde L, Becerril JM, Alkorta I, Garbisu C (2014) Adaptive long-term monitoring of soil health in metal phytostabilization: ecological attributes and ecosystem services based on soil microbial parameters. Int J Phytoremediation 16:971–981

Evangeliou N, Balkanski Y, Cozic A, Hao WM, Møller AP (2014) Wildfires in Chernobyl-contaminated forests and risks to the population and the environment: a new nuclear disaster about to happen? Environ Int 73:346–358

Fang C, Achal V (2019) The potential of microbial fuel cells for remediation of heavy metals from soil and water—review of application. Microorganisms 7:697. https://doi.org/10.3390/microorganisms7120697

FAO and ITPS (2015) Status of the world's soil resources (SWSR) – main report. Food and Agriculture Organization of the United Nations and Intergovernmental Technical Panel on Soils, Rome. Available at http://www.fao.org/3/a-i5199e.pdf

FAO and ITPS (2017) Global assessment of the impact of plant protection products on soil functions and soil ecosystems. Food and Agriculture Organization of the United Nations, Rome. 40 pp

FAO and UNEP (2021) Global assessment of soil pollution: report. FAO and UNEP, Rome. https://doi.org/10.4060/cb4894en

Faridullah F, Umar M, Alam A, Sabir MA, Khan D (2017) Assessment of heavy metals concentration in phosphate rock deposits, hazara basin, lesser himalaya Pakistan. Geosci J 21:743–752. https://doi.org/10.1007/s12303-017-0013-9

García-Mercadoa HD, Fernándezb G, Garzón-Zúñigac MA, Durán-Domínguez-de-Bazúaa MDC (2017) Remediation of mercury-polluted soils using artificial wetlands. Int J Phytoremediation 19:3–13. https://doi.org/10.1080/15226514.2016.1216074

García Frutos FJ, Escolano O, García S, Babín M, Fernández MD (2010) Bioventing remediation and ecotoxicity evaluation of phenanthrene-contaminated soil. J Hazard Mater 183(1–3):806–813. https://doi.org/10.1016/j.jhazmat.2010.07.098

Goodarzi AR, Zandi MH (2016) Assessing geo-mechanical and leaching behavior of cement–silica-fume-stabilized heavy metalcontaminated clayey soil. Environ Earth Sci 75:52. https://doi.org/10.1007/s12665-016-5730-2

Gratchev I, Towhata I (2011) Compressibility of natural soils subjected to long-term acidic contamination. Environ Earth Sci 64(1):193–200

Guo X, Zhang G, Wei Z, Zhang L, He Q, Wu Q, Qian T (2018) Mixed chelators of EDTA, GLDA, and citric acid as washing agent effectively remove Cd, Zn, Pb, and Cu from soils. J Soils Sediments 18:835–844. https://doi.org/10.1007/s11368-017-1781-6

Gupta DK, Chatterjee S, Datta S, Veer V, Walther C (2014) Role of phosphate fertilizers in heavy metal uptake and detoxification of toxic metals. Chemosphere 108:134–144. https://doi.org/10.1016/j.chemosphere.2014.01.03010

Habibul N, Hu Y, Sheng GP (2016) Microbial fuel cell driving electrokinetic remediation of toxic metal contaminated soils. J Hazard Mater 318:9–14. https://doi.org/10.1016/j.jhazmat.2016.06.041

Hasanuzzaman M, Prasad MNV (2020) Handbook of bioremediation. Academic Press. https://doi.org/10.1016/C2018-0-05109-9. ISBN 978-0-12-819382-2

Hou D (2020) Sustainable remediation of contaminated soil and groundwater: materials, processes, and assessment. Elsevier, Oxford. https://doi.org/10.1016/C2018-0-03003-0

Hou D, Gu Q, Ma F, O'Connell S (2016) Life cycle assessment comparison of thermal desorption and stabilization/solidification of mercury contaminated soil on agricultural land. J Clean Prod 139:949–956. https://doi.org/10.1016/j.jclepro.2016.08.108

Houben D, Couder E, Sonnet P (2013) Leachability of cadmium, lead, and zinc in a long-term spontaneously revegetated slag heap: implications for phytostabilization. J Soils Sediments 13:543–554. https://doi.org/10.1007/s11368-012-0546-5

Hu Z, Chu Y, Ma Y (2020) Design of a combined constructed wetland system and its application on swine wastewater treatment. J Environ Eng 146:1602. https://doi.org/10.1061/(ASCE)EE.1943-7870.0001602

Huang JW, Chen J (2003) Role of pH in phytoremediation of contaminated soils. In: Handbook of soil acidity. Marcel Dekker, New York, pp 449–472

Huang YZ, Hao XW (2012) Effect of red mud addition on the fractionation and bio-accessibility of Pb, Zn and As in combined contaminated soil. Chem Ecol 28:37–48

Indu S (2020) Bioremediation techniques for polluted environment: concept, advantages, limitations, and prospects. In: Murillo-Tovar MA, Saldarriaga-Noreña H, Saeid A (eds) Trace metals in the environment. IntechOpen, Rijeka

Jia H, Hou D, O'Connor D, Pan S, Zhu J, Bolan NS, Mulder J (2019) Exogenous phosphorus treatment facilitates chelationmediated cadmium detoxification in perennial ryegrass (Lolium perenne L.). J Hazard Mater 389:121849

Jorgensen KS (2007) In situ bioremediation. Adv Appl Microbiol 61:285–305. https://doi.org/10.1016/S0065-2164(06)61008-3. ISBN 978-0-12-002663-0

Kang CH, Kwon YJ, So JS (2016) Bioremediation of heavy metals by using bacterial mixtures. Ecol Eng 89:64–69. https://doi.org/10.1016/j.ecoleng.2016.01.023

Kapahi M, Sachdeva S (2019) Bioremediation options for heavy metal pollution. J Health Pollut 9(24):191203. https://doi.org/10.5696/2156-9614-9.24.191203

Khan MN, Mobin M, Abbas ZK, Alamri SA (2018) Fertilizers and their contaminants in soils, surface and groundwater. In: Encyclopedia of the anthropocene. Elsevier, Oxford, pp 225–240

Kim EJ, Choi SD, Chang YS (2011) Levels and patterns of polycyclic aromatic hydrocarbons (PAHs) in soils after forest fires in South Korea. Environ Sci Pollut Res 18(9):1508

Komárek M, Vaněk A, Ettler V (2013) Chemical stabilization of metals and arsenic in contaminated soils using oxides – a review. Environ Pollut 172:9–22. https://doi.org/10.1016/j.envpol.2012.07.045

Kulikowska D, Gusiatin ZM, Bulkowska K, Kierklo K (2015) Humic substances from sewage sludge compost as washing agent effectively remove Cu and Cd from soil. Chemosphere 136:42–49

Labidi S, Firmin S, Verdin A, Bidar G, Laruelle F, Douay F, Shirali P, Fontaine J, Lounès-Hadj Sahraoui A (2017) Nature of fly ash amendments differently influences oxidative stress alleviation in four forest tree species and metal trace element phytostabilization in aged contaminated soil: a long-term field experiment. Ecotoxicol Environ Saf 138:190–198

Lee RJ, Strohmeier BR, Bunker KL, Van Orden DR (2008) Naturally occurring asbestos—a recurring public policy challenge. J Hazard Mater 153(1–2):1–21

Lee DS, Lim SS, Park HJ, Yang HI, Park SI, Kwak JH, Choi WJ (2019) Fly ash and zeolite decrease metal uptake but do not improve rice growth in paddy soils contaminated with Cu and Zn. Environ Int 129:551–564

Leelarungroj K, Likitlersuang S, Chompoorat T, Janjaroen D (2018) Leaching mechanisms of heavy metals from fly ash stabilised soils. Waste Manag Res 36:616–623

Leon-Romero MA, Soto-Ríos PC, Nomura M, Nishimura O (2018) Effect of steel slag to improve soil quality of tsunami-impacted land while reducing the risk of heavy metal bioaccumulation. Water Air Soil Pollut 229:3666

Li D, Li R, Ding Z, Ruan X, Luo J, Chen J, Zheng J, Tang J (2020) Discovery of a novel native bacterium of Providencia sp. with high biosorption and oxidation ability of manganese for bioleaching of heavy metal contaminated soils. Chemosphere 241:125039

Li X, Ren Q, Liao C, Wang Q, Ren M, Zhang M, Qian X et al (2023a) Plant-microbe interactions in wheat to deal with abiotic stress. In: Khan MK, Pandey A, Hamurcu M, Gupta OP, Gezgin S (eds) Abiotic stresses in wheat. Academic Press, pp 375–391

Li C, Cui C, Zhang J, Shen J, He B, Long Y, Ye J (2023b) Biodegradation of petroleum hydrocarbons based pollutants in contaminated soil by exogenous effective microorganisms and indigenous microbiome. Ecotoxicol Environ Saf 253:114673

Liang XF, Li N, He LZ, Xu YM, Huang QQ, Xie ZL, Yang F (2019) Inhibition of Cd accumulation in winter wheat (Triticum aestivum L.) grown in alkaline soil using mercaptomodified attapulgite. Sci Total Environ 688:818–826

Lin J, He F, Su B, Sun M, Owens G, Chen Z (2019) The stabilizing mechanism of cadmium in contaminated soil using green synthesized iron oxide nanoparticles under long-term incubation. J Hazard Mater 379:120832

Liu S, Qi X, Han C, Liu J, Sheng X, Li H, Luo A, Li J (2017) Novel nano-submicron mineral-based soil conditioner for sustainable agricultural development. J Clean Prod 149:896–903

López Vizcaíno R, Yustres A, Asensio L, Saez C, Cañizares P, Rodrigo MA, Navarro V (2018) Enhanced electrokinetic remediation of polluted soils by anolyte pH conditioning. Chemosphere 199:477–485. https://doi.org/10.1016/j.chemosphere.2018.02.038

Lopez S, Benizri E, Erskine PD, Cazes Y, Morel JL, Lee G, Permana E, Echevarria G, van der Ent A (2019) Biogeochemistry of the flora of Weda Bay, Halmahera Island (Indonesia) focusing on nickel hyperaccumulation. J Geochem Explor 202:113–127

Lu G, Wang B, Zhang C, Li S, Wen J, Lu G, Zhu C, Zhou Y (2018) Heavy metals contamination and accumulation in submerged macrophytes in an urban river in China. Int J Phytoremediation 20:839–846

Mahbub KR, Krishnan K, Megharaj M, Naidu R (2016) Bioremediation potential of a highly mercury resistant bacterial strain Sphingobium SA2 isolated from contaminated soil. Chemosphere 144:330–337

Mahugija J (2013) Status and distributions of pesticides buried at five sites in arusha and mbeya regions, Tanzania. Afr J Pure Appl Chem 7(12):382–393. https://doi.org/10.5897/AJPAC2013.0531

Marchand L, Mench M, Jacob DL, Otte ML (2010) Metal and metalloid removal in constructed wetlands, with emphasis on the importance of plants and standardized measurements: a review. Environ Pollut 158:3447–3461. https://doi.org/10.1016/j.envpol.2010.08.018

Meng F, Yuan G, Wei J, Bi D, Ok YS, Wang H (2017) Humic substances as a washing agent for Cd-contaminated soils. Chemosphere 181:461–467

Mercade ME, Monleon L, De Andres C, Rodon I, Martinez E, Espuny MJ, Manresa A (1996) Screening and selection of surfactant-producing bacteria from waste lubricating soil. J Appl Bacteriol 81:161–166

Messner P, Schäffer C, Kosma P (2013) Bacterial cell-envelope glycoconjugates Elsevier 209–272

Methneni N, Morales-Gonzalez JA, Jaziri A, Mansour HB, Fernandez-Serrano M (2021) Persistent organic and inorganic pollutants in the effluents from the textile dyeing industries: ecotoxicology appraisal via a battery of biotests. Environ Res 196:110956. https://doi.org/10.1016/j.envres.2021.110956

Michelon W, Pirolli M, Soares HM, Michelon W, Mezzari MP, Busi da Silva ML (2019) Residual sugar from microalgae biomass harvested from phycoremediation of swine wastewater digestate. Water Sci Technol 79:2203–2210

Mishra R, Mohammad N, Roychudhury NVS (2016) Soil pollution: causes, effects and control. Van Sangyan 3:1–14

Mueller E (2018) The effect of acid rain on soil nutrient levels and plant growth. American Geophysical Union. 2018AGUFMED41D1217M

Muhlbachova G, Sagova M, Omelka M (2015) The influence of soil organic carbon on interactions between microbial parameters and metal concentrations at a long term contaminated site. Sci Total Environ 502:218–223

Mulligan CN (2005) Environmental applications for biosurfactants. Environ Pollut 133(2):183–198. https://doi.org/10.1016/j.envpol.2004.06.009
Naguib MM, El-Gendy AO, Khairalla AS (2018) Microbial diversity of mer operon genes and their potential rules in mercury bioremediation and resistance. Open Biotechnol J 12:56–77. https://doi.org/10.2174/1874070701812010056
Nedelkoska TV, Doran PM (2000) Characteristics of heavy metal uptake by plant species with potential for phytoremediation and phytomining. Miner Eng 13:549–561
Ning D, Liang Y, Song A, Duan A, Liu Z (2016) In situ stabilization of heavy metals in multiple-metal contaminated paddy soil using different steel slag-based silicon fertilizer. Environ Sci Pollut Res 23:23638–23647
Novo LAB, Castro PML, Alvarenga P, da Silva EF (2017) Phytomining of rare and valuable metals. In: Phytoremediation: management of environmental contaminants. Springer, Cham, pp 469–486
Nwachukwu BC, Ayangbenro AS, Babalola OO (2021) Elucidating the rhizosphere associated bacteria for environmental sustainability. Agriculture 11:75
O'Connor D, Zheng X, Hou D, Shen Z, Li G, Miao G, O'Connell S, Guo M (2019) Phytoremediation: climate change resilience and sustainability assessment at a coastal brownfield redevelopment. Environ Int 130:104945
Pacwa-Plociniczak M, Płaza GA, Piotrowska-Seget Z, Cameotra SS (2012) Environmental applications of biosurfactants: recent advances. Int J Mol Sci 12:633–654
Palansooriya KN, Yang Y, Tsang YF, Sarkar B, Hou D, Cao XD, Meers E, Rinklebe J, Kim KH, Ok YS (2019) Occurrence of contaminants in drinking water sources and the potential of biochar for water quality improvement: a review. Crit Rev Environ Sci Technol 63:1629803
Pande V, Pandey SC, Sati D, Bhatt P, Samant M (2022) Microbial interventions in bioremediation of heavy metal contaminants in agroecosystem. Front Microbiol 13:824084
Petrus AK, Rutner C, Liu S, Wang Y, Wiatrowski HA (2015) Mercury reduction and methyl mercury degradation by the soil bacterium Xanthobacter autotrophicus Py2. Appl Environ Microbiol 81:7833–7838. https://doi.org/10.1128/AEM.01982-15
Radziemska M, Bęś A, Gusiatin ZM, Cerdà A, Jeznach J, Mazur Z, Brtnický M (2020) Assisted phytostabilization of soil from a former military area with mineral amendments. Ecotoxicol Environ Saf 188:109934
Rahman KMA, Zhang D (2018) Effects of fertilizer broadcasting on the excessive use of inorganic fertilizers and environmental sustainability. Sustainability 10(3):759
Rawat G, Dhasmana A, Kumar V (2020) Biosurfactants: the next generation biomolecules for diverse applications Environmental Sustainability 3(4):353–369. https://doi.org/10.1007/s42398-020-00128-8
Robinson BH, Brooks RR, Clothier BE (1999) Soil amendments affecting nickel and cobalt uptake by Berkheya coddii: potential use for phytomining and phytoremediation. Ann Bot 84:689–694
Robl T, Oberlink A, Jones R (2017) Coal combustion products (CCPs): characteristics, utilization and beneficiation. Elsevier, Duxford
Rodríguez Eugenio N, McLaughlin NJ, Pennock D (2018) Soil pollution: a hidden reality. Food and Agriculture Organization of the United Nations, Rome. 156 pp
Ron EZ, Rosenberg E (2022) Biosurfactants and oil bioremediation. Curr Opin Biotechnol 13:249–252
Rosenkranz T, Kisser J, Wenzel WW, Puschenreiter M (2017) Waste or substrate for metal hyperaccumulating plants—the potential of phytomining on waste incineration bottom ash. Sci Total Environ 575:910–918
Saadaoui E, Ghazel N, Romdhane CB, Massoudi N (2017) Phosphogypsum: potential uses and problems – a review. Int J Environ Stud 74(4):558–567
Shayler H, Mcbride M, Harrison E (2009) Sources and impacts of contaminants in soils. Cornell Waste Management Institute, Cornell University
Shen G, Zhang S, Liu X, Jiang Q, Ding W (2018) Soil acidification amendments change the rhizosphere bacterial community of tobacco in a bacterial wilt affected field. Appl Microbiol Biotechnol 102(22):9781–9791

Shen Z, Pan S, Hou D, O'Connor D, Jin F, Mo L, Xu D, Zhang Z, Alessi DS (2019a) Temporal effect of MgO reactivity on the stabilization of lead contaminated soil. Environ Int 131:104990
Shen ZT, Jin F, O'Connor D, Hou DY (2019b) Solidification/stabilization for soil remediation: an old technology with new vitality. Environ Sci Technol 53:11615–11617. https://doi.org/10.1021/acs.est.9b04990
Sheoran V, Sheoran AS, Poonia P (2009) Phytomining: a review. Miner Eng 22:1007–1019
Shi WY, Shao HB, Li H, Shao MA, Du S (2009) Progress in the remediation of hazardous heavy metal-polluted soils by natural zeolite. J Hazard Mater 170:1–6
Shulga A, Karpenko E, Vildanova-Martsishin R, Turovsky A, Soltys M (1999) Biosurfactant enhanced remediation of oil-contaminated environments. Adsorpt Sci Technol 18(2):171–176. https://doi.org/10.1260/0263617001493369
Singh A, Agrawal S (2008) Acid rain and its ecological consequences. J Environ Biol 29(1):15–24
Singh AK, Cameotra SS (2013) Efficiency of lipopeptide biosurfactants in removal of petroleum hydrocarbons and heavy metals from contaminated soil. Environ Sci Pollut Res 20:7367–7376. https://doi.org/10.1007/s11356-013-1752-4
Smith DR, Harmel RD, Williams M, Haney R, King KW (2016) Managing acute phosphorus loss with fertilizer source and placement: proof of concept. Agric Environ Lett 1(1):150015
Sochacki A, Surmacz-Górska J, Faure O, Guy B (2014) Polishing of synthetic electroplating wastewater in microcosm upflow constructed wetlands: metals removal mechanisms. Chem Eng J 242:43–52. https://doi.org/10.1016/j.cej.2013.12.075
Song B. et al (2017) Evaluation methods for assessing effectiveness of in situ remediation of soil and sediment contaminated with organic pollutants and heavy metals Environment International 10543-55. https://doi.org/10.1016/j.envint.2017.05.001
Song TS, Zhang J, Hou S, Wang H, Zhang D, Li S, Xie J (2018) In situ electrokinetic remediation of toxic metalcontaminated soil driven by solid phase microbial fuel cells with a wheat straw addition. J Chem Technol Biotechnol 93:2860–2867. https://doi.org/10.1002/jctb.5638
Srikanth M, Sandeep TSRS, Sucharitha K, Godi S (2022) Biodegradation of plastic polymers by fungi: a brief review. Bioresour Bioprocess 9:42
Su M, Shih K (2017) Utilisation of incinerated sewage sludge ash asa matrix for cadmium stabilisation. HKIE Trans 24:35–41. https://doi.org/10.1080/1023697X.2016.1257959
Su B, Lin J, Owens G, Chen Z (2020) Impact of green synthesized iron oxide nanoparticles on the distribution and transformation of As species in contaminated soil. Environ Pollut 258:113668
Syakila A, Kroeze C (2011) The global nitrous oxide budget revisited. GHG Measure Manage 1(1):17–26. https://doi.org/10.3763/ghgmm.2010.0007
Tsadilas C, Rinklebe J, Selim M (2018) Nickel in soils and plants. CRC Press
Tahervand S, Jalali M (2017) Sorption and desorption of potentially toxic metals (Cd, Cu, Ni and Zn) by soil amended with bentonite, calcite and zeolite as a function of pH. J Geochem Explor 181:148–159
Taneez M, Hurel C (2019) A review on the potential uses of red mud as amendment for pollution control in environmental media. Environ Sci Pollut Res 26:22106–22125
Tao H, Haynes RJ (2016) Constructed wetlands: fundamental processes and mechanisms for heavy metal removal from wastewater streams. Int J Environ Eng 8:148–178
Tareen A, Saeed S, Iqbal A, Batool R, Jamil N (2022) Biodeterioration of microplastics: a promising step towards plastics waste management. Polymers 14:2275
Tognacchini A, Rosenkranz T, van der Ent A, Machinet GE, Echevarria G, Puschenreiter M (2020) Nickel phytomining from industrial wastes: growing nickel hyperaccumulator plants on galvanic sludges. J Environ Manag 254:109798
Tran HN, You SJ, Hosseini-Bandegharaei A, Chao HP (2017) Mistakes and inconsistencies regarding adsorption of contaminants from aqueous solutions: a critical review. Water Res 120(88):116. https://doi.org/10.1016/j.watres.2017.04.014
US EPA (2017a) Basics of green chemistry. Retrieved from https://www.epa.gov/greenchemistry/basics-green-chemistry
US EPA (2017b) EPA-542-R-17-001: superfund remedy report, 15th edn. US EPA

Usman S, Kundiri AM, Nzamouhe M (2017) Effects of organophosphate herbicides on biological organisms in soil medium: a mini review. J Ecol Toxicol 1:102
Vocciante M, Caretta A, Bua L, Bagatin R, Franchi E, Petruzzelli G, Ferro S (2019) Enhancements in phytoremediation technology: environmental assessment including different options of biomass disposal and comparison with a consolidated approach. J Environ Manag 237:560–568
Walaszek M, Bois P, Laurent J, Lenormand E, Wanko A (2018a) Micropollutants removal and storage efficiencies in urban stormwater constructed wetland. Sci Total Environ 645:854–864. https://doi.org/10.1016/j.scitotenv.2018.07.156
Walaszek M, Del Nero M, Bois P, Ribstein L, Courson O, Wanko A, Laurent J (2018b) Sorption behavior of copper, lead and zinc by a constructed wetland treating urban stormwater. Appl Geochem 97:167–180
Wang C, Deng H, Zhao F (2016) The remediation of chromium (VI)-contaminated soils using microbial fuel cells. Soil Sediment Contam 25:1–12. https://doi.org/10.1080/15320383.2016.1085833
Wang F, Shen Z, Al-Tabbaa A (2018a) An evaluation of stabilised/solidified contaminated model soil using PC-based and MgO-based binders under semi-dynamic leaching conditions. Environ Sci Pollut Res 25:16050–16060. https://doi.org/10.1007/s11356-018-1591-4
Wang F, Shen Z, Tabbaa AAL (2018b) PC-based and MgO-based binders stabilised/solidified heavy metal-contaminated model soil: strength and heavy metal speciation in early stage. Geotechnique 68:1025–1030. https://doi.org/10.1680/jgeot.17.P.194
Wang Y, Li F, Song J, Xiao R, Luo L, Yang Z, Chai L (2018c) Stabilization of Cd-, Pb-, Cu- and Zn-contaminated calcareous agricultural soil using red mud: a field experiment. Environ Geochem Health 40:2143–2153
Wang J, Shaheen SM, Swertz AC, Rennert T, Feng X, Rinklebe J (2019a) Sulfur-modified organoclay promotes plant uptake and affects geochemical fractionation of mercury in a polluted floodplain soil. J Hazard Mater 371:687–693
Wang L, Chen L, Cho DW, Tsang DCW, Yang J, Hou D, Baek K, Kua HW, Poon CS (2019b) Novel synergy of Si-rich minerals and reactive MgO for stabilisation/solidification ofcontaminated sediment. J Hazard Mater 365:695–706. https://doi.org/10.1016/j.jhazmat.2018.11.067
Wang H, Long X, Zhang J, Cao X, Liu S, Li X (2020a) Relationship between bioelectrochemical copper migration, reduction and electricity in a three-chamber microbial fuel cell. Chemosphere 241:125097. https://doi.org/10.1016/j.chemosphere.2019.125097
Wang L, Hou D, Cao Y, Ok YS, Tack FMG, Rinklebe J, O'Connor D (2020b) Remediation of mercury contaminated soil, water, and air: a review of emerging materials and innovative technologies. Environ Int 134:105281. https://doi.org/10.1016/j.envint.2019.105281
Wang L, Hou D, Shen Z, Zhu J, Jia X, Ok YS, Tack FMG, Rinklebe J (2020c) Field trials of phytomining and phytoremediation: a critical review of influencing factors and effects of additives. Crit Rev Environ Sci Technol 50:2724–2774
Wang L, Li X, Tsang DCW, Jin F, Hou D (2020d) Green remediation of Cd and Hg contaminated soil using humic acid modified montmorillonite: immobilization performance under accelerated ageing conditions. J Hazard Mater 387:122005. https://doi.org/10.1016/j.jhazmat.2019.122005
Wang L, O'Connor D, Rinklebe J, Ok YS, Tsang DCW, Shen Z, Hou D (2020e) Biochar aging: mechanisms, physicochemical changes, assessment, and implications for field applications. Environ Sci Technol 54:14797–14814. https://doi.org/10.1021/acs.est.0c04033
Watmough SA, Dillon PJ (2003) Base cation and nitrogen budgets for a mixed hardwood catchment's in south-central Ontario. Ecosystems 6:675–693
Weithmann N, Möller JN, Löder MGJ, Piehl S, Laforsch C, Freitag R (2018) Organic fertilizer as a vehicle for the entry of microplastic into the environment. Sci Adv 4(4):eaap8060
WHO (2009) WHO handbook on indoor radon: a public health perspective. World Health Organization, Geneva. 94 pp
Wu et al. (2011) WebMGA: a customizable web server for fast metagenomic sequence analysis Abstract BMC Genomics 12(1) https://doi.org/10.1186/1471-2164-12-444

Xiong X, Yanxia L, Wei L, Chunye L, Wei H, Ming Y (2010) Copper content in animal manures and potential risk of soil copper pollution with animal manure use in agriculture. Resour Conserv Recycl 54(11):985–990

Xu Y, Liang X, Xu Y, Qin X, Huang Q, Wang L, Sun Y (2017) Remediation of heavy metal-polluted agricultural soils using clay minerals: a review. Pedosphere 27:193–204

Xu X, Liu W, Tian S, Wang W, Qi Q, Jiang P, Gao X, Li F, Li H, Hongwen Y (2018) Petroleum hydrocarbon-degrading bacteria for the remediation of oil pollution under aerobic conditions: a perspective analysis. Front Microbiol 9:2885

Yang Z, Wu Z, Liao Y, Liao Q, Yang W, Chai L (2017) Combination of microbial oxidation and biogenic schwertmannite immobilization: a potential remediation for highly arseniccontaminated soil. Chemosphere 181:1–8

Ye X, Kang S, Wang H, Li H, Zhang Y, Wang G, Zhao H (2015) Modified natural diatomite and its enhanced immobilization of lead, copper and cadmium in simulated contaminated soils. J Hazard Mater 289:210–218

Yuvraj (2022) Microalgal bioremediation: a clean and sustainable approach for controlling environmental pollution. In: Innovations in environmental biotechnology, vol 1. Springer Singapore, Singapore, pp 305–318. https://doi.org/10.1007/978-981-16-4445-0_13

Zha F, Ji C, Xu L, Kang B, Yang C, Chu C (2019) Assessment of strength and leaching characteristics of heavy metal–contaminated soils solidified/stabilized by cement/fly ash. Environ Sci Pollut Res 26:30206–30219

Zhang X, Houzelot V, Bani A, Morel JL, Echevarria G, Simonnot MO (2014) Selection and combustion of Ni-hyperaccumulators for the phytomining process. Int J Phytoremediation 16:1058–1072

Zhang Q, Hu J, Lee DJ, Chang Y, Lee YJ (2017) Sludge treatment: current research trends. Bioresour Technol 243:1159–1172. https://doi.org/10.1016/j.biortech.2017.07.070

Zhang T, Hu L, Zhang M, Jiang M, Fiedler H, Bai W, Wang X, Zhang D, Li Z (2019) Cr(VI) removal from soils and groundwater using an integrated adsorption and microbial fuel cell (A-MFC) technology. Environ Pollut 252:1399–1405. https://doi.org/10.1016/j.envpol.2019.06.051

Zhang J, Cao X, Wang H, Long X, Li X (2020) Simultaneous enhancement of heavy metal removal and electricity generation in soil microbial fuel cell. Ecotoxicol Environ Saf 192:110314. https://doi.org/10.1016/j.ecoenv.2020.110314

Zheng C, Wang X, Liu J, Ji X, Huang B (2019) Biocharassisted phytoextraction of arsenic in soil using Pteris vittata L. Environ Sci Pollut Res 26:36688–36697

Zia-ur-Rehman M, Khalid H, Rizwan M, Ali S, Sohail MI, Usman M, Umair M (2019) Chapter 4 – Inorganic amendments for the remediation of cadmium-contaminated soils. In: Hasanuzzaman M, Vara Prasad MN, Nahar K (eds) Cadmium tolerance in plants. Academic Press, pp 113–141

Integrating Mine Rehabilitation Practices in the Context of Climate Change: A Sustainable Approach

Arby Kane

Abstract The global mining industry plays a significant role in economic development and environmental impact. As concerns about climate change intensify, there is a growing imperative to align mining practices, particularly mine rehabilitation, with sustainable and climate-resilient strategies. This chapter explores the intersection of mine rehabilitation and climate change, giving examples of some artisanal mines and industrial mines policies to emphasize the need for integrated approaches that mitigate environmental impact and contribute to climate adaptation.

Keywords Climate change · Mine industry · Mine rehabilitation · ASGM

1 Introduction

The mining industry has a long history of significant environmental impact, with activities ranging from exploration and extraction to processing and waste disposal. The environmental consequences of mining can be profound and often include deforestation, soil erosion, contamination of water sources, air pollution, and habitat destruction.

The mining industry has played a significant role in the economic development of many African countries, but it has also left a notable environmental footprint. The historical environmental impact of mining in Africa has been characterized by a range of challenges and consequences.

Mining can have numerous, primarily negative, impacts on the surrounding environment, ranging from the *exacerbation of climate change* to the *reduction of global and regional levels of biodiversity*.

A. Kane (✉)
United Nations Institute for Training and Research (UNITAR), Nouakchott, Mauritania

Council of African Youth in Minerals (CAYM), Nouakchott, Mauritania

A. A. H. Abdel Latef et al. (eds.), *Sustainable Remediation for Pollution and Climate Resilience*, https://doi.org/10.1007/978-981-96-5674-5_4

Taking examples of different zones in Africa where established mines have been established for decades, we know there is limited biodiversity, causing migration.

Significant examples include climate change, deforestation/habitat destruction, pollution, soil erosion, human-wildlife conflict, and the loss of biodiversity caused by mining and expanding explorations and exploitations of natural resources.

Climate change has emerged as a critical factor influencing global policy and resource management due to its far-reaching and potentially catastrophic impacts on the environment, economies, and societies. Recognizing climate change as a pressing global issue has led to a paradigm shift in how nations, organizations, and communities approach policymaking and resource allocation. Several key factors contribute to the prominence of climate change in global policy discussions and resource management. All these factors play a crucial role, and they are linked; managing sustainably can be done to avoid a substantial negative impact on climate change effect.

The emergence of climate change as a critical element in global policy and natural resource management reflects a growing understanding of its multifaceted impacts and the need for concerted, coordinated efforts to address this global challenge. As the urgency of climate action continues to escalate, integrating climate considerations into policy decisions and resource management practices becomes increasingly essential for building a sustainable and resilient future.

With integrating and combining small-scale mining and industrial, their impact on climate change is growing, but we can implement mitigation and adaptation strategies, and cooperate on technology transfer, avoid pollution, and management of waste and tailings.

In this chapter we will discuss the impact on climate, mine rehabilitation has its impact on climate change, different strategies on good practices and technological innovations for mine closure, starting by artisanal mines to industrials, community engagement, and how to finance the mine rehabilitation around the world, we will discuss different cases and challenges.

2 Arby Kane

Mr. Arby Kane holds a Master's degree in Natural Resources and Environmental Geology from the University of Nouakchott. He is fluent in Arabic, French, and English and has a diverse skill set encompassing administration, Environmental, Social, and Governance (ESG), and management. Notably, he has taken on pivotal roles, serving as both President and Founder of the Council Of African Youth In Minerals (CYAM) (www.caym.africa) and President of the Council of Young Mauritanian Mining Leaders (www.ymlcinmauritania.com). Through these roles, he has been advocating for compliance with Health, Safety, and Environment (HSE) standards, addressing gender diversity within the mining sector, and promoting the seamless integration of young professionals into the industry.

Actually working as a junior mining expert at Projekt-Consult GmbH since November 2024, he contributes to the development of a climate-smart mining

Fig. 1 Arby Kane during symposium Mines Mauritania 2023 (2022–2024)

roadmap for Mauritania, funded by the World Bank Group. He provides technical and local input on the industrial, artisanal, and small-scale mining sectors, as well as data analysis, coordination, translation, and communication services. He also reviews reports and strategic recommendations and networks with high-level stakeholders from the government, private sector, and civil society.

Presently, he is at the forefront of orchestrating the Symposium Mines Mauritania 2023, alongside a dedicated team from the Young Mining Leaders Council. This symposium serves as a platform for constructive dialogue and the exploration of innovative solutions within the Mauritanian mining sector. Mr. Kane's ability to spearhead such endeavors underscores his commitment to catalyzing progressive transformations within the industry. As a junior mining expert, he provides exceptional local context and showcases proactive engagement within the industry exemplifying his drive for impactful change and advancing the sustainable growth of the mining sector in Mauritania (Fig. 1).

3 Mauritania Status

It is important for all companies and industries operating in this sector in Mauritania and the world, in general, to understand the social and environmental impact of their actions and act accordingly. He explains that one of the biggest mistakes in the context of ESG approaches (social environment and good governance, a concept linked to corporate social responsibility), is to forget "to invest in young Africans, who are highly competent in this area."

Insight into the African mining sector in "Mauritania is a country very rich in natural resources", in fact, the country has significant deposits (gold, copper, iron ore) and oil reserves that have recently been discovered. The extractive sector's contribution to Mauritania's GDP in 2021 amounted to 21.8%, and represented approximately 76.9 billion international units in 2021, i.e., more than 2 billion euros, according to the Ministry of Finance. These diverse resources should benefit local youth and entrepreneurs in Mauritania.

4 Mine Rehabilitation Practices and Climate Change Impacts on Mining

Examination of traditional mine rehabilitation practices.

Traditional mine rehabilitation practices refer to the methods and strategies employed to restore a mined area's ecological and environmental condition after mining activities have ceased. The goal is to mitigate the negative impacts of mining on the landscape, biodiversity, and water resources and to promote the long-term sustainability of the ecosystem. Examining traditional mine rehabilitation practices involves assessing the various techniques and approaches commonly used in the industry.

In Africa, we face a challenge in the closure of artisanal mines; analyze the effectiveness of environmental impact assessments conducted before mining operations begin. Assess whether these assessments accurately predict the environmental impact of mining activities and if rehabilitation plans are based on thorough assessments.

Therefore, there is a different aspect to managing and adapting to artisanal systems to rehabilitate, which requires technology, sustainability, and long-term assistance to artisanal miners (Fig. 2).

Overall, examining traditional mine rehabilitation practices in Africa should consider a holistic approach that integrates environmental, social, and economic factors to ensure the long-term sustainability of rehabilitated areas. Collaboration between government agencies, mining companies, local communities, and environmental experts is crucial for developing and implementing effective rehabilitation strategies.

5 Mine Rehabilitation Impact on Deforestation

Artisanal and small-scale gold mining (ASGM) is a significant contributor to mercury (Hg) contamination and deforestation across the globe (Fig. 3). In the Colorado River watershed in Madre de Dios, Peru, mining and deforestation have increased exponentially since the 1980s, resulting in major socioeconomic shifts in the region and a national state of emergency in response to concerns for wide-scale mercury

Fig. 2 Artisanal pit non-rehabilate, Tasiast-Mauritania

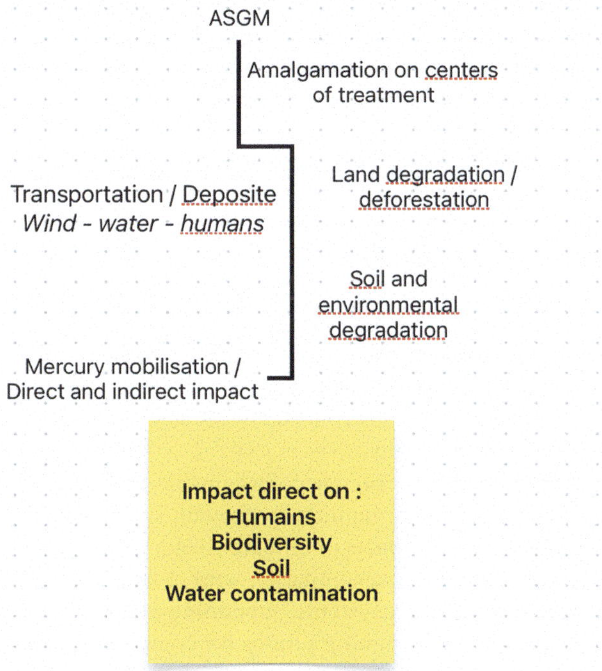

Fig. 3 ASGM impact on the environment. (Illustration done by Arby Kane) (UNEP)

poisoning by these activities. This research employed a watershed-scale soil particle detachment model and environmental field sampling to estimate the role of land cover change and soil erosion on river transport of Hg in a heavily ASGM-impacted watershed. The model estimated that observed decreases in forest cover increased soil mobilization by a factor of two in the Colorado River watershed during the 18-year period and by fourfold in the Puquiri subwatershed (the area of most concentrated ASGM activity). If deforestation continues at its current exponential rate through 2030, the annual mobilization of soil and Hg may increase by an additional 20–25% relative to 2014 levels. While the estimated total mass of Hg transported by rivers is substantially less than the estimated tons of Hg used with ASGM in Peru, this research shows that deforestation associated with ASGM is an additional mechanism for mobilizing naturally occurring and anthropogenic Hg from terrestrial landscapes to aquatic environments in the region, potentially leading to bioaccumulation in fish and exposure to communities downstream (Diringer et al. 2020).

6 Conclusion

Key Findings and Importance of Aligning Mine Rehabilitation with Climate Change Considerations

Environmental Restoration and Ecosystem Health: Effective mine rehabilitation restores ecosystems, promoting biodiversity and soil health. This is critical for mitigating climate impacts like desertification and habitat loss.

Carbon Sequestration: Rehabilitated mine lands can act as carbon sinks, particularly through reforestation and revegetation. This helps offset carbon emissions from mining operations.

Water Management: Rehabilitation plans must consider future water availability, given changing climate patterns. Properly restored sites can improve water retention and reduce the risk of floods or droughts.

Resilience to Climate Extremes: Incorporating climate change adaptation into mine rehabilitation ensures that restored lands are more resilient to extreme weather events, such as storms or rising temperatures.

Regulatory Compliance and Global Standards: Aligning rehabilitation with climate policies is becoming a regulatory requirement in many jurisdictions. Companies that integrate climate considerations into their plans show leadership and commitment to sustainability.

Community and Economic Benefits: Sustainable rehabilitation supports long-term livelihoods for local communities, improving social stability while reducing the adverse socioeconomic impacts of climate change.

In summary, aligning mine rehabilitation with climate change considerations is crucial for creating resilient ecosystems, supporting global carbon reduction goals, and ensuring sustainable post-mining futures for communities and industries alike. This alignment contributes to a future that is both environmentally and economically sustainable.

This chapter aims to contribute to the discourse on sustainable mining practices by exploring the intricate relationship between mine rehabilitation and climate change, ultimately advocating for a holistic and adaptive approach to minimize environmental impact and foster climate resilience.

References

Diringer SE, Berky AJ, Marani M, Ortiz EJ, Karatum O, Plata DL, Pan WK, Hsu-Kim H (2020) Deforestation due to artisanal and small-scale gold mining exacerbates soil and mercury mobilization in Madre de Dios, Peru. Environ Sci Technol 54(1):286–296. https://doi.org/10.1021/acs.est.9b06620

https://www.africabusinessplus.com/en/815329/mauritania-first-quantum-to-close-guelb-moghrein-in-2025/

https://www.pactworld.org/blog/supporting-responsible-gold-production-mauritania

Mine Remediation and Climate Changes

Habib Ur Rahman and Allah Ditta

Abstract Climate change and global warming are caused mainly by the buildup of greenhouse gases in the atmosphere, mostly due to human activities. Groundwater, surface water, wetlands, land, animals, and plants are all at serious risk of environmental pollution by treated and untreated mining wastes and tailings. Moreover, it is well known that substantial carbon dioxide (CO_2) emissions occur throughout the processes of ore extraction and beneficiation. However, the ability of mining waste and abandoned mines to collect carbon dioxide from the atmosphere raises the possibility that they might be a long-term solution to slow down global warming. Understanding the oxidation of sulfidic minerals, the hydrolysis of the oxidation products that lead to lower pH, SO_4^{2-}, and Fe, and hazardous metals and acid mine drainage (AMD), is necessary to facilitate remediation strategies. The effects of climate change on AMD development, water quality, and geochemical responses in semi-arid and arid mining areas are all covered in this chapter. This chapter highlights the impacts of seasonal, hydrological, and climatic changes on the composition of AMD, as well as the movement of pollutants across watersheds and the restoration of mining sites. The benefits and limits of various models vary, which affects how adaptable and appropriate these are for use in diverse mining contexts. This chapter has assessed the effects of climate change on sulfide oxidation, AMD formation from mining waste, and pollutant transmission in surface and groundwater systems by conducting a comparative review of some of the most promising and often used reactive transport models.

H. Ur Rahman
Department of Biotechnology, Shaheed Benazir Bhutto University Sheringal, Dir Upper, Khyber Pakhtunkhwa, Pakistan

A. Ditta (✉)
Department of Environmental Sciences, Shaheed Benazir Bhutto University Sheringal, Dir Upper, Khyber Pakhtunkhwa, Pakistan

School of Biological Sciences, The University of Western Australia, Perth, WA, Australia
e-mail: allah.ditta@sbbu.edu.pk

A. A. H. Abdel Latef et al. (eds.), *Sustainable Remediation for Pollution and Climate Resilience*, https://doi.org/10.1007/978-981-96-5674-5_5

Keywords Mining wastes · Tailings · Hydrolysis · Oxidation · Acid mine drainage

1 Introduction

There is ample evidence linking human activity to climate change (IPCC 2013). The future is expected to bring changes to ecosystems, manufacturing activity, and society due to increased temperatures, altered seasonal patterns, changes in precipitation, and modifications to the frequency and severity of extreme weather occurrences. It is acknowledged that anticipated future climate change will need some degree of adaptation. Recognizing this, studies have expanded to include the human aspects of changes rather than only concentrating on its origins and nature (Ford et al. 2010a). Because shifting environmental patterns may significantly impact sociocultural well-being and economic viability, it is crucial to identify and characterize impacts and vulnerabilities for specific industrial sectors in addition to the national and regional economy. Stakeholders, practitioners, and regulators may mitigate risk proactively by recognizing and anticipating weaknesses. However, much is still to be learned about how climate change may affect different industrial operations.

Primary economic activities, which underpin local economies, are particularly susceptible to the effects of climate change because of their direct reliance on the environment. Their potential susceptibility has significant regional ramifications, as many economic activities like mining, forestry, and fishing are often the main source of income for huge regions and are essential to the tertiary activities that occur both locally and globally.

The geochemical weathering and oxidation of sulfide minerals in mining waste and tailings pose a long-term environmental threat because these minerals exhibit a range of physical, hydrological, and chemical properties depending on their location and environmental setting and because they control the fate and transport of contaminants. Mining wastes create acid mine drainages (AMDs) and are dangerous for the environment, groundwater and surface waters, soil quality, and food chains because of their high sulfate and toxic metal content. Furthermore, they possess the capacity to eradicate the majority of aquatic and terrestrial species, fish, and plants. AMD generation from overburdened waste piles in open-pit lignite mines caused groundwater quality to decline to as low as 1, which also impacted surface water bodies and mine pits.

Precious natural resources include water; thus, protecting it from various pollution concerns is crucial for population health, economic development, and the sustainability of the environment. Mining activities, agriculture, and the way of life in semi-arid and arid areas will be impacted by erratic rainfall, a lack of water, an increase in the frequency of floods and droughts, and other worldwide climatic and environmental changes. Both mining waste disposal and climate change will endanger ecosystems that depend on surface and groundwater. Therefore, long-term and sustainable remediation and mitigation techniques are needed in the semi-arid and desert areas where mining is conducted to counteract the consequences of climate change and protect this vital water supply. Before implementing remediation strategies, it is essential to comprehend how climate change affects tailings dissolving, geochemical processes, AMD development, and pollutant migration in groundwater and surface water ecosystems.

Geochemical modeling is a practical and broadly applicable approach to examine the geochemical processes taking place in the AMD and the movement of contaminants in groundwater and surface water systems (Bertelli et al. 2009). Chemical speciation, ion exchange, precipitation-dissolution processes, acid-base interactions, and redox reactions are all considered in the geochemical model.

A conceptual model that is based on a range of geological and geochemical theories, data collection, and carefully controlled field observations is the first step in every geochemical modeling effort. The only things that can establish the validity of the model are the quantity and quality of the data and the coherence of the underlying assumptions. The nature of fluids interacting with rocks is a primary source of uncertainty in chemical models of geological processes. These days, the chemical system included in geochemical models may be limited using data from fluid inclusion analysis.

Predicting the consequences of mining on the environment and maintaining the site after accurate models of the dissolving of mine tailings and the development of waste in AMD can aid mining. Not many studies in the literature discuss how AMD is impacted by mining-related climate change. The natural processes that can alter the levels of acidity and metals in surface waters receiving AMD under the anticipated conditions of climatic change related to global warming (e.g., rainy events) were reviewed in a recent review by Nordstrom (2009) based on data from three mine sites in the United States. On the other hand, no research or analysis has been done on the impact of climate change on AMD development, the geochemical reactions in mining wastes, the water quality of streams affected by AMD, or the transit routes.

Few studies have specifically looked at how climate change may affect mining operations despite the significance of mining and its vulnerability to it (Ford and Pearce 2007; Ford et al. 2008). Whenever such effects have been mentioned, they have often done so subtly as a part of larger reporting on climate change (e.g., Instanes et al. 2005; Lemmen et al. 2008). Some severe climatic risks in the mining industry have been found by this inquiry. First, due to climatic variability, catastrophic weather events are happening more frequently and with increasing intensity, making transportation routes and mining infrastructure susceptible to structural failure (Auld and MacIver 2006; Infrastructure Canada 2006; Robertson 2006; Scales 2006). Depending on the type and location of the mine, thawing permafrost, rising average temperatures, stronger winds, changing water levels and ice composition, and increased intensity and frequency of precipitation can all affect containment structures, energy supply, and drainage from mine sites. These changes may compromise the structural integrity and security of ice roads, bridges, pipelines, and airstrips (Instanes et al. 2005). Furthermore, the walls of open-pit mines and polluted structures will not be able to adequately withstand exposure or melting of permafrost (Instanes et al. 2005; Furgal and Prowse 2008; Prno 2008). However, there may be advantages to climate change as well. For example, it is expected that a decrease in arctic sea ice would result in longer shipping seasons, shorter transit routes, and more exploration opportunities (Instanes et al. 2005; Comiso et al. 2008; Lemmen et al. 2008). Second, there is a chance that climate change will affect how mineral processing is conducted (UNEP-FI and SIWI 2005; Brown et al. 2006; Lemmen et al. 2008). Owing to the significant dependence on certain mining operations, such as sodium sulfate extraction, water scarcity might impact mining drainage, tailings covering, production rates, and dust control measures (Bjelkevik 2005; IPCC 2007).

It is advised that adaption strategies be created to mitigate current and projected future climate change, preserve the financial viability of mining activities, and safeguard the surrounding environment for miners. Regarding infrastructure, Auld et al. (2006) advocate for continuous education on the potential and capacities for flexibility as well as the deployment of "no regrets" measures or actions that have a net beneficial social impact. Additional experts and authorities have also urged mining companies to create long-term adaptation plans (ACIA 2005; Government of Canada 2006; Infrastructure Canada 2006; Lemmen et al. 2008). Additionally, studies on how the industry views climate change (Ford et al. 2010b, 2011) show that although there hasn't been much done to prepare for or adjust to changing conditions, climate change is becoming an issue for the mining sector.

The public's opinion of a company's commitment to solving the issue and creating laws to restrict greenhouse gas emissions are two other "policies of climate change" that mining operations are sensitive to. Implementing carbon levies and/or "cap and trade" programs, for example, may alter the financial and operational strategies chosen by some mining businesses. Businesses that are unprepared for new laws may suffer serious legal and financial consequences. Some mining industry members are especially concerned about climate change legislation, which they believe is being developed without their participation and would have severe economic consequences for the business. Large-scale businesses may be especially vulnerable due to their significant emissions and well-known public personas. The vulnerability paradigm proposed by Smit and Wandel (2006) has been employed to determine how mining operations in Canada are affected by and sensitive to climate change and evaluate how capable they are of adapting to current and projected future changes. The three case studies are presented at the beginning of the study, along with the method for assessing the sector's susceptibility. The findings section begins with discussing present vulnerabilities related to climate change, moves on to future vulnerabilities, and ends with suggestions for improving adaptation planning in the mining industry.

2 Properties of Mining Solid Waste

Around 2.6 Gt of mining tailings worldwide and 20–60 Mt are created yearly (Thom et al. 2013). In 2014, the 28 European nations' combined mine waste from quarrying and mining activities amounted to 705 million tonnes (Eurostat 2017). Ball milling quarry waste increases carbon dioxide (CO_2) sequestration, and the final product may find other uses in the building and construction industry (Rigopoulos et al. 2018b). The overburdened rocks produced might eventually serve as a carbon dioxide sink. Between 7 and 17 billion tonnes of silicate materials are produced annually worldwide by mines, igneous rock quarries, coal-fired power plants, iron and steel mills, and cement kilns, with 190–332 million tonnes of potential carbon sequestration (Renforth et al. 2011).

One crucial factor in aqueous mineral carbonation is the liquids' and solids' ratio (Huijgen et al. 2007). It has been noted that there is very little mineral carbonation in the absence of water vapor due to the gas-solid interaction of mining waste that is high in serpentinite (Veetil et al. 2015b). The combination of CO_2, water, and rock causes mineral carbonation, which is controlled by precipitation and dissolution processes (Kaszuba et al. 2013).

Particle size affects the amount of CO_2 mine waste may sequester because smaller grains provide more surface area for reactions (Wilson et al. 2009). Mineral carbonation and the viability of CO_2 sequestration are improved by mechanical activation, while waste rocks and tailings' particle sizes are decreased (Li and Hitch 2016b). It has been shown that lizardite dissolves more slowly than brucite and chrysotile, suggesting that ore heterogeneity may lessen the impact of particle size on mineral carbonation (Assima et al. 2013). The rate of weathering is greatly accelerated by the grinding of waste rocks (Renforth 2012), dissolution (Balaz et al. 2008), and thus, sequestration of CO_2 through the mineralization process (Rigopoulos et al. 2018a). As a result, tailings have a larger capacity to sequester CO_2 than overburden materials. Figure 1 explains the fate of mining waste in different environmental conditions.

Ex situ carbonation with acids like HCl utilizing the pH swing technique (House et al. 2007; Arce et al. 2017) raises the price and efficiency of mineral carbonation. Increased mineral carbonation results from releasing Mg^{2+} ions from the chrysotile tailings using biocatalysts such as *Acidithiobacillus* spp. (Power et al. 2010). Research has shown that adding ultramafic mining waste such as nickel and chrysotile to biomass torrefaction and CO_2 sequestration reduced CO_2 emissions by 62% and 22%, respectively, during the torrefaction process at room temperature (Sarvaramini et al. 2014).

Mafic and ultramafic tailings can pollute the environment since they are acidic and contain a lot of heavy metals. AMD is produced when mining waste is exposed to oxygen and water; carbonates are produced when CO_2 concentration is high. The heavy metal concentration in mining waste can be neutralized or immobilized by mineral carbonation, preventing pollution of the nearby soil, water, and plants. Furthermore, it was shown that due to weathering, twice as much CO_2 was sequestered after 4 weeks of acid leaching of mining waste (Hamilton et al. 2020). In mining tailings, CO_2 sequestration modifies the bacterial community (Li et al. 2019), suggesting that it could change in the direction of geochemical stability.

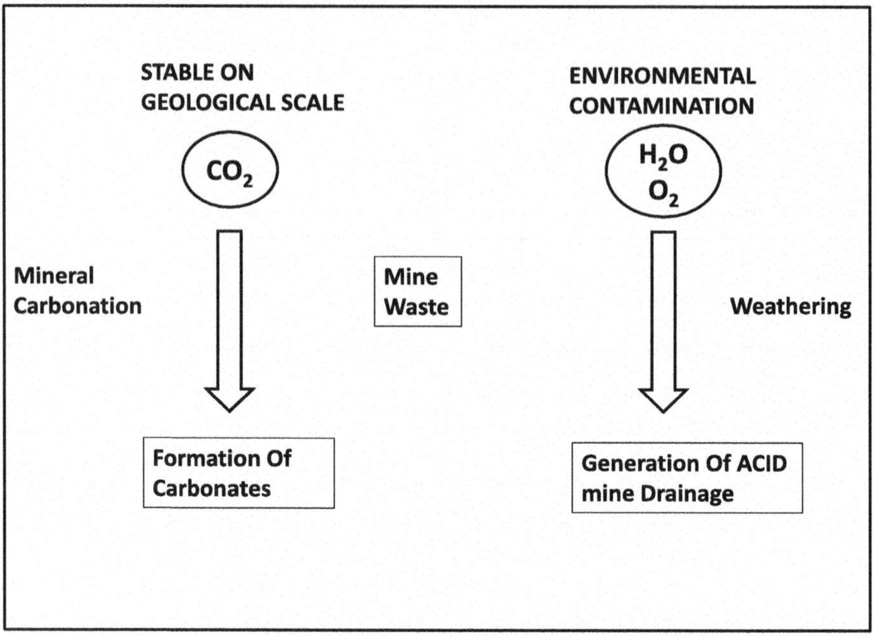

Fig. 1 The fate of mining waste in different environmental conditions

3 Properties of Mine Wastewater

Recently, research has looked into the possibility of storing CO_2 in closed open-pit mines. The mineral carbonation technique can be used to lessen the acidity of mine water. Alkalinity rises, and metal concentration falls when brucite in mine water breaks down (Bouzalakos et al. 2014). Most of the content is made up of dissolved CO_2 since carbonates dissolve in acidic pit lakes (Sánchez-España et al. 2014). Water extracted from deep coal mines contains a lot of dissolved CO_2 (Jarvis 2006). Therefore, it may be said that abandoned open pits eliminate carbon dioxide. More studies are required to determine if CO_2 sequestration by mine water is feasible (Kruse and Strosnider 2015).

Water-filled abandoned open-pit mines are a major hazard to groundwater resources and a potential source of heavy metal pollution (Abraham and Susan 2017). The exposed minerals produce carbonates, raising the water's pH and immobilizing heavy metals. During the carbonation process, it has been found that by-product minerals like nesquehonite sequester trace elements (Ni, Mn, Cr, Co, and Cu) from the solution (Hamilton et al. 2016) and stop contaminating the area. After CO_2 is sequestered, the leachate from ultramafic rocks is alkaline (pH between 9 and 10) and inhibits the mobilization of heavy metals such as Fe, Ni, Cu, and Zn (Kandji et al. 2017). Mineral carbonation may, therefore, be crucial to the reclamation of abandoned mines as it not only lowers CO_2 but also avoids heavy metal contamination. Heavy metals from the tailings may be recovered by mineral carbonation (Hamilton et al. 2018).

4 CO_2 Sequestration in the Mining Sector

Rocks and tailings are the two main waste products produced by mines. Both waste rocks (Hitch et al. 2010) and tailings (Zarandi et al. 2016) are potential materials for the sequestration of CO_2 from the atmosphere. Although mine waste has little commercial value, its profitability may increase if used to trap CO_2 (Hitch et al. 2010). It is found that carbonates are produced as a result of serpentine breakdown when a layer of carbonates builds on the upper surface of tailings due to weathering (lizardite and chrysotile) and brucite (Oskierski et al. 2013).

The kind of ore mining and the local geology determine the waste's mineral and chemical makeup. Waste rock's mineralogy and geochemical makeup determine its sequestration capability (Jacobs and Hitch 2011; Power et al. 2020) combined with the CO_2 supply (Harrison et al. 2013). Mafic and ultramafic minerals high in Ca^{2+} and Mg^{2+} react with CO_2 when they weather, making them potential materials for carbon sequestration.

4.1 CO_2 Sequestration Using Mine Waste

Lizardite, chrysotile, antigorite, and brucite-rich mining wastes undergo natural and passive mineral carbonation reactions (Lechat et al. 2016). Assima et al. (2014a) discovered that big serpentines (lizardite/antigorite) had a lower carbonation rate than mine waste that contains brucite and fibrous serpentine (chrysotile). The grain size of mining waste also significantly affects the mineral carbonation through surface area and reactivity. Fine-grained tailings are more reactive and absorb more CO_2 than waste rock (Zarandi et al. 2017).

The mineralogy and geochemistry of waste or ore dictate the mineral carbonation process. When mining waste from gold and limestone is compared, it is clear that the former can hold onto carbon dioxide longer. In addition, following mineral carbonation, waste from limestone mines produces bricks with higher compressive strength and less water absorption than waste from gold mines (Kusin et al. 2020). Therefore, after mineral carbonation, limestone mining waste might be used again as a cement-based product. The ability of mining waste to sequester CO_2 is also determined by the weight ratio of Mg^{2+} to Si^{4+}, and it has been shown that a rise in this ratio likewise increases the ability to mine waste to sequester CO_2 (Li and Hitch 2016a).

Whether wet or dry, the presence of water dictates the chemical route for mineral carbonation. Compared to the dry approach, the aqueous method is more effective for sequestering CO_2 (Revathy et al. 2015). Concrete waste samples were shown to have a greater capacity to trap CO_2 using an aqueous approach compared to anorthosite tailings (Ghacham et al. 2015). Aqueous mineral carbonation is a financially feasible method of reducing CO_2 emissions on a small to medium industrial scale (Veetil and Hitch 2020).

The upper surface layer of mining waste has a haematite coating or Fe hydroxide (III), preventing CO_2 from reacting. Better feedstock for mineral carbonation and CO_2 sequestration is provided by using magnetic separation devices to separate Fe-containing minerals from other mineral mixtures (Veetil et al. 2015a). Chelation agents, which have a greater affinity for Fe^{2+} than Mg^{2+}, enable garbage to absorb CO_2 sustainably (Assima et al. 2014a). Because magnetite precipitates iron hydroxide, it is shown to be deleterious to carbonation in chrysotile mining wastes (Assima et al. 2013).

Metals in metal mines have a strong sulfide association and are abundantly found in minerals that are richer in iron. It is questionable if waste pebbles from metal mining can effectively sequester CO_2 in ambient settings. Waste drainage pH in the Dumont deposit's ultramafic lithology varies from 9 to 10, whereas in non-ultramafic deposits, it is between 7.0 and 8.5 (Kandji et al. 2015). Thus, drainage from mine waste is greatly impacted by CO_2 sequestration.

4.2 Potential and Mechanisms for Stable C Storage in the Post-mining Soils

One of the major global concerns is what will happen to carbon in the environment. Since CO_2 is a greenhouse gas (GHG), the observed global warming and climate changes are thought to be caused by the atmosphere's dangerously high amounts of CO_2. However, as soils and forests contain the most carbon, sequestering carbon in the soil might be the primary strategy for reducing atmospheric CO_2 levels and stabilizing soil storage to mitigate or slow down climate change (Hüblová and Frouz 2021). The soil condition tightly determines how well carbon is sequestered (Kowalska et al. 2020). Regrettably, throughout the drastic environmental change, human activities significantly affected the global decline in soil quality. Specifically, mining is considered extremely harmful to the ecosystem (Placek-Lapaj et al. 2019). The loss of topsoil during opencast mining severely damages the environment, making it impossible to recover. The depletion of organic matter, which regulates fertility and, by extension, the productivity and quality of the soil, including its ability to sequester carbon, as well as the erosion of the soil microclimate and water regime, are the main causes of it. Ultimately, it results from the degradation of the health of the soil (Sarvade et al. 2019). Restoring the natural soil functioning is one of the main challenges in restoring post-mining soils. However, the growing regions with deteriorated soils compel the injured area to recover. It is crucial for the soil, but it is also extremely valuable for the ecosystem since post-mining soil remediation improves carbon sequestration, reducing climate change.

Beyond its function in reducing climate change, soil carbon has other uses. Soil carbon (C) is necessary for creating all organic compounds and influencing soil fertility. The amount of organic carbon in the soil controls every aspect of its properties, including chemical, biological, and physical ones. Additionally, carbon plays a role in the cycle of nutrients (Amoah-Antwi et al. 2020). A soil's ability to retain its mineral component is facilitated by organic carbon, which influences biodiversity and fosters the establishment and development of healthy plants (Manaye et al. 2019). The sorption and filtration capacities are two more factors that are impacted by soil organic carbon (SOC) (Yu et al. 2019), and soil structure (Roy et al. 2017), and as a result, C is crucial to the processes involved in the creation of soil (Šimanský et al. 2019). The C cycle is unaffected in healthy soil, and there are several sources of C inputs, including fertilizers, plant leftovers, secondary animal sources (like animal feces and soil fauna residues), etc. Conversely, the post-mining areas demonstrate how little the C pool is in the soil. During the early stages of restoration, soil carbon in post-mining soil is supplied by organic fertilizers and soil C sequestration. The number of C sources climbs (Placek-Lapaj et al. 2019) and C sequestration rises as cleanup proceeds (Mukhopadhyay and Masto 2022).

The soil's organic carbon concentration is heterogeneous because part of the organic carbon is stable and has a long half-life, while soil bacteria quickly mineralize other organic carbon.

According to Dignac et al. (2017), regarding their rate of deterioration, the three conceptual C pools in soil are labile, intermediate, and stable C. Both labile and intermediate C mostly come from litter and roots; turnover for labile C happens day-to-year, while for intermediate C it occurs for years to decades. The stable C, whose turnover spans decades to millennia, is obtained by extracting the labile and intermediate C. Stable C makes up most of SOC and can be absorbed on mineral surfaces or found in aggregates. The most stable components of SOC are believed to be alkyl and aromatic C, whereas O-alkyl and carbonyl C quickly mineralize. A longer-term study of manufacturing procedures should improve our comprehension of SOC sequestration mechanisms. Dignac et al. (2017) have thoroughly explained the SOC, but they have also stressed the significance of constantly monitoring the experimental sites. The methodologies and possible applications of post-mining soils for sequestering SOC have not been covered in the pertinent literature, which hampered our investigation.

Based on estimates by Blume et al. (2016), the topsoil layer of two meters may hold up to 2060 ± 215 Pg C, and the soil structure includes around 80% of the terrestrial SOC stock (Batjes 2016). Kanzler et al. (2021) show that alterations in the carbon store of the soil might significantly impact the global carbon cycle, indicating the soil's enormous potential to mitigate climate change. Fox et al. (2020) observed a significant possibility for carbon retention in post-mining sites, indicating appropriate soil regeneration. Reforestation is typically used in Poland to remediate post-mining soil, which can enhance carbon sequestration (Placek-Lapaj et al. 2019). The capacity of post-mining soil to sequester atmospheric carbon has been demonstrated, but little is understood about the mechanisms behind the permanent storage of sequestered carbon in these soils.

The growth in soil carbon storage corresponds with the cleaning of post-mining soil. Despite a plethora of research on soil carbon sequestration, little is known about the potential and strategies for carbon sequestration in post-mining environments. This long-term study uses nuclear magnetic resonance analysis to assess the development of soil organic carbon synthesis in limestone (S_1) and lignite (S_2)

post-mining soil under various restoration phases. Reforestation is the primary goal of the cleanup; however, S_2 has already added sewage sludge. The majority of the percentage in the sequestered soil for S1, according to the study, was composed of O-alkyl groups. However, S_2's remediation method revealed greater acetyl-C and aromatic groups. Improved hydrophobicity and humification were the outcomes of S_1 remediation; nonetheless, it was observed that the C/N ratio and the synthesis of aromatic groups had dropped. Sewage sludge restoration procedures have been connected to an increase in all indicators of sequestered C stability, as demonstrated by S_2. Though S_2 showed noticeably more features of stored carbon, suggesting its higher stabilization, both post-mining soils indicated considerable potential for sequestering carbon. This may suggest that it takes longer for soils not enriched with sewage sludge (S_1) to retain carbon stably (Kowalska et al. 2020).

Pretreatment (chemical, mechanical, and thermal) of mining waste is required for the slow process of mineral carbonation, which releases CO_2 and speeds up climate change. This increases implementation costs but enhances reaction kinetics. On the other hand, immobilizing heavy metals in leachate and reclaiming mining areas may also be accomplished sustainably with mineral carbonation. Only environmental hazards and consequences from mining operations are included in life cycle assessment studies; mineral carbonation has been left out. Large amounts of mining waste are produced, and its ability to absorb CO_2 from the environment should be promoted for use in large-scale industrial settings.

4.3 *Factors Influencing CO_2 Sequestration Using Mine Waste*

Up to 30 cm above the surface, mining debris sequesters CO_2 more effectively (Lechat et al. 2016). There is less CO_2 penetration at a deeper waste depth. The massive heaps of mining waste piled on top of one another minimize the amount of space needed for disposal, but they also lessen the ability to sequester CO_2. To speed up the carbonation process, trash heaps should be disposed of and stacked to improve air and water permeability (Assima et al. 2013). The technique of disposing of or piling up mining waste needs to be carefully studied for the sequestration of CO_2. The need for land is increased by the thin heaps of mining waste, which takes up a lot of space. Additionally, exposure to water and oxygen raises the risk of AMD. Because of the quick oxidation that takes place in the presence of water, the amount of AMD created would increase as mining waste expanded across bigger territories. Acid rain pollutes the surroundings and allows divalent cations, including heavy metals, to seep out. Additionally, it is shown that the amount of cations leached from tailings correlates with the precipitation of carbonates (Kemache et al. 2017; Huijgen et al. 2005).

The carbonation process is accelerated in warm climates. However, it turns out that the hurdles posed by climate change might be surmounted by mineral carbonation mediated by microorganisms (Siegrist et al. 2017). When mining waste is heated from 10 to 40 °C, its CO_2 sequestration increases tenfold (Assima et al.

2014b). At the Black Lake mine, heated air vents may be seen at the top of chrysotile waste piles, indicating that mineral carbonation is an exothermic event (Pronost et al. 2012). Solid carbonates are created when cations (such as Ca^{2+}, Mg^{2+}, etc.) leach as a result of the dissolution and precipitation process.

The rate at which mining waste weathers affects how much CO_2 is sequestered. Mineral carbonation is increased when a new surface of serpentine tailings containing brucite is exposed (Zarandi et al. 2016). One of the most important parameters when optimizing carbonation is how often mining waste is watered (Beaudoin et al. 2017). Mining waste may be a practical technique for ex situ CO_2 sequestration and storage, according to a life cycle assessment (LCA) of the complete CO_2 sequestration process (Bodénan et al. 2014). Even in frigid climates, proper air circulation in the mines improves the carbonation process' effectiveness (Beinlich and Austrheim 2012).

4.4 *Feasibility of Mining Waste for CO_2 Sequestration*

When it comes to CO_2 sequestration in laboratory settings, temperature, pressure, and the mineralogical makeup of mining waste are all extensively covered in the literature. A larger-scale implementation of CO_2 sequestration is not financially feasible, and the technology is only evaluated on a restricted basis in the lab. To encourage the CO_2 sequestration potential of mines and trash on a bigger scale, further study is required.

Various mines at various locations have varying capacities for mineral carbonation. At 39,800 t/yr of ambient CO_2, the tailings at Mount Keith can capture 11% of the mine's emissions of CO_2 (Wilson et al. 2014). 120,000 t of CO_2 might be sequestered by carbonating 0.5 weight percent brucite using 40 Mt of tailings each year (Power et al. 2020). Pan et al. (2020) calculated that 4.02 Gt of CO_2 might be fixed annually, directly and indirectly, through mineralization and use.

LCA details how a process or product affects the environment throughout its life. LCA is used to determine that CO_2 sequestration by tailings is carbon-positive, meaning that during the process, carbon is released by the mineral carbonation rather than being sequestered (Ncongwane et al. 2018). Thermal and chemical treatments that accelerate the carbonation of minerals result in net CO_2 emissions and have the largest carbon footprints. According to LCA, mechanically removing the top layer of mine waste accelerates the carbonation process and may be a better technology for the environment (Bodénan et al. 2014). When steel slag is used in mineral carbonation to produce carbonated blocks (building material), LCA reveals that the method is carbon-negative (Maria et al. 2020), indicating the technology's practicality on a ground level.

In the frigid Arctic, mining operations are also very much in operation. Colder areas have also recorded cases of mineral carbonation caused by mining waste (Wilson et al. 2011). However, because mineral carbonation also depends on mineralogy and water content, comparing mineral carbonation in warm and cold conditions is challenging. Warm, dry weather encourages Mineral carbonation and evaporation (Wilson et al. 2014). As a result, it may be said that a cold environment would have a lower capability for CO_2 sequestration than a warm environment. After a mine's sequestration capability is balanced, In comparison to a warm temperature, there would be more net CO_2 emissions in a cold environment. More research is required to comprehend the equilibrium between CO_2 emission and sequestration from a mine in a certain ecosystem completely. Most carbonate crust in semi-arid areas is created by evaporation on the upper surface of ultramafic tailings (Oskierski et al. 2016).

Another significant worry is methane emissions from coal beds, and studies have indicated that improving CO_2 injection in the coal bed improves methane recovery (Bromhal et al. 2005). Nitrogen injection further improves CO_2 sequestration and methane gas recovery in coal beds. There are 476 operational coal mines in India, 236 of which are underground mines (IBM 2019). These mines might be used as a place to sequester CO_2. In 2017, 7632 million tonnes of coal and lignite were produced worldwide (IBM 2019), indicating a massive gap in CO_2 storage. A crucial element of CO_2 sequestration is the permeability of coal, which is decreased by CO_2 injection (Vishal et al. 2013). Different coal beds react differently to CO_2 injection at different locations, necessitating unique models for effective CO_2 sequestration.

The literature contains information about CO_2 sequestration via mineable coal beds, saline aquifers, and depleted oil and gas reserves. It is best to encourage this technology's industrial application (Leung et al. 2014). Implementing CO_2 sequestration technology is hampered by high leakage and capture rates, a lack of subterranean storage, and the need for improved worldwide policy (Kalkuhl et al. 2015). Additionally, the literature ignores the possibility of employing mining waste for CO_2 sequestration; additional study is necessary before this technology can be used on a large scale.

5 Mechanisms of Sulfide Mineral Oxidation and Acid Mine Drainage

Abbassi et al. (2009) state that O_2 and water are oxidizing agents. An additional mechanism for sulfide oxidation might be an interaction with Fe(III). Both routes move quickly. They produce Fe(II) and a low pH, which can be converted to Fe(III) by O_2(aq). The conceptual model shows the successive processes of pyrite oxidation from mining waste together with the yield of metals, metalloids, acidity, and sulfate. The model continues by demonstrating how redox reactions regulate the movement of pollutants in groundwater systems, both shallow and deep. At low pH, this oxidation is very sluggish and unstable unless microorganisms are present. Bacteria may accelerate Fe(III) production by up to six orders of magnitude, and they can also increase their stability at a pH of about 8. Furthermore, chalcopyrite and sphalerite ((Zn, Fe)S), which are arbitrarily composed, exhibit acid production. Certain forms of mining contain minerals that oxidize but do not become acidic, such as chalcocite and galena (Matthews et al. 2008). As pH levels rise, so do the oxidation rates of arsenic sulfide (AsS (am), As_2S_3 (am), orpiment, and realgar).

The earlier models used first-order kinetics and Fick's first rule to assess the spoil's one-dimensional vertical oxygen transport. A two-dimensional transport model was combined with a one-dimensional oxygen diffusion model to enhance the model. In this case, the physical delivery of oxygen to the reaction site is the main rate-limiting step for the oxidation of pyrite. However, they ignored microbes' involvement and the geochemical processes' kinetics. It has been claimed that because oxidation products only showed up as source components in the aqueous transport equation and oxygen diffusion was restricted to the air phase, their model could distinguish between kinetic oxygen diffusion and aqueous phase reactive transport.

6 Geochemical Reactions and Dynamics of AMD in the Environment

The kind and concentration of metal-bearing sulfides in ore and wall rock, the rate of ore and wall rock dissolution, the permeability of the ore deposit or mine tailings, and the host rock's capacity to buffer acidity are just a few of the numerous variables that influence the chemistry of mine drainage. If buffering minerals are present in the host rock, the concentration of metals in and around the mining sites is neutralized by acid, bringing it down to background levels. According to a quantitative model, tailings leachate composition modeling requires a balance between proton-consuming processes that regulate pH and proton-producing activities that arise from pyrite oxidation. With increased pH, aqueous metal species frequently precipitate as hydroxide, oxyhydroxide, or hydroxysulfate phases (Kimball et al. 2010). Additionally, these freshly created mineral precipitates adhere to these precipitates' surfaces. Furthermore, dilution is a simple method of reducing the concentration of metal without the need for a chemical reaction.

The following is the sequence in which minerals with 100% O_2 availability can buffer AMD: minerals that are (hydro)oxide > aluminosilicate > carbonate. By prolonging the lifespan of the deposit's quick reactive buffer capacity, the aluminosilicate weathering process influences the length of time pH is buffered at relatively high values, using the conceptual model. The growth of tailings impoundments is regulated by secondary phases that precipitate into the pores and/or on the surface of waste materials. Therefore, goethite (FeO(OH)), K-jarosite ($KFe(SO_4)_2(OH)_6$), and gypsum ($CaSO_4.2H_2O$) employ the SO_4, Fe(III), and K produced during dissolution. The primary cation associations may alter if important cations like Ca^{2+}, Mg^{2+}, Cu^{2+}, Na^+, and K^+ were introduced to the clay portion of mining waste (Bea et al. 2010).

7 Influence of Climate Change on Water Quality

The makeup and operation of freshwater ecosystems are significantly impacted by climate change (Brooks et al. 2011). Climate change is causing changes in the frequency and patterns of worldwide rainfall. Climate change and drought conditions lead to higher temperatures, greater evaporation, less precipitation, and a loss in freshwater inflows to reservoirs, especially in the summer and fall. Aquatic risk evaluations of ionizable contaminants in acid mine drainage are associated with lower instream flows, diel pH variability, and the resulting uncertainty particular to each site (Brooks et al. 2005, 2011; Barnett et al. 2008; Marce et al. 2010). These elements are causing surface and groundwater quality in semi-arid and dry areas to decline. Mining waste and dump flow effluents can be the main inflows into reservoirs in semi-arid and dry areas, or they can be received simultaneously (Brooks et al. 2011). The impacts of climate change on site-specific hydrology and altered water inflows can substantially influence the physical, chemical, and biological processes in acid mine drainage, which may worsen problems with reservoir water quality (Brooks et al. 2011).

It is commonly recognized that many chemical pollutants undergo major chemical changes that affect their toxicities due to increased temperatures brought on by climate change. Because fish have greater metabolic rates and less oxygen solubility, increased temperatures, for instance, alter ventilation rates, which accelerate the pace at which pollutants are absorbed by fish (Schiedek et al. 2007; Vale et al. 2011). Climate change not only modifies the pathways that pollutants follow, but also accelerates some of these processes as temperatures rise (volatilization, adsorption, hydrolysis, biodegradation, photodegradation, absorption, and metabolism) (Schiedek et al. 2007). The pace at which trash decomposes affects the metals released and is directly influenced by temperature. As a result, metals can be created or absorbed by degrading plant tissues through adsorption on litter (Pereira et al. 2007; Vale et al. 2011).

8 Impact of Climate Change on Geochemical Reactions of AMD-Affected Streams

Acidity and dangerous components from AMD are transferred to river basins and groundwater mostly through receiving habitats. When Fe, Al, and Mn oxyhydroxides precipitate swiftly in the pH water's transition zone between acidic and nearly neutral, trace metals and metalloids are removed by sorption and/or co-precipitation (Casiot et al. 2009). Numerous biogeochemical and hydrological processes vulnerable to climate change were responsible for variations in the metal concentrations in a small Mediterranean river contaminated by AMD water from a mining stream (Casiot et al. 2009). One of the primary markers of climate change is elevated summer temperatures, which also promote the development of anoxic or suboxic conditions in sediments, boost bacterial activity, decrease oxide phases, and mobilize associated metals and metalloids (Masson et al. 2007). In alkaline water impacted by AMD, increasing the temperature and/or pH also encourages metalloid desorption (Gammons et al. 2007). Yunmei et al. (2007) discovered that the rate of arsenopyrite oxidation grows exponentially with temperature when Fe(III) concentration rises, especially when $Fe_2(SO_4)_3$ concentrations above 104 mol/kg. The results of experiments conducted at 45 degrees Celsius revealed a significantly higher value of 0.64, indicating the presence of a separate reaction mechanism at higher temperatures. The reaction order was found to be consistent between 15 and 35 °C, with a value of 0.41 (Corkhill and Vaughan 2009). Consequently, the oxidation of sulfide minerals in mining waste is expected to occur faster due to climate change, which is expected to raise world temperatures even more. A model theoretically illustrates the impacts of climate change on pyrite oxidation and the formation of acidity, sulfate, and contaminating components. The model also illustrates how hydrology and climate change impact the redox reactions, which regulate the flow of pollutants and the pollutants in groundwater systems. It has been shown that diel oscillations in element concentrations in streams impacted by AMD are caused by photoreduction of colloidal Fe oxides (Gammons et al. 2005a, 2005b).

Numerous variables, including the local temperature, climatic change, hydrology, structural geology, ore deposit composition, and host rocks, influence the amount of AMD contamination in the mining zone. Hydrodynamic and chemical processes reduce pollution concentrations in the water and sediment downstream of river systems. Solid phase precipitation, evaporation-induced efflorescent salt production, and dilution from uncontaminated surface and/or groundwater input are the main sources of natural attenuation processes at low flows. At high stream flows, rainfall dilution controls natural attenuation (Cidu et al. 2011). Chen and Jiang (2012) demonstrated that waterfall aeration and seasonal temperature changes substantially affected the iron and As attenuation rates by secondary mineral precipitate in an AMD system. One of the most effective methods for removing dangerous metals and other contaminants from mineral precipitates is adsorption onto or co-precipitation with them using several techniques, such as the adsorption onto or co-precipitation of arsenate with schwertmannite at low pH, regarding AMD (Regenspurg and Peiffer 2005; Asta et al. 2010). Due to the fluctuations in the local climate and climate change, which affect solution chemistry, temperatures, and microbe types and concentrations, iron (Fe) and other metals and metalloids measured in the field can have highly variable oxidation and precipitation rates that are subject to seasonal variations (Sánchez España et al. 2007; Chen and Jiang 2012).

9 Effect of Hydrology and Climate Change on Contaminant Transport in AMD

Crouch et al. (2013) show that modifications to the hydrologic regime may affect the biogeochemical processes related to AMD in aquatic environments. Climate change may have an impact on water quality because of the significant hydrological and watershed-scale effects. Surface and subterranean inflows carry mining debris, resulting in acid mine drainage (AMD) into streams. This leads to diffuse subsurface distribution of AMD and longitudinal alterations that long-term impact the concentration of metals in a stream environment. Watershed hydrology has a major role in controlling AMD systems' chemical, physical, and temporal variables that result in significant ion mobilization (Nordstrom 2009). Streams' chemistry and pH changes are influenced by shallow subsurface flow, groundwater table fluctuations, recharge flow paths, seasonal and annual variations in water flows, and so on (Nordstrom 2009). This highlights how crucial watershed-scale studies are in identifying the origins of AMD loads to streams.

10 Climate and Seasonal Effect on the Composition of AMD

The length and temperature of the seasons may vary due to climate change, affecting the quality of water streams affected by AMD.

10.1 Case of the Odiel River Basin

Due to many factors, including seasonal fluctuations, pollution levels, basin position, and temperature, the Odiel River Basin in Spain displays varied hydrochemical and mineralogical AMD process patterns (Sarmiento et al. 2009). For instance, the weathering of the sulfide minerals that outcrop in the zone naturally lowers pH and HCO_3 concentrations, whereas in the uncontaminated streams, SO_4^{2-} and metal concentrations rise in the autumn just in time for the first rains. Rainfall-induced dilution processes in the contaminated streams reduced SO_2, metal, and metalloid contents. However, depending on the level of stream pollution, these substances' development over time varied dramatically. As the amount of rainfall reduced, the pollution level rose. For samples that are somewhat impacted, the metal/Fe ratio rises from the wet to the dry season; for those that are severely affected, it falls. Most of the constituents demonstrated similar growth along the Odiel River. The concentrations increased in the dry season and decreased in the wet season. However, during the first several months of the dry season, the lowest section of the Odiel River had the opposite results for Fe content (Sarmiento et al. 2009). Ferrihydrite was often saturated in samples with lower impact, whereas K-jarosite was saturated in those with higher impact. For schwertmannite and ferrihydrite, the upper section of the river had higher saturation index values during the dry season, whereas the lower segment had greater values during the wet season.

10.2 Case Study of an Arctic Area in Svalbard

On Svalbard (78 N), seasonal variations in the release of AMD from waste rock related to high-arctic-sulfide coal mining were studied by Søndergaard et al. (2007). Strong connections between element concentrations indicate that substantial quantities of alkali- and alkaline earth metals, Fe, Al, Mn, Zn, and Ni, were released into the runoff by the trash pile's sulfide oxidation, ion exchange, and silicate weathering processes. Large sulfide-containing waste rock piles in the Arctic accumulated a lot of pollutants throughout the dry summer and winter months. The first week of defrosting was associated with the highest metal concentrations, the lowest pH measurements, and a significant daily acid and $SO_4{}^{2-}$ emission. The accumulation of weathering products that were produced inside the garbage pile throughout the winter and discharged as a pollution flush in the early spring is most likely the source of this. Similar instances occurred later in the summer, and significant daily metal fluxes were caused by rain events that followed very long dry spells (Søndergaard et al. 2007).

10.3 Case Study of Rum Jungle and Mt. Todd Mine in Australia

At Mt. Todd, the elemental composition of AMD changed with the year as opposed to the season. On the other hand, evaporation during the extended dry spell in Rum Jungle caused the concentration of dissolved metals in the AMD to increase throughout the dry season. Unlike Mt. Todd AMD, the bacterial population of AMD in Rum Jungle varied over the wet and dry seasons. There was a significant difference in the bacterial communities in the first and second years. The iron-oxidizing bacteria *Leptospirillum* and *Acidithiobacillus*, which are frequently seen in AMD in temperate climes, were not present in significant amounts at these two tropical mine sites (Streten-Joyce et al. 2013).

The climatic regime influences the composition of AMD in addition to geological factors such as the neutralizing capacity of the host ore, the type of rock supporting the deposit, the nature of the ore, and the concentration of trace elements in the deposit and host rock, and the amount of pyrite and iron sulfide. Mining areas located in the wet-dry tropical region of Northern Australia experience a long dry season, which runs from May to October, in contrast to AMD produced during the rainy season. During this period, AMD has a higher metal content and a lower pH (Streten-Joyce et al. 2013). Acidity, hazardous metals, and sulfate salts are produced when pyrite and sulfide minerals oxidize during the dry season. The concentration of metals in the AMD rises as the first rains fall because the salts are rinsed away. The pH falls and metal concentrations rise due to the first rain flush of AMD, which accumulates during the dry season (Nordstrom 2009).

11 Impacts of Climate Change on AMD Generation in Semi-Arid/Arid Zone

Many studies on the relationship between environmental pollution and AMD formation have previously been reported in the literature. These studies have focused largely on humid conditions with unlimited oxygen and water. On the other hand, although oxygen is not limited, semi-arid and arid regions typically experience water scarcity. Most mining operations in semi-arid and arid regions occur in isolated locations far from cities with inadequate water supply infrastructure. There are arid, drought-prone climates, water deficits, and droughts caused by evaporation that is several times higher than normal yearly precipitation, as well as sporadically significant flooding in other places. Severe weather events, such as extreme cold, floods, storms, low or irregular rainfall, droughts, high evaporative demand, high runoff potential, and extremely hot temperatures, are becoming more frequent and intense due to changes in the global climate and environment. As a result, various places receive varied quantities of rainfall. Hazardous materials at mining sites may originate from sulfidic minerals, which oxidize when exposed to air and water. While increasing rainfall intensity, water content, and flow regime would greatly impact the oxidation of pyrite/sulfidic minerals in strongly evaporative settings of semi-arid/arid areas, decreased rainfall will minimize the oxidation of sulfide minerals and the development of AMD. Aquatic habitats will experience the discharge of ions, sulfate, hazardous substances, and acidity as a result.

Because protective coatings grow on pyrite grains' surfaces, pyritic minerals are slowly weathered beneath vegetated soil/tailings coverings, along with a decreased release of contaminants and AMD. However, re-establishing plant cover (phytostabilization) on mine tailings will be more challenging due to water shortages and greater temperatures that result in lower soil moisture content. Additionally, in certain places, climate change will place additional strain on environmental mitigation efforts. On the other hand, the benefit of increased rainfall brought about by climate change may cause the re-vegetative cover on mine tailings to rise in some semi-arid or dry places. Acero et al. (2009) state that porewater growth is regulated by infiltration in humid environments with relatively high precipitation and that geochemical changes in the vadose zone can occur very slowly. On the other hand, semiarid ecosystems benefit from intermittent, heavy rainfall events that occur in short bursts, episodic reactive input, and transient water flow. These factors facilitate O_2 entry into the tailings, improved solute transport, faster rates of sulfide dissolution, and secondary phase precipitation. Thus, in semi-arid regions, the effects of climate change will hasten the geochemical reactions that lead to increased sulfide oxidation and AMD generation, lowering the quality of surface and groundwater.

12 Impact of Climate Change on Restoration of Mining Site

Mining wastes can be remedied and restored using a variety of techniques, such as geotechnical stabilization, waste pile revegetation, rainwater drainage system construction, mine adit sealing, and passive treatments like anoxic limestone drainage and anaerobic compost wetlands (Espana et al. 2005). However, chemical and climatic constraints (high acidity and metal levels and seasonal changes in water discharge) may make these restoration attempts difficult. A workable and reasonably priced alternative for the long-term management of AMD and the closure of tailings impoundments, the high water table strategy, was studied by Ouangrawa et al. (2010). Their results indicated that a higher water table might be a helpful tool for lowering the quantity of AMD generated, provided that the appropriate design parameters are selected and put into practice. With this method, the water table's position is managed to keep the reactive tailings sufficiently saturated with water through capillarity. Because saturated or nearly saturated porous media have a very low effective diffusion coefficient of oxygen, the rate at which sulfide minerals oxidize is slowed down. The water saturation, moisture content, and hydrology of mine tailings are greatly impacted by climate change and fluctuation, which will adversely affect the cleanup efforts in certain areas while helping them in others by lowering the generation of AMD.

The effects of climate change on mine infrastructure vary depending on the type and location of a mine and include permafrost thawing, rising average temperatures, stronger winds, shifting water levels and ice composition, increased intensity and frequency of precipitation, flooding from increased rainfall, and natural disasters (Pearce et al. 2011). Following will be the discharge of contaminated AMD into the adjacent watershed, a rise in environmental accountability, and consequences for public health and safety (ACIL Tasman 2007). Some abandoned mining sites will pose a major threat to the surrounding ecosystem and community since they were not constructed with the prospect of climate change in mind, especially in light of the current state of the climate. For example, because it was assumed that permafrost would always remain frozen, the Clinton Creek asbestos mine in Yukon, Northern Canada, which generated over 60 million tons of waste rock between October 1967 and August 1978, was closed without considering climate change. The disintegration of waste rock dumps and tailings piles caused by the collapse of tailings dumps ruined the fish habitat in upper Clinton Creek and Wolverine Creek, obstructing the creek's flow (Pearce et al. 2011).

13 Reactive Transport Models

For both kinds of material, a multitude of reactive transport models have been developed to simulate AMD development in both saturated and unsaturated zones. These models also include geochemical processes (e.g., mineral precipitation, reversible sorption, and the precipitation and dissolution of acid-generating salts), mineral–fluid interactions, chemical equilibrium, biogeochemistry, and groundwater chemical development. In addition, the investigation has included mineral speciation, reaction kinetics, saturation indices, and how contaminants are transported in the different flow systems observed in mining locations (Kruse 2007; Kruse and Younger 2009). Most of these models were developed without considering climate change's impact on the response mechanisms described earlier. However, a growing number of people are starting to recognize the effects of climate change. The current study aims to evaluate the potential of a few chosen modeling codes to understand how climate change affects pollutant movement, AMD formation, and water quality. The models will be compared, even though it is questionable if the current reactive transport models can effectively address physical-geochemical and even biogeochemical issues that may arise at a mining site due to climate change and unpredictability.

13.1 Conceptual Model

To develop the conceptual model, major simplifying assumptions that reduce the precision and plausibility of model responses are avoided. Most models were too complicated to consider the different ways that climate change can affect the geochemical processes and water body quality that AMD affects. The best conceptual model to evaluate the geochemical changes AMD has undergone due to mining waste in various climate change scenarios would consider these aspects. Conceptual models (a simple two-dimensional flow system) have been used for homogeneous and heterogeneous material distributions to predict the evolution of AMD resulting from sulfide oxidation inside an unsaturated overburden spoil pile at an open-pit mining site. To illustrate the mineralogy, transport processes, geochemical processes, workings, fractures, hydraulics, and airflow of abandoned underground mines, Kruse and Younger (2009) developed two purpose-built models: PLAYING POSSUM (Pollutant Loadings Above Average Pyrite Influenced Geochemistry POSSUM) and POSSUM (Pollutant Sources and Sinks in Underground Mines) (Kruse 2007).

13.2 PHREEQC Modeling

PHREEQC is a C computer software and geochemical equilibrium model. In addition to performing various aqueous geochemical calculations and basic 1D reactive-transport modeling calculations involving water-rock interaction processes, it can be used to simulate the advection of solutions through a solid column, with or without dispersion and/or diffusion effects. Inverse geochemical modeling reveals the precise dissolution and precipitation processes in mining sites, whereas forward modeling discovers the saturation states with different solid phases. Using saturation indices, element speciation, and ion activity in the water sample of sulfide-rich mine tailings, one can replicate the processes of primary mineralization interacting with the somewhat acidic and oxidizing rainfall that leaves the mine site, as well as the formation or dissolution of secondary minerals (Bhattacharya et al. 2006; Smuda et al. 2008; Arcos et al. 2008; Song and Yanful 2011). Gomez and associates created a model in 2006 to mimic the mobilization of uranium in groundwater. This model featured pitchblende as the major U ore, oxidation of pyrite and arsenopyrite, precipitation of iron oxyhydroxide, many geochemical processes, and the release of U(VI) species into groundwater as a result of oxidized fluids seeping through the cracked granite.

13.2.1 Probabilistic LaSAR-PHREEQC Model

The solutes are either immobilized or converted by a range of hydrological circumstances and (bio) geochemical processes; AMD is carried downstream from the source and splits into groundwater-surface water systems throughout the journey. The Lagrangian Stochastic Advective Reactive—PHREEQC model was used to evaluate the likelihood that AMD will seep into the groundwater near the mining sites (Malmström et al. 2008). Field data might not be sufficient for deterministic process descriptions even at well-characterized sites due to underlying variability. Nonetheless, this model may be used to examine the multicomponent reactions and transport in the heterogeneous groundwater surrounding a mining waste site, emphasizing the distribution of metals due to flow and geochemistry. The model demonstrated that distinct temporal concentration fluctuations might arise from even small amounts of flow variability due to the possibility of such changes being produced by the chemical reactions that obscure the attenuation processes. The model may be used to calculate how climate change affects hydrological processes, influencing AMD production and surface and groundwater system quality.

13.2.2 Mass-Balance Inverse Geochemical Models

The influence of climate change on AMD-polluted water bodies may be assessed using mass-balance inverse geochemical models, with the simulation base being the algorithm PHREEQC1. Climate change affects the amount and quality of water resources because it changes the length and timing of rainfall in dry and semi-arid regions. Mass-balance inverse geochemistry models may be used to assess data retrieved from the few physical measurements that are accessible in the distant semi-arid and desert zones. These models may be used in various water catchments during both rainy and dry seasons (Driscoll et al. 2011). The hydrochemistry of the start and finish catchments as well as a variety of mineral weathering processes particular to each catchment, will feed the models.

13.3 MIN3P Modeling Approach

A useful tool for examining processes, including aqueous complexation, oxidation-reduction reactions, and mineral dissolution-precipitation reactions, is MIN3P, a multicomponent reactive transport model. Large-scale mine impact studies, several kinds of mine waste systems, and laboratory column experiments have all already been conducted using and testing this 3D finite volume code (Ouangrawa et al. 2009). The geochemical development of U mine water during floods was studied by simulating the key transport and chemical reaction mechanisms. This allowed for the estimation of the risk of groundwater contamination by metals, low pH, and high dissolved solids. The MIN3P model may forecast storm, flood, and cyclone effects associated with climate change for various mining areas. These catastrophes may affect the water quality and geochemical reactivity of AMD-altered river flows and catchments at and around mining sites.

13.4 RETRASO

The most current model, known as RETRASO, was created from a multiphase perspective, accounting for the various thermal and hydraulic properties of the porous medium in its solid (mineral) phase, liquid (water and dissolved air), and gas phase (dried air and water vapor mixture). These are considered as a whole, and each component's mass balance is calculated. Acero et al. (2009) suggested a reactive transport and multiphase flow model for vadose mine tailings that can address problems with previous models (Molson et al. 2005; Molins and Mayer 2007) that treated the tailings as a single reactor and assumed time-invariant boundary conditions.

13.5 The Geochemist's Workbench

Significant changes in temperature, precipitation, hydrological processes, and the length of the wet and dry seasons are being brought about by climate change. The Geochemist's Workbench software program is used to compute the amount of chemicals needed to disinfect affected water and the seasonal variations in mine discharges brought on by climate change. Additionally, it aids in determining the chemistry of the water that results from the mixing, controlling, and diluting of various water types due to hydrological processes and rainfall. Using one or more of the two reaction-path modeling modules and the five geochemical analysis modules, the scientist may evaluate the stability of theoretical chemical processes and balance them in mining or natural contexts.

13.6 Global Modeling Based on a Network of Chemical Reactors Approach

Hydrodynamic and chemical are the two most often used approaches to mitigate groundwater damage caused by mining. A thorough grasp of a system's hydrogeological properties and structure is necessary for using distributed hydrodynamic models. More research on the second approach, which used an EPANET2 software-built physically dispersed pipe network model (PNM) to aid in water resource management, was conducted by Hamm et al. (2008). This technique views the mine as a network of pipes with certain geometric and chemical properties. Both modeling approaches need the hydrological and chemical parameters, which include the mass of materials, mine void volume, recharge flow rate, outflows, and kinetic constants of the dissolution-precipitation processes. Because of this, these models can assess the water quality of surface and groundwater impacted by AMD and take climate variables into account.

14 Likely Impacts of Climate Change and Possible Response Measures

The intricate interactions between increasing temperatures and increasingly frequent and severe floods and droughts will influence the biological processes in acid sulfate landscapes (ASLs). This chapter covers not every possibility that may arise in a more variable context. Rather, a few important ASLs are listed and explained below.

14.1 Mine Pit Lakes

When exposed, sulfide minerals oxidize, and they become a significant source of metals and acid on the pit walls. Climate change will probably increase acid production under some circumstances since acid production tends to rise with temperature. According to Castendyk and Eary (2009), the climate mostly influences the hydrological processes connected to a mining pit lake. Elevated precipitation will raise lake levels, which might contaminate shallow and surface groundwater.

Corrective action can be required to reduce acid production from the pit walls and clean up the water in mine pit lakes with a significant chance of harming the off-site environment. There are several ways to clean up mine pit lakes (Castendyk and Eary 2009). Moreover, the endeavors towards restoration may facilitate the utilization of pit lakes for advantageous purposes (Castendyk and Eary 2009). This is especially important in dry regions with limited water supplies (Kumar et al. 2009).

14.2 Underwater-Stored Mine Tailings

In actuality, potentially corrosive mining tailings have been disposed of using underwater storage techniques. Tailings Storage Facilities (TSFs) are made to store and handle tailings materials underwater to reduce the time that oxygen comes into contact with sulfide minerals. Many TSFs in mines that are still in operation were not built with the possible consequences of climate change in mind. Future extended drought occurrences are anticipated to cause a significant evaporation rate to cause the TSFs' planned layer of covering water to thin or maybe temporarily dry up, exacerbating the oxidation of the sulfide minerals found inside the TSFs. Strategies to prevent substantial oxidation of sulfide minerals before water levels rise again or the TSF is re-flooded are one way to mitigate the oxidation of sulfide minerals in TSFs caused by dryness. Possible options are as follows:

- Applying chemical agents to prevent sulfide mineral oxidation.
- Assuming no action should the dry time be too brief to allow for the emergence of sizable oxidation of sulfide minerals catalyzed by microbes.

Passivating the mineral surfaces with sodium oleate and phosphates was suggested to prevent pyrite oxidation. These methods, however, would be more suited for handling substances like sulfide ore concentrate that include a large number of sulfide minerals. It is probably not cost-effective to apply such techniques to tailings with negligible sulfide minerals.

Additionally, the proliferation of iron and sulfide-oxidizing bacteria, which are crucial to the microbially accelerated oxidation of sulfide minerals, has been suppressed by using bactericides. This technique suppresses microbial activities (Johnson and Hallberg 2005). However, given the existing circumstances, the antibacterial strategy would be appropriate for offering temporary control over the development of bacteria that oxidize sulfur and iron.

Adding acid-neutralizing components as a top dressing to the drying tailings materials to prevent pH decrease is another strategy to halt the oxidation of sulfide minerals. As a result, less aqueous Fe_3^+ is needed to catalyze the oxidation of sulfide minerals.

The following core understanding is necessary to assess these options:

- When no action is done, how long does it take for the abiotic oxidation of sulfide minerals to produce an acidic pH adequate for microbially catalyzed oxidation?
- Can oxidants other than molecular oxygen, such as H_2O_2 carried by rainfall, be involved in the oxidation of sulfide minerals?
- How do bactericides and other acid-neutralizing agents affect the tailings' ability to resist acidification?

14.3 Sulfidic Mine Spoils

Overburden materials—also known as mine spoils—are frequently deposited on land surfaces during surface mining operations. Mine spoils covered in a layer of capping or sealing materials are often held in special storage facilities to prevent moisture and air from getting to them. This is done as per environmental management plans. On the other hand, sealing layers could not hold up in places with harsh dry and wet seasons (Johnson and Hallberg 2005). Climate change is expected to exacerbate this, perhaps leading to the oxidation of sulfide minerals in the sealed structures used to store mining waste.

Untreated mine waste is often dumped on the land's surface in regions lacking an environmental management plan. This is common to many historical mines and those operating now, especially those in developing countries (Sracek et al. 2006; Lu et al. 2010). As such, there is a risk of fast oxidation and acid formation among the sulfide minerals in mine waste heaps. These metal-rich, acidic mine waste heaps are one of the leading causes of AMD. The treatment of mine water using either active or passive systems has been the primary preventive method for AMD remediation in the industrialized world due to the practical limitations in avoiding AMD development at sources (Sheoran and Sheoran 2006; Kleiv and Thornhill 2008; Johnson and Hallberg 2005). While many developing countries promptly discharge acid mine water into streams (Lin et al. 2007), climate change may cause untreated mine water to overflow mine water treatment systems' designed capacity, allowing metal and acid loadings from high flood occurrences and large amounts of mine water to infiltrate towns downstream. Mines without mine water treatment facilities are expected to cause more serious environmental impacts.

A "source control" approach is required to lessen the additional environmental damage brought about by climate change in these types of ASLs. Nonetheless, the widespread use of these technologies has yielded only patchy results thus far. Not enough thorough research has been done to identify the biogeochemical processes and different remediation techniques.

To enhance our basic comprehension of these biogeochemical processes, the following study areas might need to be investigated:

Development of the microbial community that oxidizes iron and sulfur when pH increases during acid neutralization of mine wastes.

Possible uses for bacteria that oxidize iron and sulfur-containing minerals by becoming neutrophilic.

The mechanisms and kinetics involved in acid neutralization, namely the breakdown of jarosites, the deprotonation of colloids with varying charges, and the desorption of exchangeable aluminum in alkaline environments.

14.4 Coastal Acid Sulfate Landscapes

Acid sulfate footprints that are most sensitive are thought to be found in coastal areas where pyritic soil components are immediately exposed on land surfaces. Examples of these sites include road bases, fishpond bunds, and other development sites resulting from land excavation operations. To stop climate change from causing more environmental harm in these locations, the acidity must be completely neutralized. It is not economically viable to thoroughly neutralize the acidity of the soils on broadacre agricultural farms. As a result, other approaches must be created.

Sugarcane growing seems to be the agricultural practice most vulnerable to climate change, among other practices. Many acid sulfate soils used to grow sugarcane are in a position of dynamic equilibrium concerning acidity, except for dry spells that enhance pyrite oxidation and, consequently, acid generation in the soil profiles.

Furthermore, many of the acid sulfate soils around the coast are either marginally above or below sea level. In the long term, the predicted rise in sea levels brought on by global warming may eventually render sugarcane production unprofitable due to the high expense of the extra engineering required to keep the soils' water table low. The following actions are taken into consideration to design suitable plans and economical techniques for adjusting to the potential environmental changes brought on by climate variability:

- A different cropping plan combined with a move from a deep to a shallow drainage system might allow this type of ASL to adapt to climate change. This alternative method of managing water will decrease the rate at which acid is formed during extended dry spells by delaying the lowering of the water table.
- Retention of garbage: To switch from pre-harvest sugarcane burning to green harvesting, with a minimum of 50% of the rubbish remaining in the field. Techniques for maximizing the circumstances for the sugarcane residues to break down quickly can be devised; this will raise the land's elevation and provide a healthy topsoil layer.
- Raised bed: plant the crop in single-row spacing at ground level, then mound dirt on the cane sets. Through deliberate compaction of interspaces and garbage retention, this will thicken the unsaturated topsoil layer and decrease nitrous oxide emissions.
- During periods of high rainfall and floods, surface water removal is accelerated by slot drainage and in-drain acid neutralization, which eliminate dangerous pollutants from drain water before its discharge into estuarine waterways.
- Sugarcane agriculture in acid-sulfate soils with relatively high water table conditions will be made possible by developing cultivars that are tolerant of both acid and water. These circumstances are predicted to be brought on by increased heavy rainfall events and sea level rise.

15 Research Gaps

CO_2 is one element of the atmosphere that controls both global warming and temperature variations. Mines are the source of atmospheric CO_2 emissions and atmospheric CO_2 sequestration (Fig. 2). The sequestration potential of mines does not completely offset their CO_2 emissions. One factor contributing to climate change and global warming is CO_2 emissions (Hodgkinson and Smith 2018). However, the emission of CO_2 from the environment due to mining waste (Kularatne et al. 2018) can lessen its composition and thus mitigate the effects of climate change brought on by greenhouse gas emissions. Investigating the emission and sequestration of CO_2 from mines requires an LCA study.

Research on greenhouse gas emissions and the direct or indirect consequences of mining on global warming and climate change has dominated the field. In life cycle assessment (LCA) studies on climate change assessment, greenhouse gas emissions from waste disposal to mineral extraction and beneficiation are considered. A review by Farjana et al. (2019) examines the LCA to mitigate the effects of climate change and global warming, focussing solely on emissions resulting from heavy equipment mining operations and technical shortcomings in mine operations. The sequestration of CO_2 by garbage or mines has not been included in LCA studies. The sequestration and CO_2 emission concerns are addressed independently, but their combined significance for a specific mine is not taken into consideration.

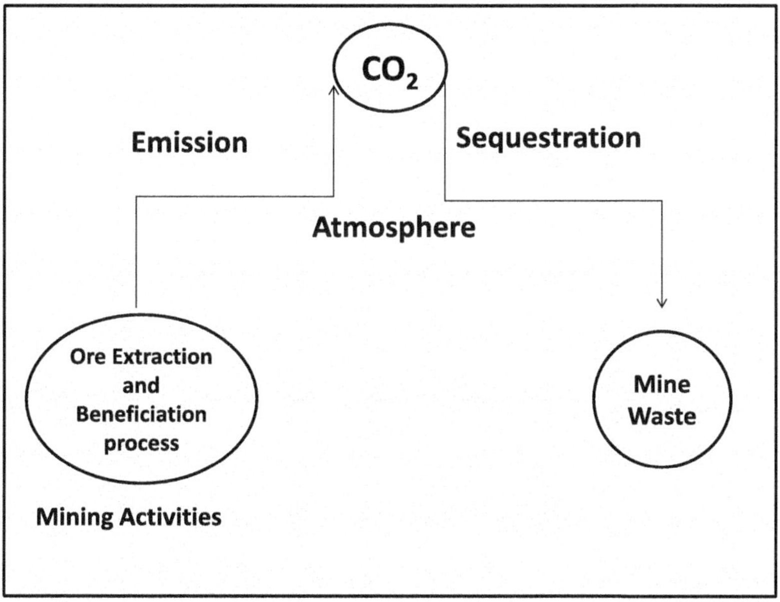

Fig. 2 Overview of the atmospheric balance of CO_2 due to mines

MgO carbonation offers a long-term option for mitigating climate change and can absorb 2–3 Gt of CO_2 year^{-1} (McQueen et al. 2020). However, after 1.4 years, the technical and financial analysis of mineral carbonation using waste from serpentinite mines demonstrates that the money raised by the carbon credit tax is advantageous (Pasquier et al. 2016). In light of greenhouse gas emissions and environmental concerns, the adoption of mining technology is also significant. In terms of the influence on climate change, conveyor belts in the Turkish surface mining industry have a greater environmental cost than vehicles (Erkayaoğlu and Demirel 2016).

The mining sector lacks research and knowledge on LCA (Awuah-Offei and Adekpedjou 2011). Zhang et al. (2018) examined the LCA of mining and land reclamation, encompassing the extraction, transportation, and processing phases, to understand the environmental risk associated with open-pit mines fully. LCA for iron (Ferreira and Leite 2015), coal (Burchart-Korol et al. 2016), copper (Memary et al. 2012), and lead-zinc (Tao et al. 2019) has taken into account the effects on the environment and its role in climate change, except for the CO_2 sequestration caused by dumping mining waste into the ecosystem. The ability of waste or abandoned mines to sequester CO_2 in natural settings needs investigation. The ability of tailings to sequester CO_2 from the atmosphere is widely established (Oskierski et al. 2013). However, LCA for tailings management (Reid et al. 2009; Adiansyah et al. 2017) has somehow ignored the tailings' ability to sequester CO_2 and their contribution to climate change.

More research is required to find out if mineral carbonation in situ is practical. LCA of ex situ mineral carbonation utilizing mafic and ultramafic mining wastes shows a viable strategy for reducing CO_2 emissions under controlled operating conditions (Bodénan et al. 2014). Ex situ CO_2 sequestration can release more CO_2 than sequestration because pretreatment and grinding need large energy or cost inputs (Wang et al. 2018) (Fig. 3). Mineral carbonation releases the majority of CO_2 during the mineralization, pretreatment, and capture stages (Khoo et al. 2011).

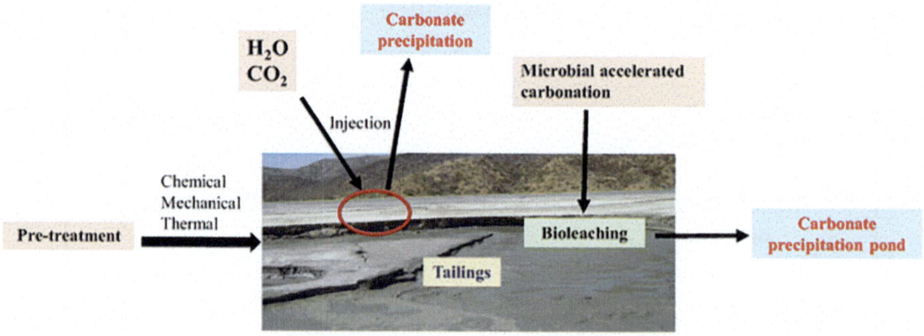

Fig. 3 Process of mineral carbonation of mining waste. (After Li et al. 2018)

16 Challenges in Industrial Applications

Sequencing man-made CO_2 from the sky is controversial since the weather of mining debris is slow. The use of heat, chemical, or mechanical pretreatment raises expenses for the environment and money. One of the main obstacles to using mineral carbonation on an industrial scale is its sluggish reaction kinetics. It has been noted that adding sodium salts improves the mineral carbonation process and chemical reaction (Wang et al. 2019). However, the process incurs additional financial expenditures due to the inclusion of chemicals and feedstock processing. Hitch and Dipple (2012) noted that integrating mineral carbonation with mining operations can provide extra carbon credits for sale and is a financially feasible industrial solution. Carbon sequestration from tailings can increase up to nine times when biosolids are added (Antonelli et al. 2018), which is cost-effective.

When applied to acidic mine tailings, steel slag and coal fly ash decrease the need for neutralization agents (NaOH) and improve mineral carbonation (Ebrahimi et al. 2018). CO_2, CH_4, and H_2S can potentially be sequestered by steel and iron slag (Reddy et al. 2019), and their inclusion in mine waste would improve the capability for sequestration. There are no reports in the literature on the sequestration of additional gases using mining waste. Thus, more investigation is required. However, one possible way to store H_2S is by using slag from the iron and steel industries (Montes-Morán et al. 2012), indicating the mining waste's likely ability to sequester additional gases. Under air circumstances, haematite (Fe_2O_3) uses sulfuration to remove H_2S when significant water content is present (Sarperi et al. 2014). The sequestration of CO_2 by Fe_2O_3 is facilitated by the injection of H_2S, indicating its potential use for geological storage in sandstone formations (Alpermann et al. 2016).

It is important to promote the identification of standard compounds that are advantageous for CO_2 sequestration as well as ore beneficiation procedures. In addition to helping further lower the amount of CO_2 in the atmosphere and lessen the burden of introducing reactive agents for the mineral carbonation process, it would also assist in keeping the process's financial costs under control. A significant obstacle to implementing and promoting CO_2 sequestration from mining waste is the absence of regulations and an insufficient legal framework.

Further study is required to enhance the grain size of tailings to ensure maximal mineral extraction during the beneficiation process and CO_2 sequestration after its discharge to the dump site. The size of grains has a major impact on CO_2 sequestration. It has been shown that following the physical and chemical activation processes, the surface area of raw serpentine grows from 8 to 330 m^2/g (Maroto-Valer et al. 2005). The energy-intensive crushing and grinding process aids in sequestering CO_2; for sequestration, the substrate, in the case of tailings, has already been crushed and pulverized. Smaller grains, however, cause less porosity, slowing the carbonation process. Therefore, more investigation is required to optimize tailings particle size for optimal mineral extraction and CO_2 sequestration.

For underground mines to sequester CO_2, CO_2 must be injected far into the ground. Compressed CO_2 injection and transportation are energy-intensive and expensive (Tayari and Blumsack 2020). The adoption of CO_2 collection and storage in natural silicates and waste is limited in scale due to its high cost (Olajire 2013). On the other hand, it appears to be a feasible way to lower atmospheric CO_2 levels and sequester them at minimal expense to the economy and environment in the upper surface of mine waste. Long-term viable options appear to be passive mineral carbonation and the development of carbonate layers on the upper surface of mine waste. Reactive surface loss and porosity are the main obstacles preventing tailings from passively mineral carbonating (Power et al. 2013b). Because the next batch of tailings is not immediately deposited in the same location, atmospheric CO_2 is exposed to freshly added material for longer. When reaction agents are added, there is a greater chance that the additional material may weather and form carbonates on the upper surface of the tailings.

17 Conclusions

Under watery mining conditions, the oxidation of iron sulfides (pyrite, arsenopyrite, and marcasite) and metal sulfide minerals converts sulfides into sulfates, produces acid, and solubilizes hazardous components. These processes are called hydration, oxidation, and hydrolysis. Significant concentrations of calcite, dolomite, and limestone, as well as the weathering of aluminosilicate minerals and the breakdown of native and rock dust gypsum, can all raise pH levels and precipitate oxides and hydroxy sulfates, which scavenge metals from solution.

This chapter aimed to determine if mineral carbonation could be utilized to reduce global warming. One of the main obstacles to solving the problem is the paucity of research on the function of mineral carbonation in this regard. Environmental pollution results from mining waste, and controlling it presents significant difficulties for the government. Using mine waste to absorb CO_2 from the atmosphere seems to be a workable and sustainable way to fight climate change and protect the environment.

The main factors influencing mineral carbonation are water availability, CO_2 atmospheric pressure, and grain size. The purpose of mineral carbonation is that the ore beneficiation process already crushes mining waste, such as tailings, into smaller particles, conserving energy. Secondly, the carbonation process is accelerated by the high water concentration in the tailings. Effluent from the mine that has been improved with magnesium and calcium ions is suitable for absorbing carbon dioxide from the atmosphere. Further research is needed to create economically viable technology for larger-scale CO_2 sequestration to mitigate global warming and climate change.

The acid sulfate landscapes are expected to be significantly impacted by the anticipated climate change in various ways. There will be an increase in the frequency and intensity of extreme weather events, such as droughts and torrential rains. This would worsen the already fragile ecosystems by hastening acid generation and increasing the frequency and amount of acid spills from acid sulfate footprints onto the adjacent regions. Reducing the ecological harm caused by climate change requires the development of affordable technologies and suitable management strategies.

Climate change will immediately affect water sources, the makeup of mine waste, and hydrogeologic and geochemical components that affect water flow, geochemical reactions, and the transport of pollutants at mine sites. Climate change-related heavy rainfall and flooding events may cause problems for the hydraulic infrastructure (dams, ditches, spillways, and holding ponds) at mine sites, thereby increasing pollution discharge or necessitating the shutdown of mining activities. It will also affect simple soil coverings, dams, store and release systems, and permafrost—all of which are essential for the long-term storage of tailings and other pollutants following mine closure. The kind of plants present and the phytoremediation of mining waste will be impacted by variations in temperature, evapotranspiration, and rainfall patterns. If mining sites are not sufficiently planned for, maintained, or controlled, they will be more vulnerable to the effects of climate change. Consequently, adopting adaptive strategies to mitigate the consequences of climate change in mining locations will be more successful and economical than just accepting relocation options.

The most successful conceptual model will be one that considers climate change and evaluates the effects of various climate change scenarios on AMD's geochemical development from mining waste.

References

Abbassi R, Khan F, Hawboldt K (2009) Prediction of minerals producing acid mine drainage using a computer-assisted thermodynamic chemical equilibrium model. Mine Water Environ 28:74–78

Abraham MR, Susan TB (2017) Water contamination with heavy metals and trace elements from Kilembe copper mine and tailing sites in Western Uganda; implications for domestic water quality. Chemosphere 169:281–287

Acero P, Ayora C, Carrera J, Saaltink MW, Olivella S (2009) Multiphase flow and reactive transport model in vadose tailings. Appl Geochem 24:1238–1250

ACIA (2005) Arctic climate impacts assessment. Cambridge University Press, Cambridge

ACIL Tasman (2007) Water reform and industry—implications of recent water initiatives for the minerals, petroleum, energy, pulp and paper industries Melbourne

Adiansyah JS, Haque N, Rosano M, Biswas W (2017) Application of a life cycle assessment to compare environmental performance in coal mine tailings management. J Environ Manag 199:181–191

Alpermann T, Dietrich M, Ostertag-Henning C (2016) Mineral trapping of a CO_2/H_2S mixture by hematite under initially dry hydrothermal conditions. Int J Greenhouse Gas Control 51:346–356

Amoah-Antwi C, Kwiatkowska-Malina J, Thornton SF, Fenton O, Malina G, Szara E (2020) Restoration of soil quality using biochar and brown coal waste: a review. Sci Total Environ 722:137852

Antonelli PM, Fraser LH, Gardner WC, Broersma K, Karakatsoulis J, Phillips ME (2018) Long-term carbon sequestration potential of biosolids-amended copper and molybdenum mine tailings following mine site reclamation. Ecol Eng 117:38–49

Arce GLAF, Neto TGS, Ávila I, Luna CMR, Carvalho JA Jr (2017) Leaching optimization of mining wastes with lizardite and brucite contents for use in indirect mineral carbonation through the pH swing method. J Clean Prod 141:1324–1336

Arcos D, Perez del Villar L, Bruno J, Domenech C (2008) Geochemical modeling of the weathering zone of the "Mina Fe" U deposit (Spain): a natural analog for nuclear-spent fuel alteration and stability processes in radwaste disposal. Appl Geochem 23:807–821

Assima GP, Larachi F, Beaudoin G, Molson J (2013) Dynamics of carbon dioxide uptake in chrysotile mining residues – effect of mineralogy and liquid saturation. Int J Greenhouse Gas Control 12:124–135

Assima GP, Larachi F, Molson J, Beaudoin G (2014a) Comparative study of five Québec ultramafic mining residues for use in direct ambient carbon dioxide mineral sequestration. Chem Eng J 245:56–64

Assima GP, Larachi F, Molson J, Beaudoin G (2014b) Impact of temperature and oxygen availability on the dynamics of ambient CO_2 mineral sequestration by nickel mining residues. Chem Eng J 240:394–403

Asta MP, Ayora C, Acero P, Cama J (2010) Field rates for natural attenuation of arsenic in Tinto Santa Rosa acid mine drainage (SW Spain). J Hazard Mater 177:1102–1111

Auld H, MacIver D (2006) Changing weather patterns, uncertainty, and infrastructure risks: emerging adaptation requirements. Paper presented at the Engineering Institute of Canada Climate Change Technology Conference, Environment Canada, Adaptation and Impacts Research Division, Ottawa, 10–12 May 2006

Auld H, MacIver D, Klaassen J (2006) Adaptation options for infrastructure under changing climate conditions. Paper presented at the Engineering Institute of Canada Climate Change Technology Conference, Environment Canada, Adaptation and Impacts Research Division, Ottawa, 10–12 May 2006

Awuah-Offei K, Adekpedjou A (2011) Application of life cycle assessment in the mining industry. Int J Life Cycle Assess 16:82–89

Balaz P, Turianicova E, Fabian M, Kleiv RA, Briancin J, Obut A (2008) Structural changes in olivine (Mg, Fe)2SiO4 mechanically activated in high-energy mills. Int J Miner Process 88:1–6

Barnett TP, Pierce DW, Hidalgo HG, Bonfils C, Santer BD, Das T, Bala G, Wood AW, Nozawa T, Mirin AA, Cayan DR, Dettinger MD (2008) Human-induced changes in the hydrology of the western United States. Science 319:1080–1083

Batjes NH (2016) Harmonized soil property values for broad-scale modeling (WISE30sec) with estimates of global soil carbon stocks. Geoderma 269:61–68

Bea SA, Ayora C, Carrera J, Saaltink MW, Dold B (2010) Geochemical and environmental controls on the genesis of soluble efflorescent salts in Coastal Mine Tailings Deposits: a discussion based on reactive transport modeling. J Contam Hydrol 111:65–82

Beaudoin G, Nowamooz A, Assima GP, Lechat K, Gras A, Entezari A, Kandji EHB, Awoh A-S, Horswill M, Turcotte S, Larachi F, Dupuis C, Molson J, Lemieux J-M, Maldague X, Plante B, Bussière B, Constantin M, Duchesne J, Therrien R, Fortier R (2017) Passive mineral carbonation of Mg-rich mine wastes by atmospheric CO_2. Energy Procedia 114:6083–6086

Beinlich A, Austrheim H (2012) In situ sequestration of atmospheric CO_2 at low temperature and surface cracking of serpentinized peridotite in mine shafts. Chem Geol 332-333:32–44

Bertelli M, Baker T, Cleverley JS, Ulrich T (2009) Geochemical modeling of a Zn–Pb skarn: constraints from LA–ICP–MS analysis of fluid inclusions. J Geochem Explor 102:13–26

Bhattacharya A, Routh J, Jacks G, Bhattacharya P, Morth M (2006) Environmental assessment of abandoned mine tailings in Adak, Vasterbotten district (Northern Sweden). Appl Geochem 21:1760–1780

Bjelkevik A (2005) Water cover closure design for tailings dams. State-of-the-Art Report

Blume H-P, Brümmer GW, Horn R, Kandeler E, Kögel-Knabner I, Kretzschmar R, Stahr K, Wilke B-M (2016) Schefer/Schachtschabel Lehrbuch der Bodenkunde, vol 16. Springer Spektrum, Berlin/Heidelberg. ISBN 978-3-662-49960-3

Bodénan F, Bourgeois F, Petiot C, Augé T, Bonfils B, Julcour-Lebigue C, Guyot F, Boukary A, Tremosa J, Lassin A, Gaucher EC, Chiquet P (2014) Ex situ mineral carbonation for CO_2 mitigation: evaluation of mining waste resources, aqueous carbonation processability and life cycle assessment (Carmex project). Miner Eng 59:52–63
Bouzalakos S, Thapa S, Timms W, Abbas A (2014) Acid mine water treatment with brucite in ultramafic tailings coupled with carbon dioxide sequestration. In: 4th international conference on water management in mining at Vina del Mar, Chile, 28–30 May 2014, pp 591–602
Bromhal GS, Sams WN, Jikich S, Ertekin T, Smith DH (2005) Simulation of CO_2 sequestration in coal beds: the effects of sorption isotherms. Chem Geol 217:201–211
Brooks BW, Chambliss CK, Stanley JK, Ramirez AJ, Banks KE, Johnson RD, Lewis RJ (2005) Determination of select antidepressants in fish from an effluent-dominated stream. Environ Toxicol Chem 24:464–469
Brooks BW, Valenti TW, Cook-Lindsay BA, Forbes MG, Doyle RD, Scott JT, Stanley JK (2011) Influence of climate change on reservoir water quality assessment and management: effects of reduced inflows on diel pH and site-specific contaminant hazards. In: Linkov I, Bridges TS (eds) Climate: global change and local adaptation, 491 NATO science for peace and security series C: environmental security. Springer Science+Business Media B.V.
Brown A, Dart J, Quesnel J (2006) Climate change and watershed management at the Hemlo Operations. Paper presented at Sudbury Restoration Workshop, Sudbury, 22 February 2006
Burchart-Korol D, Fugiel A, Czaplicka-Kolarz K, Turek M (2016) Model of environmental life cycle assessment for coal mining operations. Sci Total Environ 562:61–72
Casiot C, Egal M, Elbaz-Poulichet F, Bruneel O, Bancon-Montigny C, Cordier M-A, Gomez E, Aliaume C (2009) Hydrological and geochemical control of metals and arsenic in a Mediterranean river contaminated by acid mine drainage (the Amous River, France); preliminary assessment of impacts on fish (Leuciscus cephalus). Appl Geochem 24:787–799
Castendyk DN and LE Eary (2009) Mine pit lakes: characteristics, predictive modeling and sustainability. SME, Colorado, ISBN-13: 978-0-87335-305-2, pp: 316
Chen C-J, Jiang W-T (2012) Influence of waterfall aeration and seasonal temperature variation on the iron and arsenic attenuation rates in an acid mine drainage system. Appl Geochem 27:1966–1978
Cidu R, Frau F, Da Pelo S (2011) Drainage at abandoned mine sites: natural attenuation of contaminants in different seasons. Mine Water Environ 30:113–126
Comiso JC, Parkinson CL, Gertsen R, Stock L (2008) Accelerated decline in the Arctic sea ice cover. Geophys Res Lett 35(1):1–6
Corkhill CL, Vaughan DJ (2009) Arsenopyrite oxidation – a review. Appl Geochem 24:2342–2361
Crouch CM, McKnight DM, Todd AS (2013) Quantifying sources of increasing zinc from acid rock drainage in an alpine catchment under a changing hydrologic regime. Hydrol Process 27:721–733
Dignac MF, Derrien D, Barre P, Barot S, Cécillon L, Chenu C, Chevallier T, Freschet GT, Garnier P, Guenet B et al (2017) Increasing soil carbon storage: mechanisms, effects of agricultural practices and proxies. A review. Agron Sustain Dev 37:1–27
Driscoll JM, Meixner T, Molotch NP, Sickman JO, Williams MW, McIntosh JC, Brooks PD (2011) Inverse geochemical reaction path modeling and the impact of climate change on hydrologic structure in snowmelt-dominated catchments in the Southwestern USA. American Geophysical Union, Fall Meeting 2011, Abstract #H13C-1221
Ebrahimi A, Saffari M, Hong Y, Milani D, Montoya A, Valix M, Minett A, Abbas A (2018) Mineral sequestration of CO_2 using saprolite mine tailings in the presence of alkaline industrial wastes. J Clean Prod 188:686–697
Erkayaoğlu M, Demirel N (2016) A comparative life cycle assessment of material handling systems for sustainable mining. J Environ Manag 174:1–6
Espana JS, Pamo EL, Santofimia E, Aduvire O, Reyes J, Barettino D (2005) Acid mine drainage in the Iberian Pyrite Belt (Odiel river watershed, Huelva, SW Spain): geochemistry, mineralogy and environmental implications. Appl Geochem 20:1320–1356

Eurostat (2017) Generation of waste by waste category. http://appsso.eurostat.ec.europa.eu/nui/show.do?Dataset=env_wasgen&lang=en
Farjana SH, Huda N, Mahmud MAP, Saidur R (2019) A review on the impact of mining and mineral processing industries through life cycle assessment. J Clean Prod 231:1200–1217
Ferreira H, Leite MGP (2015) A life cycle assessment study of iron ore mining. J Clean Prod 108:1081–1091
Ford JD, Pearce T (2007) Planning for climate change in the mining sector. Can Min J
Ford JD, Pearce T, Prno J, Duerden F, Berrang-Ford L, Marshall D (2008) Stormy days ahead; weather changes pose many threats to the mining industry. Can Min J:22–24
Ford JD, Keskitalo C, Smith T, Pearce T, Berrang-Ford L, Smit B (2010a) Case study and analog methodologies in climate change vulnerability research. Wiley Interdiscip Rev 1:374. https://doi.org/10.1002/wcc.48
Ford JD, Pearce T, Prno J, Duerden F, Berrang-Ford L, Beaumier M, Smith T, Marshall D (2010b) Perceptions of climate change risks in primary resource use industries: a survey of the Canadian mining sector. Reg Environ Chang 10(1):65–81
Ford JD, Pearce T, Prno J, Duerden F, Berrang-Ford L, Smith T, Beaumier M (2011) Canary in a coal mine: perceptions of climate change risks and response options among Canadian mine operations. Clim Chang 109:399
Fox JF, Campbell JE, Acton PM (2020) Carbon sequestration by reforesting legacy grasslands on coal mining sites. Energies 13:6340
Furgal C, Prowse T (2008) Northern Canada. In: Lemmen D, Warren F, Lacroix J, Bush E (eds) From impacts to adaptation: Canada in a changing climate. Government of Canada, Ottawa, pp 57–118
Gammons CH, Nimick DA, Parker SR, Cleasby TE, McCleskey RB (2005a) Diel behavior of iron and other heavy metals in a mountain stream with acidic to neutral pH: Fisher Creek, Montana, USA. Geochim Cosmochim Acta 69:2505–2516
Gammons CH, Wood SA, Nimick DA (2005b) Diel behavior of rare earth elements in a mountain stream with acidic to neutral pH. Geochim Cosmochim Acta 69:3747–3758
Gammons CT, David G, Nimick A, Parker S, DeGrandpre M (2007) Diel changes in water chemistry in an arsenic-rich stream and treatment-pond system. Sci Total Environ 384:433–451
Ghacham AB, Cecchi E, Pasquier L-C, Blais J-F, Mercier G (2015) CO_2 sequestration using waste concrete and anorthosite tailings by direct mineral carbonation in gas-solid–liquid and gas-solid routes. J Environ Manag 163:70–77
Government of Canada (2006) Chapter 2: Adapting to the impacts of climate change. In: Report of the Commissioner of the Environment and Sustainable Development to the House of Commons. Office of the Auditor General of Canada, Ottawa, pp 1–36
Hamilton JL, Wilson SA, Morgan B, Turvey CC, Paterson DJ, Mac-Rae C, McCutcheon J, Southam G (2016) Nesquehonite sequesters transition metals and CO_2 during accelerated carbon mineralization. Int J Greenhouse Gas Control 55:73–81
Hamilton JL, Wilson SA, Morgan B, Turvey CC, Paterson DJ, Jowitt SM, McCutcheon J, Southam G (2018) Fate of transition metals during passive carbonation of ultramafic mine tailings via air capture with potential for metal resource recovery. Int J Greenhouse Gas Control 71:155–167
Hamilton JL, Wilson SA, Morgan B, Harrison AL, Turvey CC, Paterson DJ, Dipple GM, Southam G (2020) Accelerating mineral carbonation in ultramafic mine tailings via direct CO_2 reaction and heap leaching with potential for base metal enrichment and recovery. Econ Geol 115(2):303–323
Hamm V, Collon-Drouaillet P, Fabriol R (2008) Two modeling approaches to water-quality simulation in a flooded iron-ore mine (Saizerais, Lorraine, France): a semi-distributed chemical reactor model and a physically based distributed reactive transport pipe network model. J Contam Hydrol 96:97–112
Harrison AL, Power IM, Dipple GM (2013) Accelerated carbonation of brucite in mine tailings for carbon sequestration. Environ Sci Technol 47:126–134

Hitch M, Dipple GM (2012) Economic feasibility and sensitivity analysis of integrating industrial-scale mineral carbonation into mining operations. Miner Eng 39:268–275

Hitch M, Ballantyne SM, Hindle SR (2010) Revaluing mine waste rock for carbon capture and storage. Int J Min Reclam Environ 24(1):64–79

Hodgkinson JH, Smith MH (2018) Climate change and sustainability as drivers for the next mining and metals boom: the need for climate-smart mining and recycling. Resour Policy:101205. https://doi.org/10.1016/j.resourpol.2018.05.016

House KZ, House CH, Schrag DP, Aziz MJ (2007) Electrochemical acceleration of chemical weathering as an energetically feasible approach to mitigating anthropogenic climate change. Environ Sci Technol 41:8464–8470

Hüblová L, Frouz J (2021) Contrasting effect of coniferous and broadleaf trees on soil carbon storage during reforestation of forest soils and afforestation of agricultural and post-mining soils. J Environ Manag 290:112567

Huijgen WJJ, Witkamp G-J, Comans RNJ (2005) Mineral CO_2 sequestration by steel slag carbonation. Environ Sci Technol 39(24):9676–9682

Huijgen WJJ, Comans RNJ, Witkamp G-J (2007) Cost evaluation of CO_2 sequestration by aqueous mineral carbonation. Energy Convers Manag 48:1923–1935

Indian Bureau of Mines (IBM) (2019) Coal and lignite. In: Indian minerals yearbook 2018 (Part – III: mineral reviews), 57th edn. Government of India, pp 1–31

Infrastructure Canada (2006) Adapting infrastructure to climate change in Canada's cities and communities: a literature review. Research and Analysis Division, Government of Canada, Ottawa, p 23

Instanes A, Anisimov OA, Brigham L, Goering D, Khrustalev L, Ladanyi B, Larsen JO (2005) Chapter 16: Infrastructure: buildings, support systems, and industrial facilities. In: Symon C (ed) Arctic climate impact assessment scientific report. Cambridge University Press, Cambridge, pp 907–944

IPCC (2007) Climate change 2007: impacts, adaptation, and vulnerability. In: Working Group II contribution to the intergovernmental panel on climate change fourth assessment report, Geneva

IPCC (2013) Climate change, the physical science basis. In: Stocker TF, Qin D, Plattner G-K, Tignor M, Allen SK, Boschung J, Nauels A, Xia Y, Bex V, Midgley PM (eds) Contribution of Working Group I to the fifth assessment report of the intergovernmental panel on climate change. Cambridge University Press, Cambridge, UK and New York, NY

Jacobs AD, Hitch M (2011) Experimental mineral carbonation: approaches to accelerate CO_2 sequestration in mine waste materials. Int J Min Reclam Environ 25(4):321–331

Jarvis AP (2006) The role of dissolved carbon dioxide in governing deep coal mine water quality and determining treatment process selection. In: 7th international conference on acid rock drainage, pp 833–843. https://doi.org/10.21000/JASMR06020833

Johnson DB and KB Hallberg (2005) Acid mine drainage remediation options: a review. Sci. Tot. Environ., 338: 3-14. https://doi.org/10.1016/j.scitotenv.2004.09.002

Kalkuhl M, Edenhofer O, Lessmann K (2015) The role of carbon capture and sequestration policies for climate change mitigation. Environ Resour Econ 60:55–80

Kandji EHB, Plante B, Bussière B, Awoh A-S, Ibrahim Saïb BA, Beaudoin G, Dupont P-P (2015) Prediction of drainage water quality from mining waste with carbon sequestration potential. In: Proceedings of the IMWA/ICARD, Santiago, Chile, April 21–24

Kandji EHB, Plante B, Bussière B, Beaudoin G, Dupont P-P (2017) Geochemical behavior of ultramafic waste rocks with carbon sequestration potential: a case study of the Dumont nickel project, Amos, Québec. Environ Sci Pollut Res 24:11734–11751

Kanzler M, Böhm C, Freese D (2021) The development of soil organic carbon under young black locust (Robinia pseudoacacia L.) trees at a post-mining landscape in eastern Germany. New For 52:47–68

Kaszuba J, Yardley B, Andreani M (2013) Experimental perspectives of mineral dissolution and precipitation due to carbon dioxide-water-rock interactions. Rev Mineral Geochem 77(1):153–188

Kemache N, Pasquier L-C, Cecchi E, Mouedhen I, Blais J-F, Mercier G (2017) Aqueous mineral carbonation for CO_2 sequestration: from laboratory to pilot scale. Fuel Process Technol 166:209–216

Khoo HH, Bu J, Wong RL, Kuan SY, Sharratt PN (2011) Carbon capture and utilization: preliminary life cycle CO_2, energy, and cost results of potential mineral carbonation. Energy Procedia 4:2494–2501

Kimball BA, Runkel RL, Wanty RB, Verplanck PL (2010) Reactive solute transport simulation of pre-mining metal concentrations in mine-impacted catchments: Redwell Basin, Colorado, USA. Chem Geol 269:124–136

Kleiv RA , M. Thornhill (2008) Predicting the neutralisation of acid mine drainage in anoxic olivine drains. Miner Eng 21:279–287. https://doi.org/10.1016/j.mineng.2007.10.005

Kowalska A, Grobelak A, Almås ÅR, Singh BR (2020) Effect of biowastes on soil remediation, plant productivity, and soil organic carbon sequestration: a review. Energies 13:5813

Kruse NA (2007) Simulation of Hydrogeochemical processes in abandoned underground mines. Thesis. Newcastle University

Kruse NA, Strosnider WHJ (2015) Carbon dioxide dynamics and sequestration in mine water and waste. Mine Water Environ 34:3–9

Kruse NA, Younger PL (2009) Development of thermodynamically-based models for simulation of hydrogeochemical processes coupled to channel flow processes in abandoned underground mines. Appl Geochem 24:1301–1311

Kularatne K, Sissmann O, Kohler E, Chardin M, Noirez S, Martinez I (2018) Simultaneous ex-situ CO_2 mineral sequestration and hydrogen production from olivine-bearing mine tailings. Appl Geochem 95:195–205

Kumar RN, CD McCullough and MA Lund (2009) Water resources in Australian mine pit lakes. Mining Technol, 15:247–252

Kusin FM, Syed Hasan SNM, Hassim MA, Molahid VLM (2020) Mineral carbonation of sedimentary mine waste for carbon sequestration and potential reutilization as cementitious material. Environ Sci Pollut Res 27(11):12767–12780

Lechat K, Lemieux J-M, Molson J, Beaudoin G, Hébert R (2016) Field evidence of CO_2 sequestration by mineral carbonation in ultramafic milling wastes, Thetford Mines, Canada. Int J Greenhouse Gas Control 47:110–121

Lemmen D, Warren F, Lacroix J, Bush E (eds) (2008) From impacts to adaptation: Canada in a changing climate 2007. Government of Canada, Ottawa, p 448

Leung DYC, Caramanna G, Maroto-Valer MM (2014) An overview of the current status of carbon dioxide capture and storage technologies. Renew Sust Energ Rev 39:426–443

Li J, Hitch M (2016a) Carbon dioxide adsorption isotherm study on mine waste for integrated CO_2 capture and sequestration processes. Powder Technol 291:408–413

Lin C, Y Wu, W Lu, A Chen, Y Liu (2007) Water chemistry and ecotoxicity of an acid mine drainage affected stream in subtropical China during a major flood event. J Hazard Mater 142: 199–207. https://doi.org/10.1016/j.jhazmat.2006.08.006

Li J, Hitch M (2016b) Economic analysis on the application of mechanical activation in an integrated mineral carbonation process. Int Biodeterior Biodegradation 128:63–71

Li J, Hitch M, Power IM, Pan Y (2018) Integrated mineral carbonation of ultramafic mine deposits-a review. Minerals 8:147

Li Y, Wu Z, Dong X, Jia Z, Sun Q (2019) Variance in bacterial communities, potential bacterial carbon sequestration and nitrogen fixation between light and dark conditions under elevated CO_2 in mine tailings. Sci Total Environ 652:234–242

Lu W, C Lin, Y Ma, S Huang and C Si et al., (2010) Characteristics and potential environmental consequence of weathered materials in the surface layer of a spontaneously combusting mine spoil stockpile. Appl Geochem 25:496–501. https://doi.org/10.1016/j.apgeochem.2010.01.004

Malmström ME, Berglund S, Jarsjo J (2008) Combined effects of spatially variable flow and mineralogy on the attenuation of acid mine drainage in groundwater. Appl Geochem 23:1419–1436

Manaye A, Negash M, Alebachew M (2019) Effect of degraded land rehabilitation on carbon stocks and biodiversity in a semi-arid region of Northern Ethiopia. Forest Sci Technol 15:70–79

Marce R, Rodriguez-Arias MA, Garcia JC, Armengol J (2010) El Nino Southern Oscillation and climate trends impact reservoir water quality. Glob Change Biol 16:2857–2865

Maria AD, Snellings R, Alaert L, Quaghebeur M, Acker KV (2020) Environmental assessment of CO_2 mineralization for sustainable construction materials. Int J Greenhouse Gas Control 93:102882

Maroto-Valer MM, Fauth DJ, Kuchta ME, Zhang Y, Andrésen JM (2005) Activation of magnesium-rich minerals as carbonation feedstock materials for CO_2 sequestration. Fuel Process Technol 86:1627–1645

Masson M, Schafer J, Blanc G, Anschutz P (2007) Seasonal variations and annual fluxes of arsenic in the Garonne, Dordogne, and Isle Rivers, France. Sci Total Environ 373:196–207

Matthews DL, Peter JM, Scott SD, Leybourne MI (2008) Distribution, mineralogy, and geochemistry of selenium in felsic volcanic-hosted massive sulfide deposits of the Finlayson Lake District, Yukon Territory, Canada. Eco Geol 103:61–88

McQueen N, Kelemen P, Dipple G, Renforth P, Wilcox J (2020) Ambient weathering of magnesium oxide for CO_2 removal from the air. Nat Commun 11:3299

Memary R, Giurco D, Mudd G, Mason L (2012) Life cycle assessment: a time-series analysis of copper. J Clean Prod 33:97–108

Molins S, Mayer U (2007) Coupling between geochemical reactions and multicomponent gas and solute transport in unsaturated media: a reactive transport modeling study. Water Resour Res 43:W05435

Molson J, Fala O, Aubertin M, Bussiere B (2005) Numerical simulations of pyrite oxidation and acid mine drainage in unsaturated waste rock piles. J Contam Hydrol 78:343–371

Montes-Morán MA, Concheso A, Canals-Batlle C, Aguirre NV, Ania CO, Martín MJ, Masaguer V (2012) Linz-Donawitz steel slag for the removal of hydrogen sulfide at room temperature. Environ Sci Technol 46:8992–8997

Mukhopadhyay S, Masto RE (2022) Comparative evaluation of *Cassia siamea* and *Albizia lebbeck* for their potential to recover carbon and nutrient stocks in a chronosequence post-mining site. Catena 208:105726

Ncongwane MS, Broadhurst JL, Petersen J (2018) Assessment of the potential carbon footprint of engineered processes for the mineral carbonation of PGM tailings. Int J Greenhouse Gas Control 77:70–81

Nordstrom DK (2009) Acid rock drainage and climate change. J Geochem Explor 100:97–104

Olajire AA (2013) A review of mineral carbonation technology in sequestration of CO_2. J Pet Sci Eng 109:364–392

Oskierski HC, Dlugogorski BZ, Jacobsen G (2013) Sequestration of atmospheric CO_2 in chrysotile mine tailings of the Woodsreef Asbestos mine, Australia: quantitative mineralogy, isotopic fingerprinting and carbonation rates. Chem Geol 358:156–169

Oskierski HC, Dlugogorski BZ, Oliver TK, Jacobsen G (2016) Chemical and isotopic signatures of waters associated with the carbonation of ultramafic mine tailings, Woodsreef Asbestos mine, Australia. Chem Geol 436:11–23

Ouangrawa M, Molson J, Aubertin M, Bussière B, Zagury GJ (2009) Reactive transport modeling of mine tailings columns with capillarity-induced high water saturation for preventing sulfide oxidation. Appl Geochem 24:1312–1323

Ouangrawa M, Aubertin M, Molson JW, Bussière B, Zagury GJ (2010) Preventing acid mine drainage with an elevated water table: long-term column experiments and parameter analysis. Water Air Soil Pollut 213:437–458

Pan S-Y, Chen Y-H, Fan L-S, Kim H, Gao X, Ling T-C, Chiang P-C, Pei S-L, Gu G (2020) CO_2 mineralization and utilization by alkaline solid wastes for potential carbon reduction. Nat Sustain 3:399–405

Pasquier L-C, Mercier G, Blais J-F, Cecchi E, Kentish S (2016) Technical & economic evaluation of a mineral carbonation process using southern Québec mining wastes for CO_2 sequestration of raw flue gas with by-product recovery. Int J Greenhouse Gas Control 50:147–157

Pearce TD, Ford JD, Prno J, Duerden F, Pittman J, Beaumier M, Berrang-Ford L, Smit B (2011) Climate change and mining in Canada. Mitig Adapt Strateg Glob Change 16:347–368

Pereira P, Caçador I, Vale C, Caetano C, Costa A (2007) Decomposition of belowground biomass litter and metal dynamics in salt marshes (Tagus Portugal). Sci Total Environ 380:93–101

Placek-Lapaj A, Grobelak A, Fijalkowski K, Singh BR, Almås ÅR, Kacprzak M (2019) Post-mining soil as carbon storehouse under polish conditions. J Environ Manag 238:307–314

Power IM, Dipple GM, Southam G (2010) Bioleaching of ultramafic tailings by Acidithiobacillus spp. for CO_2 sequestration. Environ Sci Technol 44(1):456–462

Power IM, Wilson SA, Dipple GM (2013b) Serpentinite carbonation for CO_2 sequestration. Elements 9:115–121

Power IM, Dipple GM, Bradshaw PMD, Harrison AL (2020) Prospects for CO_2 mineralization and enhanced weathering of ultramafic mine tailings from the Baptiste nickel deposit in British Columbia, Canada. Int J Greenhouse Gas Control 94:102895

Prno J (2008) We're breaking up: what does climate change mean for the mining industry. Up Here 2008:25–26

Pronost J, Beaudoin G, Lemieux J-M, Hébert R, Constantin M, Marcouiller S, Klein M, Duchesne J, Molson JW, Larachi F, Maldague X (2012) CO_2-depleted warm air venting from chrysotile milling waste (Thetford Mines, Canada): evidence for in-situ carbon capture from the atmosphere. Geology 40(3):275–278

Reddy KR, Gopakumar A, Chetri JK (2019) Critical review of applications of iron and steel slags for carbon sequestration and environmental remediation. Rev Environ Sci Biotechnol 18:127–152

Regenspurg S, Peiffer S (2005) Arsenate and chromate incorporation in schwertmannite. Appl Geochem 20:1226–1239

Reid C, Bécaert V, Aubertin M, Rosenbaum RK, Deschênes L (2009) Life cycle assessment of mine tailings management in Canada. J Clean Prod 17:471–479

Renforth P (2012) The potential of enhanced weathering in the UK. Int J Greenhouse Gas Control 10:229–243

Renforth P, Washbourne C-L, Taylder J, Manning DAC (2011) Silicate production and availability for mineral carbonation. Environ Sci Technol 45:2035–2041

Revathy TDR, Palanivelu K, Ramachandran A (2015) Direct mineral carbonation of steelmaking slag for CO_2 sequestration at room temperature. Environ Sci Pollut Res 23:7349–7359

Rigopoulos I, Harrison AL, Delimitis A, Ioannou I, Efstathiou AM, Kyratsi T, Oelkers EH (2018a) Carbon sequestration via enhanced weathering of peridotites and basalts in seawater. Appl Geochem 91:197–207

Rigopoulos I, Török Á, Kyratsi T, Delimitis A, Ioannou I (2018b) Sustainable exploitation of mafic rock quarry waste for carbon sequestration following ball milling. Res Policy 59:24–32

Robertson R (2006) Early ice road closures could cool diamond operations. Northern Miner 92(6)

Roy ED, Willig E, Richards PD, Martinelli LA, Vazquez FF, Pegorini L, Porder S (2017) Soil phosphorus sorption capacity after three decades of intensive fertilization in Mato Grosso, Brazil. Agric Ecosyst Environ 249:206–214

Sánchez España J, López Pamo E, Santofimia E (2007) The oxidation of ferrous iron in acidic mine effluents from the Iberian Pyrite Belt (Odiel Basin, Huelva, Spain): field and laboratory rates. J Geochem Explor 92:120–132

Sánchez-España J, Boehrer B, Yusta I (2014) Extreme carbon dioxide concentrations in acidic pit lakes provoked by water/rock interaction. Environ Sci Technol 48(8):4273–4281

Sarmiento AM, Nieto JM, Olías M, Cánovas CR (2009) Hydrochemical characteristics and seasonal influence on the pollution by acid mine drainage in the Odiel River Basin (SW Spain). Appl Geochem 24:697–714

Sracek O, P. Gelinas R. Lefebvre and RV Nicholson (2006) Comparison of methods for the estimation of pyrite oxidation rate in a waste rock pile at Mine Doyon site, Quebec, Canada. J Geochem Explor 91: 99–109. https://doi.org/10.1016/j.gexplo.2006.03.002

Sarperi L, Surbrenat A, Kerihuel A, Chazarenc F (2014) The use of an industrial by-product as a sorbent to remove CO_2 and H_2S from biogas. J Environ Chem Eng 2:1207–1213

Sarvade S, Upadhyay VB, Kumar M, Imran Khan M (2019) Soil and water conservation techniques for sustainable agriculture. In: Jhariya MK, Banerjee A, Meena RS, Yadav DK (eds) Sustainable agriculture, forest and environmental management. Springer, Singapore, pp 133–188
Sarvaramini A, Assima GP, Beaudoin G, Larachi F (2014) Biomass torrefaction and CO_2 capture using mining wastes – a new approach for reducing greenhouse gas emissions of CO-firing plants. Fuel 115:749–757
Scales M (2006) The cost of less cold. Can Min J 127:5
Schiedek D, Sundelin B, Readman JW, MacDonald RW (2007) Interactions between climate change and contaminants. Mar Pollut Bull 54:1845–1856
Sheoran AS, Sheoran V (2006) Heavy metal removal mechanism of acid mine drainage in wetlands: a critical review. Miner Eng 19:105–116
Siegrist M, Southam C, Bowman G, Wilson SA, Southam G (2017) Analysis of the potential for negative CO_2 emission mine sites through bacteria-mediated carbon mineralization: evidence from Australia. Energy Procedia 114:6124–6132
Šimanský V, Juriga M, Jonczak J, Uzarowicz Ł, Stępień W (2019) How relationships between soil organic matter parameters and soil structure characteristics are affected by the long-term fertilization of a sandy soil. Geoderma 342:75–84
Smit B, Wandel J (2006) Adaptation, adaptive capacity, and vulnerability. Glob Environ Change 16:282–292
Smuda J, Dold B, Spangenberg JE, Pfeifer H-R (2008) Geochemistry and stable isotope composition of fresh alkaline porphyry copper tailings: implications on sources and mobility of elements during transport and early stages of deposition. Chem Geol 256:62–76
Søndergaard J, Elberling B, Asmund G, Gudum C, Iversen KM (2007) Temporal trends of dissolved weathering products released from a high Arctic coal mine waste rock pile in Svalbard (78 N). Appl Geochem 22:1025–1038
Song Q, Yanful EK (2011) Oxygen influx and geochemistry of percolate water from reactive mine waste rock underlying a sloping channeled soil cover. Appl Geochem 26:655–665
Streten-Joyce C, Manning J, Gibb KS, Neilan BA, Parry DL (2013) The chemical composition and bacteria communities in acid and metalliferous drainage from the wet-dry tropics are dependent on season. Sci Total Environ 443:65–79
Tao M, Zhang X, Wang S, Cao W, Jiang Y (2019) Life cycle assessment on lead-zinc ore mining and beneficiation in China. J Clean Prod 237:117833
Tayari F, Blumsack S (2020) A real options approach to production and injection timing under uncertainty for CO_2 sequestration in depleted shale gas reservoirs. Appl Energy 263:114491
Thom JGM, Dipple GM, Power IM, Harrison AL (2013) Chrysotile dissolution rates: implications for carbon sequestration. Appl Geochem 35:244–254
UNEP-FI, SIWI (2005) Challenges of water scarcity: a business case for financial institutions. UNEP Finance Initiative, Geneva, p 32
Vale C, Canário J, Caetano M, Poissant L, Ferreira AM (2011) Contaminant cycling under climate change: evidence and scenarios. In: Duarte P, SantanaCasiano JM (eds) Oceans and the atmospheric carbon content. Springer, pp 133–156
Veetil SP, Hitch M (2020) Recent developments and challenges of aqueous mineral carbonation: a review. Int J Environ Sci Technol 17:4359–4380
Veetil SP, Mercier G, Blais J-F, Cecchi E, Kentish S (2015a) Magnetic separation of serpentinite mining residue as a precursor to mineral carbonation. Int J Miner Process 140:19–25
Veetil SP, Pasquier L-C, Blais J-F, Cecchi E, Kentish S, Mercier G (2015b) Direct gas-solid carbonation of serpentinite residues in the absence and presence of water vapor: a feasibility study for carbon dioxide sequestration. Environ Sci Pollut Res 22:13486–13495
Vishal V, Ranjith PG, Singh TN (2013) CO_2 permeability of Indian bituminous coals: implications for carbon sequestration. Int J Coal Geol 105:36–47
Wang F, Dreisinger DB, Jarvis M, Hitchins T (2018) The technology of CO_2 sequestration by mineral carbonation: current status and prospects. Can Metall Q 57(1):46–58

Wang F, Dreisinger D, Jarvis M, Hitchins T (2019) Kinetics and mechanism of mineral carbonation of olivine for CO_2 sequestration. Miner Eng 131:185–197

Wilson SA, Dipple GM, Power IM, Thom JM, Anderson RG, Raudsepp M, Gabites JE, Southham G (2009) Carbon dioxide fixation within mine wastes of ultramafic-hosted ore deposits: examples from the Clinton Creek and Cassiar chrysotile deposits, Canada. Econ Geol 104:95–112

Wilson SA, Dipple GM, Power IM, Barker SLL, Fallon SJ, Southam G (2011) Subarctic weathering of mineral wastes provides a sink for atmospheric CO_2. Environ Sci Technol 45:7727–7736

Wilson SA, Harrison AL, Dipple GM, Power IM, Barker SLL, Mayer KU, Fallon SJ, Raudsepp M, Southam G (2014) Offsetting of CO_2 emissions by air capture in mine tailings at the Mount Keith nickel mine, Western Australia: rates, controls and prospects for carbon neutral mining. Int J Greenhouse Gas Control 25:121–140

Yu H, Zha T, Zhang X, Ma L (2019) Vertical distribution and influencing factors of soil organic carbon in the Loess Plateau, China. Sci Total Environ 693:133632

Yunmei Y, Yongxuan Z, Zhenmin G, Gammons CH, Dexian L (2007) Rates of arsenopyrite oxidation by oxygen and Fe(III) at pH 1.8–12.6 and 15–45 °C. Environ Sci Technol 41:6460–6464

Zarandi AE, Larachi F, Beaudoin G, Plante B, Sciortino M (2016) Multivariate study of the dynamics of CO_2 reaction with brucite-rich ultramafic mine tailings. Int J Greenhouse Gas Control 52:110–119

Zarandi AE, Larachi F, Beaudoin G, Plante B, Sciortino M (2017) Ambient mineral carbonation of different lithologies of mafic to ultramafic mining wastes/tailings – a comparative study. Int J Greenhouse Gas Control 63:392–400

Zhang L, Wang J, Feng Y (2018) Life cycle assessment of opencast coal mine production: a case study in Yimin mining area in China. Environ Sci Pollut Res 25:8475–8486

Mine Policy and Regulation

Yoshihiro Hamaguchi

Abstract The digital economy can be a substantial contributor to decarbonization. An indispensable part of this process is the production of semiconductors, which support digital goods and services. Meanwhile, seabed mineral resources are attracting attention because they contain abundant rare metals, which are raw materials for semiconductors. Consequently, the test mining of these resources has been considered or is underway in recent years. However, the impact, including irreversible ones, of resource mining on the marine environment and ecosystems remains unclear. In addition, marine ecosystem degradation may impede the climate change mitigation effects of blue carbon. Moreover, marine ecosystem changes may influence the blue economy and, thus, its associated stakeholders in the fishing, tourism, transportation, and energy industries. Seabed mineral resources can be a resource curse or blessing depending on the socioeconomic context. In particular, a better understanding of the mechanisms and interdependencies between seabed mineral resource extraction and a sustainable blue economy needs to be understood. This chapter briefly overviews the individual components necessary for future economic model analyses.

Keywords Seabed mineral resources · Blue carbon · Blue economy · Resource extraction models · Sustainable tourism

The author acknowledges financial support from the Early-Career Scientists from the Japan Society for the Promotion of Science no. 22K13409 (JSPS KAKENHI grant no. JP22K13409). The author is responsible for all errors in this chapter.

Y. Hamaguchi (✉)
Department of Economics, Faculty of Economics, Hannan University,
Matsubara-shi, Osaka, Japan
e-mail: hamaguchi@hannan-u.ac.jp

A. A. H. Abdel Latef et al. (eds.), *Sustainable Remediation for Pollution and Climate Resilience*, https://doi.org/10.1007/978-981-96-5674-5_6

1 Introduction

Mining policies have been active in the context of environmental pollution because mining processes emit various pollutants and pose serious health hazards to humans (Bridge 2004). Still, mining has also produced iron ore, coal, petroleum, and other production factors essential for economic development and has supported the world's economic growth. However, these nonrenewable resources emit large amounts of greenhouse gases, and there are increasing calls for their decarbonization. Under the Sustainable Development Goals (SDGs), new climate change mitigation measures are required in conventional mining policies to achieve sustainable development.

Furthermore, due to the artificial intelligence (AI) revolution, the importance of the digital economy in accelerating decarbonization is being increasingly recognized (Tao et al. 2023). However, the rapid shift to the digital economy has raised concerns about the shortage of semiconductors, which are essential for producing digital-related goods and services. Therefore, seabed resources have attracted attention as novel mineral resources. Besides copper, cobalt, nickel, zinc, silver, gold, and lithium, seafloor mineral resources are also rich in rare metals and raw semiconductor materials. Hence, investigations into their actual conditions and pilot mining for commercialization are underway (Levin et al. 2020a). However, in mining submarine mineral resources, as in mining resources on land, concerns remain regarding potential pollutant discharge, irreversible environmental pollution, and ecosystem destruction (Ramirez-Llodra et al. 2015). However, mining governance remains weak as facilitating resource extraction is prioritized over preserving ecosystems (Thompson et al. 2018).

To consider mining policies that mitigate the impact of pollution discharge from seabed resource mining on the environment and ecosystems, these mechanisms must first be understood. Although discussions continue on environmental assessment evaluation methods for seabed resource mining (Collins et al. 2013), the environmental impacts of seabed resource mining are still in the data collection phase, making empirical analysis difficult. Qualitative and normative analyses can provide useful insights for future empirical studies. Furthermore, analyzing the background and interrelationships will be important in modeling this new environmental issue, and then conventional model analysis can be applied. This chapter aims to provide a foundation for the development of an economic model for seabed resource extraction.

2 Mining of Submarine Resources and Environmental Impact

2.1 Distribution of Submarine Resources

The seafloor is composed of multiple oceanic plates. Seafloor mineral resources are buried in trenches and expansions on the plate boundaries. According to Hein et al. (2013) and Miller et al. (2018), the major marine resources are distributed as follows: Polymetallic nodules are mainly distributed in the western offshore areas near Guinea and Namibia, eastern offshore South Africa, southern offshore India, western offshore Australia and southern offshore Indonesia, central Pacific Ocean, offshore Peru and Chile, eastern offshore Argentina and Brazil, and eastern offshore United States. Cobalt-rich crusts are mainly distributed in western offshore Portugal, western offshore Namibia and South Africa, eastern offshore Austria and Indonesia, southeastern offshore Japan and northeastern offshore Indonesia and Philippines, western offshore Chile, western offshore Brazil, western offshore United States and Canada, and eastern offshore United States. The main distribution areas of polymetallic sulfides and polymetallic vents are the Central Indian Ocean Ridge, Nansha Islands Trench, Izu-Ogasawara Trench, area comprising the Mariana Trench, area comprising the South Solomon and Tonga Trenches, East Pacific Ocean Swell, and northern Atlantic Central Ridge.

Under the management of the United Nations Convention on the Law of the Sea (UNCLOS) and the International Seabed Authority (ISA), nearly 30 contracts have been signed since 2001 for the mining of seabed mineral resources, with pilot mining for research and preparation plans for mining underway. Contracts for the mining of polymetallic sulfide have been concluded in the Indian Ocean Central Ridge (Germany, Korea, India, and China), the northern part of the Atlantic Central Ridge (France, Poland, and Russia), the southern part of the Atlantic Central Ridge (Brazil), Izu-Ogasawara Trench for the mining of cobalt-rich crust, and the area around the Mariana Trench for the mining of cobalt-rich crust, and near the Mariana Trench (Japan, China, Russia), and the Cook Islands on the East Pacific Rise (France, Germany, the United Kingdom, Belgium, Poland, Bulgaria, Czech Republic, Slovakia, Russia, Korea, China, Japan, Singapore, Tonga, Nauru, Cuba, and Brazil) (Levin et al. 2020a).

2.2 *Mining of Submarine Resources*

Depending on the water depth, the seabed mineral resources that can be mined differ, such as cobalt crusts at water depths of 800–2500 m, polymetallic sulfides at water depths of 1000–4000 m, and polymetallic nodules at water depths of 3000–6500 m. The mining system consists of a collector of subsea resources, lifting the mined resources, beneficiation at sea, and transporting the concentrate (Miller et al. 2018). The ore collected by collectors on the seafloor is transported offshore by pipe lifting, ore beneficiated onboard is transferred to a transport vessel, and waste rock is buried on the seafloor by pipe lifting (Levin et al. 2020a, b).

UNCLOS classifies the effects on the marine environment caused by seabed mining as “harmful effects” and calls for effective protection against seabed mining (Levin et al. 2016). Specifically, it argues that seabed mining can cause irreversible damage to the marine environment and marine ecosystems; however, the actual impacts are still being studied (Thompson et al. 2018). Besides releasing sediment plumes during the mining process and affecting working equipment due to seafloor sediments, there is also concern about the adverse effects of mining operations on marine mammals due to the generated noise. However, the effects of increased water temperatures due to seawater circulation between the water surface and seafloor are unclear because of limited data (Miller et al. 2018). The interplay of these effects may have irreversible impacts on the seafloor environment, ecosystems, and deep-sea flora and fauna that have formed over tens of thousands of years.

Even in onshore resource extraction, during the production process from mining to beneficiation, the disposal of waste rock, polluted water effluents, and the generation of dust and slag result in land degradation, river acidification, air pollution, and ecosystem destruction (Bridge 2004). In particular, large amounts of greenhouse gases emitted during oil production contribute to global warming. Thus, the decarbonization of this industry needs to be promoted by switching to renewable energy sources (Wang et al. 2023). Mining seabed mineral resources may also promote global warming by undermining the blue carbon (Levin et al. 2020b). However, there are concerns about future uncertainties due to climate change's adverse effects. Considering these factors, seabed mineral mining may have more adverse impacts on the marine environment and marine ecosystems than predicted.

2.3 *Blue Carbon*

To mitigate global warming, efforts have been focused on forest conservation, such as afforestation, as it can enhance ecosystems' absorption of greenhouse gases (Forster et al. 2021). As the Amazon rainforest absorbs large amounts of carbon dioxide, reducing forest area due to slash-and-burn agriculture seriously affects global warming (Davidson et al. 2008). Recently, blue carbon, a type of carbon emissions absorbed and stored in mangrove forests, seagrass beds, wetland biomass, and soils, has attracted attention. Efforts are underway to estimate blue carbon's future carbon sequestration potential, as it is expected to play a role in mitigating global warming (Mcleod et al. 2011). According to Bertram et al. (2021), most of the global blue carbon generation occurs in Australia, China, and Cuba because of rich atolls.

The problem is the abundance of seafloor resources in these areas, including polymetallic nodules (offshore west of Australia and south of Indonesia, in the central Pacific, east of the United States, and northeast of Cuba), cobalt-rich crusts (Austria and offshore east of Indonesia), polymetallic sulfides, and polymetallic vents (the South Solomon Trench and Tonga Trench). According to Levin et al. (2020a), there are currently no plans to mine the seabed mineral resources in these areas. However, if mining begins in these areas in the future, the absorptive capacity of blue carbon could be compromised. This is because pollutants and waste rock discharged during mining can destroy marine ecosystems (Levin et al. 2020b). Blue carbon losses due to marine ecosystem degradation have been estimated to be equivalent to 3–19% of global forest loss (Pendleton et al. 2012). This estimate does not consider the impact of seabed mining, which may have an even higher impact.

A thorough analysis of its impact on blue carbon and marine ecosystems is needed to link seabed resource mining to decarbonization.

3 Blue Economy and Seabed Mining

3.1 Expectations for the Blue Economy

The ocean provides fisheries and seabed resources to humankind and, more recently, has been expected to serve as a climate adaptation and mitigation measure. In this context, there is an interdependent relationship involving various economic actors, such as fisheries, tourism, transportation, the energy industry, and mining, as well as government implementation of regulations and policies. This ocean-centered socioeconomic relationship has been examined from the perspective of sustainable development and is being conceptualized as the blue economy or marine economy (Wenhai et al. 2019). This blue economy framework is useful when considering the socioeconomic impact of seabed mineral extraction in marine environments. Rayner et al. (2019) stated that, from 2010 to 2030, the growth rates of the key industries of the blue economy were 5.69% for aquaculture, 4.10% for fishing, 6.26% for fish processing, and 3.51% for marine tourism. Offshore wind power is expected to be a renewable energy source, with a projected growth rate of 25.52%. During this period, the key industries in the blue economy are expected to grow by an average of 3.64% and generate 130% more jobs. Thus, the blue economy will play an increasingly important role in providing food, resources, and energy and in climate adaptation measures toward a sustainable society. However, the degradation of the marine environment owing to undersea mineral resources threatens to undermine its sustainability.

3.2 Impacts on Fisheries, Tourism, and Marine Energy

Heavy oil spills from shipping vessels can significantly damage the surrounding marine environment and cause severe economic losses to fisheries and tourism (Garza-Gil et al. 2006; Magris and Giarrizzo 2020). Similar economic losses can happen if contaminants are released from the seabed mineral resources and their mining. Moreover, some fishermen oppose mining plans because of the potential for reduced catches due to changes in fish habitats caused by changes in the marine environment associated with mining (Miller et al. 2018). Habitat changes can also significantly affect marine tourism. This is because if the natural observation of marine life is the purpose of tourism, the decline in marine life will reduce the attractiveness of that tourist destination (Arabadzhyan et al. 2021). In addition, island countries, such as Sri Lanka and Togo, not only have seabed mineral deposits in the surrounding waters but also rich coral reefs. Since coral reef tourism is a major contributor to these island countries' economies (Arabadzhyan et al. 2021), marine environment degradation due to resource extraction may reduce tourist numbers. Many island nations are also at risk of submergence because of rising sea levels caused by climate change (Weir et al. 2017). Lower amounts of blue carbon due to resource extraction can contribute to this crisis.

With increasing renewable energy installations for decarbonization, there are particularly high expectations for offshore wind power generation. Compared to land, large ocean areas offer potential sites for wind power generation and are free from obstacles that would impair wind power generation. However, securing locations for seabed resource mining and the increased vessel traffic associated with mining (Miller et al. 2018) may increase the difficulties in deploying offshore wind power. Besides offshore wind, renewable energy in the ocean includes coastal wave power and underwater tidal power; however, the associated power generation facilities are still being developed and commercialized (Bhuiyan et al. 2022). Mining seabed mineral resources, if not carefully considered by marine energy producers, can lead to new resource and energy conflicts. The government's ability to coordinate these is also critical. However, international interest in seabed mineral resources makes coordination difficult, and geopolitical risks exist regarding renewable energy (Vakulchuk et al. 2020). Analysis of seabed mineral resource extraction within the framework of a blue economy is necessary to enhance the sustainability of oceans.

4 Toward Economic Modeling Analysis

4.1 Resource Mining Model

Quantitative analysis is currently difficult because of limited data available on the ongoing mining of seafloor mineral resources. However, a qualitative analysis may be possible by incorporating the seabed mineral resource mining system into a blue economy. We can use economic models based on previous studies, such as those by Groth (2007), who surveyed analyses of nonrenewable resources in economic growth models. The resource curse is an important concept in the relationship between resources and the economy. Numerous studies have focused on it (Badeeb et al. 2017). Thus, an important research question is whether the traditional interpretation of the resource curse applies to submarine mineral resources. Furthermore, more advanced mining technology is required for seabed mineral resources than for coal petroleum (Guo et al. 2023). Japan and South Korea, which have traditionally been considered resource-poor, have the potential to become resource-rich countries by discovering veins of seafloor mineral resources. However, without appropriate mining technology, seafloor mineral resources cannot be extracted adequately and feasibly; therefore, for countries such as Japan and Korea, these seafloor resources could be a blessing in disguise. However, countries such as Tonga and

Nauru do not have advanced mining technology and will need financial and technical assistance from other countries. Yet, if corruption related to this aid happens, the seabed resource can instead become a political resource curse. By applying conventional resource extraction models, we can analyze whether seabed mineral resources are a resource curse or resource blessing and identify the factors that influence them.

4.2 Fishery Tourism Model

As a renewable resource, the ecological mechanisms of fisheries have been analyzed using bioeconomic models (Foley et al. 2012). These models have also been incorporated into economic models to investigate the interrelationships between ecosystems and socioeconomic factors. However, the focus has been on the impact of pollutant emissions on the ability of fisheries to regenerate, as well as pollutant emissions from factories and other production processes; few have focused on pollutant emissions from resource extraction processes. In the context of seabed mineral resource mining, it is necessary to analyze the impact of pollution from the mining of nonrenewable resources on renewable resources, such as fish catches. These impacts also affect marine tourism. Hamaguchi (2020) notes that environmental degradation can hinder tourism-led growth by reducing the tourism sector's productivity, which depends on the natural environment. Furthermore, sustainable tourism could be hampered by resource extraction. The socioeconomic mechanisms behind such a situation must first be elucidated to determine the policies that can be implemented to prevent it. This can be clarified by incorporating the seabed resource mining system into existing models, and a mining policy that contributes to sustainable tourism can be proposed.

5 Conclusion

Seabed mineral resource mining has the potential to support the digital economy and contribute significantly to decarbonization by providing raw materials for semiconductor production. However, the impact of such resource mining on the marine environment and ecosystems is still being studied and is unclear. Moreover, marine ecosystem degradation can damage blue carbon and lead to global warming. It can also affect various stakeholders in the ocean. What policies are needed to make seabed mineral resources a resource blessing rather than a resource curse and build a sustainable blue economy? A qualitative analysis using an economic model that incorporates the elements discussed in this chapter is necessary to elucidate this issue.

References

Arabadzhyan A, Figini P, García C, González MM, Lam-González YE, León CJ (2021) Climate change, coastal tourism, and impact chains–a literature review. Curr Issue Tour 24(16):2233–2268

Badeeb RA, Lean HH, Clark J (2017) The evolution of the natural resource curse thesis: a critical literature survey. Resour Policy 51:123–134

Bertram C, Quaas M, Reusch TB, Vafeidis AT, Wolff C, Rickels W (2021) The blue carbon wealth of nations. Nat Clim Chang 11(8):704–709

Bhuiyan MA, Hu P, Khare V, Hamaguchi Y, Thakur BK, Rahman MK (2022) Economic feasibility of marine renewable energy. Front Mar Sci 9:988513

Bridge G (2004) Contested terrain: mining and the environment. Annu Rev Environ Resour 29:205–259

Collins PC, Croot P, Carlsson J, Colaço A, Grehan A, Hyeong K et al (2013) A primer for the environmental impact assessment of mining at seafloor massive sulfide deposits. Mar Policy 42:198–209

Davidson EA, de Abreu Sa TD, Reis Carvalho CJ, de Oliveira Figueiredo R, Kato MDSA, Kato OR, Ishida FY (2008) An integrated greenhouse gas assessment of an alternative to slash-and-burn agriculture in eastern Amazonia. Glob Chang Biol 14(5):998–1007

Foley NS, Armstrong CW, Kahui V, Mikkelsen E, Reithe S (2012) A review of bioeconomic modelling of habitat-fisheries interactions. Int J Ecol 2012:1–11

Forster EJ, Healey JR, Dymond C, Styles D (2021) Commercial afforestation can deliver effective climate change mitigation under multiple decarbonisation pathways. Nat Commun 12(1):3831

Garza-Gil MD, Prada-Blanco A, Vázquez-Rodríguez MX (2006) Estimating the short-term economic damages from the Prestige oil spill in the Galician fisheries and tourism. Ecol Econ 58(4):842–849

Groth C (2007) A new-growth perspective on non-renewable resources. In: Bretschger L, Smulders S (eds) Sustainable resource use and economic dynamics. Springer, Dordrecht, pp 127–163

Guo X, Fan N, Liu Y, Liu X, Wang Z, Xie X, Jia Y (2023) Deep seabed mining: frontiers in engineering geology and environment. Int J Coal Sci Technol 10(1):23

Hamaguchi Y (2020) Do pollution havens restrict tourism-led growth? Achieving sustainable tourism via a mix of environmental and tourism policies. Tour Econ 26(7):1175–1196

Hein JR, Mizell K, Koschinsky A, Conrad TA (2013) Deep-ocean mineral deposits as a source of critical metals for high-and green-technology applications: comparison with land-based resources. Ore Geol Rev 51:1–14

Levin LA, Mengerink K, Gjerde KM, Rowden AA, Van Dover CL, Clark MR et al (2016) Defining "serious harm" to the marine environment in the context of deep-seabed mining. Mar Policy 74:245–259

Levin LA, Amon DJ, Lily H (2020a) Challenges to the sustainability of deep-seabed mining. Nat Sustain 3(10):784–794

Levin LA, Wei CL, Dunn DC, Amon DJ, Ashford OS, Cheung WW et al (2020b) Climate change considerations are fundamental to management of deep-sea resource extraction. Glob Chang Biol 26(9):4664–4678

Magris RA, Giarrizzo T (2020) Mysterious oil spill in the Atlantic Ocean threatens marine biodiversity and local people in Brazil. Mar Pollut Bull 153:110961

Mcleod E, Chmura GL, Bouillon S, Salm R, Björk M, Duarte CM et al (2011) A blueprint for blue carbon: toward an improved understanding of the role of vegetated coastal habitats in sequestering CO2. Front Ecol Environ 9(10):552–560

Miller KA, Thompson KF, Johnston P, Santillo D (2018) An overview of seabed mining including the current state of development, environmental impacts, and knowledge gaps. Front Mar Sci 4:312755

Pendleton L, Donato DC, Murray BC, Crooks S, Jenkins WA, Sifleet S et al (2012) Estimating global "blue carbon" emissions from conversion and degradation of vegetated coastal ecosystems. PLoS One 7(9):e43542
Ramirez-Llodra E, Trannum HC, Evenset A, Levin LA, Andersson M, Finne TE et al (2015) Submarine and deep-sea mine tailing placements: a review of current practices, environmental issues, natural analogs and knowledge gaps in Norway and internationally. Mar Pollut Bull 97(1–2):13–35
Rayner R, Jolly C, Gouldman C (2019) Ocean observing and the blue economy. Front Mar Sci 6:330
Tao M, Poletti S, Wen L, Sheng MS, Wang J, Wang G, Zheng Y (2023) Appraising the role of the digital economy in global decarbonization: a spatial non-linear perspective on globalization. J Environ Manag 347:119170
Thompson KF, Miller KA, Currie D, Johnston P, Santillo D (2018) Seabed mining and approaches to governance of the deep seabed. Front Mar Sci 5:480
Vakulchuk R, Overland I, Scholten D (2020) Renewable energy and geopolitics: a review. Renew Sust Energ Rev 122:109547
Wang Z, Li S, Jin Z, Li Z, Liu Q, Zhang K (2023) Oil and gas pathway to net-zero: review and outlook. Energ Strat Rev 45:101048
Weir T, Dovey L, Orcherton D (2017) Social and cultural issues raised by climate change in Pacific Island countries: an overview. Reg Environ Chang 17:1017–1028
Wenhai L, Cusack C, Baker M, Tao W, Mingbao C, Paige K et al (2019) Successful blue economy examples with an emphasis on international perspectives. Front Mar Sci 6:261

Strategic Conservation Plan

Gaurav Mudgal, Subhashini Sivaji, Jaspreet Kaur, Sanjoy Kumar Dhar, Vasantharekha Ramasamy, Barathi Seetharaman, Rohit Kumar Mishra, Venu Govindappa, Jagadeesh Chandra Bose, Gajendra B. Singh, Anirudh Sharma, Vani Sharma, Prabuddh Mishra, Lyudmila Asyakina, Victor Atuchin, Abhijeet Singh, Simrandeep Kaur, Harshita Thakur, Nandi LaxmiDeepika, Subhasmita Dash, Sanjana Singh, Ankit Kumar Surya, and Jasdeep Singh

Abstract Environmental pollution and climate change are two of the most pressing challenges humanity and other life forms face. They pose significant threats to health, ecosystems, and the overall sustainability of the planet. Unbelievable, but factually, nine million people die prematurely each year due to air pollution; 80% of wastewater globally is released into the environment without adequate treatment; eight million tons of plastic waste enter the oceans annually; one in four deaths worldwide are linked to environmental pollution. Other eye-openers are that global temperature is projected to rise by 1.5 °C by 2030 and 3.2 °C by 2100; sea levels are expected to rise by 26 cm by 2050 and 82 cm by 2100; climate change could lead to a 10% decline in global gross domestic product (GDP) by 2100; and that 143

Gaurav Mudgal and S. Subhashini Jr contributed equally with all other contributors.

G. Mudgal (✉)
Center for Waste Management and Renewable Energy, Saveetha Dental College and Hospitals, Saveetha Institute of Medical and Technical Sciences, Saveetha University, Chennai, India

University Institute of Biotechnology, Chandigarh University, Mohali, Punjab, India

Research and Development, Vanproz Agrovet LLP, Bagroda Industrial Area, Bhopal, Madhya Pradesh, India

S. Sivaji · V. Ramasamy · B. Seetharaman
Department of Biotechnology, School of Bioengineering, Faculty of Engineering and Technology, SRM Institute of Science and Technology, Kattankulathur, Chengalpattu District, Tamil Nadu, India
e-mail: vasanthr@srmist.edu.in; barathis@srmist.edu.in

J. Kaur · G. B. Singh · V. Sharma · A. Singh · S. Kaur · H. Thakur · N. LaxmiDeepika · S. Dash · S. Singh · A. K. Surya · J. Singh
University Institute of Biotechnology, Chandigarh University, Mohali, Punjab, India

A. A. H. Abdel Latef et al. (eds.), *Sustainable Remediation for Pollution and Climate Resilience*, https://doi.org/10.1007/978-981-96-5674-5_7

million people could be displaced due to climate change by 2050. This chapter outlines a strategic conservation plan aimed at mitigating these risks. This chapter begins by providing a comprehensive overview of the various pollutants and their detrimental effects. It then peeps into the complexities of climate change linked to these and other processes, including its causes, consequences, and future predictions. This chapter's core focuses on developing a strategic conservation plan. This plan outlines a multipronged approach encompassing (i) Pollution Control and Remediation Strategies, which explores various methods for reducing pollution at its source (prevention) and cleaning up existing contamination (remediation) across different environmental domains (air, water, land). (ii) Climate Change Mitigation and Adaptation Strategies, which addresses strategies for reducing greenhouse gas emissions and mitigating the impacts of climate change, along with adaptation strategies to build resilience in ecosystems and human societies. Beyond the conventional command and control strategies, this chapter emphasizes the importance of public awareness, education, and policy reforms in achieving successful environmental conservation. It highlights the need for international collaboration and a col-

S. K. Dhar
University Institute of Biotechnology, Chandigarh University, Mohali, Punjab, India

Banophool Adibashi Green Heart College, Mirpur, Dhaka, Bangladesh

R. K. Mishra
Centre of Science and Society, Faculty of Science, Central University of Allahabad, Prayagraj, India

V. Govindappa
Department of Zoology, Centre for Applied Genetics, Jnana Bharathi Campus, Bangalore University, Bengaluru, Karnataka, India

J. C. Bose
Pallavi International School, Moinabad Mandal, Hyderabad, Telangana, India

A. Sharma
Department of biotechnology, Jaypee Institute of Information Technology, Noida, Uttar Pradesh, India

P. Mishra
Research and Development, Vanproz Agrovet LLP, Bagroda Industrial Area, Bhopal, Madhya Pradesh, India

L. Asyakina
Laboratory for Phytoremediation of Technogenically Disturbed Ecosystems, Kemerovo State University, Kemerovo, Russia

V. Atuchin
Laboratory of Optical Materials and Structures, Institute of Semiconductor Physics, SB RAS, Novosibirsk, Russia

Research and Development Department, Kemerovo State University, Kemerovo, Russia

Department of Industrial Machinery Design, Novosibirsk State Technical University, Novosibirsk, Russia

R&D Center "Advanced Electronic Technologies", Tomsk State University, Tomsk, Russia

lective global effort toward a sustainable future. Given that pollution and climate change are the major impactors to productivity multifacetedly, this chapter highlights the measures and initiatives toward realizing the UN Sustainable Development Goals.

Keywords Command-and-control · Eutrophic · Health · Public · Resources · Nanotechnology · Bioremediation · 3Rs · Technogenic waste

1 Introduction

Our planet, a vibrant tapestry of life, faces a critical juncture. Human activity, fueled by innovation and progress, has cast a long shadow, manifesting in environmental pollution and climatic changes. These challenges threaten the delicate balance of ecosystems and the very fabric of human civilization. According to the World Health Organization (WHO), almost 99% of the global population breathes polluted air, leading to an estimated seven million premature deaths annually (WHO 2024a). Water is not spared either, with contaminated sources contributing to diarrheal diseases that claim a sizeable count of lives of children under five every year (WHO 2024b). With climate change, the global average temperatures have risen by 1 °C since pre-industrial times, with projections indicating a further increase if left unchecked. This warming translates into extreme weather events—more frequent and intense heatwaves, droughts, floods, and storms—displacing millions and jeopardizing food security (Gul et al. 2020; O'Sullivan 2015; Kumar et al. 2023a). In effect, in both conditions, the impacts are severe and multifaceted. For example, the rate of species extinction is currently estimated to be hundreds to thousands of times higher than the natural background rate (Maximillian et al. 2019; Schiedek et al. 2007; Shivanna 2022). This decline disrupts ecosystems, weakens their resilience, and threatens the critical services they provide, such as pollination and climate regulation. The effects of pollution are not merely environmental; they directly impact human health. Air pollution can lead to respiratory illnesses, heart disease, and even cancer (Schraufnagel et al. 2019; Kim et al. 2018; Hoek et al. 2013). Water pollution contributes to waterborne diseases, including diarrheal diseases, cholera, and typhoid (Shayo et al. 2023; Sharma et al. 2023, 2024; Adegoke et al. 2023). Chemical pollution can disrupt hormones, impair cognitive development, and increase the risk of birth defects (Fucic et al. 2023; Ramírez et al. 2023; Luhach et al. 2023). The economic burden of pollution is vast, with healthcare costs, lost productivity, and damage to infrastructure reaching trillions of dollars annually. The consequences of climate change are already being felt worldwide. Rising sea levels threaten coastal communities, displacing millions and submerging vital infrastructure. Extreme weather events inflict widespread damage on agriculture, leading to food shortages and malnutrition. The disruption of weather patterns also risks water availability, impacting sanitation and public health. Furthermore, climate change acts as a threat multiplier, exacerbating existing inequalities and conflicts.

Historically, environmental protection and climate change mitigation efforts have focused on regulations and command-and-control (C&C) approaches, such as strict emission regulations on specific industries or bans on certain harmful chemicals. These strategies have achieved only some success in reducing pollution levels. For instance, emission regulations might reduce factory pollution, but they may not incentivize the development of cleaner production methods. Similarly, chemical bans might eliminate a specific chemical problem but could lead to using less-understood alternatives. Additionally, these approaches often face challenges with enforcement and compliance. Companies may find ways to circumvent regulations or enforcement agencies may lack the resources to monitor compliance. Furthermore, past efforts have sometimes been limited in their effectiveness due to limited geographic scope. Regulations or initiatives might be implemented in one country or region but fail to address the global nature of environmental problems. For example, a country might enact stricter air pollution standards, but this would not address the contribution of international shipping or long-range air pollution transport. There had been more focus on end-of-pipe solutions. For example, past efforts might have focused on cleaning up pollution after it has been created rather than preventing it in the first place. This could include building expensive pollution scrubbers in factories instead of encouraging cleaner production processes. Other than this, in many countries, subsidies for unsustainable practices have been the routine regime. In some cases, government subsidies might unintentionally support activities that harm the environment. This could include subsidies for fossil fuel production or unsustainable agricultural practices. Yet, in addition to these, there was a lack of public awareness and participation. Environmental protection efforts may not have effectively engaged the public in understanding the issues and taking action. This can lead to a lack of public pressure for stricter regulations or resistance to change. By acknowledging these limitations, we can explore new approaches that are more comprehensive, preventative, and involve collaborative efforts from governments, businesses, and individuals.

Despite the daunting challenges, the surge of innovative solutions offers a glimmer of hope in the face of environmental challenges. The renewable energy evolution champions cleaner alternatives like solar, wind, and geothermal power, with advancements driving their efficiency and affordability. The circular economy promotes minimizing waste and maximizing resource use by designing durable, recyclable products. Bioremediation harnesses nature's power through bacteria, fungi, and plants to break down or neutralize pollutants (Fotina et al. 2024). Precision agriculture utilizes data analysis to optimize agricultural practices, reducing water consumption, fertilizer use, and pesticide application while boosting yields. Finally, climate-smart agriculture focuses on improving food security in the face of climate change by utilizing drought-resistant crops, improved soil management, and efficient irrigation systems.

At the outset, there is a highly sought-after need to think and apply the new normal to mitigate these challenges, and the fight for a sustainable future demands a concerted effort on multiple fronts. Besides the above, governments must implement robust environmental policies, invest in clean technologies, and foster international cooperation. Businesses must embrace sustainable practices throughout their entire supply chains. Individually, we can play a role by adopting environmentally conscious lifestyles, reducing our carbon footprint, and supporting sustainable products and services. The road ahead is undoubtedly challenging. However, the human spirit of innovation and collaboration can pave the way for a more sustainable future. By acknowledging the gravity of environmental threats, embracing innovative solutions, and working collectively, we can safeguard our planet and ensure a healthy and prosperous future for future generations.

In this chapter, we embark on a granular exploration of these nascent approaches, dissecting the conservation and mitigation paradigms at the core of a sustainable Anthropocene. Through the harnessing of renewable energy sources, the implementation of a closed-loop circular economic model, and the strategic deployment of bioremediation techniques, precision agriculture methodologies, and climate-resilient agricultural practices, we can collectively orchestrate a transformative shift toward a more salubrious planetary ecosystem for the benefit of future generations.

2 Understanding Air Pollutants Hurdling Mitigation Regimes

2.1 The Invisible Threat: Unveiling Challenges in Air Pollution Mitigation

A complex cocktail of pollutants hangs heavy in many parts of the world, posing a significant threat to human health and environmental well-being (Xu et al. 2023a; Sicard et al. 2023; Shi et al. 2023; Li et al. 2023a). While regulations have aimed to curb this invisible threat, effective mitigation remains a complex challenge due to the nature of the pollutants and the limitations of existing strategies. Air pollution is not a single entity; it is a deadly concoction of various gases and particulate matter, but it encompasses composites of an incredible and dreadful list of key culprits and the challenges they pose.

Some of these entities have been discussed here, along with multifaceted impacts on climate, environment, and public health (and others are inclusively listed in Table 1).

2.1.1 Particulate Matter (PM2.5)

These are fine microscopic particles, less than 2.5 microns in diameter, which penetrate deep into the lungs and enter the bloodstream, causing respiratory illnesses, heart disease, and even cancer (Poulsen et al. 2023; Xu et al. 2023b; Raaschou-Nielsen et al. 2023; Li et al. 2023b, c; Yang et al. 2023; Craver et al. 2024; Wang

Table 1 A comprehensive list of air pollutants, sources, mitigation challenges, and impacts

Pollutant	Sources	Mitigation challenges	Health impacts	References
Fine particulate matter (PM2.5)	Vehicle emissions, industrial processes, biomass burning	Difficult to monitor due to small size Diverse sources Evolving technologies	Respiratory illnesses Heart disease Cancer	Wyer et al. (2022), Maciejczyk et al. (2021), and University, N. T (2024)
Ground-level ozone	Reaction of sunlight with pollutants	Not directly emitted Weather dependent	Respiratory problems Particularly harmful to children and asthmatics	Rosser and Balmes (2023), Cheesman et al. (2023), Winiwarter and Klimont (2011), Sharma et al. (2021)
Nitrogen oxides (NOx)	Vehicle exhaust, industrial processes	Contributes to smog and acid rain Precursor to ozone formation	Respiratory problems Cardiovascular issues	Hoek et al. (2013), Hui et al. (2023), Winiwarter and Klimont (2011), Wade and Farrauto (2012)
Volatile organic compounds (VOCs)	Vehicles, paints, solvents	Diverse sources across sectors Difficult to comprehensively control	Smog formation Ozone formation Respiratory irritation Potential cancer links	Hui et al. (2023), Singh et al. (2023), Zheng et al. (2016), Chen et al. (2018), and Kim (2011)
Sulfur oxides (SOx)	Burning fossil fuels (coal)	Success stories in regulation Challenges in regions reliant on coal	Acid rain Respiratory problems Cardiovascular diseases	Fazakas et al. (2023), Rodrigues and Joekes (2011), Jyethi (2016), Dignon (1992), and Shelyapina et al. (2016)
Ammonia (NH_3)	Agricultural practices, animal waste	Contributes to PM2.5 formation Requires targeting agricultural practices	Respiratory problems	Winiwarter and Klimont (2011), Wyer et al. (2022)
Black carbon (BC)	Incomplete combustion of fossil fuels & biomass	Major climate change contributor Requires cleaner burning practices	Respiratory problems Accelerates climate change	Tiwari et al. (2021) and Venkatraman et al. (2021)
Heavy Metals (Lead, Mercury, Arsenic)	Industrial processes, Contaminated sites	Persistent pollutants Difficult and expensive remediation	Cancer Developmental problems Nervous system damage	Patel et al. (2021), Rahman and Singh (2019), Mishra et al. (2019), and Kumar et al. (2019)
Persistent organic pollutants (POPs)	Pesticides, industrial solvents	Banned in many countries Potential for illegal production	Cancer Developmental problems Bioaccumulation in the food chain	Miglioranza et al. (2023), Nieder et al. (2018a), Mohapatra and Basu (2023)

Carbon Monoxide (CO)	Incomplete combustion of fuels	Disrupts oxygen transport in blood	Headaches Dizziness Death in high concentrations	Bleecker (2015), Kleinman (2020), Savioli et al. (2024), and Raub (1999)
Chlorofluorocarbons (CFCs)	Refrigerants, aerosol propellants (banned)	Ozone depletion (historical) Potential for illegal production	Increased UV radiation exposure Skin cancer	Egelston (2022), Parker (2021), and Downie (2015)
PM10	Dust, construction activities, some industrial processes	Inadequate monitoring networks, limited public awareness, difficulty in controlling emissions, high costs of mitigation technologies, regulatory challenges	Irritates respiratory system Worsens chronic respiratory conditions	Williams et al. (2017) and Ćurić et al. (2022)
Coarse particulate matter (PM10-PM2.5)	Mechanical processes (grinding, crushing)	Similar health impacts as PM10	Respiratory irritation Worsens chronic respiratory conditions	Khan and Srivastava (2024), Corona Sánchez et al. (2021)
Bioaerosols	Pollen, mold spores, bacteria	Lack of standardization in measurement methods, limited understanding of sources and fate, difficulty in controlling emissions, high costs of mitigation technologies, regulatory challenges	Triggers allergic reactions Worsens respiratory problems	Alam et al. (2022), Atta (2023), Fromme (2023), and Nag (2019)
Dioxins	Industrial processes, waste incineration	Highly toxic and persistent Difficult to eliminate from the environment	Cancer Developmental problems	Fiedler (2003), Mudhoo et al. (2013), and Mandal (2005)
Polycyclic aromatic hydrocarbons (PAHs)	Vehicle exhaust, cigarette smoke, grilled foods	Persistence in environment, bioaccumulation, limited regulations, difficulty in remediation, high costs of cleanup, soil and groundwater contamination	Irritate respiratory system Potential cancer link	Abdel-Shafy and Mansour (2016), Yang et al. (2021), Kumar et al. (2016), and Kim et al. (2013)
Microplastics	Synthetic fibers, car tire wear, plastic waste incineration	Emerging concern Health impacts under investigation	Potential respiratory problems (research ongoing)	Prata et al. (2022), Bhat et al. (2023), Mamun et al. (2023), and Osman et al. (2023)

(continued)

Table 1 (continued)

Pollutant	Sources	Mitigation challenges	Health impacts	References
Nanoparticles	Engineered products	Emerging concern Potential health impacts under investigation	Potential respiratory problems (research ongoing)	Hoyt and Mason (2008), Borm et al. (2006), Kreyling et al. (2006), and Hoet et al. (2004)
Ozone-depleting substances (ODS)	Industrial processes, improper disposal of appliances	Depletes ozone layer Allows harmful UV radiation to reach Earth	Increased skin cancer risk Impacts ecosystems	Weber et al. (2022), Murphy et al. (2002), Singh and Bhargawa (2019), Jana et al. (2013)
Endocrine-disrupting chemicals (EDCs)	Personal care products, pharmaceuticals	Potential for airborne pathways Health impacts under investigation	Potential hormonal disruption Reproductive health concerns (research ongoing)	Dagar and Bagchi (2020), Xiao et al. (2023), Annamalai and Namasivayam (2015)
Hexavalent chromium (Cr(VI))	Chrome plating, metal finishing, leather tanning	Highly toxic Difficult to control emissions	Lung cancer	Baral and Engelken (2002), Junaid et al. (2016), Laxmi and Kaushik (2020), Das and Mathew (2011)
Polychlorinated biphenyls (PCBs)	Banned electrical equipment & coolants	Persistent organic pollutants Difficult to remediate	Cancer Developmental problems	Akhtar et al. (2021), Qi et al. (2014), and Iqbal et al. (2022)
Radon	Naturally occurring radioactive gas	Seeps from the ground into buildings Requires detection and mitigation measures	Lung cancer	Yoon et al. (2016), Pawade and Charhate (2024)
Indoor air pollutants (various)	Building materials, furnishings, cleaning products	Diverse sources within buildings Requires proper ventilation and source control	Respiratory irritation Allergies	Carrer et al. (2001), Levin (1992) and Nandan et al. (2021)

et al. 2023; Krittanawong et al. 2023; Yuan et al. 2023). Monitoring and controlling PM2.5 are difficult due to its minuscule size and diverse sources, including vehicle emissions, industrial processes, and biomass burning. Limited research and documentation hinder our understanding of the cost-effectiveness of PM2.5 prevention and control technologies. To achieve optimal control, strategies will likely integrate a diverse set of options (Tucker 2000). Extensive research confirms a significant correlation between fine particle pollutants (PM2.5) and respiratory problems, including increased morbidity and mortality. European studies link PM2.5 to shorter lifespans and higher respiratory illness rates (Zanobetti et al. 2009; Analitis et al. 2006; Dominici et al. 2006; Huynh et al. 2006; de Oliveira et al. 2012). Research conducted in the USA shows that PM2.5 exposure increases cardiopulmonary problems and lung cancer deaths. A large American Cancer Society study demonstrates PM2.5's link to higher overall mortality, heart disease deaths, and lung cancer deaths (Schwartz 2000; Franklin et al. 2008). Vulnerable populations are at greater risk from PM2.5, including the elderly, pregnant women, adolescents, Infants, and people regardless of age but with preexisting respiratory or heart conditions. In countries like China, recent research highlights PM2.5's impact on public health, with a high PM2.5 concentration significantly exceeding safety standards in major cities. Studies show a strong correlation between PM2.5 levels and daily mortality rates.

2.1.2 Ground-Level Ozone

This gas, formed by the reaction of sunlight with pollutants, attacks the lungs, triggering respiratory problems, particularly in children and asthmatics (Rosser and Balmes 2023; Lu and Yao 2023). This, in many ways, also affects crop productivity (Pui Kin and Yang 2023; Cheesman et al. 2023). Controlling ozone is complex because it is not directly emitted but forms in the atmosphere. Additionally, weather patterns significantly influence ozone levels, making it challenging to predict and mitigate (Hui et al. 2023).

2.1.3 Ammonia (NH_3) and Nitrogen Oxides (NOx) Gases

Atmospheric nitrogen oxides (NOx) and ammonia (NH_3) are primary contributors to environmental challenges. These gases, primarily emitted from vehicle exhaust and industrial processes, form a complex air pollution web. They are key precursors to ground-level ozone and particulate matter, contributing to smog formation and impairing human health. Additionally, NOx and NH_3 deposition acidifies terrestrial and aquatic ecosystems, while nitrogen enrichment from ammonia exacerbates water quality issues through eutrophication (Jonson et al. 2017). The synergistic effects of these pollutants complicate mitigation strategies, necessitating comprehensive and integrated pollution control measures (Winiwarter and Klimont 2011). Ammonia is released routinely from agricultural practices, like fertilizers and animal waste, and reacts with other pollutants to form PM2.5 (Wyer et al. 2022), further compromising respiratory health.

2.1.4 Volatile Organic Compounds (VOCs)

These contagious compounds are emitted from vehicles, paints, and solvents, contributing to smog and ozone formation (Fazakas et al. 2023). Their diverse sources across various sectors, from industrial facilities to household products, make comprehensive control challenging and continue health effects on the public and other organismic health (Singh et al. 2023; Ma and Li 2024). To move ahead within the list is SOx emissions, contributing to acid rain, respiratory problems, and cardiovascular diseases.

2.1.5 SOx Emissions

Regulations have successfully curbed sulfur oxides (SOx) from burning fossil fuels. For example, regulations such as implementing air quality acts in many countries, like the Clean Air Act in the United States, have mandated stricter emission standards for industries and power plants. These regulations limit the amount of sulfur content allowed in fuels and require using pollution control technologies like scrubbers to capture SOx emissions before they reach the atmosphere. Similarly, regulations have encouraged a shift toward cleaner-burning fuels with lower sulfur content, such as natural gas or low-sulfur coal. This switch has significantly reduced SOx emissions from burning fossil fuels. Some regions have implemented market-based solutions like cap-and-trade programs for SOx emissions. These programs create a market for emission allowances, incentivizing industries to reduce their emissions or purchase allowances from others who emit less. The success of these regulations can be measured by reduced SOx levels, improved air quality, and public health benefits. Lower SOx levels translate to a reduction in respiratory illnesses, cardiovascular diseases, and other health problems associated with air pollution. It is important to acknowledge that while regulations have succeeded, SOx challenges remain. For example, implementation and enforcement of regulations might vary significantly across regions. Developed countries may have stricter regulations and see greater progress than developing ones. Some regions still heavily rely on coal for energy production. Transitioning to cleaner energy sources or implementing advanced emission control technologies remains crucial. Moreover, rapid industrialization in developing economies can increase SOx emissions if proper regulations and emission control measures are not implemented. So, regulations play a significant role in curbing SOx emissions and improving air quality. However, continued efforts are necessary to ensure consistent enforcement, promote cleaner energy sources, and address challenges in regions still reliant on coal.

2.1.6 Black Carbon (BC)

BC, a component of PM2.5 from incomplete combustion, is a major climate concern. It absorbs sunlight and warms the atmosphere, accelerating climate change. Black carbon comes from burning fossil fuels and biomass. Heavy metals like lead, mercury, and arsenic persist in the environment, posing long-term health risks like cancer and developmental problems due to their presence in industrial processes and contaminated sites. While black carbon remains a significant climate and health concern, regulations and technological advancements have successfully reduced its presence in some regions. For example, stricter vehicle emission standards have mandated cleaner burning engines and the use of particulate filters in diesel vehicles. These filters capture significant black carbon before it can be released into the atmosphere. Secondly, regulations have driven a shift toward cleaner-burning fuels with lower sulfur content. This is because complete combustion of lower-sulfur fuels results in less black carbon formation. Another success has been with improved industrial practices. Industries like brick kilns and steel mills have adopted cleaner technologies to reduce black carbon emissions. This may involve switching fuels, implementing improved combustion techniques, and employing filters to capture black carbon particles. These strategies have succeeded despite reduced black carbon levels, improved air quality, and climate benefits. Studies show a decline in black carbon concentrations in some regions, particularly in developed countries with stricter regulations. Lower black carbon levels contribute to better air quality, decreasing respiratory illnesses and other health problems associated with air pollution. Reduced black carbon emissions contribute to mitigating climate change by limiting atmospheric warming caused by this pollutant. Despite these successes, challenges remain in effectively tackling black carbon in the context of incomplete regulatory compliance and/or implementation, regions with developing economies, and open burning practices. Stricter regulations might be in place, but consistent and effective enforcement across regions is crucial. Rapid industrialization and urbanization in developing countries can become significant black carbon emissions sources if cleaner technologies and regulations are not implemented. Open burning of agricultural waste and household trash remains a major source of black carbon in many regions. Addressing this requires awareness campaigns, promoting sustainable waste management practices, and providing alternative solutions. While significant progress has been made, efforts are essential to reduce black carbon emissions. This includes promoting stricter regulations globally, supporting technological advancements in cleaner fuels and industrial processes, and encouraging a shift toward sustainable practices that minimize open burning.

2.1.7 Microplastics

Emerging threats like microplastics from synthetic fibers, car tire wear, and plastic incineration raise concerns about potential respiratory problems, with research ongoing to determine the exact health impacts. Similarly, nanoparticles used in various products have unknown health effects that require further investigation.

2.1.8 Endocrine-Disrupting Chemicals (EDCs)

These are found in personal care products and pharmaceuticals and might have airborne pathways and pose potential health risks like hormonal disruption and reproductive problems, although research is still underway. Indoor air quality is also a concern, with pollutants like building materials, furnishings, and cleaning products contributing to respiratory irritation, allergies, headaches, dizziness, and fatigue. Effective mitigation strategies require a multipronged approach, addressing regulation limitations, static approaches that do not adapt to change, enforcement challenges, and the need for innovation under stricter regulations. By understanding the diverse sources of air pollutants and their health impacts, we can work toward cleaner air for a healthier planet.

Well, the list of culprits will not end just with these many entities in the air, and an expansive tabulation to suffice for this has been shown in Table 1.

2.2 Beyond Command-and-Control: Hurdles in Mitigation

Traditional C&C regulations, which set specific emission limits or mandate specific control technologies, have achieved some success. However, they encounter limitations when tackling the diverse nature of air pollutants.

2.3 Limited Scope

C&C regulations often target specific pollutants or industries, neglecting other contributing factors. For example, controlling NOx emissions from vehicles might not address VOCs from paints and solvents, leaving a gap in mitigation efforts (Wade and Farrauto 2012; Zheng et al. 2016; Kim 2011; Brezinski et al. 2012; Klimont et al. 2002).

2.4 Static Approach

C&C regulations might struggle to adapt to evolving pollution sources and technologies. New industrial processes or emerging pollutants may not be adequately addressed by existing regulations (Geissen et al. 2015; Rizzo et al. 2019; Tickner et al. 2019).

2.5 *Enforcement Challenges*

Monitoring and enforcing C&C regulations can be resource-intensive, especially for geographically dispersed sources like agricultural burning or small-scale industries (Twerefou et al. 2015; Feng and Liao 2016; Peng et al. 2020; Zeng et al. 2019; Heyes 2000).

2.6 *Innovation Disincentive*

Rigid C&C regulations might discourage investments in cleaner technologies if companies comply with existing standards (Song et al. 2023; Zhang and Yu 2023). Companies may prioritize meeting current regulations rather than exploring innovative solutions for deeper emission reductions (Wang et al. 2022; Blackman et al. 2018; List et al. 2004; Popp 2003).

2.7 *Multipronged Approach: The Way Forward*

To effectively combat air pollution, a multipronged approach is crucial. This includes the following:

2.8 *Market-Based Instruments*

Strategies like emissions trading can create economic incentives for reducing pollution. Companies can buy and sell pollution permits, encouraging innovation in cleaner technologies to reduce their overall emission footprint (Yan et al. 2020; Atkinson and Tietenberg 1991).

2.9 *Investing in Cleaner Technologies*

Promoting research and development of cleaner energy sources like renewables and energy efficiency can significantly reduce air pollution at its source.

2.10 Public Awareness and Behavior Change

Empowering the public with information on air pollution sources and health impacts can lead to individual actions that reduce emissions, such as opting for public transportation or energy-efficient appliances.

2.11 International Cooperation

Air pollution respects no borders. International collaboration on regulations, pollution monitoring, technology transfer, and data sharing is crucial for addressing the global challenge of air pollution (Bernauer and Moser 1996; Calvillo 2018; de Coninck et al. 2008).

By understanding the complex nature of air pollutants and moving beyond the limitations of traditional mitigation strategies, we can pave the way for cleaner air and a healthier future for generations to come.

3 Breathing Easy: Emerging Strategies to Combat Air Pollution

Air pollution, a global health crisis, demands innovative solutions. While existing regulations have yielded some success, the fight continues. Here's a glimpse into some of the newest and most promising strategies emerging on the scene:

3.1 Cutting-Edge Technologies

- *Electrostatic Precipitators (ESPs) and Scrubbers*: These advanced filtration systems, already used in some industries, can be further optimized for wider application in power plants and factories to capture even finer particulate matter.
- *Bioremediation*: Harnessing the power of nature, this technique utilizes specially engineered microbes to break down pollutants like nitrogen oxides and sulfur dioxide released from industrial processes. While still under development, it holds promise for a sustainable approach.
- *Photocatalysis*: This technology utilizes sunlight or ultraviolet light to activate catalysts that break down pollutants like nitrogen oxides and volatile organic compounds (VOCs) into harmless substances. Pilot projects show promise, but more comprehensive implementation requires further research and cost reductions.

3.2 Shifting Gears in Transportation

- *Electric Vehicles (EVs)*: The rapid growth of EVs, powered by batteries and not fossil fuels, significantly reduces tailpipe emissions, particularly in urban areas. Continued advancements in battery technology, charging infrastructure, and affordability will be crucial for widespread adoption.
- *Hydrogen Fuel Cell Vehicles*: These vehicles produce only water vapor as a by-product, offering a clean alternative to gasoline and diesel-powered vehicles. However, challenges remain in terms of hydrogen production and infrastructure development.
- *Urban Mobility Solutions*: Promoting cycling, walking, and public transportation can significantly reduce reliance on personal vehicles, particularly in densely populated areas. Integrating these options with smart traffic management systems can optimize efficiency and reduce emissions.

3.3 Nature-Based Solutions

- *Urban Green Infrastructure*: Planting trees and creating green spaces in cities is a natural air filter, capturing pollutants and improving air quality. Strategic placement of green corridors and vertical gardens in urban landscapes can maximize benefits.
- *Precision Agriculture*: Optimizing fertilizer application and irrigation practices in agriculture can minimize ammonia emissions, a key contributor to particulate matter formation. This requires advancements in soil testing technologies and farmer education.

3.4 The Power of Policy and Public Participation

- *Carbon Pricing*: Implementing carbon pricing mechanisms like carbon taxes or cap-and-trade systems can incentivize industries and consumers to reduce emissions. Revenue generated can be used to fund clean energy initiatives.
- *Community-Based Monitoring*: Empowering citizens with low-cost air quality monitoring tools can raise awareness about pollution levels and encourage advocacy for cleaner air policies.

3.5 Additional Ideas and Opinions

- *Regional Solutions*: Developing strategies tailored to address specific air pollution sources and meteorological conditions in different regions is crucial. For instance, tackling agricultural burning in Southeast Asia may require a different approach than controlling European industrial emissions.

- *Investing in Research & Development*: Continued funding for research on cleaner technologies, alternative fuels, and innovative air pollution control methods is essential for long-term progress.
- *Public Education and Behavior Change*: Raising public awareness about the health impacts of air pollution and promoting individual actions like carpooling and responsible waste disposal can contribute significantly to curbing emissions.

Combining these emerging strategies with existing regulations and a renewed commitment to clean air can create a healthier future for all. The key lies in fostering innovation, promoting regional solutions, and empowering communities to actively participate in the fight for clean air. This is a collective battle, and our collective action holds the power to make a difference. Remember, clean air is not a luxury; it is necessary for a healthy planet and a thriving society.

4 Understanding the Murky Threat: Critical Issues and Challenges of Water Pollution

Water, the elixir of life, is under siege. Unlike air, which readily disperses pollutants, water concentrates them, posing a significant threat to the health of ecosystems and human populations worldwide. While vast oceans dominate the Earth's surface, freshwater resources, crucial for terrestrial life, are a finite and increasingly vulnerable treasure. Freshwater sources, including rivers, lakes, and aquifers, serve as a source of drinking water and a vital component of the global drainage system, carrying away household and industrial waste. While many urban centers have implemented robust sewage treatment infrastructure, a significant portion of the world's wastewater remains untreated or inadequately managed. This influx of contaminants, ranging from microscopic pathogens to persistent chemicals, disrupts the delicate balance of freshwater ecosystems. The consequences of this water pollution are far-reaching and often severe. This can lead to a cascade of consequences, firstly, waterborne diseases. Untreated sewage introduces a multitude of pathogens like bacteria, viruses, and parasites into water bodies. These pathogens can cause a spectrum of waterborne diseases, ranging from mild diarrheal illnesses to life-threatening cholera and dysentery. Secondly, algal blooms and eutrophication. Excess nutrients like nitrates and phosphates from agricultural runoff and fertilizers act as superchargers for algae. These excessive algal blooms (eutrophication) deplete oxygen levels in the water, creating dead zones devoid of aquatic life. Next, pollutants like heavy metals and persistent organic chemicals bioaccumulate the food chain, reaching dangerous concentrations in fish and shellfish consumed by humans. This bioaccumulation poses a significant threat to human health. Further, contaminated irrigation water can percolate through the soil, polluting agricultural land and impacting crop quality. The consequences extend beyond reduced yields, potentially introducing contaminants into the food chain through plant uptake. The impact of water pollution is not uniform. Developing countries often lack the infrastructure and resources for effective wastewater treatment, leading to a higher burden of waterborne diseases. However, developed nations are not immune. Industrial waste, agricultural practices, and emerging contaminants like pharmaceuticals pose challenges even in regions with advanced sanitation systems.

4.1 Pollutants in Global Water: Diverse Culprits, Devastating Impacts

Some have been discussed here, while others are listed in Table 2.

4.1.1 Pathogens

Bacteria, viruses, and parasites from untreated sewage contaminate water sources, causing waterborne diseases like cholera, typhoid, and dysentery (Hamner et al. 2006; Kristanti et al. 2022). These diseases lead to significant illness, mortality, and strain on healthcare systems, particularly in developing countries.

4.1.2 Nutrient Pollution

Excessive fertilizers and agricultural runoff introduce nitrates and phosphates into water bodies. This nutrient overload triggers eutrophication, where algae blooms deplete oxygen levels, suffocating aquatic life and creating dead zones (Bailey et al. 2020; Devlin and Brodie 2023).

4.1.3 Industrial Pollutants

Heavy metals, toxic chemicals, and pharmaceuticals from industrial processes contaminate water sources (Wang and Li 2019; Xia et al. 2020; Michael Beman et al. 2005; Sharpley et al. 2001; Smith et al. 1999). These pollutants can cause cancer, and developmental problems and disrupt ecosystems.

4.1.4 Microplastics

Tiny plastic fragments from various sources find their way into water bodies. Microplastics' ecological and health impacts are still under investigation, but they pose a potential threat to aquatic life and potentially even health (Winiwarter and Klimont 2011; Karbalaei et al. 2018).

Table 2 A comprehensive table detailing the major culprit pollutants in global water, their sources, health impacts, environmental effects, and mitigation challenges

Pollutant category	Specific pollutants	Sources	Health impacts	Environmental effects	Mitigation challenges	References
Pathogens	Bacteria (e.g., *E. coli*), viruses (e.g., Hepatitis A), parasites (e.g., Giardia)	Untreated sewage, animal waste runoff, poor sanitation practices	Diarrhoeal diseases (cholera, typhoid, dysentery), hepatitis, gastrointestinal illness	Disrupts aquatic ecosystems, reduces biodiversity, creates unsafe water for recreation	Lack of sanitation infrastructure in developing countries, defective or overwhelmed sewage systems, inadequate wastewater treatment	Pandey et al. (2014), Baig et al. (2021), and Kumar et al. (2021a)
Nutrient pollution	Nitrates, phosphates	Agricultural fertilizers, animal manure runoff, lawns & gardens (excessive fertilization)	Methemoglobinemia (blue baby syndrome) in infants, cancers (colorectal, stomach), eutrophication (algal blooms, oxygen depletion)	Creates dead zones in water bodies, harms fish and other aquatic life, disrupts food webs	Excessive use of fertilizers in agriculture, urban runoff carrying lawn & garden chemicals, manure management practices in factory farms	Segerson and Walker (2002), Sakadevan and Nguyen (2017), Grizzetti et al. (2021), Mosheim and Sickles (2021)
Industrial pollutants	Heavy metals (lead, mercury, arsenic), organic solvents, pharmaceuticals, endocrine-disrupting chemicals (EDCs), textile dyes	Industrial discharges, mining operations, pharmaceutical manufacturing, personal care products, textile dyeing facilities	Cancers, nervous system disorders, developmental problems, hormonal disruption	Bioaccumulation in aquatic organisms, contaminates fish and shellfish, reduces biodiversity	Ineffective or lax industrial waste treatment, limited regulations on specific pollutants, illegal dumping practices	Ahamad et al. (2020), Pereira et al. (2022), and Madhav et al. (2020)

Microplastics	Fragments of larger plastic debris, synthetic fibers (clothing), microbeads (in cosmetics)	Plastic breakdown in the environment, washing clothes containing synthetic fibers, microbeads in personal care products	Potential health impacts (research ongoing, endocrine disruption, digestive issues), disrupts feeding habits of aquatic life, can block digestive systems	Lack of regulations on microplastics in products, limited infrastructure for capturing microplastics in wastewater treatment, incomplete understanding of long-term health effects	Inadequate waste management, lack of regulations, limited public awareness, difficulty in detection and removal, harm to marine life, contamination of food chain, persistence in environment, impact on human health, limited technological solutions, high cleanup costs, lack of international cooperation, insufficient research on effects	Mamun et al. 2023, Osman et al. (2023), and Chen et al. (2021)
Emerging contaminants	Pharmaceuticals (antibiotics, hormones), personal care products (triclosan, sunscreen chemicals), endocrine-disrupting chemicals (EDCs)	Improper disposal of unused medications, household waste runoff, industrial discharges	Antibiotic resistance in bacteria, hormonal disruption, developmental problems (research ongoing)	Limited data on long-term health and ecological effects, difficulties in removing these contaminants from wastewater treatment, incomplete regulations on emerging pollutants	Various as above and below	Lofrano et al. (2020), Kumar et al. (2023b), Dhir (2022), and Chaturvedi et al. (2021)

(continued)

Table 2 (continued)

Pollutant category	Specific pollutants	Sources	Health impacts	Environmental effects	Mitigation challenges	References
Salinity (ocean water intrusion)	Rising sea levels, over-extraction of freshwater	Climate change, agricultural water use, deforestation	Salinization of freshwater sources, reduces the availability of freshwater for drinking & agriculture, disrupts coastal ecosystems	Curbing climate change (mitigation & adaptation), sustainable water management practices, protecting coastal ecosystems	Aquifer contamination, saltwater corrosion of infrastructure, agricultural land degradation, freshwater scarcity, ecosystem disruption, increased energy costs for desalination, reduced crop yields, soil erosion, decreased water quality, impacts on human health, economic losses to industries, difficulty in monitoring and predicting intrusion	Artiola et al. (2019), Haq et al. (2018), and Kaushal et al. (2021)
Sedimentation	Soil particles	Deforestation, Poor land management practices	Reduced water storage capacity, Harms aquatic life by clogging gills and smothering eggs	Damages aquatic habitats, Reduces light penetration, Impacts fish spawning grounds	Sustainable forestry practices, Soil conservation techniques	Issaka and Ashraf (2017), Rashmi et al. (2022)
Thermal Pollution	Hot water discharges	Industrial processes (power plants, factories)	Disrupts ecosystems, Reduces dissolved oxygen levels, harms sensitive species	Impacts fish migration patterns, reduces biodiversity, creates unsuitable habitats	Regulations on thermal discharge limits, cooling water treatment technologies	Raptis et al. (2016), Naimaee et al. (2024), De Jonckheere et al. (1975), and Ma et al. (2019)

Acidification	Sulfuric & nitric acid (air & water pollution)	Fossil fuel combustion, acid mine drainage	Harms aquatic life (especially shellfish), makes it difficult for some organisms to form shells or skeletons	Disrupts aquatic food webs, reduces biodiversity, damages coral reefs	Curbing air and water pollution from fossil fuels, Limestone treatment of acidic water bodies	Khemani et al. (1994) and Galloway (2001)
Textile dyes	Azodyes, Reactive dyes, disperse dyes	Textile dyeing facilities	Skin irritation, respiratory problems, potential carcinogenicity (some dyes)	Harms aquatic life, disrupts ecosystems, reduces water clarity	Toxicity, color persistence, soil contamination, scalability, cost, color consistency, regulatory frameworks, public awareness, technological limitations	Kaur et al. (2023) and Zhang and Wu (2014)

4.1.5 Emerging Contaminants

Pharmaceuticals, personal care products, and endocrine-disrupting chemicals (EDCs) are emerging as new water pollutants (Ahamad et al. 2020; Kasonga et al. 2021). Their long-term effects on human health and ecosystems are still being studied.

4.1.6 Azodye Effluents

Azodyes, a vibrant class of synthetic dyes used extensively in textiles, plastics, and other industries, cast a long shadow on water quality. Discharged as effluents from dyeing facilities, these colorful contaminants pose a significant threat to aquatic ecosystems and human health. Azodyes qualify as major culprits in water pollution. This is because, unlike natural dyes, azodyes are complex molecules highly resistant to biodegradation (Copaciu et al. 2013). They persist in water bodies for extended periods, accumulating in sediments and the food chain. The intense color of azodyes reduces light penetration in water, disrupting photosynthesis by aquatic plants and phytoplankton, the base of the aquatic food chain. This disrupts the entire ecosystem. The decomposition of organic matter, including azodyes, consumes dissolved oxygen in water. This oxygen depletion creates hypoxic zones, areas devoid of aquatic life. Certain azodyes and their breakdown products can be toxic to aquatic organisms. This can directly harm fish and other aquatic life, impacting their survival and reproduction. Some azodyes can bioaccumulate in fish and shellfish, concentrating up the food chain (Maheshwari et al. 2021). This poses a potential health risk to humans who consume contaminated seafood. The widespread use of azodyes and their persistence and potential toxicity necessitates stricter regulations and the adoption of sustainable dyeing practices. By promoting alternative, biodegradable dyes and implementing advanced wastewater treatment technologies, we can minimize the environmental impact of these colorful contaminants and safeguard our precious water resources.

5 The Road to Recovery

Combating water pollution requires a multipronged approach, and understanding the scope of the issue is crucial. Currently, 80% of global wastewater is released into the environment without adequate treatment. Unsustainable farming practices contribute significantly to the problem, with 40% of agricultural pollutants in water bodies coming from these sources. Stricter regulations and enforcement could reduce industrial waste disposal by 20%. If conservation efforts are not implemented, 60% of the global population will face water scarcity by 2030. However, public education campaigns and awareness programs can prevent 25% of water pollution, and technological advancements in wastewater treatment and contaminant detection can mitigate an additional 30%. Without action, 50% of the world's population will depend on imported water by 2025. Upgrading wastewater treatment facilities, promoting sustainable agricultural practices, and developing stricter regulations on industrial waste disposal are crucial steps. Public education campaigns can raise awareness about water conservation and responsible waste disposal practices. Technological advancements in areas like wastewater treatment and contaminant detection offer promising solutions for mitigating water pollution. Water pollution is a global crisis demanding immediate and collective action. By acknowledging the severity of the problem, implementing effective solutions, and fostering international cooperation, we can safeguard this vital resource for future generations. This scientific examination of water pollution highlights the urgency of protecting this essential element for the well-being of our planet and all its inhabitants.

5.1 *Command and Control Strategies: A Double-Edged Sword in Water Pollution Control*

C&C strategies have played a significant role in mitigating water pollution, but their effectiveness has limitations. While they have succeeded in specific regions, their rigid, top-down approach struggles to address the full complexity of the problem. Here's a deeper look at C&C strategies, their successes, and limitations, supported by data and case studies.

5.1.1 C&C Strategies in Action

- *Regulations*: The Clean Water Act (CWA) in the USA and the Water Framework Directive (WFD) in the European Union are prime examples of C&C strategies. These regulations set specific pollution discharge limits for industries and municipalities.
- *Technology Mandates*: C&C policies can mandate specific technologies like advanced wastewater treatment plants to reduce pollutant discharge.
- *Permits and Enforcement*: Permits with strict compliance requirements and penalties for violations are key tools for enforcing regulations.

5.1.2 Success Stories

- *Reduced Point Source Pollution*: C&C regulations have significantly reduced point source pollution from industrial facilities in developed countries. Studies by the US Environmental Protection Agency (EPA) show a 65% decline in industrial wastewater discharges since the C&C implementation of the CWA.

- *Improved Water Quality*: The WFD, focusing on achieving "good ecological status" in water bodies, has successfully improved water quality in some European rivers and lakes.

5.1.3 Challenges and Limitations

- *High Implementation Costs*: Upgrading infrastructure and implementing new technologies to comply with C&C regulations can be expensive, particularly for developing countries.
- *Limited Scope*: C&C strategies primarily focus on point source pollution, neglecting nonpoint sources like agricultural runoff and urban stormwater.
- *Lack of Flexibility*: Rigid regulations may not adapt to changing environmental conditions or address emerging contaminants effectively.
- *Enforcement Challenges*: Enforcing C&C regulations can be complicated, especially in regions with limited resources or weak governance.

5.1.4 Statistical Evidence

- *Uneven Progress*: A 2020 UN Environment Programme report suggests that 80% of the global wastewater remains untreated, highlighting the limitations of C&C strategies in achieving universal success.
- *Persistent Issues*: Despite regulations, agricultural runoff remains a major source of water pollution. The US EPA estimates that agriculture contributes to over half of the water quality impairments in US rivers and streams.

5.1.5 Case Studies

- *Success in the USA*: The CWA in the USA has seen significant declines in point-source pollution, but challenges remain in addressing nonpoint-source pollution and emerging contaminants.
- *Challenges in China*: China's rapid industrialization has led to severe water pollution. While C&C regulations have been implemented, weak enforcement and lack of investment in wastewater treatment infrastructure hinder progress.

5.1.6 Moving Forward

C&C strategies have played a valuable role, but a more holistic approach is needed. Integrating C&C with market-based instruments like pollution trading and economic incentives for sustainable practices can improve effectiveness. Additionally, fostering innovation in wastewater treatment technologies and promoting public participation in water conservation efforts are crucial steps toward a cleaner future.

C&C strategies have yielded undeniable success in specific regions. However, their limitations in addressing the full complexity of water pollution are becoming increasingly evident. A multifaceted approach combining C&C with market incentives and technological advancements is needed to combat this global challenge effectively. By acknowledging the limitations of C&C and exploring innovative solutions, we can ensure the protection of this vital resource for generations to come.

5.1.7 The Road to Cleaner Water

Despite these challenges, various strategies offer hope for a cleaner future:

- *Investing in Infrastructure*: Building and upgrading sanitation systems and wastewater treatment plants is crucial for preventing untreated sewage from entering water bodies.
- *Sustainable Agriculture*: Promoting sustainable agricultural practices that minimize fertilizer and pesticide use can significantly reduce nutrient pollution.
- *Industrial Regulations*: Stricter regulations and enforcement regarding industrial waste disposal and treatment are essential to curb the release of toxic pollutants.
- *Public Education*: Raising public awareness about water pollution and its impacts can encourage water conservation and responsible waste disposal practices.
- *Technological Advancements*: New technologies for wastewater treatment, pollutant detection, and remediation offer promising solutions for mitigating water pollution.

6 The Silent Threat: Unveiling the Challenges of Land Pollution

While air and water pollution often grab headlines, land pollution is a more insidious and frequently overlooked environmental crisis. A recent geographically mapped impact on the global land area shows various areas affected by human-led disturbances (Theobald et al. 2020). Accordingly, the study indicates that mangroves, tropical and subtropical moist and dry broadleaf forests suffered the greatest losses between 1990 and 2015, with Oceania, Asia, and Europe being the most affected regions (Theobald et al. 2020). Numerous pollutants emanate from activities on land and effectively contribute to the above-discussed air and water pollutants. Sources other than agriculture, industries, and household processes are also added with least considered metals, battery electrolytes, acids, dyes, and other technogenic factors, and there are growing concerns about not undermining electronic wastes as well. The advent of use and throw plastics and packaging materials, bottling composites, rubber, etc., make up only a small fraction of the culprits. To say more truly, land pollution can be considered more detrimental. Firstly, because it is cumulative and Long-lasting, in the view that, unlike air and water pollution, which can disperse or be remediated to some extent, land pollution, especially from hazardous waste and persistent pollutants, tends to accumulate and persist in the soil for decades or even centuries (Ijaz et al. 2024; Pasetto et al. 2024). Secondly, it poses a direct impact on food chains as it directly contaminates the soil, which is the foundation of our food production systems. Toxic substances can bioaccumulate in the food chain, posing significant risks to human health (Saeedi and Technology 2024). The third reason should be the affected habitat destruction and biodiversity loss through deforestation, urbanization, and improper waste disposal, all of which destroy critical habitats, leading to loss of biodiversity and ecosystem imbalance (Singh 2024). The fourth reason would be land pollution caused by groundwater contamination, as pollutants on the top layer of soil seep into groundwater, contaminating a vital resource for drinking and irrigation (Artiola et al. 2019; Burmaster 1982; Lerner and Harris 2009). Last but not least, there can be aesthetic damage, such as littering and illegal dumping, negatively impacting the visual appeal of landscapes and reducing the quality of life (Damineva et al. 2019; Kapper 2004). Hence, while air and water pollution undoubtedly pose severe threats, the cumulative, long-term, and often irreversible effects of land pollution make it a particularly challenging and insidious environmental problem.

6.1 A Multitude of Culprits of Land Pollution

Several factors interact in complex ways, often exacerbating the land pollution problem. We discuss some of them here while inclusively others are tabulated ahead (Table 4).

6.1.1 Municipal Solid Waste (MSW)

Improper waste disposal, including plastics, food scraps, and hazardous materials, creates overflowing landfills and pollutes surrounding areas through leachate—toxic liquids that seep from decomposing waste (Pillai et al. 2014). While landfilling is the cheapest and most common waste disposal method around the globe, this requires an unimaginably large area and proper drainage regimes (Pheakdey et al. 2022). If unchecked, it is probable to create potential groundwater contamination and air pollutant densities to surpass the thresholds stipulated in and around the sites. Again, the leachates may grow sizeable in places with high moisture and precipitation. Municipal and household waste consists of a variety of materials (Table 3), but the major components are organic waste, which includes food scraps and yard trimmings; then paper and cardboard products such as newspapers and cardboard boxes; thereafter, the plastics from packaging, bottles, and other containers; and then followed with a variety of different materials, including glass, metal, textiles, and electronics (Kaza et al. 2018). The composition of municipal and household waste can vary depending on factors such as location, income levels, and

Table 3 General composition of municipal and household waste

Material	Percentage (%)
Organic waste	44
Paper and cardboard	17
Plastics	12
Glass	5
Metal	4
Other waste	18

Data adapted from published sources (USEPA 2024a, b)

recycling programs. However, organic waste, paper, cardboard, and plastics are the major components of municipal and household waste streams worldwide. Medical waste also needs to be considered well. Although 85% of it continues to be general, the remaining 15% constitutes the contagious portion, which is rightly disposed of in bigger population-dense countries (WHO 2018). Low-income countries face a significant challenge in managing healthcare waste. Proper segregation of hazardous and nonhazardous medical waste is often neglected, drastically underestimating the true volume of hazardous materials (Ferronato and Torretta 2019).

6.1.2 Industrial Waste

Industrial processes generate diverse pollutants, including heavy metals, chemicals, and radioactive materials. These can contaminate soil and groundwater, posing health risks and impacting agricultural productivity (Sharma et al. 2021). Surprisingly, industrial waste is a massive problem, accounting for at least half of all global waste (Pratt 2021). Manufacturing is the leading culprit, rapidly expanding as our population grows. With billions more people expected in the coming decades, we cannot simply build more factories without catastrophic consequences for our planet. Crucially, much of this waste is preventable. Poor-quality products are a major source of manufacturing waste, but this issue can be tackled with data-driven solutions. Technological advances, like artificial intelligence, offer new hope for reducing waste and creating a more sustainable future. Industrial processes generate a staggering 50% of global waste. This includes contaminants like heavy metals, chemicals, and radioactive materials, which have profound impacts on ecosystems, soil health, and human well-being. For instance, it is estimated that around 400 million tons of hazardous waste are produced globally every year, a large portion of which comes from manufacturing and industrial activities. Heavy metals like lead and mercury from industrial waste have long-lasting effects on both land and water, with up to 60% of groundwater contamination in some regions attributed to industrial pollutants. Manufacturing alone is a significant contributor to the waste problem, producing over 2 billion tons of solid waste globally each year. This sector is responsible for roughly 30% of global greenhouse gas emissions, further exacerbating climate change. Poor product quality and inefficiencies in

Table 4 Culprit pollutants of land globally

Category	Pollutant	Sources	Impacts	Examples	References
Industrial waste	Heavy metals (e.g., lead, mercury, arsenic)	Mining operations, metal processing facilities, battery manufacturing, electronics waste (e-waste)	Bioaccumulation in the food chain, posing health risks for humans and wildlife Soil contamination, impacting plant growth and agricultural productivity Water pollution through leaching	Lead contamination from abandoned mines, mercury contamination from chloralkali plants, arsenic contamination from legacy pesticide use	Gaur et al. (2020), Hogland and Stenis (2000), Freitas et al. (2021), Misra and Pandey (2005)
	Industrial chemicals	Chemical manufacturing, pharmaceutical production, solvents, paints, dyes	Soil and groundwater contamination Disruption of soil microbial communities Potential human health risks through exposure or bioaccumulation	Soil contamination from leaking underground storage tanks (USTs) at industrial facilities, endocrine disruption from pharmaceuticals in soil	
	Hazardous materials	Industrial waste disposal sites, spills during transportation, or accidents	Acute and chronic health risks for humans and wildlife Soil sterilization, rendering land unusable Groundwater contamination	Soil and groundwater contamination from abandoned hazardous waste landfills, dioxin contamination from industrial accidents	
Agricultural practices	Excess fertilizers and pesticides	Intensive farming practices, reliance on chemical inputs	Eutrophication of water bodies through runoff Soil acidification and salinization Disruption of soil microbial communities Loss of biodiversity	Nitrate leaching from excessive fertilizer use causes algal blooms in water bodies, salinization of agricultural land due to irrigation with saline water	Baweja et al. (2020), Lal et al. (1988), and De Haan (1987)
Laboratory and pharma giants	Antibiotics	Use of antibiotics in veterinary medicine	Development of antibiotic-resistant bacteria in soil Potential human health risks through exposure or bioaccumulation in the food chain	Antibiotic resistance in soil bacteria from manure application in agriculture	Obimakinde et al. (2017), Singh et al. (2024), Nieder et al. (2018b), and Kotwani et al. (2021)

Municipal waste	Landfills	Improper waste disposal practices, limited recycling, and composting infrastructure	Landfilling consumes valuable land space Methane emissions from landfill decomposition contribute to climate change Leachate generation contaminates soil and groundwater	Overcrowded landfills in developing countries, methane emissions from municipal solid waste landfills	Chen et al. (2021), Redko et al. (2024), Idumah and Nwuzor (2019), Shahnawaz et al. (2019), and Idris et al. (2004)
	Plastic waste	Single-use plastics, packaging waste, improper waste management	Fragmentation into microplastics, harming soil biota and potentially entering the food chain Reduced soil fertility Visual pollution	Microplastic contamination in agricultural soils, plastic pollution harming wildlife	
Other sources	Salinization	Overuse of irrigation water with high salt content, seawater intrusion	Reduction of soil fertility and agricultural productivity Loss of arable land	Salinization of agricultural land in coastal areas	Artiola et al. (2019), Haq et al. (2018), Kaushal et al. (2021), Li et al. (2014), Van Meter et al. (2011), and Kazemi et al. (2018)
	Acid rain	Emissions of sulfur and nitrogen oxides from industrial activities	Soil acidification, impacting plant growth Mobilization of heavy metals in soil	Forest dieback in areas impacted by acid rain	
	Urban runoff	Stormwater runoff from urban areas carrying pollutants	Contamination of soil with pollutants like oil, grease, and heavy metals Increased pollution of waterways receiving urban runoff	Soil contamination near busy roads due to runoff carrying pollutants from vehicles	

manufacturing processes lead to vast amounts of waste, but technological solutions like artificial intelligence (AI) are now offering ways to optimize production, improve efficiency, and reduce waste. For example, data-driven AI solutions have the potential to reduce industrial waste by up to 30%, transforming the sector into a more sustainable and eco-friendly industry. A strategic conservation plan is vital to address these growing challenges. Such a plan should focus on integrating advanced technologies like AI to improve waste management, enforce stricter environmental regulations, and encourage industries to adopt circular economy principles. These actions could dramatically reduce industrial waste and minimize its devastating environmental impacts.

6.1.3 Agricultural Runoff

Agriculture harms the environment through pollution, climate change, and health risks (Abbasi et al. 2014). The excessive use of fertilizers, pesticides, and herbicides in modern agriculture poses a significant threat to land and water quality (Li and Zhang 1999). While designed to boost crop yields, these chemicals have far-reaching negative consequences. Fertilizers, primarily composed of nitrogen and phosphorus, are essential for plant growth. However, their overuse can lead to a phenomenon known as nutrient runoff. When excess fertilizers are washed into water bodies, they cause algal blooms, depleting oxygen levels and harming aquatic life (Baweja et al. 2020). Moreover, nitrates from fertilizers can seep into groundwater, contaminating drinking water sources. This contamination is hazardous for infants, as high nitrate levels in the blood can prevent oxygen from being transported effectively, a condition known as methemoglobinemia or "blue baby syndrome." Pesticides are designed to control pests, but their indiscriminate use can devastate the environment. These chemicals can kill beneficial insects, disrupt delicate ecosystems, and contaminate soil and water bodies. Pesticides can also accumulate in the food chain, posing risks to human health (Baweja et al. 2020; Srivastav 2020). Herbicides, used to control weeds, can also contaminate soil and water. They can harm nontarget plants, including crops, and affect soil microorganisms essential for maintaining soil fertility. It is crucial to adopt sustainable agricultural practices that minimize the use of these chemicals while ensuring food security. This includes precision agriculture, crop rotation, integrated pest management, and organic farming methods.

6.1.4 Electronic Waste (e-Waste)

E-waste, the electronic detritus of our digital age, has emerged as a formidable environmental challenge. This burgeoning waste stream, composed of discarded computers, smartphones, televisions, and other electronic devices, is a toxic cocktail of hazardous substances, including lead, mercury, cadmium, and beryllium. Improper disposal, a prevalent practice in many parts of the world, allows these pollutants to seep into the soil and groundwater, contaminating ecosystems and posing severe health risks to humans and wildlife. As per the Global e-Waste Monitor (ITU 2024), the scale of the e-waste crisis is staggering. I reported a global generation of 53.6 million metric tons of e-waste in 2019, with only 17.4% formally documented as collected and recycled. This implies that the vast majority of e-waste ends up in landfills or is illegally dumped, often in developing countries with lax environmental regulations. A significant portion of e-waste is shipped to countries with less stringent environmental laws for informal recycling. This practice, often called "e-waste dumping," involves hazardous processes like acid baths and open burning to extract valuable metals. These activities release toxic fumes into the atmosphere and contaminate soil and water bodies, disproportionately affecting the health of local communities. Beyond the immediate environmental impact, e-waste also contributes to climate change. The energy-intensive processes involved in producing, using, and disposing of electronic devices generate substantial greenhouse gas emissions. Additionally, the incineration of e-waste releases harmful pollutants into the atmosphere. To address the e-waste crisis, a comprehensive approach is required. This includes extended producer responsibility, where manufacturers are held accountable for the end-of-life management of their products, investments in e-waste collection and recycling infrastructure; and promoting consumer awareness about the importance of responsible e-waste disposal. By adopting sustainable practices and policies, we can mitigate the environmental and health risks associated with e-waste and transition to a more circular economy.

6.2 Hurdles in Mitigating Land Pollution

Limited public awareness about land pollution and its consequences hinders responsible waste disposal practices. Many developing and least developing countries lack proper waste management facilities and infrastructure for safe disposal and treatment of industrial waste (Bundhoo 2018). Implementing sustainable waste management practices and clean technologies can be expensive, posing a challenge for some regions. Other than these, ineffective enforcement of regulations on waste disposal and industrial practices allows for continued pollution. Additionally, regulations may not address emerging contaminants. Also, growing population densities in countries like India and China increase waste, which outpaces management regimes (Kumar et al. 2017). The Ministry of Environment and Forests established India's waste management regulations, but their implementation is inconsistent and incomplete. India generates a massive amount of waste, producing around 133,760 tons daily. However, the country struggles to manage this waste effectively, with only about 68% of it collected and a mere 19% treated (Kumar et al. 2017). Similarly, the Philippines is emerging as a major contributor to plastic pollution in Southeast Asia, generating approximately 133,760 tons of municipal solid waste daily (Manas 2023). While around 91,152 tons are collected, only 25,884 tons are treated. This, coupled with a low recycling rate of just 9% for high-value recyclables, highlights the country's significant plastic waste management challenge. Filipinos consume an average of 20 kg of plastic packaging annually, with 15.43 kg ending up as waste (Manas 2023). Similarly, China's rapid development has led to a severe shortage of waste disposal facilities due to local resistance, outdated infrastructure, and increasing waste generation, despite a centralized planning system (Chung et al. 2002).

To further emphasize the urgency of addressing waste management challenges, it is important to consider global waste generation trends. The world currently produces around 2.01 billion tons of municipal solid waste annually, with 33% of it being mismanaged in ways that harm the environment. In India, as highlighted, the nation generates 133,760 tons of waste each day. Projections show that by 2041, urban areas in India will produce 543 million tons annually, placing immense strain on the existing waste management infrastructure. Similarly, Southeast Asia, particularly the Philippines, faces a plastic waste crisis, contributing 2.7 million tons of plastic waste annually, with 20% of its population lacking access to waste collection services. China, the world's largest waste producer, generated 235 million tons of municipal solid waste in 2020, a figure expected to rise to 500 million tons annually by 2030. These alarming statistics underscore the urgent need for a strategic conservation plan to address infrastructure gaps, improve public awareness, and strengthen enforcement of waste management regulations.

To mitigate these growing challenges, a strategic conservation plan is essential. This plan should prioritize public education on land pollution, emphasize the adoption of sustainable waste management technologies, and advocate for stronger enforcement of environmental regulations. By addressing infrastructure gaps and focusing on emerging contaminants, countries can better align waste management practices with conservation goals. Engaging communities and fostering partnerships between governments, industries, and international organizations will be key to ensuring that both developing and developed nations can sustainably manage their waste and protect natural resources for future generations.

6.3 *Navigating the Path Forward: Collective Responsibility Toward Mitigating Land Pollution*

Land pollution is a complex issue demanding collective action from individuals, corporations, and governments. Efforts to combat land pollution require a multi-pronged approach. By promoting sustainable practices, enforcing regulations, and fostering public awareness, we can safeguard this vital resource and protect the health of our planet for generations to come. This fight against land pollution requires a shift toward a more responsible relationship with the land, ensuring a cleaner and healthier future for all. This may need the promotion of the 3Rs, reduce, reuse, and recycle initiatives, which can significantly reduce waste generation and conserve resources (Pariatamby and Fauziah 2014; Wichai-utcha and Chavalparit 2019; Shah et al. 2024). In addition, a lean thinking management model has recently been proposed for operationalizing the circular economy in companies (Junior and

Nagai 2024). Next, sustainable practices, like encouraging organic farming, responsible use of fertilizers and pesticides, and soil conservation techniques, can minimize agricultural pollution. Improved waste management regimes, such as investing in proper waste collection, treatment, and disposal facilities, are crucial for preventing land contamination. Incentivization for proper waste recovery and enhanced regulations would further strengthen compliance with waste disposal, industrial practices, and e-waste management, ensuring responsible waste handling. Another way is to mass-initiate public education programs, whereby raising awareness about the dangers of land pollution and promoting responsible waste disposal habits would be crucial for long-term success.

C&C regulatory policies include the Clean Air Act (CAA) and Resource Conservation and Recovery Act (RCRA) in the USA, which set limits on pollutants released from industries and mandate proper waste disposal practices (Sampson et al. 2024; Nunez et al. 2024). Similarly, a recent review summarizes similar success with the National Clean Air Programme (NCAP) in Indian cities and assesses the impact on surface ozone from 2018 to 2022 (Gopikrishnan and Kuttippurath 2024). Other than these, certain landfill restrictions on the types of waste allowed in landfills and mandate-specific liner systems to prevent leachate contamination have also been implemented as the norms. In addition, permits with strict compliance requirements and penalties for violations are key tools for enforcing regulations. Undoubtedly, various success ratings have been witnessed with time, leading to decreased industrial waste generation and improved disposal practices in developed countries. The US EPA reports a 30% decline in industrial hazardous waste generation from 1991 to 2018 (USEPA 2024a). Regulations have spurred investments in waste collection and treatment infrastructure in some regions. Similarly, The European Union's Landfill Directive has increased waste diversion from landfills and a rise in recycling rates in some member states (EEA 2024). In China, rapid industrialization has led to severe land contamination (Yuan et al. 2020). While C&C regulations exist, weak enforcement and lack of investment in cleaner technologies hinder progress. Although various C&C regulations have been the mainstay in combating land pollution, their effectiveness is limited. While they have succeeded in specific areas, their rigid, top-down approach struggles to address the problem's vast scope and evolving nature of the problem. Hence, the C&C approaches are double-edged swords in land pollution control. C&C policies often focus on specific pollutants or waste types, neglecting emerging contaminants and diffuse sources like agricultural runoff. Implementation and enforcement can be weak in regions with limited resources or weak governance (Turner and Kim 2024). Moreover, rigid regulations do not adapt easily to new technologies or address unique regional challenges (Fogleman 2024). Ultimately, the bigger challenge has been the high costs of compliance. Industries might resist adopting cleaner technologies due to the associated financial burden (Brooks and Simon 2024; Ogieriakhi et al. 2024). Despite regulations, the UN Environment Programme estimates that over 2 billion hectares of land are degraded globally (Khan et al. 2024; Dinerstein et al. 2024). The World Bank estimates that only around 13.5% of global waste is recycled (Ellis 2018). Hence, no doubt, C&C approaches do show merits, but they need to be complex with newer, more real-time developments backed by newer technological, economic, and biotechnological interventions.

6.3.1 Rejuvenating the Land: Emerging Strategies to Combat Land Pollution

We highlighted the need to incorporate land pollution management, mitigation, and redressal with the new normal. It is imperative to discuss these emerging approaches. These may include efforts to detect, recover, neutralize, remediate, convert waste to valorization, and/or implement new management policies and regulatory frameworks to deal with the pollutants. While many new innovative strategies have been devised, their implementation has only been in the testing phase or only available sporadically in some global regions, and many are undergoing in-depth research to seek translational merits.

6.3.2 Conversion Techs and Tactics

Solid waste materials are being converted to other products like base and composites and/or other alternate products to manage waste or their ready recyclability. For example, plastic trash is being turned into ropes, fibers for clothing, mops, brooms, baskets, handbags, etc. Such wastes are also seen in their use in making road materials and building constructions as fillers (Wilson et al. 2013). Municipal solid wastes produced by industries releasing biomass slurries and cakes have also been seen as potential sources to produce biogas via thermochemical, hydrothermal, and biological approaches and for recyclization to manures through biomethanation and composting (Parashar et al. 2020; Nanda and Berruti 2021; Jeevahan et al. 2021; Sudalai et al. 2024; Vinayak et al. 2020). The realization of an old proverb is being seen as "one man's trash is another man's treasure." In a revolutionary work by Karthikeyan Kandan and a group from the De Montfort University, Leicester, UK (Wood 2019), they developed prosthetic limbs from waste plastics. The group believes these new prosthetics are significantly cheaper, costing around $12 per socket compared to the usual $6000. This affordability will primarily benefit amputees in developing countries who previously could not afford quality limbs. For centuries, waste was carelessly discarded without considering its potential value. Today, this mindset is shifting dramatically, particularly in food waste. Once considered a disposal challenge, used cooking oil is now transformed into valuable resources like renewable diesel, sustainable aviation fuel, and even renewable plastics. This transformation is driven by the urgent need to reduce greenhouse gas emissions and dependence on fossil fuels. The transportation sector, responsible for nearly a quarter of global carbon dioxide emissions in 2019, is at the forefront of this change. The global used cooking oil market is experiencing significant growth, reaching USD 7.09 billion in 2023. Analysts project further expansion, with the market value expected to climb to USD 13.96 billion by 2032, representing a compound annual growth rate (CAGR) of 8.00%. Notably, North America currently leads the market, holding a dominant share of 42.17% in 2023 (FBI 2024). Cities like Bournemouth in the UK are leading the way by converting waste cooking oil into hydrotreated vegetable oil (HVO) to power municipal vehicles (Amsen 2021). Major corporations, such as McDonald's, are also joining the effort by partnering with companies to recycle used cooking oil into fuel for their delivery fleets (Amsen 2021). These initiatives highlight the growing recognition of waste as a potential resource and the potential for circular economy models to combat climate change. In this way, Dubai pioneered sustainable transportation by becoming the world's first city to exclusively use biodiesel, produced entirely from locally sourced waste cooking oil, to fuel its municipal vehicles (Thorpe 2021). Biodiesel produced from used cooking oil offers a significant environmental advantage. Compared to traditional fossil diesel, it substantially reduces carbon emissions by 84%. This makes it a cleaner and more sustainable fuel option without requiring vehicle modifications, and additionally, the process allows for promising sources of biolubricants (Joshi et al. 2023). While other countries are expected to follow here, in India, most of the cooking oil used is resold to small food outlets and joints (Srivastava 2021). India generates a substantial amount of used cooking oil—approximately three million metric tons annually. However, due to policy gaps and insufficient implementation, a significant portion of this valuable resource is not converted into biodiesel (Srivastava 2021). This missed opportunity hinders the country's efforts to reduce reliance on fossil fuels and mitigate environmental impact.

6.3.3 Nature-Inspired Solutions

The inherent ability of certain life forms, their body, tissue or cell structures, and/or, for that matter, their secretions' chemistry may allow capturing, converting, and metabolizing certain entities that might be deemed pollutants to us. Phytoremediation leverages the natural ability of specific native plants to absorb and break down contaminants like heavy metals and organic pollutants from soil (Sarkar and Sadhukhan 2022). Using native species ensures adaptation to local conditions and minimizes disruption to existing ecosystems (Dhar et al. 2021; Ramamoorthy et al. 2009). Green waste—comprising food, agricultural, and forest residues—is a rich source of polyphenols with potent environmental remediation properties. These compounds remove heavy metals, bacteria, and dyes from contaminated soil and water through adsorption, chelation, and biodegradation (Kuppusamy et al. 2015). Moreover, polyphenols serve as precursors for green nanoparticles, enhancing pollutant detoxification. While bioremediation techniques harness some of these benefits, the full potential of polyphenols remains untapped. Other than this, certain plant species can absorb, accumulate, and detoxify harmful substances (Erakhrumen 2011). This eco-friendly approach offers a cost-effective and sustainable solution to environmental pollution. A recent work uncovers that wheat straw, corncobs, and sugarcane bagasse are valuable resources teeming with hydrocarbon-degrading, nitrogen-fixing (diazotrophic), and heavy metal-resistant microorganisms (Dashti et al. 2019). These plant-based materials physically absorb oil spills and harbor microbial communities that actively break down hydrocarbons. Remarkably, these consortia, consisting of just a few species, can thrive in oily environments, fixing nitrogen and mitigating the toxic effects of heavy metals. This synergistic relationship between plant waste and microbes offers a promising approach to bioremediation, transforming a waste product into a potent tool for environmental cleanup. Scientists are engineering microbial strains to enhance their ability to degrade specific pollutants in soil. These can tackle complex contamination challenges by targeting specific contaminants more effectively (Kaur et al. 2023; Zhang et al. 2011). Many bacteria and fungi can degrade pollutants that contaminate both soil and public waters (Kulshreshtha et al. 2013; Ghosh et al. 2023; Shrivastava 2020; Awasthi et al. 2017). At the same time, they effectively maintain the productive fertility of the soils, as well as many crop growth-promoting microbes (Rajhans et al. 2021; Kaur et al. 2022; Dhar et al. 2024; Parashar et al. 2023; Parashar and Mudgal 2024). Polychlorinated dibenzo-p-dioxins and dibenzofurans (PCDD/Fs) are highly toxic compounds often found in contaminated fly ash. Due to their persistent nature and harmful effects, effective remediation strategies are crucial. Microbial biocatalysts offer a promising approach to tackle this challenge (Nam et al. 2008). These biological catalysts, composed of specifically selected microorganisms, can degrade PCDD/Fs into less harmful substances (Kaewlaoyoong et al. 2020; Cheng et al. 2023). By introducing these biocatalysts into contaminated fly ash, researchers have demonstrated significant reductions in PCDD/F levels. Another approach is vermicomposting with engineered worms. Research is underway to develop worm species with enhanced capabilities to break down certain pollutants and improve soil quality. This can accelerate the breakdown of organic contaminants and enrich the soil for agricultural use.

6.3.4 Technological Advancements in Land Remediation

We need new normal like the Superfund Innovative Technology Evaluation (SITE) program as a key initiative to develop innovative solutions for hazardous waste management (Cabrera-Guzmán et al. 1990).

- *Electrokinetic Remediation (EKR)*
  This innovative technique uses low-voltage electricity to mobilize contaminants in the soil, making them easier to extract or break down through other treatment methods. This approach promises to remediate sites contaminated with heavy metals and inorganic pollutants (Akemoto et al. 2023). Municipal solid waste incineration (MSWI) generates substantial amounts of bottom ash, a by-product often landfilled due to concerns about pollutant leaching. While encapsulation with cement can stabilize this waste, it is costly and may not fully address the issue. Electrokinetic remediation (EKR) offers a promising alternative, which is removing heavy metals and salts from bottom ash. EKR involves applying a low-level direct current to the ash, causing contaminants to migrate toward the electrodes. Combined with electro-osmosis (fluid movement due to electrical

potential), this process effectively extracts pollutants. Unlike traditional EKR, electrodialytic remediation (EDR) uses ion exchange membranes to separate the ash from electrolyte solutions, enhancing contaminant removal efficiency (Traina et al. 2007). This innovative approach holds significant potential for treating MSWI bottom ash and recovering valuable resources. By coupling EKR or EDR with other waste management strategies, we can move closer to a circular economy for waste management. Halogenated persistent organic pollutants (POPs) in e-waste present a potential challenge. These compounds, including PCBs, PBDEs, PAHs, and others, are released into the environment through e-waste burning. EKR is presented as a potential solution for cleaning up contaminated soil and water from e-waste. However, the low solubility of POPs in water hinders the effectiveness of traditional EKR methods (Liyanage and Walpita 2020). To effectively remove organic contaminants from soil using electrokinetics, either the contaminants' solubility must be increased, or the technology must be combined with other remediation methods. Complementary techniques like bioremediation, Fenton oxidation, or ultrasonic remediation can enhance contaminant removal (Liyanage and Walpita 2020; Han et al. 2021).

- *Thermal Desorption*
  This technology uses controlled heating to vaporize contaminants from soil, which can then be captured and treated. While effective for certain pollutants, it requires significant energy input and proper management of vapor emissions. By applying heat, it effectively removes various pollutants. While efficient, it generates off-gases requiring further treatment (Zhao et al. 2019). Research is focused on optimizing thermal desorption processes and developing effective off-gas management strategies. Low-temperature thermal desorption (LTTD) is an effective method for cleaning up soil contaminated with diesel (Zivdar et al. 2019). By applying heat, volatile pollutants are evaporated and captured. Factors such as temperature, residence time, and soil composition influence the process's efficiency. While LTTD offers advantages like rapid treatment and high removal rates, managing the resulting off-gases is crucial (Zivdar et al. 2019). This technology holds promise for remediating diesel-contaminated sites, especially when combined with waste heat recovery to reduce energy consumption. A study investigated the thermal remediation of soil contaminated with hexachlorocyclohexane (HCH) (Araújo et al. 2016). By subjecting soil samples to varying temperatures and exposure times, researchers analyzed the effectiveness of this method. Thermogravimetric analysis of pure HCH provided insights into its thermal behavior, while soil samples were heated in an oven to assess contaminant removal. The results demonstrated that HCH can be effectively removed from soil at temperatures below its boiling point. However, the process generates gaseous by-products, including HCH isomers, chlorine derivatives, and benzene compounds, necessitating proper off-gas treatment (Araújo et al. 2016). The study highlights the importance of optimizing parameters like temperature, heating time, and airflow for successful soil remediation. Besides these, a novel method for measuring volatile organic compounds (VOCs) at specific aerial

locations has been developed using multicopters equipped with sorbent tubes (Chen et al. 2018). This cost-effective approach enables simultaneous sampling at multiple heights, increasing efficiency compared to traditional methods. By studying VOC concentrations around various emission sources, including industrial sites and forests, the research demonstrated the effectiveness of multicopters for capturing spatial variations in air quality. This technology holds promise for future environmental monitoring and research. Thermal desorption can effectively remove polychlorinated biphenyls (PCBs) from contaminated soil. A study found that higher temperatures and finer soil particles increased PCB removal efficiency (Qi et al. 2014). While thermal desorption primarily removes PCBs through vaporization, dechlorination and decomposition also occur at higher temperatures. These findings highlight the potential of thermal desorption as a viable remediation technique for PCB-contaminated soil, though careful management of off-gases remains crucial.

- *Advanced Nanotechnologies*
  Researchers are exploring the potential of nanomaterials like nanoparticles to degrade contaminants or immobilize them within the soil, preventing their spread (Kumar et al. 2021b; Chaudhary et al. 2019). By combining nanomaterials with biological processes, nanobioremediation effectively tackles contaminants like heavy metals, hydrocarbons, and organic pollutants (Devi et al. 2021). This approach holds the potential for more efficient, cost-effective, and environmentally friendly cleanup compared to conventional methods (Kharangate-Lad and D'Souza 2021). This is a rapidly developing field with ongoing research to ensure the safety and effectiveness of these technologies. Also, nanomaterials can be extracted from solid waste for further applications. Transforming solid waste into valuable carbon-based nanomaterials like graphene and carbon quantum dots offers a promising approach. Many recent reviews explore the potential of these materials in energy storage applications, addressing the critical need to balance waste management with resource recovery and sustainability (Tatrari et al. 2021; Samaddar et al. 2018). Nanotechnology offers a promising approach to reclaiming degraded mine sites. By incorporating nanoparticles into traditional soil amendments, it is possible to enhance soil quality, mitigate contamination, and promote ecosystem recovery (Liu and Lal 2012). Materials like zeolites, iron oxides, and carbon nanotubes show potential for improving soil structure, retaining nutrients, and remediating pollutants. While research is ongoing, combining nanoparticles with conventional methods could provide a cost-effective and sustainable solution for mine land reclamation.

6.4 A Focus on Prevention and Circular Economy Principles

Planting fast-growing, nonfood crops on contaminated land can prevent soil erosion, suppress weeds, and help stabilize the soil while it undergoes remediation. These cover crops can also contribute organic matter to the soil, enhancing its fertility over time. Hence, photo management presents a holistic solution for addressing metal(loid) contamination while providing ecological and economic benefits (Moreira et al. 2021). Other than this, developing technologies to transform industrial waste streams into usable resources can significantly reduce landfilling and create opportunities for land rehabilitation. Examples include using construction waste to create new building materials or processing organic waste into compost for soil enrichment, as highlighted before. Next, we need to promote sustainable land management. Encouraging practices like crop rotation, reduced tillage, and responsible use of fertilizers and pesticides can minimize soil degradation and promote long-term soil health (Böttcher et al. 2012; Fleskens et al. 2014). This is particularly crucial in agricultural regions to ensure land productivity and reduce the need for land expansion into natural ecosystems.

6.5 The Power of Policy and Public Participation

We need to initiate brownfield redevelopment programs186 substantially. Offering incentives and facilitating the process of redeveloping abandoned or contaminated land for productive use can revitalize degraded areas and create economic opportunities (Cutts et al. 2024; Chen et al. 2024a; Becerra 2024; Tendero and Bazart 2024). Other than this, community-based land remediation projects will also foster the management of waste in lands. Empowering local communities to participate in land remediation efforts can foster a sense of ownership and encourage responsible land management practices (Kusumadewi et al. 2024; Sam and Zibima 2024; Ullah et al. 2024). Engaging citizen science initiatives can also provide valuable data for monitoring land health. Besides these, there is a need for regional focus, such that strategies must be tailored to address the specific types of land pollution prevalent in different regions. For example, tackling salinization in agricultural areas requires solutions different from those in managing industrial waste in urban centers. Similarly required is investing in research & development. Continued funding for research on bioremediation techniques, advanced soil treatment technologies, and sustainable land management practices is essential to develop more effective and affordable solutions. Yet ahead, public-private partnerships are highly sought need at the time. Collaboration between governments, industries, research institutions, and NGOs can accelerate land remediation efforts, share costs and expertise, and ensure sustainable land management practices.

We can work toward a future where our land thrives by embracing these emerging strategies alongside existing regulations. This requires a multipronged approach emphasizing prevention, innovative remediation technologies, responsible waste management, and promoting sustainable land use practices (Yang and Koondhar 2024; Sovacool et al. 2024). By fostering regional adaptation, empowering communities, and investing in cutting-edge solutions, we can reclaim contaminated land, rejuvenate our soil, and ensure a healthy and productive land base for future generations. Remember, healthy land is the foundation for a healthy planet and a healthy society.

6.6 Technogenic Waste: A Multi-pollutant Menace

Technogenic waste, the unwanted byproducts of human industrial and technological activities, casts a long shadow on our environment (Sharma et al. 2021). Globally, industries produce over 13 billion tons of industrial waste annually, contributing significantly to land, water, and air pollution. For example, in India, around 90 million tons of hazardous waste are generated annually from industrial activities, such as mining, manufacturing, and chemical processing. Additionally, China, being one of the largest industrial hubs, generated approximately 4.5 billion tons of solid industrial waste in 2020 alone, with significant portions still being inadequately managed.

This vast amount of technogenic waste not only contaminates ecosystems but also introduces hazardous materials like heavy metals and toxic chemicals, which further complicate waste treatment efforts. In many developing countries, improper handling of technogenic waste has led to severe health problems, affecting both human populations and biodiversity. Given the scale of these challenges, a strategic conservation plan is essential to address the environmental and health risks posed by technogenic waste. Strengthening regulatory frameworks, investing in waste treatment technologies, and promoting sustainable industrial practices will be crucial steps in mitigating the impact of this growing waste stream.

Its diverse composition wreaks havoc on air, land, and water, posing a significant threat to ecosystems and human health. Burning fossil fuels in power plants and industrial facilities releases fine particulate matter (PM) into the air. These tiny particles lodge deep in the lungs, causing respiratory problems like asthma and chronic obstructive pulmonary disease (COPD). Industrial processes release harmful gases like sulfur dioxide, nitrogen oxides, and volatile organic compounds (VOCs). These gases contribute to smog formation, acid rain, and ozone depletion, impacting human health and ecosystems. Waste incineration and mining operations can release heavy metals like lead, mercury, and arsenic into the air. These metals can bioaccumulate in the food chain, posing a risk to human health, especially for pregnant women and children. The effects of these pollutants have been covered before. Yet, it is important to reiterate a few. Landfills containing industrial waste can leach heavy metals into the soil and groundwater, contaminating agricultural land and drinking water sources. Mining operations often generate acidic wastewater laden with heavy metals. This acidic runoff contaminates soil and water bodies, impacting plant growth and aquatic life. Improper e-waste disposal releases toxic substances like lead, mercury, and flame retardants into the soil. These toxins can enter the food chain and harm human health. Discarded plastic waste fragments into microplastics, polluting soil and impacting plant growth and soil organisms. Microplastics can also contaminate food products grown in polluted soil. This can follow up with effects on water and air purity as well. Untreated industrial wastewater containing heavy metals, organic pollutants, and other toxins can contaminate rivers, lakes, and groundwater. This disrupts aquatic ecosystems, harms fish and other wildlife, and renders water unsuitable for drinking or irrigation. Fertilizers, pesticides, and herbicides used in agriculture can contaminate waterways. Nitrates from fertilizers can cause eutrophication, leading to algal blooms and oxygen depletion in water bodies. Pesticides can harm aquatic organisms and contaminate the food chain. Microplastics released from clothing, cosmetics, and plastic break into waterways, harming aquatic life and potentially entering the human food chain through consuming contaminated fish.

Hence, technogenic waste is a complex environmental threat with far-reaching consequences. Its impact on air, land, and water necessitates a multipronged approach. Implementing stricter regulations, promoting cleaner technologies, and fostering responsible waste management practices are crucial steps toward mitigating the detrimental effects of technogenic waste and ensuring a healthier planet for all. Absolutely! Various guidelines and regulatory policies have been established at both national and international levels to address the challenges posed by technogenic waste. Among these, the Basel Convention stands out as a pivotal international treaty. This agreement regulates the transboundary movement of hazardous waste and establishes strict guidelines for its environmentally sound disposal, thereby aiming to protect human health and the environment. It aims to minimize the generation of hazardous waste and ensure its environmentally sound management (Ummik et al. 2024; Khan and Ewe 2024; Puckett 2024). The Stockholm Convention focuses on persistent organic pollutants (POPs), a specific subset of technogenic waste known for their persistence and bioaccumulation in the environment (Wania and McLachlan 2024; Biggi 2024). It aims to eliminate or restrict the production and use of POPs. The Rotterdam Convention promotes informed decision-making concerning the import of hazardous chemicals and pesticides (Halkos and Aslanidis 2024; Chen et al. 2024b). It ensures countries have the right to know about and potentially ban the import of hazardous substances. Many developed countries have established comprehensive national legislation to manage various aspects of technogenic waste (Sampson et al. 2024; Gopikrishnan and Kuttippurath 2024). The US Clean Air Act (CAA) sets limits on air pollutants released from various sources. The Clean Water Act (CWA) in the USA, regulates the discharge of pollutants into waterways and aims to restore and maintain the quality of the nation's waters. Resource Conservation and Recovery Act (RCRA) in the USA focuses on the proper management of hazardous waste from its generation to disposal. The Waste Framework Directive (WFD) in the European Union establishes a framework for waste management within the EU, promoting waste reduction, reuse, and recycling. Many developing countries are still in the process of establishing robust waste management regulations. International assistance and capacity building are crucial to improve their waste management infrastructure and enforcement capabilities. Many industry-led initiatives like the Responsible Care program also exist that encourage companies to adopt environmental management practices that minimize waste generation and pollution. Extended Producer Responsibility (EPR) policies hold manufacturers responsible for the entire lifecycle of their products, including their end-of-life management and proper disposal.

Despite these regulations and initiatives, significant challenges remain. Effective enforcement, particularly in developing countries, is crucial. Additionally, addressing emerging technogenic waste streams like e-waste requires innovation and adaptation of existing regulations. We can move toward a future with less technogenic waste and a healthier environment by strengthening national and international frameworks, fostering technological advancements in waste treatment, and promoting responsible production and consumption practices. The battle against technogenic waste requires a multipronged approach, building upon the successes of existing regulations while embracing innovative solutions. We can create a future with less waste and a healthier planet by prioritizing waste prevention, adopting circular economy principles, and investing in new technologies. Remember, this continuous process requires ongoing adaptation to address emerging challenges.

7 Unveiling Climate Change: A Threat Magnified by Pollution

Our planet's climate is not static but a dynamic system accustomed to fluctuations over vast timescales. Many factors, like volcanic eruptions, solar activity variations, and Earth orbit shifts, have continually reshaped climatic conditions throughout history. These natural oscillations played a crucial role in shaping the evolution of life, influencing the development of diverse species and ecosystems that adapted to thrive in specific climatic niches. The climate is accustomed to changes on the globe due to a variety of other factors. It is a phenomenon that framed the evolution of life and its nuances over the earth and should not be surprising. What is more surprising to gather is the pace with which this climate change is advancing. One needs to provoke thoughts over the physics of climate change (Ghil and Lucarini 2020).

Undoubtedly, the earth's climate, a constantly changing and uneven mix of forces, presents challenges in understanding and predicting its future due to its inherent complexities (Ghil and Lucarini 2020). A new World Economic Forum report (Eitelwein 2024) highlights the potential consequences of climate change by 2050; it estimates an additional 14.5 million deaths and $12.5 trillion in economic losses worldwide. However, the report also emphasizes there's still time for global stakeholders to act. Through decisive and strategic measures, we can still mitigate the health impacts of climate change worldwide. Climate change is not an isolated issue; it is intricately linked to and exacerbated by pollution. Understanding these connections is crucial for tackling both challenges effectively.

7.1 Causes of Climate Change

Pollution (covered before in this chapter) is a major contributor to climate change, but several other human activities also play a significant role. Deforestation, for example, releases large amounts of stored carbon dioxide into the atmosphere. It can worsen climate change and lead to even more carbon loss from remaining forests (Li et al. 2022). Previous studies suggested deforestation warms things up in most areas except high latitudes. However, a new research (Longobardi et al. 2016) digs deeper. It simulated removing different percentages of forests in different regions and found that at high latitudes, removing forests always leads to cooling, both globally and in those areas; in mid-latitudes, the effect is mixed, with some areas warming and others cooling, while at low latitudes, removing forests always leads to warming, both globally and in those areas. Interestingly, in high latitudes, removing a large portion of the forest (over 45%) can lead to a decrease in atmospheric CO2. This happens because even though deforestation releases carbon, the colder and drier conditions that follow allow the soil to store more carbon, offsetting some of the initial increase (Longobardi et al. 2016). Truly speaking, a comprehensive analysis of landscape ecology requires acknowledging the interconnectedness of the land use-climate change-energy nexus (Dale et al. 2011). Ignoring these interactions can lead to a fragmented understanding of how ecological patterns and processes evolve within landscapes. There is a greater need to delve into the importance of modeling these changes to understand their broader environmental impact (Riebsame et al. 1994). A more historical perspective and emphasis on the human dimension in driving land-use changes has also been discussed often (Dale 1997).

To continue, converting natural landscapes to agriculture or urban areas can alter local climate patterns. Intensive farming practices, particularly the use of nitrogen-based fertilizers, contribute to nitrous oxide emissions, a potent greenhouse gas. A lot has been talked about how converting forests to farmland can significantly reduce the amount of carbon stored in the soil (Murty et al. 2002). Murty and coworkers highlight that on average, converting forests to cropland led to a loss of about 22% of soil carbon (after correcting for changes in soil density). However, converting forests to pastureland (grazing) did not significantly change soil carbon levels (Murty et al. 2002). A study by Silvério and coworkers (Silvério et al. 2015) emphasizes that deforestation for agriculture significantly impacts the Amazon's climate beyond just greenhouse gas emissions. The study details that it reduces the amount of water returned to the atmosphere (evapotranspiration); increases the amount of heat released from the land surface; and raises the overall land surface temperature (Silvério et al. 2015). Additionally, methane emissions from livestock like cows and sheep are significant contributors to global warming (Moumen et al. 2016; Johnson and Johnson 1995; Vaghar Seyedin et al. 2022; Sejian et al. 2015).

Many industrial processes, such as cement production and steel manufacturing, release significant amounts of carbon dioxide into the atmosphere (Rodrigues and Joekes 2011; Gao et al. 2015; Barcelo et al. 2014), and these emissions occur throughout the production process, from obtaining raw materials to the final product. Even if the primary process does not directly emit CO_2, the electricity used to power machinery and equipment often comes from fossil fuel-burning power plants, indirectly contributing to atmospheric CO_2. Burning fossil fuels like coal, oil, and natural gas is the primary driver of human-caused climate change. When burned, these fuels release large amounts of carbon dioxide, trapping heat in the atmosphere. These activities, along with pollution, act synergistically to accelerate climate change. For instance, deforestation can worsen the effects of rising temperatures by reducing the Earth's ability to absorb carbon dioxide. It is important to note that the relative impact of these activities varies depending on the region and specific practices employed. However, addressing all these factors is crucial to mitigate climate change and create a more sustainable future. When food decomposes in landfills, it releases methane, with a warming impact 25 times stronger than carbon dioxide over 100 years. According to the UN estimates, it accounts for a third of all greenhouse gas emissions, a surprisingly large chunk (Wasserman 2024). To detail, from a 2021 report on the World Economic Forum (Marchant 2021), surprisingly, a staggering 931 million tons of food go to waste each year globally, according to the UN, and this unconsumed produce contributes to a hefty 8–10% of global greenhouse gas emissions and believe it or not roughly 17% of all food produced is wasted, with most of it (61%) coming from our households. Food service establishments (26%) and retailers (13%) contribute significantly to this global food waste problem. This waste happens throughout the food chain, from inefficiencies in local agriculture to consumer discards in affluent regions. Hilarious would be to read, but seriously, in fact, while ten million pumpkins are grown annually in the UK, 95% are used solely for Halloween decorations before being discarded, generating 18,000 tons of wasted food potential (WEF 2019). Similar is the case with aquatic and seafood waste. Fish and seafood are vital for global nutrition, but a growing problem is how much is wasted. A study reveals that in 2021, a significant 14.8% (23.8 million tons) of all fish intended for people never reached them (Giron and Pickerel 2021). The biggest losses occur during processing (39%) and wild fish discards (35%). The study highlights specific areas needing improvement, like minimizing waste of bottom-dwelling fish and frozen products, especially in Asian and European production and processing (Giron and Pickerel 2021). Scorching heatwaves and droughts are causing a surprising consequence, misshapen fruits and vegetables. Farmers are pleading with consumers to embrace these "wonky" crops to avoid food waste and potential price hikes. The extreme weather has left many of the farmlands parched, leading to stunted or oddly shaped produce. Farmers warn that supermarkets rejecting these "ugly" options could lead to higher food prices, urging consumers to be more accepting of imperfect shapes and sizes. After all, tossing out good food simply for aesthetics contributes to global waste and environmental concerns. All this is worth reading for more brainstorming facts and figures with this on a World Economic Forum's report of 2022 by Simon Read (2022). Adding to this, climate change in the context of beef production is also emerging as a hot topic, with discussions often becoming polarized. While animal-based foods offer valuable nutrition and are crucial for many diets, especially in developing countries, they also have a large environmental footprint (Waite et al. 2019). Beef production is improving efficiency, but deforestation for new pastures continues. Consumers express interest in plant-based diets, yet meat consumption keeps rising. These seemingly contradictory trends highlight the beef and sustainability conversation's complex and often contentious nature. Other than these, leaky or improperly disposed-of refrigerators and air conditioners release harmful coolants like chlorofluorocarbons (CFCs) and hydrofluorocarbons (HFCs). These gases are potent greenhouse gases, contributing significantly to global warming (di Filippo et al. 2022; Ko et al. 1999; Seidel et al. 2016; Xu et al. 2013). The fast-fashion industry, churning out trendy clothing at breakneck speed and rock-bottom prices, leaves a significant environmental footprint. A recent discussion at the World Economic Forum shed more light on this concern (McFall-Johnsen 2020). Production involves energy-intensive steps, reliance on synthetic materials derived from fossil fuels, and frequent garment disposal in landfills. The affordability of fast fashion comes at a steep environmental cost. As a major polluter, the industry is responsible for 10% of global carbon emissions and is a massive water consumer. Our desire for ever-cheaper and more clothes has fueled a near-doubling of clothing production since 2000. This translates to a 60% increase in clothing purchases between 2000 and 2014, with garments being discarded twice as fast (McFall-Johnsen 2020). This unsustainable cycle, driven by fast fashion, demands changing our consumption habits to protect the environment. While not directly contributing to global warming, excessive artificial lighting disrupts natural light cycles (Davies and Smyth 2018). This can have cascading effects on ecosystems, impacting wildlife behavior and potentially influencing climate patterns in localized areas. A recent review explores how artificial light disrupts natural light cycles across the globe (Gaston and Sánchez de Miguel 2022). Artificial light at night (ALAN) offers advantages but carries environmental costs. For example, while light-emitting diodes (LEDs) are often seen as a lighting revolution, their blue light emissions pose new environmental concerns for wildlife and ecosystems (Gaston and Sánchez de Miguel 2022). Another work reveals they reduce activity in some bat species, potentially disrupting their flight paths (Stone et al. 2012). In addition, blue light disrupts our sleep cycles by suppressing melatonin (sleep hormone) production, potentially leading to issues like cancer, heart disease, and diabetes (Benke and Benke 2013). Artificial light has multifaceted effects on the growth patterns of species in the global flora and fauna (Macgregor et al. 2017; Falcón et al. 2020; Bennie et al. 2018).

As global temperatures rise, permafrost, permanently frozen ground in polar regions, begins to thaw. These releases trapped methane gas (Byun et al. 2017; Walter Anthony et al. 2012; Anisimov 2007), further accelerating warming. While a natural process, human activities like burning fossil fuels exacerbate permafrost thaw. These are just a few examples, and research is ongoing to identify and quantify the impact of various human activities on climate change. It highlights the interconnectedness of our actions and the complex nature of this global challenge (Hope and Schaefer 2016).

7.2 *Moving Forward: A Holistic Approach*

Addressing climate change requires a multi-pronged approach that tackles pollution at its source. Strategic actions may include initiatives such as the steady but quick transition to shifting dependence on (i) clean energy sources, (ii) development of better policies and regulations, (iii) paradigm shifts in market mechanisms and investments, and finally, (iv) improving public awareness, perception, and acceptability. Let us discuss these further comprehensively:

7.3 Transitioning to Clean Energy

Shifting from fossil fuels to renewable energy sources like solar, wind, and geothermal can significantly reduce CO_2 emissions and improve air quality (Cha 2017). However, to achieve these, many technological advancements are sought to encourage continued development and cost reduction of solar, wind, geothermal, and tidal energy technologies. Similarly required are advancements in battery storage technology to capture and store excess renewable energy for later use and the development of intelligent electricity grids that can integrate and manage diverse energy sources efficiently. Other than this, a major emphasis needs to be placed on carbon capture, utilization, and storage (CCUS) technologies to capture carbon emissions from power plants and industrial facilities, potentially storing them underground or utilizing them for other purposes. A recent release from the World Economic Forum (Ellerbeck 2023), talks about the astounding new emerging trends such as 35% of global electricity will be from renewables (up from 29%) by 2025; China is leading the way with nearly half the additional renewable capacity; coal and gas-fired generation expected to decrease; nuclear power output to increase by 3.6% annually; solar to dominate new US electricity capacity, exceeding half of all additions.

7.4 Policy and Regulations

Transitioning away from fossil fuels and embracing clean energy sources is crucial. Several policy instruments have emerged to incentivize this shift, such as carbon pricing: implementing carbon pricing mechanisms (e.g., carbon tax) to incentivize transition away from fossil fuels (Gorbach et al. 2022; Dominioni 2022; Timilsina 2018) and setting mandates for utilities to generate a certain percentage of their electricity from renewable sources. Newer programs also offer financial incentives to individuals and businesse generating their own renewable energy resources (Qadir et al. 2021; Gouchoe et al. 2002; Dóci and Vasileiadou 2015; Wilson 2022). Government support for research and development of clean energy technologies and tax breaks for businesses and individuals who invest in clean energy solutions (Bose et al. 2019; Bürer and Wüstenhagen 2009; Norberg-Bohm 2000; Yan et al. 2023).

7.5 Market Mechanisms and Investments

- *Green Bonds*: Issuing bonds specifically designated for financing clean energy projects.
- *Environmental, Social, and Governance (ESG) Investing*: Investments considering a company's environmental practices and sustainability efforts.
- *Decentralized Energy Systems*: Promoting electricity generation closer to where it is used, potentially through rooftop solar panels and microgrids.
- *Energy Efficiency Improvements*: Investing in technologies and practices that reduce overall energy consumption across sectors (buildings, transportation, industry).

7.6 Behavioral Shifts and Public Awareness

The current education system is a major bottleneck in the clean energy transition. Investing in renewable energy education and refocusing universities can bridge the skills gap and accelerate the path toward a sustainable future (Vakulchuk 2024). According to the recent release from the WEF (Vakulchuk 2024), (i) Demand for green jobs like wind technician and solar installer is surging. (ii) However, the number of workers with relevant skills is not keeping pace, with only one in eight people qualified for climate change mitigation jobs. (iii) Universities still produce more graduates for fossil fuel industries (68%) compared to renewables (32%). (iv) It is estimated that this rate, universities will not offer a majority of degrees in renewables until 2107. (v) Asia Pacific still leads in clean energy education, while developing regions like Africa and South America lag way behind. (vi) A challenge due to "carbon lock-in" is from past fossil fuel funding and ingrained infrastructure. (vii) Private universities are quicker to adopt renewable energy education due to less reliance on fossil fuel support. (viii) Renewable energy education receives less investment compared to fossil fuels. (ix) Urgent need to increase funding and prioritize renewable energy education globally. (x) Redirect public and private funding from fossil fuel education to renewables. (xi) Encouraging signs with programs like the World Economic Forum's Reskilling Revolution aiming to equip millions for green jobs. Similar initiatives may include investing in green technologies, sustainable land management, circular economy principles, education and awareness campaigns, consumer-driven demand, and community-owned renewable energy projects. These approaches all work together to create a multifaceted strategy to accelerate the clean energy transition. We can move toward a more sustainable energy future by combining technological advancements, supportive policies, market forces, and behavioral changes. Adopting practices like crop rotation, reduced tillage, and responsible use of fertilizers can minimize soil erosion and enhance its ability to store carbon. Additionally, promoting reforestation and sustainable forestry practices can increase carbon capture. This would encourage waste reduction, reuse, and recycling, can significantly reduce landfilling, and create opportunities for waste-to-resource strategies. This reduces reliance on virgin resources and minimizes resource extraction and processing pollution. Supporting research and development of innovative technologies for capturing and utilizing carbon emissions and cleaner industrial processes can play a crucial role in combating climate change. We can create a cleaner, healthier planet by integrating pollution control strategies into climate change mitigation efforts. The solutions are interconnected, and a holistic approach is essential for long-term environmental sustainability.

7.7 Adaptation Strategies

While mitigation strategies are crucial to addressing the root causes of climate change, adaptation strategies are essential for preparing for the inevitable impacts we are already experiencing and will face. Healthy ecosystems can be vital in mitigating climate change and adapting to its effects. This includes conservation efforts. Protecting and restoring natural ecosystems like forests, mangroves, and wetlands can enhance their ability to absorb carbon dioxide and buffer against the impacts of extreme weather events. Next should be sustainable land management. Promoting practices that improve soil health, reduce deforestation, and conserve water resources can make ecosystems more resilient to climate change impacts like drought and floods. Further research is on climate-smart agriculture. Adapting agricultural practices to cope with a changing climate is essential for ensuring food security. This could involve, for example, developing drought-resistant crops. Research and development of crop varieties more resistant to drought and heat stress can help ensure food production in a changing climate. Next in the lineup of adaptation strategies can be water conservation techniques. Implementing irrigation practices that minimize water waste and utilizing drought-tolerant crops can reduce agriculture's water footprint and adapt to changing precipitation patterns. Therefore, encouraging a wider variety of crops can improve agricultural resilience by reducing reliance on crops more susceptible to specific climate change impacts. Disaster preparedness and risk reduction would be the other key needful to consider. Climate change is expected to increase the frequency and intensity of extreme weather events. Preparing for these events can save lives and minimize damage. This would include early warning systems. Investing in and strengthening early warning systems for extreme weather events like floods, storms, and heatwaves can provide communities with crucial lead time to prepare and evacuate if necessary. Building seawalls, levees, and other infrastructure that can withstand extreme weather events can protect communities and critical infrastructure from flooding and storm (Fig. 1).

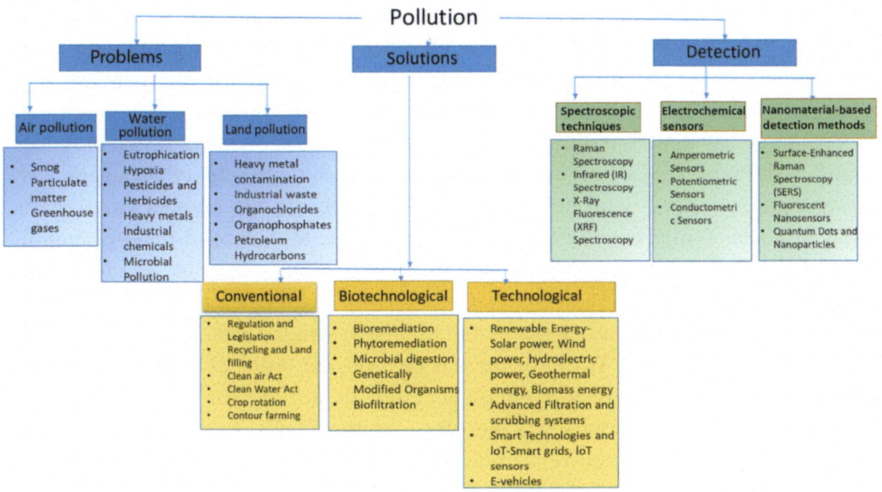

Fig. 1 A compelling view of various pollution mitigation strategies

8 Conclusion

Our planet's current condition presents a sobering reality. The indelible impact of human activities is evident in the escalating pollution levels and the manifestations of climate change, including extreme weather events, rising seas, and ecological imbalances. These environmental challenges extend beyond mere data points, posing significant threats to societal stability and the prospects of future generations. This dire situation necessitates a paradigm shift from environmental exploitation to conscious stewardship. The United Nations' Sustainable Development Goals (SDGs) for 2030 provide a comprehensive blueprint for a sustainable future. These 17 interlinked objectives address crucial issues ranging from poverty alleviation and access to clean water to developing sustainable urban environments and tackling climate change. Achieving these ambitious targets requires concerted efforts from individuals, corporations, and governments alike. On a personal level, adopting an environmentally conscious lifestyle is crucial. This involves making deliberate choices in our daily lives, such as reducing energy consumption, minimizing waste production, and embracing sustainable practices like using public transit, selecting energy-efficient products, and supporting eco-friendly local businesses. Moreover, we must champion environmental responsibility by holding ourselves and others accountable, backing businesses committed to sustainable practices, and actively participating in environmental advocacy and civic engagement.

The business sector must integrate sustainability as a core principle, moving beyond profit-centric models to those that balance environmental and social impacts with financial gains. This includes investing in renewable energy, implementing pollution control measures, designing eco-friendly products, and fostering a culture of sustainability within their organizations.

Governments play a pivotal role in setting sustainability standards. This involves implementing robust environmental policies, such as carbon pricing and regulations that promote cleaner production and waste reduction. Additionally, investing in clean energy research and development, supporting green infrastructure projects, and promoting sustainable agricultural practices are essential for a systemic shift toward sustainability. Collaboration between governments, businesses, and civil society organizations is vital in creating a unified approach to combat environmental degradation. While the challenges are formidable, they are not insurmountable. The SDGs offer a clear action plan, and through collective efforts, we can forge a more sustainable future.

It is important to remember that small changes, when adopted by millions, can yield significant impacts. We have moved beyond the point of complacency. Our shared responsibility is to restore the planet's health, not just for our benefit, but for future generations. By embracing the ethos of the SDGs and taking purposeful action, we can reverse the current trends and construct a sustainable future for all. Let us choose a path of responsible action, guided by a shared vision of humanity thriving in harmony with nature.

References

Abbasi A et al (2014) Agricultural Pollution: An Emerging Issue, In: Ahmad P, Wani MR, Azooz MM, Tran L-SP (eds) Improvement of crops in the era of climatic changes: volume 1. Springer, New York, pp 347–387. https://doi.org/10.1007/978-1-4614-8830-9_13

Abdel-Shafy HI, Mansour MSM (2016) A review on polycyclic aromatic hydrocarbons: source, environmental impact, effect on human health and remediation. Egypt J Pet 25:107–123. https://doi.org/10.1016/j.ejpe.2015.03.011

Adegoke HA, Solihu H, Bilewu SO (2023) Analysis of sanitation and waterborne disease occurrence in Ondo State, Nigeria. Environ Dev Sustain 25:11885–11903. https://doi.org/10.1007/s10668-022-02558-2

Ahamad A, Madhav S, Singh AK, Kumar A, Singh P (2020) Types of Water Pollutants: Conventional and Emerging. In: Pooja D, Kumar P, Singh P, Patil S (eds) Sensors in water pollutants monitoring: role of material. Springer, Singapore, pp 21–41. https://doi.org/10.1007/978-981-15-0671-0_3

Akemoto Y, Putra RS, Tanaka S (2023) Electrokinetic Remediation. In: Tanaka S, Kurasaki M, Morikawa M, Kamiya Y (eds) Design of materials and technologies for environmental remediation. Springer Nature Singapore, Singapore, pp 147–197

Akhtar ABT, Naseem S, Yasar A, Naseem Z (2021) Persistent Organic Pollutants (POPs): Sources, Types, Impacts, and Their Remediation. In: Prasad R (ed) Environmental pollution and remediation. Springer, Singapore, pp 213–246. https://doi.org/10.1007/978-981-15-5499-5_8

Alam S, Nisar M, Bano SA, Ahmad T (2022) Chapter 10. Impact of Aerial Fungal Spores on Human Health. In: Ahmed T, Hashmi MZ (eds) Hazardous environmental micro-pollutants, health impacts and allied treatment technologies. Springer International Publishing, Cham, pp 219–240. https://doi.org/10.1007/978-3-030-96523-5_10

Amsen E (2021) The end of waste as we know it? 4 ways to turn waste into treasure. https://www.weforum.org/agenda/2021/05/the-end-of-waste-as-we-know-it-5-ways-to-turn-waste-into-treasure/

Analitis A et al (2006) Short-term effects of ambient particles on cardiovascular and respiratory mortality. Epidemiology 17:230–233. https://doi.org/10.1097/01.ede.0000199439.57655.6b

Anisimov OA (2007) Potential feedback of thawing permafrost to the global climate system through methane emission. Environ Res Lett 2:045016. https://doi.org/10.1088/1748-9326/2/4/045016

Annamalai J, Namasivayam V (2015) Endocrine disrupting chemicals in the atmosphere: their effects on humans and wildlife. Environ Int 76:78–97. https://doi.org/10.1016/j.envint.2014.12.006

Araújo MM et al (2016) Thermal desorption of HCH. J Therm Anal Calorim 123:1019–1029. https://doi.org/10.1007/s10973-015-4810-y

Artiola JF, Walworth JL, Musil SA, Crimmins MA (2019)Chapter 10 - The Role of Environmental Monitoring in Pollution Science. In: Brusseau ML, Pepper IL, Gerba CP (eds) Environmental and pollution science, 3rd edn. Academic Press, London, pp 219–235. https://doi.org/10.1016/B978-0-12-814719-1.00010-0

Atkinson S, Tietenberg T (1991) Market failure in incentive-based regulation: the case of emissions trading. J Environ Econ Manag 21:17–31. https://doi.org/10.1016/0095-0696(91)90002-Z

Atta HI (2023) Chapter 2 - Aerosolization and bioaerosols. In: Ilori MO, Obayori OS, Salam LB (eds) Aeromicrobiology. Academic Press, London, pp 17–37. https://doi.org/10.1016/B978-0-323-96122-6.00008-6

Awasthi AK, Pandey AK, Khan J (2017) Biosorption an innovative tool for bioremediation of metal-contaminated municipal solid waste leachate: optimization and mechanisms exploration. Int J Environ Sci Technol 14:729–742. https://doi.org/10.1007/s13762-016-1173-2

Baig U, Faizan M, Sajid M (2021) Effective removal of hazardous pollutants from water and deactivation of water-borne pathogens using multifunctional synthetic adsorbent materials: a review. J Clean Prod 302:126735. https://doi.org/10.1016/j.jclepro.2021.126735

Bailey A, Meyer L, Pettingell N, Macie M, Korstad J (2020) Agricultural Practices Contributing to Aquatic Dead Zones. In: Bauddh K, Kumar S, Singh RP, Korstad J (eds) Ecological and practical applications for sustainable agriculture. Springer, Singapore, pp 373–393. https://doi.org/10.1007/978-981-15-3372-3_17

Baral A, Engelken RD (2002) Chromium-based regulations and greening in metal finishing industries in the USA. Environ Sci Pol 5:121–133. https://doi.org/10.1016/S1462-9011(02)00028-X

Barcelo L, Kline J, Walenta G, Gartner E (2014) Cement and carbon emissions. Mater Struct 47:1055–1065. https://doi.org/10.1617/s11527-013-0114-5

Baweja P, Kumar S, Kumar G (2020) Fertilizers and Pesticides: Their Impact on Soil Health and Environment. In: Giri B, Varma A (eds) Soil health. Springer International Publishing, Cham, pp 265–285. https://doi.org/10.1007/978-3-030-44364-1_15

Becerra M (2024) Gentrifying force or a force for environmental justice? A national assessment of brownfield redevelopment and gentrification in the United States from 2006 to 2015. Am Behav Sci 68:486–502. https://doi.org/10.1177/00027642221140839

Benke KK, Benke KE (2013) Uncertainty in health risks from artificial lighting due to disruption of circadian rhythm and melatonin secretion: a review. Hum Ecol Risk Assess Int J 19:916–929. https://doi.org/10.1080/10807039.2012.702608

Bennie J, Davies TW, Cruse D, Bell F, Gaston KJ (2018) Artificial light at night alters grassland vegetation species composition and phenology. J Appl Ecol 55:442–450. https://doi.org/10.1111/1365-2664.12927

Bernauer T, Moser P (1996) Reducing pollution of the river Rhine: the influence of international cooperation. J Environ Dev 5:389–415. https://doi.org/10.1177/107049659600500402

Bhat MA, Gedik K, Gaga EO (2023) Atmospheric micro (nano) plastics: future growing concerns for human health. Air Qual Atmos Health 16:233–262. https://doi.org/10.1007/s11869-022-01272-2

Biggi G (2024) Industrial dynamics and firm strategies in the agrochemical industry. Springer Nature Switzerland, Cham, pp 19–37

Eitelwein O, Fricker R, Green A, Racloz V (2024) Insight report January 2024, World economic Forum, pp 2–48 https://www.weforum.org/publications/quantifying-the-impact-of-climate-change-on-human-health/

Blackman, A., & Harrington, W. (2018). The use of economic incentives in developing countries: Lessons from international experience with industrial air pollution. The Theory and Practice of Command and Control in Environmental Policy, 199-238. https://doi.org/10.4324/9781315197296

Bleecker ML (2015) Chapter 12 - Carbon monoxide intoxication. In: Lotti M, Bleecker ML (eds) Handbook of clinical neurology, vol 131. Elsevier, Amsterdam, pp 191–203. https://doi.org/10.1016/B978-0-444-62627-1.00024-X

Borm PJA et al (2006) The potential risks of nanomaterials: a review carried out for ECETOC. Part Fibre Toxicol 3:11. https://doi.org/10.1186/1743-8977-3-11

Bose S, Dong G, Simpson A (2019) The financial ecosystem: the role of finance in achieving sustainability. Springer International Publishing, Cham, pp 339–368

Böttcher H et al (2012) Setting priorities for land management to mitigate climate change. Carbon Balance Manag 7:5. https://doi.org/10.1186/1750-0680-7-5

Brezinski JJ, Litton RK, Koleske JV (2012) Paint and coating testing manual: 15th edition of the Gardner-Sward handbook, vol MNL17-2ND-EB 0. ASTM International, West Conshohocken

Brooks KB, Simon D-C (2024) Sustainability: business and investment implications. World Scientific, New Jersey, pp 411–437

Bundhoo ZMA (2018) Solid waste management in least developed countries: current status and challenges faced. J Mater Cycles Waste Manag 20:1867–1877. https://doi.org/10.1007/s10163-018-0728-3

Bürer MJ, Wüstenhagen R (2009) Which renewable energy policy is a venture capitalist's best friend? Empirical evidence from a survey of international cleantech investors. Energy Policy 37:4997–5006. https://doi.org/10.1016/j.enpol.2009.06.071

Burmaster DE (1982) The new pollution. Environ Sci Policy Sustain Dev 24:6–36. https://doi.org/10.1080/00139157.1982.9932469

Byun E, Yang J-W, Kim Y, Ahn J (2017) Trapped greenhouse gases in the permafrost active layer: preliminary results for methane peaks in vertical profiles of frozen alaskan soil cores. Permafr Periglac Process 28:477–484. https://doi.org/10.1002/ppp.1935

Cabrera-Guzmán D, Swartzbaugh JT, Weisman AW (1990) The use of electrokinetics for hazardous waste site remediation. J Air Waste Manage Assoc 40:1670–1676. https://doi.org/10.1080/10473289.1990.10466815

Calvillo N (2018) Political airs: from monitoring to attuned sensing air pollution. Soc Stud Sci 48:372–388. https://doi.org/10.1177/0306312718784656

Carrer P, Maroni M, Alcini D, Cavallo D (2001) Allergens in indoor air: environmental assessment and health effects. Sci Total Environ 270:33–42. https://doi.org/10.1016/S0048-9697(00)00791-9

Cha JM (2017) A just transition: why transitioning workers into a new clean energy economy should be at the center of climate change policies. Fordham Environ Law Rev 29:196–220

Chaturvedi P et al (2021) Prevalence and hazardous impact of pharmaceutical and personal care products and antibiotics in environment: a review on emerging contaminants. Environ Res 194:110664. https://doi.org/10.1016/j.envres.2020.110664

Chaudhary S, Sharma P, Chauhan P, Kumar R, Umar A (2019) Functionalized nanomaterials: a new avenue for mitigating environmental problems. Int J Environ Sci Technol 16:5331–5358. https://doi.org/10.1007/s13762-019-02253-2

Cheesman AW et al (2023) Impacts of ground-level ozone on sugarcane production. Sci Total Environ 904:166817. https://doi.org/10.1016/j.scitotenv.2023.166817

Chen J et al (2018) On the use of multicopters for sampling and analysis of volatile organic compounds in the air by adsorption/thermal desorption GC-MS. Air Qual Atmos Health 11:835–842. https://doi.org/10.1007/s11869-018-0588-y

Chen HL et al (2021) The plastic waste problem in Malaysia: management, recycling and disposal of local and global plastic waste. SN Appl Sci 3:437. https://doi.org/10.1007/s42452-021-04234-y

Chen Z, Yang W, Yang Y, Ding H (2024a) Decision-making modeling of brownfield redevelopment in the context of dual carbon: the case of brownfield in Chengdu industrial zone. Acad J Sci Technol 10:81–88. https://doi.org/10.54097/rdnp9303

Chen J et al (2024b) Exploring long-term global environmental impacts of chlorinated paraffins (CPs) in waste: implications for the Stockholm and Basel Conventions and the global plastic treaty. Environ Int 185:108527. https://doi.org/10.1016/j.envint.2024.108527

Cheng C-Y et al (2023) Improved bioremediation of PCDD/Fs contaminated soil by mycelium-free liquids induced by agro-industrial residues. Bioresour Technol Rep 22:101435. https://doi.org/10.1016/j.biteb.2023.101435

Chung S-S, Lo CWH, Poon C-S (2002) Factors affecting waste disposal facilities siting in Southern China. JEAPM 4:241–262 https://doi.org/10.1016/S1464-3332(02)00100-5

Copaciu F et al (2013) Diffuse water pollution by anthraquinone and azo dyes in environment importantly alters foliage volatiles, carotenoids and physiology in wheat (Triticum aestivum). Water Air Soil Pollut 224:1478. https://doi.org/10.1007/s11270-013-1478-4

Corona Sánchez JE et al (2021) Metal(loid) bioaccessibility of atmospheric particulate matter from mine tailings at Zimapan, Mexico. Environ Sci Pollut Res 28:19458–19472. https://doi.org/10.1007/s11356-020-11887-6

Craver A et al (2024) Air quality and cancer risk in the All of Us Research Program. Cancer Causes Control 35:749–760. https://doi.org/10.1007/s10552-023-01823-7

Ćurić M, Zafirovski O, Spiridonov V (2022) Essentials of medical meteorology. Springer International Publishing, Cham, pp 143–182

Cutts S, Fricano R, Peters R (2024) Environmental justice for whom? Citizen participation and brownfield redevelopment in downtown Birmingham, Alabama. Environ Plan E 7:583–608. https://doi.org/10.1177/25148486231199330

Dagar G, Bagchi G (2020) Nanoparticles as Potential Endocrine Disruptive Chemicals. In: Saxena SK, Paul Khurana SM (eds) NanoBioMedicine. Springer Singapore, Cham, pp 411–429. https://doi.org/10.1007/978-981-32-9898-9_17

Dale VH (1997) The relationship between land-use change and climate change. Ecol Appl 7:753–769. https://doi.org/10.1890/1051-0761(1997)007[0753:TRBLUC]2.0.CO;2

Dale VH, Efroymson RA, Kline KL (2011) The land use–climate change–energy nexus. Landsc Ecol 26:755–773. https://doi.org/10.1007/s10980-011-9606-2

Damineva R, Shulaev N, Pryanichnikova V, Kadyrov R, Bykovsky NJSWC (2019) Economic assessment of the effects of pollution of land resources. SHS Web Conf 69:00145

Das N, Mathew L (2011) Chromium Pollution and Bioremediation: An Overview. In: Khan MS, Zaidi A, Goel R, Musarrat J (eds) Biomanagement of metal-contaminated soils. Springer Netherlands, Dordrecht, pp 297–321. https://doi.org/10.1007/978-94-007-1914-9_13

Dashti N, Ali N, Khanafer M, Radwan SS (2019) Plant-based oil-sorbents harbor native microbial communities effective in spilled oil-bioremediation under nitrogen starvation and heavy metal-stresses. Ecotoxicol Environ Saf 181:78–88. https://doi.org/10.1016/j.ecoenv.2019.05.072

Davies TW, Smyth T (2018) Why artificial light at night should be a focus for global change research in the 21st century. Glob Chang Biol 24:872–882. https://doi.org/10.1111/gcb.13927

de Coninck H, Fischer C, Newell RG, Ueno T (2008) International technology-oriented agreements to address climate change. Energy Policy 36:335–356. https://doi.org/10.1016/j.enpol.2007.09.030

De Haan FAM (1987) Effects of Agricultural Practices on the Physical, Chemical and Biological Properties of Soils: Part III—Chemical Degradation of Soil as the Result of the Use of Mineral Fertilizers and Pesticides: Aspects of Soil Quality Evaluation. In: Barth H, L'Hermite P (eds) Scientific basis for soil protection in the European Community. Springer Netherlands, Dordrecht, pp 211–236. https://doi.org/10.1007/978-94-009-3451-1_11

De Jonckheere J, Van Dijck P, van de Voorde H (1975) The effect of thermal pollution on the distribution of Naegleria fowleri. J Hyg 75:7–13. https://doi.org/10.1017/S0022172400047021

de Oliveira BF et al (2012) Risk assessment of PM(2.5) to child residents in Brazilian Amazon region with biofuel production. Environ Health 11:64. https://doi.org/10.1186/1476-069x-11-64

Devi A, Mudgal G, Khan ZA (2021) Adversities of Nanoparticles in Elderly Populations. In: Kesari KK, Jha NK (eds) Free radical biology and environmental toxicity. Springer International Publishing, Cham, pp 115–130. https://doi.org/10.1007/978-3-030-83446-3_5

Devlin M, Brodie J (2023) Nutrients and Eutrophication. In: Reichelt-Brushett A (ed) Marine pollution – monitoring, management and mitigation. Springer Nature Switzerland, Cham, pp 75–100. https://doi.org/10.1007/978-3-031-10127-4_4

Dhar, S., Kaur, J., & Mudgal, G. (2021). Unveiling the multifaceted exploration from genomic insights to functional applications of the Agave genus: A comprehensive review. Natl. Volatiles Essent. Oils, 8, 605-6628. https://www.nveo.org/index.php/journal/article/view/5623

Dhar SK et al (2024) Novel Bacillus and Prestia isolates from Dwarf century plant enhance crop yield and salinity tolerance. Sci Rep 14:14645. https://doi.org/10.1038/s41598-024-65632-x

Dhir B (2022) In: Sarma H, Dominguez DC, Lee W-Y (eds) Emerging contaminants in the environment. Elsevier, Amsterdam, pp 159–182

di Filippo R, Bursi OS, di Maggio R (2022) Global warming and ozone depletion potentials caused by emissions from HFC and CFC banks due to structural damage. Energ Buildings 273:112385. https://doi.org/10.1016/j.enbuild.2022.112385

Dignon J (1992) NOx and SOx emissions from fossil fuels: a global distribution. Atmos Environ Part A 26:1157–1163. https://doi.org/10.1016/0960-1686(92)90047-O

Dinerstein, E., Joshi, A. R., Hahn, N. R., Lee, A. T., Vynne, C., Burkart, K., ... & Zolli, A. (2024). Conservation Imperatives: securing the last unprotected terrestrial sites harboring irreplaceable biodiversity. Frontiers in Science, 2, 1349350. https://doi.org/10.3389/fsci.2024.1349350

Dóci G, Vasileiadou E (2015) "Let′s do it ourselves" Individual motivations for investing in renewables at community level. Renew Sust Energ Rev 49:41–50. https://doi.org/10.1016/j.rser.2015.04.051

Dominici F et al (2006) Fine particulate air pollution and hospital admission for cardiovascular and respiratory diseases. JAMA 295:1127–1134. https://doi.org/10.1001/jama.295.10.1127

Dominioni G (2022) Pricing carbon effectively: a pathway for higher climate change ambition. Clim Pol 22:897–905. https://doi.org/10.1080/14693062.2022.2042177

Downie D (2015) Still no time for complacency: evaluating the ongoing success and continued challenge of global ozone policy. J Environ Stud Sci 5:187–194. https://doi.org/10.1007/s13412-014-0199-3

EEA (2024) Diversion of waste from landfill in Europe. https://www.eea.europa.eu/en/analysis/indicators/diversion-of-waste-from-landfill#:~:text=In%20line%20with%20the%20EU,municipal%20waste%20generated%20by%202035

Egelston A (2022) Worth saving: international diplomacy to protect the environment. Springer International Publishing, Cham, pp 69–84

Ellerbeck S (2023) IEA: more than a third of the world's electricity will come from renewables in 2025. https://www.weforum.org/agenda/2023/03/electricity-generation-renewables-power-iea/

Ellis C (2018) World Bank: global waste generation could increase 70% by 2050. https://www.wastedive.com/news/world-bank-global-waste-generation-2050/533031/#:~:text=The%20World%20Bank%20estimates%20overall,instead%20dumped%20or%20openly%20burned

Erakhrumen AA (2011) Research Advances in Bioremediation of Soils and Groundwater Using Plant-Based Systems: A Case for Enlarging and Updating Information and Knowledge in Environmental Pollution Management in Developing Countries. In: Khan MS, Zaidi A, Goel R, Musarrat J (eds) Biomanagement of metal-contaminated soils. Springer Netherlands, Dordrecht, pp 143–166. https://doi.org/10.1007/978-94-007-1914-9_6

Falcón J et al (2020) Exposure to artificial light at night and the consequences for flora, fauna, and ecosystems. Front Neurosci 14. https://doi.org/10.3389/fnins.2020.602796

Fazakas E, Neamtiu IA, Gurzau ES (2023) Health effects of air pollutant mixtures (volatile organic compounds, particulate matter, sulfur and nitrogen oxides) – a review of the literature. https://doi.org/10.1515/reveh-2022-0252

FBI (2024) Used cooking oil market size, share & industry analysis, by source (food services and households), applications (industrial usage, animal feed, and others), and regional forecast, 2024–2032. Report No. Report ID: FBI103665

Feng L, Liao W (2016) Legislation, plans, and policies for prevention and control of air pollution in China: achievements, challenges, and improvements. J Clean Prod 112:1549–1558. https://doi.org/10.1016/j.jclepro.2015.08.013
Ferronato N, Torretta V (2019) Waste mismanagement in developing countries: a review of global issues. Int J Environ Res Public Health 16. https://doi.org/10.3390/ijerph16061060
Fiedler H (2003) Dioxins and Furans. (PCDD/PCDF) In: Fiedler H (ed) Persistent organic pollutants. Springer, Berlin Heidelberg, pp 123–201. https://doi.org/10.1007/10751132_6
Fleskens L, Nainggolan D, Stringer LC (2014) An exploration of scenarios to support sustainable land management using integrated environmental socio-economic models. Environ Manag 54:1005–1021. https://doi.org/10.1007/s00267-013-0202-x
Fogleman V (2024) Managing environmental risks through insurance: legal and economic aspects. Springer, Cham, pp 169–216
Fotina NV et al (2024) Enhancement of wheat growth by plant growth-stimulating bacteria during phytopathogenic inhibition. Biocatal Agric Biotechnol 60:103294. https://doi.org/10.1016/j.bcab.2024.103294
Franklin M, Koutrakis P, Schwartz P (2008) The role of particle composition on the association between PM2.5 and mortality. Epidemiology 19:680–689. https://doi.org/10.1097/ede.0b013e3181812bb7
Freitas LC et al (2021) From waste to sustainable industry: how can agro-industrial wastes help in the development of new products? Resour Conserv Recycl 169:105466. https://doi.org/10.1016/j.resconrec.2021.105466
Fromme H (2023) Indoor air quality: occurrence and health effects of contaminants. Springer Nature Switzerland, Cham, pp 391–442
Fucic A et al (2023) Impact of endocrine disruptors from mother's diet on immuno-hormonal orchestration of brain development and introduction of the virtual human twin tool. Reprod Toxicol 117:108357. https://doi.org/10.1016/j.reprotox.2023.108357
Galloway JN (2001) Acidification of the world: natural and anthropogenic. Water Air Soil Pollut 130:17–24. https://doi.org/10.1023/A:1012272431583
Gao T et al (2015) Analysis on differences of carbon dioxide emission from cement production and their major determinants. J Clean Prod 103:160–170. https://doi.org/10.1016/j.jclepro.2014.11.026
Gaston KJ, Sánchez de Miguel A (2022) Environmental impacts of artificial light at night. Annu Rev Environ Resour 47:373–398. https://doi.org/10.1146/annurev-environ-112420-014438
Gaur VK et al (2020) Assessing the impact of industrial waste on environment and mitigation strategies: a comprehensive review. J Hazard Mater 398:123019. https://doi.org/10.1016/j.jhazmat.2020.123019
Geissen V et al (2015) Emerging pollutants in the environment: a challenge for water resource management. Int Soil Water Conserv Res 3:57–65. https://doi.org/10.1016/j.iswcr.2015.03.002
Ghil M, Lucarini V (2020) The physics of climate variability and climate change. Rev Mod Phys 92:035002. https://doi.org/10.1103/RevModPhys.92.035002
Ghosh D et al (2023) Microbial assemblage for solid waste bioremediation and valorization with an essence of bioengineering. Environ Sci Pollut Res 30:16797–16816. https://doi.org/10.1007/s11356-022-24849-x
Giron A, Pickerel T (2021) Investigating global aquatic food loss and waste World Economic forum: Whitepapers, October 2024. https://www3.weforum.org/docs/WEF_Investigating_Global_Aquatic_Food_Loss_and_Waste_2024.pdf
Gopikrishnan GS, Kuttippurath J (2024) Four years of National Clean Air Programme (NCAP) in Indian cities: assessment of the impact on surface ozone during the period 2018–2022. Sustain Cities Soc 101:105207. https://doi.org/10.1016/j.scs.2024.105207
Gorbach OG, Kost C, Pickett C (2022) Review of internal carbon pricing and the development of a decision process for the identification of promising Internal Pricing Methods for an Organisation. Renew Sust Energ Rev 154:111745. https://doi.org/10.1016/j.rser.2021.111745

Gouchoe S, Everette V, Haynes R (2002) Case studies on the effectiveness of state financial incentives for renewable energy. United States. Natural Renewable Energy Laboratory. https://ncsolarcen-prod.s3.amazonaws.com/wp-content/uploads/2015/01/2002-Gouchoe-Case-Studies....pdf

Grizzetti B et al (2021) How EU policies could reduce nutrient pollution in European inland and coastal waters. Glob Environ Chang 69:102281. https://doi.org/10.1016/j.gloenvcha.2021.102281

Gul, A., Jamil, M., Hussain, T., Hafeez, Z., Masood, U., Mazhar, A., ... & Asad, R. M. (2020). The economic aspects of climate risks and food insecurity. In Climate Change and Food Security with Emphasis on Wheat (pp. 347-355). Academic Press. https://doi.org/10.1016/B978-0-12-819527-7.00025-X

Halkos GE, Aslanidis P-SC (2024) Reviewing the integrated institutional waste-related framework for circular economy in the European Union. Waste Manag Bull 2:28–35. https://doi.org/10.1016/j.wmb.2024.06.002

Hamner S et al (2006) The role of water use patterns and sewage pollution in incidence of waterborne/enteric diseases along the Ganges river in Varanasi, India. Int J Environ Health Res 16:113–132. https://doi.org/10.1080/09603120500538226

Han D et al (2021) Critical review of electro-kinetic remediation of contaminated soils and sediments: mechanisms, performances and technologies. Water Air Soil Pollut 232:335. https://doi.org/10.1007/s11270-021-05182-4

Haq S, Kaushal SS, Duan S (2018) Episodic salinization and freshwater salinization syndrome mobilize base cations, carbon, and nutrients to streams across urban regions. Biogeochemistry 141:463–486. https://doi.org/10.1007/s10533-018-0514-2

Heyes A (2000) Implementing environmental regulation: enforcement and compliance. J Regul Econ 17:107–129. https://doi.org/10.1023/A:1008157410380

Hoek G et al (2013) Long-term air pollution exposure and cardio- respiratory mortality: a review. Environ Health 12:43. https://doi.org/10.1186/1476-069X-12-43

Hoet PHM, Brüske-Hohlfeld I, Salata OV (2004) Nanoparticles – known and unknown health risks. J Nanobiotechnol 2:12. https://doi.org/10.1186/1477-3155-2-12

Hogland W, Stenis J (2000) Assessment and system analysis of industrial waste management. Waste Manag 20:537–543. https://doi.org/10.1016/S0956-053X(00)00020-9

Hope C, Schaefer K (2016) Economic impacts of carbon dioxide and methane released from thawing permafrost. Nat Clim Chang 6:56–59. https://doi.org/10.1038/nclimate2807

Hoyt VW, Mason E (2008) Nanotechnology. J Chem Health Saf 15:10–15. https://doi.org/10.1016/j.jchas.2007.07.015

Hui K, Yuan Y, Xi B, Tan W (2023) A review of the factors affecting the emission of the ozone chemical precursors VOCs and NOx from the soil. Environ Int 172:107799. https://doi.org/10.1016/j.envint.2023.107799

Huynh M, Woodruff TJ, Parker JD, Schoendorf KC (2006) Relationships between air pollution and preterm birth in California. Paediatr Perinat Epidemiol 20:454–461. https://doi.org/10.1111/j.1365-3016.2006.00759.x

Idris A, Inanc B, Hassan MN (2004) Overview of waste disposal and landfills/dumps in Asian countries. J Mater Cycles Waste Manag 6:104–110. https://doi.org/10.1007/s10163-004-0117-y

Idumah CI, Nwuzor IC (2019) Novel trends in plastic waste management. SN Appl Sci 1:1402. https://doi.org/10.1007/s42452-019-1468-2

Ijaz S et al (2024) Chapter 3 - Heavy metal–polluted arable land and its consequences: a global scenario. In: Fahad S et al (eds) Biochar-assisted remediation of contaminated soils under changing climate. Elsevier, Amsterdam, pp 71–99. https://doi.org/10.1016/B978-0-443-21562-9.00003-7

Iqbal N et al (2022) Effects of Polychlorinated Biphenyls on Plant Growth. In: Mahmood Q (ed) Sustainable plant nutrition under contaminated environments. Springer International Publishing, Cham, pp 187–208. https://doi.org/10.1007/978-3-030-91499-8_10

Issaka S, Ashraf MA (2017) Impact of soil erosion and degradation on water quality: a review. Geol Ecol Landsc 1:1–11. https://doi.org/10.1080/24749508.2017.1301053

ITU (2024) International Telecommunication Union. https://www.itu.int/en/ITU-D/Environment/Pages/Spotlight/Global-Ewaste-Monitor-2020.aspx
Jana PK, Saha DK, Sarkar D (2013) Contribution of some ozone depleting substances (ODS) and greenhouse gases (GHGs) on total column ozone growth at Srinagar (34°N, 74.8°E), India. J Earth Syst Sci 122:239–252. https://doi.org/10.1007/s12040-012-0246-y
Jeevahan J et al (2021) Waste into energy conversion technologies and conversion of food wastes into the potential products: a review. Int J Ambient Energy 42:1083–1101. https://doi.org/10.1080/01430750.2018.1537939
Johnson KA, Johnson DE (1995) Methane emissions from cattle. J Anim Sci 73:2483–2492. https://doi.org/10.2527/1995.7382483x
Jonson JE et al (2017) Impact of excess NOx emissions from diesel cars on air quality, public health and eutrophication in Europe. Environ Res Lett 12:094017. https://doi.org/10.1088/1748-9326/aa8850
Joshi JR, Bhanderi KK, Patel JV (2023) Waste cooking oil as a promising source for bio lubricants- a review. J Indian Chem Soc 100:100820. https://doi.org/10.1016/j.jics.2022.100820
Junaid M, Hashmi MZ, Malik RN, Pei D-S (2016) Toxicity and oxidative stress induced by chromium in workers exposed from different occupational settings around the globe: a review. Environ Sci Pollut Res 23:20151–20167. https://doi.org/10.1007/s11356-016-7463-x
Junior AST, Nagai RA (2024) Lean and Green: 3Ms Meets 5Rs Creating Environmental Value for the Planet. In: Rego C et al (eds) Entrepreneurship, technological change and circular economy for a green transition: research contributions for a more productive environment. Springer Nature Switzerland, Cham, pp 99–119. https://doi.org/10.1007/978-3-031-48079-9_6
Jyethi, D.S. (2016). Air Quality: Global and Regional Emissions of Particulate Matter, SOx, and NOx. In: Kulshrestha, U., Saxena, P. (eds) Plant Responses to Air Pollution. Springer, Singapore. https://doi.org/10.1007/978-981-10-1201-3_2
Kaewlaoyoong A et al (2020) White rot fungus Pleurotus pulmonarius enhanced bioremediation of highly PCDD/F-contaminated field soil via solid state fermentation. Sci Total Environ 738:139670. https://doi.org/10.1016/j.scitotenv.2020.139670
Kapper T (2004) Bringing beauty to account in the environmental impact statement: the contingent valuation of landscape aesthetics. Environ Pract 6:296–305. https://doi.org/10.1017/S146604660400047X
Karbalaei S, Hanachi P, Walker TR, Cole M (2018) Occurrence, sources, human health impacts and mitigation of microplastic pollution. Environ Sci Pollut Res 25:36046–36063. https://doi.org/10.1007/s11356-018-3508-7
Kasonga TK, Coetzee MAA, Kamika I, Ngole-Jeme VM, Benteke Momba MN (2021) Endocrine-disruptive chemicals as contaminants of emerging concern in wastewater and surface water: a review. J Environ Manag 277:111485. https://doi.org/10.1016/j.jenvman.2020.111485
Kaur J et al (2022) An exopolysaccharide-producing novel Agrobacterium pusense strain JAS1 isolated from snake plant enhances plant growth and soil water retention. Sci Rep 12:21330. https://doi.org/10.1038/s41598-022-25225-y
Kaur J et al (2023) Reactive black-5, congo red and methyl orange: chemical degradation of azo-dyes by Agrobacterium. Water 15:1664
Kaushal SS et al (2021) Freshwater salinization syndrome: from emerging global problem to managing risks. Biogeochemistry 154:255–292. https://doi.org/10.1007/s10533-021-00784-w
Kaza S, Yao L, Bhada-Tata P, Van Woerden F (2018) What a waste 2.0: a global snapshot of solid waste management to 2050. World Bank Publications, Washington, DC
Kazemi F, Golzarian MR, Myers B (2018) Potential of combined Water Sensitive Urban Design systems for salinity treatment in urban environments. J Environ Manag 209:169–175. https://doi.org/10.1016/j.jenvman.2017.12.046
Khan SA, Ewe SY (2024) Case 10: Heng Hiap Industries—Pioneer in the Plastic Recycling Industry. In: Mutum DS, Ewe SY, Ghazali EM, Vincent RL (eds) Marketing case studies in emerging markets: contemporary multi-industry issues and best-practices. Springer Nature Switzerland, Cham, pp 135–145 https://www.springerprofessional.de/en/case-10-heng-hiap-industries-pioneer-in-the-plastic-recycling-in/26929874

Khan AS, Srivastava P (2024) Particle Pollution and Health – Risk and Resilience Evaluation. In: Ghosh S, Kumari M, Mishra VN (eds) Geospatial technology to support communities and policy: pathways to resiliency. Springer Nature Switzerland, Cham, pp 305–326. https://doi.org/10.1007/978-3-031-52561-2_17

Khan S et al (2024) Ecosystem Degradation to Restoration: A Challenge. In: Chatterjee, U., Shaw, R., Kumar, S., Raj, A.D., Das, S. (eds) Climate crisis: adaptive approaches and sustainability. Springer, Cham, pp 19–33. https://doi.org/10.1007/978-3-031-44397-8_2

Kharangate-Lad A, D'Souza NC (2021) Remediation of Toxic Environmental Pollutants Using Nanoparticles and Integrated Nano-Bio Systems. In: Kumar V, Prasad R, Kumar M (eds) Rhizobiont in bioremediation of hazardous waste. Springer, Singapore, pp 443–482. https://doi.org/10.1007/978-981-16-0602-1_19

Khemani LT et al (1994) Atmospheric pollutants and their influence on acidification of rain water at an industrial location on the West Coast of India. Atmos Environ 28:3145–3154. https://doi.org/10.1016/1352-2310(94)00148-E

Kim BR (2011) VOC emissions from automotive painting and their control: a review. Environ Eng Res 16:1–9. https://doi.org/10.4491/eer.2011.16.1.001

Kim K-H, Jahan SA, Kabir E, Brown RJC (2013) A review of airborne polycyclic aromatic hydrocarbons (PAHs) and their human health effects. Environ Int 60:71–80. https://doi.org/10.1016/j.envint.2013.07.019

Kim D, Chen Z, Zhou L-F, Huang S-X (2018) Air pollutants and early origins of respiratory diseases. Chronic Dis Transl Med 4:75–94. https://doi.org/10.1016/j.cdtm.2018.03.003

Kleinman MT (2020) Environmental toxicants, pp 455–486. https://doi.org/10.1002/9781119438922.ch12

Klimont Z et al (2002) Anthropogenic emissions of non-methane volatile organic compounds in China. Atmos Environ 36:1309–1322. https://doi.org/10.1016/S1352-2310(01)00529-5

Ko M, Shia R-L, Sze N-D, Magid H, Bray RG (1999) Atmospheric lifetime and global warming potential of HFC-245fa. J Geophys Res 104:8173–8181. https://doi.org/10.1029/1998JD100097

Kotwani A, Joshi J, Kaloni D (2021) Pharmaceutical effluent: a critical link in the interconnected ecosystem promoting antimicrobial resistance. Environ Sci Pollut Res 28:32111–32124. https://doi.org/10.1007/s11356-021-14178-w

Kreyling WG, Semmler-Behnke M, Möller W (2006) Health implications of nanoparticles. J Nanopart Res 8:543–562. https://doi.org/10.1007/s11051-005-9068-z

Kristanti RA, Hadibarata T, Syafrudin M, Yılmaz M, Abdullah S (2022) Microbiological contaminants in drinking water: current status and challenges. Water Air Soil Pollut 233:299. https://doi.org/10.1007/s11270-022-05698-3

Krittanawong C et al (2023) PM2.5 and cardiovascular diseases: State-of-the-Art review. Int J Cardiol Cardiovasc Risk Prev 19:200217. https://doi.org/10.1016/j.ijcrp.2023.200217

Kulshreshtha S, Mathur N, Bhatnagar P (2013) Mycoremediation of Paper, Pulp and Cardboard Industrial Wastes and Pollutants. In: Goltapeh EM, Danesh YR, Varma A (eds) Fungi as bioremediators. Springer Berlin Heidelberg, pp 77–116. https://doi.org/10.1007/978-3-642-33811-3_4

Kumar V, Kothiyal NC, Saruchi, Vikas P, Sharma R (2016) Sources, distribution, and health effect of carcinogenic polycyclic aromatic hydrocarbons (PAHs) – current knowledge and future directions. J Chin Adv Mater Soc 4:302–321. https://doi.org/10.1080/22243682.2016.1230475

Kumar S et al (2017) Challenges and opportunities associated with waste management in India. 4:160764. https://doi.org/10.1098/rsos.160764

Kumar S et al (2019) Hazardous heavy metals contamination of vegetables and food chain: role of sustainable remediation approaches – a review. Environ Res 179:108792. https://doi.org/10.1016/j.envres.2019.108792

Kumar S, Anwer R, Sehrawat A, Yadav M, Sehrawat N (2021a) Assessment of bacterial pathogens in drinking water: a serious safety concern. Curr Pharmacol Rep 7:206–212. https://doi.org/10.1007/s40495-021-00263-8

Kumar L, Ragunathan V, Chugh M, Bharadvaja N (2021b) Nanomaterials for remediation of contaminants: a review. Environ Chem Lett 19:3139–3163. https://doi.org/10.1007/s10311-021-01212-z

Kumar, S., Chatterjee, U., David Raj, A., Sooryamol, K.R. (2023a). Global Warming and Climate Crisis/Extreme Events. In: Chatterjee, U., Shaw, R., Kumar, S., Raj, A.D., Das, S. (eds) Climate Crisis: Adaptive Approaches and Sustainability. Sustainable Development Goals Series. Springer, Cham. https://doi.org/10.1007/978-3-031-44397-8_1

Kumar M et al (2023b) Current research trends on emerging contaminants pharmaceutical and personal care products (PPCPs): a comprehensive review. Sci Total Environ 859:160031. https://doi.org/10.1016/j.scitotenv.2022.160031

Kuppusamy S, Thavamani P, Megharaj M, Naidu R (2015) Bioremediation potential of natural polyphenol rich green wastes: a review of current research and recommendations for future directions. Environ Technol Innov 4:17–28. https://doi.org/10.1016/j.eti.2015.04.001

Kusumadewi SD, Purnomo H, Nadhira S, Putrizulfan R (2024) Systematic review on the implementation of mangrove community-based restoration in Indonesia and beyond. IOP Conf Ser Earth Environ Sci 1315:012052. https://doi.org/10.1088/1755-1315/1315/1/012052

Lal R, Miller FP, Logan TJ (1988) Are intensive agricultural practices environmentally and ethically sound? J Agric Ethics 1:193–210. https://doi.org/10.1007/BF01833409

Laxmi V, Kaushik G (2020) Toxicity of Hexavalent Chromium in Environment, Health Threats, and Its Bioremediation and Detoxification from Tannery Wastewater for Environmental Safety. In: Saxena G, Bharagava RM (eds) Bioremediation of industrial waste for environmental safety: volume I: industrial waste and its management. Springer, Singapore, pp 223–243. https://doi.org/10.1007/978-981-13-1891-7_11

Lerner DN, Harris B (2009) The relationship between land use and groundwater resources and quality. Land Use Policy 26:S265–S273. https://doi.org/10.1016/j.landusepol.2009.09.005

Levin H (1992) Controlling Sources of Indoor Air Pollution. In: Knöppel H, Wolkoff P (eds) Chemical, microbiological, health and comfort aspects of indoor air quality — state of the art in SBS. Springer Netherlands, Dordrecht, pp 321–341. https://doi.org/10.1007/978-94-015-8088-5_22

Li Y, Zhang J (1999) Agricultural diffuse pollution from fertilisers and pesticides in China. Water Sci Technol 39:25–32. https://doi.org/10.1016/S0273-1223(99)00029-3

Li Z et al (2014) Osmotically driven membrane process for the management of urban runoff in coastal regions. Water Res 48:200–209. https://doi.org/10.1016/j.watres.2013.09.028

Li Y et al (2022) Deforestation-induced climate change reduces carbon storage in remaining tropical forests. Nat Commun 13:1964. https://doi.org/10.1038/s41467-022-29601-0

Li C et al (2023a) Reversal of trends in global fine particulate matter air pollution. Nat Commun 14:5349. https://doi.org/10.1038/s41467-023-41086-z

Li N et al (2023b) Secondary PM2.5 dominates aerosol pollution in the Yangtze River Delta region: environmental and health effects of the Clean air Plan. Environ Int 171:107725. https://doi.org/10.1016/j.envint.2022.107725

Li B, Ma Y, Zhou Y, Chai E (2023c) Research progress of different components of PM2.5 and ischemic stroke. Sci Rep 13:15965. https://doi.org/10.1038/s41598-023-43119-5

List JA, Millimet DL, McHone W (2004) The unintended disincentive in the Clean Air Act 4. https://doi.org/10.2202/1538-0637.1204

Liu R, Lal R (2012) Nanoenhanced materials for reclamation of mine lands and other degraded soils: a review. J Nanotechnol 2012:461468. https://doi.org/10.1155/2012/461468

Liyanage D, Walpita J (2020) 8 - Organic pollutants from E-waste and their electrokinetic remediation. In: Prasad MNV, Vithanage M, Borthakur A (eds) Handbook of electronic waste management. Butterworth-Heinemann, Oxford, pp 171–189. https://doi.org/10.1016/B978-0-12-817030-4.00006-1

Lofrano G et al (2020) 1 - Occurrence and potential risks of emerging contaminants in water. In: Sacco O, Vaiano V (eds) Visible light active structured photocatalysts for the removal of emerging contaminants. Elsevier, Amsterdam, pp 1–25. https://doi.org/10.1016/B978-0-12-818334-2.00001-8

Longobardi P, Montenegro A, Beltrami H, Eby M (2016) Deforestation induced climate change: effects of spatial scale. PLoS One 11:e0153357. https://doi.org/10.1371/journal.pone.0153357

Lu J, Yao L (2023) Observational evidence for detrimental impact of inhaled ozone on human respiratory system. BMC Public Health 23:929. https://doi.org/10.1186/s12889-023-15902-6

Luhach, K., Sharma, P., Sharma, N., Dhiman, N., Kharkwal, H., Sharma, B. (2023). Prenatal and Early Life Environmental Stressors: Chemical Moieties Responsible for the Development of Autism Spectrum Disorder. In: El Idrissi, A., McCloskey, D. (eds) Neurobiology of Autism Spectrum Disorders. Springer, Cham. https://doi.org/10.1007/978-3-031-42383-3_3

Ma J, Li L (2024) VOC emitted by biopharmaceutical industries: source profiles, health risks, and secondary pollution. J Environ Sci 135:570–584. https://doi.org/10.1016/j.jes.2022.10.022

Ma S et al (2019) Advanced treatment technology for FGD wastewater in coal-fired power plants: current situation and future prospects. Desalin Water Treat 167:122–132. https://doi.org/10.5004/dwt.2019.24630

Macgregor CJ, Evans DM, Fox R, Pocock MJO (2017) The dark side of street lighting: impacts on moths and evidence for the disruption of nocturnal pollen transport. Glob Chang Biol 23:697–707. https://doi.org/10.1111/gcb.13371

Maciejczyk P, Chen L-C, Thurston G (2021) The role of fossil fuel combustion metals in PM25 air pollution health associations. Atmosphere 12:1086

Madhav S et al (2020) Water Pollutants: Sources and Impact on the Environment and Human Health. In: Pooja D, Kumar P, Singh P, Patil S (eds) Sensors in water pollutants monitoring: role of material. Springer, Singapore, pp 43–62. https://doi.org/10.1007/978-981-15-0671-0_4

Maheshwari K, Agrawal M, Gupta AB (2021) Dye Pollution in Water and Wastewater. In: Muthu SS, Khadir A (eds) Novel materials for dye-containing wastewater treatment. Springer, Singapore, pp 1–25. https://doi.org/10.1007/978-981-16-2892-4_1

Mamun AA, Prasetya TAE, Dewi IR, Ahmad M (2023) Microplastics in human food chains: food becoming a threat to health safety. Sci Total Environ 858:159834. https://doi.org/10.1016/j.scitotenv.2022.159834

Manas JJI (2023) Top 5 waste management challenges in the Philippines and how to solve them. Plasticbank, Blog. https://plasticbank.com/blog/top-5-waste-management-challenges-in-the-philippines-and-how-to-solve-them/

Mandal PK (2005) Dioxin: a review of its environmental effects and its aryl hydrocarbon receptor biology. J Comp Physiol B 175:221–230. https://doi.org/10.1007/s00360-005-0483-3

Marchant N (2021) The world's food waste problem is bigger than we thought – here's what we can do about it. World Economic Forum. https://www.visionofhumanity.org/global-food-waste-problem-is-bigger-than-we-thought/

Maximillian J, Brusseau ML, Glenn EP, Matthias AD (2019) Pollution and Environmental Perturbations in the Global System. In: Brusseau ML, Pepper IL, Gerba CP (eds) Environmental and pollution science, 3rd edn. Academic Press, London, pp 457–476. https://doi.org/10.1016/B978-0-12-814719-1.00025-2

McFall-Johnsen M (2020) These facts show how unsustainable the fashion industry is. https://www.weforum.org/agenda/2020/01/fashion-industry-carbon-unsustainable-environment-pollution/

Michael Beman J, Arrigo KR, Matson PA (2005) Agricultural runoff fuels large phytoplankton blooms in vulnerable areas of the ocean. Nature 434:211–214. https://doi.org/10.1038/nature03370

Miglioranza KSB, Ondarza PM, Grondona SI, Scenna LB (2023) Persistent Organic Contaminants. In: Blasco, J., Tovar-Sánchez, A. In: Blasco J, Tovar-Sánchez A (eds) Marine analytical chemistry. Springer International Publishing, Cham, pp 275–306. https://doi.org/10.1007/978-3-031-14486-8_5

Mishra, S. et al. (2019). Heavy Metal Contamination: An Alarming Threat to Environment and Human Health. In: Sobti, R., Arora, N., Kothari, R. (eds) Environmental Biotechnology: For Sustainable Future. Springer, Singapore. https://doi.org/10.1007/978-981-10-7284-0_5

Misra V, Pandey SD (2005) Hazardous waste, impact on health and environment for development of better waste management strategies in future in India. Environ Int 31:417–431. https://doi.org/10.1016/j.envint.2004.08.005

Mohapatra P, Basu AR (2023) Status of Persistent Organic Pollutants Pollution in India. In: Chakraborty P et al (eds) Managing persistent organic pollutants in India: case studies on Vapi and Surat, Gujarat. Springer Nature Switzerland, Cham, pp 1–18. https://doi.org/10.1007/978-3-031-31311-0_1
Moreira H et al (2021) Phytomanagement of metal(loid)-contaminated soils: options, efficiency and value. Front Environ Sci 9. https://doi.org/10.3389/fenvs.2021.661423
Mosheim R, Sickles RC (2021) Spatial effects of nutrient pollution on drinking water production. Empir Econ 60:2741–2764. https://doi.org/10.1007/s00181-021-02019-1
Moumen A, Azizi G, Chekroun KB, Baghour M (2016) The effects of livestock methane emission on the global warming: a review. International Journal of Global Warming 9:229–253. https://doi.org/10.1504/ijgw.2016.074956
Mudhoo A et al (2013) Dioxins and Furans: Sources, Impacts and Remediation. In: Lichtfouse E, Schwarzbauer J, Robert D (eds) Pollutant diseases, remediation and recycling. Springer International Publishing, Wien, pp 479–541. https://doi.org/10.1007/978-3-319-02387-8_10
Murphy AB, Farmer AJD, Horrigan EC, McAllister T (2002) Plasma destruction of ozone depleting substances. Plasma Chem Plasma Process 22:371–385. https://doi.org/10.1023/A:1015365032020
Murty D, Kirschbaum MUF, Mcmurtrie RE, Mcgilvray H (2002) Does conversion of forest to agricultural land change soil carbon and nitrogen? A review of the literature. Global Change Biol 8:105–123. https://doi.org/10.1046/j.1354-1013.2001.00459.x
Nag PK (2019) Office buildings: health, safety and environment. Springer, Singapore, pp 53–103
Naimaee R et al (2024) Long-term water quality monitoring: using satellite images for temporal and spatial monitoring of thermal pollution in water resources. Sustainability 16:646
Nam I-H et al (2008) Bioremediation of PCDD/Fs-contaminated municipal solid waste incinerator fly ash by a potent microbial biocatalyst. J Hazard Mater 157:114–121. https://doi.org/10.1016/j.jhazmat.2007.12.086
Nanda S, Berruti F (2021) A technical review of bioenergy and resource recovery from municipal solid waste. J Hazard Mater 403:123970. https://doi.org/10.1016/j.jhazmat.2020.123970
Nandan A, Siddiqui NA, Singh C, Aeri A (2021) Occupational and environmental impacts of indoor air pollutant for different occupancy: a review. Toxicol Environ Heal Sci 13:303–322. https://doi.org/10.1007/s13530-021-00102-9
Nieder R, Benbi DK, Reichl FX (2018a) Soil components and human health. Springer Netherlands, Dordrecht, pp 575–657
Nieder R, Benbi DK, Reichl FX (2018b) Soil components and human health. Springer Netherlands, Dordrecht, pp 659–721
Norberg-Bohm V (2000) Creating incentives for environmentally enhancing technological change: lessons from 30 years of U.S. Energy Technology Policy. Technol Forecast Soc Chang 65:125–148. https://doi.org/10.1016/S0040-1625(00)00076-7
Nunez Y et al (2024) An environmental justice analysis of air pollution emissions in the United States from 1970 to 2010. Nat Commun 15:268. https://doi.org/10.1038/s41467-023-43492-9
O'Sullivan TM (2015) Environmental security is homeland security: climate disruption as the ultimate disaster risk multiplier 6:183–222. https://doi.org/10.1002/rhc3.12084
Obimakinde S, Fatoki O, Opeolu B, Olatunji O (2017) Veterinary pharmaceuticals in aqueous systems and associated effects: an update. Environ Sci Pollut Res 24:3274–3297. https://doi.org/10.1007/s11356-016-7757-z
Ogieriakhi MO, Wang XJEA, Policy. (2024) Do mandatory environmental policies really work? A case study of California's mandatory commercial recycling law. Econ Anal Policy 81:915–930
Osman AI et al (2023) Microplastic sources, formation, toxicity and remediation: a review. Environ Chem Lett 21:2129–2169. https://doi.org/10.1007/s10311-023-01593-3
Pandey PK, Kass PH, Soupir ML, Biswas S, Singh VP (2014) Contamination of water resources by pathogenic bacteria. AMB Express 4:51. https://doi.org/10.1186/s13568-014-0051-x
Parashar, M., & Mudgal, G. (2024). The Prostrate Spurge-isolated PGPB Endophytes, EP1-AS, and EP1-BM That Can Tolerate High Levels of Salinity and Heavy Metals and Allow Wheat

Growth Under These Stressors. Nature Environment & Pollution Technology, 23(2). https://neptjournal.com/upload-images/(12)B-4110.pdf

Parashar CK, Das P, Samanta S, Ganguly A, Chatterjee PK (2020) Municipal Solid Wastes—A Promising Sustainable Source of Energy: A Review on Different Waste-to-Energy Conversion Technologies. In: Ghosh SK (ed) Energy recovery processes from wastes. Springer, Singapore, pp 151–163. https://doi.org/10.1007/978-981-32-9228-4_13

Parashar M et al (2023) Two novel plant-growth-promoting Lelliottia amnigena isolates from Euphorbia prostrata aiton enhance the overall productivity of wheat and tomato. Plants (Basel, Switzerland) 12. https://doi.org/10.3390/plants12173081

Pariatamby A, Fauziah SH (2014) Sustainable 3R Practice in the Asia and Pacific Regions: The Challenges and Issues. In: Pariatamby A, Tanaka M (eds) Municipal solid waste management in asia and the pacific islands: challenges and strategic solutions. Springer, Singapore, pp 15–40. https://doi.org/10.1007/978-981-4451-73-4_2

Parker ER (2021) The influence of climate change on skin cancer incidence – a review of the evidence. Int J Women's Dermatol 7:17–27. https://doi.org/10.1016/j.ijwd.2020.07.003

Pasetto R et al (2024) Promotion of environmental public health and environmental justice in communities affected by large and long lasting industrial contamination: methods applied and lessons learned from the case study of Porto Torres (Italy). Front Public Health 12. https://doi.org/10.3389/fpubh.2024.1408127

Patel, N., Chauhan, D., Shahane, S., Rai, D., Ali Khan, M. Z., Mishra, U., & Chaudhary, V. K. (2021). Contamination and health impact of heavy metals. Water pollution and remediation: Heavy metals, 259–280. https://www.springerprofessional.de/contamination-and-health-impact-of-heavy-metals/18599326

Pawade, A., & Charhate, D. S. Radon (2024) Transport & Analysis of Radon Level in Existing Buildings Structures in India. SSRG International Journal of Civil Engineering. https://www.internationaljournalssrg.org/IJCE/paper-details?Id=670

Peng W et al (2020) The critical role of policy enforcement in achieving health, air quality, and climate benefits from India's clean electricity transition. Environ Sci Technol 54:11720–11731. https://doi.org/10.1021/acs.est.0c01622

Pereira CP, Goldenstein JPN, Bassin JP (2022) Biosorption for wastewater contaminants:1–22. https://doi.org/10.1002/9781119737629

Pheakdey DV, Quan NV, Khanh TD, Xuan TD (2022) Challenges and priorities of municipal solid waste management in Cambodia. Int J Environ Res Public Health 19:8458

Pillai S, Peter AE, Sunil B, Shrihari S (2014) In: International conference on biological, civil and environmental engineering (BCEE-2014), March, Dubai

Popp D (2003) Pollution control innovations and the Clean Air Act of 1990. J Policy Anal Manage 22:641–660. https://doi.org/10.1002/pam.10159

Poulsen AH et al (2023) Air pollution with NO2, PM2.5, and elemental carbon in relation to risk of breast cancer–a nationwide case-control study from Denmark. Environ Res 216:114740. https://doi.org/10.1016/j.envres.2022.114740

Prata JC et al (2022) Airborne Microplastics. In: Rocha-Santos T, Costa MF, Mouneyrac C (eds) Handbook of microplastics in the environment. Springer International Publishing, Cham, pp 177–201. https://doi.org/10.1007/978-3-030-10618-8_37-2

Pratt S (2021) How AI can solve manufacturing's waste problem. https://www.weforum.org/agenda/2021/04/how-ai-can-cut-waste-in-manufacturing/

Puckett J (2024) Global and National Instruments to Stop the Export and Dumping of Plastic Wastes. In: Gündoğdu S (ed) Plastic waste trade: a new colonialist means of pollution transfer. Springer Nature Switzerland, Cham, pp 67–103. https://doi.org/10.1007/978-3-031-51358-9_4

Pui Kin FL, Yang H (2023) Introduction of modelling the impact of ground-level ozone on crops at a local and global scale. In: Pui Kin FL (ed) Advances in botanical research, vol 108. Academic Press, pp 1–45. https://doi.org/10.1016/bs.abr.2023.04.001

Qadir SA, Al-Motairi H, Tahir F, Al-Fagih L (2021) Incentives and strategies for financing the renewable energy transition: a review. Energy Rep 7:3590–3606. https://doi.org/10.1016/j.egyr.2021.06.041
Qi Z et al (2014) Effect of temperature and particle size on the thermal desorption of PCBs from contaminated soil. Environ Sci Pollut Res 21:4697–4704. https://doi.org/10.1007/s11356-013-2392-4
Raaschou-Nielsen O et al (2023) PM2.5 air pollution components and mortality in Denmark. Environ Int 171:107685. https://doi.org/10.1016/j.envint.2022.107685
Rahman Z, Singh VP (2019) The relative impact of toxic heavy metals (THMs) (arsenic (As), cadmium (Cd), chromium (Cr)(VI), mercury (Hg), and lead (Pb)) on the total environment: an overview. Environ Monit Assess 191:419. https://doi.org/10.1007/s10661-019-7528-7
Rajhans G, Barik A, Sen SK, Raut S (2021) Degradation of dyes by fungi: an insight into mycoremediation. Biotechnologia 102:445–455. https://doi.org/10.5114/bta.2021.111109
Ramamoorthy S et al (2009) Characterisation of novel pH indicator of natural dye Oldenlandia umbellata L. Nat Prod Res 23:1210–1217. https://doi.org/10.1080/14786410802696635
Ramírez V et al (2023) Developmental programming in animal models: critical evidence of current environmental negative changes. Reprod Sci 30:442–463. https://doi.org/10.1007/s43032-022-00999-8
Raptis CE, van Vliet MTH, Pfister S (2016) Global thermal pollution of rivers from thermoelectric power plants. Environ Res Lett 11:104011. https://doi.org/10.1088/1748-9326/11/10/104011
Rashmi I et al (2022) Soil Erosion and Sediments: A Source of Contamination and Impact on Agriculture Productivity. In: Naeem M, Bremont JFJ, Ansari AA, Gill SS (eds) Agrochemicals in soil and environment: impacts and remediation. Springer Nature, Singapore, pp 313–345. https://doi.org/10.1007/978-981-16-9310-6_14
Raub JA (1999) Health effects of exposure to ambient carbon monoxide. Chemosphere Global Change Sci 1:331–351. https://doi.org/10.1016/S1465-9972(99)00005-7
Read S (2022) Vegetables are changing shape because of the climate crisis. World Economic Forum: Food and Water. https://www.weforum.org/stories/2022/09/food-waste-climate-drought-farmers/
Redko V et al (2024) Environmental impacts of 5-year plastic waste deposition on municipal waste landfills: a follow-up study. Sci Total Environ 906:167710. https://doi.org/10.1016/j.scitotenv.2023.167710
Riebsame WE, Meyer WB, Turner BL (1994) Modeling land use and cover as part of global environmental change. Clim Chang 28:45–64. https://doi.org/10.1007/BF01094100
Rizzo L et al (2019) Consolidated vs new advanced treatment methods for the removal of contaminants of emerging concern from urban wastewater. Sci Total Environ 655:986–1008. https://doi.org/10.1016/j.scitotenv.2018.11.265
Rodrigues FA, Joekes I (2011) Cement industry: sustainability, challenges and perspectives. Environ Chem Lett 9:151–166. https://doi.org/10.1007/s10311-010-0302-2
Rosser F, Balmes J (2023) Ozone and childhood respiratory health: a primer for US pediatric providers and a call for a more protective standard. Pediatr Pulmonol 58:1355–1366. https://doi.org/10.1002/ppul.26368
Saeedi MJJOFS, Technology (2024) How microplastics interact with food chain: a short overview of fate and impacts. J Food Sci Technol 61:403–413
Sakadevan K, Nguyen ML (2017) Chapter Four - Livestock Production and Its Impact on Nutrient Pollution and Greenhouse Gas Emissions. In: Sparks DL (ed) Advances in agronomy, vol 141. Academic Press, San Diego, pp 147–184. https://doi.org/10.1016/bs.agron.2016.10.002
Sam K, Zibima T (2024) Inclusive environmental decision-making in a developing nation: insights from the Ogoni Remediation Project, Niger Delta, Nigeria. Environ Manag 73:323–337. https://doi.org/10.1007/s00267-023-01885-y
Samaddar P, Ok YS, Kim K-H, Kwon EE, Tsang DCW (2018) Synthesis of nanomaterials from various wastes and their new age applications. J Clean Prod 197:1190–1209. https://doi.org/10.1016/j.jclepro.2018.06.262

Sampson N, Price C, Sampson M, Bradshaw M, Freeman B (2024) Lessons from a plain language analysis: U.S. Clean Air Act Title V public notices as barriers to environmental justice. Environ Sci Pol 151:103604. https://doi.org/10.1016/j.envsci.2023.103604

Sarkar AK, Sadhukhan S (2022) Bioremediation of Salt-Affected Soil Through Plant-Based Strategies. In: Malik JA (ed) Advances in bioremediation and phytoremediation for sustainable soil management: principles, monitoring and remediation. Springer International Publishing, Cham, pp 81–100. https://doi.org/10.1007/978-3-030-89984-4_5

Savioli G et al (2024) Carbon monoxide poisoning: from occupational health to emergency medicine. J Clin Med 13:2466

Schiedek D, Sundelin B, Readman JW, Macdonald RW (2007) Interactions between climate change and contaminants. Mar Pollut Bull 54:1845–1856. https://doi.org/10.1016/j.marpolbul.2007.09.020

Schraufnagel DE et al (2019) Air pollution and noncommunicable diseases: a review by the forum of international respiratory societies' environmental committee, part 2: air pollution and organ systems. Chest 155:417–426. https://doi.org/10.1016/j.chest.2018.10.041

Schwartz J (2000) Harvesting and long term exposure effects in the relation between air pollution and mortality. Am J Epidemiol 151:440–448. https://doi.org/10.1093/oxfordjournals.aje.a010228

Segerson K, Walker D (2002) Nutrient pollution: an economic perspective. Estuaries 25:797–808. https://doi.org/10.1007/BF02804906

Seidel S, Ye J, Andersen SO, Hillbrand A. Not-in-kind alternatives to high global warming HFCs. 27 (2016) center for climate and energy solutions (C2ES). https://www.c2es.org/wp-content/uploads/2016/10/not-in-kind-alternatives-high-global-warming-hfcs.pdf

Sejian V et al (2015) Global Warming: Role of Livestock. In: Sejian V, Gaughan J, Baumgard L, Prasad C (eds) Climate change impact on livestock: adaptation and mitigation. Springer India, New Delhi, pp 141–169. https://doi.org/10.1007/978-81-322-2265-1_10

Shah A, Che Mat CR, Ibrahim A, Zhang Y, Muzammil S (2024) Industrial ecology: a fusion of material and energy in green supply chain context. Springer Nature, Singapore, pp 177–196

Shahnawaz M, Sangale MK, Ade AB (2019) Bioremediation technology for plastic waste. Springer, Singapore, pp 21–30

Sharma, A., Chand, K., Singh, G.B., Mudgal, G. (2021). Toxicity with Waste-Generated Ionizing Radiations: Blunders Behind the Scenes. In: Kesari, K.K., Jha, N.K. (eds) Free Radical Biology and Environmental Toxicity. Molecular and Integrative Toxicology. Springer, Cham. https://doi.org/10.1007/978-3-030-83446-3_15

Sharma MD et al (2023) Microbial waterborne diseases in India: status, interventions, and future perspectives. Curr Microbiol 80:400. https://doi.org/10.1007/s00284-023-03462-2

Sharma K, Rajan S, Nayak SK (2024) Chapter 1 - Water pollution: Primary sources and associated human health hazards with special emphasis on rural areas. In: Madhav S, Srivastav AL, Izah SC, van Hullebusch E (eds) Water resources management for rural development. Elsevier, Amsterdam, pp 3–14. https://doi.org/10.1016/B978-0-443-18778-0.00014-3

Sharpley AN, McDowell RW, Kleinman PJA (2001) Phosphorus loss from land to water: integrating agricultural and environmental management. Plant Soil 237:287–307. https://doi.org/10.1023/A:1013335814593

Shayo GM, Elimbinzi E, Shao GN, Fabian C (2023) Severity of waterborne diseases in developing countries and the effectiveness of ceramic filters for improving water quality. Bull Natl Res Cent 47:113. https://doi.org/10.1186/s42269-023-01088-9

Jyethi, D.S. (2016). Air Quality: Global and Regional Emissions of Particulate Matter, SOx, and NOx. In: Kulshrestha, U., Saxena, P. (eds) Plant Responses to Air Pollution. Springer, Singapore. pp 2429–2458. https://doi.org/10.1007/978-981-10-1201-3_2

Shi L et al (2023) Incident dementia and long-term exposure to constituents of fine particle air pollution: a national cohort study in the United States. Proc Natl Acad Sci 120:e2211282119. https://doi.org/10.1073/pnas.2211282119

Shivanna KR (2022) Climate change and its impact on biodiversity and human welfare. Proc Indian Natl Sci Acad 88:160–171. https://doi.org/10.1007/s43538-022-00073-6
Shrivastava NG (2020) Bioremediation of Solid Waste Management. In: Baskar C et al (eds) Handbook of solid waste management: sustainability through circular economy. Springer, Singapore, pp 1–51. https://doi.org/10.1007/978-981-15-7525-9_48-1
Sicard P et al (2023) Trends in urban air pollution over the last two decades: a global perspective. Sci Total Environ 858:160064. https://doi.org/10.1016/j.scitotenv.2022.160064
Silvério DV et al (2015) Agricultural expansion dominates climate changes in southeastern Amazonia: the overlooked non-GHG forcing. Environ Res Lett 10:104015. https://doi.org/10.1088/1748-9326/10/10/104015
Singh V (2024) Textbook of environment and ecology. Springer, Singapore, pp 217–224
Singh AK, Bhargawa A (2019) Atmospheric burden of ozone depleting substances (ODSs) and forecasting ozone layer recovery. Atmos Pollut Res 10:802–807. https://doi.org/10.1016/j.apr.2018.12.008
Singh BP et al (2023) Substantial changes in selected volatile organic compounds (VOCs) and associations with health risk assessments in industrial areas during the COVID-19 pandemic. Toxics 11:165
Singh AP et al (2024) Chapter 6 - Pharmaceutical compounds: a recent threat to agricultural soils (sources, transport, and negative impacts on the crop's quality and other adjacent ecosystems). In Kumar P et al (eds) Bioremediation of emerging contaminants from soils. Elsevier, Amsterdam, pp 107–134. https://doi.org/10.1016/B978-0-443-13993-2.00006-2
Smith VH, Tilman GD, Nekola JC (1999) Eutrophication: impacts of excess nutrient inputs on freshwater, marine, and terrestrial ecosystems. Environ Pollut 100:179–196. https://doi.org/10.1016/S0269-7491(99)00091-3
Song Y, Yue Q, Zhu J, Zhang M (2023) Air pollution, human capital, and urban innovation in China. Environ Sci Pollut Res 30:38031–38051. https://doi.org/10.1007/s11356-022-25002-4
Sovacool BK, Baum CM, Low S, Fritz L (2024) The sociotechnical dynamics of blue carbon management: testing typologies of ideographs, innovation, and co-impacts for marine carbon removal. Environ Sci Pol 155:103730. https://doi.org/10.1016/j.envsci.2024.103730
Srivastav AL (2020) Chapter 6 - Chemical fertilizers and pesticides: role in groundwater contamination. In: Prasad MNV (ed) Agrochemicals detection, treatment and remediation. Butterworth-Heinemann, Oxford, pp 143–159. https://doi.org/10.1016/B978-0-08-103017-2.00006-4
Srivastava K (2021) Used cooking oil' in india diverted to food chain; public health and biodiesel sectors suffer. https://earthjournalism.net/stories/used-cooking-oil-in-india-diverted-to-food-chain-public-health-and-biodiesel-sectors-suffer
Stone EL, Jones G, Harris S (2012) Conserving energy at a cost to biodiversity? Impacts of LED lighting on bats. Glob Change Biol 18:2458–2465. https://doi.org/10.1111/j.1365-2486.2012.02705.x
Sudalai S et al (2024) A review on oilcake biomass waste into biofuels: current conversion techniques, sustainable applications, and challenges: waste to energy approach (WtE). Energy Convers Manag 314:118724. https://doi.org/10.1016/j.enconman.2024.118724
Tatrari G et al (2021) Solid waste-derived carbon nanomaterials for supercapacitor applications: a recent overview. Mater Adv 2:1454–1484. https://doi.org/10.1039/D0MA00871K
Tendero M, Bazart C (2024) "I can't get no satisfaction": soil contaminated brownfield redevelopment in France. Cities 145:104719. https://doi.org/10.1016/j.cities.2023.104719
Theobald DM et al (2020) Earth transformed: detailed mapping of global human modification from 1990 to 2017. Earth Syst Sci Data 12:1953–1972. https://doi.org/10.5194/essd-12-1953-2020
Thorpe D (2021) The first city running its vehicles on biodiesel from waste cooking oil. World Economic Forum. https://www.weforum.org/stories/2015/02/the-first-city-running-its--vehicles-on-biodiesel-waste-cooking-oil/#:~:text=Dubai%20is%20to%20become%20the,use%20in%20its%20municipal%20vehicles

Tickner J, Jacobs MM, Mack NB (2019) Alternatives assessment and informed substitution: a global landscape assessment of drivers, methods, policies and needs. Sustain Chem Pharm 13:100161. https://doi.org/10.1016/j.scp.2019.100161

Timilsina G (2018) Carbon pricing for climate change mitigation:1–13. https://doi.org/10.1079/pavsnnr201813008

Tiwari S, Chen B, Singh S, Singh AK, Srivastava AK (2021) chapter 5 - Impacts of black carbon on environment and health In: Hussain CM, Shukla SK, Joshi GM (eds) Functionalized nanomaterials based devices for environmental applications. Elsevier, San Diego, 107–125. https://doi.org/10.1016/B978-0-12-822245-4.00007-6

Traina G, Morselli L, Adorno GP (2007) Electrokinetic remediation of bottom ash from municipal solid waste incinerator. Electrochim Acta 52:3380–3385. https://doi.org/10.1016/j.electacta.2006.05.067

Tucker WG (2000) An overview of PM2.5 sources and control strategies. Fuel Process Technol 65–66:379–392. https://doi.org/10.1016/S0378-3820(99)00105-8

Turner K, Kim YJS (2024) Problems of the US recycling programs: what experienced recycling program managers tell. Sustainability 16:3539

Twerefou DK, Tutu K, Owusu-Afriyie J, Adjei-Mantey KJ (2015) Attitudes of local people to mining policies and interventions. International Growth Centre, pp 1–92. https://www.theigc.org/sites/default/files/2015/08/Twerefou-Tutu-2015-Policy-brief.pdf

Ullah A, Zeb A, Shah AA, Bavorova M (2024) From education to rehabilitation: empowering farming communities through extension services for landscape restoration. Environ Dev Sustain 26:11681–11701. https://doi.org/10.1007/s10668-023-03452-1

Ummik M-L, Järvik O, Reinik J, Konist A (2024) Ecotoxicity assessment of ashes from calcium-rich fuel combustion: contrasting results and regulatory implications. Environ Sci Pollut Res 31:48523–48533. https://doi.org/10.1007/s11356-024-34387-3

University, N. T (2024) Study estimates that between 1980 and 2020, 135 million premature deaths could be linked to fine particulate matter pollution. https://www.sciencedaily.com/releases/2024/06/240610140259.htm

USEPA (2024a) National overview: facts and figures on materials, wastes and recycling. https://www.epa.gov/facts-and-figures-about-materials-waste-and-recycling/national-overview-facts-and-figures-materials

USEPA (2024b) Guide to the facts and figures report about materials, waste and recycling. https://www.epa.gov/facts-and-figures-about-materials-waste-and-recycling/guide-facts-and-figures-report-about

Vaghar Seyedin SM, Zeidi A, Chamanehpour E, Nasri MHF, Vargas-Bello-Pérez E (2022) Methane emission: strategies to reduce global warming in relation to animal husbandry units with emphasis on ruminants. Sustainability 14:16897

Vakulchuk R (2024) The green skills gap: educational reform in favour of renewable energy is now urgent. https://www.weforum.org/agenda/2024/04/green-skills-gap-educational-reform-renewable-energy/

Van Meter RJ, Swan CM, Snodgrass JW (2011) Salinization alters ecosystem structure in urban stormwater detention ponds. Urban Ecosyst 14:723–736. https://doi.org/10.1007/s11252-011-0180-9

Venkatraman K, Bhaskar V, Kesarkar AP (2021) A systematic approach to comprehend the role of atmospheric black carbon in different environmental segments. Aerosol Sci Eng 5:253–274. https://doi.org/10.1007/s41810-021-00100-x

Vinayak A, Mudgal G, Singh GJPCB, Biology, M (2020) Waste to energy: sustainable approach for renewable energy. Plant Cell Biotechnol Mol Biol 21:292–302

Wade J, Farrauto RJ (2012) 12 - Controlling emissions of pollutants in urban areas. In: Zeman F (ed) Metropolitan sustainability. Woodhead Publishing, Cambridge, pp260–291. https://doi.org/10.1533/9780857096463.3.260

Waite R, Searchinger T, Ranganathan J (2019) 6 pressing questions about beef and climate change, answered. World Resources Institute. updated March 20022 issue. https://www.wri.org/insights/6-pressing-questions-about-beef-and-climate-change-answered
Walter Anthony KM, Anthony P, Grosse G, Chanton J (2012) Geologic methane seeps along boundaries of Arctic permafrost thaw and melting glaciers. Nat Geosci 5:419–426. https://doi.org/10.1038/ngeo1480
Wang Z-H, Li S-X (2019) Chapter Three - Nitrate N loss by leaching and surface runoff in agricultural land: A global issue (a review). In Sparks DL (ed) Advances in agronomy, vol 156. Academic Press, San Diego, pp 159–217. https://doi.org/10.1016/bs.agron.2019.01.007
Wang Y, Woodward RT, Liu J-Y (2022) The impact of exogenous pollution on green innovation. Environ Resour Econ 81:1–24. https://doi.org/10.1007/s10640-021-00614-5
Wang TH et al (2023) PM2.5 promotes lung cancer progression through activation of the AhR-TMPRSS2-IL18 pathway. EMBO Mol Med 15:e17014. https://doi.org/10.15252/emmm.202217014
Wania F, McLachlan MS (2024) The Stockholm Convention at a crossroads: questionable nominations and inadequate compliance threaten its acceptance and utility. Environ Sci Technol. https://doi.org/10.1021/acs.est.4c06775
Wasserman A (2024) Food waste is a huge problem. Here's how new innovations can help prevent it. World Economic Forum May 2024. https://www.weforum.org/stories/2024/05/food-waste-is-a-huge-problem-but-new-innovations-can-help-prevent-it/#:~:text=A%20fresh%20take%20on%20food%20preservation&text=New%20innovations%20by%20companies%2C%20such,also%20preserve%20more%20nutritional%20value
Weber M et al (2022) Global total ozone recovery trends attributed to ozone-depleting substance (ODS) changes derived from five merged ozone datasets. Atmos Chem Phys 22:6843–6859. https://doi.org/10.5194/acp-22-6843-2022
WEF (2019) Are Halloween pumpkins a problem for the planet? World Economic Forum. October, 2019. https://www.weforum.org/stories/2019/10/halloween-pumpkins-food-waste-energy/
WHO (2018) Health-care waste. https://www.who.int/news-room/fact-sheets/detail/health-care-waste
WHO (2024a) Air pollution. https://www.who.int/health-topics/air-pollution#tab=tab_1
WHO (2024b) Diarrhoeal disease. https://www.who.int/news-room/fact-sheets/detail/diarrhoeal-disease#:~:text=Key%20facts,aged%205%20to%209%20years
Wichai-utcha N, Chavalparit O (2019) 3Rs Policy and plastic waste management in Thailand. J Mater Cycles Waste Manag 21:10–22. https://doi.org/10.1007/s10163-018-0781-y
Williams LJ, Chen L, Zosky GR (2017) The respiratory health effects of geogenic (earth derived) PM10. Inhal Toxicol 29:342–355. https://doi.org/10.1080/08958378.2017.1367054
Wilson J (2022) Understanding the impact of state-level financial incentives on the deployment of renewable energy at colleges and universities. Electr J 35:107060. https://doi.org/10.1016/j.tej.2021.107060
Wilson B, Williams N, Liss B, Wilson BJERBR FL, USA (2013) A comparative assessment of commercial technologies for conversion of solid waste to energy. https://sustainablelibrary.org/wp-content/uploads/2023/08/00015-Comparative_WTE-Technologies-1.pdf
Winiwarter W, Klimont Z (2011) The role of N-gases (N2O, NOx, NH3) in cost-effective strategies to reduce greenhouse gas emissions and air pollution in Europe. Curr Opin Environ Sustain 3:438–445. https://doi.org/10.1016/j.cosust.2011.08.003
Wood J (2019) These researchers are turning plastic bottles into prosthetic limbs. https://www.weforum.org/agenda/2019/10/plastic-bottles-waste-prosthetic-limbs/
Wyer KE, Kelleghan DB, Blanes-Vidal V, Schauberger G, Curran TP (2022) Ammonia emissions from agriculture and their contribution to fine particulate matter: a review of implications for human health. J Environ Manag 323:116285. https://doi.org/10.1016/j.jenvman.2022.116285
Xia Y et al (2020) Recent advances in control technologies for non-point source pollution with nitrogen and phosphorous from agricultural runoff: current practices and future prospects. Appl Biol Chem 63:8. https://doi.org/10.1186/s13765-020-0493-6

Xiao Y, Han D, Currell M, Song X, Zhang Y (2023) Review of Endocrine Disrupting Compounds (EDCs) in China's water environments: implications for environmental fate, transport and health risks. Water Res 245:120645. https://doi.org/10.1016/j.watres.2023.120645
Xu Y, Zaelke D, Velders GJM, Ramanathan V (2013) The role of HFCs in mitigating 21st century climate change. Atmos Chem Phys 13:6083–6089. https://doi.org/10.5194/acp-13-6083-2013
Xu R et al (2023a) Global population exposure to landscape fire air pollution from 2000 to 2019. Nature 621:521–529. https://doi.org/10.1038/s41586-023-06398-6
Xu F et al (2023b) The challenge of population aging for mitigating deaths from PM2.5 air pollution in China. Nat Commun 14:5222. https://doi.org/10.1038/s41467-023-40908-4
Yan Y, Zhang X, Zhang J, Li K (2020) Emissions trading system (ETS) implementation and its collaborative governance effects on air pollution: the China story. Energy Policy 138:111282. https://doi.org/10.1016/j.enpol.2020.111282
Yan H, Qamruzzaman M, Kor S (2023) Nexus between green investment, fiscal policy, environmental tax, energy price, natural resources, and clean energy—a step towards sustainable development by fostering clean energy inclusion. Sustainability 15:13591
Yang J, Koondhar MA (2024) Impact of mineral based social equity on unearthing inequities: a longitudinal analysis of environmental and social equity of mineral exploitation in the global south. J Environ Manag 365:121492. https://doi.org/10.1016/j.jenvman.2024.121492
Yang L et al (2021) Exposure to atmospheric particulate matter-bound polycyclic aromatic hydrocarbons and their health effects: a review. Int J Environ Res Public Health 18:2177
Yang L et al (2023) The PM2.5 concentration reduction improves survival rate of lung cancer in Beijing. Sci Total Environ 858:159857. https://doi.org/10.1016/j.scitotenv.2022.159857
Yoon JY, Lee J-D, Joo SW, Kang DR (2016) Indoor radon exposure and lung cancer: a review of ecological studies. Ann Occup Environ Med 28:15. https://doi.org/10.1186/s40557-016-0098-z
Yuan J et al (2020) Ecology of industrial pollution in China. Ecosyst Health Sustain 6:1779010. https://doi.org/10.1080/20964129.2020.1779010
Yuan X et al (2023) Maternal exposure to PM2.5 and the risk of congenital heart defects in 1.4 million births: a nationwide surveillance-based study. Circulation 147:565–574. https://doi.org/10.1161/CIRCULATIONAHA.122.061245
Zanobetti A, Franklin M, Koutrakis P, Schwartz J (2009) Fine particulate air pollution and its components in association with cause-specific emergency admissions. Environ Health 8:58. https://doi.org/10.1186/1476-069x-8-58
Zeng Y, Cao Y, Qiao X, Seyler BC, Tang Y (2019) Air pollution reduction in China: recent success but great challenge for the future. Sci Total Environ 663:329–337. https://doi.org/10.1016/j.scitotenv.2019.01.262
Zhang W, Wu CW (2014) Dyeing of multiple types of fabrics with a single reactive azo disperse dye. Chem Pap 68:330–335. https://doi.org/10.2478/s11696-013-0444-3
Zhang Z, Yu D (2023) Unintended consequences of heavy air pollution control: efficiency losses, resource misallocation, and firm innovation in China. Environ Sci Pollut Res 30:20203–20222. https://doi.org/10.1007/s11356-022-23623-3
Zhang S-Y, Wang Q-F, Wan R, Xie S-G (2011) Changes in bacterial community of anthracene bioremediation in municipal solid waste composting soil. J Zhejiang Univ Sci B 12:760–768. https://doi.org/10.1631/jzus.B1000440
Zhao C, Dong Y, Feng Y, Li Y, Dong Y (2019) Thermal desorption for remediation of contaminated soil: a review. Chemosphere 221:841–855. https://doi.org/10.1016/j.chemosphere.2019.01.079
Zheng J, Chang M, Xie H, Guo P (2016) Exploring the spatiotemporal characteristics and control strategies for volatile organic compound emissions in Jiangsu, China. J Clean Prod 127:249–261. https://doi.org/10.1016/j.jclepro.2016.03.076
Zivdar Z, Heidarzadeh N, Asadollahfardi G (2019) Remediation of diesel-contaminated soil by low-temperature thermal desorption. Int J Environ Sci Technol 16:6113–6124. https://doi.org/10.1007/s13762-018-2020-4

Coastal and Marine Management

Yoshihiro Hamaguchi

Abstract From food supply, marine tourism, resource extraction, and blue carbon, the ocean provides various benefits to the social economy. Hence, the blue economy, which consists of economic actors, policy instruments, and regulatory systems related to the ocean, is attracting increasing efforts to conceptualize it. However, its definition is not univocal. Further, qualitative analysis is needed to elicit useful policy recommendations. Through a literature analysis, this chapter analyzes the blue economy framework, considering the impact of wastewater from the fisheries, fish processing, and tourism industries on fishery resources. Further, global warming due to the blue carbon degradation may have undesirable effects on the blue economy itself. Meanwhile, transferable individual quota (ITQ) policies indirectly negatively impact fishery resources through contaminated water. Furthermore, rent-seeking for IQTs may also degrade the blue carbon. Hence, an environmental policy dimension is added to fisheries policy, thereby linking the environmental Kuznets curve and the pollution haven hypothesis. This new concept implies that international environmental conflicts reduce blue carbon, owing to inappropriate resource management. Theoretical and empirical analyses supporting these interdependencies are expected to assist in identifying fisheries management and climate change mitigation measures to mitigate this conflict.

Keywords Blue economy · Blue carbon · Blue tourism · Marine environmental Kuznets curve · Water pollution haven hypothesis

The author acknowledges the financial support from the Early-Career Scientists from the Japan Society for the Promotion of Science no. 22K13409 (JSPS KAKENHI grant no. JP22K13409). The author is responsible for all errors in this paper.

Y. Hamaguchi (✉)
Department of Economics, Faculty of Economics, Hannan University,
Matsubara-shi, Osaka, Japan
e-mail: hamaguchi@hannan-u.ac.jp

A. A. H. Abdel Latef et al. (eds.), *Sustainable Remediation for Pollution and Climate Resilience*, https://doi.org/10.1007/978-981-96-5674-5_8

1 Introduction

The ocean is attracting global attention as a space for solving the challenges faced by humanity. Amid concerns about food crises caused by global population growth, oceans provide an abundance of marine products. Moreover, as global warming becomes more severe, seaweed's absorption of carbon dioxide is expected to contribute to decarbonization as blue carbon. The ocean also provides valuable leisure opportunities for the tourism industry, which accounts for nearly 10% of the world's gross domestic product (GDP). Furthermore, the practical application of seabed mining is underway because rare metals—which are raw materials for semiconductors indispensable for the development of information and artificial intelligence technologies—lie beneath the seafloor. Although the ocean provides various benefits to humankind, concerns about resource degradation and depletion due to overfishing and polluted water discharge make ocean conservation a major issue. The ocean comprises various economic entities, including fisheries, fish processing, tourism, ship transportation, consumers, and governments. This complex mix of interests, combined with those among nations, makes implementing measures to conserve marine resources difficult. To develop effective solutions, understanding the ocean-based socioeconomic interrelationship is important. This relationship is recognized as the blue economy and has attracted extensive research recently (Martínez-Vázquez et al. 2021).

Klinger et al. (2018) analyzed the interrelationships between basic (fisheries, transportation, and offshore power generation) and emerging sectors (aquaculture, tourism, and seabed mining), which are considered the main economic actors comprising the blue economy. With the development of these sectors, the global value-added generated by the blue economy is projected to reach US$1.5 trillion by 2030 (Rayner et al. 2019). By examining various case analyses, Wenhai et al. (2019) presented a successful case study of the blue economy from an international perspective. At the 2012 UN Conference on Sustainable Development (Rio+20), the blue economy was recognized as the new arena of global environmental governance. However, even if we view the ocean only from the perspective of natural resources, trade in goods, marine island states, and small-scale fisheries, the concept is not univocal (Silver et al. 2015). This is also true for the geographic dimension of the blue economy, which Garland et al. (2019) attempted to specify. Besides clarifying the concept of the blue economy, identifying its mechanisms is important to promote the appropriate use of marine resources while preserving them through efficient environmental policies.

While case studies and empirical surveys are useful for analyzing the concept of the blue economy, qualitative analyses using theoretical models are necessary to elucidate its mechanisms systematically. To proceed with this analysis, it is necessary to analyze the interrelationships among the various components of the blue economy after limiting the scope of the analysis to a limited number of components. In this chapter, we limit the sectors of the blue economy to fishing, fisheries, and tourism and analyze the socioeconomic impacts of pollution emitted by these industries through a literature review. The interrelationships presented in this chapter will be useful for unraveling through theoretical analyses and the effective environmental policies to conserve marine resources under a blue economy.

2 Blue Economy Framework

2.1 Economic Actors and Sectors

Referring to Hamaguchi's (2024) economic model, we describe the blue economy's assumed economic agents and industrial sectors in this chapter. Households provide labor to each sector and use their wage income to purchase goods and services produced by each sector. The economy includes fishing, fisheries, tourism, and research and development (R&D) sectors. The fishing industry sells the catch obtained from labor input to the fisheries industry. The fisheries industry produces processed seafood products with the inputs of labor and fish catches. The tourism sector produces tourism goods and services using only labor input. Fisheries and tourism receive positive externalities from marine resources. Improvements in the environmental quality increase production through higher productivity in these sectors. However, each of these sectors emits pollution, which reduces household welfare and productivity through environmental and resource degradation. Islam et al. (2004) comprehensively surveyed the impacts of pollution emissions from various economic agents on coastal and marine ecosystems.

In Matulich et al. (1996), Hackett et al. (2005), and McEvoy et al. (2009), the fisheries industry earns monopolistic profits in imperfectly competitive markets. Hamaguchi (2024) linked the development of the fisheries industry to economic growth by assuming that the fisheries industry has patents on the processed seafood products it manufactures. To enter the seafood market, a new entrepreneur requires a blueprint for the newly processed seafood product. Entrepreneurs finance their R&D activities by selling shares of their company to households. By investing in labor, R&D firms develop blueprints for new seafood products. The entrepreneur enters the market by applying for a patent based on a blueprint purchased with the R&D funds. Monopoly profits earned by entrepreneurs are distributed to households as dividends, which can be used for spending on goods and services. The production of processed seafood products has also increased because of their expanded variety. The analytical framework for economic growth through R&D activities relies on the R&D-based growth model of Grossman and Helpman (1991) and Romer (1990).

2.2 Pollution from Fisheries, Fish Processing Industry, and Tourism

Overfishing degrades global marine resources (Pauly and Zeller 2016). Additionally, plastics from fishing gear and heavy oil spills from hulls cause environmental pollution (Garza-Gil et al. 2006; Dowarah and Devipriya 2019). The aquaculture industry is of particular concern. As non-native species, fish that escape from aquaculture farms negatively affect the surrounding ecosystems and reduce the productivity of fisheries and aquaculture (Belias et al. 2003; Dempster et al. 2009; Lorenzen et al. 2012; Liu et al. 2014). Analyses of the interrelationships between aquaculture and fisheries include Hoagland et al. (2003), Mikkelsen (2007), and Foley et al. (2012). In addition, the large amount of excess food from fish farms and feces from farmed fish degrades the surrounding water quality (Watson et al. 2016). Bergland et al. (2020) characterized this conceptual relationship through a model analysis and argued for sustainable fishery management. Catches from fisheries and aquaculture farms are marketed as fresh fish and processed seafood products. Polluted water containing organic lipids and solids generated while producing processed seafood products degrades the surrounding waters' quality and affects fishery resources (Islam and Tanaka 2004; Chowdhury et al. 2010). In a blue economy, the extraction and processing of raw materials also cause pollution.

If marine resorts are located in the vicinity of fishing and fish processing industries, such pollution can negatively affect tourism. Initially, few scholarly discussions considered this. For instance, in an overview of marine tourism by Orams (2002), there was no separate chapter on marine pollution from tourism. However, in the 2000s, discussions gradually emerged on the relationship between marine pollution and marine tourism (Selvaduray et al. 2023). Scholars have also focused on the impact of marine tourism on coastal pollution and resources (Kocasoy 1995; Burak et al. 2004; Lin 2022). According to Qiang et al. (2020), marine pollution in coastal areas reduces the number of days tourists spend, thus hampering the development of tourism in the region owing to reduced tourism income. However, when the increase in the number of visitors and number of days spent in the area exceeds the tourism-carrying capacity of the area, it causes environmental degradation of the surrounding environment via tourism pollution. Using a model-based analysis, Metilelu et al. (2022) argued that the number of tourists should be kept within the carrying capacity for sustainable tourism. The number of tourists does not just cause marine pollution from tourism; as noted by Gola et al. (2021), the negative impacts of microplastics on marine resources and the marine environment are becoming increasingly severe. For instance, in Indian beaches and surrounding marine areas, microplastics discharged from fishing and tourism can negatively impact fishing and tourism resources (Dowarah and Devipriya 2019). Thus, the impact and role of tourism on the economy and environment are significant in the blue economy.

To understand how pollution emitted by the fishing, fish processing, and tourism industries affects them and the environment through their interaction, a theoretical analysis using a mathematical model should be conducted, followed by verification through empirical analysis. Hamaguchi (2024) had attempted such an analysis. In the logistic function of the present model, the growth rate of fishery resources increases in accordance with the amount of resources at a resource level below the carrying capacity of the environment, but the growth rate decreases at a resource level above the carrying capacity. At a resource level below the carrying capacity of the environment, the decrease in fishery resources leads to the depletion of resources through the decrease in the growth rate of resources. On the other hand, at a resource level above the carrying capacity, the increase in resources leads to the expansion of resources, although it also leads to a decrease in the growth rate. Assuming this logistic function for fishery resources and resource decline due to fish catch (fishing effect), resource deterioration due to wastewater from the fish processing industry has been analyzed by Islam and Tanaka (2004) and Chowdhury et al. (2010), resource degradation due to tourism in Cerina (2007), and the habitat effect in Foley et al. (2012). The fisheries' catch provides the raw materials for the fish processing industry, but its abundance declines with increased fishing (the fishing effect). Wastewater from the seafood processing industry reduces fishery resources due to water quality degradation, which is exacerbated by marine tourism (water pollution effect). The reduction in fish numbers owing to water pollution makes reproductive activities more difficult, leading to the deterioration of fishery resources (habitat effect). In this setting, whether an increase in the number of fish caught leads to stock recovery or stock depletion depends on the relative magnitudes of the three effects. Therefore, further verification through empirical analysis is required to determine which case is more likely.

2.3 Global Warming and the Blue Economy

In addition to the negative impacts of various pollutants on fisheries and tourism resources, climate change may also significantly impact the blue economy. Due to the impacts such as increased seawater temperatures, climate change will have significant impacts on marine ecosystems. It may not only lead to declines in populations and fish species but also reduced farmed fish production due to deteriorating water quality and degradation of coastal tourism facilities and resources due to storms and hurricanes (Weatherdon et al. 2016). According to Arabadzhyan et al. (2021), although certain caveats are necessary to the results of previous studies, which show markedly heterogeneous impacts, climate change significantly affects both the demand and supply of tourism. Specifically, if the purpose of tourism is to observe nature, such as whale watching, a decrease in marine life will reduce the attractiveness of that tourist destination. In island countries such as Sri Lanka and Togo, where coral reef-related tourism accounts for a large share of the GDP, its deterioration will lead to a decline in the number of tourists. Rising sea levels due to climate change will also reduce the number of tourists because it will reduce the number of available beaches at tourist destinations. Mitigating climate change is an important policy issue in the pursuit of a sustainable blue economy.

Moreover, blue carbon is gaining attention as a climate change mitigation measure. Blue carbon is carbon dioxide (CO_2), absorbed and stored in mangrove forests, seagrass beds, wetland biomass, and soils. Various estimates of its absorption capacity are being vigorously undertaken, such as Mcleod et al. (2011). Bertram et al. (2021) estimated the global blue carbon wealth; the authors found that marine ecosystems generate a mean ± s.e.m. of US\$190.67 ± 30 billion per year of blue carbon, with the majority from Australia, China, and Cuba. Pendleton et al. (2012) estimated that the loss of blue carbon due to marine ecosystem degradation is equivalent to 3–19% of global forest loss, or US\$6–42 billion annually. Blue economies will benefit greatly from blue carbon adaptation measures, as they are the most adversely affected by climate change. Therefore, understanding the relationship between the blue economy and blue carbon and its mechanisms will contribute to effective blue carbon management.

3 Marine Environmental Policy and Emerging Environmental Issues

3.1 Fisheries Policy and Rent-Seeking

Traditionally, total allowable catch (TAC) policies have been introduced worldwide to ensure the sustainable conservation of fishery resources. More recently, transferable individual quota (ITQ) policies have been introduced by applying market mechanisms to promote more efficient resource management (Newell et al. 2005). However, tradable quota allocation is a potential rent. Some have expressed concerns that rent-seeking to acquire quotas may reduce the conservation effectiveness of fishery resources (Grainger and Parker 2013). Therefore, scholars have investigated the mechanisms (Edwards 2001; Foley et al. 2015) and economic scale of rent-seeking in fisheries (Flaaten et al. 2017; Gunnlaugsson et al. 2020). Hilborn et al. (2005) noted that whether fishery management systems lead to the efficient conservation of fishery resources depends on the maturity of private property rights systems. Bertram et al. (2021) found that Australian coastal countries with abundant coral reefs, as well as Southeast Asian and Caribbean coastal countries, contribute to blue carbon production. However, many of these regions are developing countries that do not have well-developed private property rights systems. Rent-seeking for fishing quotas encourages the extraction of fishery resources (Sutinen and Soboil 2003), and this quota system itself significantly affects marine ecosystems (Branch 2009). Hence, rent-seeking in ITQ policies can undermine the efficiency of blue carbon. Even if a country implements resource management through ITQ policies, overfishing due to rent-seeking in other countries may cause international conflicts over resource management because international fishery resources will deteriorate (Hamaguchi 2022). Thus, indirect blue carbon inhibition through rent-seeking may lead to conflict over climate-change mitigation measures.

3.2 Fisheries and Marine Policy as Environmental Policy

When rent-seeking for ITQs impedes blue carbon efficiency, fishery policies that regulate rent-seeking will indirectly reduce greenhouse gases. Considering this blue economy framework, the fisheries policy mimics an environmental policy that aims to reduce pollutants and greenhouse gases. This point is analyzed by Hamaguchi (2024), who showed the growth effects of fisheries policy: When ITQ prices rise due to a reduction in the ITQ, fisheries prices rise. Faced with higher raw material prices, fishery processing firms pass on the price to their processed fishery products. This price shift increases the firms' profits. Seeing this increase in profit, the entrepreneur attempts to enter the market using the R&D funds obtained from selling their company's shares. The resulting expanded variety of new seafood products will lead to economic growth. A blue economy framework is important for clarifying these combined effects. Hamaguchi (2024) interpreted blue growth simply as resource recovery and economic growth through ITQ policies. However, this win-win relationship may not hold when certain components of the blue economy are considered. As Klinger et al. (2018) noted, the definition of blue growth is not univocal, and the debate continues. If we can identify the characteristics of fishery policies that lead to blue growth, a pathway to blue growth could be proposed.

3.3 Marine Environmental Kuznets Curve and Water Pollution Haven Hypothesis

Since Grossman and Krueger (1995) first demonstrated an inverse U-shaped relationship between economic growth and the natural environment as an environmental Kuznets curve, numerous empirical and theoretical analyses have validated it (Stern 2017; Shahbaz and Sinha 2019). However, these studies have focused on air pollution and greenhouse gases, while fishery resources and water pollution have not been considered. Considering the environmental Kuznets curve is important for conceptualizing and defining blue growth. Solarin et al. (2021) and Yilanci et al. (2022b) presented the concept of fish footprint by applying the idea of ecological footprint to fishery resources. Yilanci et al. (2022a) were the first to relate this fish footprint to the environmental Kuznets curve.

In the context of sustainable tourism, the tourism carbon Kuznets curve has been analyzed as the relationship between the carbon emissions of the tourism industry and economic growth (Papavasileiou and Tzouvanas 2021). Revisiting this hypothesis from a blue economy perspective, blue carbon may significantly affect this hypothesis. The deterioration of fishery resources associated with economic growth can boost carbon emissions associated with economic growth, owing to the lower absorption of blue carbon. However, if technological innovation improves fishery resources due to further growth, it will boost carbon dioxide emissions due to the recovery of blue carbon sinks. Thus, through the lens of the environmental Kuznets curve, marine tourism can play a larger role in the blue economy.

Hamaguchi (2022) pointed out that overfishing due to rent-seeking can lead to international conflicts over fishery resources. However, in the blue economy framework discussed in this chapter, polluted water from production activities leads to the deterioration of fishery resources. Then, fish processing enterprises may relocate to areas with better water quality and stable catches. Moreover, if the government seeks to attract fish processing companies and raises the catch quota to expand the catch, polluted water from relocated factories may degrade local fishery and tourism resources. The possibility that environmental regulations in developing countries, which are more lenient than those in developed countries, may increase pollutant emissions through company relocation continues to be studied as the pollution haven hypothesis (Dechezleprêtre and Sato 2017; Zeng and Zhao 2009; Zheng and Shi 2017). Hamaguchi (2024) proposed a water pollution haven hypothesis by considering contaminated water from fish processing companies and bilateral ITQ policies. Yilanci et al. (2023) used the Indonesian fish footprint to show the pollution halo effect and environmental Kuznets curve empirically. Although these relationships require further analysis, they suggest that fishery policies may create international firestorms concerning environmental pollution.

4 Conclusion

Through the concept of the blue economy, the ocean is increasingly attracting attention because of its potential to bring various benefits to the social economy. However, owing to the various interrelationships among the diverse actors involved in the blue economy, debates continue over its definition. In this chapter, to present a brief overview of the qualitative analysis of the blue economy, we have limited the economic actors to the fishing, fish processing, and tourism industries, and described the blue economy through the impact of polluted water from these industries on fishery resources. Referring to the literature supporting this interrelationship, we point out that new interpretation of the environmental Kuznets curve and pollution haven hypothesis may be generated. Blue carbon in the ocean is expected to serve as a promising measure for mitigating global warming. However, declining global blue carbon emissions due to inappropriate resource management could lead to international conflict. To link blue carbon to decarbonization and mitigate international conflicts over resource management and climate change, efforts to identify the underlying socioeconomic mechanisms should continue.

References

Arabadzhyan A, Figini P, García C, González MM, Lam-González YE, León CJ (2021) Climate change, coastal tourism, and impact chains–a literature review. Curr Issue Tour 24(16):2233–2268

Belias CV, Bikas VG, Dassenakis MJ, Scoullos MJ (2003) Environmental impacts of coastal aquaculture in eastern Mediterranean bays the case of Astakos Gulf, Greece. Environ Sci Pollut Res 10:287–295

Bergland H, Burlakov E, Pedersen PA, Wyller J (2020) Aquaculture, pollution and fishery – dynamics of marine industrial interactions. Ecol Complex 43:100853

Bertram C, Quaas M, Reusch TB, Vafeidis AT, Wolff C, Rickels W (2021) The blue carbon wealth of nations. Nat Clim Chang 11(8):704–709

Branch TA (2009) How do individual transferable quotas affect marine ecosystems? Fish Fish 10(1):39–57

Burak S, Dogan E, Gazioglu C (2004) Impact of urbanization and tourism on coastal environment. Ocean Coast Manag 47(9–10):515–527
Cerina F (2007) Tourism specialization and environmental sustainability in a dynamic economy. Tour Econ 13(4):553–582
Chowdhury P, Viraraghavan T, Srinivasan A (2010) Biological treatment processes for fish processing wastewater–a review. Bioresour Technol 101(2):439–449
Dechezleprêtre A, Sato M (2017) The impacts of environmental regulations on competitiveness. Rev Environ Econ Policy 11(2):183–206
Dempster T, Uglem I, Sanchez-Jerez P, Fernandez-Jover D, Bayle-Sempere J, Nilsen R, Bjørn PA (2009) Coastal salmon farms attract large and persistent aggregations of wild fish: an ecosystem effect. Mar Ecol Prog Ser 385:1–14
Dowarah K, Devipriya SP (2019) Microplastic prevalence in the beaches of Puducherry, India and its correlation with fishing and tourism/recreational activities. Mar Pollut Bull 148:123–133
Edwards SE (2001) Rent-seeking and property rights formation in the US Atlantic sea scallop fishery. Mar Resour Econ 16(4):263–275
Flaaten O, Heen K, Matthíasson T (2017) Profit and resource rent in fisheries. Mar Resour Econ 32(3):311–328
Foley NS, Armstrong CW, Kahui V, Mikkelsen E, Reithe S (2012) A review of bioeconomic modelling of habitat-fisheries interactions. Int J Ecol 2012:1–11
Foley P, Mather C, Neis B (2015) Governing enclosure for coastal communities: social embeddedness in a Canadian shrimp fishery. Mar Policy 61:390–400
Garland M, Axon S, Graziano M, Morrissey J, Heidkamp CP (2019) The blue economy: identifying geographic concepts and sensitivities. Geogr Compass 13(7):e12445
Garza-Gil MD, Prada-Blanco A, Vázquez-Rodríguez MX (2006) Estimating the short-term economic damages from the Prestige oil spill in the Galician fisheries and tourism. Ecol Econ 58(4):842–849
Gola D, Tyagi PK, Arya A, Chauhan N, Agarwal M, Singh SK, Gola S (2021) The impact of microplastics on marine environment: a review. Environ Nanotechnol Monit Manag 16:100552
Grainger CA, Parker DP (2013) The political economy of fishery reform. Annu Rev Resour Econ 5(1):369–386
Grossman G, Helpman E (1991) Innovation and growth in the global economy. MIT Press, Cambridge
Grossman GM, Krueger AB (1995) Economic growth and the environment. Q J Econ 110(2):353–377
Gunnlaugsson SB, Saevaldsson H, Kristofersson DM, Agnarsson S (2020) Resource rent and its distribution in Iceland's fisheries. Mar Resour Econ 35(2):113–135
Hackett SC, Krachey MJ, Brown S, Hankin D (2005) Derby fisheries, individual quotas, and transition in the fish processing industry. Mar Resour Econ 20(1):47–60
Hamaguchi Y (2022) Welfare effect of rent-seeking activities under international management of fishery resources. Fish Res 246:106170
Hamaguchi Y (2024) A water pollution haven hypothesis in a dynamic agglomeration model with fisheries resource management. Environ Dev Sustain. https://doi.org/10.1007/s10668-024-04788-y
Hilborn R, Orensanz JM, Parma AM (2005) Institutions, incentives and the future of fisheries. Philos Trans R Soc B 360(1453):47–57
Hoagland P, Jin D, Kite-Powell H (2003) The optimal allocation of ocean space: aquaculture and wild-harvest fisheries. Mar Resour Econ 18(2):129–147
Islam MS, Tanaka M (2004) Impacts of pollution on coastal and marine ecosystems including coastal and marine fisheries and approach for management: a review and synthesis. Mar Pollut Bull 48(7–8):624–649
Islam MS, Khan S, Tanaka M (2004) Waste loading in shrimp and fish processing effluents: potential source of hazards to the coastal and nearshore environments. Mar Pollut Bull 49(1–2):103–110

Klinger DH, Eikeset AM, Davíðsdóttir B, Winter AM, Watson JR (2018) The mechanics of blue growth: management of oceanic natural resource use with multiple, interacting sectors. Mar Policy 87:356–362
Kocasoy G (1995) Effects of tourist population pressure on pollution of coastal seas. Environ Manag 19:75–79
Lin Z (2022) The impact of marine tourism behavior on the ecological effect of marine benthos Community in the South China Sea. J Coast Conserv 26(3):17
Liu Y, Olaussen JO, Skonhoft A (2014) Fishy fish? The economic impacts of escaped farmed fish. Aquac Econ Manag 18(3):273–302
Lorenzen K, Beveridge MC, Mangel M (2012) Cultured fish: integrative biology and management of domestication and interactions with wild fish. Biol Rev Camb Philos Soc 87(3):639–660
Martínez-Vázquez RM, Milán-García J, de Pablo Valenciano J (2021) Challenges of the Blue Economy: evidence and research trends. Environ Sci Eur 33(1):61
Matulich SC, Mittelhammer RC, Reberte C (1996) Toward a more complete model of individual transferable fishing quotas: implications of incorporating the processing sector. J Environ Econ Manag 31(1):112–128
McEvoy DM, Brandt S, Lavoie N, Anders S (2009) The effects of ITQ management on fishermen's welfare when the processing sector is imperfectly competitive. Land Econ 85(3):470–484
Mcleod E, Chmura GL, Bouillon S, Salm R, Björk M, Duarte CM et al (2011) A blueprint for blue carbon: toward an improved understanding of the role of vegetated coastal habitats in sequestering CO2. Front Ecol Environ 9(10):552–560
Metilelu OO, Adeniyi MO, Ekum MI (2022) Modelling the dynamic effect of environmental pollution on coastal tourism. Sci Afr 17:e01364
Mikkelsen E (2007) Aquaculture-fisheries interactions. Mar Resour Econ 22(3):287–303
Newell RG, Sanchirico JN, Kerr S (2005) Fishing quota markets. J Environ Econ Manag 49(3):437–462
Orams M (2002) Marine tourism: development, impacts and management. Routledge, London
Papavasileiou EF, Tzouvanas P (2021) Tourism carbon Kuznets-curve hypothesis: a systematic literature review and a paradigm shift to a corporation-performance perspective. J Travel Res 60(4):896–911
Pauly D, Zeller D (2016) Catch reconstructions reveal that global marine fisheries catches are higher than reported and declining. Nat Commun 7(1):10244
Pendleton L, Donato DC, Murray BC, Crooks S, Jenkins WA, Sifleet S et al (2012) Estimating global "blue carbon" emissions from conversion and degradation of vegetated coastal ecosystems. PLoS One 7(9):e43542
Qiang M, Shen M, Xie H (2020) Loss of tourism revenue induced by coastal environmental pollution: a length-of-stay perspective. J Sustain Tour 28(4):550–567
Rayner R, Jolly C, Gouldman C (2019) Ocean observing and the blue economy. Front Mar Sci 6:330
Romer P (1990) Endogenous technical change. J Polit Econ 98(5, Part 2):S71–S102
Selvaduray M, Bandara YM, Zain RM, Ramli A, Mohd Zain MZ (2023) Bibliometric analysis of maritime tourism research. Aust J Marit Ocean Aff 15(3):330–356
Shahbaz M, Sinha A (2019) Environmental Kuznets curve for CO2 emissions: a literature survey. J Econ Stud 46(1):106–168
Silver JJ, Gray NJ, Campbell LM, Fairbanks LW, Gruby RL (2015) Blue economy and competing discourses in international oceans governance. J Environ Dev 24(2):135–160
Solarin SA, Gil-Alana LA, Lafuente C (2021) Persistence and sustainability of fishing grounds footprint: evidence from 89 countries. Sci Total Environ 751:141594
Stern DI (2017) The environmental Kuznets curve after 25 years. J Bioecon 19(1):7–28
Sutinen JG, Soboil M (2003) The performance of fisheries management systems and the ecosystem challenge. In: Sinclair M, Valdimarsson G (eds) Responsible fisheries in the marine ecosystem. CABI Publisher, Wallingford, pp 291–309

Watson SCL, Paterson DM, Queirós AM, Rees AP, Stephens N, Widdicombe S, Beaumont NJ (2016) A conceptual framework for assessing the ecosystem service of waste remediation: in the marine environment. Ecosyst Serv 20:69–81

Weatherdon LV, Magnan AK, Rogers AD, Sumaila UR, Cheung WW (2016) Observed and projected impacts of climate change on marine fisheries, aquaculture, coastal tourism, and human health: an update. Front Mar Sci 3:48

Wenhai L, Cusack C, Baker M, Tao W, Mingbao C, Paige K et al (2019) Successful blue economy examples with an emphasis on international perspectives. Front Mar Sci 6:261

Yilanci V, Cutcu I, Cayir B (2022a) Is the environmental Kuznets curve related to the fishing footprint? Evidence from China. Fish Res 254:106392

Yilanci V, Pata UK, Cutcu I (2022b) Testing the persistence of shocks on ecological footprint and sub-accounts: evidence from the big ten emerging markets. Int J Environ Res 16(1):10

Yilanci V, Cutcu I, Cayir B, Saglam MS (2023) Pollution haven or pollution halo in the fishing footprint: evidence from Indonesia. Mar Pollut Bull 188:114626

Zeng D-Z, Zhao L (2009) Pollution havens and industrial agglomeration. J Environ Econ Manag 58(2):141–153

Zheng D, Shi M (2017) Multiple environmental policies and pollution haven hypothesis: evidence from China's polluting industries. J Clean Prod 141:295–304

Landscape-Safe Environment and Remediation

Meltem Erdem Kaya

Abstract Today, 53% of the world's population lives in urban areas, and according to the United Nations, this number is expected to rise to 68% by the year 2050. The increase in urban populations and the concentration of their activities signal a growing impact on the environment. One indicator of human impact on the environment is the amount of hazardous pollutants produced by human activities, mainly from industry and related forms of production. It is estimated that, globally, the number of potentially contaminated sites exceeds five million. Governments have taken action to mitigate the negative environmental impact as part of the Sustainable Development Goals. "*Make cities and human settlements inclusive, safe, resilient and sustainable*" is one of the goals (SDG 11) of the United Nations 2030 Agenda for Sustainable Development. In this context, transforming brownfield sites into clean and safe environments has become one of the most important issues for governments, involving two separate but related fields of brownfield regeneration: landscape design and remediation. Spatial planners and designers have an ethical responsibility in the regeneration process of contaminated environments, such as postindustrial landscapes. Poor design choices can spread contamination through multiple pathways and pose threats to human and environmental health, or create spaces where people may be easily exposed to contaminants.

Designers can employ precautionary design strategies and tactics that integrate appropriate remediation methods to provide safe environments and protect human and ecosystem health. As remediation methods have evolved, new approaches in landscape design have also developed. In addition to conventional remediation methods, there is a growing interest in incorporating process-based remediation techniques in the form of soft remediation options and/or phytoremediation or nature-based solutions to create a continuous green infrastructure system with economic, environmental, and social outcomes. In this way, isolated contaminated sites

M. Erdem Kaya
Faculty of Architecture, Department of Landscape Architecture,
Istanbul Technical University, İstanbul, Turkey
e-mail: erdemme@itu.edu.tr

A. A. H. Abdel Latef et al. (eds.), *Sustainable Remediation for Pollution and Climate Resilience*, https://doi.org/10.1007/978-981-96-5674-5_9

can become integral parts of the human-dominated environment and contribute to the development of green infrastructure systems with associated environmental functions such as mitigating air pollution, regulating water regimes, creating habitats for urban fauna, contributing to biological diversity, and providing aesthetically pleasing and attractive spaces.

This chapter uses case studies to explore the relationship between landscape design and remediation, demonstrating various ways to integrate remediation for a landscape-safe environment. "Remediated Landscapes" is offered as a conceptual framework to express the multipurpose structure of remediated sites with four integration modes: ecological, spatial, historical-cultural, and socioeconomic.

Keywords Landscape design · Remediation · Postindustrial landscape · Phytoremediation · Gentle remediation options · Integration

9.1 Introduction

Since the industrial era, especially beginning with the second industrial era, which covers the years between the mid-nineteenth century and the early twentieth century, the scale and scope of human impact on the environment have increased with advanced transport and communication networks, turning isolated urban centers into the planet's large urban systems. Large urban systems refer to both the continuing population growth in urban areas and the increasing interconnectedness of urban regions with massive and pervasive infrastructure systems. Today, 53% of the world's population lives in urban areas, and that number is expected to rise to 68% by 2050 (United Nations 2018, 2016). The increase in urban populations and the concentration of these populations' activities are leading to a growing environmental impact. One of the indicators of human impact on the environment is the amount of hazardous pollutants produced by human activities, mainly based on industry and related forms of production. The metabolism of urban systems, with their diverse urban activities and processes, produces numerous toxic substances in the air, water, and soil, including various organic compounds and heavy metals. Both human health and ecosystem health are seriously threatened by these dangerous pollutants (Etim 2012; Smith 2015). Today, these hazardous substances' rates exceed planetary boundaries, where crossing a boundary can cause irreversible environmental changes (Rockström et al. 2009).

In recent decades, governments have taken action and provided strategies to mitigate negative environmental impacts as part of the Sustainable Development Goals. "*Make cities and human settlements inclusive, safe, resilient and sustainable*" is one of the goals (SDG 11) of the United Nations 2030 Agenda for Sustainable Development (UN 2018, 2016). Providing healthy and sustainable urban living is one of the goals in focus, calling for novel strategies to address environmental pressures at the spatial scale. In this context, transforming brownfield sites into clean and safe environments has become one of the most critical issues for governments.

The relocation of old industrial sites within the city has left significant areas of contaminated land and infrastructure near community facilities (Kennen and Kirkwood 2015). These sites, which vary in size, are suitable for redevelopment and offer opportunities to be reused as a public asset with a new postindustrial identity. The term postindustrial is used here to refer to the spaces left behind by industry and includes contamination. Although the term refers to contaminated landscapes, including former factory buildings and associated infrastructure and spaces, postindustrial sites are now valued for their heritage values (Historic England n.d.; Kapp 2019; Agnes (2001); Loures 2008; Swensen and Granberg 2024) and have become the subject of large-scale transformation projects with cultural, historical, social, and economic outcomes as catalysts for urban and regional regeneration. Emscher Park Duisburg Nord, Ruhr, Germany, is a prime example of postindustrial transformation that preserves industrial heritage and employs adaptive reuse strategies to support diverse programs (sports, culture, tourism, historical significance) as a part of a large-scale regional green network. The Neuoerlikon district of Zurich is another case of postindustrial transformation at the neighborhood level. The former site of a weapons manufacturing complex, the district has been transformed into a new urban neighborhood supporting 5000 new residents and 12,000 new jobs in a new spatial setting (Matthews and Winter 2010). One of the core concepts of the redevelopment project was the creation of a series of neighborhood parks as a central figure of neighborhood identity.

Postindustrial sites, with their capacity to host mega-events such as the Olympic Games and Expos, can be catalysts for urban and regional regeneration at various levels. As mega-events require large tracts of land near cities, postindustrial sites at the periphery of the city offer strategic locations. The overall impact of transforming derelict land into an Olympic Park as a part of mega-event strategies on cities can be multifaceted, such as local economic growth, improved city image, social development, and environmental upgrading (Chen 2012). Sydney Olympic Park, Sydney, Australia; Queen Elizabeth Olympic Park, London, England; and Vancouver Olympic Village, Canada, are some representative cases underlying the positive impacts of Olympic Parks developed on large tracts of derelict land.

Their urban location near critical infrastructures and urban amenities, existing spatial frameworks and structures, postindustrial ecologies, and sublime aesthetic character are some of the characteristics of postindustrial sites. On the other hand, most of these brownfield sites have the problem of known and perceived contamination, which is one of the main obstacles to developing these sites with new programs (De Sousa 2008). Remediation of contaminated sites and the regeneration of inner-city sites are priorities for urban management and urban planning initiatives to provide a safe environment for districts and neighborhoods (Kennen and Kirkwood 2015). In line with this priority, there is an urgent need to develop novel strategies and a holistic framework that integrates the remediation process and spatial design. Remediation methods and design processes are two different but related fields in postindustrial regeneration. Remediation methods define a clear set of methods to clean up contaminated soil, and design includes a creative process to transform the site into a functional, ecological, and aesthetically pleasing environment (Erdem Kaya 2023).

The context of postindustrial land transformation projects ranges from residential to commercial, retail, light industrial, recreational, and open space. In particular, sites affected by former industrial uses in urban centers are being transformed into recreational areas, sports fields, and open spaces, ranging from waterfront promenades on former docklands and harbors to cycling greenways. One of the growing trends is using brownfield sites for community gardens, providing interim or longer-term urban agricultural uses adjacent to residential neighborhoods (Hollander et al. 2010; De Sousa 2008). Some sites require more scientific and sensitive approaches to ecological quality, such as wetland restoration, open parkland, and habitat restoration in areas of environmental sensitivity and scientific interest (Hollander et al. 2010), while other sites require strategies that consider dynamic urban life and public use. For these complex projects, landscape design is key in ensuring a safe environment with different spatial design strategies integrated with specific remediation methods. On the other hand, beyond individual sites, the potential of transformative remediation can be used as a holistic framework to build urban form as part of a large-scale landscape network (Sleegers 2010). In this case, the environmental impact of remediation can extend beyond the spatial boundary when brownfield development is integrated with greenfield development. With this method, brownfield redevelopment can contribute to the environment by reducing greenhouse gas emissions at the city level, as shown in research conducted in the San Francisco Bay Area of California, USA. The research found that greenhouse gas emissions were potentially reduced by 14% over a 70-year period (Hou et al. 2018, 2023).

This chapter aims to explain the role of landscape design in providing safe environments with appropriate remediation methods in postindustrial settings. It offers four different approaches to creating landscape-safe environments.

9.2 Landscape Design and Remediation for Landscape-Safe Environment

As an integral part of the regeneration process with different solutions, landscape design becomes imperative to make sites safe and usable for the public. Landscape design is the decision to use land regarding function, spatial organization, and movement for a specific program related to ecological, aesthetic, sociocultural, and economic insights (Dee 2005; Nassauer 2002; Nijhuis and de Vries 2019). It is a cultural act toward the environment that addresses a creative process involving both human and natural systems (Nassauer 2002).

Designers can adopt precautionary design strategies and design tactics that integrate appropriate remediation methods to create multifunctional landscape spaces. Today, the integration mode between spatial design and remediation offers more diverse spatial settings as site technologies have advanced and produced different landscape typologies, as presented in the work of Erdem Kaya 2023. Here, remediation methods can be considered a critical element of the design process that leads to the emergence of diverse spatial settings with different design tactics such as topographical manipulations, surface formation, living systems articulation, and thematic configurations (Erdem Kaya 2023). On the other hand, landscape designers have an ethical responsibility to protect human and ecosystem health through their design choices when designing contaminated sites. Since the effects of contamination can spread across spatial boundaries, remediation methods can also affect the broader environmental context. Here, how design strategies respond to mitigate the negative environmental impacts or how to integrate a broader spatial perspective to remediation becomes a critical issue.

Remediation can be defined in many different ways. A common understanding of remediation is to remove or contain contaminants from the environment and limit their adverse effects to ensure ecosystem and human health. Depending on the context of remediation, it also aims to restore degraded ecological and hydrological systems and prepare sites for new land uses (Cambridge Dictionary n.d.; Diamond et al. 1999; Ridsdale and Noble 2016; US Environmental Protection Agency (USEPA)). Although there are various remediation methods that fall into three main categories, namely, physical, chemical, and biological, a combination of these methods usually provides the most effective solutions for site remediation (US Environmental Protection Agency (USEPA) 2022; Koal and Taak 2018; Bradl and Xenidis 2005). In their study, Lee et al. (2024) provide a comparative study on the effectiveness of different remediation methods. According to this study, combined approaches appear as one of the most effective and low-impact methods (Lee et al. 2024). On the other hand, the environmental impact of selected remediation methods is another issue being addressed by researchers. In their research, Hou et al. (2023) discuss the potential environmental impacts of traditional remediation methods as a basis for developing more sustainable remediation processes for brownfields. It is emphasized that the environmental impact of the remediation can extend the spatial boundaries of the contaminated sites. Because environmental systems are highly interconnected, the remediation method and the processes used to make it available, such as providing an energy source by burning fuel off-site or disposing of hazardous waste off-site, may cause potential pollution in the surrounding environment (Hou et al. 2023).

For example, a sustainability assessment study of a single brownfield site shows that the dig & haul method will release 2.7 million tons of CO_2 emissions, which is approximately 2% of the annual CO_2 emissions for the entire state (Hou et al. 2023). In this case, designers should understand the possible impacts of remediation on the environment to provide a landscape-safe environment. Wrong design decisions can lead to the proliferation of contamination through multiple pathways and threaten human and environmental health or create spaces where people can easily expose themselves to contaminants (Hollander et al. 2010; Erdem and Nassauer 2013).

According to Young et al. (1997), remediation methods fall into three broad categories: risk avoidance, engineering, and process-based. Within these three methods, risk avoidance methods can be used to a limited extent as they involve steps that need to be considered at the planning stage of development, such as land use, site layout, location of services, and other necessary infrastructure to cut off potential pathways for contaminants. Engineering and process-based methods are the most commonly used remediation techniques. Engineering methods include civil engineering techniques such as excavation (cut and haul), in-ground barriers such as cut-off/vent trenches and slurry walls, capping and covering systems, and hydraulic measures such as pumping to control contaminated groundwater. From a landscape architectural perspective, engineering methods and integrated spatial settings create more formal and structured spaces in the landscape that provide room for diverse activities and add dynamism to the space (Erdem Kaya 2023). Several design cases use engineering methods to stabilize contaminants on-site. Designed by Richard Haag, Gas Works Park, Seattle, USA, is one of the first examples of the transformation of an industrial site into an urban park. For more than 50 years, the 7.97 ha site was operated by the Seattle Gas Company as a coal and oil gasification plant to provide energy to the city of Seattle on the shores of Lake Union. Over the years, the site was contaminated with benzene, polyaromatic hydrocarbons (PAHs), arsenic, and volatile organic compounds (VOCs) (Lyall 1991). Remediation efforts included bioremediation by injecting oil-eating bacteria into the soil. However, with the limited technology available at that time, it appeared that this method could not clean up the soil as further soil testing was approved. As part of the in situ remediation process, some parts of the contaminated site were excavated, retained on-site, and redesigned to create a large mound as one of the iconic landscape elements of this landscape design project. Another prime example is the Crissy Field project in San Francisco, USA. Designed by Hargreaves Associates, the 40.4 ha public park was formerly used as a military airfield with associated facilities and light industry. The back-dune tidal marsh ecosystem was one of the ecologically sensitive zones within the site boundary that was filled during airfield operations (Rieder 2001). More than half of the project site was covered with impervious asphalt and hard-packed dirt. 1.8 km of shoreline was reconstructed with gravel. Some areas within the project site were contaminated with heavy metals, PAHs, VOCs, and pesticides (Erler and Kalinowski 2004). As a part of the remediation plan, the asphalt surface was removed, crushed, and recycled to be used as a structural element for the new topography. The asphalt surface was transformed into a grassy field by keeping the contours of the former surface. In this way, the design gave reference to the former structure of the site. Contaminated sediments were excavated from the coastal shoreline and filled marshland to be used in the grassy field and to restore a 7.2 ha marshland ecosystem.

On-site groundwater remediation was used to protect the restored marsh ecosystem (Erler and Kalinowski 2004). In its current form, the wetland ecosystem provides a habitat for more than 110 native plant species and 135 animal species (Golden Gate National Park Conservancy 2025; Ward and Ablog 2006).

Process-based methods include operations to remove, stabilize, or destroy contaminants. These methods include processes such as physical, biological, chemical, stabilization/solidification, and thermal processes (Young et al. 1997).

The process-based approach to remediation has become very popular in the last year, as these methods activate living systems and provide aesthetically pleasing and biologically diverse spatial settings in the landscape (Erdem Kaya 2023). There are prime examples that successfully address how process-based remediation methods can be integrated into landscape design. Especially, it is an effective method for cleaning up contaminants from water as a part of stormwater management. Here, rain gardens, biological swales, and constructed wetlands are some design elements that integrate process-based remediation with spatial design.

Postindustrial sites can be used as a laboratory to test different remediation methods combined with different spatial elements where ecological knowledge, engineering knowledge, and design come together. This hybrid nature of postindustrial projects requires comprehensive design solutions that use the remediation process as a framework to generate new design tactics that combine engineering solutions and spatial configuration with an aesthetic dimension (Erdem Kaya 2023).

West Soul Lake Park and Seonyudo Park from Seoul, South Korea, are among the best examples that address postindustrial transformation from aesthetic, ecologic, social, and cultural aspects. The award-winning projects include strategies for the protection of industrial legacy as a part of urban memory, adaptive reuse of industrial structures and infrastructures, and process-based remediation methods to clean up contaminants from water and soil (Lee 2019).

9.3 From Site-Based Approach to System-Based Approach

Traditional remediation methods include excavation and off-site disposal, on-site treatment, or containment through impermeable layers of materials such as plastic, concrete, or clean soil (Young et al. 1997). These technologies can be evaluated as on-site solutions, usually in site-specific solutions such as traditional park designs (Sleegers 2010). Today, there is a growing tendency to integrate remediation, spatial planning, and design through a systemic approach with socioeconomic, environmental, and aesthetic objectives. Some groups of studies address transformative restoration as part of the green infrastructure approach, while others discuss the potential of a nature-based approach to create healthy urban contexts through community planning (Smith 2015).

As part of process-based methods, phytoremediation or gentle remediation options (GROs) are promising methods for the systematic remediation of contaminants in landscapes. GRO is defined as a risk management strategy or technologies involving plant (phyto-), fungal (myco-), and/or bacterial-based methods that result in a net gain in soil function as well as effective risk management (Cundy et al. 2016). In this method, the main objective is to reduce the transfer of contaminants within the environment through extraction, transformation, or degradation of contaminants, or through in situ stabilization such as biological and/or chemical processes (Cundy et al. 2016). GROs stand out as an effective remediation method, especially for postindustrial transformation projects that envisage the use of green areas in the form of urban parks, gardens, urban agriculture, etc.

Nature-based solution is another term used to describe the integration of natural systems into the remediation process. As a cost-effective solution, this approach supports the use of natural features and processes to remove and/or reduce contaminants with systematic interventions to create more diverse and resilient cities, landscapes, and seascapes (European Commission (2025)). Both methods share the principle of activating natural systems as part of a sustainable redevelopment strategy and provide positive contributions to the urban environment from economic, ecologic, and social perspectives. Biomass production, restoration of ecological communities, stormwater management, amenity, and recreation are some of the desired end uses that can be achieved in the projects (Vangronsveld et al. 1995, 2009; Witters et al. 2012; Cundy et al. 2016). Both approaches highlight the potential of phytoremediation, GROs, and nature-based solutions to remove contaminants from soil and water and evaluate these methods to integrate remediation into the urban context.

Phytoremediation-by-design is the name Smith (2015) used to describe "*the symbiotic relationship between site remediation and long-term planning.*" This approach provides three interdependent stages to holistically complete the phytoremediation process from start to finish. (Smith 2015). The first stage starts with inventory studies to analyze and design the contaminated site within the community scale contexts before implementing the phytoremediation method. The second stage includes the implementation of phytoremediation according to site design and development goals. The final stage includes post-phytoremediation design at the site and community scale. The main objective here is to provide a coordinated post-remediation site development in a holistic manner, where natural elements used in the remediation process as well as built elements such as buildings, roads, and open spaces are utilized (Smith 2015).

Since the 1970s, phytoremediation has been used as a process-based method to remediate contaminated sites, using specific plant species to absorb, remove, and release contaminants from soil and water (Smith 2015). On the other hand, Kennen and Kirkwood (2015) separate two related terms, phytoremediation and phytotechnologies. According to Kennen and Kirkwood (2015), phytoremediation is the "*degradation and/or removal of a specific contaminant at a contaminated site by a specific plant or group of plants. In addition to the degradation and/or removal of contaminants, phytotechnology also includes techniques such as the stabilization of contaminants with the surrounding soil or root structure of a plant and the preventative installation of plant-based approaches to treat a contaminant or mitigate an environmental problem before it occurs.*"

A study focusing on the potential environmental impact of different remediation methods found that microbial bioremediation can reduce the global warming potential by 50%–90% compared to the dig & haul method, while phytoremediation can reduce the environmental impact by up to 250%. The same research also showed that in a case study in Denmark, the potential impact of bioremediation on human health was 76% lower than thermal desorption and 92% lower than the dig and haul option (Hou et al. 2023).

Phytotechnologies can also include measures to prevent contamination that may occur in the future. Phytoremediation involves planting to clean up soil and groundwater. On the other hand, phytotechnologies include a wide range of options such as constructed wetlands, bioswales, green roofs, green walls, and planted landfill caps

(Kennen and Kirkwood 2015). It is evident that for contaminated sites as part of ecological design strategies, phytoremediation and phytotechnologies are becoming an integral part of spatial design, especially in landscape design projects developed for parks, community gardens, green spaces, riparian corridors, and vegetated filter strips as a response to multiple environmental problems (Kennen and Kirkwood 2015). Phytoremediation and phytotechnologies also add social value to contaminated sites by creating attractive public spaces and aesthetically pleasing environments essential for urban regeneration projects (Kirkwood 2001). Although the time frame required to remediate a site with plants has been a limiting factor in the use of phytoremediation, a project can successfully manage this time limitation by adapting the entire program over time as part of process-based design approaches. In this way, a dynamic urban environment can be achieved as the project site grows and changes over time to reflect the character and sense of place (Sleegers 2010).

As conventional remediation methods are expensive, phytoremediation and phytotechnologies offer more cost-effective and ecologically sound in situ remediation options (Smith 2015; Kennen and Kirkwood 2015; Compernolle et al. 2012; Smith 2015; Susarla et al. 2002) that can be easily linked to urban green infrastructure. Once isolated, contaminated land can be transformed into a continuous network on a regional or local scale (Sleegers 2010). In addition to providing spatial settings for diverse outdoor activities, urban landscapes have an operational role in relation to remediation when considered with a system approach. Green infrastructure, one of the latest spatial planning methods to create landscape systems, provides a continuous network of open and green spaces at multiple scales to protect natural systems and agricultural landscapes and provide an integrated framework for cultural, social, and ecosystem benefits (Benedict and McMahon 2002). Elements of green infrastructure can include a wide range of spatial components, such as individual home gardens, parcels, neighborhood parks, playgrounds, streets, urban parks, plazas, squares, large-scale peripheral parks, and rural zones, which can be classified as corridors as the connecting elements of the system and nodes as the homogeneous areas that are distinct from the surrounding landscape (Pozoukidou 2020).

The functions of green infrastructure can be thought of in three broad categories, namely, abiotic (hydrological regulation, soil development, carbon sequestration, regulating climatic extremes, etc.), biotic (providing habitat for species, protecting genetic reserves, biomass production, etc.), and cultural (recreation, social interactions, environmental education, etc.) (Ahern 2007).

Therefore, as a contemporary approach, when considered as a part of a large-scale green infrastructure system, the positive outcomes of remediation of a single site increase. Detroit, Michigan, USA, is another successful example that shows how a holistic framework for site remediation can be a catalyst of social, economic, and ecologic regeneration at the city-wide scale. Stoss Landscape Urbanism provides a green infrastructure plan for the whole city that includes urban vacant lots, abandoned industrial sites, existing green fields, and open areas. Stoss's proposal redefines the role of landscape apart from traditional notions by showing various ways that landscape can provide a healthy urban environment. Departing from the operational notion of the landscape, they incorporate strategies by using the productivity of the landscape to generate resources and to heal the environment through natural processes. Here they set up a detailed program to clean up air, water, and soil by applying green infrastructure concepts to improve the health of citizens and ecosystems. The team provides various types of blue/green infrastructure, including stormwater management, productive landscapes (food and energy) that foster public engagement, and research opportunities (Stoss Landscape Urbanism).

9.4 Cases: Four Approaches to Landscape-Safe Environment

Today, most contemporary landscape design projects occupy former industrial sites. Many include soil and water remediation measures as part of the spatial design. This section examines four projects to illustrate different approaches to providing a landscape-safe environment and to highlight landscape design strategies. The cases were studied according to their former land use, contamination history, program, and remediation methods. Each case addresses critical remediation issues and demonstrates how remediation can be part of landscape design scenarios. The cases studied in the research are classified according to different modes of integration.

9.4.1 Ecological Integration Approach

Any remediation has a very important role in making the contaminated site safe for future use. Some projects may prioritize supporting existing ecological systems depending on the preferred remediation method. James Corner and Field Operations's Freshkills Staten Island Lifescape project exemplifies design strategies and tactics to integrate the contaminated site into a broader ecological context. As the largest sanitary landfill in the world, the 890-acre site has been transformed into a public park with diverse programmatic organizations, from education to housing and conservation to recreation. The Freshkills project addresses several important aspects of remediation and landscape design. The former sanitary landfill was close to the island's wetland ecosystem. Here, one of the main objectives of the remediation was to regenerate the site and ecologically integrate it into the existing wetland ecosystem. The second important aspect is that the design strategies developed for the 890 ha site were based on process-based strategies. The complexity of the site in terms of size and contamination requires a sensitive approach to programming and timing.

The strategies offered by the project span the long term (30 years) by placing natural processes, agricultural practices, and plant life cycles at the center of this large-scale rehabilitation work (NYC Parks 2015).

The process-based design strategies include several phases to manage the restoration and remediation work and provide access to the site. As part of a long-term development plan, visitors will be able to have different experiences that change

Fig. 1 Freshkills Staten Island, New York, USA. (Source: Images courtesy of Freshkills Park and the city of New York. Source: https://www.flickr.com/photos/138495278@N07/38534564492/in/album-72157689892407705)

over time as nature grows and regenerates over time. Once inaccessible, degraded landscapes and brownfield sites will be completely accessible with the pathways and new programs as nature heals itself over time.

The primary strategy of the project was to rehabilitate the site through its natural cycles, using the mechanism of phytoremediation and successive growth dynamics, activating different landscape features that operate at different scales and contexts (NYC Parks 2015). The Freshkills project is also a very good example of using the idea of remediation as part of a broader environmental context and systems (Fig. 1).

9.4.2 Spatial Integration Approach

Contaminated sites usually have boundaries that separate the land from its surroundings. After remediation, the site becomes safe and needs to be spatially integrated into the existing urban context. The Sculpture Olympic Park in Seattle is a prime example of design strategies to integrate this former industrial site into the urban fabric. The 3.5-hectare site was used as a transfer and distribution area for petroleum products by the Unicol plant between the years 1900 and 1975. The site was occupied by related infrastructure and structures such as storage tanks, pipelines, loading racks, and a tanker loading dock, leaving a degraded site contaminated with gasoline, diesel fuel, lubricating and heating oils, and petroleum-based solvents after the facility closed.

As a result of heavy industrial use, the site was contaminated with gasoline, diesel fuel, lubricating and heating oils, and petroleum-based solvents. As part of the remediation process, 65,000 tons of petroleum-contaminated soil was removed, a new layer of soil was placed, a layer of reduced permeability was installed within the clean fill, stormwater drainage pipes were installed, and erosion control measures were implemented to stabilize the site (Herman and Heffner 2004). In parallel with these remedial actions, a highly manipulated topographical design was introduced, and a structured landscape appearance was created, culminating in a zigzag shape that crosses rail and road. This zigzag landform is proposed as a unifying element, providing the spatial setting for different activities and new vegetation communities while providing continuous pedestrian movement across the railways and roads. This particular landform design is supported by a system of mechanically stabilized earth, which is safer and more economical than a concrete retaining wall (Margolis and Robinson 2007). A sculpture pavilion is located at the top of the park. The park provides a setting for art events as well as a series of outdoor spaces for different native ecosystems. Here, the park itself becomes a spatial link between the coastline and the city. Another important environmental outcome of this remediation project is the increased number of salmon fish in the coastal zone of the project through the integration of a series of new tidal terraces for salmon habitat and saltwater vegetation. The number of epibenthic invertebrates, an important food source for juvenile salmon, increased by 20% in the first 3 years. According to shoreline monitoring within the first 3 years, the population of the juvenile fish increased from 500 to over 265.000 along the park's pocket beach and habitat bench. Another environmental dimension of the project was the strategies incorporated into the site design to manage stormwater. As part of the spatial design, the new landform of the site and planting beds regulate the flow of surface water, collecting and cleaning stormwater before it reaches Elliot Bay. Stormwater management and drought-tolerant plantation strategies have dramatically reduced the amount of water needed to maintain the plantation to nearly 95,000 gallons per week, saving the coast nearly $7900 annually. The social impact of the project can be seen in the increased number of visitors. This brownfield transformation project hosts an average of over 452,000 visitors annually by providing free access to world-renowned sculpture and Puget Sound (Nancy et al. 2012) (Fig. 2).

9.4.3 Historic-Cultural Integration Approach

Most postindustrial landscape designs can have the main objective of preserving industrial heritage, and some projects become symbols of this heritage. This is true for projects that include industrial structures as the main element of the design. Instead of demolishing the industrial structures and the spaces between them, the

Fig. 2 Olympic Sculpture Park, Seattle, Washington. Before and after construction of the park. (Image credit: © Benjamin Benschneider Photography. Source: https://www.weissmanfredi.com/projects/386-seattle-art-museum-olympic-sculpture-park)

design treats these structures as symbols of industrial culture and reuses them with new programs. The adaptive use of the existing structures provides a wide range of new activities. Emscher Park Duisburg Nord is one of the earliest examples of the postindustrial transformation of industry into a park. The site was redeveloped as part of the larger Emscher Regional Park Plan, which aimed to restore the River Rhine and create an ecological corridor with economic, social, and cultural objectives (Internationale Bauausstellungen 2024a, b). The Internationale Bauausstellung Emscher Park (IBA) was a project organized by the state of North Rhine-Westphalia, the federal government, and local authorities between 1989 and 1999 for the ecological, economic, and social renewal of the Ruhr area. As one of Europe's largest centers of coal mining and steel production, the Emscher region experienced the decline of heavy industry, leaving behind a degraded landscape of polluted environment and industrial ruins (Latz + Partners 2005; Shaw 2002). The project includes a total of 120 separate projects to create economic, social, and cultural impetus as well as to promote green and open spaces through the ecological development program (Shaw 2002). As part of this comprehensive plan, the 223-hectare site, formerly a steel mill, was transformed into a park designed by Latz+Partners and built between 1989 and 2002. The soil and water were contaminated with pollutants such as cyanide, arsenic, heavy metals, and PAHs. The project represents important aspects of postindustrial landscape design, such as the adaptive reuse of existing industrial structures, infrastructures, and materials for new purposes (gasometer as a diving center, blast furnace as a venue for cultural events, a steel plaza, etc.), and the integration of phytoremediation with spatial design. The site includes several layers of design elements within the context of different park systems to make the site legible for visitors at the human scale, such as a railway park, a sinter park, a ballast furnace park, and an ore bunker gallery (Internationale Bauausstellungen 2024a; Latz 2001; Weilacher 2007).

Highly contaminated areas were cut and hauled away, and new clean soil was introduced. In order to reduce remediation costs, 10,000 ha, including some of the most contaminated sites in the corridor, are being treated by slow remediation through natural attenuation, which was managed by the forest administration (Dettmar 2005; Hamm 2006). The Emscher Park, with its historical and cultural context, demonstrates how brownfield redevelopment can be an integral part of regional development with economic, social, and environmental objectives. As one of the region's most successful and iconic projects, Emscher Park's design strategies integrate the social, cultural, and historical aspects of the site through the protection and adaptive reuse of the structures and ecological aspects of redevelopment. (Network Nature).

Coming 25 years after the earlier work of Richard Haag's Gas Works Park in Seattle, Washington, Emscher Park addresses key points of integrating restoration and design and provides a robust framework for understanding how scientific and design aspects of degraded landscapes can be managed through innovative approaches (Erdem and Nassauer 2013). The site was transformed into a revolutionary new landscape type with a successful remediation program and attracted more than 500.000 visitors per year (Latz + Partners 2005). The most successful aspects of the project are ecological preservation, keeping a sense of place, adaptive reuse of materials, and industrial structure with precautionary measures. As a concrete result of the remediation, the site is now home to more than 700 different species of plants, 45 species of birds, and 100 species of beetles (Network Nature) (Fig. 3).

9.4.4 Socioeconomic Integration Approach

Postindustrial sites carry the stigma of contamination, which can be a barrier to development. However, with the right remediation methods and design strategies, these problematic brownfield sites can catalyze new development, bringing new social and economic life back to the surrounding area. The High Line Project, in New York, is an example of postindustrial transformation, demonstrating how spatial design can transform an abandoned industrial structure into one of the city's most touristic and popular sites. The High Line, built in 1930, is an elevated railway that runs through Manhattan's industrial district 30 feet above street level to take freight trains off the street. This elevated train line was closed after 50 years of operation and has been abandoned for more than two decades. The Friends of High Line, a nongovernmental organization formed to preserve the historical memory of

Fig. 3 Emscher Park, Duisburg Nord credit # Christa Panick. (Source: https://landezine.com/post-industrial-landscape-architecture/)

the High Line and to prevent the structure from falling into disrepair, together with the City of New York, organized an international design competition to find the best alternatives to transform this elevated structure into an urban space (The High Line 2011). The winning proposal was submitted by landscape architect James Corner and his firm Field Operation, in collaboration with architecture firm Diller Scofidio + Renfro and Piet Oudolf, who used the entire structure as the main scaffolding of the design, integrating different surface strategies to develop functional, ecological, and aesthetic urban space. High Line's remediation work included the removal of all existing debris, steel structures, gravel, contaminated soil, and ruderal vegetation that had occupied the site for more than 20 years. Steel structures were replaced, and a clean layer of soil was added as a growing medium for new plant communities. The wild appearance of the existing plant communities was used as a design inspiration for the planting scheme, which includes a rich combination of native plants from the region. The High Line project reconnects New Yorkers to their city, but the project has also significantly increased economic activity as a result of its redevelopment and design. The project has significantly impacted property values, transforming the neighborhood into one of the "*hottest markets for upscale residential, retail, and office center development*" (Hassel 2024). From this perspective, the High Line project can be presented as a case to show how postindustrial transformation changes the socioeconomic and sociocultural structure of urban neighborhoods through strong design considerations (Fig. 4).

Fig. 4 The high line, New York. (Source: Author)

9.5 Discussion

Pollution is one of the natural consequences of industrialization and urban development. This problem is even more critical for cities with an industrial past. With advanced manufacturing technologies, old industries have relocated, leaving behind large contaminated areas. In the past, these problematic landscapes were left as they were. Still, today there is a growing interest in cleaning up these contaminated sites and transforming the land for new uses to provide a safe environment for both public and ecosystem health.

Remediation and landscape design are closely related, and the success of the design project depends on the integration of these two fields. The context of "from industry to urban park projects" can show a variety of large-scale urban regeneration projects to small-scale individual projects. Whatever the scale, planners and designers have an ethical responsibility to carry out these complex projects. The site can pose serious risks to public and ecosystem health without a precautionary approach. A designed park can provide clues to the site's past land use or a completely new setting and appearance to hide the contamination.

Historically, contaminated landscapes have been isolated from public and urban life simply because of the threat of contamination. Known and/or perceived contamination from both the public and private sectors can be a barrier to these problematic sites. However, in recent decades, with the advancement of site technologies in remediation, these sites provide suitable ground for testing new techniques and design methods. Especially with the growing demand for land in densely populated urban areas, it is imperative to transform these sites with appropriate remediation methods. As remediation methods vary, so do the ways of integrating these problematic sites. Today, with the availability of process-based methods such as gentle remediation methods or phytoremediation, phytotechnologies, or nature-based solutions, these isolated sites can be an integral part of a more comprehensive environmental system. There is a growing trend to integrate remediation into the urban form as part of green infrastructure to achieve multiple positive economic, environmental, and social outcomes.

Investigated case projects reflect different contamination histories and different integration modes concerning design strategies and remediation methods. Although their design approaches can show differences, each project has environmental, social, and economic benefits. This situation addresses the multidimensional aspects of postindustrial transformation projects (Table 1). According to the investigation, these high-profile landscape design projects can successfully become a part of the surrounding urban landscape. Some projects have a sensitive approach to industrial heritage through precautionary approaches such as Emscher Park and High Line, whereas others have efforts to transform the contaminated land with structural and engineering solutions. Each project has significant environmental benefits such as restoring existing ecosystems, stormwater management, carbon sequestration, and supporting biodiversity as well as social and economic benefits. Aesthetic value and active spatial configuration add value to these derelict lands through creative design solutions that reference remediation methods.

"Remediated Landscapes" is proposed as a conceptual approach to representing the complex sites with contaminated histories that have been integrated into environmental systems with an appropriate remediation method and spatial design choices (Fig. 5). Remediated Landscapes not only have the intention to provide a better environmental quality from a physical point of view, but these projects also have an agenda to become a full part of urban life. Four approaches derived from case studies show the extent to which remediated landscapes provide ecological, spatial, historical-cultural, and socioeconomic integration to create safe environments. Using different remediation methods, each integration mode helps designers create an interconnected system from the local to the regional level with multiple outcomes.

9.6 Conclusion and Future Outlook

As urbanization keeps going with a growing magnitude, demand for new land will increase as well. The transformation of postindustrial sites has become one of the urgent topics on the agenda of the governments as a part of the Sustainable Development Goals. On a global scale, the number of recorded transformation projects is increasing each day. Especially, it becomes imperative for the cities of the postindustrial era to remake and redefine open spaces to cope with contamination. As Peter Reed (2005) underlines, the fact that today most of the remarkable new

Table 1 Comparison of case projects in terms of environmental, social, and economic benefits

Project name	Location	Size (ha-km)	Former land-use	Landscape typology	Environmental benefits	Social benefits	Economic benefits	Integration mode
Freshkills Staten Island Lifescape	Staten Island, New York, USA	890	Sanitary Landfill	Parkland	Enhancement and restoration of wetlands Remediation of sanitary landfill Increased biodiversity Enhanced flood control Carbon sequestration	Recreational areas Social and cultural amenities Educational activities	Natural gas collection from the decomposing waste will be used to heat approximately 22,000 homes, and photo-voltaic cells, wind turbines, and geothermic heating and cooling will be contribute cities sustainable energy	Ecological Integration
Sculpture Olympic Park	Seattle, Washington, USA	3.4	Transfer and distribution center for petroleum products	Sculpture Park	Salmon habitat restoration Number of Juvenile fishes increased from 500 to 265.000 Increased biodiversity Storm-water management	Open-access sculpture park and museum Hosts more than 452.000 visitors annually, The project connects the shoreline with the neighborhoods	Annual cost savings with this water conservation total nearly $7900.	Spatial Integration

Emscher Park Duisburg Nord	Duisburg-Nord, Germany	223	Steal Manufacturing Facility	Urban Park	Remediation of contaminated soil, New habitat creation Habitat remediation by natural attenuation Re-use of materials (Steel Plaza) Reduced load to sewer system Reducing temperature at meso or micro scale Greater ecological connectivity across urban regenerated sites Improved connectivity and functionality of green and blue infrastructure Carbon sequestration Urban Forest as Nature Base Solutions Enhance ecological preservation (700 different species of plant, 45 species of birds, 100 species beetles recorded)	New recreational spaces by adaptive re-use of old industrial structures, 500.000 visitors per year Keeping the industrial memory and becoming a landmark for the region Heritage protection Increase accessibility to green open space, increased amount of green open spaces for residents Increase social interaction Increase well-being Social learning Keeping sense of place	As a part of large-scale regeneration project-new housing, new jobs were created with the idea of "Work in The Park"	Historic-cultural integration

(continued)

Table 1 (continued)

Project name	Location	Size (ha-km)	Former land-use	Landscape typology	Environmental benefits	Social benefits	Economic benefits	Integration mode
High Line-New York	Manhattan, New York, USA	2 km.	Elevated Railroad built to transport good within meatpacking district	Urban Promanade	Strom-water management Integration of native vegetation and support biological diversity The ecology of the area has been enhanced with over 100,000 plants, comprising over 1500 species of plants, birds and insects.	Become a ist place, Providing new public space supported with cultural programs Aesthetic value	Increased economic activity and real estate values	Socio-economic integration

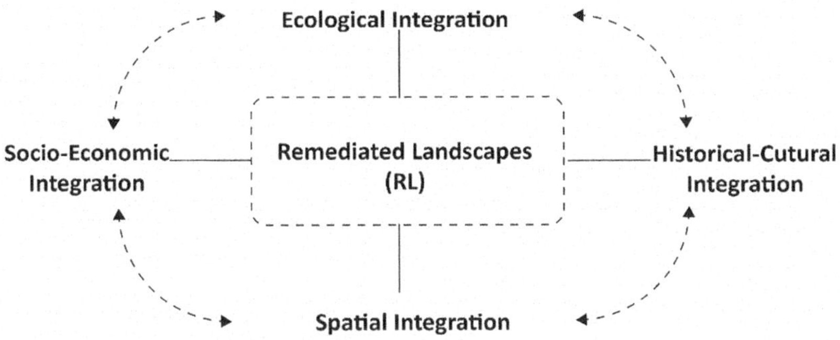

Fig. 5 Conceptual diagram of remediated landscape and integration modes

landscape architecture projects occupy sites that have been reinvented and reclaimed from degradation (Reed 2005). Contemporary urbanism requires novel design strategies to cope with contamination and to provide diverse, rich, and landscape-safe environments. Postindustrial transformation projects have multiple layers to be considered. These projects have ecological, economic, and social impacts on the environment, which leads designers to search for different integration modes with the surrounding environment to reflect this multilayered structure.

The relation between the remediation process and spatial design should be considered at the beginning of the transformation process. Instead of evaluating contaminated sites isolated from surrounding environments, today's system-based approaches, such as green infrastructure, offer a more plausible and effective solution that maximizes the positive environmental impact of remediation at multiple scales.

Large-scale transformation projects can be a catalyst for urban regeneration, which brings social, economic, and environmental change. As a part of the transformation of contaminated sites, site history and remedial action become two of the parameters that define the success of the project. However, the design process and its relation with design and remediation become as crucial as upper-scale planning projects.

Here the role of landscape architecture becomes crucial, and designers have an ethical responsibility to cope with contamination for public and ecosystem health. As remediation technologies advance, how spatial design responds to these technologies becomes an important question for research. These projects are complex in nature, and multidisciplinary teams can provide the best solutions to create landscape-safe environments. Therefore, dialogue between different disciplines and the translation of scientific information into design language is an important aspect of postindustrial landscape design projects that is ready to be discovered.

References

Ahern J (2007) Green infrastructure for cities: the spatial dimension. In: Novotny V, Brown P (eds) Cities of the future towards integrated sustainable water and landscape management. IWA Publishing, London, UKSBN: 1843391368

Agnes B (2001) Industrial Heritage: Rusty Memories, UNESCO Source 131:4–5

Benedict MA, McMahon ET (2002) Green infrastructure: smart conservation for the 21st century, Sprawlwatch clearinghouse monograph series. The Conservation Fund, Washington DC

Bradl H, Xenidis A (2005) Remediation techniques. In: Bradl HL (ed) Heavy metals in the environments. Elsevier, London

Cambridge Dictionary (n.d.). Accessed from http://dictionary.cambridge.org/dictionary/business-english/remediation

Chen Y (2012) Urban strategies and post-event legacy: the cases of summer Olympic cities, AESOP 26th annual congress, 11–15 METU Ankara

Compernolle T, Passel SV, Weyens N, Vangronsveld J, Lebbe L, Thewys T (2012) Groundwater remediation and the cost-effectiveness of phytoremediation. Int J Phytoremediation 14(9):861–877. https://doi.org/10.1080/15226514.2011.628879

Cundy AB, Bardos RP, Puschenreiter M, Mench M, Bert V, Friesl-Hanl W, Müller I, Li XN, Weyens N, Witters N, Vangronsveld J (2016) Brownfields to green fields. Realising wider benefits from practical contaminant phytomanagement strategies. J Environ Manag 184:67–77

De Sousa C (2008) Brownfields redevelopment and the quest for sustainability. Curr Res Urban Reg Stud 3:1–25

Dee C (2005) Form and fabric in landscape architecture: a visual introduction. Spon Press, London/New York

Dettmar J (2005) Nature-dominated development in urban landscapes. In: Hubertus EA (ed) Landscape architecture in mutation: essays on urban landscape. Institute for Landscape Architecture, ETH Zurich, pp 79–98

Erdem Kaya M (2023) From remediation to landscape design: design tactics and landscape typologies derived from post-industrial experiences. Landsc Res 48(3):331–353. https://doi.org/10.1080/01426397.2022.2150754

Erdem M, Nassauer IJ (2013) Design of brownfield landscapes under different contaminant remediation Policies in Europe and the United States. Landsc J 32(2):277–292

Erler and Kalinowski Inc (2004) Crissy field operable unit 4 implementation report, Presidio San Francisco. Accessed from: http://www.Archieve.org/strean/CrissyField0perableUnit4ImplementationReportPresidioOfSanFrancisco/CrissyFieldOperableUnit4lmplementationReportclosureReport#page/n13/mode/1up[October10

European Commission, Research and Innovation, (2025) https://research-and-innovation.ec.europa.eu/research-area/environment/nature-based-solutions_en

Etim EE (2012) Phytoremediation and its mechanisms: a review. Int J Environ Bioenergy 2:120–136

Golden Gate National Park Concervancy- Crissy Field, (2025) https://www.parksconservancy.org/parks/crissy-field

Hamm A (2006) A landscape laboratory in Germany—reaching out for new landscape concepts (Master dissertation). Swedish University of Agricultural Sciences

Hassel A (2024) How the high line reshaped New York City, in Rhetorikos. Accessed from: https://rhetorikos.blog.fordham.edu/?p=1313, Accessed 02 June 2024

Herman L, Heffner D (2004) Engineering design report for phase I remediation Olympic Sculpture Park, Aspect Consulting, LLC, Project No 020118-001-04 April 14

Hester, Harrison RM (1997) Contaminated land and its reclamation, issues in environmental science and technology, vol 7. The Royal Society of Chemistry, Great Britain

Historic England (n.d.) Understanding Industrial and Infrastructure Heritage. Retrieved from https://historicengland.org.uk/advice/heritage-at-risk/industrial-heritage/how-are-we-classifying-industrial/

Hollander JB, Kirkwood N, Gold JL (2010) Principles of Brownfield regeneration: cleanup, design, and reuse of derelict land, 2nd edn. Island Press, Washington DC

Hou D, Song Y, Zhang J, Hou M, O'Connor D, Harclerode M (2018) Climate change mitigation potential of contaminated land redevelopment: a city-level assessment method. J Clean Prod 171:1396–1406

Hou D, Al-Tabbaa A, O'Connor D, Hu Q, Zhu Y, Wang L, Kirkwood N, Ok YS, Tsang DCW, Bolan NS, Rinklebe J (2023) Sustainable remediation and redevelopment of brownfield sites. Nat Rev Earth Environ 4:271–286. https://doi.org/10.1038/s43017-023-00404-1

Internationale Bauausstellungen (2024a) Duisburg-Nord Landscape Park, Duisburg: A New Type of Park and a Part of the Emscher Landscape Park, Accessed from: https://www.internationale-bauausstellungen.de/en/history/1989-1999-iba-emscher-park-a-future-for-an-industrial-region/duisburg-nord-landscape-park-duisburg-a-new-type-of-park-and-a-part-of-the-emscher-landscape-park/, Access date: 060624

Internationale Bauausstellungen (2024b) IBA Emscher Park: future for an Industrial Region. Accessed from: https://www.internationale-bauausstellungen.de/en/history/1989-1999-iba-emscher-park-a-future-for-an-industrial-region/, Access date: 060624

Kapp PH (2019) How intangible cultural heritage can make historic preservation and post-industrial regeneration more sustainable. J Archit Plan Res 36:1

Kennen K, Kirkwood N (2015) Phyto: principles and resources for site remediation and landscape design. Routledge

Kirkwood N (2001) Manufactured sites: Integrating technology and design in reclaimed landscapes. In: Kirkwood EN (ed) Manufactured sites: rethinking the post-industrial landscape. Taylor and Francis, New York, NY, pp 3–11

Landschaft Park Duisburg Nord. Internation Building Exhibition. Accesed from https://www.landschaftspark.de/en/background-knowledge/international-building-exhibition/. Accessed in 25 July

Latz P (2001) Landscape park duisburg-nord: the metamorphosis of an industrial site. In: Kirkwood EN (ed) Manufactured sites: rethinking the post-industrial landscape. Taylor & Francis, New York, pp 150–161

Latz +Partners (2005) Landschaftspark Duisburg Nord—Duisburg, Germany. *Places* 17:3

Lee M-J (2019) Transforming post-industrial landscapes into urban parks: design strategies and theory in Seoul: 1998–present, Habitat Int, 91, 102023, ISSN 0197-3975, https://doi.org/10.1016/j.habitatint.2019.102023

Lee H, Sam K, Coulon F, De Gisi S, Notarnicola M, Labianca C (2024) Recent developments and prospects of sustainable remediation treatments for major contaminants in soil: a review. Sci Total Environ 912:168769

Loures L (2008) Industrial Heritage: the past in the future of the city. WSEAS Trans Environ Dev 4(8):687–696

Lyall S (1991) Richard Haag Associates, gas works park, Seattle, Washington. In: Designing the new landscape. Thames and Hudson, London, pp 204–207

Margolis L, Robinson A (2007) Living systems: innovative materials and technologies for landscape architecture. Birkhauser

Matthews J, Winter M (2010) MFO Park. Multi-Tiered Vine Park, Zurich. Retrieved from http://courses.washington.edu/gehlstud/WEBSITE_CONTENT/Winter2010/MFOpark.pdf

Diamond ML, Page CA, Campbell M, McKenna S, Lall R (1999) Life-framework for assessment of site remediation options: method and generic survey. Environ. Toxicol. Chem 18:788–800

Nancy R, Emerson P, Lacson D (2012) Olympic Sculpture Park, Landscape Performance Seriesa. Landscape Architecture Foundation. https://doi.org/10.31353/cs0380

Nassauer JI (2002) Ecological science & landscape design: a necessary relationship in changing. In: Johnson KBR, Hill (eds) Ecology and design: frameworks for learning. Island Press, pp 217–230

Network Nature, Landscape Park Duisburg Nord-Germany. Accessed from https://networknature.eu/casestudy/22582. Accessed in 25 July

Nijhuis S, de Vries J (2019) Design as research in landscape architecture. Landsc J 38(1–2):87–103. https://doi.org/10.3368/lj.38.1-2.87
NYC Parks (2015) Freshkills Park. Accessed from http://www.nycgovparks.org/park-features/freshkills-park/aboutthe-site
Pozoukidou G (2020) Designing a green infrastructure network for metropolitan areas: a spatial planning approach. Euro-Mediterr J Environ Integr 5:40
Reed P (2005) Groundswell: constructing the contemporary landscape. The Museum of Modern Art
Ridsdale DR, Noble BF (2016) Assessing sustainable remediation frameworks using sustainability principles, Journal of Environmental Management 184:36–44
Rieder K (2001) Crissy field: Tidal marsh restoration and form. In: Kirkwood N (ed) Manufactured Sites: Rethinking the Post-Industrial Landscape. Taylor and Francis, New York, pp 193–205
Rockström J, Steffen W, Noone K, Persson A, Chapin FS III, Lambin E, Lenton TM, Scheffer M, Folke C, Schellnhuber HJ, Nykvist B, de Wit CA, Hughes T, van der Leeuw S, Rodhe H, Sörlin S, Snyder PK, Costanza R, Svedin U, Falkenmark M, Karlberg L, Corell RW, Fabry VJ, Hansen J, Walker B, Liver D (2009) Planetary boundaries: exploring the safe operating space for humanity. Ecol Soc 14(2):32
Shaw R (2002) The international building exhibition (IBA) Emscher Park, Germany: a model for sustainable restructuring? Eur Plan Stud 10(1):77–97. Carfax Publishing, Taylor and Francis
Sleegers F (2010) Phytoremediation as green infrastructure and a landscape of experiences. In Proceedings of the Annual International Conference on Soils, Sediments, Water and Energy: vol. 15, Article 13. Available at: http://scholarworks.umass.edu/soilsproceedings/vol15/iss1/13
Smith GR (2015) Phytoremediation-by-design: community-scale landscape systems design for healthy communities. Int J Sustain Dev World Ecol 22(5):1–7. https://doi.org/10.1080/13504509.2015.1079276
Song Y, Kirkwood N, Maksimovic C, Zheng X, O'Conner D, Jin Y, Hou D (2019) Nature base solutions for contaminated land remediation and brownfield redevelopment in cities: a review. Sci Total Environ 663:568–579
Stoss Landscape Urbanism, Detroit Future City. Accessed from https://www.stoss.net/projects/planning-urbanism/detroit-future-city. Accessed in 25 July
Susarla S, Medina VF, McCutcheon SC (2002) Phytoremediation: an ecological solution to organic chemical contamination. Ecol Eng 18(5):647–658. https://doi.org/10.1016/S0925-8574(02)00026-5
Swensen G, Granberg M (2024) The impact of images on the adaptive reuse of post-industrial site. Hist Environ: Policy Prac 15(1):101–129. https://doi.org/10.1080/17567505.2024.2311005
The High Line (2011) High Line. Accessed from https://www.thehighline.org
United Nations (2018) Department of Economic and Social Affairs, Retrieved from: https://www.un.org/development/desa/en/news/population/2018-revision-of-world-urbanization-prospects.html#:~:text=News-,68%25%20of%20the%20world%20population%20projected%20to%20live%20in,areas%20by%202050%2C%20says%20UN&text=Today%2C%2055%25%20of%20the%20world's,increase%20to%2068%25%20by%202050. Accessed 6 June 2024
United Nations (2016) The New Urban Agenda: Key Commitments. Retrieved From https://www.un.org/sustainabledevelopment/blog/2016/10/newurbanagenda/
US Environmental Protection Agency (USEPA) (2022) Remediation technology descriptions for cleaning up contaminated sites. Retrieved from https://www.epa.gov/remedytech/remediation-technology-descriptions-cleaning-contaminated-sites
US Environmental Protection Agency (USEPA). Ecological risk assessment glossary of terms. Retireved from https://sor.epa.gov/sor_internet/registry/termreg/searchandretrieve/glossariesandkeywordlists/search.do?details=&vocabName=Eco%20Risk%20Assessment%20Glossary&filterTerm=remediation&checkedAcronym=false&checkedTerm=false&hasDefinitions=false&filterTerm=remediation&filterMatchCriteria=Contains, Accessed in 11.06.2024
Vangronsveld J, Colpaert J, Van Tichelen K, Clijsters H (1995) Reclamation of a bare industrial area contaminated by non-ferrous metals: physico-chemical and biological evaluation of the durability of soil treatment and revegetation. Environ Pollut 94:131e140

Vangronsveld J, Herzig R, Weyens N, Boulet J, Adriaensen K, Ruttens A, Thewys T, Vassilev A, Meers E, Nehnevajova E, van der Lelie D, Mench M (2009) Phytoremediation of contaminated soils and groundwater: lessons from the field. Environ Sci Pollut Res Int 16(7):765e794. https://doi.org/10.1007/s11356-009-0213-6

Ward Kr, Myla A (2006) Crissy Field Restoration Project Summary of Monitoring Data 2000-2004. San Francisco: National Park Service, Golden Gate National Recreation Area

Weilacher U (2007) Duisburg-Nord Landscape Park. In Syntax of Landscape, The Landscape Architecture of Peter Latz and Partner, ed. Udo Weilacher, 102–132. Basel, Switzerland: Birkhausen

Witters N, Mendelsohn RO, Van Slycken S, Weyens N, Schreurs E, Meers E, Tack F, Carleer R, Vangronsveld J (2012) Phytoremediation, a sustainable remediation technology? Conclusions from a case study. I: energy production and carbon dioxide abatement. Biomass Bioenergy 39:454e469. https://doi.org/10.1016/j.biombioe.2011.08.016

Young PJ, Pollard S, Crowcroft P (1997) Overview: context, calculating risk and using consultants. In: Hester ER, Harrison RM (eds) The Royal Society of Chemistry, Cambridge, pp 1–24

Zheng X, Kirkwood NG (2020) Landscape architecture and sustainable remediation. In: Hou D (ed) Sustainable remediation of contaminated soil and groundwater. (Chapter 12. Elsevier Inc., Butterworth-Heinemann, pp 301–324

New Cities Based on Safe Climate and Remediation

Mohamed Saleh Mohamed Esmail

Abstract This chapter examines the concept of creating new climate-safe cities focused on ecological remediation. It opens with a discussion on how global and groundbreaking the movement for sustainability in urban development has become. It highlights examples from countries leading this change, like the Netherlands, the United States, Germany, and the United Arab Emirates (UAE). Each of these cases offers a model for cities as resilient, regenerative places that function as ecosystems within nature. This chapter then moves on to Egypt and the important projects in building new cities based on climate change. This chapter provides an overview of how sustainable principles are being incorporated into urban planning through projects like the New Administrative Capital (NAC), Al Alamein City, Galala City, and the Toshka Project in Egypt. These cities are designed with low-energy buildings, water management systems, and green infrastructure, setting a template for new development in arid regions. As such, this conclusion of the chapter discloses an analysis related to how such cities can be sustainably and affordably built, and the projects detailed are case studies for other regions under such environmental pressures. This chapter argues that such an exercise indicates that Egypt has not only secured local sustainability through urban planning but also illustrates critical lessons for international efforts in climate-safe urbanism.

Keywords Climate-safe cities · Sustainability · Urban development · Renewable energy · Environmental remediation · Smart cities · Green infrastructure

M. S. M. Esmail (✉)
Basic Science Department, Faculty of Engineering, Misr University for Science and Technology, Giza, Egypt
e-mail: mohamed.esmail@must.edu.eg

A. A. H. Abdel Latef et al. (eds.), *Sustainable Remediation for Pollution and Climate Resilience*, https://doi.org/10.1007/978-981-96-5674-5_10

1 Introduction

The cities are mostly focused on sustainability, resiliency, and environmental remediation. These new urban areas are not accommodations for increased population on the ground but for growth with the environment in harmony (Rockström et al. 2013). The traditional cities, wriggling and wrestling with overcrowding, pollution, and resource unavailability, contrast with these innovative developments, constructed with the future in mind (Carmin et al. 2009).

This chapter debates the creation of new cities according to climate safety and environmental sustainability. It stops presenting international models in indeed pioneering countries of sustainable urban development: Germany, the Netherlands, and the United States. Subsequently, it focuses on Egypt, a country where extraordinary efforts are taken to create cities appropriate to this specific climatic and environmental demand. These examples, among other illustrations, show various ways of being employed worldwide in developing cities that are resilient but, importantly, restorative to the environment.

2 Global Perspectives: Leading Examples of Sustainable Urban Development

2.1 Houten, Netherlands

Located near Utrecht, Netherlands, the small town of Houten has been internationally recognized for its innovation. The town has a cyclist and pedestrian focus, and the architectural design for this is completely different from that of towns that have developed from road building. Houten has a circular road system around the town center and is enclosed by cycle paths, so walking from one location to another (e.g., residential areas, schools, and city center) becomes safer, thus reducing reliance on automobiles for intra-town commutes (Lux 2007).

As shown in the map in Fig. 1, Houten was a bold plan with large portions of energy-efficient housing and expansive green areas (for regulating the microclimate) for residents to recreate. These steps have now enabled the town to claim victory in shrinking its carbon footprint, making it a great example of suburban sustainability.

2.2 Portland, Oregon, USA

Portland, Oregon, is often refuted as one of the greenest cities in America (Lundqvist 2007; Goodling et al. 2015). In this respect, it has long been a leader in green urban planning and allows for the smallest emission of greenhouse gases, thus

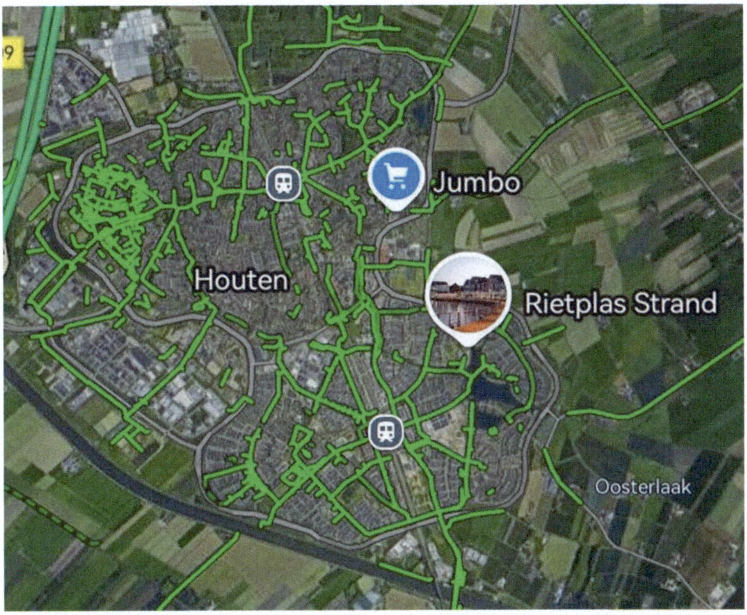

Fig. 1 Street Layout in Houten. (Green line represents the bicycling paths: Google Maps)

Fig. 2 Transportation in Portland, Oregon (AI Image)

contributing to sustainable transport modes. During the 1970s, the City of Portland established an urban growth boundary, which has greatly decreased urban sprawl and preserved the city's natural landscape (Huber and Currie 2007).

Besides the transportation systems (shown in Fig. 2) and green buildings, Portland has been synonymous with sustainability. The same thing goes for the city, which has made huge investments in hydroelectric power and other renewable energy. This level of sustainability has carried over to waste management, with Portland recycling and composting 70% or more in their comprehensive programs, all designed around keeping as much waste out of the landfill (Richardson et al. 2012).

2.3 Masdar City, United Arab Emirates

Classified as the template of a sustainable model city, Masdar City in Abu Dhabi, United Arab Emirates (UAE), is probably one of the most iconic ever conceived and halfway accomplished. A city that produces zero carbon emissions and no waste, with solar and wind energy sources for an infrastructure concept. The buildings are highly energy efficient. In fact, many people here cool their homes passively and hardly run the air conditioning at all (Fig. 3). Masdar is also not skimping on water conservation, reclaiming greywater with a recycling system, and using treated sewage effluent for landscaping (Reiche 2010).

2.4 Freiburg, Germany

As Germany's "Green City," as shown in Fig. 4, Freiburg stands as a great example of such sustainable urban development (Newman et al. 2017). The city has an ambitious program of energy transition, doing particularly well in microgeneration and

Fig. 3 Masdar City, Abu Dhabi, UAE (AI Image)

Fig. 4 Green City: Freiburg, Germany (AI Image)

solar power with a high-level usage per capita. It also encourages people to live without cars thanks to its very efficient public transport network and extensive cycle paths. Community participation is significant in Freiburg, and its success lies primarily in its culture (Rohracher and Späth 2014).

2.5 *Lessons for Egypt*

The examples of Houten, Portland, Masdar City, and Freiburg hold important lessons for the new cities that Egypt is building. Making walkability the cornerstone of designs for urban real estate, with schools, hospitals, and public greenery scattered throughout their development pattern, as in Houten's case, is one way to address both unbearable congestion and CO_2 emissions while increasing human happiness. If Portland offers a model to limit the encroaching development of sprawl and encourage infrastructure investment in other forms of commuting, this comprehensive strategy could be adapted for Egypt's cities' greener spaces to improve healthy living. While ideas on how to reduce waste and boost recycling from sources such as Portland's Department of Solid Waste have been proven popular, less is said about the need to manage expanding cities like Egypt. In theory, the advanced thinking behind renewable energy and water viability solutions coming out of Masdar City would be a great reference for how Egypt could survive its desert climate environment while having a decaying infrastructure. The new cities developed in Egypt could be made much more self-reliant by utilizing solar power, and advanced water management systems would improve their sustainability. Furthermore, the involvement of different stakeholders in Freiburg's sustainable planning indicates a strong sense of ownership among residents on greening initiatives, which is the key to their sustainability. Freiburg is a lesson that applies to larger urban settings, especially in places like Egypt, where sustainable transport use and renewables may be more challenging to implement than it was for the small German town. This model allows Egypt to build newer, modern, and sustainable cities.

3 Egyptian Cities: Leading the Way in Climate-Safe Development

Egypt is intentionally pushing this idea by preempting the establishment of climate change-induced shocks to its economy, and with rapid urbanization and increased environmental fragility crisscrossing this region, Egypt is advancing the logic further by planning economic cities as shock absorbers (Smith et al. 2013; Ali 2022). The ambition of these new urban projects tells us how well their architects know the big forces vying for control and also how widely they can imagine resilient, climate-safe cities (Smith et al. 2014). The pattern of building new smart cities has become one of the urban development strategies for spatial development under Egypt's vision (Hamza 2016; Mohamed Hassanein 2017; Ali 2022).

3.1 The New Administrative Capital (NAC)

The New Administrative Capital, which anchors an urban development about 45 km east of Cairo, has the ambitious mission of relieving a metropolis that has struggled with infrastructure and environmental woes for years (Fig. 5). The New Administrative Capital (NAC) spans an area of 700 km^2 and boasts a population of over six million residents, government buildings, foreign embassies, residential areas, commercial areas, and recreational spots (Hakeh n.d.).

Fig. 5 The New Administrative Capital (NAC), Egypt. (AI Image)

Fig. 6 Green River in NAC, Egypt. (AI Image)

3.1.1 Climate-Safe Features

Sustainability is a key focus in the development of the NAC. Solar farms also play an important part in the city's energy strategy, with extensive plans for huge tracts of land to be carpeted with arrays supplying large percentages of the electricity sold within Vegas. Its building materials restrict the necessity for mechanical cooling and heating from stemming energy-efficient design in buildings. As shown in Fig. 6, green spaces in such large areas of the city, like interconnected parks and lakes in Green River (it regulates microclimate and reduces the urban heat island effect significantly), provide the population with places for relaxation.

3.1.2 Environmental Remediation Efforts

Environmental remediation forms part of the NAC design. The waste-to-energy (W2E) plant is one example of the smart solution by the city of Bengaluru, where garbage can be burned and used as an incinerator so that electricity production occurs, curtailing methane emissions and reducing landfills (Ali 2022; Metwally et al. 2023; Hasan and Mohmed 2024). The river walk becomes a teaching tool for local plants to filter stormwater using low-intensity techniques, and greywater recycling systems are utilized on-site, cleaning and reusing wastewater within the property boundaries for landscaping and non-potable uses. Another way to reduce reliance on the water outside the region is rainwater harvesting (Hamed and Ramadan 2022; Selim et al. 2024). The transport network in the NAC is designed to have minimal environmental impact. The city is made to accommodate pedestrians and cyclists with many walkways and bike paths. Moreover, the NAC will be connected to Cairo and its surrounding cities via a sustainable public transport network of electric buses and a monorail system (Fig. 7) that ensures lower vehicular demand while putting more power into our hands.

Fig. 7 Monorail system, which connects NAC to Cairo City, Egypt. (AI Image)

3.2 Al Alamein City

Situated on the Mediterranean coast, Al Alamein City is one of Egypt's most important urban development projects (Fig. 8). Branded for its World War II history as a battleground, Al Alamein is witnessing a transformation into what can only be described as an astonishing combination of a modern city living with a green urban lifestyle and the preservation of heritage.

3.2.1 Climate-Safe Features

Coastal geophysical and climatological stresses: The city is located along the coastline, which poses challenges that are distinct for building climate resilience. Similarly, coastal defenses at Al Alamein comprise sea walls, breakwaters, and the restoration of natural barriers, including dunes to absorb wind-borne storms (reconstructing back earthen-created heights to recede storm impacts). In contrast, wetlands are reconstructed to absorb rising floodwater (Saeed et al. 2023). The city is built on renewable energy, substantially activated by wind and solar power activities as part of the overall system (Attia 2019).

3.2.2 Environmental Remediation Efforts

The environmental remediation at Al Alamein is being completed with a focus on the significant coastal ecology in this region necessary for its health and longevity. Seagrass beds, coral reefs, and other key habitats supporting the ecological services that underpin human existence are being rehabilitated. To combat water shortages, desalination plants were also developed in the city, converting salt-contaminated seawater to fresh water for consumption by residents, laying pressure on Egypt's fresher varieties (Attia 2019; Fahmy and Kamiya 2019).

Fig. 8 Al Alamein City, Egypt. (Google Maps)

3.3 Galala City

Galala City on the Red Sea coast is part of a shift in Egypt toward developing new city areas that are further away from densely populated locations. The city was envisioned as an international tourism place, educational center, and green belt, which can be replicated in future planned cities with climate-resilient areas along these lines where environment-friendly activities will attract others to the place (Mahgoub 2022, 2024).

3.3.1 Climate-Safe Features

There are various solar and wind energies in Galala City, located on an elevated plateau overlooking the Red Sea. The city has made the most of renewable energy, utilizing solar panels and wind turbines to power many areas, as shown in the map of Fig. 9. In addition to that, the city focuses on water management, using sophisticated systems to purify wastewater and use it for irrigation in a non-drinkable way, which also helps reduce the total amount of consumed potable water, as presented in Fig. 10.

3.3.2 Environmental Remediation Efforts

The city will give room to the attention paid to environmental protection in Galala City. The city has implemented erosion control measures to prevent wind- and water-based erosion from harming the surrounding landscape. The marine protected areas are established along the coast to conserve special environments from the Red Sea that are important in attracting and maintaining the tourism sector. Greenery and sustainability beyond environmental protection, Galala City is taking on the challenge of desertification head-on with landscaping that includes vegetative cover

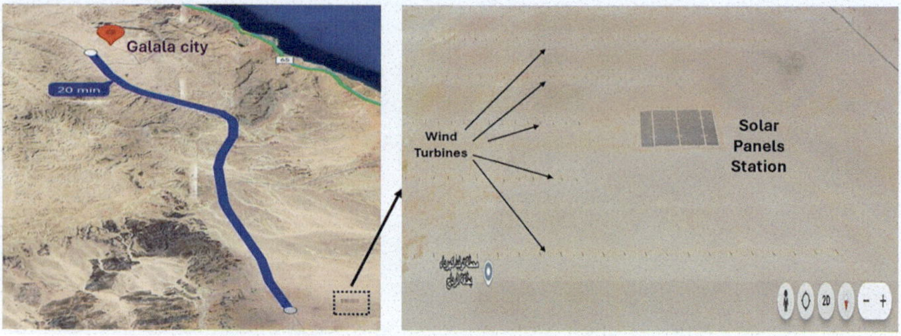

Fig. 9 Solar panel station and wind turbines at 33 km from Galala City. (Google Maps)

Fig. 10 Desalination station of seawater for Galala City. (Google Maps)

using an innovative approach (Siahaya and Hermana 2013). These trees mitigate desertification from sand encroachment, reduce the phenomenon of urban heat islands in our towns, and provide good places for residents' recreation opportunities and areas where visitors can stay.

3.4 Toshka Project (Toshka Lakes)

One of the largest new land reclamation projects is Egypt's Toshka Project (or New Valley Project), an extremely ambitious initiative to increase the livable area in areas outside the old-fashioned Nile Valley, thus diminishing population density during part of it. The first of these projects is the transformation from desert land into arable agriculture, as in Fig. 11, with new urban areas being developed to help support it in a vast region in southern Egypt (Abo-Khalil and Ahmed 2016; Soliman et al. 2019; Abd Ellah 2021).

3.4.1 Climate-Safe Features

The main interest area for the Toshka Project is water management, utilizing excess water from Lake Nasser to irrigate desert regions and create new agricultural zones (Abou Samra 2022). This is of core interest to the Egyptian government in its pursuit of food security, as a large portion of imported products involves the agricultural sector. Many modern irrigation systems have methods in place so that water can be used in the most efficient way possible; drip irrigation and soil moisture monitoring try to maximize the use of available water (Iwasaki et al. 2021). The new urban areas that will be built for the Toshka Project are a hallmark of sustainability. These will also include renewable energy resources such as solar and wind, which shall reduce the carbon footprints, so that they are independent of any fossil fuels (Shin et al. 2022). The cities will be built using energy-efficient materials and methods of construction, minimizing their environmental impact.

3.4.2 Environmental Remediation Efforts

One of the main concerns regarding Toshka is its environmental implications, one being that it seems essential they have some serious remediation processes to maintain and restore soil fertility. It enforces agricultural practices that help prevent soil degradation, including crop rotation and organic fertilizers, to ensure the sustainability of farmlands long after reclaiming. Green corridors are also being developed

Fig. 11 Toshka Lakes and agriculture regions. (Google Maps)

to link areas of the city with existing natural habitats, promoting biodiversity and offering cooler environments in a desert environment (Al-Soghir et al. 2022; Abou-Shady et al. 2023; Aly et al. 2023).

4 Challenges and Future View

4.1 Challenges Faced

Contrary to the attendant progress in creating new sustainable cities in Egypt, several obstacles have occurred. Of these, the most basic barrier may be related to the high building and maintenance costs of each facility that is required for renewable energies, water management, and waste management systems. This technology requires big investment upfront and operational costs, thus stressing public and private resources. Another important challenge pertains to ensuring that such new cities are affordable and accessible to a broad part of the population. Such developments are possibly highly exclusionary in that they cater only to higher-income groups, and people experiencing poverty remain marginal or have minimal access to sustainable urban living. Most importantly, however, social equity in the planning and construction of such cities is critical. It cannot just be an afterthought with existing inequalities being played out on bigger screens than their current location. Apart from this, developing a sense of community in newly developed areas is also hard to find, especially if residents must be relocated from elsewhere or if the cities are mainly designed for tourism or business. Setting up local governance structures that involve residents in decision-making is important in ensuring cohesive community support and sustainability initiatives.

4.2 Future View

However, the future of Egypt's new cities appears bright despite these challenges. It remains very promising that continued investment in smart cities and Internet of Things (IoT) solutions, such as smart grids with a broad set of sensors and data analytics tools, could drastically reduce the energy sources used for urban operations. Supporting the development and deployment of these technologies may enable better resource utilization and reduce waste, thereby elevating the overall quality of life for people in cities and reducing vulnerability to some climate change impacts.

These cities represent future opportunities with Egypt poised for renewable energy capacity growth: providing proof of concept in new technologies and methodologies that will, in due course, quicken innovation and set the benchmark in sustainable urban development. Experience is likely to influence the building of future cities, whether in Egypt or elsewhere, as humanity wakes up to environmental challenges.

In addition, they could provide an example of sustainable urban development for cities in other areas that are struggling with similar issues. Egypt can support the global conversation on building more climate-safe cities. It also has an opportunity to add a new strand of knowledge economy by showing how sustainable city building can look in arid and semiarid environments.

5 Conclusion

These are not only much-needed answers to the century's challenges but also bold projects by Egypt to reinvent towns that can reshape its future. Whether through the New Administrative Capital, Al Alamein City, Galala City, or even the Toshka Project, it is definite that Egypt is showing how urban planning provides an answer to mitigate climate challenges while creating healthier environments for residents.

With their use of renewable energy, sustainable infrastructure, and extensive environmental restoration strategies, these cities are defining new expectations for a functionally sustainable city in the twenty-first century. As they advance the projects from blueprint to reality, they improve the quality of life for their citizens and provide valuable lessons learned for any other nation or region in similar dire straits.

What makes Egypt's new cities great laboratories is that, with further investment in innovation and the active engagement of the community, they can rise to become a leading force around the globe regarding sustainable and resilient urban construction. These cities will not only be a place to live but also templates showing how urban living can be rewritten in the future—this time, together with nature.

References

Abd Ellah RG (2021) Morphometric analysis of Toshka Lakes in Egypt: a succinct review of geographic information systems & remote sensing based techniques. Egypt J Aquat Res. Elsevier 47(2):215–221

Abo-Khalil AG, Ahmed SS (2016) Water pumping using powered solar system—more than an environmentally alternative: the case of Toshka, Egypt. J Energy Nat Resour 5(1–1):19–25

Abou Samra RM (2022) Dynamics of human-induced lakes and their impact on land surface temperature in Toshka Depression, Western Desert, Egypt. Environ Sci Pollut Res. Springer 29(14):20892–20905

Abou-Shady A, Siddique MS, Yu W (2023) A critical review of innovations and perspectives for providing adequate water for sustainable irrigation. Water. MDPI 15(17):3023

Ali MA (2022) Smart city policy in developing countries: case study of the new administrative capital in Egypt. J Public Aff. Wiley Online Library 22:e2774

Al-Soghir MMA et al (2022) Comprehensive assessment of soil chemical properties for land reclamation purposes in the Toshka area, Egypt. Sustainability. MDPI 14(23):15611

Aly MM et al (2023) Integrated management and environmental impact assessment of sustainable groundwater-dependent development in Toshka district, Egypt. Water. MDPI 15(12):2183

Attia S (2019) Al Alamein New City, a sustainability battle to win. In: New cities and community extensions in Egypt and the Middle East: visions and challenges. Springer, pp 1–18

Carmin J, Roberts D, Anguelovski I (2009) Planning climate resilient cities: early lessons from early adapters. In: Fifth urban research symposium, cities and climate change: responding to an urgent agenda, pp 28–30

Fahmy B, Kamiya M (2019) Productive urban development: linking planning and economy in Al-Alamein New City, Egypt. In: New cities and community extensions in Egypt and the Middle East: visions and challenges. Springer, pp 19–34

Goodling E, Green J, McClintock N (2015) Uneven development of the sustainable city: shifting capital in Portland, Oregon. Urban Geogr. Taylor & Francis 36(4):504–527

Hakeh AE (n.d.) The new administrative capital from an inclusive perspective. Available at https://aps.aucegypt.edu/en/articles/666/the-new-administrative-capital-from-an-inclusive-perspective

Hamed RE, Ramadan AM (2022) Collaborate to generate a more sustainable and smart new capital city. Int J Mech Eng 7:396–416

Hamza K (2016) Smart city implementation framework for developing countries: the case of Egypt. In: Smarter as the new urban agenda: a comprehensive view of the 21st century city. Springer, pp 171–187

Hasan S, Mohmed M (2024) Orientation towards using smart technical solutions in new cities, case study: Noor-city, new administrative capital in Egypt. Int Des J. Scientific Society for Egyptians Designers 14(2):185–193

Huber MT, Currie TM (2007) The urbanization of an idea: imagining nature through urban growth boundary policy in Portland, Oregon. Urban Geogr. Taylor & Francis 28(8):705–731

Iwasaki E, Negm AM, Elbeih SF (2021) Update, conclusions, and recommendations of "sustainable water solutions in the western desert, Egypt: Dakhla Oasis". In: Sustainable water solutions in the western desert, Egypt: Dakhla Oasis. Springer, pp 287–298

Lundqvist M (2007) Sustainable cities in theory and practice: a comparative study of Curitiba and Portland. Master thesis, University in Karlstad, Karlstad, Sweden

Lux A (2007) Towards a sustainable suburb: lessons from the Netherlands

Mahgoub Y (2022) Sustainability of tourism development in the city of Ain-Sukhna, Egypt. J Contemp Urban Aff 6(1):13–22

Mahgoub Y (2024) Assessing the efficacy of spatial planning and architecture in new cities: a case study of Galala City and its implications for responsible urban development. In: IOP conference series: earth and environmental science. IOP Publishing, p 12016

Metwally E, Samir E, Ayman M (2023) The indicators of the smartness to assess Egypt's New Administrative Capital as a smart city. Mansoura Eng J 48(6):12

Mohamed Hassanein H (2017) Smart technical street furniture design: case study of "New Cairo Administrative Capital". The Academic Research Community Publication

Newman P, Beatley T, Boyer H (2017) Resilient cities: overcoming fossil fuel dependence. Island Press

Reiche D (2010) Renewable energy policies in the Gulf countries: a case study of the carbon-neutral "Masdar City" in Abu Dhabi. Energy Policy. Elsevier 38(1):378–382

Richardson EA et al (2012) Green cities and health: a question of scale? J Epidemiol Community Health. BMJ Publishing Group Ltd 66(2):160–165

Rockström J et al (2013) Sustainable development and planetary boundaries. JSTOR

Rohracher H, Späth P (2014) The interplay of urban energy policy and socio-technical transitions: the eco-cities of Graz and Freiburg in retrospect. Urban Stud. Sage Publications 51(7):1415–1431

Saeed EH, Shenouda YA, Omar NA (2023) An integrated approach for sustainable development cities on the developed urban coastal corridors. J Eng Res. Tanta University, Faculty of Engineering 7(4):183–194

Selim HS, Mayhoub MS, Abuzaid A (2024) A comprehensive model to assess sustainable architecture in emerged megacities: a closer look at Cairo's New Administrative Capital (NAC). Sustainability. Multidisciplinary Digital Publishing Institute 16(12):5046

Shin S et al (2022) Systems thinking for planning sustainable desert agriculture systems with saline groundwater irrigation: a review. Water. Multidisciplinary Digital Publishing Institute 14(20):3343

Siahaya FZ, Hermana J (2013) Environmental impact assessment of land use due to the change of water resources potential in urban watershed: a review. J Basic Appl Sci Res 3:420–425

Smith J et al (2013) Potential impacts of climate change on the Egyptian economy. Report of a study implemented under the UN Climate Change Risk Management Joint Programme funded by the UN MDG Fund and the Finnish Government, prepared for the United Nations Development Programme (UNDP) with the Government of Egypt, UNDP, Cairo, Egypt

Smith JB et al (2014) Egypt's economic vulnerability to climate change. Clim Res 62(1):59–70

Soliman AH, El Zawahry A, Bekhit H (2019) GERD failure analysis and the impacts on downstream countries. In: Grand Ethiopian Renaissance Dam versus Aswan High Dam: a view from Egypt. Springer, pp 149–171

Phycoremediation Using Freshwater Algae and Safe Ecology

Sergei Yazvenko

Abstract Phycoremediation, the application of algae for the extraction and biotransformation of contaminants from aquatic environments, has surfaced as a viable sustainable method for water treatment and environmental remediation. This chapter reviews existing information on phycoremediation utilizing freshwater algae, examining its core concepts, frequently utilized species, and practical applications. It discusses the concepts of using freshwater algae for phycoremediation and assesses the species most commonly used and real-world applications of phycoremediation. The research investigates the processes involved in phycoremediation, such as biosorption, bioaccumulation, biotransformation, and biological volatilization. Certain species of freshwater algae, e.g., *Chlorella, Scenedesmus, Spirulina, Chlamydomonas,* and *Botryococcus*, have demonstrated their effectiveness in removing various pollutants, such as heavy metals, nutrients, and organic contaminants from water sources. These algae possess qualities that make them suitable for phycoremediation processes due to their high growth rates and efficiency in nutrient removal. Though there are benefits to using these algae for remediation, there are also challenges to consider, such as species selection, optimization of environmental conditions, difficulties in harvesting the algae, ensuring cost-effectiveness, and expanding applications to larger industrial-scale systems. Progress made in designing photobioreactors and applying modern engineering has significantly improved the efficiency of algae-based environmental remediation systems. The research underscores the importance of continuing and expanding research to address challenges and enhance the effectiveness of phycoremediation methods. This chapter gives an up-to-date overview of the status of using freshwater algae for environmental cleanup purposes, points to future areas of study, and informs strategies for managing the environment in the dynamic realm of sustainable water treatment.

S. Yazvenko (✉)
Faculty of Agro-Science and Business, International Agriculture University,
Tashkent, Uzbekistan
e-mail: sergei.yazvenko@rau.ac.uk

A. A. H. Abdel Latef et al. (eds.), *Sustainable Remediation for Pollution and Climate Resilience*, https://doi.org/10.1007/978-981-96-5674-5_11

Keywords Phycoremediation · Freshwater algae · Contaminants · Water treatment · Heavy metals · Biofuels · Wastewater

1 Introduction

Phycoremediation, the use of algae for the elimination or biotransformation of contaminants, has been attracting growing interest as a sustainable method for water treatment and environmental remediation (Guleri et al. 2020; Krishnamoorthy and Premalatha 2021; Covaliu et al. 2020). Freshwater algae have been identified as potentially highly suitable candidates for this technology based on their unique characteristics and versatility (Zainith et al. 2021; Pacheco et al. 2020; Jabri et al. 2020). The objective of this chapter is to summarize the current state of knowledge on phycoremediation using freshwater algae and to identify the directions of further development of the discipline, which is rapidly evolving at present.

Early research on phycoremediation can be traced back to the mid-twentieth century with groundbreaking studies of the treatment of human wastewater using algae. In the 1950s and 1960s, scientists initiated the studies of utilizing algae for nutrient removal from wastewater, recognizing algae's capacity to absorb nitrogen and phosphorus effectively (Emparan et al. 2019; Sunday 2018; Pacheco et al. 2020). William J. Oswald and his colleagues from the University of California, Berkeley, made great contributions in laying the fundamental principles for algal wastewater treatment (Oswald and Gotaas 1957; Oswald 2003; Covaliu et al. 2020; Porto et al. 2020).

Responding to increasing environmental concerns during the later part of the twentieth century, the research on algal-based remediation became more extensive in scope. It focused on a growing variety of pollutants. During the 1970s and 1980s, a surge of interest in employing algae (Wong and Tam 1998) to eliminate heavy metals occurred (Wilde and Benemann 1993; Pascucci and Kowalak 1996; Bulgariu and Gavrilescu 2015). Numerous studies documented the biosorption and bioaccumulation capabilities of various types of algae (Zeraatkar et al. 2016). In the 1990s, the scope of phycoremediation studies broadened to focus on the ability of algae to break down organic contaminants, such as pesticides and industrial toxicants (Sunday 2018; Pacheco et al. 2020; Wong and Tam 1998).

New scientific breakthroughs in biotechnology, material science, and environmental engineering have stimulated phycoremediation research to a new generation of technology and practical applications. The enhanced designs of photobioreactors, advances in genetic engineering methods, and sophisticated analytical procedures opened new opportunities for remediation using algae-based systems (Bansal et al. 2018; Sivasubramanian 2016; Emparan et al. 2019). The interest in phycoremediation has been consistently growing due to the ever-increasing focus on sustainable development, the application of circular economy concepts, and the ability of phycoremediation to combat environmental problems while generating valuable biomass (Emparan et al. 2019).

This chapter discusses the fundamental principles of phycoremediation, the most commonly used freshwater algae species, the types of contaminants that can be treated using freshwater algae, existing technologies, and practical applications, limiting factors to the effectiveness of phycoremediation, methods of analyzing the process, and potential areas for future research. It aims to provide academics, environmental managers, and policymakers valuable guidance on where and how further efforts in this promising sector should be directed, identifying major trends and gaps in current knowledge.

2 Essential Principles of Phycoremediation

The essential biological principles forming the foundation of phycoremediation are based on the inherent metabolic processes of algae. Algae encompass a heterogeneous collection of nonvascular photosynthetic organisms that display a wide range of morphological complexity (Büdel and Friedl 2024). Their structural organization varies from single-celled microalgae to large multicellular macroalgae reaching the size of a tree, e.g., giant kelp or bull kelp. The bulk of algal taxa lack the specialized tissues typical of vascular plants, but their distinct feature is that they all contain chlorophyll and are capable of photosynthesis (Anbuchezhian et al. 2015). Phycoremediation, the use of algae to remove contaminants, has become increasingly prevalent as a novel method for mitigating heavy metal contamination in water-based ecosystems.

The advantages of phycoremediation include relatively low cost, long-term biomass utility (algal biomass can be used for several years), the ability to sequester a broad spectrum of contaminants, high performance in diluted effluents due to high surface area-to-volume ratios of microscopic algal cells, superior heavy metal sequestration efficiency, resilience even in wastewaters with high concentrations of contaminants (e.g., heavy metals), viability in both aerobic and anaerobic treatment systems, and scalability from laboratory to industrial-scale implementations.

Algae play a critical role in aquatic environments as primary producers of organic matter and, thus, the main source of energy in aquatic food webs (Stevenson et al. 1996). This organic matter is created out of atmospheric CO_2 and water through the process of photosynthesis, which converts sun energy into the chemical energy of organic molecules and, as a byproduct, produces molecular oxygen O_2, which has become an essential element that sustains aerobic life on Earth, both aquatic and terrestrial. As critical constituents in biogeochemical cycles, algae impact nutrient dynamics and energy transfer throughout aquatic food webs.

Algal species are crucial in regulating water quality, actively participating in wastewater treatment processes, and producing substantial biomass. Their various ecological services include the sequestration of atmospheric CO_2 through photosynthesis, rehabilitation of polluted ecosystems, their possible effect on global warming, and many others.

Algae exhibit promise as a sustainable energy resource due to their high growth rates and ability to flourish even in wastewater (Alam et al. 2015). Algal species have higher biomass production per unit area than land-based crops. Lipids derived from algae provide a promising raw material for the development of biofuels, as they have a considerable potential to sequester CO_2 without requiring much of the arable land (Energy 2021). These microorganisms have developed robust mechanisms for absorbing nutrients (primarily nitrogen and phosphorus). These mechanisms can be used to remove pollutants from the ambient aquatic environment. This characteristic makes various algae suitable for treating wastewater and other eutrophic waters laden with nutrients. A better understanding of these processes provides a basis for optimizing phycoremediation processes and selecting appropriate algae species for particular applications.

1. Biological sorption, or biosorption, is a passive process in which the pollutants are adsorbed onto the surface of algal cells. Heavy metals present in wastewater and some organic contaminants are removed by biosorption (Derco and Vrana 2018; Srivastava et al. 2023). The algal cell wall contains polysaccharides, proteins, and lipids and includes functional groups, i.e., carboxyl, hydroxyl, sulfate, and phosphate, which can bind contaminants (Na et al. 2020). The factors that determine the extent of biosorption include pH, temperature, ionic strength, and the presence of competing ions (El-Ghany et al. 2023).
2. Bioaccumulation, in contrast, is a dynamic process in which contaminants are absorbed into the cell and subsequently concentrated within certain cellular organelles. This mechanism facilitates the removal of nutrients, some metals, and various other chemicals. Bioaccumulation frequently involves particular transport proteins and metabolic activity and, therefore, is a more selective process than biosorption (Chen et al. 2021; Chugh et al. 2022).
3. Biotransformation means altering the chemical structure of pollutants into less hazardous forms through a range of enzymatic reactions of algae. This is particularly relevant to organic contaminants, which can undergo chemical degradation into simpler compounds or complete mineralization (conversion into simple inorganic molecules; Zainith et al. 2021; Kandasamy et al. 2021). Enzymes present in some algae species can degrade complex organic molecules, including insecticides and medicinal substances (Casanova et al. 2023; Touliabah et al. 2022).
4. Biological volatilization is a process through which algae transform a pollutant into a volatile reactant released into the atmosphere (Touliabah et al. 2022; Saha and Fink 2022). Although less common, this mechanism has been reported for bioremediation of some metals, such as mercury, and certain chemicals (Feldmann 2002; Saxena et al. 2023).
5. Photodegradation (chemical degradation of organic compounds by the energy of the UV light): A high-density algal population will enhance the photodegradation of pollutants by increasing the level of dissolved oxygen in water and the production of reactive oxygen species (ROS) through photosynthesis.

Environmental parameters like pH, temperature, oxygen level, nutrients, presence of other ions, and light intensity exert a strong influence on the functioning of various mechanisms of phycoremediation (Ankit and Korstad 2022; Yuliasni et al. 2023; Hajipour et al. 2023).

The synergistic nature of these processes improves the efficiency of phycoremediation. The relative significance of each process varies concerning particular algae species, types of pollutants, and environmental factors (Redha 2020; Laraib et al. 2020; Zeraatkar et al. 2016; Dönmez et al. 1999).

The utilization of freshwater algae for phycoremediation presents numerous benefits:

1. Under optimal conditions, many algae species can achieve high growth rates, sometimes doubling their biomass every 24 hours (Carvalho et al. 2021; Almeida et al. 2023). This phenomenon enables the rapid uptake of pollutants and the generation of valuable algal biomass (Singh et al. 2021; Bulynina et al. 2023; Baldisserotto et al. 2023).
2. Nutrient removal efficiency: Algae are highly effective in the removal of excess nutrients from water, especially nitrogen and phosphorus (Razaviarani et al. 2022; Zhang et al. 2023a, b). This characteristic makes them ideal for the treatment of eutrophic water as well as nutrient-rich wastewater.
3. Photosynthesis by algae contributes to CO_2 sequestration, adding to the system's environmental benefits. Recent estimates suggest that algae can fix CO_2 at rates that are 10–50 times greater than those of terrestrial plants, per unit mass of the plant (Ralph and Pernice 2023; Zieliński et al. 2023; Onyeaka et al. 2021).
4. Algal species show high adaptability and flexibility and are able to adequately respond to diverse and fluctuating environmental conditions and varying levels of pollutants (Vale et al. 2020; Even et al. 2022; Barati et al. 2021).
5. Biomass valorization is defined as the use of the biomass obtained from the phycoremediation process for the production of biofuels, fertilizers, or other high-value chemicals (Martins et al. 2022; Amulya et al. 2023; Calvo-Flores and Martín-Martínez 2022). Valorization helps make phycoremediation more economically viable by creating additional products.
6. Algae produce oxygen through photosynthesis, which is beneficial to aerobic bacteria involved in the degradation of organic contaminants in so-called mixed treatment systems. Algae and aerobic bacteria can operate synergistically to increase the overall effectiveness of remediation (Kilbane 2022; Zheng et al. 2023; Arango et al. 2022).
7. Last but not least, freshwater algae live and thrive in fresh water, which means that they could be naturally used to decontaminate and clean up drinking supply, irrigation water, and fresh water used by humans for a variety of other purposes (Radziff et al. 2021; Salman et al. 2023; Baldisserotto et al. 2023; Touliabah et al. 2022).

Despite these benefits, phycoremediation with freshwater algae presents certain obstacles and limitations that need to be addressed:

1. The efficiency of phycoremediation greatly depends upon the algal species composition; considerable knowledge is necessary to select algae strains for particular applications (Carvalho et al. 2021; Abdelfattah et al. 2023).
2. Environmental parameters such as light intensity, temperature, and pH may significantly affect the process of algae growth and their effectiveness in removing pollutants (Dhiab 2012; Hanifzade et al. 2017; Tarlan et al. 2002; Zhang et al. 2023a, b). Fine-tuning of conditions in systems at an industrial scale offers enormous challenges.
3. Some of the challenges that are associated with harvesting microalgae include the microscopic size of the unicellular algae, which makes the process of collecting and recovering the biomass technically complex, energy-intensive, and therefore costly (Muylaert et al. 2017; Uduman et al. 2010; Singh and Patidar 2018).
4. In open phycoremediation systems, there is a chance of algal escape and ensuing ecological effects on the adjacent natural aquatic ecosystems (Saxena et al. 2019; Zaimes and Khanna 2013).
5. Seasonal variations: Outdoor phycoremediation systems are exposed to changes in weather conditions such as temperature and light intensity; these fluctuations can impact their operational efficiency over the year (Xu et al. 2021; Lukács et al. 2021).
6. Potential contamination: Open algal growth methods are vulnerable to introducing undesirable microorganisms, which may reduce the overall effectiveness of the entire phycoremediation system (Lam et al. 2018; Narala et al. 2016).

Finding solutions to these challenges is one of the major goals of modern research in phycoremediation (Kandasamy et al. 2021).

3 Species of Freshwater Algae Frequently Employed in Phycoremediation

A large number of species of freshwater algae have been investigated for their capacity to be used in phycoremediation (Guleri et al. 2020). Some freshwater algal taxa are especially promising for phycoremediation due to their high tolerance, rapid growth rates, and efficacy in removal, degrading, and immobilizing pollutants.

1. Some species of Chlorella, such as *Chlorella vulgaris, C. Sorokiniana,* and *C. pyrenoidosa (Condylium pyrenoisodes)*, were among the algae that have been extensively researched for phycoremediation (Saxena et al. 2019; Doshi et al. 2008; Takáčová et al. 2021; Bulynina et al. 2023). Their rapid growth, high biomass productivity, and ability to adapt to a wide range of environments are some useful characteristics of these unicellular green microalgae. Chlorellas have demonstrated high efficiency in removing nutrients, heavy metals, and various forms of organic pollutants in wastewater (CORONADO-REYES et al. 2022;

Abreu et al. 2023). Ahmad et al. (2019) have used *C. vulgaris* to treat the palm oil mill effluent (POME). The investigations indicated these algae could remove 60.1% of the chemical oxygen demand (COD), 57.8% of the ammoniacal nitrogen, and 84.1% of the orthophosphate from the POME within 12 days. In addition, the algal biomass with a lipid content of 32.3% of its dry weight could be used in biodiesel production.

2. Certain species of *Scenedesmus*, e.g., *S. obliquus* and *S. quadricauda,* are widely used in phycoremediation studies (Santo et al. 2023; Wang et al. 2023; Shen et al. 2020). The high efficiency of *Scenedesmus* species in nutrient removal, biodiesel production, and heavy metal biosorption has been widely reported (Abomohra et al. 2016; Msanne et al. 2020). These colonial algae show high viability in wastewater; they contain high lipid content, making them suitable for biofuel synthesis as an additional benefit. He et al. (2023) evaluated the capability of *S. obliquus* in removing piggery-related effluent. The study found elimination rates of 98.6% for total nitrogen, 98.4% for total phosphorus, and 86.5% for COD over 10 days of treatment (Ji et al. 2013; He et al. 2023; Tan et al. 2024).

   The researchers also noted that the algal biomass created had substantial protein content, making it a good potential source for livestock feed production.
3. Species of spirulina (genus *Arthrospira*), including *Arthrospira platensis* and *A. maxima,* are filamentous cyanobacteria (Cyanobacteria, Prokaryota) known for their abundant protein content and capacity to thrive in alkaline environments. They have shown promise in the bioremediation of industrial effluents and the removal of heavy metals from polluted watercourses (Lisboa et al. 2018; Tabagari et al. 2019; Papadopoulos et al. 2022; Abreu et al. 2023). Markou et al. (2012) studied the effectiveness of *A. platensis* in treating effluent from olive oil mills. After 24 days of treatment, a 79% reduction in phenolic chemicals and an 80% reduction of COD were observed. Due to its high nitrogen content, the biomass produced was considered a potential source of biofertilizer.
4. The ability of *Chlamydomonas* species to degrade organic pollutants and due to their genetic malleability, these species are invaluable both for the applied and theoretical research of phycoremediation processes (Bellido-Pedraza et al. 2024). *Chlamydomonas reinhardtii* is a single-cell green alga that has become a model organism for studying algal systems and has significant potential for phycoremediation applications (Souza and Silva 2019; Tatarová et al. 2021). Its well-studied genome and relative ease of genetic technologies make it a suitable candidate for creating genetically modified strains with enhanced remediation capabilities. A study by Yu et al. (2022) discussed using *C. reinhardtii* to eliminate tetracycline from water. It has been found that *C. reinhardtii* can remove up to 75% of tetracycline from the aquatic environment within one week; both biosorption and biodegradation played a major role.
5. The remarkable ability of *Botryococcus* species to absorb large amounts of hydrocarbons, including oil and oil products, makes species of *Botryococcus* highly suitable candidates for the coupled processes of bioremediation and biofuel production (Lozoya-Gloria et al. 2020). Research carried out by Órpez et al. (2009) assessed the ability of *B. braunii* to treat urban wastewater. The result was

that nitrogen removal efficiency was 79.6% and phosphorus was 96.8%. In addition, the algal biomass produced had up to 45% hydrocarbon by dry weight, which showed the ability of algae to manufacture biofuels.
6. Studies have demonstrated that *Ankistrodesmus* species have a high capacity to remove pharmaceuticals and personal care products (PPCPs) from water and have the potential to remediate effluents from the textile industry (Goswami et al. 2022). Santos et al. (2023) assessed the suitability of *A. falcatus* for removing pharmaceuticals from the effluent of the wastewater treatment plants. The study found that within one week, *A. falcatus* could eliminate over 80% of diclofenac and over 95% of metoprolol from the effluent.
7. The reference strain PCC 6803 of *Synechocystis*, a cyanobacterium, has been examined in detail for its application in phycoremediation and biotechnologies (Wu et al. 2023). Marchetto et al. (2021) focused on the potential of genetically modified *Synechocystis* sp. PCC 6803 to eliminate industrial pollutants such as per- and poly-fluoroalkyl substances (PFAS). The strain has partially removed 88% and 37% of perfluorooctanoic acid (PFOA) and perfluorooctane sulfonate (PFOS), respectively. This was an indication of its relative efficiency in treating industrial wastewater.
8. In the removal of medicines, *Chlorococcum* species have shown exceptional efficiency (Chandel et al. 2022).

The variation of abilities and traits found in various algal species implies that careful consideration is needed when deciding on the exact strains of algae to be used in phycoremediation processes. Critical considerations for selecting algae species for certain purposes include the following:

1. Environmental pollution mitigation effectiveness.
2. Growth rate and biomass yield.
3. Tolerance of organisms to stress factors and concentrations of pollutants in the environment.
4. Facilitation of cultivation and harvesting.
5. Biomass valorization potential.
6. Designation of native species status (to minimize ecological risks in open systems).

Phycoremediation application of algae in a specific environment starts with the identification of the most appropriate algae species that are suitable for a given task, which includes aspects such as the type of pollutants to be treated, the geographical region and environmental conditions, and the intended use of the accumulated algal biomass (Bhatt et al. 2022). Recent studies are identifying new algae species that can be utilized in phycoremediation, thus increasing the arsenal of tools that can be used in environmental remediation systems. Effective techniques, including selective breeding of current strains, genetic engineering, and adaptive laboratory evolution, are expanding the arsenal of methods available for algal-based remediation.

4 Categories of Contaminants Remediated by Freshwater Algae

The versatility of freshwater algae used in water treatment can be attributed to their ability to remove or immobilize a diverse array of contaminants that can be classified into different categories, with nutrients and heavy metals being the most significant among them.

4.1 Nutrients, Namely, Nitrogen and Phosphorus

Removal of excess nutrients, particularly nitrogen and phosphorus, can be effectively done by algal phycoremediation (Bansal et al. 2018; Emparan et al. 2019). Freshwater algae can metabolize nitrogen and phosphorus compounds, including nitrates, nitrites, ammonium, and various phosphates. Due to this trait, they are useful in treating municipal wastewater, agricultural runoff waters, and various eutrophic water bodies. Removing excess nutrients enhances water quality and mitigates or even eliminates adverse effects like eutrophication, algal blooms, and hypoxia in the water that receives it. Abolhasani et al. (2018) conducted a detailed study using *Chlorella vulgaris* and *Scenedesmus obliquus* to treat municipal wastewater in an open-air pond system with high algae growth rates. Over two weeks, the system demonstrated mean removal rates of 90% for total nitrogen and almost 100% for total phosphorus.

4.2 Heavy Metals

Research has indicated that freshwater algae can be highly efficient in heavy metal remediation (Ahmad et al. 2020). These metals include lead (Pb), cadmium (Cd), copper (Cu), zinc (Zn), chromium (Cr), and arsenic (As), while the processes that enhance their removal by algae include biosorption on the cell surface and intracellular accumulation. The efficiency of the process depends on such factors as pH, concentration of metals, and species of algae used. Some of the algae can hyperaccumulate (i.e., selectively accumulate) certain metals, concentrating them to higher levels than those found in the surrounding aquatic system.

Numerous researchers examined the use of the freshwater alga *Spirogyra* sp. for the removal of toxic heavy metals from industrial wastewater (Nimisha and Joseph 2020; Almeida et al. 2023). Removal efficiencies of >90% for lead and copper and >70% for zinc were reported in a number of studies.

4.3 Organic Pollutants

Organic pollutants cover a range of contaminants that can be effectively treated using phycoremediation with freshwater algae. These pollutants include pesticides, herbicides, polycyclic aromatic hydrocarbons (PAHs), and pharmaceutical and personal care products (PPCPs). Certain species of algae possess enzyme systems capable of breaking down these chemicals via enzymatic transformations or converting them into less harmful metabolites. For example, studies have documented that several species of green algae, *Chlorella* and *Scenedesmus*, can break down pesticides and PAHs effectively (Touliabah et al. 2022; Sharma et al. 2024). Lu et al. (2023) explored how effective various microalgae are in removing antibiotics from the environment. The review concluded that algae were efficient in removing antibiotics from the environment.

4.4 Persistent Organic Pollutants (POPs)

Some algae species have demonstrated promise in the removal of persistent organic pollutants, which is significant because of their recalcitrance to conventional treatment methods. Luo et al. (2014) focused on the effectiveness of using the green alga *Selenastrum capricornutum* to remove polycyclic aromatic hydrocarbons (PAHs). The study found that *S. capricornutum* could remove 96% of phenanthrene and 89% of pyrene from water within 12 days. Both bioaccumulation and metabolic transformation have been confirmed as pathways of the elimination process.

4.5 Emerging Contaminants

There are growing concerns regarding emerging pollutants such as microplastics, nanoparticles, antibiotic-resistant bacteria (ARB), and endocrine-disrupting compounds (EDCs) in aquatic environments. Although current studies in this field are still limited, some research has identified positive results in using algae to tackle these pollutants. Some species of algae can absorb microplastics (Manzi et al. 2022), thus presenting the potential for removing these particles from water. In addition, certain algae have been shown to possess the ability to degrade or sequester some of the substances that may disrupt the normal functioning of the endocrine system. Cunha et al. (2020) analyzed the process of microplastics' interaction with the freshwater algae *Chlorella* sp. The study found that the algae could absorb and agglomerate the microplastic particles and, therefore, can help in the removal of such pollutants from water systems. Algae cannot be considered thus far as a complete solution for the solution of the problem of microplastic contamination, but this work shows that algae can contribute to the reduction of microplastic contamination.

4.6 Chemical Dyes and Pigments

Algae have been identified to have the ability to remove chemical dyes and effluents from wastewater of textile and other industries (Adane et al. 2021; Khan et al. 2022). Hernández-Zamora et al. (2015) investigated the usability of *Chlorella vulgaris* in the removal of textile dye, Reactive Red 272. The study revealed the maximum decolorization efficiency of 94.5% after 96 hours of treatment. *Chlorella* was found to have the ability to retain its viability after being subjected to dye, thus showing the possibility of continued treatment.

4.7 Persistent Radioactive Pollutants

Some algae species have been found to have the ability to absorb radioactive materials, thus having the capacity to treat nuclear waste or contaminated sites (Kim et al. 2020). Microalgae, among other organisms, were evaluated as potential agents for treating radioactive materials (Vanhoudt et al. 2018). Over 20 organisms/biosorbents were identified that showed high potential to clean waters contaminated with (mixtures of) radionuclides, several species of microalgae among them. The main mode of cesium (Cs^+) elimination has probably been linked to potassium (K^+) uptake because the two ions share comparable chemical characteristics (Lee et al. 2018).

4.8 Combustion and Refined Petroleum Products

Certain algae can decompose or trap hydrocarbons, and this makes them valuable in the process of cleaning up oil spills or petroleum pollution in water bodies. A study that assessed the use of *Scenedesmus obliquus* for the biodegradation of crude oil (El-Sheekh et al. 2013) showed that *S. obliquus* can degrade up to 89% of the components of crude oil within a 15-day treatment period. It was also noticed that the algal biomass increased during the process, suggesting that the algae might use oil as a source of carbon.

4.9 Pathogens

Some of the species of algae have been shown to help reduce the presence of diseases and bacterial pollution in water (Moondra et al. 2020; Kumar et al. 2021). Water quality can be improved in terms of its microbiological quality by algal bioremediation because some algae species synthesize antimicrobial substances as well as create adverse conditions for pathogens, such as high pH and increased levels of dissolved oxygen.

The high number of contaminants that can be treated by freshwater algae proves the potential and versatility of phycoremediation for wastewater treatment and ecological rehabilitation. The efficiency of phycoremediation depends on the choice of specific microorganisms (algae species), as well as on characteristics of pollution and its concentration and environmental conditions (Guleri et al. 2020). These parameters are important for the enhancement of the use of phycoremediation systems in real-world situations.

5 Phycoremediation Systems and Technical Advances

Optimizing algal phycoremediation requires carefully engineered systems and advanced protocols that facilitate algal growth and concurrently facilitate the efficient removal of pollutants. Several technologies have also been designed and optimized for this purpose, including open pond systems, closed photobioreactors, and immobilized algal systems.

1. Open pond systems include raceway ponds and high-rate algal ponds (HRAPs) that are widely used for the cultivation of algae and wastewater treatment (Kumar et al. 2021). These systems have a straightforward design and function that includes shallow channels in which the algae culture is cultivated under natural sunlight, and water flow is managed by paddlewheels. The advantages of open ponds are that they are cost-effective to construct and maintain, easy to scale up, and have low energy requirements. Almost 90% of microalgae cultivation occurs globally in open surface systems (Acién Fernández et al. 2017). The disadvantages of open systems are that they are vulnerable to pollution and have limited environmental control; the productivity per unit volume is lower compared to closed systems; therefore, larger areas of land are required for open systems. Craggs et al. (2012) evaluated a pilot-scale high-rate algal pond system in New Zealand that was designed to treat municipal wastewater. The culture used was a mixed culture, mostly of *Pediastrum* sp. and *Micractinium* sp. The system achieved mean removal efficiencies of 82.8% for nitrogen and 80.1% for phosphorus. The study's findings pointed to the feasibility of algal ponds for the treatment of wastewater on a year-round basis in a temperate climate with mild winters.
2. Closed photobioreactor systems have better control over the conditions of growth and can achieve higher biomass production rates than open ponds (Shen et al. 2009). These systems are available in different designs, such as a tube, flat panel, and column photobioreactors. Tubular photobioreactors are made up of transparent tubes where the movement of the algal culture is enhanced by the flow of the culture medium to optimize light exposure. Flat panel photobioreactors with thin, long, and flat panels have a high surface area-to-volume ratio for optimal

algal growth. Vertical column photobioreactors are cylinder-shaped systems designed for efficient aeration and thorough mixing of the culture. Each layout has its advantages and disadvantages as to light distribution, the maximization of the surface area, and space management. Some advantages of using photobioreactors include increased biomass production rates, better control of culture conditions, and reduced chances of contamination. These designs improve the effectiveness and reliability of the bioremediation processes.

Ting et al. (2017) compared the performance of tubular and flat panel photobioreactors in treating secondary effluent of municipal wastewater treatment plants using *Chlorella zofingiensis*. Compared with the tubular system, the flat panel design had 20% higher nitrogen removal efficiency and 15% higher phosphorus removal efficiency. This work emphasized the significance of photobioreactor design in nutrient removal efficiency and biomass production.

Compared to open systems, photobioreactors require greater capital investment and are more expensive to operate. The systems are complex, thus demanding specialized expertise to operate and maintain them. Furthermore, there are significant technical challenges in increasing the size of photobioreactors to industrial magnitudes.

Notwithstanding, photobioreactors have a high potential to be an effective tool for bioremediation using algae. These systems are especially useful in processes that demand exceptional purity and reliable performance, as they can maintain ideal growing conditions and reduce extraneous influences (Acién Fernández et al. 2017). However, more research has to be conducted on the scaling issues and increasing operational effectiveness.

3. Immobilized algal systems are a new concept of phycoremediation, where algae cells are attached to a support matrix (Wang et al. 2023). These systems employ different immobilization techniques, including encapsulation in polymeric matrices like alginate beads, adsorption on porous materials, and covalent bonding to activated surfaces.

Advantages of immobilized systems include streamlined biomass separation and reduced operational expenses for subsequent handling. This technology can achieve cell densities greater than that of suspended cultures and, thus, has the potential to increase productivity per unit volume of the system. The support matrix acts as a protective microenvironment that shields algal cells from toxic substances and other stress factors in the environment, thereby increasing the life span of the bioremediation system and ensuring uninterrupted functioning. Using algal beads in packed bed bioreactors for wastewater treatment and growing algal biofilms on various surfaces in contact with polluted water are both viable applications (Zainith et al. 2021). The immobilized systems can achieve higher rates of nutrient and pollutant removal compared to free cell systems (Pant et al. 2024) and can be reused for several treatment cycles, thus reducing operational costs.

However, immobilized systems are not without challenges. Mass transfer constraints are possible, especially in thick matrices where the uptake rate of pollutants may decline. The support matrix may develop cell leakage, which may require frequent system maintenance. The initial costs of the immobilization materials and procedures are generally higher compared to suspended systems (Mollamohammada et al. 2022).

Despite these difficulties, immobilized algal systems have the potential for targeted phycoremediation applications, primarily when biomass conservation is needed or when toxic components hinder the growth of the algal cells. Current research efforts are directed toward the development of innovative immobilization matrices and the optimization of the mass transfer to enhance the efficiency of these systems.

4. The bioremediation systems may involve algal–bacterial consortia (Muñoz and Guieysse 2006) in which both photoautotrophic algae and heterotrophic bacteria work to increase the efficiency of pollutant removal through synergistic interaction. These systems take advantage of complementary metabolic processes of both organisms since algae produce oxygen and organic compounds for the bacteria through photosynthesis, while the bacteria release CO_2 and nutrients that are used by algae to produce more oxygen and organic compounds through photosynthesis.

There are many advantages of using algal–bacterial consortia in bioremediation systems, such as enhanced process efficiency due to the combined metabolic capabilities, the ability to treat a broader spectrum of contaminants, including nutrients, organic compounds, and certain xenobiotics, and improved system stability achieved by increased biodiversity (Guo et al. 2023). Such a system that combined algal and bacterial components was studied by Mu et al. (2021) to determine its effectiveness in the treatment of high concentrations of agricultural wastewater. The system dominated by *Chlorella vulgaris* and heterotrophic bacteria achieved the removal efficiency of 98% for ammonia nitrogen, 96% for total phosphorus, and 95% for COD. The study emphasized the synergistic effect of algal–bacterial interaction on increasing pollutant removal.

Like any other system, algal–bacterial consortia also have constraints. Algae and bacteria are known to interact in a highly complex manner, making the system's behavior difficult to predict and control. Certain circumstances may lead to competition for resources such as nutrients and light. Also, it may be challenging to maintain the intended population composition in the system, especially in open systems that are susceptible to environmental variations and contamination.

However, algal–bacterial consortia show potential for use in some bioremediation activities, especially in cases when single-species systems may be ineffective. Current research is focused on improving species selection, better understanding of species interactions, and developing strategies for maintaining a stable community structure within the system.

5. The algal turf scrubber (ATS) system is an innovative bioremediation technique that utilizes benthic algae grown on a specific surface to eliminate contaminants from flowing water (Adey et al. 2011). These systems consist of a sloping

textured surface on which water is either pumped or flows naturally, thus allowing the growth of filamentous algae, microalgae and other microorganisms.

ATS has the benefit of high efficiency in treating large volumes of water, especially in open systems such as rivers or coastal water. The relatively shallow layer of the flowing water, which can be sustained, allows enough light to reach the bottom and high oxygen concentration, thus promoting algal growth. The algal community has a natural propensity to grow attached to surfaces, making it easier and more cost-effective to harvest biomass through simple mechanical means such as scraping. ATS can be incorporated into natural water systems; hence, it can treat water in situ. Also, ATS enhances CO_2 sequestration by encouraging its uptake via algal photosynthesis. Also, they contribute to the sustainable cultivation of biomass by using the harvested algae for biofuels, animal feed, and other high-value products. Generally, these systems use less energy than conventional wastewater treatment systems, making them more eco-friendly. In addition, ATS systems can provide habitat for various aquatic organisms, increasing the biodiversity of the water systems (Higgins and Kendall 2012).

Adey et al. (2013) evaluated a large-scale ATS system to handle agricultural runoff in the Chesapeake Bay watershed. The system dominated by filamentous green algae achieved a removal efficiency of 20–40% for total nitrogen and 40–60% for total phosphorus from the influent water. The study's findings showed that ATS systems have the potential to improve water quality on a large scale and in a cost-effective manner.

However, there are limitations associated with ATS systems. In some applications, it will be necessary to pump water to maintain the flow of nutrients needed for algae growth and to prevent desiccation. The disadvantage of ATS systems is that the initial investment costs are comparatively high compared to conventional systems. Although most of them need minimal maintenance relative to other technologies, periodic maintenance is essential to check for the state of the algae and potential blockages (Adey et al. 2013). The growth of algae is temperature-dependent, thus making the system vulnerable to changes in weather and climatic conditions. Currently, there is still limited data on the efficiency of this technology in the long run, as well as on best practices for its application. Mismanagement of the ATS systems can lead to the occurrence of adverse algal blooms, which will affect water quality and aquatic ecosystems.

Nevertheless, ATS systems have demonstrated potential in many fields, including the treatment of agricultural runoff, the polishing of urban wastewater, and as part of environmental restoration projects. Current research efforts are directed toward the optimization of algal community structure, improvement of nutrient removal efficiency, and the development of more efficient harvesting algorithms to increase the overall performance and economic feasibility of these systems.

6. The use of membrane pressure bed reactors (MPBRs) that combine algae cultivation and membrane filtration technologies (Castro-Muñoz and García-Depraect 2021) is a sophisticated approach aimed at increasing biomass attachment and improving the quality of the effluent in bioremediation systems.

Usually, these systems consist of a photobioreactor that enables algae cultivation and a membrane filtration unit that predominantly utilizes microfiltration or ultrafiltration membranes.

One of the main benefits of MPBRs is their ability to retain suspended particles and microbes, thus producing purified effluent of superior quality. This filtration technique helps in the efficient segregation of biomass, which in turn helps with processing and may reduce harvesting costs. MPBR allows for the maintenance of high cell density within the reactor, thus enhancing the volumetric productivity and rates of pollutant removal. In recent work, Gao et al. (2018) assessed a membrane photobioreactor with *Chlorella vulgaris* to treat tertiary wastewater. The removal efficiency of the system was 98% for ammonia nitrogen, 97% for total phosphorus, and 95% for COD. This work highlighted the potential of membrane photobioreactors for the production of purified water of superior quality.

The critical challenge of using MPBRs includes biofilm fouling, resulting from the accumulation of algal cells, extracellular polymeric substances, and other organic matter on the surface of the membrane. Fouling can reduce filtration efficiency; therefore, the membrane must be cleaned or changed regularly. The energy consumption of MPBRs is generally higher than that of conventional systems because of the energy needed for the operation and maintenance of the membrane. Furthermore, the application of membrane technology complicates the operations, thus requiring staff with specialized expertise in system management and performance enhancement.

However, studies indicate the potential of using MPBRs in situations that require high-quality effluents, such as water reuse systems or sensitive receiving environments (Gonzalez-Camejo et al. 2020). Recent studies are directed toward developing membranes that are resistant to fouling, improving hydrodynamics to minimize fouling, and improving energy efficiency to make these systems more economical and environmentally sustainable.

7. Algal-based constructed wetlands are human-made systems that incorporate basic principles of constructed wetlands with biological mechanisms of algae control. Such systems usually consist of shallow basins containing aquatic plants and substrate. They are engineered to encourage the growth of different types of algal species together with other microorganisms, and phytobenthos (Gupta et al. 2021).

The benefits of algal-based built wetlands are their minimal energy demands because they mostly depend on natural processes in their operations. The constructed wetlands can create habitats and, as a result, enhance biodiversity and provide ecosystem services that go beyond water treatment. They are effective in removing a wide range of pollutants—nutrients, organic matter, pathogens, and some new pollutants—through biosorption, biodegradation, and phytoremediation. Patro et al. (2023) investigated the applicability of an algal-based built wetland system in China for the treatment of rural household sewage. A green alga *Cladophora* sp. and duckweed (Lemnaceae, Anthophyta) were the dominant species in the system. Duckweed and other plants remove up to 88% of the total nitrogen, 94% of the total phosphorus, and 93% of the COD. The work highlighted the efficacy of algal-based constructed wetlands for decentralized wastewater treatment in rural areas.

Algal-based built wetlands have some constraints. A majority of such systems require large areas to provide treatment; hence they are unsuitable for urban areas or limited spaces. Seasonal variations in performance can occur due to changes in temperature, radiation, and plant growth and development, which may require substantial knowledge of ecosystem functioning and the application of adaptive management techniques. The wetlands can potentially breed mosquitoes, particularly in areas with stagnant water, and hence require other vector control measures (Patro et al. 2023).

Despite these challenges, manmade wetlands based on algae can be considered an efficient, sustainable solution for water treatment systems, particularly in rural or peri-urban areas, where the availability of land is not a limiting factor. Most of the current research is directed toward a deeper understanding of algal community structure, increasing the rate of pollution removal, and providing solutions to the problems of seasonality and issues with mosquitoes. Adopting these systems in the landscape design and ecosystem restoration programs presents opportunities for establishing multifunctional green infrastructure.

8. A new generation of bioremediation systems includes the revolving algal biofilm reactors (RABRs; Jones et al. 2023) that use discs or drums that are partially submerged in wastewater to support algal growth on their surface. The systems also consist of a series of circular semi-submerged discs or cylindrical drums affixed to a horizontal shaft, which rotates the discs with biofilm in an alternating pattern that makes them in contact with both air and wastewater.

A key advantage of RABRs is the effective use of light, because the system's rotation provides uniform access of the whole biofilm to full sunlight and, therefore, to photosynthesis without desiccation (Jones et al. 2023). Biofilm biomass is easily collected by mechanically scraping the discs. RABRs can attain higher treatment capacity within a small area compared to conventional pond systems. Christenson and Sims (2012) looked into a pilot-scale rotating algal biofilm reactor that is intended to treat municipal wastewater. The system hosted a heterogeneous algal community and achieved nitrogen removal efficiency of 92% and phosphorus of 85%. The study indicated that rotating biofilm systems had the potential for wastewater treatment and algal biomass production.

Nevertheless, RABRs are not without limitations. The complex mechanical design of the rotating RABRs requires skilled maintenance and may require the periodic intervention of a specialist. Rotational motion requires energy, making it more costly to operate compared to passive systems. Also, there may be different rates of biofilm formation and development in different parts of the system where light exposure or hydrodynamic circumstances may vary, which can affect the system's effectiveness.

However, RABRs have possible uses in scenarios requiring high biomass production or situations with limited space (Gupta et al. 2015). Presently, research is focused on the optimization of rotation rates and patterns, enhancement of attachment and uniformity of biofilms, and energy efficiency. Potential exists for making water treatment systems more sustainable and resource-efficient by integrating RABRs with other processes, such as anaerobic fermentation of harvested biomass.

There is ongoing research for the development of novel and efficient methods of phycoremediation. One area of concern is the enhancement of the remediation potential of algae through genetic engineering, e.g., by increasing their metal absorption or their ability to degrade specific organic pollutants. Another emerging field is the application of nanotechnology in algal systems to augment algal growth or improve pollutant removal effectiveness.

Factors that affect the choice of a phycoremediation system include the type and quantity of the wastewater to be treated, climate, size of the area used, and cost. Several systems that have been developed possess distinct benefits and constraints; hence, ongoing research aims to refine and optimize these systems for various applications and operational scales.

6 Possible Further Developments in Phycoremediation Systems

1. The use of sophisticated materials, including nanocomposites and smart polymers, to enhance the growth of algae and achieve efficient removal of pollutants.
2. Expansion of hybrid systems that integrate phycoremediation with other treatment systems, including enhanced oxidation processes and electrochemical treatment methods.
3. The use of artificial intelligence and machine learning to improve the system's performance in real time.
4. Exploration of the options for the vertical and urban cultivation of algae to address space issues in densely populated areas.
5. Improvement of portable and modular phycoremediation systems that can be readily deployed in the case of emergencies or remote areas.

As these technologies continue to advance, phycoremediation systems will increase in their application to water quality issues in various economic sectors and geographic areas.

7 Factors Influencing the Efficiency of Phycoremediation

The effectiveness of phycoremediation procedures depends on the complex interaction of technological, environmental, and biological factors. Therefore, a deep understanding and effective optimization of these parameters are vital in efficiently applying water treatment systems that contain freshwater algae.

Environmental parameters affect algae's growth and efficiency in removing pollutants. Light intensity and photoperiod influence photosynthesis directly, thus affecting the growth and metabolism of algae. Even though most algae grow best in high light intensities, too much light exposure may result in photoinhibition. Temperature is another factor, as the majority of freshwater algae thrive within the temperature range of +20–30 °C. Algal growth, as well as the mobility of certain pollutants, especially heavy metals, is also dependent on the pH. The algal respiration rate and organic contaminants' degradation rate depend on dissolved oxygen concentration (Almeida et al. 2023).

Nitrogen and phosphorus are the key nutrients critical for the functioning of algae (Bulynina et al. 2023); therefore, their availability and equilibrium affect algal growth and community composition. Even though algae are widely used for nutrient removal, it is imperative to foster nutrient balance to ensure the optimal growth of the algal community and the elimination of pollutants. Carbon concentration (dissolved CO_2 and/or bicarbonate) also plays a role in algal productivity.

The choice of algae species or their combination is essential for the success of the phycoremediation process (Mustafa et al. 2021). Individual species vary in their tolerance to environmental factors and the ability to remove environmental pollutants. Under some circumstances, using a mix of different algae species is more advantageous and has greater resilience and efficacy than a monoculture.

Remediation efficiency depends on the concentration and the type of contaminants in the water to be treated (Almeida et al. 2023). High concentrations of pollutants may be detrimental to algae, while low levels may not present enough concentration for adsorption or degradation. A combination of contaminants may bring about competitive reactions or synergistic effects on removal efficiency.

Hydraulic retention time is one of the operational parameters that affect the performance of phycoremediation systems since it determines the length of time algae will contact pollutants. Longer retention times generally lead to better pollutant removal, but they reduce the overall treatment rate. To ensure that the algae get the necessary light, nutrients, and contaminants, they have to be thoroughly mixed and have an efficient mass transfer. It will also help to avoid sedimentation of algal biomass at the bottom of the culture and ensure uniform distribution of contaminants in the culture system.

Other factors that may significantly influence the success of algal colonies in an open environment include competition with other microorganisms and predation by zooplankton (Ullmann and Grimm 2021). These interactions can impact the rates of algal growth and the composition of the algal community, which may, in turn, influence the efficiency of the phycoremediation process.

These elements usually have intertwined relationships in which the best conditions for algal growth have to be balanced with the practicalities of water treatment. Current research efforts are directed toward creating robust algae strains and efficient cultivation protocols that could achieve high remediation efficiency across various environmental settings.

8 Contemporary Applications: Examples and Case Studies

Freshwater algae phycoremediation has been implemented in many industries to test and confirm its versatility and potential as a sustainable water treatment solution. Numerous pilot research and case studies have been conducted to prove this approach is suitable for application in practice.

Algae-based systems have been utilized to treat municipal wastewater (Nandeshwar and Satpute 2014), particularly as a tertiary treatment to eliminate remaining nutrients and polish effluents before discharge.

The second major area of freshwater algae phycoremediation is the treatment of industrial wastewater. In the textile sector, where wastewater frequently contains elevated concentrations of dyes and heavy metals, *Chlorella vulgaris* and *Scenedesmus obliquus* have been identified as promising candidates for decolorizing wastewater and removing hazardous metals. A series of studies carried out in India demonstrated the effectiveness of *Chlorella*-based and other algal systems in treating textile industry effluents with high efficiency in color and heavy metals elimination (Bhattacharya et al. 2021; Verma et al. 2024; Ravishankar et al. 2024).

Algae systems have been used in food processing to treat wastewater laden with nutrients (Covaliu et al. 2020). Algal systems have been identified to be efficient in treating effluents from mining and metallurgical industries that contain elevated levels of metal contaminants and high acidity. In a pilot work performed in Canada, an algal bioreactor was successfully applied for the treatment of acid mine drainage containing sulfate and metals such as Fe and Al in acid mine drainage tailings.

Algal systems have been applied to the treatment of agricultural nonpoint source pollution (Li et al. 2020), a major source of nutrient pollution in numerous watersheds. Full-scale algal turf scrubber (ATS) systems have been installed in the Midwestern United States to remove nutrients from nutrient-enhanced agricultural drainage water, which has significantly reduced nitrogen and phosphorus loads.

There is a growing interest in the use of algal systems for the remediation of eutrophic water bodies. In China, a system that used floating photobioreactors with microalgae was applied to remediate a eutrophic lake, resulting in improved water quality and reduced frequency of algal blooms.

Another interesting development in the area involves a combination of biofuel production with phycoremediation (Amit and Ghosh 2018). Several examples have been described where the algae cultivated on the treated wastewater can be used to produce biofuels, making it possible to create a mutually beneficial interaction between water treatment and biofuel production.

The discussed applications demonstrate that algae phycoremediation has practical potential for implementation in various industries. Nonetheless, one has to recognize that the usefulness of these systems is contingent upon meticulous planning and fine-tuning to match specific regional contexts and treatment objectives. With the ongoing research and technological progress, it is foreseeable that large-scale applications of freshwater algae phycoremediation will emerge in the next few years.

9 Methods for Quantitative Assessment of the Parameters of Phycoremediation

Rigorous monitoring of the phycoremediation process is crucial for evaluating treatment effectiveness, optimizing the system, and ensuring that water quality standards are maintained (Shah et al. 2021). Various analytical methods are employed to assess aspects of the process, including water quality, algal and biomass development, and the dynamics of microbial communities.

An essential component of monitoring phycoremediation is the observation of water quality. Spectrophotometric methods are widely used to determine parameters such as nutrition contents, chemical oxygen demand, and color levels. Such methods enable a rapid and straightforward assessment, which makes them suitable for routine use. A more detailed analysis of the organic contaminants is conducted using chromatographic methods such as high-performance liquid chromatography (HPLC) and gas chromatography-mass spectrometry (GC-MS; Rontani 2022). These techniques help identify and quantify specific organic compounds in water samples. Atomic absorption spectroscopy (AAS) or inductively coupled plasma mass spectrometry (ICP-MS) is widely used to determine heavy metal concentration due to its exceptional sensitivity and specificity.

Effective monitoring of algal growth and biomass is crucial for evaluating the health and productivity of the treatment system. Conventional methods include counting by microscopy or automated cell counters and determining the optical density by spectrophotometry. Chlorophyll fluorescence analysis is used to determine the photosynthetic activity of the algal culture; it can detect either stress or optimal growth of the culture. Dry weight is widely used as a direct measure of biomass, particularly when evaluating the possibility of biomass valorization.

Molecular techniques have gained significance in assessing microbial community dynamics within the phycoremediation systems. Polymerase chain reaction (PCR) techniques, including quantitative PCR (qPCR) and denaturing gradient gel electrophoresis (DGGE), are used to detect and quantify certain algae species or functional genes (Leong and Chang 2020). Advanced next-generation sequencing techniques provide a detailed view of the composition of microbial communities and allow for tracking changes in the structure of the community over time or in response to various treatment regimes.

Phycoremediation systems may impact the environment, and biomarkers and ecotoxicological assessments are effective tools for evaluating the possible environmental effects of phycoremediation processes. Such measurements may include quantifying enzyme activity and tracking stress proteins or other cellular contents in the algae or organisms used as indicators. Such assessments may help identify system load conditions or potential environmental impacts at an early stage.

Applying these multiple techniques for analysis enhances our understanding of the phycoremediation process, which helps researchers and operators optimize the system and achieve the objectives of effectively eliminating pollutants. Over the years this field has shown more and more emphasis on developing real-time monitoring systems and the application of big data analytics to the management of monitoring information (Leong and Chang 2020). This has the potential to lead to the development of more flexible, efficient, and cost-effective phycoremediation systems.

10 Areas of Future Research and Associated Challenges

The development of phycoremediation using freshwater algae has produced crucial research questions that will shape the future of this technology.

Advances in the selection and genetic manipulation of algae strains are among the most important areas of research (Srimongkol et al. 2022). This includes the development of stress-tolerant strains capable of thriving in diverse environmental conditions with elevated levels of pollutants. Research into the techniques of genetic engineering is being carried out to improve algae species' ability to absorb or metabolize pollutants, increase their growth rates, and provide resistance to predators or competing organisms. However, using genetically modified organisms in environmental applications presents regulatory and public acceptance challenges that must be resolved.

Maximizing the performance within large-scale phycoremediation systems, i.e., scaling the systems up to industrial levels, has remained a challenge (Yuliasni et al. 2023). This requires an increase in the size of photobioreactor designs to take advantage of the economy of scale by treating a large volume while maintaining high efficiency. Another key area is enhancing biomass collection techniques because the current methods are relatively energy-intensive and expensive. Research on new techniques of harvesting, including bio-flocculation or magnetic separation, has the potential to improve the economic viability of large-scale phycoremediation systems significantly.

Understanding the integrated effects of algal–microbial consortia offers opportunities for enhancing the efficiency of cleanup (Kandasamy et al. 2021). Investigating and enhancing the interactions between algae and beneficial bacteria in the system may lead to the development of more stable, resilient, and effective treatment systems. This includes the assessment of co-culture approaches and the design of new microbial communities customized for specific remedial purposes.

To create truly sustainable phycoremediation systems, it is important to trace the fate of the accumulated contaminants in algal biomass. This includes the identification of potential conversion products of organic pollutants and the mobility of incorporated heavy metals. Such a study is vital in searching for safe ways to dispose of or utilize the biomass produced.

The issue of developing successful and environmentally friendly uses for algal biomass after the remediation process is thus of great economic and environmental significance. Biofuel generation from algal biomass has been extensively researched, but there is an increasing focus on other high-value marketable by-products, such as bioplastics, bio-based products, biofertilizers, and various supplements for animal feed (Chandra et al. 2019; Syrpas and Venskutonis 2020). Further developing the methods for efficient and cost-effective extraction of valuable chemicals from algal biomass can help enhance the economic feasibility of phycoremediation systems.

Mitigating possible impacts of climate change on phycoremediation processes is a matter of increasing concern (Sahu et al. 2019). This entails assessing the effects that changed temperature patterns, extreme weather events, and changes in the water chemistry may have on the performance of algal bioremediation systems. Therefore, future research must focus on developing adaptive management strategies and designing resilient systems that will ensure the sustainability of this technology.

Integrating phycoremediation into the existing treatment technologies offers an opportunity to develop complete and effective water treatment solutions. One of the possible avenues includes linking algal systems with the existing treatment facilities, constructed wetlands, or other biological treatment systems. More work has to be done to establish the best pathways for integrating these systems and designing the treatment systems to ensure that high-efficiency standards are attained at minimum costs.

A new direction is to facilitate the removal of emerging pollutants, such as microplastics, pharmaceuticals, and personal care products (Ricky and Shanthakumar 2022). With the improvement of analytical techniques used to detect these pollutants in the environment, it is possible to identify them at low concentrations. Thus, there is an acute need to engineer algae strains and treatment systems that can effectively remove or transform these pollutants.

There is a need for standardization in assessing and comparing phycoremediation systems for this technology to be more widely accepted. This means that standardized techniques should be implemented to quantify the effectiveness of the remediation process and assess the impact of algal systems on the environment (Di Caprio 2020), including detailed life cycle assessments. Adopting such standardization will streamline regulatory approval procedures and allow for more rigorous yet cost-effective comparisons of the various phycoremediation methods.

Solving these issues and challenges will require collaborative efforts among experts in algae biology, environmental engineering, chemistry, and other fields (Miksch et al. 2015). As the knowledge in algal biology and phycoremediation improves and advanced technologies emerge, the potential of freshwater algal systems to contribute to sustainable water treatment and environmental remediation is progressively improving.

11 Conclusion

Phycoremediation using freshwater algae has emerged as an efficient and sustainable approach toward the treatment of water and removing contaminants from the environment. This chapter presents an account of the historical development of the subject, the fundamental principles involved, and the state of the art in this rapidly evolving field. Freshwater algae such as *Chlorella, Scenedesmus, Spirulina, Botryococcus,* and *Chlamydomonas* have been found to be highly efficient in the removal of a wide range of contaminants, including nutrients, heavy metals, and organic pollutants present in aquatic environments. Multiple mechanisms of phycoremediation, such as biosorption, bioaccumulation, and biotransformation, comprise an array of versatile and sustainable solutions to diverse environmental challenges.

Though phycoremediation has multiple advantages, including high growth rates of many algae, efficient CO_2 sequestration, and valorization of biomass produced, it has a number of challenges. These include optimization needing to be species-specific, challenges to maintain control of environmental parameters, and efficient biomass-harvesting techniques. It remains one of the most important avenues for future research to try and overcome such limitations by using an interdisciplinary methodology that includes advances in algal ecology, biotechnology, genetic engineering, and process optimization. Scaling up from laboratory to industrial applications and potential ecological impacts of algal escape from open systems are all areas that need further investigation. In view of increased global awareness of environmental issues, phycoremediation with freshwater algae is well positioned to play an increasingly important role in sustainable water management and pollution control.

References

Abdelfattah A, Ali SS, Ramadan H, El-Aswar EI, Eltawab R, Ho S, Elsamahy T, Li S, El-Sheekh MM, Schagerl M, Kornaros M, Sun J (2023) Microalgae-based wastewater treatment: mechanisms, challenges, recent advances, and future prospects. Environ Sci Ecotechnol (Elsevier BV) 13:100205–100205

Abolhasani MH, Pirestani N, Nehbandani A, Sanatinia B (2018) Nutrient removal from municipal wastewater using mixture of two algae, *Scenedesmus obliquus* and *Chlorella vulgaris*. Int J Aquatic Sci 9(1):44–50

Abomohra AE, Jin W, El-Sheekh MM (2016) Enhancement of lipid extraction for improved biodiesel recovery from the biodiesel promising microalga *Scenedesmus obliquus*. Energy Convers Manage (Elsevier BV) 108:23–29. https://doi.org/10.1016/j.enconman.2015.11.007

Abreu AP, Martins R, Nunes J (2023) Emerging applications of *Chlorella* sp. and Spirulina (*Arthrospira*) sp. Bioengineering (Basel) (Multidisciplinary Digital Publishing Institute) 10(8):955–955. https://doi.org/10.3390/bioengineering10080955

Acién Fernández FG, Fernández Sevilla JM, Molina Grima E (2017) Microalgae: the basis of mankind sustainability. InTechOpen, pp 123–140. https://doi.org/10.5772/67930

Adane T, Adugna AT, Alemayehu E (2021) Textile industry effluent treatment techniques. J Chem 2021(1):5314404

Adey WH, Laughinghouse HD IV, Miller JB, Hayek LAC, Thompson JG, Bertman S, Puvanendran S (2013) Algal turf scrubber (ATS) floways on the Great Wicomico River, Chesapeake Bay: productivity, algal community structure, substrate and chemistry. J Phycol 49(3):489–501

Ahmad A, Bhat AH, Buang A, Shah SM, Afzal M (2019) Biotechnological application of microalgae for integrated palm oil mill effluent (POME) remediation: a review. Int J Env Sci Technol 16:1763–1788

Ahmad S, Pandey A, Pathak VV, Tyagi VV, Kothari R (2020) Phycoremediation: algae as eco-friendly tools for the removal of heavy metals from wastewaters. In: bioremediation of industrial waste for environmental safety: volume II: biological agents and methods for industrial waste management. Springer, Singapore, pp 53–76

Alam F, Mobin S, Chowdhury H (2015) Third generation biofuel from algae. Procedia Eng 105:763–768

Almeida Â, Cotas J, Pereira L, Carvalho PCS (2023) Potential role of *Spirogyra* sp. and *Chlorella* sp. in bioremediation of mine drainage: a review. Phycology 3(1):186–201. https://doi.org/10.3390/phycology3010012

Amit, Ghosh UK (2018) An approach for phycoremediation of different wastewaters and biodiesel production using microalgae. Environ Sci Pollut Res 25:18673–18681

Amulya K, Morris S, Lens PN (2023) Aquatic biomass as sustainable feedstock for biorefineries. Biofuel Bioprod Biorefin (Wiley) 17(4):1012–1029. https://doi.org/10.1002/bbb.2471

Anbuchezhian R, Karuppiah V, Li Z (2015) Prospect of marine algae for production of industrially important chemicals. In: Algal biorefinery: an integrated approach. Springer, Cham, pp 195–217

Ankit BK, Korstad J (2022) Phycoremediation: use of algae to sequester heavy metals. Hydrobiology 1(3):288–303. https://doi.org/10.3390/hydrobiology1030021

Arango J, Muñoz R, Donoso-Bravo A, Bernard O, Casagli F, Jeison D (2022) A half-century of research on microalgae-bacteria for wastewater treatment. Algal Res (Elsevier BV) 67:102828–102828. https://doi.org/10.1016/j.algal.2022.102828

Baldisserotto C, Demaria S, Arcidiacono M, Benà E, Giacò P, Marchesini R, Ferroni L, Benetti L, Zanella M, Benini A, Pancaldi S (2023) Enhancing urban wastewater treatment through isolated *Chlorella* strain-based phytoremediation in concentrated stream: an analysis of Algae Morpho-physiology and nutrients removal efficiency. Plants (Multidisciplinary Digital Publishing Institute) 12(5):1027. https://doi.org/10.3390/plants12051027

Bansal A, Shinde O, Sarkar S (2018) Industrial wastewater treatment using Phycoremediation technologies and co-production of value-added products. Int J Biodegrad Bioremediat (OMICS Publishing Group) 9(01). https://doi.org/10.4172/2155-6199.1000428

Barati B, Zeng K, Baeyens J, Wang S, Addy M, Gan SY, Abomohra AE (2021) Recent progress in genetically modified microalgae for enhanced carbon dioxide sequestration. Biomass Bioenerg (Elsevier BV) 145:105927. https://doi.org/10.1016/j.biombioe.2020.105927

Bellido-Pedraza CM, Torres MJ, Llamas A (2024) The microalgae *Chlamydomonas* for bioremediation and bioproduct production. Cells 13(13):1137

Bhatt P, Bhandari G, Bhatt K, Simsek H (2022) Microalgae-based removal of pollutants from wastewaters: occurrence, toxicity and circular economy. Chemosphere 306:135576

Bhattacharya S, Rohilla P, Joshi VC, Gupta AR, Prajapati S, Patel S, Sharma S (2021) Textile wastewater treatment. In: The future of effluent treatment plants. Elsevier, pp 551–568

Büdel B, Friedl T (2024) Introduction: the new cryptogams. In: Biology of Algae, Lichens and Bryophytes. Springer, Berlin, Heidelberg, pp 1–8

Bulgariu L, Gavrilescu M (2015) Bioremediation of heavy metals by microalgae. In: Handbook of marine microalgae. Academic Press, pp 457–469. https://www.sciencedirect.com/science/article/pii/B9780128007761000303

Bulynina SS, Ziganshina EE, Ziganshin AM (2023) Growth efficiency of *Chlorella sorokiniana* in synthetic media and unsterilized domestic wastewater. BioTech (Basel) (Multidisciplinary Digital Publishing Institute) 12(3):53. https://doi.org/10.3390/biotech12030053

Calvo-Flores FG, Martín-Martínez FJ (2022) Biorefineries: achievements and challenges for a bio-based economy. Front Media 10. https://doi.org/10.3389/fchem.2022.973417

Carvalho EMD, Santos CRD, Ansilago M, Menegazzo ML, Nunes NSP (2021) Phycoremediation of fish farm effluents using *Chlorella vulgaris*: evaluation of the removal of nutrients and emerging pollutants. Environ Sci Pollut Res 28:19940–19950. https://doi.org/10.33448/rsd-v10i13.20723

Casanova LM, Macrae A, Souza JED, Neves A, Vermelho AB (2023) The potential of allelochemicals from microalgae for biopesticides. Plan Theory 12:1896. https://doi.org/10.3390/plants12091896

Castro-Muñoz R, García-Depraect O (2021) Membrane-based harvesting processes for microalgae and their valuable-related molecules: a review. Membranes 11:585

Chandel N, Ahuja V, Gurav R, Kumar V, Tyagi VK, Pugazhendhi A, Bhatia SK (2022) Progress in microalgal mediated bioremediation systems for the removal of antibiotics and pharmaceuticals from wastewater. Sci Total Environ 825:153895

Chandra R, Iqbal HMN, Vishal G, Lee HS, Nagra S (2019) Algal biorefinery: a sustainable approach to valorize algal-based biomass towards multiple product recovery. Bioresour Technol 278:346–359

Chen X, Tian Z, Cheng H, Xu G, Zhou H (2021) Adsorption process and mechanism of heavy metal ions by different components of cells, using yeast (*Pichia pastoris*) and Cu2+ as biosorption models. RSC Adv 11:7075–7084. https://pubs.rsc.org/en/content/articlepdf/2021/ra/d0ra09744f

Christenson LB, Sims RC (2012) Rotating algal biofilm reactor and spool harvester for wastewater treatment with biofuels by-products. Biotechnol Bioeng 109:1674–1684

Chugh M, Kumar L, Bhardwaj D, Bharadvaja N (2022) Bioaccumulation and detoxification of heavy metals: an insight into the mechanism. In: Saxena G, Kumar V, Shah MP (eds) Bioremediation for environmental sustainability. Elsevier, Amsterdam, pp 345–366. https://www.sciencedirect.com/science/article/pii/B9780323856577000134

Coronado-Reyes JA, Salazar-Torres JA, Juárez-Campos B, González-Hernández JC (2022) *Chlorella vulgaris*, a microalgae important to be used in biotechnology: a review. Food Sci Technol 42:e37320. https://doi.org/10.1590/fst.37320

Covaliu CI, Nedelcu A, Deák G, Stoian O, Noor NM (2020) Biotechnology of water treatment based on algae cultures. IOP Conf Ser Earth Environ Sci 616:012081. https://doi.org/10.1088/1755-1315/616/1/012081

Craggs R, Sutherland D, Campbell H (2012) Hectare-scale demonstration of high rate algal ponds for enhanced wastewater treatment and biofuel production. J Appl Phycol 24:329–337

Cunha C, Lopes J, Paulo J, Faria M, Kaufmann M, Nogueira N, Cordeiro N (2020) The effect of microplastics pollution in microalgal biomass production: a biochemical study. Water Res 186:116370

Derco J, Vrana B (2018) Introductory chapter: biosorption. In: Derco J, Vrana B (eds) Biosorption. IntechOpen, London. https://doi.org/10.5772/intechopen.78961

Dhiab B (2012) Photosynthetic behavior of microalgae in response to environmental factors. In: Najafpour M (ed) Advances in photosynthesis – fundamental aspects. IntechOpen, London. https://doi.org/10.5772/27944

Di Caprio F (2020) Methods to quantify biological contaminants in microalgae cultures. Algal Res 49:101943

Dönmez G, Aksu Z, Öztürk A, Kutsal T (1999) A comparative study on heavy metal biosorption characteristics of some algae. Process Biochem 34:885–892. https://doi.org/10.1016/s0032-9592(99)00005-9

Doshi H, Ray A, Kothari IL (2008) Bioremediation potential of *Chlorella*: spectroscopic, kinetics, and SEM studies. Int J Phytoremediation 10:264–277. https://www.tandfonline.com/doi/abs/10.1080/15226510802096028

El-Ghany NAA, Elella MHA, Abdallah HM, Mostafa MS, Samy M (2023) Recent advances in various starch formulation for wastewater purification via adsorption technique: a review. J Polym Environ 31:2792–2825. https://doi.org/10.1007/s10924-023-02798-x

El-Sheekh MM, Hamouda RA, Nizam AA (2013) Biodegradation of crude oil by *Scenedesmus obliquus* and *Chlorella vulgaris* growing under heterotrophic conditions. Int Biodeterior Biodegradation 82:67–72. https://doi.org/10.1016/j.ibiod.2012.12.015

Emparan Q, Harun R, Danquah MK (2019) Role of phycoremediation for nutrient removal from wastewaters: a review. Appl Ecol Environ Res 17:889–915. https://doi.org/10.15666/aeer/1701_889915

Energy JOR (2021) Retracted: microalgae as a renewable source of energy: a niche opportunity. J Renew Energy 2021:9813285

Even C, Hadroug D, Boumlaik Y, Simon G (2022) Microalgae-based bioenergy with carbon capture and storage quantified as a negative emissions technology. Energy Nexus 7:100117. https://doi.org/10.1016/j.nexus.2022.100117

Feldmann J (2002) Volatilization of metals from a landfill site. In: Cai Y, Braids OC (eds) Biogeochemistry of environmentally important trace elements, ACS symposium series, vol 835. American Chemical Society, Washington, DC, pp 128–140. https://doi.org/10.1021/bk-2003-0835.ch010

Gao F, Peng YY, Li C, Cui W, Yang ZH, Zeng GM (2018) Coupled nutrient removal from secondary effluent and algal biomass production in membrane photobioreactor (MPBR): effect of HRT and long-term operation. Chem Eng J 335:169–175

Gonzalez-Camejo J, Aparicio S, Jiménez-Benítez A, Pachés M, Ruano MV, Borrás L, Seco A (2020) Improving membrane photobioreactor performance by reducing light path: operating conditions and key performance indicators. Water Res 172:115518

Goswami RK, Agrawal K, Verma P (2022) An exploration of natural synergy using microalgae for the remediation of pharmaceuticals and xenobiotics in wastewater. Algal Res 64:102703

Guleri S, Singh K, Kaushik R, Dhanker R, Tiwari A (2020) Phycoremediation: a novel and synergistic approach in wastewater remediation. J Microbiol Biotechnol Food Sci 10:98–106. https://doi.org/10.15414/jmbfs.2020.10.1.98-106

Guo Q, Qi F, Mu R, Yu G, Ma G, Meng Q (2023) Advances in sustainable wastewater treatment: microalgal–bacterial consortia process, greenhouse gas reduction and energy recovery technologies. Water Environ J 37:192–205

Gupta PL, Lee SM, Choi HJ (2015) A mini review: photobioreactors for large scale algal cultivation. World J Microbiol Biotechnol 31(9):1409–1417

Gupta S, Nayak A, Roy C, Yadav AK (2021) An algal assisted constructed wetland-microbial fuel cell integrated with sand filter for efficient wastewater treatment and electricity production. Chemosphere 263:128132

Hajipour S, Alinia-Ahandani E, Behnaz S, Zahra A, Selamoğlu Z, Akram M (2023) Phytoremediation potential and its methods- a review. Bioeng Stud 4:1–8. https://doi.org/10.37868/bes.v4i1.id246

Hanifzade MM, Nabati Z, Tavakoli O, Sarrafzadeh M (2017) Waste to Energy from flue gas of industrial plants to biodiesel: effect of CO2 on microalgae growth. J Bioremed Biodegr 7:300. https://doi.org/10.4172/2572-0805.1000300

He Z, Fan X, Qu L, Zhou X, Jin W, Hatshan MR, Wang Q (2023) Cultivation of *Chlorella pyrenoidosa* and *Scenedesmus obliquus* in swine wastewater: nitrogen and phosphorus removal and microalgal growth. Process Saf Environ Prot 179:887–895

Hernández-Zamora M, Cristiani-Urbina E, Martínez-Jerónimo F, Perales-Vela HV, Ponce-Noyola T, Montes-Horcasitas MDC, Cañizares-Villanueva RO (2015) Bioremoval of the azo dye Congo red by the microalga *Chlorella vulgaris*. Environ Sci Pollut Res 22:10811–10823

Higgins BT, Kendall A (2012) Life cycle environmental and cost impacts of using an algal turf scrubber to treat dairy wastewater. J Ind Ecol 16:436–447

Jabri HA, Das P, Khan S, Thaher M, AbdulQuadir M (2020) Treatment of wastewaters by microalgae and the potential applications of the produced biomass—a review. Water 13:27. https://doi.org/10.3390/w13010027

Ji M, Abou-Shanab RAI, Hwang J, Timmes TC, Kim H, Oh Y, Jeon B (2013) Removal of nitrogen and phosphorus from piggery wastewater effluent using the green microalga *Scenedesmus obliquus*. J Environ Eng 139:1198–1205. https://ascelibrary.org/doi/10.1061/%28ASCE%29EE.1943-7870.0000726

Jones GB, Sims RC, Zhao J (2023) Experimental and theoretical investigations of rotating algae biofilm reactors (RABRs): areal productivity, nutrient recovery, and energy efficiency. Biotechnol Bioeng 120(10):2865–2879

Kandasamy S, Narayanan M, He Z, Liu G, Ramakrishnan M, Thangavel P, Pugazhendhi A, Raja R, Carvalho IS (2021) Current strategies and prospects in algae for remediation and biofuels: an overview. Biocatal Agric Biotechnol 35:102045. https://doi.org/10.1016/j.bcab.2021.102045

Khan AA, Gul J, Naqvi SR, Ali I, Farooq W, Liaqat R, Juchelková D (2022) Recent progress in microalgae-derived biochar for the treatment of textile industry wastewater. Chemosphere 306:135565

Kilbane JJ (2022) Shining a light on wastewater treatment with microalgae. Arab J Sci Eng 47:45–56. https://doi.org/10.1007/s13369-021-06444-3

Kim TY, Hong JE, Park HM, Lee UJ, Lee SY (2020) Decontamination of low-level contaminated water from radioactive cesium and cobalt using microalgae. J Radioanal Nucl Chem 323:903–908

Krishnamoorthy S, Premalatha M (2021) Phycoremediation of industrial wastewater: challenges and prospects. In: Yadav AN, Singh J, Singh C, Mishra N (eds) Current developments in biotechnology and bioengineering. Elsevier, Amsterdam, pp 99–123. https://doi.org/10.1016/b978-0-12-820318-7.00005-8

Kumar V, Jaiswal KK, Verma M, Vlaskin MS, Nanda M, Chauhan PK, Kim H (2021) Algae-based sustainable approach for simultaneous removal of micropollutants, and bacteria from urban wastewater and its real-time reuse for aquaculture. Sci Total Environ 774:145556

Lam TP, Lee T, Chen C, Chang J (2018) Strategies to control biological contaminants during microalgal cultivation in open ponds. Bioresour Technol 252:180–187. https://doi.org/10.1016/j.biortech.2017.12.088

Laraib N, Hussain A, Javid A, Hafeez-ur-Rehman M, Bukhari SM, Rashid MI, Ali W (2020) Recent advancements in microalgal-induced remediation of wastewaters. In: Yadav AN, Singh J, Rastegari AA, Yadav N (eds) Algal biotechnology. Elsevier, Amsterdam, pp 205–217. https://doi.org/10.1016/b978-0-12-819001-2.00010-3

Lee K, Lee S, Lee JE, Lee S (2018) Biosorption of radioactive cesium from contaminated water by microalgae *Haematococcus pluvialis* and *Chlorella vulgaris*. J Environ Manag 233:83–88. https://doi.org/10.1016/j.jenvman.2018.12.022

Leong YK, Chang JS (2020) Bioremediation of heavy metals using microalgae: recent advances and mechanisms. Bioresour Technol 303:122886

Li X, Wu S, Yang C, Zeng G (2020) Microalgal and duckweed based constructed wetlands for swine wastewater treatment: a review. Bioresour Technol 318:123858

Lisboa FC, Okura MH, Barboza M, Pelizer LH (2018) Comparative analysis of three growth medium for *Arthrospira platensis* cultivation based on lab-scale results. Int J Adv Eng Res Sci 5:200–207. https://doi.org/10.22161/ijaers.5.6.33

Lozoya-Gloria E, Cruz XML, Ozawa-Uyeda TA (2020) The colonial microalgae *Botryococcus braunii* as biorefinery. In: Yousuf A (ed) Microalgae cultivation for biofuels production. Academic Press, London, pp 161–172. https://web.archive.org/web/20200209154818/https:/cdn.intechopen.com/pdfs/69185.pdf

Lu W, Xu C, Liu F, Su M, Cheng S, Zhang Y (2023) Antibiotic removal efficiency by microalgae: a systematic analysis combined with meta-analysis. Process Saf Environ Prot 174:912–920

Lukács Á, Bácsi I, Kókai Z, Borics G, Várbíró G, T-Krasznai E, B-Béres V (2021) Strong influence of climatic extremes on diversity of benthic algae and cyanobacteria in a lowland intermittent stream. Ecohydrology 14:e2286. https://doi.org/10.1002/eco.2286

Luo L, Wang P, Lin L, Luan T, Ke L, Tam NFY (2014) Removal and transformation of high molecular weight polycyclic aromatic hydrocarbons in water by live and dead microalgae. Process Biochem 49:1723–1732

Manzi HP, Abou-Shanab RA, Jeon B, Wang J, Salama E (2022) Algae: a frontline photosynthetic organism in the microplastic catastrophe. Trends Plant Sci 27:1159–1172

Marchetto F, Roverso M, Righetti D, Bogialli S, Filippini F, Bergantino E, Sforza E (2021) Bioremediation of per-and poly-fluoroalkyl substances (PFAS) by *Synechocystis* sp. PCC 6803: a chassis for a synthetic biology approach. Life 11:1300

Markou G, Chatzipavlidis I, Georgakakis D (2012) Cultivation of *Arthrospira* (Spirulina) *platensis* in olive-oil mill wastewater treated with sodium hypochlorite. Bioresour Technol 112:234–241

Martins M, Carvalheiro F, Gírio F (2022) An overview of lignin pathways of valorization: from isolation to refining and conversion into value-added products. Biomass Convers Biorefinery 14:3183–3207

Miksch K, Cema G, Corvini PFX, Felis E, Sochacki A, Surmacz-Górska J, Żabczynski S (2015) R&D priorities in the field of sustainable remediation and purification of agro-industrial and municipal wastewater. New Biotechnol 32:128–132

Mollamohammada S, Hassan AA, Dahab M (2022) Application of immobilized algae in water and wastewater treatment. In: Alam MA, Wang Z (eds) Algal biofuel. CRC Press, Boca Raton, pp 237–257

Moondra N, Jariwala N, Christian RA (2020) Sustainable treatment of domestic wastewater through microalgae. Int J Phytoremediation 22:1480–1486

Msanne J, Polle J, Starkenburg SR (2020) An assessment of heterotrophy and mixotrophy in *Scenedesmus* and its utilization in wastewater treatment. Algal Res 48:101911

Mu R, Jia Y, Ma G, Liu L, Hao K, Qi F, Shao Y (2021) Advances in the use of microalgal–bacterial consortia for wastewater treatment: community structures, interactions, economic resource reclamation, and study techniques. Water Environ Res 93:1217–1230

Muñoz R, Guieysse B (2006) Algal–bacterial processes for the treatment of hazardous contaminants: a review. Water Res 40:2799–2815

Mustafa S, Bhatti HN, Maqbool M, Iqbal M (2021) Microalgae biosorption, bioaccumulation and biodegradation efficiency for the remediation of wastewater and carbon dioxide mitigation: prospects, challenges and opportunities. J Water Process Eng 41:102009

Muylaert K, Bastiaens L, Vandamme D, Gouveia L (2017) Harvesting of microalgae: overview of process options and their strengths and drawbacks. In: Gonzalez-Fernandez C, Muñoz R (eds) Microalgae-based biofuels and bioproducts. Woodhead Publishing, Sawston, pp 113–132

Na Y, Lee J, Lee S, Kumar P, Kim JH, Patel R (2020) Removal of heavy metals by polysaccharide: a review. Environ Sci Pollut Res 59:1770–1790

Nandeshwar SN, Satpute GD (2014) Green technical methods for treatment of waste water using microalgae and its application in the management of natural water resources—a review. Curr World Environ 9:837

Narala RR, Garg S, Sharma K, Thomas-Hall SR, Deme M, Li Y, Schenk PM (2016) Comparison of microalgae cultivation in photobioreactor, open raceway pond, and a two-stage hybrid system. Front Energy Res 4. https://doi.org/10.3389/fenrg.2016.00029

Nimisha P, Joseph S (2020) Assessment of phycoremediation efficiency of *Spirogyra maxima* by using heavy metals manganese and lead. Int J Environ Agric Biotech 5(3):616–620. https://doi.org/10.22161/ijeab.53.13

Onyeaka H, Miri T, Obileke K, Hart A, Anumudu C, Al-Sharify ZT (2021) Minimizing carbon footprint via microalgae as a biological capture. Carbon Capture Sci Technol 1:100007. https://doi.org/10.1016/j.ccst.2021.100007

Órpez R, Martínez ME, Hodaifa G, El Yousfi F, Jbari N, Sánchez S (2009) Growth of the microalga *Botryococcus braunii* in secondarily treated sewage. Desalination 246(1–3):625–630

Oswald WJ (2003) My sixty years in applied algology. J Appl Phycol 15:99–106

Oswald WJ, Gotaas HB (1957) Photosynthesis in sewage treatment. Trans Am Soc Civil Eng 122:73–105

Pacheco D, Rocha ACS, Pereira L, Verdelhos T (2020) Microalgae water bioremediation: trends and hot topics. Appl Sci 10:1886. https://doi.org/10.3390/app10051886

Pant A, Singh J, Tewari K, Meehnian H, Jana AK (2024) Immobilization of microalgae for bioremediation of wastewater. In: Ravishankar GA, Rao AR, Kim SK (eds) Algae mediated bioremediation: industrial prospectives, vol 1. Wiley, Hoboken, pp 43–61

Papadopoulos KP, Economou CN, Μάρκου Γ, Nicodemou A, Koutinas M, Tekerlekopoulou AG, Vayenas D (2022) Cultivation of *Arthrospira platensis* in brewery wastewater. Water 14(10):1547. https://doi.org/10.3390/w14101547

Pascucci PR, Kowalak AD (1996) Public health benefits of using algae for simultaneous multiple metal extraction from waters. Rev Environ Health 11(4):205. https://doi.org/10.1515/reveh.1996.11.4.205

Patro A, Dwivedi S, Panja R, Saket P, Gupta S, Mittal Y, Yadav AK (2023) Constructed wetlands for wastewater management: basic design, abiotic and biotic components, and their interactive functions. In: Material-microbes interactions. Academic Press, Cambridge, pp 315–348

Porto B, Gonçalves AL, Esteves AF, Souza SMDAGU, Souza AAU, Vilar VJ, Pires JC (2020) Microalgal growth in paper industry effluent: coupling biomass production with nutrients removal. Appl Sci 10(9):3009. https://doi.org/10.3390/app10093009

Radziff SBM, Ahmad SA, Shaharuddin NA, Merican F, Kok Y, Zulkharnain A, Gómez-Fuentes C, Wong C (2021) Potential application of algae in biodegradation of phenol: a review and bibliometric study. Plan Theory 10(12):2677. https://doi.org/10.3390/plants10122677

Ralph PJ, Pernice M (2023) Save the planet with green industries using algae. PLoS Biol 21(3):e3002061. https://doi.org/10.1371/journal.pbio.3002061

Ravishankar GA, Rao AR, Kim SK (eds) (2024) Algae mediated bioremediation: industrial prospectives, volumes 1–2. Wiley, Hoboken

Razaviarani V, Arab G, Lerdwanawattana N, Gadia Y (2022) Algal biomass dual roles in phycoremediation of wastewater and production of bioenergy and value-added products. Int J Environ Sci Technol 20(7):8199–8216. https://doi.org/10.1007/s13762-022-04696-6

Redha AA (2020) Removal of heavy metals from aqueous media by biosorption. Arab J Basic Appl Sci 27(1):183–193. https://doi.org/10.1080/25765299.2020.1756177

Ricky R, Shanthakumar S (2022) Phycoremediation integrated approach for the removal of pharmaceuticals and personal care products from wastewater – a review. J Environ Manag 302:113998

Rontani JF (2022) Use of gas chromatography-mass spectrometry techniques (GC-MS, GC-MS/MS and GC-QTOF) for the characterization of photooxidation and autoxidation products of lipids of autotrophic organisms in environmental samples. Molecules 27:1629

Saha M, Fink P (2022) Algal volatiles – the overlooked chemical language of aquatic primary producers. Biol Rev 97:2162–2173

Sahu SN, Sahoo NK, Naik SN (2019) Phycoremediation technology: a global prospective. In: Application of microalgae in wastewater treatment: volume 1: domestic and industrial wastewater treatment. Springer, pp 1–18

Salman JM, Salman JM, Kadhim NF (2023) Review on toxicity and removal of pharmaceutical pollutants using immobilised microalgae. Ecol Eng Environ Technol 24:44–60

Santo GE, Barros A, Costa M, Pereira H, Trovão M, Cardoso H et al (2023) *Scenedesmus rubescens* heterotrophic production strategies for added value biomass. Mar Drugs 21:411

Santos MDJ, de Souza CO, Marcelino HR (2023) Blue technology for a sustainable pharmaceutical industry: microalgae for bioremediation and pharmaceutical production. Algal Res 69:102931

Saxena G, Singh R, Singh JS, Singh DP (2019) Microalgae-assisted phycoremediation and energy crisis solution: challenges and opportunity. In: Bioremediation of industrial waste for environmental safety. Elsevier, pp 295–307

Saxena G, Mallick S, Singh R, Singh L, Singh N, Mandotra SK et al (2023) Novel cost-effective design for bio-volatilization studies in photosynthetic microalgae exposed to arsenic with emphasis on growth and glutathione modulation. Front Microbiol 14:1170740

Shah MP, Rodriguez-Couto S, De La Cruz CBV, Biswas JK (eds) (2021) An integration of Phycoremediation processes in wastewater treatment. Elsevier, Amsterdam

Sharma M, Sharma A, Batra N, Pareek R, Patel S (2024) Persistent organic pollutants in water and their microalgae based bioremediation. Biosci Biotechnol Res Asia 21:3

Shen X, Gao L, Zhou S, Huang J, Wu C, Qin Q et al (2020) High fatty acid productivity from *Scenedesmus obliquus* in heterotrophic cultivation with glucose and soybean processing wastewater via nitrogen and phosphorus regulation. Sci Total Environ 708:134596

Shen Y, Yuan W, Pei ZJ, Wu Q, Mao E (2009) Microalgae mass production methods. Am Soc Agri Biol Eng 52(4):1275–1287

Singh G, Patidar SK (2018) Microalgae harvesting techniques: a review. J Environ Manag 217:499–508

Singh DV, Bhat RA, Saxena G, Singh R, Singh D (2021) Microalgae in aquatic environs: a sustainable approach for remediation of heavy metals and emerging contaminants. Environ Technol Innov 21:101340

Sivasubramanian V (2016) Phycoremediation and business prospects. In: Algae biotechnology. Springer, pp 421–456
Souza CBD, Silva GR (2019) Phytoremediation of effluents contaminated with heavy metals by floating aquatic macrophytes species. In: Heavy metals. IntechOpen
Srimongkol P, Sangtanoo P, Songserm P, Watsuntorn W, Karnchanatat A (2022) Microalgae-based wastewater treatment for developing economic and environmental sustainability: current status and future prospects. Front Bioeng Biotechnol 10:904046
Srivastava I, Pal A, Sahu S, Singh KR (2023) A study of the suitability of microbial cells for the biosorption and bioaccumulation of heavy metal removal. Int J Curr Sci Res Rev 6:5
Stevenson RJ, Bothwell ML, Lowe RL, Thorp JH (1996) Algal ecology: freshwater benthic ecosystem. Academic Press, San Diego
Sunday ER (2018) Phycoremediation: an eco-solution to environmental protection and sustainable remediation. J Chem Environ Biol Eng 2:5–5
Syrpas M, Venskutonis PR (2020) Algae for the production of bio-based products. In: Biobased products and industries. Elsevier, pp 203–243
Tabagari I, Kurashvili M, Varazi T, Adamia G, Gigolashvili G, Pruidze M et al (2019) Application of *Arthrospira (Spirulina) platensis* against chemical pollution of water. Water 11:1759
Takáčová A, Bajuszová M, Kohanová J, Lux A, Valent P, Kokavcová A et al (2021) Biosorption of chromium by dry algae Chlorella kessleri. IOP Conf Ser Earth Environ Sci 900:012046
Tan XB, Zhao ZY, Gong H, Jiang T, Liu XP, Liao JY et al (2024) Growth of *Scenedesmus obliquus* in anaerobically digested swine wastewater from different cleaning processes for pollutants removal and biomass production. Chemosphere 352:141515
Tarlan E, Dilek FB, Yetiş Ü (2002) Effectiveness of algae in the treatment of a wood-based pulp and paper industry wastewater. Bioresour Technol 84:1–5
Tatarová D, Galanda D, Kuruc J, Gaálová B (2021) Phytoremediation of 137Cs, 60Co, 241Am, and 239Pu from aquatic solutions using *Chlamydomonas reinhardtii*, *Scenedesmus obliquus*, and *Chlorella vulgaris*. Int J Phytoremediation 23:1376–1381
Ting H, Haifeng L, Shanshan M, Zhang Y, Zhidan L, Na D (2017) Progress in microalgae cultivation photobioreactors and applications in wastewater treatment: a review. Int J Agric Biol Eng 10:1–29
Touliabah HE, El-Sheekh MM, Ismail MM, El-Kassas HY (2022) A review of microalgae- and cyanobacteria-based biodegradation of organic pollutants. Molecules 27:1141
Uduman N, Qi Y, Danquah MK, Forde GM, Hoadley A (2010) Dewatering of microalgal cultures: a major bottleneck to algae-based fuels. J Renew Sustain Energy 2:012701
Ullmann J, Grimm D (2021) Algae and their potential for a future bioeconomy, landless food production, and the socio-economic impact of an algae industry. Org Agric 11:261–267
Vale MA, Ferreira A, Pires JC, Gonçalves AL (2020) CO2 capture using microalgae. In: Biomass, biofuels and biochemicals. Elsevier, pp 381–405
Vanhoudt N, Vandenhove H, Leys N, Janssen P (2018) Potential of higher plants, algae, and cyanobacteria for remediation of radioactively contaminated waters. Chemosphere 207:239–254
Verma P, Sirotiya V, Rathore R, Kumar A, Rai A, Soni U et al (2024) A comprehensive review on microalgal chromium detoxification in tannery wastewater: paving the way for biobased products. Process Saf Environ Prot 183:409–428
Wang S, Zhao Q, Yu H, Du X, Zhang T, Sun T et al (2023) Assessing the potential of *Chlorella* sp. phycoremediation liquid digestates from brewery wastes mixture integrated with bioproduct production. Front Bioeng Biotechnol 11:1199472
Wilde E, Benemann JR (1993) Bioremoval of heavy metals by the use of microalgae. Biotechnol Adv 11:781–812
Wong YS, Tam N (eds) (1998) Wastewater treatment with algae. Springer, Berlin
Wu P, Hu T, Sun L, Fan J (2023) Efficient degradation of sulfadiazine by photosynthetic cyanobacteria Synechocystis sp. PCC 6803 coupling with recombinant laccase strategy. Chem Eng J 474:145974

Xu K, Zou X, Xue Y, Qu Y, Li Y (2021) The impact of seasonal variations about temperature and photoperiod on the treatment of municipal wastewater by algae-bacteria system in lab-scale. Algal Res 54:102175

Yu C, Pang H, Wang JH, Chi ZY, Zhang Q, Kong FT et al (2022) Occurrence of antibiotics in waters, removal by microalgae-based systems, and their toxicological effects: a review. Sci Total Environ 813:151891

Yuliasni R, Kurniawan SB, Marlena B, Hidayat MR, Kadier A, Ma P et al (2023) Recent progress of phytoremediation-based technologies for industrial wastewater treatment. J Ecol Eng 24:208–220

Zaimes GG, Khanna V (2013) Microalgal biomass production pathways: evaluation of life cycle environmental impacts. Biotechnol Biofuels 6:88

Zainith S, Saxena G, Kishor R, Bharagava RN (2021) Application of microalgae in industrial effluent treatment, contaminants removal, and biodiesel production: opportunities, challenges, and future prospects. In: Microalgae cultivation for biofuels production. Elsevier, pp 481–517

Zeraatkar AK, Ahmadzadeh H, Talebi AF, Moheimani NR, McHenry MP (2016) Potential use of algae for heavy metal bioremediation, a critical review. J Environ Manag 181:817–831

Zhang L, Han J, Ma S, Zhang Y, Wang Y, Xu J (2023a) Comprehensive evaluation of growth characteristics, nitrogen removal capacity, and nutritional properties of three diet microalgae. Front Mar Sci 10:1117043

Zhang M, Ning R, Zheng Q, Gao K (2023b) Microalgae-based biotechnology as a promising strategy for removing antibiotics from wastewater: opportunities, challenges and future directions. Front Bioeng Biotechnol 11:1248765

Zheng Q, Ning R, Zhang M, Deng X (2023) Biofuel production as a promising way to utilize microalgae biomass derived from wastewater: progress, technical barriers, and potential solutions. Front Bioeng Biotechnol 11:1250407

Zieliński M, Kazimierowicz J, Dębowski M (2023) The possibility of deploying CO_2 from biogas combustion to improve the productivity of a periodical *Chlorella vulgaris* culture. Front Biosci Eng 15:3–3

Phycoremediation Algae Marine and Its Role in Climate Changes

Alyan Ashraf, Tanzeela Basheer, Usman Khalid Chaudhary, and Arslan Asim

Abstract Marine ecosystems support various organisms, with phycoremediation algae standing out as key participants in the global carbon cycle and climate management. Phycoremediation uses algae's fundamental abilities to absorb and neutralize contaminants in aquatic environments, providing an eco-friendly solution to environmental degradation. Marine algae, which include diverse types of seaweed, are particularly good at removing heavy metals, excess nutrients, and organic pollutants from water bodies. Seaweeds significantly contribute to climate change mitigation. Phycoremediation algae are effective carbon sequesters due to photosynthesis, which converts atmospheric carbon dioxide into organic biomass. The biomass produced by seaweeds can be utilized for biofuel production, providing a renewable energy source that can help reduce reliance on fossil fuels. Their ability to develop in various marine settings, from coastal areas to wide oceans, makes them able for carbon capture and storage. Furthermore, phycoremediation algae improve marine ecosystem health by reducing nutrient pollution. They absorb excess nutrients like nitrogen and phosphorus from agricultural runoff and wastewater, reducing hazardous algal blooms and increasing water quality. Phycoremediation algae contribute significantly to climate change adaptation and resilience. They contribute to biodiversity and environmental stability by buffering against the effects of climate-induced disturbances such as ocean acidification and temperature changes. Moreover, the carbon stored in seaweed biomass can be transmitted to the deep ocean, providing long-term carbon storage. Large-scale algae cultivation projects, such as seaweed farms and algae-based carbon capture initiatives, are being

A. Ashraf (✉) · U. K. Chaudhary · A. Asim
Pakistan Environmental Protection Agency, Islamabad, Pakistan

Ministry of Climate Change and Environmental Coordination, Islamabad, Pakistan
e-mail: alyanashraf383@gmail.com; usmanchaudhry455@gmail.com; arslanasim92@gmail.com

T. Basheer
Department of Environmental Science, COMSATS University Islamabad, Vehari, Pakistan
e-mail: tanzeela.bashir555@gmail.com

A. A. H. Abdel Latef et al. (eds.), *Sustainable Remediation for Pollution and Climate Resilience*, https://doi.org/10.1007/978-981-96-5674-5_12

implemented worldwide. These projects often target both environmental restoration and commercial interests, highlighting the practical and economic benefits of algae. Moreover, algae research is increasingly focusing on developing innovative biotechnological solutions, including enhanced carbon capture technologies and pharmaceuticals with climate-related applications. Understanding the ecological significance and potential applications of phytoremediation algae is crucial for advancing climate change mitigation efforts. Through algae farms in diverse environments, including deserts, CO_2 can be absorbed from the atmosphere and oxygen can be produced, potentially transforming marine ecosystems into powerful allies in combating climate change and fostering a more sustainable future.

Keywords Phycoremediation algae · Marine ecosystems · Climate change · Carbon sequestration · Nutrient pollution · Sustainability

1 Introduction

Bioremediation involves using organisms to treat pollutants in various environmental mediums such as soil, groundwater, sludge, and solids, ultimately reducing their toxicity before releasing them into the environment. Utilizing the nutrients found in industrial wastewater is crucial for fostering the growth of microalgae cultures. This approach not only facilitates the production of biomass from desirable species but also aids in removing contaminants, thereby establishing an economically viable process. The efficiency of phycoremediation depends on several factors, including the species of microalgae, the design of the bioreactor, growth conditions, and the composition of residual wastewater. This chapter emphasizes the key considerations for achieving effectiveness in both phycoremediation and microalgae production processes.

Phycoremediation, also referred to as algal bioremediation, stands out as a highly effective biological process for treating wastewater. Algae and cyanobacteria thrive in effluents containing high concentrations of inorganic and organic compounds. These effluents originate from various industries including poultry processing, slaughterhouses, paper production, textiles, distilleries, and supercritical water gasification processes, as well as from produced water during oil and gas extraction, and landfill leachate. Integrating biomass for bioenergy with phycoremediation enhances the feasibility of practical commercialization and sustainable energy generation, thereby aiding in mitigating climate change and reducing competition for resources such as crop irrigation, carbon, and water. Moreover, phycoremediation facilitates bioenergy production with carbon capture and storage, consequently lessening the water footprint. In the context of a biorefinery approach, recycled wastewater used for cultivating algae serves as a medium for developing biomass for biofuel and co-products, further emphasizing the versatility and efficiency of phycoremediation as a water treatment method (Mathias et al. 2023). Climate change induces significant challenges in marine environments, endangering biodiversity, ecosystem equilibrium, and human sustenance (Onoh et al. 2024). To address this issue, creative strategies are being investigated to alleviate these effects and rejuvenate the well-being of our oceans (Borja et al. 2024).

Efforts are underway to explore innovative solutions to reduce these impacts and rejuvenate our oceans' health (Reisoglu and Aydin 2023). The algae can absorb carbon dioxide from saltwater through photosynthesis, effectively reducing greenhouse gas (GHG) emissions and mitigating ocean acidification (Kumar 2023). These pollutants may consist of heavy metals, nutrients like nitrogen and phosphorus, organic toxins, and oil spills (Sengupta et al. 2023). Algae in phycoremediation employ their inherent capability to absorb, metabolize, and transform pollutants into less harmful forms. By means of mechanisms like adsorption, absorption, and bioaccumulation, algae incorporate toxins into their biomass, efficiently purifying the water (Kumar et al. 2023). This method presents a feasible tactic for tackling water contamination while also offering additional advantages like carbon capture, habitat rehabilitation, and potential biofuel generation (Varshney 2024). Algae, a greatly diverse collection of photosynthetic organisms, can be found in a variety of aquatic environments, which are usually classified as freshwater or saltwater. Freshwater algae, which grow in rivers, lakes, and ponds, play an important part in ecosystem function by producing oxygen and cycling nutrients. They have practical applications, such as biofertilizers in agriculture, where species like Azolla and Spirulina increase soil fertility and crop productivity. Furthermore, freshwater algae, such as Chlorella, play an important role in wastewater treatment procedures by removing contaminants and heavy metals (Mahapatra et al. 2018). Saltwater algae, which live in oceans and seas, are essential to marine ecosystems and are used for a variety of industrial applications. Compounds originating from saltwater algae, like carrageenan from red algae and alginates from brown algae, are used in pharmaceuticals and cosmetics to gel, thicken, and stabilize (Lomartire and Gonçalves 2023). Recognizing the intricate mechanisms and possible applications of phycoremediation algae is essential for fully harnessing their potential in climate change mitigation endeavors (Danouche et al. 2021). Algae, with their huge variety and adaptability, play an important part in environmental treatments for water, air, and land, an area known as phycoremediation (Sengupta et al. 2023). Chlorella, Spirulina, and other marine macroalgae are among the most important algae employed in these applications. Chlorella, a freshwater microalga, is well-known for its effectiveness in wastewater treatment, where it eliminates contaminants, heavy metals, and excess nutrients, reducing eutrophication (Geremia et al. 2021). Spirulina, another freshwater alga, is used for soil remediation, enhancing soil quality and fertility due to its high nitrogen content. Macroalgae such as kelp and seaweed are vital in marine habitats for regulating ocean acidification and sequestering carbon dioxide, which helps to counteract climate change. These sea algae also help to purify the air by absorbing CO2 and releasing oxygen (Islam et al. 2021). The research explores the significance of phycoremediation algae in marine environments, examining their techniques, challenges, and potential benefits, while underscoring their crucial role in shaping a more sustainable future for our oceans and planet. Incorporating phycoremediation algae into comprehensive climate strategies allows us to safeguard the resilience and well-being of marine ecosystems for generations to come (Yong et al. 2021).

2 Role of Phycoremediation Algae in Climate Change

Algae involved in phycoremediation play a crucial role in alleviating the impacts of climate change by capturing CO_2 from both the atmosphere and aquatic environments (Sahu et al. 2019). Through photosynthesis, algae can assimilate CO_2 and transform it into organic biomass, effectively reducing the presence of greenhouse gases in the environment (Tsai et al. 2023). Anthropogenic climate change, stemming from greenhouse gases such as carbon dioxide, methane, and nitrous oxide, seems to be inducing notable environmental disturbances (Sahu et al. 2019). This mechanism aids in decreasing CO_2 gas levels in the atmosphere, thereby mitigating the greenhouse effect and the phenomenon of global warming (Ouedraogo et al. 2022). Algae and higher plants naturally absorb inorganic carbon through the process of photosynthesis. Algae, a term encompassing various types, have been categorized over the years by many indigenous civilizations into three groups: macroalgae, microalgae, and cyanobacteria. Macroalgae, often likened to aquatic moss because of their fast growth and widespread presence, are a common example (Blessing and Pisciotta 2023). Algae farming presents an opportunity for the production of biofuel, a sustainable energy alternative capable of mitigating carbon emissions from fossil fuels (Chia et al. 2022).

One approach to address climate change and move closer to a sustainable future involves utilizing the carbon sequestration potential of phycoremediation algae (Mehta et al. 2023). Their rapid expansion rates and ability to flourish in diverse conditions render them highly significant in sustainable biofuel production, offering a viable alternative to fossil fuels (Kour et al. 2019). Phycoremediation has found wide-ranging applications, including (i) utilizing algae-derived sorbents to eliminate xenobiotic compounds and nutrients, (ii) extracting nutrients from organic matter-rich wastewater, (iii) treating heavy metal-rich wastewater, (iv) facilitating the transformation and degradation of xenobiotics, (v) capturing CO_2, and (vi) employing algae-based biosensors for detecting hazardous substances (Emparan et al. 2019). Algae utilized in phycoremediation can combat climate change by serving as natural agents for capturing carbon (Sengupta et al. 2023). These algae absorb carbon dioxide from the environment through photosynthesis, thus reducing greenhouse gas emissions. Their capacity to transform carbon dioxide into biomass makes them valuable assets for mitigating carbon emissions generated from manufacturing processes and fossil fuel combustion (Budzianowski 2012).

This fuel gas emits high levels of carbon dioxide directly into the environment. Utilizing algae for phycoremediation could address additional environmental concerns, like nutrient pollution in aquatic systems (Priyadharshini et al. 2021). As research advances in this field, utilizing phycoremediation algae holds promise for accomplishing a deal with carbon management and ecological restoration (Li et al. 2023).

3 Effectiveness Over Other Traditional Methods and Division of Macroalgae

Macroalgae hold significant economic value in phycoremediation compared to traditional methods due to their cost-effectiveness, sustainability, and versatility. Unlike chemical and mechanical treatments that often require high energy inputs and produce secondary pollutants, macroalgae-based phycoremediation operates through natural processes such as photosynthesis and biochemical reactions, which are energy-efficient (Chung et al. 2011). This method not only reduces operational costs but also minimizes environmental impact, making it a preferred choice in coastal and marine ecosystems (Mata et al. 2010). Moreover, the biomass generated from macroalgae can be utilized for various value-added products like biofuels, fertilizers, and animal feed, providing additional economic benefits (Milledge and Harvey 2016). These advantages underscore macroalgae's potential to revolutionize environmental remediation practices while promoting sustainable economic development.

Macroalgae, commonly referred to as seaweeds, are divided into three main groups: Chlorophyta (green algae), Phaeophyceae (brown algae), and Rhodophyta (red algae). Each division possesses unique characteristics and plays a significant role in environmental remediation, particularly in marine and coastal ecosystems. Chlorophyta are efficient in nutrient removal and heavy metal adsorption, making them valuable for managing water quality (Mata et al. 2010). Phaeophyceae are known for their capacity to sequester carbon dioxide and remove heavy metals, aiding in both pollution control and climate change mitigation (Chung et al. 2011). Rhodophyta excels in nutrient cycling and bioremediation of aquaculture effluents, enhancing ecosystem health and sustainability (Schiewer and Patil 2008). Various countries have become pioneers in the application of macroalgae for phycoremediation. China, Japan, and South Korea are noted for their extensive use of Chlorophyta, while the United States and Norway are leading in the use of Phaeophyceae (Park et al. 2021). Additionally, China, South Korea, and Norway also excel in utilizing Rhodophyta for environmental remediation (Chopin 2014). Table 1 outlines the diverse applications of these macroalgae divisions in phycoremediation, highlighting their specific uses, examples of species involved, the mechanisms by which they achieve environmental cleanup, and the countries renowned for their application.

4 Assessing the Effectiveness of Phycoremediation Algae in Mitigating Climate Change

Phycoremediation plays a crucial role in mitigating the effects of climate change, encompassing a broad array of essential ecological processes and potential solutions that hold promise for addressing the challenges posed by a warming planet (Svenning et al. 2024; Kaur et al. 2023).

Table 1 Division of macroalgae along with its uses

Sr. no.	Division	Common uses in phycoremediation	Examples of species	Mechanisms	Famous countries
1	*Chlorophyta* (Green algae)	Nutrient removal (nitrogen, phosphorus)	*Ulva lactuca*, *Cladophora* spp.	Uptake of nutrients through surface area	China, Japan, South Korea
		Heavy metal adsorption	*Ulva lactuca*	Binding of metals to cell walls	
		Organic pollutant degradation	*Enteromorpha* spp.	Biodegradation by metabolic processes	
2	*Phaeophyceae* (Brown algae)	*CO_2 sequestration*	*Macrocystis pyrifera*, *Sargassum* spp.	Photosynthesis and carbon fixation	United States, Norway, Japan
		Removal of heavy metals (e.g., arsenic, mercury)	*Laminaria digitata*	Adsorption to algal surfaces	
		Eutrophication control	*Ascophyllum nodosum*	Uptake of excess nutrients	
3	*Rhodophyta* (Red algae)	*Nutrient cycling (nitrogen, phosphorus)*	*Gracilaria* spp., *Porphyra* spp.	Assimilation and conversion	China, South Korea, Norway
		Removal of heavy metals (e.g., cadmium, lead)	*Gelidium* spp., *Palmaria palmata*	Ion exchange and chelation	
		Bioremediation of aquaculture effluents	*Gracilaria* spp.	Filtration and nutrient uptake	

Mata et al. (2010); Schiewer and Patil (2008); Chung et al. (2011); Park et al. (2021); Chopin (2014)

Their significant contribution is centered on their remarkable capacity for carbon sequestration, an essential process in reducing atmospheric carbon dioxide levels, which serves as the main catalyst for global warming (Karduri and Ananth 2021). Phycoremediation harnesses photosynthesis to draw carbon dioxide from both the atmosphere and oceans, converting it into organic matter (Kaur et al. 2023). This decreases the concentration of greenhouse gases in the atmosphere and offers a route for the sequestration of carbon in algal biomass, eliminating carbon from the carbon cycle (van den Burg et al. 2024). This decreases the concentration of greenhouse gases in the atmosphere and offers a route for the sequestration of carbon in algal biomass, eliminating carbon from the carbon cycle (Elhabodi et al. 2023).

By enhancing the reflectivity of surfaces, such as water bodies where they thrive, phycoremediation can assist in cooling the nearby environment, consequently alleviating the effects of increasing temperatures. This approach may also have wider implications for regional climate dynamics, affecting rainfall, wind patterns, and cloud formation (Boza et al. 2024). Phycoremediation plays a crucial role in the atmosphere and serves as a vital tool in combating ocean acidification, a consequence of seawater absorbing excessive carbon dioxide (Webb 2024). Ocean acidification presents significant dangers to marine life, such as coral reefs and shell-forming organisms. Phycoremediation's capacity to absorb carbon dioxide from seawater serves as a buffer against these detrimental impacts. This process aids in preserving the delicate equilibrium of ocean chemistry crucial for the health and resilience of marine ecosystems (Althea 2023; Yuan 2024).

Furthermore, phycoremediation contributes to nutrient cycling and water quality upkeep in aquatic settings. It effectively absorbs nutrients like nitrogen and phosphorus, thereby reducing nutrient pollution and the probability of detrimental algal blooms (Ankit et al. 2022). By rectifying nutrient cycles and enhancing water quality, phycoremediation aids in maintaining the health and sustainability of marine and freshwater ecosystems, which are progressively susceptible to the impacts of climate change (Nguyen et al. 2024). Also, these algae provide answers for coastal preservation and habitat rehabilitation (Schoonees et al. 2019). Phycoremediation enhances coastal resilience against the escalating threats of rising sea levels and intensified weather events triggered by climate change, accomplishing this by stabilizing coastlines, mitigating erosion, and fostering diverse marine habitats (Ghani and Hidayati 2024).

Moreover, phycoremediation shows potential for promoting sustainable aquaculture and ensuring food security. Serving as a nutrient-dense food and feed option, it presents an alternative to conventional agricultural methods, alleviating strain on land-based ecosystems and offering a robust nutritional source for human communities (Mrabet 2023). Despite the vast potential of phycoremediation in addressing the impacts of climate change, numerous obstacles persist, such as scalability, technological advancement, and economic viability (Almaraz et al. 2023). Algae, observed for their diversity and adaptability, play an important role in environmental treatments such as phytoremediation, which includes water, air, and land applications. In marine habitats, macroalgae like kelp and seaweed play an important role in carbon sequestration and ocean acidification mitigation (Islam et al. 2021). Countries with large coastlines, such as Norway, the United States, and China, are at the forefront of using marine algae for these reasons. These macroalgae not only sequester substantial amounts of CO_2 but also release oxygen, which helps to purify the air. Globally, the algal biomass production sector is expanding, with predictions indicating that by 2025, the market could exceed $1 billion, reflecting greater recognition of algae's potential for environmental sustainability and climate change mitigation (Sedighi et al. 2023). Through persistent research, innovation, and investment, these obstacles can be tackled, thereby unleashing the complete potential of phycoremediation as an effective solution for battling climate change and constructing a sustainable future for future generations (Xie et al. 2022).

Macroalgae, also known as seaweeds, play a crucial role in environmental remediation due to their diverse applications in nutrient removal, heavy metal adsorption, and organic pollutant degradation. Studies indicate that macroalgae can effectively remove nitrogen and phosphorus from wastewater, with some species capable of eliminating up to 90% of nitrogen and 70% of phosphorus content (Xu et al. 2017). They also exhibit high adsorption capacities for heavy metals like arsenic, mercury, and cadmium, contributing significantly to water pollution control efforts. For example, certain brown algae species have demonstrated the ability to adsorb as much as 80% of arsenic from contaminated water sources (Papazi et al. 2018). Moreover, macroalgae possess enzymatic capabilities that facilitate the degradation of organic pollutants, thereby enhancing water quality in marine ecosystems (Munoz et al. 2009). These findings highlight the potential of macroalgae not only as a sustainable solution for environmental remediation but also as a valuable resource for sustainable development initiatives globally.

5 The Carbon Sink Potential of Algae

The carbon recycling process has a limited capacity to absorb carbon dioxide, whereas algae possess significant potential for carbon dioxide reduction across the atmosphere. The risk of compressing excessive greenhouse gases into the seafloor is evident. Algae can naturally sink to the seafloor as organic remnants, avoiding conversion into carbon dioxide. Furthermore, the high costs associated with cultivating Tung trees in coastal deserts hinder the feasibility of this solution. Conversely, cultivating algae proves cost-effective. Land-based plants currently absorb 52% of the total carbon dioxide in the Earth's biosphere, while oceanic algae contribute 45–50%, highlighting algae's efficient carbon dioxide absorption despite their small size and short life cycles. With suitable environmental support—such as temperature, nutrients (including crucial iron), sunlight, and controlling predator populations—algae can further mitigate atmospheric carbon dioxide levels. Hence, algae cultivation is a promising solution for combating greenhouse effects (Haoyang 2018).

Algae, comprising a significant portion of biomass in marine and freshwater environments, play a paramount role as CO_2 fixers within aquatic ecosystems. As primary producers, they utilize photosynthesis to convert carbon into biomass, occupying a critical niche in the ecosystem. Despite their significance, the ecological role of algae remains incompletely elucidated. Algal carbon uptake potentially serves as an unrecognized carbon sink, given that the comprehension of carbon storage within algae lags behind that of physical and chemical processes. The emulation of natural environments offers a promising avenue for enhancing our understanding of algae's contribution to the carbon cycle (Tsai et al. 2017). Efficiency and rate of CO_2 absorption by algae in aquatic ecosystems are vital subjects of research, especially concerning global carbon cycling and mitigating climate change. Algae in freshwater and marine habitats play a crucial role in absorbing CO_2 from the atmosphere, functioning as natural carbon sinks. The rate at which algae absorb CO_2 varies significantly among different species and environmental conditions. Factors such as light intensity, water temperature, nutrient availability, and the specific algal species present influence this absorption rate in aquatic ecosystems. Generally, algae demonstrate higher CO_2 absorption rates compared to terrestrial plants, particularly under optimal conditions. This heightened efficiency is primarily attributed to their simple cellular structures and growth mechanisms, which facilitate more effective photosynthesis (Tsai et al. 2023).

It is reported that globally, marine macroalgae alone sequester approximately 173 million tons of carbon annually (Chung et al. 2011). This carbon is stored in biomass and in the form of dissolved organic carbon in seawater, contributing to the reduction of atmospheric CO_2 levels.

5.1 Efficiency of CO_2 Bio-Fixation

The effectiveness of algae in converting CO_2 into biomass (bio-fixation) is a pivotal aspect. This efficiency not only gauges CO_2 absorption but also signifies the transformation of CO_2 into algal biomass. A high bio-fixation efficiency implies that more carbon gets stored as organic matter, playing a substantial role in carbon cycling within aquatic ecosystems (Tsai et al. 2023). Specific statistics on the efficiency of CO2 bio-fixation by macroalgae are less commonly reported compared to microalgae. However, studies have indicated that macroalgae, particularly large brown algae such as kelps (e.g., Macrocystis, Laminaria species), can absorb substantial amounts of CO_2 during their growth cycles. For instance, large kelp forests have been estimated to sequester carbon dioxide at rates comparable to terrestrial forests, with some estimates suggesting that kelp forests can capture hundreds of kilograms of carbon per hectare annually (Wernberg et al. 2020).

5.2 Role in Aquatic Ecosystems

Algae serve as the foundation of the food web in aquatic settings. Their capacity for carbon fixation holds profound implications for ecosystem vitality and carbon cycling. Through CO_2 absorption, algae aid in regulating carbon levels in water bodies, which is vital for sustaining pH balance and overall ecosystem well-being (Tsai et al. 2023). Macroalgae contribute significantly to nutrient cycling by absorbing excess nutrients like nitrogen and phosphorus from the water, thereby helping to maintain water quality and prevent eutrophication (Lavery et al. 2012).

5.3 Environmental Implications

Investigating algal CO_2 uptake is vital for comprehending and tackling global climate change. Augmenting algal proliferation in aquatic ecosystems can be a natural means to enhance carbon sequestration. Moreover, the cultivation of algae for biofuel production is garnering increased attention. Algal biofuels offer a more sustainable alternative to fossil fuels, given their lower carbon footprint and noncompetition with arable land required for food production (Tsai et al. 2023). Macroalgae also serve as carbon sinks, sequestering carbon dioxide from the atmosphere and storing it in their biomass and surrounding sediments, which helps mitigate climate change (Wernberg et al. 2020). Moreover, their ability to provide habitats and food for diverse marine organisms supports biodiversity in coastal areas, enhancing ecosystem resilience and productivity (Steneck et al. 2002). Additionally, macroalgae contribute to shoreline stabilization by reducing wave energy and preventing coastal erosion, making them integral components of coastal management strategies (Wernberg et al. 2020). These environmental roles highlight the importance of macroalgae in maintaining marine ecosystem balance and addressing global environmental challenges.

5.4 Challenges and Future Research

One challenge in leveraging algae for CO_2 uptake is the management of algal blooms, which can adversely affect water quality and marine biodiversity. Furthermore, there is a necessity for further research to optimize conditions for maximal CO_2 absorption and to grasp the enduring consequences of algal biofixation across varied aquatic environments. To sum up, the efficiency and pace of algal CO_2 uptake in aquatic ecosystems represent a critical domain of study with far-reaching implications for environmental sustainability, carbon cycling, and strategies for mitigating climate change. Progress in this realm could yield innovative approaches to environmental stewardship and renewable energy generation (Tsai et al. 2023).

6 Overview of Macroalgae Applications in Environmental Remediation

Macroalgae, commonly known as seaweeds, are increasingly recognized for their multifaceted roles in environmental remediation. These versatile organisms have been utilized for various applications due to their efficient nutrient absorption, heavy metal adsorption, carbon sequestration, and more. Each type of macroalgae offers unique benefits that make them suitable for specific environmental applications. Table 2 provides a detailed overview of different macroalgae species, their first recorded uses, geographical locations of use, and notable characteristics. This comprehensive summary highlights the global impact and potential of macroalgae in addressing various environmental challenges.

Table 2 First-time use of macroalgae in environmental remediation application

Sr. no.	Use	Algae	First use	Locations	Characteristics
1	Nutrient removal (Nitrogen, Phosphorus)	Ulva (Green algae)	1970s	China, South Korea, Japan	High nutrient absorption capacity
2	Heavy metal adsorption	Sargassum (Brown algae)	1980s	China, Japan, United States	High adsorption capacity for heavy metals
3	Organic pollutant degradation	Gracilaria (Red algae)	1990s	South Korea, Norway, United States	Biochemical degradation of organic pollutants
4	Carbon sequestration	Macrocystis (Brown algae, Giant kelp)	Early 2000s	United States, Australia, Chile	High biomass, effective carbon sink
5	Biofuel production	Laminaria (Brown algae)	2000s	Norway, United Kingdom, Japan	High carbohydrate content, suitable for bioethanol
6	Food and nutraceuticals	Porphyra (Red algae, Nori)	Centuries ago, commercial use in the twentieth century	Japan, China, South Korea	High nutritional value, rich in bioactive compounds
7	Fertilizer and soil conditioner	Ascophyllum (Brown algae)	Nineteenth century	United Kingdom, Ireland, France	Rich in minerals, growth-promoting properties
8	Cosmetic and pharmaceutical ingredients	Fucus (Brown algae, Bladderwrack)	Twentieth century	France, United States, United Kingdom	Anti-inflammatory, antioxidant properties

Lavery et al. (2012); Milledge and Harvey (2016); Steneck et al. (2002); Wernberg et al. (2020); Xu et al. (2017)

7 Algal Biofuel Production

Renewed interest in various renewable biofuels stems from fluctuating petroleum fuel costs, impending peak oil scenarios, dependence on unstable foreign petroleum sources, and the escalating threat of rising atmospheric CO_2 levels. Photosynthetic algae, encompassing both microalgae and macroalgae (such as seaweeds), have garnered significant attention as potential biofuel reservoirs for many years. Numerous species exhibit biomass production rates surpassing those of terrestrial plants. At the same time, many eukaryotic microalgae possess the capacity to store substantial quantities of energy-rich compounds like triacylglycerol (TAG) and starch, viable for diverse biofuel production, including biodiesel and ethanol. Given their lipid profiles and productivity, there is a prevailing belief that a substantial portion of crude oil originates from microalgae, particularly diatoms. Recognizing this, exploring the biofuel potential of living microalgae becomes imperative. Microalgae offer environmental advantages as fuel sources, consuming carbon dioxide and thriving on marginal land with waste or saltwater. Furthermore, leveraging microalgae's metabolic pathways may yield a broad spectrum of biofuels (Radakovits et al. 2010).

Unlike corn-based ethanol or soy/palm-based biodiesel, biofuels sourced from microalgal feedstocks will not directly vie with the resources required for agricultural food production, provided that inorganic components can be reused and cultivation systems based on saltwater are established (Radakovits et al. 2010).

While microalgae primarily engage in oxygenic photosynthesis—breaking down water to generate O_2 and converting CO_2 into biomass through sunlight—many can also utilize organic substrates like glucose and acetic acid, enabling mixotrophic or heterotrophic growth in light or darkness, respectively. Furthermore, certain species have adapted, or reverted, to a non-photosynthetic state, lacking pigmentation and residing permanently in darkness (Lundquist et al. 2010).

As the worldwide need for energy escalates, the utilization of fossil fuels is expected to grow persistently. Although currently abundant and relatively inexpensive, the future outlook suggests a potential shift in their availability and cost. However, the fundamental issue lies in the sustainability of escalating fossil fuel consumption over the long haul, primarily due to the associated rise in greenhouse gas (GHG) emissions and their detrimental impact on global warming (Hill et al. 2006).

Hence, considerable attention is being focused on discovering alternative renewable fuel sources that can be carbon neutral (Demirbas 2009; Hill et al. 2006; Rittmann 2008).

The majority of the biofuels currently on the market are derived from sugar cane or corn starch for bioethanol and oil crops like soybean and oilseed rape for biodiesel. While biofuels offer potential environmental benefits over fossil fuels, there is ongoing debate regarding their economic competitiveness. Additionally, there are concerns about the potential impact on food availability when these crops are diverted for biofuel production (Demirbas 2009; Hill et al. 2006).

Algae-derived biofuels have thus emerged as a viable alternative, offering a solution that avoids encroaching on agricultural resources.

Algae, especially green unicellular microalgae, have long been suggested as a promising source of renewable fuel (Benemann et al. 1977; Oswald and Golueke 1960). Microalgae possess the capability to produce substantial amounts of biomass and oil suitable for biodiesel conversion. Studies suggest that microalgae exhibit higher biomass productivity per unit of land area compared to traditional plant crops, potentially resulting in a lower cost per yield. Furthermore, the utilization of microalgae has the potential to mitigate greenhouse gas emissions by substituting fossil fuels.

The drive to produce algal biofuel stems from the aim to replace conventional petroleum-based fuels, which are finite; generate fuels domestically to lessen energy imports; and decrease greenhouse gas emissions by cultivating algae that recycle carbon dioxide emitted from industrial facilities. Theoretically, algae possess the capability to yield substantial substitutes for petroleum fuel, all while bypassing the necessity for significant quantities of fresh water and arable land. These characteristics have sparked extensive interest in algal biofuels. Nonetheless, profitable algal biofuel production encounters several significant challenges (Beal et al. 2012).

The optimal microalgal biofuel process may involve a growth system prioritizing ease of harvesting over maximal yield. Additionally, the energy extraction process could be designed to minimize the concentration required by the microalgal harvesting method. Many researchers recognize efficient harvesting as a primary obstacle in commercializing microalgal biofuel, which significantly impacts the design and operation of both upstream and downstream processes in the overall production process (Milledge and Heaven 2013).

Algae, as a sustainable energy source, and the feedstocks they generate hold immense promise for replacing petroleum-based fuels. The adaptability of algae in generating lipids, carbohydrates, and protein is crucial for developing various products across different markets to meet economic demands effectively. Biotechnology firms and the algae industry prioritize manufacturing small quantities of high-value items like pharmaceuticals or dietary supplements. These sectors need to shift their focus toward producing large volumes of biofuels at competitive prices and utilizing byproducts such as protein for distiller's grains and carbohydrates for ethanol. It is essential to utilize postextraction byproducts efficiently and completely (Menetrez 2012).

7.1 Algal Biofuel Production: Process, Mechanics, and Implementation

Algal biofuel production has gained significant importance as a sustainable and environmentally friendly alternative to fossil fuels. This process involves cultivating algae, harvesting their biomass, and converting it into various forms of biofuels such as biodiesel, bioethanol, and biogas. The following sections provide a detailed explanation of the process, the mechanics involved, and how different countries can scientifically benefit from this technology, particularly as a potential project for unemployed youth.

7.1.1 Important Algae Species for Biofuel Production

- *Laminaria* (*Brown algae*): Known for its high carbohydrate content, suitable for bioethanol production.
- *Ulva* (*Green algae*): Contains high levels of lipids and carbohydrates, making it ideal for biodiesel and bioethanol.
- *Nannochloropsis* (*Microalgae*): Rich in lipids, widely used for biodiesel production.
- *Chlorella* (*Microalgae*): High lipid content, suitable for biodiesel and biogas production.

7.1.2 The Process of Algal Biofuel Production

1. *Cultivation*:

   - *Open Ponds*: Open pond production is a method of cultivating algae in large, shallow ponds exposed to natural environmental conditions. These ponds are typically constructed in a circular or raceway design to facilitate water circulation and nutrient distribution. Algae in open ponds grow using sunlight, carbon dioxide, and nutrients, which can be supplied through fertilizers or wastewater. This method is cost-effective and suitable for large-scale biomass production, but it can be susceptible to contamination and environmental fluctuations. Open pond systems are commonly used for macroalgae like Ulva and for microalgae species that thrive in outdoor conditions.
   - *Photobioreactors (PBRs)*: Photobioreactors (PBRs) for macroalgae cultivation are closed systems designed to provide a controlled environment for optimal algae growth. These reactors use transparent tubes or panels to allow maximum light penetration, which is essential for photosynthesis. PBRs can regulate temperature, nutrient levels, and carbon dioxide supply, creating ideal conditions for macroalgae growth. This controlled environment minimizes the risk of contamination and maximizes biomass yield. PBRs are more efficient than open ponds in terms of space utilization and productivity, making them suitable for high-value macroalgae species and locations with limited space or harsher environmental conditions.
   - *Raceway Ponds*: Raceway ponds are a type of open pond system designed for the large-scale cultivation of algae, including macroalgae. These ponds are typically shallow, elongated basins with a continuous loop, resembling a racetrack. A paddlewheel continuously circulates the water, ensuring even distribution of nutrients, carbon dioxide, and light exposure throughout the pond. Raceway ponds are cost-effective and relatively easy to construct and operate, making them suitable for producing large volumes of algal biomass. However, they can be susceptible to environmental fluctuations and contamination, requiring careful management to maintain optimal growth conditions.

2. *Harvesting*:

   - *Centrifugation*: Algae are separated from the water using high-speed centrifuges. This method is efficient but energy-intensive.
   - *Flocculation*: Chemicals are added to the algae culture to aggregate the cells, making them easier to separate.
   - *Filtration*: Physical filters are used to separate algae from the water.

3. *Biomass Processing*:

   - *Drying*: Algal biomass is dried to reduce water content. This can be done using solar drying, drum drying, or freeze-drying.
   - *Extraction*: Lipids are extracted from the dried biomass using solvents like hexane or through mechanical pressing.

4. *Biofuel Conversion*:
   - *Transesterification*: The extracted lipids are converted into biodiesel by reacting with alcohol (usually methanol) in the presence of a catalyst.
   - *Fermentation*: Carbohydrates from algae are fermented to produce bioethanol. This involves enzymatic hydrolysis to break down polysaccharides into fermentable sugars.
   - *Anaerobic Digestion*: Algal biomass is broken down by microorganisms in the absence of oxygen to produce biogas, primarily composed of methane and carbon dioxide.

7.1.3 Mechanics of the Process

- *Light and CO_2 Supply*: Optimal growth of algae requires sufficient light and carbon dioxide. Photobioreactors can be equipped with artificial light sources and CO_2 injection systems to enhance productivity.
- *Nutrient Management*: Algae require nutrients like nitrogen, phosphorus, and trace elements. These can be supplied through fertilizers or wastewater, which also helps in wastewater treatment.
- *Temperature Control*: Maintaining optimal temperature is crucial for algae growth. This can be managed through environmental control systems in photobioreactors.

8 Enhanced Marine Productivity and Ecosystem Resilience

Biotic degradation is a frequently observed phenomenon utilized to eliminate various organic pollutants. Phycoremediation, employing live algae to enhance effluent quality, is widely recognized as a promising, efficient, and economically viable approach to wastewater treatment. This method is advantageous because the organic and inorganic compounds present in aquaculture systems largely fulfill the nutrient requirements of algae (Turcios and Papenbrock 2014). Algae play a crucial role in treating various elements present in wastewater, including nutrients like dissolved nitrogen and dissolved phosphorus, biological oxygen demand (BOD), coliforms, metals, minerals, and CO_2. Additionally, in the phycoremediation process, organic nutrients undergo uptake, transformation, and assimilation into nucleic acids and proteins within algal cells, consequently promoting algae biomass growth (Whitton et al. 2015). Ammonium derived from the breakdown of organic residues or nitrate reduction is assimilated into amino acids to facilitate protein synthesis. Meanwhile, inorganic phosphorus serves either to synthesize ribosomal RNA or is stored as polyphosphate (Rawat et al. 2011; Wu et al. 2012; Beuckels et al. 2015).

Microalgae have the capacity to endure challenging environmental circumstances (Weissman and Goebel 1987). Achieving an optimal balance of nutrients

(N/P ratio) is essential for the proper development of microalgae. Various methods have been utilized to enhance microalgae growth, involving adjustments to both chemical and physical parameters. In addition to nutrient levels within the cultivation medium, factors such as pH, temperature, and light intensity also significantly promote biomass growth (Oh et al. 2009). Uncovering the phycoremediation potential and suitability of diverse microalgae for nutrient removal across various wastewater types. Abomohra et al. (2017) discovered that among 21 halophilic microalgae, *Tetraselmis elliptica* exhibited the most significant biomass growth. Mathimani et al. (2017) demonstrated the suitability of marine *Chlorella vulgaris* for biodiesel production. Aravantinou et al. (2013) found that marine microalgae displayed a higher growth rate in synthetic wastewater than freshwater microalgae. However, there is currently no available information in the open literature regarding the biodiesel production potential of marine Tetraselmis indica. This study highlights the potential of *T. indica* for phycoremediation of secondary treated domestic sewage (STDS) and biodiesel production from harvested algal biomass. Phycoprospecting, which entails evaluating potential native microalgae possessing inherent traits for bioresource production and waste management, is widely regarded as the most sustainable approach for advancing algal biotechnology development (Wilkie et al. 2011). For phycoremediation (utilizing algae for waste management), its effectiveness relies on various factors such as achievable algal productivity, efficiency in nutrient and pollutant elimination, and the expenses associated with harvesting (Olguín 2003). Furthermore, considering algal strains possessing unique attributes such as resilience to extreme temperatures, a chemical composition rich in high-value products, rapid sedimentation behavior, and the ability to grow mixotrophically is imperative. Another avenue gaining significant attention is the enhancement of strain performance through genetic modification technology, aimed at augmenting microalgae's suitability for bioremediation, thanks to advancements in molecular and functional genomic tools. The utilization of filamentous microalgae with a high capacity for autoflocculation and the exploration of immobilized cells have also been explored for phycoremediation purposes. Additionally, the immobilization of both cyanobacteria and eukaryotic algae in diverse materials like alginate, glass beads, loofa sponge, carrageenan, and silica gel has demonstrated efficacy in removing nutrients such as nitrogen and phosphorus, as well as heavy metals (Hameed and Ebrahim 2007).

In today's context, the scarcity of adequate water resources for drinking, domestic, and industrial use presents a significant global challenge (Chiu et al. 2015). Treated wastewater offers cleaning and gardening opportunities, thus preserving fresh municipal water for essential purposes like drinking, cooking, and bathing. However, conventional wastewater treatment methods are costly, involve the use of hazardous chemicals, and consume significant energy. Therefore, water treatment methods must prioritize cost-effectiveness and environmental sustainability (Piao et al. 2016). The limitations outlined above in the currently employed methods have prompted scholars and industries to innovate a cost-efficient and eco-friendly approach for wastewater treatment. Kelp (Laminaria) is important because of its ability to absorb carbon, create marine habitats, and protect coastlines from erosion.

Kelp is produced mostly in China, South Korea, Japan, and the United States, and its ecological value has been recognized since the mid-twentieth century. Kelp is worth billions of dollars per year and is mainly grown in coastal seas using ropes or nets. It is also utilized in food, pharmaceuticals, and cosmetics (Wright et al. 2022). Red algae (Rhodophyta) is also important for carbon sequestration, marine animal feed, and the production of agar and carrageenan, with the Philippines, Indonesia, Malaysia, and Chile leading the way in cultivation. It is worth hundreds of millions of dollars due to its industrial and food applications, with coastal farming serving as the principal cultivation method (Wertz and Perez 2024). Bioremediation involves utilizing naturally occurring biological agents or processes to eradicate environmental contaminants within a specific location (Rodrigues and Silva 2016). Green algae (Chlorophyta) is known for its potential in biofuel generation, nutrient bioextraction, and biofertilization, with significant contributions from the United States, China, India, and Australia. Green algae farming, which uses bioreactors or open ponds, was recognized for its benefits in the early twenty-first century (Choudhary et al. 2021). Bioremediation has been utilized to purify water, soil, atmosphere, or any other polluted area in the environment. Microbial, algal, or fungal sources are typically employed as biological agents for environmental remediation (El-Sheekh and Mahmoud 2017). To develop appropriate genetically engineered algal strains capable of effectively increasing nitrogen and phosphorus uptake, it is necessary to explore genome databases such as the National Center for Biotechnology Information (NCBI) and GenBank to identify genes with enhanced nitrogen and phosphorus uptake capacity. Upon identification, these genes can be overexpressed using a suitable promoter system, resulting in heightened nitrogen and phosphorus uptake in the modified algal strain and subsequent removal of organic compounds from wastewater. Consequently, creating genetically modified algal strains with improved absorption and metabolic traits can aid in addressing wastewater treatment challenges. This analysis will enable the identification of effective algal species and genes for wastewater treatment, thereby enhancing overall treatment efficacy. Such an approach will facilitate the strategic planning of future wastewater treatment endeavors.

9 Future Prospects: Implementation and Benefits for Different Countries, Project for Unemployed Youth, and Steps to Initiate the Project

Implementing algal biofuel production offers significant future benefits for different countries, particularly those with varying climatic conditions and available resources. Coastal nations with access to seawater can harness the potential of macroalgae cultivation, while countries with limited water resources can focus on microalgae using photobioreactors (Xu et al. 2017). This approach reduces reliance on fossil fuels, mitigates climate change through carbon sequestration, and aids in wastewater treatment (Milledge and Harvey 2016). Additionally, it supports economic diversification and environmental sustainability, integrating well with national energy strategies and global climate goals.

Future prospects for algal biofuel production are promising, as technological advancements continue to improve efficiency and cost-effectiveness. Countries investing in this field can achieve energy independence and enhance their economic resilience. For example, nations like the United States, China, and Japan are already leading in algae research and cultivation techniques, paving the way for broader adoption and innovation (Wernberg et al. 2020).

Algal biofuel production also presents a significant opportunity as a project for unemployed youth, providing meaningful employment and skill development. Establishing training programs to teach algae cultivation, harvesting, and biofuel conversion can empower youth, foster entrepreneurship, and reduce unemployment. Community-based algae farms can be set up with support from government and nongovernmental organizations, ensuring sustainability and scalability. This initiative not only creates jobs but also promotes sustainable energy production and environmental conservation, addressing multiple socioeconomic and environmental challenges simultaneously.

To initiate such a project, a feasibility study should be conducted to assess local conditions for algae cultivation, considering factors like climate, water availability, and potential sites. Following this, training and capacity-building programs can be developed to equip youth with the necessary skills. Infrastructure development involves establishing algae cultivation systems like raceway ponds or photobioreactors and necessary processing facilities. Forming partnerships with research institutions, government agencies, and private sector stakeholders is crucial for securing technical and financial support. Finally, implementing regular monitoring and evaluation ensures the project's success, allowing for adjustments and improvements as needed (Xu et al. 2017).

Investing in algal biofuel production promises a sustainable future by integrating renewable energy solutions with socioeconomic development. By following these steps, countries can harness the potential of algal biofuel production, driving sustainable development and innovation while creating significant socioeconomic benefits (Milledge and Harvey 2016).

10 Summary/Conclusion

The pivotal role of phycoremediation in marine ecosystems has profound implications for climate change mitigation. Despite facing challenges such as space constraints, nutrient requirements, and contamination risks, phycoremediation emerges as a sustainable solution for wastewater treatment, offering benefits ranging from environmental preservation to industrial applications. By harnessing algae's natural capabilities, phycoremediation improves water quality and facilitates carbon sequestration, biofuel production, and ecosystem resilience. The versatility of algal systems, both microalgae and macroalgae, underscores their potential as invaluable resources for bioremediation and biofuel generation. Moreover, the consortial approach, harnessing synergistic interactions among different algae species, holds promise for enhancing the efficiency and effectiveness of phycoremediation processes. As we navigate the challenges of climate change and environmental degradation, embracing phycoremediation as a viable and sustainable solution is imperative for safeguarding the health and integrity of marine ecosystems and ensuring a more sustainable future for generations to come.

References

Abomohra AE-F, El-Sheekh M, Hanelt D (2017) Screening of marine microalgae isolated from the hypersaline Bardawil lagoon for biodiesel feedstock. Renew Energy 101:1266–1272

Almaraz M, Simmonds M, Boudinot FG, Di Vittorio AV, Bingham N, Khalsa SDS, Ostoja S, Scow K, Jones A, Holzer IJ (2023) Soil carbon sequestration in global working lands as a gateway for negative emission technologies. Glob Chang Biol 29(21):5988–5998

Althea AJ (2023) Impact of ocean acidification on coral reefs and the marine ecosystems in Phillipines. Int J Nat Sci 3(2):48–60

Ankit, Bauddh K, Korstad J (2022) Phycoremediation: use of algae to sequester heavy metals. Hydrobiology 1(3):288–303

Aravantinou AF, Theodorakopoulos MA, Manariotis ID (2013) Selection of microalgae for wastewater treatment and potential lipids production. Bioresour Technol 147:130–134

Beal CM, Hebner RE, Webber ME, Ruoff RS, Seibert AF, King CW (2012) Comprehensive evaluation of algal biofuel production: experimental and target results. Energies 5(6):1943–1981

Benemann JR, Weissman JC, Koopman BL, Oswald WJ (1977) Energy production by microbial photosynthesis. Nature 268(5615):19–23

Beuckels A, Smolders E, Muylaert K (2015) Nitrogen availability influences phosphorus removal in microalgae-based wastewater treatment. Water Res 77:98–106

Blessing S, Pisciotta JM (2023) Carbon sequestration: principle and recent advances. In: Clean technologies toward the development of a sustainable environment and future. IWA Publishing, London, pp 43–74

Borja A, Elliott M, Teixeira H, Stelzenmüller V, Katsanevakis S, Coll M, Galparsoro I, Fraschetti S, Papadopoulou N, Lynam C (2024) Addressing the cumulative impacts of multiple human pressures in marine systems, for the sustainable use of the seas. Front Ocean Sustain 1:1308125

Boza B, Herrington AR, Ilicak M, Danabasoglu G, Sen OL (2024) Exploring regional climate of the Euro-Mediterranean using a variable-resolution configuration of the global Community Earth System Model (VR-CESM). J Geophys Res Atmos 130:e2024JD041300

Budzianowski WM (2012) Negative carbon intensity of renewable energy technologies involving biomass or carbon dioxide as inputs. Renew Sust Energ Rev 16(9):6507–6521

Chia SR, Nomanbhay SBHM, Chew KW, Munawaroh HSH, Show PL (2022) Algae as potential feedstock for various bioenergy production. Chemosphere 287:131944

Chiu S-Y, Kao C-Y, Chen T-Y, Chang Y-B, Kuo C-M, Lin C-S (2015) Cultivation of microalgal Chlorella for biomass and lipid production using wastewater as nutrient resource. Bioresour Technol 184:179–189

Chopin T (2014) Seaweeds: top mariculture crop, ecosystem service provider. World Aquacult 45(1):50–54

Choudhary P, Khade M, Savant S, Musale A, Chelliah MS, Dasgupta S (2021) Empowering blue economy: from underrated ecosystem to sustainable industry. J Environ Manag 291:112697

Chung IK, Beardall J, Mehta S, Sahoo D, Stojkovic S (2011) Using marine macroalgae for carbon sequestration: a critical appraisal. J Appl Phycol 23(5):877–886

Danouche M, El Ghachtouli N, El Arroussi H (2021) Phycoremediation mechanisms of heavy metals using living green microalgae: physicochemical and molecular approaches for enhancing selectivity and removal capacity. Heliyon 7(7):e07609
Demirbas A (2009) Biofuels securing the planet's future energy needs. Energy Convers Manag 50(9):2239–2249
Elhabodi TS, Yang S, Parker J, Khattak S, He B-J, Attia S (2023) A review on BIPV-induced temperature effects on urban heat islands. Urban Clim 50:101592
El-Sheekh MM, Mahmoud YA (2017) Technological approach of bioremediation using microbial tools: bacteria, fungi, and algae. In: Handbook of research on inventive bioremediation techniques. IGI Global, pp 134–154
Emparan Q, Harun R, Danquah M (2019) Role of phycoremediation for nutrient removal from wastewaters: a review. Appl Ecol Environ Res 17(1):889–915
Geremia E, Ripa M, Catone CM, Ulgiati S (2021) A review about microalgae wastewater treatment for bioremediation and biomass production—a new challenge for Europe. Environments 8(12):136
Ghani M, Hidayati I (2024) Adapting to sea level rise: migration decisions and disaster mitigation strategies in small-scale fishing communities. IOP Conf Ser Earth Environ Sci 1314:012013
Hameed MA, Ebrahim OH (2007) Biotechnological potential uses of immobilized algae. Int J Agric Biol 9:183–192
Haoyang C (2018) Algae-based carbon sequestration. IOP Conf Ser Earth Environ Sci 120:012011
Hill J, Nelson E, Tilman D, Polasky S, Tiffany D (2006) Environmental, economic, and energetic costs and benefits of biodiesel and ethanol biofuels. Proc Natl Acad Sci 103(30):11206–11210
Islam MN, van Amstel A, Bhuyan MS, Islam MN, Sharif ASM, Islam MM et al (2021) Seaweed: a powerful tool for climate change mitigation that provides various ecological services. In: Bangladesh II: climate change impacts, mitigation and adaptation in developing countries. Springer, pp 159–192
Karduri RKR, Ananth C (2021) Bioenergy with carbon capture and storage (BECCS): viability and potential. Int J Adv Res Basic Eng Sci Technol 7:16–27
Kaur R, Gupta S, Tripathi V, Chauhan A, Parashar D, Shankar P, Kashyap V (2023) Microbiome based approaches for the degradation of polycyclic aromatic hydrocarbons (PAHs): a current perception. Chemosphere 341:139951
Kour D, Rana KL, Yadav N, Yadav AN, Rastegari AA, Singh C, Negi P, Singh K, Saxena AK (2019) Technologies for biofuel production: current development, challenges, and future prospects. In: Prospects of renewable bioprocessing in future energy systems. Springer, Cham, pp 1–50
Kumar A (2023) The role of algae in environmental remediation. In: Building a blueprint for progress. BFC Publications, pp 74–81
Kumar A, Ponmani S, Sharma G, Sangavi P, Chaturvedi A, Singh A, Malyan S, Khan S, Shabnam AA, Jigyasu D (2023) Plummeting toxic contaminates from water through phycoremediation: mechanism, influencing factors and future outlook to enhance the capacity of living and non-living algae. Environ Res 239:117381
Lavery PS, Mateo MA, Serrano O, Rozaimi M (2012) Variability in the carbon storage of seagrass habitats and its implications for global estimates of blue carbon ecosystem service. PLoS One 7(9):e43550
Li S-N, Zhang C, Li F, Ren N-Q, Ho S-H (2023) Recent advances of algae-bacteria consortia in aquatic remediation. Crit Rev Environ Sci Technol 53(3):315–339
Lomartire S, Gonçalves AM (2023) Algal phycocolloids: bioactivities and pharmaceutical applications. Mar Drugs 21(7):384
Lundquist TJ, Woertz IC, Quinn N, Benemann JR (2010) A realistic technology and engineering assessment of algae biofuel production. Energy Biosciences Institute, Berkeley, pp 1–178
Mahapatra DM, Chanakya H, Joshi N, Ramachandra T, Murthy G (2018) Algae-based biofertilizers: a biorefinery approach. In: Microorganisms for green revolution. Springer Nature, Singapore, pp 177–196
Mata TM, Martins AA, Caetano NS (2010) Microalgae for biodiesel production and other applications: a review. Renew Sust Energ Rev 14(1):217–232

Mathias RC, Kushalan S, Hegde H, Jathanna NN, Hegde S (2023) Phycoremediation: algae-based bioremediation. In: Algae materials. Elsevier, pp 451–469
Mathimani T, Kumar TS, Chandrasekar M, Uma L, Prabaharan D (2017) Assessment of fuel properties, engine performance and emission characteristics of outdoor grown marine Chlorella vulgaris BDUG 91771 biodiesel. Renew Energy 105:637–646
Mehta P, Sahil K, Sarao LK, Jangra M, Bhardwaj S (2023) Algal biofuels: clean energy to combat the climate change. In: Basic research advancement for algal biofuels production. Springer, pp 187–210
Menetrez MY (2012) An overview of algae biofuel production and potential environmental impact. Environ Sci Technol 46(13):7073–7085
Milledge JJ, Harvey PJ (2016) Potential process "hot spots" in algal biofuel production – understanding costs, energy and greenhouse gas emissions. Energy 116(Part 1):1037–1045
Milledge JJ, Heaven S (2013) A review of the harvesting of micro-algae for biofuel production. Rev Environ Sci Biotechnol 12:165–178
Mrabet R (2023) Sustainable agriculture for food and nutritional security. In: Sustainable agriculture and the environment. Elsevier, pp 25–90
Munoz R, Guieysse B, Mattiasson B (2009) Phenanthrene biodegradation by the microalga Chlorella vulgaris. Biotechnol Bioeng 102(2):331–337
Nguyen XC, Bui VKH, Cho KH, Hur J (2024) Practical application of machine learning for organic matter and harmful algal blooms in freshwater systems: a review. Crit Rev Environ Sci Technol 54(12):953–975
Oh SH, Han JG, Kim Y, Ha JH, Kim SS, Jeong MH, Jeong HS, Kim NY, Cho JS, Yoon WB (2009) Lipid production in Porphyridium cruentum grown under different culture conditions. J Biosci Bioeng 108(5):429–434
Olguín EJ (2003) Phycoremediation: key issues for cost-effective nutrient removal processes. Biotechnol Adv 22(1–2):81–91
Onoh UC, Ogunade J, Owoeye E, Awakessien S, Asomah JK (2024) Impact of climate change on biodiversity and ecosystems services. Int J Geogr Environ Manag 10(1):77–93
Oswald WJ, Golueke CG (1960) Biological transformation of solar energy. In: Advances in applied microbiology, vol 2. Elsevier, pp 223–262
Ouedraogo M, Peng D, Chen X, Hashmi SH (2022) Testing the mineral resources-induced environmental Kuznets curve hypothesis in Africa. Nat Resour Res 31(5):2435–2459
Papazi A, Kotzabasis K, Fakiri EM (2018) Marine macroalgae: an untapped resource for producing green chemicals for sustainable development. Diversity 10(3):68
Park J, Jin HF, Hwang SJ (2021) Current status and prospects of marine biotechnology in Korea. J Mar Sci Eng 9(3):278
Piao W, Kim Y, Kim H, Kim M, Kim C (2016) Life cycle assessment and economic efficiency analysis of integrated management of wastewater treatment plants. J Clean Prod 113:325–337
Priyadharshini SD, Babu PS, Manikandan S, Subbaiya R, Govarthanan M, Karmegam N (2021) Phycoremediation of wastewater for pollutant removal: a green approach to environmental protection and long-term remediation. Environ Pollut 290:117989
Radakovits R, Jinkerson RE, Darzins A, Posewitz MC (2010) Genetic engineering of algae for enhanced biofuel production. Eukaryot Cell 9(4):486–501
Rawat I, Kumar RR, Mutanda T, Bux F (2011) Dual role of microalgae: phycoremediation of domestic wastewater and biomass production for sustainable biofuels production. Appl Energy 88(10):3411–3424
Reisoglu Ş, Aydin S (2023) Microalgae as a promising candidate for fighting climate change and biodiversity loss. In: Microalgae – current and potential applications. IntechOpen
Rittmann BE (2008) Opportunities for renewable bioenergy using microorganisms. Biotechnol Bioeng 100(2):203–212
Rodrigues TB, Silva AET (2016) Molecular diversity of environmental prokaryotes. CRC Press
Sahu SN, Sahoo NK, Naik SN (2019) Phycoremediation technology: a global prospective. In: Application of microalgae in wastewater treatment. Springer, Cham, pp 1–18

Schiewer S, Patil SB (2008) Pectin-rich fruit wastes as biosorbents for heavy metal removal: equilibrium and kinetics. Bioresour Technol 99(6):1896–1903
Schoonees T, Gijón Mancheño A, Scheres B, Bouma T, Silva R, Schlurmann T, Schüttrumpf H (2019) Hard structures for coastal protection, towards greener designs. Estuar Coasts 42:1709–1729
Sedighi M, Pourmoghaddam Qhazvini P, Amidpour M (2023) Algae-powered buildings: a review of an innovative, sustainable approach in the built environment. Sustainability 15(4):3729
Sengupta SL, Chaudhuri RG, Dutta S (2023) A critical review on phycoremediation of pollutants from wastewater—a novel algae-based secondary treatment with the opportunities of production of value-added products. Environ Sci Pollut Res Int 30(54):114844–114872
Steneck RS, Graham MH, Bourque BJ, Corbett D, Erlandson JM, Estes JA, Tegner MJ (2002) Kelp forest ecosystems: biodiversity, stability, resilience and future. Environ Conserv 29(4):436–459
Svenning J-C, McGeoch MA, Normand S, Ordonez A, Riede F (2024) Navigating ecological novelty towards planetary stewardship: challenges and opportunities in biodiversity dynamics in a transforming biosphere. Philos Trans R Soc B 379(1902):20230008
Tsai DD-W, Chen PH, Ramaraj R (2017) The potential of carbon dioxide capture and sequestration with algae. Ecol Eng 98:17–23
Tsai DD, Ramaraj R, Unpaprom Y, Chen PH (2023) Exploring algae role in global carbon dynamics and CO2 biofixation. In: The 3rd international conference on science technology & innovation (3rd ICSTI) – Maejo University, Chiang Mai, Thailand
Turcios AE, Papenbrock J (2014) Sustainable treatment of aquaculture effluents—what can we learn from the past for the future? Sustainability 6(2):836–856
van den Burg SW, Koch SJ, Poelman M, Veraart J, Selnes T, Foekema EM, Lansbergen R (2024) Seaweed as climate mitigation solution: categorizing and reflecting on four climate mitigation pathways. Wiley Interdiscip Rev Clim Chang 15(2):e868
Varshney K (2024) Regenerative architecture: carbon sequestration and habitat provisioning through building design. Te Herenga Waka-Victoria University of Wellington
Webb RM (2024) International governance of ocean-based carbon dioxide removal: recent developments and future directions
Weissman JC, Goebel R (1987) Design and analysis of microalgal open pond systems for the purpose of producing fuels: a subcontract report
Wernberg T, Krumhansl KA, Filbee-Dexter K, Pedersen MF (2020) Status and trends for the world's kelp forests. In: Duarte CM (ed) World seas: an environmental evaluation, 2nd edn. Academic Press, pp 375–398
Wertz J-L, Perez S (2024) Algae in the bioeconomy. CRC Press
Whitton R, Ometto F, Pidou M, Jarvis P, Villa R, Jefferson B (2015) Microalgae for municipal wastewater nutrient remediation: mechanisms, reactors and outlook for tertiary treatment. Environ Technol Rev 4(1):133–148
Wilkie AC, Edmundson SJ, Duncan JG (2011) Indigenous algae for local bioresource production: phycoprospecting. Energy Sustain Dev 15(4):365–371
Wright LS, Pessarrodona A, Foggo A (2022) Climate-driven shifts in kelp forest composition reduce carbon sequestration potential. Glob Chang Biol 28(18):5514–5531
Wu Y-H, Yu Y, Li X, Hu H-Y, Su Z-F (2012) Biomass production of a Scenedesmus sp. under phosphorous-starvation cultivation condition. Bioresour Technol 112:193–198
Xie L, Bulkeley H, Tozer L (2022) Mainstreaming sustainable innovation: unlocking the potential of nature-based solutions for climate change and biodiversity. Environ Sci Pol 132:119–130
Xu G, Zhao L, Guo X, Shen H, Han H, Jin P (2017) Macroalgae cultivation for nutrient recovery from wastewater and energy production. Chemosphere 174:492–498
Yong JJJY, Chew KW, Khoo KS, Show PL, Chang J-S (2021) Prospects and development of algal-bacterial biotechnology in environmental management and protection. Biotechnol Adv 47:107684
Yuan CJ (2024) Climate change, ocean pollution, and acidification: the application of integrated management strategies within the framework of the United Nations Decade of Ocean Science. Int J Mar Sci 14:83–93

Mobilizing Microbes for Bioremediation Strategies in the Context of Climate Change

Md Abdur Rahman, Biplob Dey, Md Abdul Halim, and Romel Ahmed

Abstract Microbial remediation has emerged as a paramount strategy in addressing environmental pollution and mitigating climate change impacts. This chapter explores the dynamic interplay between pollutants, microbial communities, and climate change, emphasizing the critical role of microbial remediation in restoring ecosystems. Environmental pollutants, including heavy metals, microplastics, and volatile organic compounds (VOCs), not only degrade ecosystems but also exacerbate climate change by altering atmospheric and terrestrial processes. Microbial remediation uses the natural metabolic capabilities of diverse microorganisms—bacteria, fungi, archaea, and algae—to degrade, detoxify, or immobilize these contaminants, offering a sustainable and cost-effective alternative to conventional methods. This chapter delves into mechanisms of microbial action, such as biodegradation, bioaccumulation, and enzyme transformation, highlighting how these processes can be optimized under varying environmental conditions. Factors such as temperature, pH, moisture, and nutrient availability critically influence microbial

Md A. Rahman
Department of Forestry & Environmental Science, Shahjalal University of Science & Technology, Sylhet, Bangladesh

B. Dey
Center for Research in Environment, iGen and Livelihood (CREGL), Sylhet, Bangladesh

Department of Tropical and International Forestry, Faculty of Forest Sciences and Forest Ecology, Georg-August-University Göttingen, Göttingen, Germany

Md A. Halim
Center for Research in Environment, iGen and Livelihood (CREGL), Sylhet, Bangladesh

Institute of Forestry and Conservation, John H Daniels Faculty of Architecture Landscape and Design, University of Toronto, Toronto, ON, Canada

R. Ahmed (✉)
Department of Forestry & Environmental Science, Shahjalal University of Science & Technology, Sylhet, Bangladesh

Center for Research in Environment, iGen and Livelihood (CREGL), Sylhet, Bangladesh
e-mail: romel-fes@sust.edu

A. A. H. Abdel Latef et al. (eds.), *Sustainable Remediation for Pollution and Climate Resilience*, https://doi.org/10.1007/978-981-96-5674-5_13

efficacy, necessitating tailored interventions to enhance performance. Advanced techniques, including bioaugmentation, biostimulation, and genetically engineered microbes, are discussed as promising approaches to improve resilience and efficiency against complex environmental pollutants. However, the effectiveness of microbial remediation is intricately linked to abrupt climate changes, which can alter microbial community dynamics and disrupt bioremediation processes. This chapter examines how climate-induced shifts in temperature, precipitation, and extreme weather events affect pollutant bioavailability and microbial activity, posing both challenges and opportunities. It also highlights the potential risks associated with genetically modified microorganisms and incomplete degradation pathways, underscoring the need for comprehensive risk assessments and adaptive management strategies. Looking ahead, integrating microbial remediation into broader climate change mitigation frameworks is crucial. Innovative solutions, such as microbial carbon capture cells (MCCs) and synthetic biology applications, offer new avenues for enhancing carbon sequestration and reducing greenhouse gas emissions. This chapter calls for a multidisciplinary approach, combining scientific research, technological innovation, and policy support, to fully unlock the potential of microbial remediation. By harnessing the power of microbes, we can build a more sustainable and resilient future, transforming environmental challenges into opportunities for ecosystem restoration and stability in the face of climate change.

Keywords Bioremediation · Heavy metals · Microplastics · Atmospheric pollution · Environmental pollution · Volatile organic compounds · Phytoremediation · Bioaugmentation · Biostimulation · Synthetic biology · Pollution mitigation

1 Introduction

Climate change is intricately related to environmental pollution and bioremediation (Alkorta et al. 2017). An undeniable harsh reality is that the changing climate casts a long shadow over our planet, aggressively fading habitats of all living organisms, including humans. The changes in temperature, precipitation, and extreme weather events due to climate change have cascading effects on migration, extinction, and biodiversity of microbial populations. Mitigating climate change by reducing pollution is one of the major concerns of local and international bodies. Numerous policies, actions, and technologies are employed to abate and mitigate the pollutants and ameliorate the environment across the globe (Sofia et al. 2020). Examples of technologies employed are particulate filters (Tsai et al. 2011) and activated carbon for removing flue gas (Wen et al. 2024), catalytic oxidization, and chemisorption formaldehyde removal (Wang et al. 2013) and photocatalysis for reducing volatile organic compounds (VOCs) (Shayegan et al. 2019). Besides, using plants to reduce pollutants as a phytoremediation process is being practiced for a longer time with modifications of phyto-mechanisms for enhancing the efficacy (Kafle et al. 2022). This economical and environmentally friendly technique utilizes suitable plants to retain, sequester, and detoxify environmental contaminants (Ashraf et al. 2019).

Bioremediation of ambient atmospheric pollutants involving plants and microorganisms to absorb, degrade, or transform environmental pollutants has emerged as a more practical approach to cleaning up contaminated environments (Maurya et al. 2023). Unlike traditional methods that can be expensive and leave harmful residues, microbial remediation offers a sustainable and eco-friendly approach (Ayilara and Babalola 2023; Maurya et al. 2023). In microbial remediation techniques, the natural capabilities of microorganisms such as archaea, bacteria, and fungi are exploited to break down pollutants and contaminants in and from soil, water, and air. By harnessing the natural abilities of microorganisms, it is being used to restore contaminated environments and mitigate the environmental impacts of pollutants. For example, microbe-assisted green technology was employed in a project in Maharashtra to restore 6.3 ha of mine area contaminated with manganese, and after 8 years of bioremediation process, the land was successfully reclaimed (Juwarkar et al. 2009; Juwarkar and Jambhulkar 2008). Furthermore, such approaches can mitigate climate change by promoting carbon sequestration and aiding in ecosystem restoration. However, the efficacy of bioremediation in remediating the contaminants in the environment is adversely affected by climate change (Alkorta et al. 2017; ASM 2022; Saha et al. 2021). So, as we navigate the challenges of a changing climate, understanding the role of microbial remediation becomes crucial.

Biodegradation, bioaccumulation, bioabsorption, and enzyme transformation are fundamental processes underpinning the effectiveness of microbial remediation strategies. By understanding and optimizing these mechanisms, we can enhance the efficacy of bioremediation efforts, ensuring cleaner and safer environments. Continued research and innovation in this field will further enhance the effectiveness of microbial bioremediation strategies. Extensive literature reviews have been accomplished in this field in recent years, discussing the progress, challenges, and prospects (for example, Maurya et al. 2023; Ayilara and Babalola 2023; Grifoni et al. 2022). Nevertheless, considering microbial remediation a large and time-demanding environmental issue, understanding the state of knowledge on this requires a synthesis of a comprehensive write-up comprising the crucial aspects, including the subtopic that was not discussed adequately. More specifically, the interaction of climate change and microbial remediation is not sufficiently discussed; the assisted microbial remediation for enhancing the efficacy of the microbes is not well addressed; and remediation in the aquatic and terrestrial ecosystems is not discussed eloquently in a single article so far. This chapter explores the mechanisms microbes employ to clean up contaminated environments, how pollutants affect climate change, which in turn adversely affects the efficacy of bioremediation by affecting the microbes and indirectly affecting the plant growth, and the recent progress of potential microbial remediation as a powerful tool for building a more sustainable future.

2 Pollutants, Climate Change, and Microbial Remediation: Interplay the Game

How pollution and climate change are interrelated and interplayed is a complex issue that warrants a discussion for a deep understanding in the context of global climate change. Climate change is the flip side of air pollution, as coined by the United Nations Environmental Program (UNEP). For instance, particulate matter (PM), if transported over a long distance and falling on ice or snow, causes a slight dark effect on the surface of snow and ice that causes less light to reflect back and leads to an increase in the temperature (Chen et al. 2021). Increased PM in the air also affects the global radiative balance (He et al. 2018). Equally, climate change has a tremendous effect on increasing air pollutants, such as the accumulation of PM in the air (He et al. 2018; Shi et al. 2021). It is more complicated when such pollutants can also modulate the number and distribution of microbial populations, particularly plant microbial community (PMC). This proposition is supported by a study that showed that PM highly affects the abundance of taxa belonging to PMC (Postiglione et al. 2022). Like PM, hydrocarbons can change the community structure of bacteria when they accumulate in the leaves of *Platanus* sp. (Espenshade et al. 2019).

Microbes have their realm of actions governed by many environmental parameters such as temperature, moisture, humidity, and nutrient availability, which are drastically modulated by climate change (Fig. 1). The mounting concerns about the effects of climate change on bioremediation have resulted in many theoretical propositions and experimental explorations (Mishra et al. 2022; Naorem 2022; Zak et al. 2011). Climate change-induced temperature, moisture, and precipitation abruptly change a microbial community's population, biomass, reproduction, structure, and functions, which are tightly linked with the degradation and reduction of soil pollutants (Alkorta et al. 2017; Burns et al. 2013). The example of climate change-induced global warming increases the root exudation of many chemicals affecting the rhizospheric microbiota and increases the volatilization of many persistent organic pollutants. The volatility and longer transportation of pollutants due to global warming are also reported in Lamon et al. (2009) and Miraglia et al. (2009). Pollutant-induced changes in the structure and function of bioremediate microbial communities ultimately affect the process of climate change (Abhilash et al. 2013).

To some extent, global warming aids bioremediation; for example, elevated temperature is often reported to accelerate the microbial activity in breaking down organic pollutants (Gouin et al. 2013), which might have a strong link with the upregulation of functional genes associated with soil carbon degradation under warm climatic environments (Wu et al. 2011; Zhou et al. 2011). However, such findings pose the question of whether this is a general trend or applicable only to specific environments of temperate climate. Nevertheless, an increase in temperature within a specific range is reported to increase the production of some vital molecules, such as enzymes, extracellular polysaccharides, and biosurfactants which augment the remediation process (Wang et al. 2022). Many other findings refuted

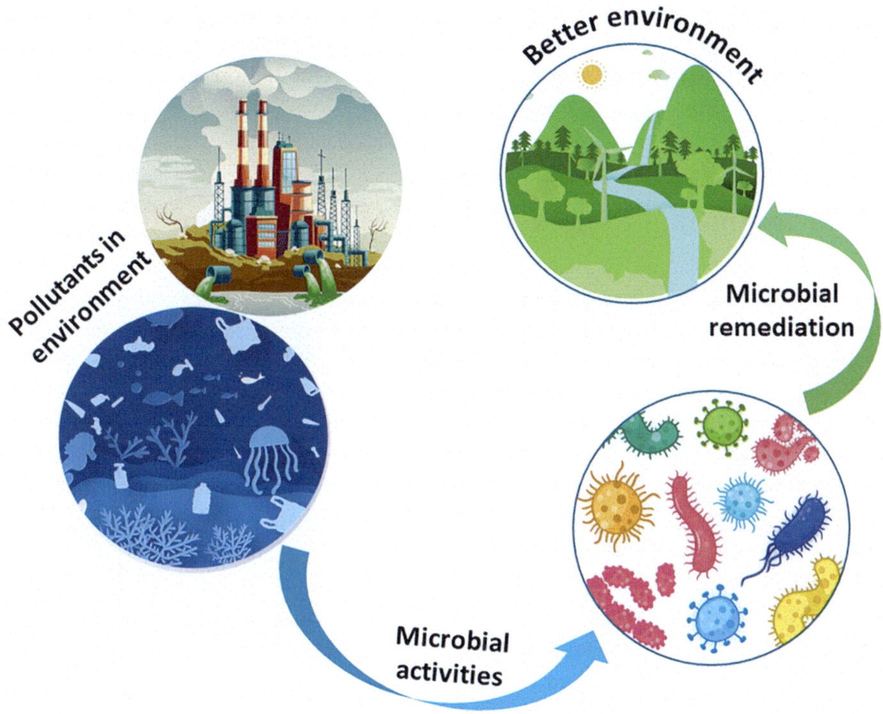

Fig. 1 Remediation of environmental pollutants through microbial actions makes the environment better and safer

the positive warming on microbial activity; this stimulation is only factual for the time being, but prolonged warming declined the overall degradation process (Frey et al. 2008; Steinweg et al. 2008). While a moderate increase in temperature (1–3 °C) may not severely impact bioremediation, more extreme heat events can have variable effects depending on moisture levels and the specific microbial community present (Alkorta et al. 2017). Higher temperatures can increase the toxicity of certain pollutants, complicating remediation efforts. Climate change can also alter the chemical state of contaminants, affecting their bioavailability. For example, some pollutants may become more soluble or mobile under changing environmental conditions, impacting microbial degradation rates. Many studies have shown short-term warming effects (for instance, Hartley et al. 2008) on bioremediation that hardly echo the actual field conditions that prevail, as climate change is a long-term phenomenon and microbial activity is a highly complex process that involves many interacting factors, including acclimation, adaptation, and evolution of microbes (T Classen et al. 2015). Adapting to the situation and developing resilience strategies are crucial. Microbial communities are generally resilient and can adapt to changing environmental conditions. However, the speed and effectiveness of these adaptations can vary widely among different microbial species and communities (ASM

2022). Continuous monitoring of environmental conditions and microbial communities is essential to adapt remediation strategies to changing circumstances. Adaptive management practices can help optimize remediation efforts in response to real-time environmental changes.

3 Environmental Factors Affecting Microbial Remediation

Microbial remediation is a complex process that is governed by several biophysical and chemical factors (Fig. 2). The acting microbes in the remedial process are not the only type in the environment; they must survive through diverse interactions with other microbes, such as competition, predation, and succession. Their population numbers, strains, and type of enzyme secretions are some of the biological factors that determine the speed of the degradation and detoxification process (Sodhi et al. 2021). While single microbial species could be specialized for specific pollutants, a consortium has a multifunctionality, is resistant to some toxic substances, and thus is more effective than single species in degrading and/or detoxifying all the substrates effectively using different metabolic pathways (Bhatt et al. 2021). The presence of oxygen, particularly for aerobic bacteria, is a condition for accelerating the remediation process (Mazumder et al. 2021). Like oxygen, optimum moisture conditions depending on the types of microbes are also required; the wet soil facilitates less degradation, while the same is true for too-dry soil. Several experimental evidence suggest keeping the optimum nutrient conditions in the site to maintain the proper carbon, nitrogen, and phosphorus (C: N: P) ratio in bacterial cells for enhancing their metabolic activity that harnesses the remediation process (Bala et al. 2022). Soil chemistry, especially the soil pH, has a profound effect; even a small shift in pH

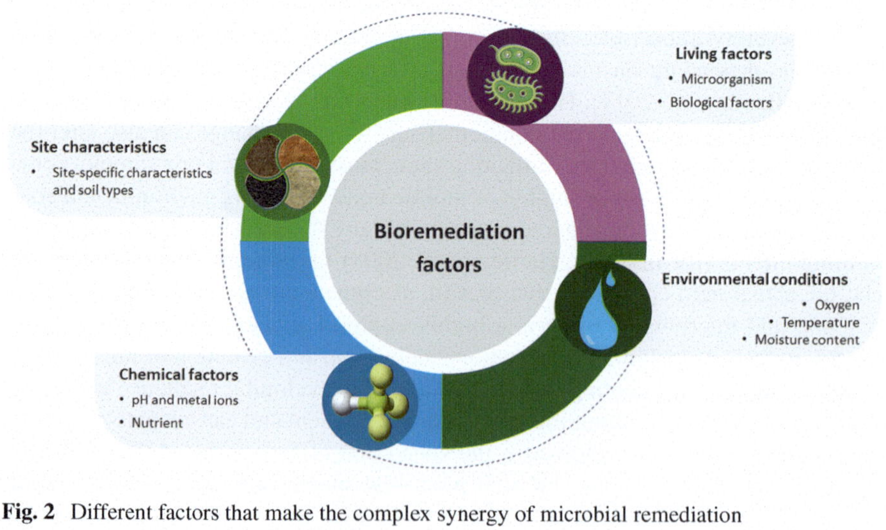

Fig. 2 Different factors that make the complex synergy of microbial remediation

can change the activity and distribution of microbial populations, ultimately affecting the bioremediation process (Rajkumar and Kurinjimalar 2021). The other soil chemistry is the presence of metal ions that affect the degradation process directly and indirectly (Padhan et al. 2021). Metal ions are required for the microbes—bacteria and fungi—but if there is an excess amount, it directly affects the metabolism of microbes. Of the physical factors, the temperature is the most critical variable that determines the metabolic turnover of the enzymes required for degradation, in addition to the survival and reproduction of microbes (Ren et al. 2018).

4 Microbial Remediation Techniques

Microbes use their inherent abilities to degrade, detoxify, or immobilize pollutants.

This process harnesses the metabolic capabilities of microbes to convert toxic substances into harmless byproducts, such as water, carbon dioxide, and biomass, making them an effective choice for mitigating pollution in both terrestrial and aquatic ecosystems. The fundamental principles of microbial remediation involve understanding the types of microorganisms used, their mechanisms of action, and the conditions that optimize their activity.

4.1 Microorganisms Involved in Bioremediation and Mechanism

(a) *Bacteria:* Bacteria are the most commonly used microorganisms in bioremediation due to their diverse metabolic pathways, enabling them to degrade a wide range of organic and inorganic pollutants, including hydrocarbons, heavy metals, and pesticides. For instance, *Pseudomonas putida* is a well-known bacterium frequently employed in bioremediation, particularly for the degradation of hydrocarbons found in oil spills. This bacterium metabolizes various organic pollutants, including aromatic hydrocarbons and alkanes, using specific enzymes that break down these compounds into less harmful substances (Margesin and Schinner 2001).

(b) *Fungi:* Fungi, especially white-rot fungi, are effective in degrading complex organic pollutants like lignin, polycyclic aromatic hydrocarbons (PAHs), and dyes. Their extracellular enzymes, including lignin peroxidases and manganese peroxidases, play a crucial role in breaking down these compounds. *Phanerochaete chrysosporium*, a widely studied white-rot fungus, is renowned for its ability to degrade lignin and various environmental pollutants, including PAHs and dyes, through its robust enzymatic system (Janusz et al. 2017).

(c) *Archaea:* Archaea are known for their ability to thrive in extreme environments, making them particularly useful in the bioremediation of heavy metals and hydrocarbons in extreme conditions, such as high salinity or temperature. *Haloferax mediterranei*, an extremophilic archaeon, has demonstrated the capacity to reduce heavy metals like chromium (VI) in high-salinity environments, demonstrating its potential in bioremediation applications under such conditions (Bonfá et al. 2011).

(d) *Algae:* Algae contribute to bioremediation primarily through their ability to absorb heavy metals and other pollutants from water. They can also produce biomass that can be used as a renewable resource. Green algae, in particular, have shown promise in removing nutrients and toxins from wastewater. For example, *Chlorella vulgaris* has been effectively used to remove heavy metals such as cadmium (Cd) and lead (Pb) from wastewater, demonstrating its potential for environmental remediation (Gupta et al. 2006).

(e) *Engineered microbes:* Genetically engineered microorganisms (GEMs) are increasingly recognized for their potential in bioremediation. Strains of *Pseudomonas putida, Escherichia coli, Bacillus subtilis, Alcaligenes eutrophus*, and microalgae are genetically modified to absorb heavy metals and nutrients from wastewater. These engineered microbes are designed to degrade various contaminants more effectively than their natural counterparts. For example, a genetically engineered strain of *Pseudomonas putida* has been modified to express pathways for the degradation of toluene and other hydrocarbons, significantly enhancing its bioremediation efficiency (Cases and de Lorenzo 2005).

4.2 Mechanisms of Action

Microbial remediation is a multifaceted process that employs various mechanisms to detoxify, degrade, and immobilize environmental pollutants. Understanding these mechanisms is crucial for optimizing bioremediation strategies and enhancing their effectiveness in different environments. The mechanisms include enzymatic breakdown, cometabolism, and immobilization. These processes can be illustrated through specific biochemical pathways and reactions.

(a) *Enzymatic breakdown:* Microorganisms produce enzymes that catalyze the breakdown of pollutants. For example, bacteria like *Pseudomonas putida* produce enzymes such as oxygenases and dehydrogenases to degrade hydrocarbons (Juwarkar et al. 2010). The general reaction for hydrocarbon degradation can be represented as follows:

$$R-CH_2-CH_2-R'+O_2 \rightarrow R-CHO+R'-CH_2-OH \rightarrow R-COOH+H_2O$$

(b) *Cometabolism:* In cometabolism, microorganisms degrade pollutants incidentally while metabolizing another primary substrate. For instance, methanotrophic bacteria degrade trichloroethylene (TCE) while oxidizing methane:

$$CH_4 + TCE + O_2 \rightarrow CO_2 + HCl + \text{intermediates}$$

(c) *Immobilization:* Certain microorganisms can immobilize heavy metals by transforming them into less soluble forms. For example, sulfate-reducing bacteria convert soluble metals into insoluble metal sulfides:

$$Me^{2+} + H_2S \rightarrow MeS + 2H^+.$$

In a broader spectrum, the mechanisms of remediation include biodegradation, bioaccumulation, bioabsorption, and enzyme transformation.

4.3 Biodegradation: Microbial Breakdown of Contaminants

Biodegradation is the process by which microorganisms convert complex organic contaminants into simpler, nontoxic compounds through metabolic processes. This mechanism is central to microbial remediation, as it directly reduces the toxicity and persistence of pollutants in the environment.

4.3.1 Mechanism and Examples

Microorganisms such as bacteria and fungi produce specific enzymes that catalyze the breakdown of contaminants. For instance, the bacterium *Pseudomonas putida* can degrade petroleum hydrocarbons by producing oxygenases and dehydrogenases, which sequentially oxidize hydrocarbons to less harmful compounds (Rojo 2009).

$$R - CH_2 - CH2 - R' + O_2 \rightarrow R - CHO + R' - CH_2 - OH \rightarrow R - COOH + H_2O.$$

In this pathway, hydrocarbons (R-CH_2-CH_2-R′) are first oxidized to alcohols (R′-CH_2-OH), then to aldehydes (R-CHO), and finally to carboxylic acids (R-COOH), which are further mineralized to carbon dioxide and water.

Another example is the degradation of chlorinated solvents by methanotrophic bacteria. These bacteria utilize methane as a primary substrate and cometabolize trichloroethylene (TCE) during methane oxidation.

$$CH_4 + TCE + O_2 \rightarrow CO_2 + HCl + \text{intermediates}$$

Methanotrophs oxidize methane to methanol, formaldehyde, and formic acid, which incidentally degrades TCE into less harmful intermediates and ultimately carbon dioxide and hydrochloric acid.

4.4 Bioaccumulation: Microbial Uptake and Concentration of Contaminants

Bioaccumulation refers to the process by which microorganisms take up and concentrate contaminants from their environment into their biomass. This mechanism is particularly important for the remediation of heavy metals and other inorganic pollutants.

4.4.1 Mechanism and Examples

Certain microorganisms can accumulate heavy metals through active and passive transport mechanisms. For example, the alga *Chlorella vulgaris* can accumulate cadmium (Cd) and lead (Pb) ions from contaminated water. This accumulation occurs via metal-binding proteins and peptides such as phytochelatins and metallothioneins, which sequester metals within the cells (Edris et al. 2014).

$$\mathrm{Cd}^{2+} + \mathrm{Phytochelatin} \rightarrow \mathrm{Cd} - \mathrm{Phytochelatin\,Complex}.$$

In this reaction, cadmium ions bind to phytochelatins, forming stable complexes that are sequestered within the cell, thereby reducing the bioavailability and toxicity of the metal in the environment.

Another example is the accumulation of arsenic by the fungus *Aspergillus niger*. This fungus can bioaccumulate arsenic by converting it into less toxic forms, primarily through the production of acidic metabolites that facilitate the conversion of arsenic into biogenic mineral forms. This bioaccumulation process involves the conversion of arsenic within the fungal hyphal structures, effectively reducing its bioavailability and environmental toxicity (Farkas et al. 2020).

4.5 Bioabsorption: Microbial Surface Uptake of Contaminants

Bioabsorption involves the adherence of contaminants onto the surface of microbial cells. This mechanism is often the first step before internalization and further degradation or immobilization of the pollutants.

4.5.1 Mechanism and Examples

Microbial cell walls possess functional groups such as carboxyl, hydroxyl, and amino groups that can bind to contaminants. For instance, *Bacillus subtilis* can bioabsorb heavy metals like zinc (Zn) and nickel (Ni) through its cell wall structures. The biosorption process involves interactions between the metal ions and the cell wall's active sites, making *Bacillus subtilis* an effective tool for removing heavy metals from contaminated environments (Wróbel et al. 2023).

$$Zn^{2+} + \text{Cell Wall} \rightarrow Zn - \text{Cell Wall Complex}.$$

In this process, zinc ions interact with carboxyl groups on the bacterial cell wall, forming stable complexes that reduce the mobility of the metal.

Fungi such as *Penicillium chrysogenum* also exhibit bioabsorption capabilities. The cell walls of these fungi contain chitin and glucans, which can adsorb dyes and heavy metals from wastewater. Studies have shown that *P. chrysogenum* can effectively bind metals such as cadmium (Cd), zinc (Zn), copper (Cu), and lead (Pb) through these mechanisms. For example, the biosorption process with *P. chrysogenum* biomass operates rapidly, with heavy metals binding within minutes, and the process being saturable, indicating high efficiency of adsorption (Skowro'nski et al. 2001).

4.6 *Enzyme Transformation: Role of Microbial Enzymes in Contaminant Breakdown*

Enzyme transformation is a critical aspect of microbial remediation, where specific microbial enzymes catalyze the conversion of contaminants into less toxic or more easily degradable forms. This mechanism underlies many bioremediation processes.

4.6.1 Mechanism and Examples

Microbial enzymes such as peroxidases, oxygenases, and reductases play vital roles in transforming pollutants. For instance, the white-rot fungus *Phanerochaete chrysosporium* produces lignin peroxidase and manganese peroxidase, which degrade lignin and other complex organic pollutants. These enzymes break down the lignin structure by oxidizing lignin polymers, allowing the fungus to access and decompose these otherwise recalcitrant materials. This process is crucial not only for lignin degradation but also for breaking down various environmental pollutants, making *P. chrysosporium* highly valuable in bioremediation (Coconi-Linares et al. 2014).

$$\text{Lignin} + \text{Lignin Peroxidase} \rightarrow \text{Phenolic Compounds}.$$

In this reaction, lignin peroxidase breaks down the complex structure of lignin into smaller phenolic compounds that are further degraded by other microbial enzymes.

Bacterial oxygenases, such as toluene dioxygenase produced by *Pseudomonas putida*, catalyze the initial oxidation of aromatic hydrocarbons, transforming them into catechols that are further degraded through ring-cleavage pathways (Spain et al. 1989).

$$\text{Toluene} + O_2 + \text{NADH} \rightarrow \text{Catechol} + \text{NAD}^+ + H_2O.$$

This reaction highlights the role of dioxygenase in adding oxygen to the aromatic ring of toluene, converting it into catechol, which is a more easily degradable compound.

5 Microbial Remediation of Soil Pollution

Non-contaminated soil is increasingly rare due to widespread organic and inorganic pollutants, and their levels are rising alarmingly. Despite various physical and chemical remediation methods, outcomes often fail to adequately address soil pollution, sometimes compromising the integrity of soil ecosystems. Restoration of soil ecosystem services alongside reducing soil pollution is critical. Microbes play a pivotal role in degrading or transforming pollutants, particularly effective for organic contaminants, heavy metals, and other hazardous substances. The efficacy is often enhanced by employing specific microbial strains or genetically modified organisms.

Organic contaminants, such as hydrocarbons and organometallic substances, are the major challenges due to their broad persistence and accumulation (Table 1). Different classes of microbes, such as *Fusarium*, *Cunninghamella*, *Graphium*, *Aspergillus*, and *Penicillium*, can reduce complex organic pollutants to less harmful substances (Al-Hawash et al. 2018). While petroleum plays a pivotal role in the world economy, its complex structural properties (aliphatic, aromatic, and heterocyclic carbons) make it prevalent in the environment. Microbes like algae, fungi, and

Table 1 Example of microbes in degrading antibiotics in the environment

Antibiotics	Microbes	References
Sulfonamides	*Geobacillus thermoleovorans* *Achromobacter denitrificans*	Reis et al. (2014) Jiang et al. (2014)
Tetracyclines	*Burkholderia cepacia*	Hong et al. (2020)
Fluoroquinolones	*Microbacterium* sp. *Labrys portucalensis* *Rhodococcus* sp.	Kim et al. (2011) Amorim et al. (2014)
Gentamicin	*Brevundimonas diminuta* *Providencia vermicola, Alcaligenes* sp., and *Acinetobacter*	Nzila (2013) Liu et al. (2017)
Sulfamethoxazole	*Bjerkandera adusta*	Eibes et al. (2011)
Pradofloxacin	*Gleophyllum striatum*	Wetzstein et al. (2012)
Ofloxacin	*Trametes versicolor*	Gros et al. (2014)

bacteria, particularly *Bacillus*, efficiently degrade these hydrocarbons into simpler compounds, reducing them to biomass, CO_2, and H_2O (Zhang et al. 2021).

The rapid growth of the agricultural sector has led to increased pesticide use, posing severe environmental and health risks (Silva et al. 2019). The serious concern is their entrance into the food chain, making the public health system vulnerable. Pesticide residues can disperse vertically and horizontally in soils, impacting several microbial communities (Tarla et al. 2020). Fortunately, there are microbes that are tolerant and even can use the pesticides as energy and carbon sources, degrading them by catalytic enzymes secreted outside of the cells (Roy et al. 2022). In a recent study, Kumar et al. (2023) reported that many microbes, such as *Flammulina velutipes, Pandoraea* sp., *Auricularia auricular, Avatha discolor, Pseudomonas* sp., *Mycobacterium* sp., *Stereum hirsutum, Bacillus* sp., *Phanerochaete chrysosporium, Klebsiella* sp., *Dichomitus squalene, Agrocybe semiorbicularis, Coriolus versicolor, Hypholoma fasciculare,* and *Pleurotus ostreatus*, can use pesticides as energy and carbon sources, breaking them down into less harmful molecules through extracellular enzymes.

Antibiotic pollution, particularly from pharmaceutical waste, has been a health and environmental concern since 1982, killing beneficial microorganisms. Microbes such as *Geobacillus thermoleovorans* and *Burkholderia cepacia* effectively degrade various antibiotics, as outlined in Table 1. Moreover, microbial consortia often outperform individual strains in degrading pollutants like antibiotics (Wang and Wang 2018), as demonstrated by the effective consortium of *Aquamicrobium defluvium, Alcaligenes* sp., and *Pseudomonas putida* in sulfonamide degradation from soils (Islas-Espinoza et al. 2012).

Similar to organic pollutants, inorganic pollutants such as heavy metals and metalloids pose severe risks to soil ecosystems. *Pseudomonas putida* and *Bacillus subtilis* are known to bioaccumulate metals like lead, cadmium, and arsenic, reducing their bioavailability in soil (Kafle et al. 2022). Genetic modifications have further enhanced these strains for better remediation capabilities, making them valuable tools in the sustainable restoration of contaminated soils.

6 Bioremediation of Ambient Pollutants

Clean air is the lifeblood, and the presence of hazardous contaminants in it modulates the climate in the long run and poses a serious threat to lives. Air pollutants are gaseous and nongaseous substances generated from various sources since the Anthropocene period. Of the gaseous pollutants, the most notorious are the oxides of carbon, nitrogen, and sulfur (CO, CO_2, NO_x, SO_x); ozone (O_3); methane (CH_4); and VOCs such as benzene, toluene, ethylbenzene, xylene, polycyclic aromatic hydrocarbons (PAHs), and formaldehyde. Besides, the most common nongaseous pollutants, particulate matter, have profound effects on human health and climate change. Among various efforts and techniques employed to clean the air, the microbial role in detoxifying and degrading the contaminants is found to be an effective way of the natural remediation process. Numerous species of microbes belonging to bacteria, fungi, and archaea act as scavengers of these pollutants by using them as sources of their major nutrition and energy sources.

Microbes associated with plant roots can help degrade pollutants and improve plant health, leading to better pollutant removal rates (Li et al. 2023; Maurya et al. 2023). Bacteria, fungi, and other microorganisms living on the leaf surfaces called phyllosphere degrade or transform pollutants. This process, also known as phyllo remediation, shows promise in addressing various atmospheric contaminants (Maurya et al. 2023). Specific microbes can be isolated and used to target particular pollutants, such as polycyclic aromatic hydrocarbons (PAHs) and VOCs. Recent studies highlight the importance of selecting appropriate plant species and microbial communities for effective bioremediation. Research emphasizes the need for targeted approaches that consider the specific types of pollutants present in the environment (Li et al. 2023; Maurya et al. 2023). The effectiveness of indoor plants in improving indoor air quality (IAQ) through phytoremediation has gained attention. Various studies have assessed the potential of specific indoor plants to absorb indoor air pollutants, contributing to healthier living environments (Ravindra and Mor 2022).

6.1 Volatile Organic Compounds

VOCs are potential contributors to global warming by changing ozone concentration, which can originate from indoor and outdoor environments from various sources, including cooking, cleaning, and washing products, industrial emissions, vehicle exhaust, furniture, and natural processes. To reduce the effects of VOC emissions, the inclusion of biological agents such as various bacteria, fungi, and algae could be an effective and sustainable solution. Recently, *Aglaonema commutatum, A. rotundum*, and aloe vera have been used to absorb (60–65 mmol d^{-1} cm^{-2}-leaf area) (Sriprapat et al. 2014) indoor BTEX VOCs (benzene, toluene, ethylbenzene, and xylene). *Epipremnum aureum* is efficient at absorbing formaldehyde (Feng et al. 2021) because of its leaf area and high stomata density. Also, Soreanu et al. (2020) recommended using *Arthrospira platensis*, a microalga in a photobioreactor, and reported that this microalga grown under LED light could remove more than 90% of acetic acid vapor at 1 ppm. Besides, *Botrychium ternatum* and *Asparagus densiflorus* ferns uptake α-pinene, octane, and toluene. In parallel, some Capnodiales (*Cladosporium cladosporioides* and *Neohortaea acidophila*) and Chaetothyriales (*Cladophialophora psammophila*) fungi are also reported (Prenafeta-Boldú et al. 2019) to be used as remediators of indoor VOCs.

Biofilters associated with microorganisms are used to degrade outdoor VOCs. These biofilters are designed to treat air contaminated with VOCs by passing it through a medium where microbial populations can metabolize the pollutants. The advantages of biofilters include their cost-effectiveness and the fact that they do not produce secondary pollutants, as the VOCs are converted into biodegradable waste that can decompose naturally (Gopinath et al. 2018). Various microorganisms have been identified that can degrade specific VOCs, including chlorinated compounds and BTEX. The biodegradation process can occur under both aerobic and anaerobic conditions, depending on the type of VOC and the microbial community involved (Yoshikawa et al. 2017). Cometabolism, where one VOC enhances the degradation of another, is a critical factor in the biodegradation of multiple VOCs. However, competition among VOCs can also hinder degradation processes, highlighting the need for tailored bioremediation strategies that consider the specific contaminants present in a given environment. Certain plant species have shown significant potential in removing VOCs through their root systems and associated microbial communities. For example, plants can enhance the degradation of VOCs by promoting the growth of specific bacteria that metabolize these compounds (Kim et al. 2018; Lee et al. 2021). The big challenge remains in degrading multiple VOCs present in a specific site (Vidali 2001) because of their unknown nature and interaction with each other. However, progress is there in the complete degradation of VOCs with integrated anaerobic and aerobic biodegradation demonstrated in a case study (Yoshikawa et al. 2017).

6.2 Particulate Matter

Airborne PM, mainly originating from anthropogenic sources, is categorized based on size primarily into $PM_{2.5}$ and PM_{10}. Though relatively nonreactive, PM often carries biologically active compounds, such as metals, organic compounds, and free radicals, which are adsorbed onto PM surfaces, increasing their toxicity (Weyens et al. 2015). PM also acts as a carrier for airborne microbes (AMs), forming aerosol complexes that influence the biogeochemical cycles of terrestrial and oceanic ecosystems (Mandal et al. 2023). These PM–AM interactions can benefit plant growth by acting as proxy nutrient sources.

Bioremediation has emerged as a promising strategy for improving air quality by mitigating PM. This process utilizes the natural metabolic capabilities of microorganisms and plants to degrade or transform pollutants, effectively reducing their concentration in the atmosphere. Microorganisms, particularly epiphytic bacteria residing on plant surfaces, play a key role by metabolizing airborne contaminants, including PM and PAHs (Gandolfi et al. 2017; Ismail Soifiata et al. 2022; Wei et al. 2017). The phyllosphere—the plant's aerial parts—acts as a critical bioremediation interface, hosting diverse microbes capable of biotransformation and biodegradation of airborne contaminants (Gandolfi et al. 2017; Wei et al. 2017).

Plants also play a crucial role in bioremediation through processes such as phytoremediation, where they absorb, accumulate, and detoxify pollutants from the air and soil (Putri et al. 2023). For instance, certain tree species have shown physiological responses that enhance their ability to mitigate air pollution, effectively lowering the levels of particulate matter and other harmful substances in urban environments (Putri et al. 2023). The synergistic action of plants and microorganisms enhances bioremediation efficiency, reducing pollutants via bioaccumulation and biotransformation (Mandal et al. 2022; Putri et al. 2023; Wei et al. 2017).

Moreover, the use of specific bioremediation techniques, such as bioaugmentation and biostimulation, can further enhance the degradation of PM in the atmosphere. Bioaugmentation involves the introduction of specialized microbial strains that can effectively degrade specific pollutants, while biostimulation enhances the growth and activity of indigenous microbial populations (Mani and Kumar 2014; Mitra 2024; Tekere and Tekere 2019). These approaches can be particularly effective in urban areas where particulate matter concentrations are high due to industrial activities and vehicular emissions (Sonavane and Pinjari 2024). The application of bioremediation not only addresses air quality issues but also contributes to the restoration of ecological balance by promoting the health of plant and microbial communities (Azubuike et al. 2016; Wei et al. 2017).

7 Microbial Remediation of Polluted Sea and Freshwater

The marine ecosystem is polluted directly by effluent discharge containing many organic and inorganic pollutants and indirectly by washing away the pollutants from the terrestrial ecosystem. Climate change exacerbates marine pollution by driving numerous physical and chemical changes in oceanic conditions, including surface temperature, pH, salinity, oxygen levels, ice coverage, and currents (Alava et al. 2017; Cabral et al. 2019). These changes are impacting the behavior of contaminants in the marine environment, affecting their accumulation, distribution, bioavailability, and toxicity. The interplay between climate change and pollutants can be categorized as either climate change-dominated, where climate change enhances contaminant exposure or organism susceptibility to toxicity, or contaminant-dominated, where chemical exposure increases susceptibility to climate change effects. These interactions can be antagonistic, additive, or synergistic, with synergistic effects often being the most prevalent (Cabral et al. 2019), especially in coastal ecosystems facing multiple stressors like sedimentation, nutrient loadings, and metal contamination. These ecosystems are particularly vulnerable due to high levels of anthropogenic pollution and pronounced environmental changes (Islam et al. 2023), such as hypoxia and acidification from increased freshwater runoff and nutrient discharge.

7.1 Hydrocarbon Remediations

Changes in ocean currents due to climate change influence oil spill transport, potentially affecting larger areas and marine habitats. Higher temperatures increase volatility, leading to faster evaporation of lighter hydrocarbons while reducing viscosity and enhancing their infiltration into aquatic ecosystems. Extreme weather events like hurricanes can exacerbate spills by dispersing oil widely and hindering cleanup efforts. Over the past two decades, there has been significant progress in utilizing biological processes to mitigate hazardous pollutants in marine and freshwater environments, breaking them into less toxic substances or reducing them to levels below regulatory thresholds.

Studies have identified hydrocarbon-degrading bacteria in various marine habitats, such as deep-sea sediments, coastal seawater, and sediments near industrial sites (Table 2). These bacteria, including genera like *Pseudomonas*, *Acinetobacter*, and *Marinobacter*, play key roles in hydrocarbon bioremediation. In coastal anoxic sediments, microbial communities like Desulfosarcina/Desulfococcus are crucial for anaerobic hydrocarbon degradation. Following oil spills, ecological succession involves oil-degrading genera like *Hyphomonas*, *Pseudomonas*, and *Methylophaga*. Polar regions harbor unique oil-degrading genera such as *Oleispira* and *Psychrobacter*. A recent study reported several bacteria like *Marinomonas*

Table 2 Oil-degrading marine bacteria with their media composition for screening and culturing

Compounds	Marine bacteria	Media composition (lab scale) *, α	References
Hydrocarbon	Erythrobacter, *Halomonas* spp.*, *Pseudomonas* spp., *Vibrio gazogene**, Xanthomarina, *Marinobacter hydrocarbonoclasticus* SP17*, Brevibacterium	*Marine broth 2216 with 375 mg/L oil	Abbasian et al. (2015)
Crude oil	*Bacillus atrophaeus*$^{\alpha}$, *Oleispira*, *Pseudomonas aeruginosa* Dutlxm0725, 〖Pseudomonas resinovorans$^{\alpha}$, *Streptomyces parvus* B7*, *Microbacterium esteraromaticum*, *Plantibacter auratus*$^{\alpha}$, Luteibacter, *Rhodococcus erythropolis* Dut-lxm1018 strains, *Stenotrophomonas maltophilia*, Thalassolituus, Parvibaculum, *Bacillus subtilis*$^{\alpha}$	$^{\alpha}$ Salt medium with 1% crude oil *Bushnell has medium with 1% crude oil	Gao et al. (2019); Kumari et al. (2018); Murphy et al. (2021)
Diesel	*Flavobacterium petrolei* sp., Micrococcus*, *Psychrophilic strains*, *Staphylococcus Genii**, *Flavobacterium psychrolimnae* LMG22018	*Nutrient medium with different diesel concentrations	Chaudhary et al. (2019); Sulistiyaning Titah et al. (2018)
Polycyclic aromatic hydrocarbon	*Ochrobactrum anthropi* IITR07, *Cobetia marina**, *Microbacterium esteraromaticum* IITR47, *Rhodococcus soli**, *Pseudomonas mendocina* IITR46, *Pseudoalteromonas agarivorans** *Ochrobactrum anthropi* IITR07	*Modified minimal basal medium containing 20 mg/100 ml PAHs	Lee et al. (2018)
Marine oil	Alteromonas*, Thalassospira*	*Bushnell has medium prepared in seawater with marine oil	Godini et al. (2018)

*, α indicate standardized media composition for the respective strains.

communis, Marinobacter santoriniensis, Pseudomonas pseudoalcaligenes, Exiguobacterium indicum, Rhodobium marinum, Rhodobacter sphaeroides, Alteromonas haloplanktis, and *Marinobacter hydrocarbonoclasticus* remediate Ar, Cr, Cu, Fe, Cd, Pb, Hg, Ni, Se, and Co from marine ecosystems (Ghosh et al. 2022). Additionally, recent studies have highlighted previously unnoticed oil-degrading taxa like Acinetobacter and Methylococcales in tropical waters.

By considering the role of microorganisms, the European Union has recently implemented the Kill.Spill project, aimed at enhancing bioremediation in marine environments, mainly focusing on the biodegradation of oil for all types of aquatic environments from the Eastern Mediterranean to Disko Bay in Greenland. The Kill. Spill technology provides essential nutrients along with hydrocarbon-degrading microbial consortia, primarily based on Pseudomonas spp. and *Burkholderia* spp. biosurfactants to accelerate and optimize bioremediation rates. Recently, in 2021, a major oil company in Chennai, India, carried out ex situ bioremediation at Ennore Port with 180 tons of oil sludge using oilivorous (bacterial consortium) to assess the efficacy of bioremediation (Tamizhdurai et al. 2022). They observed higher weight hydrocarbon transformation after six months, and total petroleum hydrocarbon concentration was decreased.

7.2 *Microplastic Bioremediations*

Millions of metric tons of plastic waste enter the ocean annually, predominantly from coastal countries, influencing marine environments, and their combined impact with global warming significantly affects marine primary production. While studying the triangle interaction between marine microbes, phytoplankton, and plastic pollution, still immediate stages (Ford et al. 2022; Sjollema et al. 2016), it reveals that this microplastic pollution can disrupt essential biogeochemical cycles in marine ecosystems (biological carbon pumps). Some studies have shown that microplastics can disrupt microalgal growth at high concentrations, though they may not significantly impact photosynthetic rates. However, climate change is expected to further impact plastic pollution fluxes and concentrations globally, particularly in regions like the Arctic (Obbard et al. 2014), where melting sea ice releases trapped microplastics into the marine environment. In 2016, Yang et al. (2016) discovered that microbes have a potential role in breaking down plastic using bacteria. Not only that, but recently, Rüthi et al. (2023) identified that even at low temperatures, microbes can degrade plastic.

For marine microplastic bioremediation, various fungal species, notably *Cladosporium, Verticillium, Aureobasidium, Emericellopsis, Debaryomyces, Mucor, Acremonium, Chaetomium, Paecilomyces, Cryptococcus, Rhodosporidium, Eupenicillium, Curvularia,* and *Fusarium*, are known for their ability to biodegrade plastics (Narayanan 2023), particularly low-density polyethylene (LDPE). One dominant fungal isolate in the sea, *Zalerion maritimum*, has shown potential for

LDPE degradation (Zhang et al. 2020). These fungi typically attach to the plastic surface, form biofilms, and produce specialized enzymes that break down the polymer bonds (C–C). These enzymes accelerate oxidative degradation, converting microplastics into smaller compounds like monomers, oligomers, and dimers. Some marine fungi, such as *Fusarium oxysporum* and *Penicillium citrinum*, can degrade polyethylene terephthalate (PET) (Ben Ali et al. 2020), while others, like *Trichoderma harzianum*, produce enzymes like laccase, manganese peroxidase, and lignin peroxidase, enabling them to degrade polyurethane (PU) and polyethylene (PE).

Besides those fungal species, various bacterial species commonly found in marine aquatic systems, including *Schlegelella* sp., *Bacillus* sp., *Amycolatopsis* sp., *Methanosarcina barkei, Brevibacillus* sp., *Alcanivorax* sp., *Hyphomonas* sp., *Cycloclasticus* sp., *Streptomyces* sp., *Clostridium* sp., and *Pseudomonas aeruginosa,* have been identified as capable of synthesizing extracellular hydrolytic enzymes (Belahmadi et al. 2023) such as xylanase, chitinase, protease, lipase, and keratinase, facilitating microplastic biodegradation. Specifically, *Methanosarcina barkei* has demonstrated the ability to degrade PVC, while *Rhodococcus ruber* secretes laccase enzymes capable of degrading PE. Additionally, *Azotobacter* sp. can degrade PS plastics using hydroquinone peroxidase, while species of *Cycloclasticus* sp., *Alcanivorax* sp., and *Hyphomonas* sp. degrade PET through ester bond hydrolysis (Othman et al. 2021). Gram-negative bacilli, particularly *Pseudomonas* sp., have shown significant potential for degrading PE (Narayanan 2023), with strains isolated from polluted environments and municipal sludge dumps proving effective. *Pseudomonas* sp. forms more viscous and flocculent biofilms on plastic surfaces than other bacterial species within a short timeframe.

7.3 Heavy Metals Bioremediations

Climate change affects the hydrological cycle globally and regionally (Dey et al. 2023), altering water flow patterns and temperatures. Reduced water flow and higher temperatures in summer and autumn can elevate pollutant levels in water and fish biomass due to limited dilution (Lazăr et al. 2024). Conversely, intensified rainfall and runoff can transport heavy metals from terrestrial sources into freshwater ecosystems (Hong et al. 2022). Pb and As concentrations increase in water with lower pH and temperature, while sediment Hg concentrations rise with higher pH and temperature (Mok et al. 2023). Climate change-induced ocean acidification can worsen heavy metal contamination in coastal waters by hindering photosynthesis and reducing CO_2 absorption efficiency. For instance, Cu, Cd, and Pb are highly toxic to marine life, with even slight increases in their concentrations causing harmful effects on organisms unaccustomed to such changes.

In such a scenario, various marine fungi, including Ascomycetes, have shown potential in heavy metal biosorption. For instance, *Penicillium fellutanum* and bentonite bio-composite exhibited promising nickel and zinc adsorption capabilities (Rashid et al. 2016). *Aspergillus niger,* among other Aspergillus species, demonstrated effective Hg biosorption and was able to eliminate 40.53 mg/g^{-1} under favorable conditions (Khambhaty et al. 2008). Additionally, *A. candidus* (isolated from different Indian coastlines) displayed resistance to arsenic, while *A. favus* and *Rhizopus* sp. exhibited varying degrees of arsenic bioaccumulation (Vala 2010). *Trichoderma viride* from the Mediterranean Sea successfully removed (4.66 mg/g^{-1}) chromium (VI), and *Cryptococcus* sp. from deep-sea environments showed resistance and removal of various heavy metals (El-Kassas and El-Taher 2009). *Yarrowia* sp. was found effective in mercury removal from Hg-contaminated sediments, with growing cells serving as bio-reduction agents and leftover cells as reusable bio-adsorbents (Oyetibo et al. 2015). *Monodictys pelagica* exhibited differential responses to cadmium (exceeding 60 mg/g^{-1}) and lead (surpassing 6 mg/g^{-1}) exposure, with significant bioaccumulation observed in both species (Taboski et al. 2005). *Dendryphiella salina* demonstrated high mercury absorption potential (80–92%) (Mendoza et al. 2012), highlighting the suitability of marine fungi in bioremediation efforts.

Besides fungi, marine or freshwater microalgae species show different levels of arsenic tolerance based on their growth rates and As accumulations. For instance, Morelli et al. (2005) identified that the growth rate of marine microalgae *Phaeodactylum tricornutum* gradually decreased when exposed to arsenate concentrations exceeding 0.1 μm, experiencing a notable 35% inhibition at 1.0 μm As.

Most bacteria in seawater belong to the viable but unculturable group. Bacterial bio-blocs, with active functional groups like aldehydes and carboxyl groups in their cell walls, serve as effective adsorbents for heavy metals. Moreover, studies have identified significant microbial community variations at different ocean depths, influenced by factors such as temperature and phosphate levels. Various bacterial strains, including *Pseudomonas* sp. and *Bacillus firmus*, have demonstrated heavy metal resistance and efficient metal sorption capacities (Alireza Safahieh 2012; Camacho-Chab et al. 2018), with some studies indicating the superiority of dead biomass over live cells in heavy metal removal processes (El-Gendy and El-Bondkly 2016). Additionally, halophilic actinomycetes like *Nocardiopsis halophila* and *Nocardiopsis rosea* exhibit high efficiency in heavy metal removal from industrial effluent samples (Alabssawy and Hashem 2024; El-Gendy and El-Bondkly 2016). Furthermore, marine bacteria such as *Shewanella* sp. show promising results in bioremediating metal-contaminated environments, especially when supplemented with cysteine (Staicu et al. 2022). However, challenges remain in efficiently removing specific elements like Cu from polluted grounds using microbial bioemulsifiers (Alabssawy and Hashem 2024). Nonetheless, bacterial strains like Streptomyces (separated from the Indian shore) demonstrate potential as biosurfactant producers and heavy metal-resistant organisms (Lakshmipathy and Kannabiran 2010).

8 Integrating Microbial Remediation into Climate Change Mitigation

Microbial carbon capture cells (MCCs) are an approach that combines carbon sequestration with algal biomass production and has two purposes: wastewater treatments and clean energy production. In MCCs, an algal biocathode generates the necessary O_2 for cathodic reactions, eliminating the need for external aeration. The algal photosynthesis process converts CO_2, produced during organic matter oxidation in the anodic chamber, into biomass while simultaneously generating electricity, treating wastewater, and sequestering CO_2. Moreover, MCCs advance microbial fuel cell technology by using algal biomass to supply O_2 for cathodic reduction, with the harvested algae serving as an organic substrate in the anodic chamber, potentially without additional oil and acting as a methanogen inhibitor (Yaashikaa et al. 2024).

In addition to natural microbial processes, the field of synthetic biology also showed microbial efficacy in carbon capture and sequestration. Companies like LanzaTech exemplify how genetically modified microorganisms can be utilized to convert industrial emissions into valuable chemicals. By engineering *Clostridium autoethanogenum* to metabolize waste gases such as CO_2 and CO, LanzaTech has achieved the production of chemicals like ethanol and acetone, which have a wide range of industrial applications. Not only that, but they are also working to capture one million tons of CO_2 annually and convert it to produce around seven million tons of useful chemicals like ethanol and acetone.

Moreover, cyanobacteria were also explored for their ability to produce biofuels and biomass directly from CO_2. Another company, CyanoCapture, for instance, focuses on optimizing cyanobacteria for bio-oil production, an approach that could provide sustainable alternatives to fossil fuels while simultaneously sequestering CO_2. The scalability of such solutions will be crucial for their success, and further genetic engineering advancements may enhance the carbon-fixing abilities of cyanobacteria and other photosynthetic microbes.

Microbial engineering is another potential option for facilitating climate change mitigation by modulating microbial-regulated ecosystem functions. A current example is the innovation of PivotBio (Pivot Bio 2010), which has come up with innovative and sustainable solutions to replace synthetic nitrogen soil fertilizers. In 2021, they applied their engineered soil bacteria on corn fields over 293,000 ha across the USA, which reduced approximately 80,000 metric tons of CO_2 emissions. Microbial solutions can stabilize carbon levels directly and indirectly to deal with the greenhouse effect by targeting carbon use efficiency, nitrogen cycling, and waste bioremediation. Effective inoculum design, which considers functional and establishment traits such as diversity and metabolic interconnectedness, is crucial for maximizing the efficacy of these microbial interventions.

Beyond carbon capture and emissions reduction, microbes can also facilitate the remediation of the growing plastic pollution problem. Plastics are derived from petrochemicals, and their degradation in the environment releases both CO_2 and methane. The MIX-UP consortium (MixUp-EU 2024), a collaboration between the EU and China, is using microbes to depolymerize plastics into their monomers. By using enzymes produced by specialized microbes, this approach offers a sustainable alternative to traditional plastic recycling methods, which are often energy-intensive and inefficient.

9 Future Directions

Despite significant advances in microbial remediation, several challenges remain that require focused research and development and the efficacy of large-scale applications in the real field. The future of microbial remediation hinges on addressing fundamental limitations and optimizing the environmental, technical, and genetic aspects of microbial processes.

A central challenge in microbial remediation is the influence of abiotic factors such as pH, temperature, moisture, and nutrient availability. These variables directly affect microbial activity, growth, and the overall effectiveness of pollutant degradation. Future research needs to focus on optimizing these abiotic factors in diverse environmental settings to ensure that microbial remediation can be applied on a large scale with consistent success. For example, nutrient levels in contaminated environments can significantly impact the ability of microbial communities to degrade pollutants. Understanding the ideal concentrations of macronutrients (such as N and P) and micronutrients (like trace metals) will be essential in maximizing microbial efficiency. Also need to focus on pH regulation, as different microbes have specific pH ranges within which they are most effective. Further, abiotic factors are not static; they can change with seasons, industrial activity, or climate change, all of which can impact microbial efficacy. Adaptive strategies, like dynamic nutrient dosing or temperature regulation in bioremediation systems, must be explored, which will increase the success stories of the natural environment.

Many microbes are not inherently equipped to degrade complex or recalcitrant pollutants, and even when they can, their efficiency might not be optimal for large-scale applications. By focusing on enhancing the production of enzymes that break down toxic substances such as PAHs, polychlorinated biphenyls, and heavy metals, GEMs can be developed to accelerate pollutant degradation. Similarly, the creation of microbial consortia, which involve multiple strains of bacteria, fungi, or algae working synergistically, could enhance the breakdown of complex pollutants that no single strain could tackle alone. These consortia could be genetically modified to ensure compatibility and optimization for a specific polluted environment. However, the release of genetically modified organisms (GMOs) into the environment raises ethical and ecological concerns. Future research must include rigorous risk assessment and the development of safety mechanisms, such as built-in gene circuits that trigger the self-destruction of GMOs once their task is completed.

High-throughput sequencing technologies, metagenomics, and metatranscriptomics can be used to reveal which microbes are actively degrading pollutants under specific conditions. Machine learning algorithms could be applied to these datasets to predict the most promising microbial candidates for bioremediation in different environments, streamlining the process of selecting effective microbial strains or consortia. Besides, integrated technologies, such as nano-bioremediation and electrokinetic bioremediation, have already shown promise in coupling microbial degradation with other remediation strategies to enhance efficiency. The use of electrokinetic techniques, where electrical currents are applied to contaminated soils or water, can enhance microbial mobility and pollutant accessibility, further increasing the efficiency of microbial remediation.

10 Conclusion

Microbial remediation stands at the forefront of sustainable environmental management, offering a powerful, cost-effective, and eco-friendly approach to mitigate pollution in the face of climate change. Microbes, with their unparalleled ability to degrade, detoxify, and transform contaminants, provide a natural solution that not only cleans polluted environments but also enhances ecosystem resilience. Their unique capacity to thrive across diverse habitats, from extreme Arctic regions to deep-sea thermal vents, positions them as essential agents in our efforts to restore and maintain ecological balance in the face of climate change.

Compared to conventional remediation techniques, microbial approaches offer distinct advantages: they are cost-effective, minimize harmful residues, and promote carbon sequestration, thereby aiding in climate change mitigation. The versatility of microbial remediation extends across various environments, including water, soil, and air, making it an invaluable tool for restoring polluted ecosystems while supporting biodiversity.

However, the success of microbial remediation depends on optimizing microbial activity through targeted interventions such as nutrient supplementation, pH regulation, and strategic bioaugmentation. While the potential of GMOs holds promise for enhancing remediation efficacy, these approaches must be carefully managed to address potential ecological and health risks. Ethical considerations, including the spread of antibiotic-resistant genes and incomplete degradation leading to more toxic byproducts, necessitate rigorous risk assessment and monitoring.

To fully utilize the potential of microbial remediation, it is important to advance our understanding through continued research and innovation. Future efforts should focus on integrating advanced technologies, such as metagenomics, machine learning, and synthetic biology, to refine and scale up remediation strategies. Collaborative efforts between scientists, policymakers, industry leaders, and communities will be vital in developing and deploying these solutions responsibly.

As we confront the unprecedented challenges posed by climate change, embracing microbial remediation offers a path toward a cleaner, healthier future. Using the power of the microscopic world, we can drive transformative changes in environmental management, creating resilient ecosystems and safeguarding our planet from the changing climate for generations to come.

References

Abbasian F, Lockington R, Mallavarapu M, Naidu R (2015) A comprehensive review of aliphatic hydrocarbon biodegradation by bacteria. Appl Biochem Biotechnol 176(3):670–699. https://doi.org/10.1007/S12010-015-1603-5

Abhilash PC, Dubey RK, Tripathi V, Srivastava P, Verma JP, Singh HB (2013) Remediation and management of POPs-contaminated soils in a warming climate: Challenges and perspectives. Environ Sci Pollut Res 20:5879–5885. https://doi.org/10.1007/S11356-013-1808-5

Alabssawy AN, Hashem AH (2024) Bioremediation of hazardous heavy metals by marine microorganisms: a recent review. Arch Microbiol 206(3):103. https://doi.org/10.1007/S00203-023-03793-5

Alava JJ, Cheung WWL, Ross PS, Sumaila UR (2017) Climate change–contaminant interactions in marine food webs: toward a conceptual framework. Glob Chang Biol 23:3984–4001. https://doi.org/10.1111/GCB.13667

Al-Hawash AB, Dragh MA, Li S, Alhujaily A, Abbood HA, Zhang X, Ma F (2018) Principles of microbial degradation of petroleum hydrocarbons in the environment. Egypt J Aquat Res 44:71–76. https://doi.org/10.1016/J.EJAR.2018.06.001

Alkorta I, Epelde L, Garbisu C (2017) Environmental parameters altered by climate change affect the activity of soil microorganisms involved in bioremediation. FEMS Microbiol Lett 364:200. https://doi.org/10.1093/FEMSLE/FNX200

Amorim CL, Moreira IS, Maia AS, Tiritan ME, Castro PML (2014) Biodegradation of ofloxacin, norfloxacin, and ciprofloxacin as single and mixed substrates by Labrys portucalensis F11. Appl Microbiol Biotechnol 98:3181–3190. https://doi.org/10.1007/S00253-013-5333-8

Ashraf S, Ali Q, Zahir ZA, Ashraf S, Asghar HN (2019) Phytoremediation: Environmentally sustainable way for reclamation of heavy metal polluted soils. Ecotoxicol Environ Saf 174:714–727. https://doi.org/10.1016/J.ECOENV.2019.02.068

ASM (2022) Microbes and climate change – science, people & impacts. American Society for Microbiology, Washington, DC

Ayilara MS, Babalola OO (2023) Bioremediation of environmental wastes: the role of microorganisms. Front Agron 5:1183691. https://doi.org/10.3389/FAGRO.2023.1183691

Azubuike CC, Chikere CB, Okpokwasili GC (2016) Bioremediation techniques–classification based on site of application: principles, advantages, limitations and prospects. World J Microbiol Biotechnol 32(11):180. https://doi.org/10.1007/S11274-016-2137-X

Bala S, Garg D, Thirumalesh BV, Sharma M, Sridhar K, Inbaraj BS, Tripathi M (2022) Recent strategies for bioremediation of emerging pollutants: a review for a green and sustainable environment. Toxics 2022(10):484. https://doi.org/10.3390/TOXICS10080484

Belahmadi MSO, Charchar N, Abdessemed A, Gherib A (2023) Impact of petroleum refinery on aquatic ecosystem of Skikda Bay (Algeria): Diversity and abundance of viable bacterial strains. Mar Pollut Bull 188:114704. https://doi.org/10.1016/J.MARPOLBUL.2023.114704

Ben Ali W, Chaduli D, Navarro D, Lechat C, Turbé-Doan A, Bertrand E, Faulds CB, Sciara G, Lesage-Meessen L, Record E, Mechichi T (2020) Screening of five marine-derived fungal strains for their potential to produce oxidases with laccase activities suitable for biotechnological applications. BMC Biotechnol 20:1–13. https://doi.org/10.1186/S12896-020-00617-Y

Bhatt P, Gangola S, Bhandari G, Zhang W, Maithani D, Mishra S, Chen S (2021) New insights into the degradation of synthetic pollutants in contaminated environments. Chemosphere 268:128827. https://doi.org/10.1016/J.CHEMOSPHERE.2020.128827

Bonfá MRL, Grossman MJ, Mellado E, Durrant LR (2011) Biodegradation of aromatic hydrocarbons by Haloarchaea and their use for the reduction of the chemical oxygen demand of hypersaline petroleum produced water. Chemosphere 84:1671–1676. https://doi.org/10.1016/J.CHEMOSPHERE.2011.05.005

Burns RG, DeForest JL, Marxsen J, Sinsabaugh RL, Stromberger ME, Wallenstein MD, Weintraub MN, Zoppini A (2013) Soil enzymes in a changing environment: Current knowledge and future directions. Soil Biol Biochem 58:216–234. https://doi.org/10.1016/J.SOILBIO.2012.11.009

Cabral H, Fonseca V, Sousa T, Leal MC (2019) Synergistic effects of climate change and marine pollution: an overlooked interaction in Coastal and Estuarine areas. Int J Environ Res Public Health 16. https://doi.org/10.3390/IJERPH16152737

Camacho-Chab JC, del Castañeda-Chávez MR, Chan-Bacab MJ, Aguila-Ramírez RN, Galaviz-Villa I, Bartolo-Pérez P, Lango-Reynoso F, Tabasco-Novelo C, Gaylarde C, Ortega-Morales BO (2018) Biosorption of cadmium by non-toxic Extracellular Polymeric Substances (EPS) synthesized by bacteria from marine intertidal biofilms. Int J Environ Res Public Health 15. https://doi.org/10.3390/IJERPH15020314

Cases I, de Lorenzo V (2005) Genetically modified organisms for the environment: stories of success and failure and what we have learned from them. Int Microbiol 8:213–222

Chaudhary DK, Kim D-U, Kim D, Kim J (2019) Flavobacterium petrolei sp. nov., a novel psychrophilic, diesel-degrading bacterium isolated from oil-contaminated Arctic soil. Sci Rep 9:4134. https://doi.org/10.1038/s41598-019-40667-7

Chen SL, Chang SW, Chen YJ, Chen HL (2021) Possible warming effect of fine particulate matter in the atmosphere. Commun Earth Environ 2(1):1–9. https://doi.org/10.1038/s43247-021-00278-5

Classen AT, Sundqvist MK, Henning JA, Newman GS, Moore JAM, Cregger MA, Moorhead LC, Patterson CM (2015) Direct and indirect effects of climate change on soil microbial and soil microbial-plant interactions: What lies ahead? Ecosphere 6:1–21. https://doi.org/10.1890/ES15-00217.1

Coconi-Linares N, Magaña-Ortíz D, Guzmán-Ortiz DA, Fernández F, Loske AM, Gómez-Lim MA (2014) High-yield production of manganese peroxidase, lignin peroxidase, and versatile peroxidase in Phanerochaete chrysosporium. Appl Microbiol Biotechnol 98:9283–9294. https://doi.org/10.1007/S00253-014-6105-9

Dey B, Abir KAM, Ahmed R, Salam MA, Redowan M, Miah MD, Iqbal MA (2023) Monitoring groundwater potential dynamics of north-eastern Bengal Basin in Bangladesh using AHP-Machine learning approaches. Ecol Indic 154:110886. https://doi.org/10.1016/j.ecolind.2023.110886

Edris G, Alhamed Y, Alzahrani A (2014) Biosorption of cadmium and lead from aqueous solutions by Chlorella vulgaris biomass: equilibrium and kinetic study. Arab J Sci Eng 39:87–93. https://doi.org/10.1007/S13369-013-0820-X/METRICS

Eibes G, Debernardi G, Feijoo G, Moreira MT, Lema JM (2011) Oxidation of pharmaceutically active compounds by a ligninolytic fungal peroxidase. Biodegradation 22:539–550. https://doi.org/10.1007/S10532-010-9426-0

El-Gendy MMAA, El-Bondkly AMA (2016) Evaluation and enhancement of heavy metals bioremediation in aqueous solutions by Nocardiopsis sp. MORSY1948, and Nocardia sp. MORSY2014. Braz J Microbiol 47:571–586. https://doi.org/10.1016/J.BJM.2016.04.029

El-Kassas HY, El-Taher EM (2009) Optimization of batch process parameters by response surface methodology for mycoremediation of Chrome-VI by a chromium resistant strain of marine Trichoderma viride. J Agric Environ Sci 5:676–681

Espenshade J, Thijs S, Gawronski S, Bové H, Weyens N, Vangronsveld J (2019) Influence of urbanization on epiphytic bacterial communities of the platanus × hispanica tree leaves in a biennial study. Front Microbiol 10:439804. https://doi.org/10.3389/FMICB.2019.00675

Farkas B, Kolenčík M, Hain M, Dobročka E, Kratošová G, Bujdoš M, Feng H, Deng Y, Yu Q, Illa R, Ratna Sunil B, Kim H, Matúš P, Urík M (2020) Aspergillus Niger decreases bioavailability of Arsenic(V) via biotransformation of manganese oxide into biogenic oxalate minerals. J Fungi 6:270. https://doi.org/10.3390/JOF6040270

Feng L, He S, Wei L, Zhang J, Wu H (2021) Impacts of aeration and biochar on physiological characteristics of plants and microbial communities and metabolites in constructed wetland microcosms for treating swine wastewater. Environ Res 200:111415. https://doi.org/10.1016/J.ENVRES.2021.111415

Ford HV, Jones NH, Davies AJ, Godley BJ, Jambeck JR, Napper IE, Suckling CC, Williams GJ, Woodall LC, Koldewey HJ (2022) The fundamental links between climate change and marine plastic pollution. Sci Total Environ 806:150392. https://doi.org/10.1016/J.SCITOTENV.2021.150392

Frey SD, Drijber R, Smith H, Melillo J (2008) Microbial biomass, functional capacity, and community structure after 12 years of soil warming. Soil Biol Biochem 40:2904–2907. https://doi.org/10.1016/J.SOILBIO.2008.07.020

Gandolfi I, Canedoli C, Imperato V, Tagliaferri I, Gkorezis P, Vangronsveld J, Padoa Schioppa E, Papacchini M, Bestetti G, Franzetti A (2017) Diversity and hydrocarbon-degrading potential of epiphytic microbial communities on Platanus x acerifolia leaves in an urban area. Environ Pollut 220:650–658. https://doi.org/10.1016/J.ENVPOL.2016.10.022

Gao LL, Lu YC, Zhang JL, Li J, Zhang JD (2019) Biotreatment of restaurant wastewater with an oily high concentration by newly isolated bacteria from oily sludge. World J Microbiol Biotechnol 35:1–11. https://doi.org/10.1007/S11274-019-2760-4

Ghosh S, Bhattacharya J, Nitnavare R, Webster TJ (2022) Microbial remediation of metals by marine bacteria. In: Development in wastewater treatment research and processes: microbial degradation of xenobiotics through bacterial and fungal approach. Elsevier, pp 131–158. https://doi.org/10.1016/B978-0-323-85839-7.00011-6

Godini K, Samarghandi MR, Zafari D, Rahmani AR, Afkhami A, Arabestani MR (2018) Isolation and identification of new strains of crude oil degrading bacteria from Kharg Island, Iran. Pet Sci Technol 36:869–874. https://doi.org/10.1080/10916466.2018.1447961

Gopinath M, Pulla RH, Rajmohan KS, Vijay P, Muthukumaran C, Gurunathan B (2018) Bioremediation of volatile organic compounds in biofilters. In: Energy, environment, and sustainability. Springer, Singapore, pp 301–330. https://doi.org/10.1007/978-981-10-7485-1_15

Gouin T, Armitage JM, Cousins IT, Muir DCG, Ng CA, Reid L, Tao S (2013) Influence of global climate change on chemical fate and bioaccumulation: the role of multimedia models. Environ Toxicol Chem 32:20–31. https://doi.org/10.1002/ETC.2044

Grifoni M, Franchi E, Fusini D, Vocciante M, Barbafieri M, Pedron F, Rosellini I, Petruzzelli G (2022) Soil remediation: towards a resilient and adaptive approach to deal with the ever-changing environmental challenges. Environments 2022(9):18. https://doi.org/10.3390/ENVIRONMENTS9020018

Gros M, Cruz-Morato C, Marco-Urrea E, Longrée P, Singer H, Sarrà M, Hollender J, Vicent T, Rodriguez-Mozaz S, Barceló D (2014) Biodegradation of the X-ray contrast agent iopromide and the fluoroquinolone antibiotic ofloxacin by the white rot fungus Trametes versicolor in hospital wastewaters and identification of degradation products. Water Res 60:228–241. https://doi.org/10.1016/J.WATRES.2014.04.042

Gupta VK, Rastogi A, Saini VK, Jain N (2006) Biosorption of copper(II) from aqueous solutions by Spirogyra species. J Colloid Interface Sci 296:59–63. https://doi.org/10.1016/J.JCIS.2005.08.033

Hartley IP, Hopkins DW, Garnett MH, Sommerkorn M, Wookey PA (2008) Soil microbial respiration in arctic soil does not acclimate to temperature. Ecol Lett 11:1092–1100. https://doi.org/10.1111/j.1461-0248.2008.01223.x

He H, Liang XZ, Wuebbles DJ (2018) Effects of emissions change, climate change and long-range transport on regional modeling of future U.S. particulate matter pollution and speciation. Atmos Environ 179:166–176. https://doi.org/10.1016/J.ATMOSENV.2018.02.020

Hong X, Zhao Y, Zhuang R, Liu J, Guo G, Chen J, Yao Y (2020) Bioremediation of tetracycline antibiotics-contaminated soil by bioaugmentation. RSC Adv 10:33086–33102. https://doi.org/10.1039/D0RA04705H

Hong J, Zhang J, Song Y, Cao X (2022) Spatial and temporal distribution characteristics of nutrient elements and heavy metals in surface water of Tibet, China and their pollution assessment. Water 14:3664. https://doi.org/10.3390/W14223664

Islam R, Ahmed R, Dey B, Haque S, Aktar S, Bhuiyan MS, Arif MS, Habib Ador MA, Ul Haque MM, Saha N (2023) Salinity hazard drives the alteration of occupation, land use and ecosystem service in the coastal areas: evidence from the south-western coastal region of Bangladesh. Heliyon 9:e18512. https://doi.org/10.1016/j.heliyon.2023.e18512

Islas-Espinoza M, Reid BJ, Wexler M, Bond PL (2012) Soil bacterial consortia and previous exposure enhance the biodegradation of sulfonamides from pig manure. Microb Ecol 64:140–151. https://doi.org/10.1007/S00248-012-0010-5

Ismail Soifiata S, Abdou Karima Y, Asnat S, Hassane Fahimat S, Allaoui Said Ahmed A, Mohamed A, Mohamed Abdou N, Sailine R, Hamdane Aladine B, Ahmed Mohamed N, Mohamed Elyamine A (2022) Polycyclic aromatic hydrocarbons effect on the phyllosphere bacterial community of Gliricidia sepium leaves. Int J Adv Eng Res Sci 9:2022. https://doi.org/10.22161/ijaers.98.9

Janusz G, Pawlik A, Sulej J, Świderska-Burek U, Jarosz-Wilkolazka A, Paszczyński A (2017) Lignin degradation: microorganisms, enzymes involved, genomes analysis and evolution. FEMS Microbiol Rev 41:941–962. https://doi.org/10.1093/FEMSRE/FUX049

Jiang B, Li A, Cui D, Cai R, Ma F, Wang Y (2014) Biodegradation and metabolic pathway of sulfamethoxazole by Pseudomonas psychrophila HA-4, a newly isolated cold-adapted sulfamethoxazole-degrading bacterium. Appl Microbiol Biotechnol 98:4671–4681. https://doi.org/10.1007/S00253-013-5488-3

Juwarkar AA, Jambhulkar HP (2008) Phytoremediation of coal mine spoil dump through integrated biotechnological approach. Bioresour Technol 99:4732–4741. https://doi.org/10.1016/J.BIORTECH.2007.09.060

Juwarkar AA, Yadav SK, Thawale PR, Kumar P, Singh SK, Chakrabarti T (2009) Developmental strategies for sustainable ecosystem on mine spoil dumps: a case of study. Environ Monit Assess 157:471–481. https://doi.org/10.1007/S10661-008-0549-2/METRICS

Juwarkar AA, Singh SK, Mudhoo A (2010) A comprehensive overview of elements in bioremediation. Rev Environ Sci Biotechnol 9:215–288. https://doi.org/10.1007/S11157-010-9215-6

Kafle A, Timilsina A, Gautam A, Adhikari K, Bhattarai A, Aryal N (2022) Phytoremediation: mechanisms, plant selection and enhancement by natural and synthetic agents. Environ Adv 8:100203. https://doi.org/10.1016/J.ENVADV.2022.100203

Khambhaty Y, Mody K, Basha S, Jha B (2008) Hg(II) removal from aqueous solution by dead fungal biomass of marine Aspergillus Niger: kinetic studies. Sep Sci Technol 43:1221–1238. https://doi.org/10.1080/01496390801888235

Kim DW, Heinze TM, Kim BS, Schnackenberg LK, Woodling KA, Sutherland JB (2011) Modification of Norfloxacin by a Microbacterium sp. strain isolated from a wastewater treatment plant. Appl Environ Microbiol 77:6100. https://doi.org/10.1128/AEM.00545-11

Kim KJ, Khalekuzzaman M, Suh JN, Kim HJ, Shagol C, Kim HH, Kim HJ (2018) Phytoremediation of volatile organic compounds by indoor plants: a review. Hortic Environ Biotechnol 59:143–157. https://doi.org/10.1007/S13580-018-0032-0

Kumar P, Arshad M, Gacem A, Soni S, Singh S, Kumar M, Yadav VK, Tariq M, Kumar R, Shah D, Wanale SG, Al Mesfer MKM, Bhutto JK, Yadav KK (2023) Insight into the environmental fate, hazard, detection, and sustainable degradation technologies of chlorpyrifos—an organophosphorus pesticide. Environ Sci Pollut Res 30(50):108347–108369. https://doi.org/10.1007/S11356-023-30049-Y

Kumari S, Regar RK, Manickam N (2018) Improved polycyclic aromatic hydrocarbon degradation in a crude oil by individual and a consortium of bacteria. Bioresour Technol 254:174–179. https://doi.org/10.1016/J.BIORTECH.2018.01.075

Lakshmipathy T, Kannabiran K (2010) Production of biosurfactant and heavy metal resistance activity of Streptomyces Sp. VITDDK3-a novel Halo Tolerant Actinomycetes Isolated from Saltpan Soil. Biology, Chemistry, Environmental Science

Lamon L, Dalla Valle M, Critto A, Marcomini A (2009) Introducing an integrated climate change perspective in POPs modelling, monitoring and regulation. Environ Pollut 157:1971–1980. https://doi.org/10.1016/J.ENVPOL.2009.02.016

Lazăr NN, Simionov IA, Petrea ŞM, Iticescu C, Georgescu PL, Dima F, Antache A (2024) The influence of climate changes on heavy metals accumulation in Alosa immaculata from the Danube River Basin. Mar Pollut Bull 200:116145. https://doi.org/10.1016/J.MARPOLBUL.2024.116145

Lee DW, Lee H, Lee AH, Kwon BO, Khim JS, Yim UH, Kim BS, Kim JJ (2018) Microbial community composition and PAHs removal potential of indigenous bacteria in oil contaminated sediment of Taean coast, Korea. Environ Pollut (Barking, Essex: 1987) 234:503–512. https://doi.org/10.1016/J.ENVPOL.2017.11.097

Lee H, Jun Z, Zahra Z (2021) Phytoremediation: the sustainable strategy for improving indoor and outdoor air quality. Environments 2021(8):118. https://doi.org/10.3390/ENVIRONMENTS8110118

Li Y, Chen X, Sonne C, Lam SS, Yang Y, Ma NL, Peng W (2023) Reduction and control of air pollution: based on plant-microbe interactions. Environ Pollut Bioavailab 35. https://doi.org/10.1080/26395940.2023.2173657

Liu Y, Chang H, Li Z, Feng Y, Cheng D, Xue J (2017) Biodegradation of gentamicin by bacterial consortia AMQD4 in synthetic medium and raw gentamicin sewage. Sci Rep 7(1):1–11. https://doi.org/10.1038/s41598-017-11529-x

Mandal D, Aghababaei M, Das SK, Majumder S, Chatterjee D, Basu A (2022) Isolation and identification of arsenic hyper-tolerant bacterium with potential plant growth promoting properties from soil. Minerals 2022(12):1452. https://doi.org/10.3390/MIN12111452

Mandal M, Das S, Roy A, Rakwal R, Jones OAH, Popek R, Agrawal GK, Sarkar A (2023) Interactive relations between plants, the phyllosphere microbial community, and particulate matter pollution. Sci Total Environ 890:164352. https://doi.org/10.1016/J.SCITOTENV.2023.164352

Mani D, Kumar C (2014) Biotechnological advances in bioremediation of heavy metals contaminated ecosystems: an overview with special reference to phytoremediation. Int J Environ Sci Technol 11:843–872. https://doi.org/10.1007/S13762-013-0299-8

Margesin R, Schinner F (2001) Biodegradation and bioremediation of hydrocarbons in extreme environments. Appl Microbiol Biotechnol 56:650–663. https://doi.org/10.1007/S002530100701

Maurya A, Sharma D, Partap M, Kumar R, Bhargava B (2023) Microbially-assisted phytoremediation toward air pollutants: current trends and future directions. Environ Technol Innov 31:103140. https://doi.org/10.1016/J.ETI.2023.103140

Mazumder MAR, Jubayer MF, Ranganathan TV (2021) Biodegradation of plastics by microorganisms. In: Biotechnology for zero waste: emerging waste management techniques. Wiley, pp 123–141. https://doi.org/10.1002/9783527832064.CH9

Mendoza RAJS, Estanislao KB, Aninipot JFP, Dahonog RA, De Guzman JA, Torres J, Cruz TD (2012) Biosorption of mercury by the marine fungus Dendryphiella salina. Environmental Science, Biology

Miraglia M, Marvin HJP, Kleter GA, Battilani P, Brera C, Coni E, Cubadda F, Croci L, De Santis B, Dekkers S, Filippi L, Hutjes RWA, Noordam MY, Pisante M, Piva G, Prandini A, Toti L, van den Born GJ, Vespermann A (2009) Climate change and food safety: an emerging issue with special focus on Europe. Food Chem Toxicol 47:1009–1021. https://doi.org/10.1016/J.FCT.2009.02.005

Mishra S, Hora S, Mishra R, Kanaujia PK (2022) Bioremediation-based microorganisms to break down pollutants decelerate due to climate change. In: Plant stress mitigators: action and application. Springer, Singapore, pp 125–143. https://doi.org/10.1007/978-981-16-7759-5_7

Mitra AK (2024) Current trends in green bioremediation of environmental organic pollutants. In: Green solutions for degradation of pollutants, pp 65–95. https://doi.org/10.2174/9789815238969124010007

MixUp-EU (2024) Mix-up: home [WWW Document]. URL https://www.mix-up.eu/. Accessed 4.9.24

Mok WJ, Ghaffar MA, Noor MIM, Lananan F, Azra MN (2023) Understanding climate change and heavy metals in coastal areas: a Macroanalysis assessment. Water 15:891. https://doi.org/10.3390/W15050891

Morelli E, Mascherpa MC, Scarano G (2005) Biosynthesis of phytochelatins and arsenic accumulation in the marine microalga Phaeodactylum tricornutum in response to arsenate exposure. Biometals 18:587–593. https://doi.org/10.1007/S10534-005-2998-1

Murphy SMC, Bautista MA, Cramm MA, Hubert CRJ (2021) Diesel and crude oil biodegradation by cold-adapted microbial communities in the Labrador Sea. Appl Environ Microbiol 87:1–21. https://doi.org/10.1128/AEM.00800-21/SUPPL_FILE/AEM.00800-21-S0005.XLSX

Naorem A (2022) Will climate change alter the efficiency of bioremediation? In: Advances in bioremediation and phytoremediation for sustainable soil management: principles, monitoring and remediation, pp 429–435. https://doi.org/10.1007/978-3-030-89984-4_27

Narayanan M (2023) Origination, fate, accumulation, and impact, of microplastics in a marine ecosystem and bio/technological approach for remediation: a review. Process Saf Environ Prot 177:472–485. https://doi.org/10.1016/J.PSEP.2023.07.013

Nzila A (2013) Update on the cometabolism of organic pollutants by bacteria. Environ Pollut (Barking, Essex: 1987) 178:474–482. https://doi.org/10.1016/J.ENVPOL.2013.03.042

Obbard RW, Sadri S, Wong YQ, Khitun AA, Baker I, Thompson RC (2014) Global warming releases microplastic legacy frozen in Arctic Sea ice. Earth's Future 2:315–320. https://doi.org/10.1002/2014EF000240

Othman AR, Hasan HA, Muhamad MH, Ismail NI, Abdullah SRS (2021) Microbial degradation of microplastics by enzymatic processes: a review. Environ Chem Lett 19(4):3057–3073. https://doi.org/10.1007/S10311-021-01197-9

Oyetibo GO, Ishola ST, Ikeda-Ohtsubo W, Miyauchi K, Ilori MO, Endo G (2015) Mercury bioremoval by Yarrowia strains isolated from sediments of mercury-polluted estuarine water. Appl Microbiol Biotechnol 99:3651–3657. https://doi.org/10.1007/S00253-014-6279-1

Padhan D, Rout PP, Kundu R, Adhikary S, Padh PP (2021) Bioremediation of heavy metals and other toxic substances by microorganisms. In: Soil bioremediation: an approach towards sustainable technology, pp 285–329. https://doi.org/10.1002/9781119547976.CH12

Pivot Bio (2010) Home | Pivot Bio [WWW Document]. URL https://www.pivotbio.com/. Accessed 9.4.24

Postiglione A, Prigioniero A, Zuzolo D, Tartaglia M, Scarano P, Maisto M, Ranauda MA, Sciarrillo R, Thijs S, Vangronsveld J, Guarino C (2022) Quercus ilex Phyllosphere microbiome environmental-driven structure and composition shifts in a Mediterranean Contex. Plan Theory 11:3528. https://doi.org/10.3390/PLANTS11243528/S1

Prenafeta-Boldú FX, Roca N, Villatoro C, Vera L, de Hoog GS (2019) Prospective application of melanized fungi for the biofiltration of indoor air in closed bioregenerative systems. J Hazard Mater 361:1–9. https://doi.org/10.1016/J.JHAZMAT.2018.08.059

Putri A, Yuniati R, Putrika A (2023) Six tree species physiological responses to air pollution in Pulogadung Industrial Estate, East Jakarta, Indonesia and Universitas Indonesia Campus, Depok, Indonesia. Int J Adv Appl Sci 12:152–162. https://doi.org/10.11591/IJAAS.V12.I2.PP152-162

Rajkumar R, Kurinjimalar C (2021) Microbes and plant mineral nutrition. In: Microbiological activity for soil and plant health management, pp 111–132. https://doi.org/10.1007/978-981-16-2922-8_5

Rashid A, Bhatti HN, Iqbal M, Noreen S (2016) Fungal biomass composite with bentonite efficiency for nickel and zinc adsorption: a mechanistic study. Ecol Eng 91:459–471. https://doi.org/10.1016/J.ECOLENG.2016.03.014

Ravindra K, Mor S (2022) Phytoremediation potential of indoor plants in reducing air pollutants. Front Sustain Cities 4:1039710. https://doi.org/10.3389/FRSC.2022.1039710

Reis PJM, Reis AC, Ricken B, Kolvenbach BA, Manaia CM, Corvini PFX, Nunes OC (2014) Biodegradation of sulfamethoxazole and other sulfonamides by Achromobacter denitrificans PR1. J Hazard Mater 280:741–749. https://doi.org/10.1016/J.JHAZMAT.2014.08.039

Ren X, Zeng G, Tang L, Wang J, Wan J, Wang J, Deng Y, Liu Y, Peng B (2018) The potential impact on the biodegradation of organic pollutants from composting technology for soil remediation. In: Waste Manag (New York, N.Y.), vol 72, pp 138–149. https://doi.org/10.1016/J.WASMAN.2017.11.032

Rojo F (2009) Degradation of alkanes by bacteria. Environ Microbiol 11:2477–2490. https://doi.org/10.1111/j.1462-2920.2009.01948.x

Roy A, Roy M, Alghamdi S, Dablool AS, Almakki AA, Ali IH, Yadav KK, Islam MR, Cabral-Pinto MMS (2022) Role of microbes and nanomaterials in the removal of pesticides from wastewater. Int J Photoenergy 2022:1. https://doi.org/10.1155/2022/2131583

Rüthi J, Cerri M, Brunner I, Stierli B, Sander M, Frey B (2023) Discovery of plastic-degrading microbial strains isolated from the alpine and Arctic terrestrial plastisphere. Front Microbiol 14:1178474. https://doi.org/10.3389/FMICB.2023.1178474

Safahieh A (2012) Isolation and characterization of Pseudomonas resistant to heavy metals and poly aromatics hydrocarbons (PAHs) from Persian Gulf sediments. Afr J Biotechnol 11:4418–4423. https://doi.org/10.5897/ajb11.3450

Saha L, Tiwari J, Bauddh K, Ma Y (2021) Recent developments in microbe–plant-based bioremediation for tackling heavy metal-polluted soils. Front Microbiol 12:731723. https://doi.org/10.3389/FMICB.2021.731723

Shayegan Z, Haghighat F, Lee CS (2019) Photocatalytic oxidation of volatile organic compounds for indoor environment applications: three different scaled setups. Chem Eng J 357:533–546. https://doi.org/10.1016/J.CEJ.2018.09.167

Shi L, Zhu A, Huang L, Yaluk E, Gu Y, Wang Y, Wang S, Chan A, Li L (2021) Impact of the planetary boundary layer on air quality simulations over the Yangtze River Delta region. China Atmos Environ 263:118685. https://doi.org/10.1016/J.ATMOSENV.2021.118685

Silva V, Mol HGJ, Zomer P, Tienstra M, Ritsema CJ, Geissen V (2019) Pesticide residues in European agricultural soils – a hidden reality unfolded. Sci Total Environ 653:1532–1545. https://doi.org/10.1016/J.SCITOTENV.2018.10.441

Sjollema SB, Redondo-Hasselerharm P, Leslie HA, Kraak MHS, Vethaak AD (2016) Do plastic particles affect microalgal photosynthesis and growth? Aquat Toxicol 170:259–261. https://doi.org/10.1016/J.AQUATOX.2015.12.002

Skowro'nski T, Skowro'nski S, Pirszel J, Pawlik-Skowro'nska B, Skowro'nska S (2001) Heavy metal removal by the waste biomass of Penicillium chrysogenum. Water Qual Res J 36:793–803. https://doi.org/10.2166/WQRJ.2001.042

Sodhi KK, Kumar M, Singh DK (2021) Insight into the amoxicillin resistance, ecotoxicity, and remediation strategies. J Water Proc Eng 39:101858. https://doi.org/10.1016/J.JWPE.2020.101858

Sofia D, Gioiella F, Lotrecchiano N, Giuliano A (2020) Mitigation strategies for reducing air pollution. Environ Sci Pollut Res 27:19226–19235. https://doi.org/10.1007/S11356-020-08647-X

Sonavane P, Pinjari S (2024) Assessment of the Ambient Air Quality of Sangli City, Maharashtra, vis-à-vis National Ambient Air Quality Standards (NAAQS). Int Res J Adv Eng Hub 2:182–189. https://doi.org/10.47392/IRJAEH.2024.0031

Soreanu G, Cretescu I, Diaconu M, Cojocaru C, Ignat M, Samoila P, Harabagiu V (2020) Investigation of a biosystem based on Arthrospira platensis for air revitalisation in spacecrafts: Performance evaluation through response surface methodology. Chemosphere 264:128465–128465. https://doi.org/10.1016/J.CHEMOSPHERE.2020.128465

Spain JC, Zylstra GJ, Blake CK, Gibson DT (1989) Monohydroxylation of phenol and 2,5-dichlorophenol by toluene dioxygenase in Pseudomonas putida F1. Appl Environ Microbiol 55:2648–2652. https://doi.org/10.1128/AEM.55.10.2648-2652.1989

Sriprapat W, Boraphech P, Thiravetyan P (2014) Factors affecting xylene-contaminated air removal by the ornamental plant Zamioculcas zamiifolia. Environ Sci Pollut Res 21:2603–2610. https://doi.org/10.1007/S11356-013-2175-Y

Staicu LC, Wójtowicz PJ, Baragaño D, Pósfai M, Molnár Z, Ruiz-Agudo E, Gallego JLR (2022) Bioremediation of a polymetallic, arsenic-dominated reverse osmosis reject stream. Lett Appl Microbiol 75:1084–1092. https://doi.org/10.1111/lam.13578

Steinweg JM, Plante AF, Conant RT, Paul EA, Tanaka DL (2008) Patterns of substrate utilization during long-term incubations at different temperatures. Soil Biol Biochem 40:2722–2728. https://doi.org/10.1016/J.SOILBIO.2008.07.002

Sulistiyaning Titah H, Pratikno H, Moesriati A, Fauzul Imron M, Putera RI (2018) Isolation and screening of diesel degrading bacteria from ship dismantling facility at Tanjungjati, Madura, Indonesia. J Eng Technol Sci 50:99. https://doi.org/10.5614/j.eng.technol.sci.2018.50.1.7

Taboski MAS, Rand TG, Piórko A (2005) Lead and cadmium uptake in the marine fungi Corollospora lacera and Monodictys pelagica. FEMS Microbiol Ecol 53:445–453. https://doi.org/10.1016/J.FEMSEC.2005.02.009

Tamizhdurai P, Sakthipriya N, Sivagami K, Rajasekhar B, Nambi IM (2022) Field studies on monitoring the marine oil spill bioremediation site in Chennai. Process Saf Environ Prot 163:227–235. https://doi.org/10.1016/J.PSEP.2022.05.005

Tarla DN, Erickson LE, Hettiarachchi GM, Amadi SI, Galkaduwa M, Davis LC, Nurzhanova A, Pidlisnyuk V (2020) Phytoremediation and bioremediation of pesticide-contaminated soil. Appl Sci 2020(10):1217. https://doi.org/10.3390/APP10041217

Tekere M, Tekere M (2019) Microbial bioremediation and different bioreactors designs applied. Biotechnol Bioeng. https://doi.org/10.5772/INTECHOPEN.83661

Tsai YI, Yang HH, Wang LC, Huan JL, Young LH, Cheng MT, Chiang PC (2011) The influences of diesel particulate filter installation on air pollutant emissions for used vehicles. Aerosol Air Qual Res 11:578–583. https://doi.org/10.4209/AAQR.2011.05.0066

Vala AK (2010) Tolerance and removal of arsenic by a facultative marine fungus Aspergillus candidus. Bioresour Technol 101:2565–2567. https://doi.org/10.1016/J.BIORTECH.2009.11.084

Vidali M (2001) Bioremediation. An overview. Pure Appl Chem 73:1163–1172. https://doi.org/10.1351/PAC200173071163/MACHINEREADABLECITATION/RIS

Wang S, Wang J (2018) Degradation of emerging contaminants by acclimated activated sludge. Environ Technol 39:1985–1993. https://doi.org/10.1080/09593330.2017.1345989

Wang Z, Pei J, Zhang J (2013) Catalytic oxidization of indoor formaldehyde at room temperature – effect of operation conditions. Build Environ 65:49–57. https://doi.org/10.1016/J.BUILDENV.2013.03.007

Wang Q, Guo S, Ali M, Song X, Tang Z, Zhang Z, Zhang M, Luo Y (2022) Thermally enhanced bioremediation: a review of the fundamentals and applications in soil and groundwater remediation. J Hazard Mater 433:128749. https://doi.org/10.1016/J.JHAZMAT.2022.128749

Wei X, Lyu S, Yu Y, Wang Z, Liu H, Pan D, Chen J (2017) Phylloremediation of air pollutants: exploiting the potential of plant leaves and leaf-associated microbes. Front Plant Sci 8:270745. https://doi.org/10.3389/FPLS.2017.01318

Wen W, Wen C, Wang D, Zhu G, Yu J, Ling P, Xu M, Liu T (2024) A review on activated coke for removing flue gas pollutants (SO2, NOx, Hg0, and VOCs): preparation, activation, modification, and engineering applications. J Environ Chem Eng 12:111964. https://doi.org/10.1016/J.JECE.2024.111964

Wetzstein HG, Schneider J, Karl W (2012) Metabolite proving fungal cleavage of the aromatic core part of a fluoroquinolone antibiotic. AMB Express 2:1–7. https://doi.org/10.1186/2191-0855-2-3

Weyens N, Thijs S, Popek R, Witters N, Przybysz A, Espenshade J, Gawronska H, Vangronsveld J, Gawronski SW (2015) The role of plant–microbe interactions and their exploitation for phytoremediation of air pollutants. Int J Mol Sci 16:25576. https://doi.org/10.3390/IJMS161025576

Wróbel M, Śliwakowski W, Kowalczyk P, Kramkowski K, Dobrzyński J (2023) Bioremediation of heavy metals by the genus Bacillus. Int J Environ Res Public Health 20:4964. https://doi.org/10.3390/IJERPH20064964

Wu Z, Dijkstra P, Koch GW, Peñuelas J, Hungate BA (2011) Responses of terrestrial ecosystems to temperature and precipitation change: a meta-analysis of experimental manipulation. Glob Chang Biol 17:927–942. https://doi.org/10.1111/J.1365-2486.2010.02302.X

Yaashikaa PR, Saravanan A, Kumar PS, Thamarai P, Rangasamy G (2024) Role of microbial carbon capture cells in carbon sequestration and energy generation during wastewater treatment: a sustainable solution for cleaner environment. Int J Hydrog Energy 52:799–820. https://doi.org/10.1016/j.ijhydene.2023.05.307

Yang Y, Yang J, Jiang L (2016) Comment on "a bacterium that degrades and assimilates poly(ethylene terephthalate)". Science 353:759. https://doi.org/10.1126/science.aaf8305

Yoshikawa M, Zhang M, Toyota K (2017) Integrated anaerobic-aerobic biodegradation of multiple contaminants including chlorinated ethylenes, benzene, toluene, and dichloromethane. Water Air Soil Pollut 228:25. https://doi.org/10.1007/S11270-016-3216-1

Zak DR, Pregitzer KS, Burton AJ, Edwards IP, Kellner H (2011) Microbial responses to a changing environment: implications for the future functioning of terrestrial ecosystems. Fungal Ecol 4:386–395. https://doi.org/10.1016/J.FUNECO.2011.04.001

Zhang J, Gao D, Li Q, Zhao Y, Li L, Lin H, Bi Q, Zhao Y (2020) Biodegradation of polyethylene microplastic particles by the fungus Aspergillus flavus from the guts of wax moth Galleria mellonella. Sci Total Environ 704:135931. https://doi.org/10.1016/J.SCITOTENV.2019.135931

Zhang X, Li R, Song J, Ren Y, Luo X, Li Y, Li X, Li T, Wang X, Zhou Q (2021) Combined phyto-microbial-electrochemical system enhanced the removal of petroleum hydrocarbons from soil: a profundity remediation strategy. J Hazard Mater 420:126592. https://doi.org/10.1016/J.JHAZMAT.2021.126592

Zhou J, Xue K, Xie J, Deng Y, Wu L, Cheng X, Fei S, Deng S, He Z, Van Nostrand JD, Luo Y (2011) Microbial mediation of carbon-cycle feedbacks to climate warming. Nat Clim Chang 2(2):106–110. https://doi.org/10.1038/nclimate1331

Ecological Importance of Bacteria in Bioremediation of Contaminated Sites

Kamal A. M. Abo-Elyousr, Suleiman K. Bello, Hadeel M. Khalil Bagy, and Nashwa M. A. Sallam

Abstract Every day, pollution levels in the environment rise due to the increase in the number of people and industrialization. To abate the presence of contaminants in the environment, bacterial bioremediation is a low-cost process that removes contaminants by utilizing bacterial metabolism. The different types of bioremediation processes include bioaugmentation (introduction of a sustainable population artificially), biostimulation (promoting the survival of a viable native microbial population), biosorption (microbiological biomass that has died), bioaccumulation (live cells), and rhizoremediation (interaction between phytoremediation (plants) and microbes). In recent years, rapid advancements have helped humanity in its knowledge of the bioremediation process. The application of molecular techniques that are not dependent on culture has aided us in comprehending the mechanics and dynamics of microbial community structure in bioremediation. This has undoubtedly aided humanity in making bioremediation safer and more reliable.

Keywords Pollution · Bioremediation · Phytoremediation · Biostimulation · Bioaugmentation · Bioaccumulation · Biosorption · Rhizoremediation

---

K. A. M. Abo-Elyousr (✉)
King Abdulaziz University, Department of Agriculture, Faculty of Environmental Sciences, Jeddah, Saudi Arabia

University of Assiut, Faculty of Agriculture, Department of Plant Pathology, Assiut, Egypt
e-mail: kaaboelyousr@agr.au.edu.eg

S. K. Bello
King Abdulaziz University, Department of Agriculture, Faculty of Environmental Sciences, Jeddah, Saudi Arabia

Department of Soil Science, Faculty of Agriculture/Institute for Agricultural Research, Ahmadu Bello University, Zaria, Kaduna State, Nigeria
e-mail: bellosuleimank@gmail.com

H. M. Khalil Bagy · N. M. A. Sallam
University of Assiut, Faculty of Agriculture, Department of Plant Pathology, Assiut, Egypt
e-mail: hadel_magdy@aun.edu.eg; nashwasallam@aun.edu.eg

A. A. H. Abdel Latef et al. (eds.), *Sustainable Remediation for Pollution and Climate Resilience*, https://doi.org/10.1007/978-981-96-5674-5_14

1 Introduction

Pollution is a significant issue worldwide. Pollution is of two types: pollution with organic compounds and pollution with inorganic compounds. Both types of pollution affect the health of all living organisms in the ecosystem. Chemical and physical processes of removing pollutants, such as sedimentation, ion exchange, electrochemical treatment, evaporation, osmosis, and absorption, are not suitable for the environment (Mulligan et al. 2001; Kadirvelu et al. 2002). The biological treatment technique is also based on its high efficiency in removing heavy metals from contaminated locations. Bacteria can be considered a biological tool to remove heavy metals from contaminated aquatic ecosystems (Riggle and Kumamoto 2000). Therefore, as an alternative to typical treatment methods, microorganisms are used to absorb heavy metals in polluted water (Tsezos and Volesky 1981; Gadd and White 1993; Texier et al. 1999). Bacterial remediation of pollutants is advantageous because bacteria act on contaminants even when they are found in highly diluted solutions, and they can also adapt to harsh environments, and their choice is economically feasible.

Contaminated sites greatly threaten the ecosystem, including the living and nonliving things in the environment. Due to the hazardous nature of contaminated sites, safe and eco-friendly methods of remediating contaminated sites must be deeply researched and developed for wide use. Organic compounds, herbicides, and heavy metals rank foremost on the list of contaminants that could cause serious damage to the environment and human health. Apart from the environmental consequences of contaminants, they also cause a great setback in the attainment of global food security as they render a lot of arable land uncultivable. Thus, it is highly important to find environmentally friendly ways of removing or eliminating these contaminants from our environment. Biotechnological strategies using bacteria to remediate contaminated sites are gaining wide attention globally. Generally, bioremediation is achieved through microorganisms' metabolic activities, which are regulated via enzyme activities (Pileggi et al. 2020).

Bioremediation is the ecological use of microbes, including bacteria, to decontaminate, decrease, or eliminate contaminants from the ecosystem (Vidali 2001). Bioremediation of contaminated sites using bacteria can eliminate contaminants, pollutants, and toxins from soil, water, and the environment at large. Bioremediation relies on natural processes, and it is a relatively green method that minimizes damage to the ecosystem compared to other methods of removing contaminants from the environment. Most often, contaminated sites are remediated by excavation and transportation to a landfill, far away from human activities and population, to contain the contaminants to a particular point and avoid the spread of the contaminants. However, this method is costly, and the site for final disposal of the contaminants is sometimes challenging to find. Rather than excavating and transporting the contaminants, a safer approach is to decontaminate the affected site in situ using bioremediation techniques. Organic pollutants include polychlorinated biphenyls (4,4-dichloroethylenel and 4-chloroethylenel), chlorinated solvents (perchloroethylene and trichloroethylene (TCE)), BTEX (xylene, toluene, ethylbenzene, and benzene), chlorinated phenol (pentachlorophenol), polyaromaticatrazine, and carbaryl (Kang 2014). Some of these organic compounds are very dangerous to human health, and they pose health hazards all over the world (Kang 2014). Inorganic pollutants include cadmium, copper, chromium, arsenic, nickel, mercury, and lead. Furthermore, these heavy metals are unable to biodegrade in the environment. Many varieties of bioremediation approaches are applied to remove heavy metals from industrial wastewater (Fomina and Gadd 2014). Techniques such as chemical, physical, and biological processes have been deployed in removing contaminants from the environment. Humans produce the bulk of heavy metal pollution, and it accumulates over time. Bioremediators are bacteria that perform the function of bioremediation (Terra Nova's Environmental Remediation Resources 2009). Bioremediation is a technique that can be used on contamination that is to be removed from the original site (in situ) or based on contaminants that have been taken down from the original site (ex situ). When it comes to contaminated soil, we must plow the land to provide more nutrients and oxygen for microorganisms. It may require land tilling to increase the availability of nutrients and oxygen to microorganisms. As a result, it employs low-cost techniques that are widely accepted and may often be deployed on-site. Through a full site assessment, significant experience may be necessary to plan and conduct an effective bioremediation program.

In the bioremediation of polluted sites, bacteria can perform the following tasks: (1) Microbial detoxification and the exclusion of heavy metal ions. (2) Bacteria destroy heavy metals by changing their chemical properties through their natural ability to absorb biologically toxic heavy metal ions. (3) Bacteria can absorb heavy metals at cellular binding sites (Dixit et al. 2015). Bacteria can change heavy metal oxidation states in extensively contaminated soil by immobilizing heavy metals (Dixit et al. 2015), allowing their easy removal. Bacterial bioremediation strategies may rely on bacterial consortia present at contaminated sites that have a tremendous metabolic capacity.

2 Bioremediation: Features, Strategies, and Applications

Features

The phrase "bioremediation" refers to the biological process of eliminating contaminants from the environment. In the environment, bioremediation is a technique that uses biological methods that act as active biological inhibitors of pollutants. It is a natural process that takes some time to treat polluted materials such as soil. In this regard, bacteria used in bioremediation are highly specialized and can mineralize organic and inorganic pollutants and biodegrade them into biologically reduced pollutants. Bioremediation is a process that uses microorganisms or their enzymes to remove or neutralize contaminants in the environment, converting them into less harmful forms. Many harmful substances in the environment are converted into harmless products, and no hazardous chemicals should be used in biological treatment. Fertilizers with nutrients added to help bacteria grow better are routinely used on lawns and gardens. This is because bioremediation converts hazardous chemicals into harmless water and gases, where they are eliminated.

Strategies
Biosparging, bioventing, phytoremediation, landfarming, bioleaching, bioreactor, composting, bioaugmentation, biopiling, rhizofiltration, and biostimulation are some examples of bioremediation technologies that can be deployed in situ or ex situ. Brief descriptions of some of these technologies are given below.

2.1 *Bioremediation (In Situ)*

In situ bioremediation is the application of biological treatment to remove harmful compounds from the subsurface at the contamination site. The optimization and control of microbial transformations of organic contaminants in situ necessitate the combination of numerous scientific and engineering fields. Some applicable strategies under in situ bioremediation include the following:

1. Biosparging

   Biosparging is a technique for increasing oxygen levels in groundwater, which increases the rate at which bacteria degrade contaminants. Biosparging is simple and inexpensive.
2. Bioventing

   Bioventing is a new technology for promoting natural in situ biodegradation of any biodegradable molecule in the air and inside the soil by supplying oxygen to soil bacteria. The most common way of delivering oxygen is through direct air injection into pollutants in the soil. This technology is used to biodegrade fuel wastes and volatile compounds (Lee et al. 2006). Plants can also be used to deliver contaminants degrading bacteria into contaminated soils during bioventing.
3. Bioaugmentation

   Bioaugmentation is the use of a variety of consortia of natural microbial strains or genetically modified bacteria to remediate contaminated soil or water. It is extensively used in the treatment of wastewater.
4. Biopiling

   Biopile treatment is a large-scale process that generates energy from excavated soils. Forced aeration is used in this type of bioremediation in addition to blending soil additives into the soil pile before treatment. The basic biopile system consists of a treatment bed, an aeration system, an irrigation/nutrient system, and a leach-away collection system. Biodegradation is aided by moisture, heat, nutrients, oxygen, and pH levels. The irrigation/nutrient system is buried beneath the soil and uses vacuum or positive pressure to move air and nutrients. Soil mounds can be up to 20 feet high and covered with plastic to prevent runoff, evaporation, and volatilization while also increasing solar heating. If volatile organic compounds (VOCs) in the soil volatilize into the air, the air leaving the soil could be treated to remove or eliminate the VOCs before they reach the atmosphere. Treatment lasts 3–6 months on average.

2.2 Bioremediation (Ex Situ)

This is the bioremediation of contaminants away from where the contamination occurred. Some applicable strategies under in situ bioremediation include:

1. Composting

   Composting is the decomposition of organic wastes by microbes at high temperatures. The compost is kept at a temperature of 55–65 °C. Temperatures rise due to heat created by microbes during the decomposition of organic material in garbage. The following basic steps have been used to demonstrate windrow composting: First, to remove huge boulders and debris from polluted soils, they are dug and screened (Antizar-Ladislao et al. 2007, 2008).
2. Bioreactors

   Slurry reactors or aqueous reactors are used to treat soil and wastewater off-site. In reactors, bioremediation entails treating contaminated particles such as sediment, soil, or wastewater using a specially designed containment system. A slurry bioreactor is a vessel and system for containing slurry that produces a three-phase (solid, liquid, and gas) mixing condition to boost the bioremediation rate of soil-bound and water-soluble pollutants as a water slurry of contaminated soil and biomass (typically indigenous microorganisms) capable of decomposing target contaminants as a water slurry of contaminated soil and biomass (usually native microorganisms) capable of degrading the target.

Applications

1. In Situ

   This is capable of native microorganism biodegradation, metal and inorganic compound decomposition, and pollutant biodegradation (Sei et al. 2001; Niu et al. 2009).
2. Ex Situ

   To make plants healthier, surface application, aerobic process, and application of organic materials to natural soils, followed by irrigation and tilling; surface application, agricultural to urban garbage, is viable and convenient as a substitute for landfilling or incineration (Antizar-Ladislao et al. 2007, 2008).
3. Microfiltration

   More than 90% of the original wastewater is treated, recovered, and reused.
4. Electrodialysis

   Effective removal of dissolved solids using cation and anion exchange membrane pairs.

3 Factors Limiting Bioremediation Technologies

Bioremediation optimization is a complex, multi-factored system. Biological remediation is affected by various factors, including the presence of a microbial community capable of breaking down contaminants, environmental factors, and the action of bacteria in detoxifying various physical and chemical pollutants from the environment. Bacteria act as biocatalysts through the enzymatic pathways triggered by the bacteria, thus limiting contamination. Bacteria have access to a wide range of resources, chemicals, and compounds that enable them to have access to energy and nutrients for the construction of additional cells, which can allow bacteria to play a significant role in battling pollution. The effectiveness of bioremediation is determined by a variety of physical and chemical parameters, including pollutant concentrations, environmental physical and chemical qualities, and microbe availability (El Fantroussi and Agathos 2005).

3.1 Environmental Factors

1. Nutrients:

   Carbon is the most abundant element in living things, outnumbering other elements like nitrogen, oxygen, and hydrogen. The addition of nutrients balances out the key nutrients required for microbial proliferation while also speeding up and improving biodegradation. To boost the rate of biodegradation in the environment, microorganisms require a variety of nutrients such as nitrogen, carbon, and phosphorus to increase their metabolic activity (Couto et al. 2014). It was also shown that the availability of nutrients constrains wastewater biodegradation. Oil-eating bacteria also need nutrients to thrive and develop properly.
2. Available Soil Moisture:

   Bacteria need enough water to grow. For microbial activity to occur, the ratio must not range out from 25% to 28% of the soil water storage capacity.
3. Oxygen:

   Bacteria require oxygen to facilitate biodegradation more efficiently (Macaulay 2014).
4. pH:

   pH influences bacteria's metabolic activity and can boost or reduce biological processing (Asira 2013).
5. Temperature:

   Temperature is one of the most important factors to consider. Das and Chandran (2011) identified elements that influence microbe survival and hydrocarbon production. Temperature depends on how quickly biodegradation occurs in the vast majority of aquatic and terrestrial systems. In the Arctic, oil degradation through biodegradation processes has been found to occur at a slower rate and puts bacteria under greater pressure to clean up spilled oil. When the water temperature is below zero in the Arctic, the entire cytoplasm of the bacteria freezes, making the metabolism of most oleic bacteria inactive (Macaulay 2014; Yang et al. 2009).

6. Contaminants:

   Optimum conditions: hydrocarbons account for about 15% of the dry weight of the soil. Not too toxic conditions are required for microbial activity.

7. Heavy Metals:

   It was found that heavy metals are important for bacteria, but in small quantities, and when these heavy metals are present in large quantities, they inhibit the metabolic activity of bacteria. These heavy metals affect wastewater's biological degradation rate in both direct and indirect ways.

8. Site Characterization and Selection:

   Contaminated sites must be selected and identified to carry out biodegradation processes and determine the contamination's size and extent. Site characterization and selection include several important points, such as outlining the horizontal and vertical range of contamination sites, publishing enough information about the sampled sites and the rationale for their selection, and explaining the sample acquisition and analysis procedures.

9. Toxic Compounds:

   Hazardous chemicals found in high concentrations in some contaminants can have toxic effects on microorganisms, delaying disinfection and biological treatment processes. Poisonous chemicals vary in their degree and mechanism of toxicity according to the toxic substances, their concentration, and the microorganisms.

3.2 Natural Biological Processes of Microorganisms

Biological factors influence the degradation of organic and inorganic waste materials via competition for carbon sources among microbes. The contamination rate depends on the pollutant quantity and the number of stimuli present among the microorganisms. The number of these catalysts corresponds to the total number of organisms with the ability to metabolize contaminants. In addition, the number of enzymes each bacterial cell produces is an important factor. These bacterial cell enzymes can either speed up or slow down the breakdown of pollutants. Here, enzyme activity, gene transfer, mutation, inter-organism interaction, and bacterial growth until they achieve critical biomass are all considered biological factors (Madhavi and Mohini 2012; Boopathy 2000). Bacteria can be isolated from any type of environment. These bacteria can be divided into two groups:

1. Aerobic Bacteria:

   They thrive in an oxygen-rich environment. *Alcaligenes*, *Pseudomonas*, *Rhodococcus*, *Mycobacterium*, and *Sphingomonas* are the most well-known aerobic bacteria with the ability to digest organic and inorganic compounds as well as heavy metals. These microorganisms degrade pesticides and hydrocarbons. Many of these bacteria rely only on pollution for carbon and energy.

2. Anaerobic Bacteria:

   They can survive in the absence of oxygen. Anaerobic bacteria are not used nearly as often as aerobic bacteria. The employment of anaerobic bacteria in the bioremediation of polychlorinated biphenyls (PCBs) in river sediments, the dechlorination of the solvent trichloroethylene (TCE), and the dechlorination of chloroform is gaining popularity. They are aerobic bacteria that evolve by consuming methane as a source of carbon and energy. Monooxygenated methane and chlorinated aliphatics, such as trichloroethylene and 1,2-dichloroethane, are the most common examples.

4 Mechanisms of Bacterial in Bioremediation

Bacteria have evolved several strategies to eradicate destructive substances from the environment. The cytoplasmic membrane of bacterial cells is the first point of interaction between a cell and a contaminant. The most critical factor determining cell viability is membrane fluidity. Depending on the external conditions, bacterial cells can vary the fluidity of this surface component. This can be achieved by changing fatty acids' structure, chain length, and phospholipid synthesis (Jurkiewicz et al. 2012). Changes in the cytoplasmic membrane structure are important for adjusting to high levels of hazardous pollutants. The most essential responses of bacteria toward membrane-active chemicals are modifications in the fatty acid content of cell membrane lipids. Most adaptation mechanisms have only been characterized for Gram-negative (G-) aerobic bacteria (Pepi et al. 2008; Zorádová et al. 2011), although some studies report on anaerobic (Duldhardt et al. 2010; Murínová and Dercová 2014) and Gram-positive (G+) aerobic bacteria (Fernandes et al. 2011; Torres et al. 2011). Duldhardt et al. (2010) found that anaerobic bacteria were three times more sensitive to various organic chemicals than aerobic bacteria. This is because anaerobic bacteria develop at a slower pace than aerobic bacteria.

4.1 Modifications in Fatty Acid Structure

The most adaptable strategies are preservation of cell membrane flexibility and lipid-phase consistency (Weber et al. 1994). Fluidity is a significant aspect of the cell membrane structure, characterized as the reverse value of its viscosity. It may be affected by changes in the fatty acids that make up the membrane phospholipids (Katarína et al. 2018). Contaminants containing hydrophobic organics alter the flexibility of bacterial membranes, causing physiological instability and mortality. That is why membrane elasticity is so important to a cell's survival (Čertík et al. 2003; Denich et al. 2003). Bacterial cell membrane mitigates this impact by phospholipids, which stack fatty acid alkyl chains to increase membrane rigidity and avoid solvent buildup (Šajbidor 1997; Murínová and Dercová 2014). According to earlier studies, aromatic chemicals like benzene, biphenyl, phenol, PCBs, and toluene have been observed to increase in the phospholipid bilayer between the acyl bonds of fatty acids (Murínová and Dercová 2014). Increased saturated membrane lipids are one of the important processes of bacterial cell membranes. The closely packed linear acyl chains of saturated fatty acids reduce fluidity and combat the fluidizing effects of hazardous chemical substances (Duldhardt et al. 2010). A similar process was found in *Pseudomonas stutzeri* in the presence of naphthalene. Even though bacterial cells try to improve membrane stiffness to oppose the impacts of pollutants, the membrane should still be responsible for executing its physiological activities. Hence, a proportion of the membrane should be in the fluid-crystalline case.

The energy-based de novo generation of saturated fatty acids enables bacteria to enhance their saturation levels, which could explain why changes in saturation levels were only detected in developing cells (Heipieper and de Bont 1994). Also, the strategy for increasing saturation levels has limitations because of the saturated fatty acid composition of the environment. In *Pseudomonas putida* P8, a relationship was discovered between higher fatty acid saturation of the membrane and enhanced resistance to hazardous chemicals (Heipieper et al. 1992; Murínová and Dercová 2014). This is regarded to be one of the most important long-term adaptation processes in microorganisms that have been exposed to hazardous aromatic chemicals. As a result of this, bacterial membranes are becoming more immune to the fluidizing effects of aromatic chemicals, allowing the cells to exist in hydrocarbon-contaminated environments (Mrozik et al. 2010).

4.2 *Enzymatic Characterization*

Bioremediation relies on microorganisms that destroy contaminants enzymatically and transform them into harmless compounds (Karigar and Rao 2011). In certain situations, several enzymatic outputs have the potential to be substantially more harmful than the original pollutant. Enzymes that are important in the degradation of various pollutants can be targeted in this approach.

4.2.1 Hydrolytic Enzymes

Proteases, esterases, lipases, phosphatases, carbohydrates (amylases, cellulases, xylanases, etc.), and phytases are among the enzymes that belong to this category. These enzymes have weak substrate specificity, which is important in the degradation of various pollutants, including unsolvable wastes, oil spills, cellulose substances, organophosphate and carbamate pesticides, kraft pulp, chitin, keratin, starch materials, wastewater sludge, polyurethane, and polyacrylate (Gianfreda and

Rao 2004). The toxicity of poisonous compounds is reduced as a result of the destruction of chemical bonds by these enzymes (Karigar and Rao 2011).

4.2.2 Oxidoreductase and Dehydrogenase

These enzymes play a role in microbial metabolism (both catabolic and anabolic processes) by converting pairs of electrons from a substrate to nicotinamide adenine dinucleotide phosphate (NADP) or nicotinamide adenine dinucleotide (NAD+), as well as renewing electron transporters during anabolism. Various studies have demonstrated that oxidoreductase enzymes are essential in removing excessive hazardous metals such as copper and chromium (Irazusta et al. 2018). Investigations using cell-free mitochondrial extracts and extracellular yeast extracts from *Cyberlindnera jadinii* (M9) and *Wickerhamomyces anomalus* (M10) demonstrated that Cr (VI) elimination is controlled by intracellular types of enzymes such as oxidoreductase and dehydrogenase.

4.2.3 Laccases

Laccases can be found in a wide range of bacteria, such as *Escherichia coli*, *Pseudomonas syringae*, *Mycobacterium tuberculosis*, *Bacillus subtilis*, *Yersinia pestis*, *Caulobacter crescentus*, *Pseudomonas aeruginosa*, and *Bordetella campestris* (Bains et al. 2003). It is an extracellular multi-copper oxidase isozyme with a broad substrate selectivity used for bioremediation of various industrial pollutants. Karigar and Rao (2011) noted that these enzymes are a kind of polyphenol oxidases, ligninolytic enzymes that use molecular oxygen as an electron acceptor to catalyze the oxidation of different aromatic compounds, notably those with electron-contributing groups, including anilines ($-NH_2$) and phenols ($-OH$). Due to their relatively low specificity, these enzymes are suitable for bioremediation of wastes from the paper, pulp, textile, and distillery sectors as well as industrial contaminants and effluents. In addition, these enzymes, specifically in soils, in association with peroxidases, improve natural xenobiotic and organochloride decomposition, oxidation of pentachlorophenol and aromatic hydrocarbons, and degradation of chemical pesticides like bentazon and diuron, among other things (Duran and Esposito 2000; Chandra and Chowdhary 2015).

4.2.4 Peroxidases

Peroxidases work on a wide range of organic and inorganic substances and are identified as oxidoreductases. They can activate processes in the existence of hydrogen peroxide (Duran and Esposito 2000). They can be used to decompose pollutants such as textile dye regression, sewage treatment, and lignin degeneration (Bansal and Kanwar 2013). Three essential peroxidases, versatile peroxidase (VP), lignin peroxidase (LiP), and manganese-dependent peroxidase (MnP), have been investigated in terms of bioremediation importance, owing to their strong potential to destroy harmful compounds in nature (Karigar and Rao 2011).

4.3 Biosurfactant Production

Microorganisms with significant surface and emulsifying properties produced biosurfactants known as amphiphilic chemicals (Singh et al. 2007). Pacwa-Płociniczak et al. (2011) mentioned that biosurfactants' hydrophilic and hydrophobic groups can raise the surface area of unsolvable materials. Their value in bioremediation derives from the fact that a contaminant's biodegeneration is determined by its bioavailability to metabolizing organisms, and they can improve the water bioavailability of such compounds as well as alter the characteristics of bacterial membranes (Pacwa-Płociniczak et al. 2011). Ahmad Khan et al. (2015) mentioned that biosurfactants perform two basic effects in this regard: on the one hand, they enhance substrate availability, and on the other hand, they improve biodegeneration by solubilizing, mobilizing, and emulsifying the substrate. These compounds are able to decompose hydrocarbons (polynuclear aromatic hydrocarbons) and heavy metals (Rangarajan and Narayanan 2018).

4.4 Siderophore Synthesis

These substances are beneficial for heavy metal purification because they create a stabilized synthesis of metal-ligand and affect the movement of metal in the environment (Naik and Dubey 2013).

5 Bacterial Degradation of Fuels

Hydrocarbons are considered one of the most harmful pollutants worldwide. Bioremediation of a hydrocarbon-polluted environment by natural microorganisms is considered one of the cheapest pollutant removal technologies; nevertheless, it might take a long time, practically in hard situations, like non-suitable average temperatures (Horel and Schiewer 2020). They have emerged as a viable technique for dealing with environmental pollution and are environmentally friendly (Guerra et al. 2018).

Bacteria have been evaluated and used to decompose waste materials from food, agricultural, chemical, and medicinal origins. In addition, they can use organic compounds from the fuel spill as a source of energy by decomposing hydrocarbons, creating CO_2 and water (Kleindienst et al. 2015).

A hydrocarbon degradation process (e.g., polyaromatics and aliphatics) has been proven to utilize oxidizing mechanisms; nevertheless, these mechanisms differ substantially due to the diverse oxygenases in different bacterial species. Certain bacteria can destroy aromatic or resin portions of hydrocarbons, whereas others can degrade specific alkanes. About 79 genera of bacteria have been recognized with the ability to degrade hydrocarbons (Tremblay et al. 2017), such as *Burkholderia*, *Marinobacter*, *Achromobacter*, *Pandoraea*, *Acinetobacter*, *Alkanindiges*, *Arthrobacter*, *Kocuria*, *Mycobacterium*, *Alteromonas*, *Rhodococcus*, *Pseudomonas*, *Dietzia*, *Staphylococcus*, and *Enterobacter* (Varjani and Upasani 2016; Sarkar et al. 2017; Xu et al. 2017). Fuentes et al. (2015) reported that in soil, bacteria *Alkanindiges* sp. displayed scarce-to-predominant due to the influence of environmental constraints like petroleum pollution. Likewise, several hydrocarbonoclastic bacteria that are obligatory, such as *Oleispira*, *Marinobacter*, *Alcanivorax*, *Thallassolituus*, and *Cycloclasticus*, had low frequency or were invisible before the contamination but became dominant following petrol oil pollution (Yakimov et al. 2007). These findings indicate that these microorganisms are important in the decomposition of hydrocarbons and considerably impact their conversion and destiny in the environment (Xu et al. 2018).

The secretion and activity of hydrocarbon-decomposing bacterial enzymes are directly tied to the physiological function of the bacterium (Mukherjee et al. 2017; Song et al. 2017). Wang et al. (2011) found that certain bacteria can be responsible for breaking down a wide spectrum of petroleum, such as *Dietzia* sp. (DQ12-45-1b). It utilizes n-alkanes (C6–C40) and other compounds as carbon sources. The bacteria *Achromobacter xylosoxidans* (DN002) is effective against various mono- and polyaromatic hydrocarbons (Ma et al. 2015). Almost no microorganism is capable of destroying the hydrocarbon part of petroleum. This is due to the reality that various regional bacteria contain diverse catalytic enzymes; therefore, their breakdown techniques in oil-polluted areas differ drastically. To get the optimum environmental purity impact, the cleanup of petroleum hydrocarbon pollution necessitates the cooperation of various functional bacteria (Dombrowski et al. 2016). Bacterial isolates such as *Ochrobactrum* sp., *Stenotrophomonas maltophilia*, and *Pseudomonas aeruginosa* were used to break down crude oil, with a decomposition percentage of up to 83.49% (Varjani et al. 2015).

The bioremediation rate for most petroleum hydrocarbons is often restricted in the environment due to their hydrophobicity and poor water solubility. The major phase in the breakdown of crude oil frequently involves bacterial cell membrane-bound oxygenases, which require effective association among bacterial membranes and hydrocarbon sources (Xu et al. 2018). Bacteria can develop anti-petrochemical defenses, including changing cell membrane components to improve adherence and secreting bioemulsifiers to improve reaching the target hydrocarbon substrates (Zhang et al. 2015). Hydrophobic fimbriae, exterior proteins, fibrils, lipids, gramicidin S, and prodigiosin are the primary mediators of bacterial cell adhesion with hydrophobic pollutants (Ron and Rosenberg 2014). Bacteria exhibiting these characteristics are frequently evaluated for utilization as bioremediation agents, which can help to speed up the cleanup of hydrocarbon contaminants from the environment (Krasowska and Sigler 2014).

6 Role of Bacteria in Agricultural Application

Due to intensive anthropogenic activities, there is a continuous buildup of inorganic and organic pollutants in the environment that needs to be mitigated before they cause more danger to both the living and nonliving things in the ecosystem. Regarding pollutant degradation, ecosystem sustainability depends on the critical roles that bacteria and other microbes play in the decontamination of pollutants. Since bacteria are ubiquitous in the ecosystem, they have developed the capability of adapting swiftly to environmental changes. Thus, several investigations have exploited their genetic composition, phenotypic expression, biochemical synthesis, and environmental proliferation to improve agricultural production and alleviate the impact of contaminants in the ecosystem.

6.1 *Bioremediation of Organic Compounds in the Soil*

Organic wastes from intensive agricultural and industrial activities that are toxic and recalcitrant in the environment are called persistent organic pollutants (POPs). Some of the chemicals that have been tagged as POPs in the Stockholm Convention include toxaphene, mirex, dieldrin, dichlorodiphenyltrichloroethane (DDT), chlordane, aldrin, heptachlor, hexachlorobenzenes (HCBs), endrin, furans, polychlorinated biphenyls (PCBs) and dioxins (Mondal et al. 2019). To alleviate the effect of organic contaminants on agriculture and the environment, bacteria can transform chemical contaminants into less or nontoxic substances. Organic pollutants like hydrocarbons can be transformed by bioremediation into compounds that are less toxic or harmful to the environment. Through the production of catalytic enzymes, bacteria can biodegrade organic contaminants into non-harmful compounds (Orji et al. 2012). To remediate polluted sites, indigenous bacteria adapted to such contaminated sites are isolated, grown for higher efficiency, and reintroduced to the contaminated sites to effect bioremediation. Isolation of indigenous bacteria for bioremediation is more effective than introducing bacteria.

6.2 *Bioremediation of Herbicides*

Herbicides are chemical compounds or agents that are used to control or eliminate weeds or unwanted plants in the environment, mostly on agricultural fields to increase crop production. Currently, there are about 261 herbicides in existence globally (www.hracglobal.com). Examples of herbicides that are used widely include diazinon, metolachlor, butachlor, simazine, azimsulfron, atrazine, paraquat, 2,4-dichlorophenoxy acetic acid (2,4-D), trifluralin, dicamba, imazethapyr, glyphosate, alachlor, and acetochlor (Beffa et al. 2019; Maggi et al. 2019; Pileggi et al. 2020). Once applied to the target plants, herbicides eliminate weeds by obstructing their metabolism, e.g., synthesis of lipids, carotenoids, amino acids, and photosynthetic pathways (Pileggi et al. 2020). However, herbicide application may also negatively affect nontarget plants and contaminate soil, groundwater, and the entire ecosystem (Pileggi et al. 2020). Herbicides such as atrazine and phenylurea have low biodegradability, high persistence in the environment, and are easily leached to the groundwater if improperly applied or disposed of. With excessive application, herbicide residues may also get into the human system through the consumption of crops sprayed with herbicides during cultivation. Several studies have demonstrated that the metabolites derived from the breakdown of herbicides could contaminate groundwater, surface and drinking water, watercourses, plants, and farmers' urine (Li et al. 2018a, b; Rendón-von Osten and Dzul-Caamal 2017; Székács et al. 2015). Thus, there is an urgent need to find ways of getting rid of herbicide metabolites from the ecosystem.

Factors that determine the degradability of herbicides at the site of application include their efficacy rate, leaching potential, and their carry-over effect on subsequent crops (Kanissery and Sims 2011). Soil microorganisms play significant roles in the fate of any chemical compound applied to the soil; however, the biodegradability of herbicides by bacteria has been explored less. Likewise, herbicides also negatively affect microorganisms in the environment, thus hindering them from carrying out their ecological functions and services. One of the most significant effects of herbicides is the production of reactive oxygen species (ROS), which greatly disrupts the survivability of soil microbes (Pileggi et al. 2020). However, in adapting to the continuous application of herbicides, several soil microbes have acquired or developed genes that enable them to degrade herbicides and sometimes use herbicide degradation byproducts as nourishment (Pileggi et al. 2020). Moreover, despite the potential abundance of indigenous microorganisms in contaminated sites, factors such as the presence of bioavailable phosphorus (P) and nitrogen (N), electron donors and acceptors, and the upregulation of toxic substances degradation metabolic pathways affect the biodegradation of herbicides by microorganisms (Kanissery and Sims 2011).

Microbial activities have been shown to enhance the degradation of different herbicides, as evident from increased microbial biomass, leading to higher availability of N, P, and carbon (Barman and Das 2015). *Cupriavidus pinatubonensis* AEO106 biodegraded phenoxy acid by upregulating the peroxiredoxins protein family, which included osmotically inducible protein C (OsmC) and organic hydroperoxide resistance protein (Ohr) (Svenningsen et al. 2017). Through metabolic activities, *Sphingobium* sp. MEA3-1 can degrade chloroacetamide into simpler chemical compounds that can be utilized as a source of nourishment (Dong et al. 2015). By sharing common metabolic pathways and biochemical mechanisms involved in herbicide degradation, the simultaneous use of groups of bacteria can potentially improve the efficiency of herbicide bioremediation. A consortium of *Advenella* sp., *Variovorax soli*, and *Anthrobacter sulfonivorans* has also been demonstrated to degrade diuron in contaminated soils (Villaverde et al. 2017). Likewise, a consortium of bacteria in different genera, including *Variovorax*, *Rhodococcus*, *Pseudomonas*, and *Bacillus*, degraded azimsulfron into simpler chemical compounds that were utilized as sources of nourishment for their growth and survival (Valle et al. 2006). Other herbicides that consortia of bacteria have degraded include thiobencarb, glyphosate, and phenylurea (Sviridov et al. 2015; Chu et al. 2017; Albers et al. 2018).

Bacteria can also cope with herbicide residues by producing appropriate enzymes that confer tolerance to oxidative stress induced by the presence of herbicides in their system. For example, through genetic mutations in superoxide dismutase (SOD) genes, *Escherichia coli* adapted to oxidative stress induced by paraquat application (Gravina et al. 2017). On application of mesotrione, the production of catalase enzymes by *Pantoea ananatis* allowed this bacterium to tolerate the induced chemical toxicity by enhancing the production of glutathione-s-transferase complexes with mesotrione and increasing saturation of lipid membranes (Prione et al. 2016). Evolved changes in the saturation of lipid membranes can potentially prevent the herbicide from permeating through the membrane. Further investigations revealed that *Pantoea ananatis* can degrade mesotrione into a supplementary energy source in the presence of glucose (Pileggi et al. 2012).

6.3 Bioremediation of Heavy Metals Using Bacteria

Contaminated sites are mostly ladened with heavy metals such as arsenic (As), lead (Pb), zinc (Zn), chromium (Cr), copper (Cu), nickel (Ni), mercury (Hg), vanadium (V), and cadmium (Cd), which are nonbiodegradable, toxic, and persistent in the environment (Gupta and Diwan 2017). Environmental pollution caused by heavy metals has been classified as one of the major problems affecting the quality of life and the ecosystem on a global scale due to their high persistence level (Beiyuan et al. 2020). From time immemorial, various dead and living organisms have been implicated in the remediation of toxic elements and substances from the environment. Among all organisms living in the ecosystem, microorganisms, including bacteria, are key players in the alleviation of heavy metal contamination in the environment (Table 1).

Heavy metals can be taken up by bacteria passively or actively. Passive uptake involves the biosorption of metals on the cell membrane and is independent of metabolic activities. The active uptake involves the transport of metals across the cell membrane and requires metabolic activities (Malik 2004; Kumar and Oommen 2012). Some of the effective strategies to remediate heavy metals from contaminated sites are those that reduce their bioavailability, toxicity, and mobility (Mohammed et al. 2011). Some of the specific mechanisms by which bacteria bioremediate heavy metals include (i) intracellular accretion that renders heavy metals

Table 1 Various bacterial strains used for bioremediation of heavy metals

Accession no.	Strains	Characteristics	References
DQ205432	*Proteus vulgaris* 16S rRNA	Pb and Cd resistant	Rani et al. (2008)
AY970345	*Bacillus cereus* 16S rRNA	Arsenic resistant	Satlewal et al. (2010)
EU512943	*Pseudomonas monteilli*	Cd resistant	Rani et al. (2009)
EU512944	*Pseudomonas putida*	Cd resistant	Rani et al. (2009)
EF207716	*Comamonas aquatica*	Cd resistant	Rani et al. (2009)

redundant through microbial metabolism, (ii) extracellular accretion or precipitation of heavy metals by living bacterial cells, and (iii) adsorption or complexation of heavy metals on the membranes of living and dead bacterial cells (Mohammed et al. 2011).

One of the major causes of heavy metal toxicity in the environment is ion imbalance, which could also lead to oxidative stress. Viable bacteria cells can chemically reduce heavy metal ions to nontoxic forms or use the ions in anaerobic respiration as terminal electron acceptors (Mohammed et al. 2011). *Synechocystis* sp., a cyanobacterium, has been demonstrated to tolerate arsenate (AsO_4) and arsenite (AsO_3) toxicity by evolving different stress responses, which included the downregulation of genes regulating growth and photosynthesis, induction of chaperones, and upregulation of redox scavenging systems (Sánchez-Riego et al. 2014). In *Cupriavidus pinatubonensis*, there was an upregulation of proteins regulating responses to high ROS concentration, which was induced as a consequence of exposure to moderate copper concentrations (Svenningsen et al. 2017).

Overall, the ability of bacteria to tolerate and bioremediate heavy metal contaminants is dependent on their genetic constituents, such as transposons, which can transfer genetic materials from one part of the bacterial cell to the other (Mazhar et al. 2020). Some highly heavy metal-resistant bacterial strains usually have heavy metal efflux proteins, resistance pumps, and cation diffusion facilitators in their genetic makeup, allowing them enough time to survive in contaminated sites while developing a detoxification mechanism (Mazhar et al. 2020). Furthermore, introducing inorganic nutrients such as phosphorus, nitrogen, and organic carbon to heavy metal-contaminated soils increases microbial metabolism, which could lead to a higher bioremediation rate (Xu et al. 2021).

7 Bioremediation of Polluted Water Using Bacteria

Pollution of water by organic and inorganic toxic substances is a major problem worldwide and deserves a wide array of attention from everyone on Earth. Organic toxic substances that can cause water pollution include hydrocarbons, oils, pesticides, dyes, herbicides, and POPs, while inorganic water pollutants are majorly heavy metals. In some cases, a heavy load of harmful microorganisms in water bodies can also lead to water pollution. The main objectives in polluted water remediation are the removal or reduction of excess inorganic nutrients such as nitrate, nitrite, ammonium, and phosphate; chemical and biological oxygen demands; organic pollutants; and general toxicity (El-Sheekh et al. 2016). Thus, there have been research studies geared toward the decontamination of water since all living things depend on water for survival. Among the several strategies that have been developed to ameliorate water pollution, bioremediation seems to be one of the most effective strategies in addition to being environmentally friendly. Bioremediation of polluted water is achieved by using microbial biomass or molecules to fixate or alter toxic compounds or ions in water. Compared to physicochemical methods of remediating water pollution, bioremediation can recover or remove toxic substances from water even when available in minute quantities (Coelho et al. 2015).

Several microorganisms, including yeast, fungi, algae, and bacteria, can be deployed for the bioremediation of water pollution, but this review focuses on the use of bacteria in remediating polluted water. Bacterial remediation of polluted water is regulated by the metabolic activities of different bacteria, which can render water pollutants redundant, nontoxic, or as sources of energy and nutrition for their survival. In the absence of the appropriate bacterial enzymatic activities that can enhance the detoxification of pollutants, these pollutants may cause the death of the bacteria present in the water. Thus, pollutant-degrading bacteria can attract or search for pollutants in their environment by moving toward the area with a high concentration gradient of the pollutants (Thapa et al. 2012). Bacterial remediation of hydrocarbon-polluted water is proceeded by the release of molecular oxygen which causes the breakdown of hydrocarbons into smaller compounds that bacteria can use to activate specific metabolic pathways leading to bioremediation. For oil-contaminated water, some bacteria can produce chemical compounds that stay afloat on the surface of water to emulsify the oil, thereby enhancing the removal of the oil easily from the surface (Kumar et al. 2011). Functional groups, such as aldehydes, ketones, galactans, xylans, and carboxyl, present on the cell walls of bacteria can adsorb heavy metals from wastewater, thereby reducing their bioavailability in the water (Qu et al. 2014).

Several bacteria species, including *Staphylococcus aureus*, *Enterococcus faecium*, *Bacillus subtilis*, and *Escherichia coli*, have been identified with heavy metals decontamination from water (Bhakta et al. 2014). Due to the presence of peptides (e.g., histidines- and cysteines-rich phytochelatins and metallothioneins) that can bind different heavy metals, *Escherichia coli* have been shown to remediate Cd contamination (Mejáre and Bülow 2001). Chromium resistance has also been found in *Pseudomonas aeruginosa* and was aided by genes upregulating enzymatic reductions (Thatoi et al. 2014). *Pseudomonas aeruginosa* has also been used to significantly bioremediate hydrocarbon-polluted water (Efeovbokhan et al. 2014). Bioremediation of uranium from groundwater and water drainages has been demonstrated by *Geobacter metallireducens* (Kumar et al. 2011). In the remediation of polluted river water, three floated bacteria strains, including *Pseudomonas gessardii*, *Bacillus cereus*, and *Aeromonas salmonicida*, reduced the water's total organic carbon content, phosphorus, nitrate, total phosphorus, and biological and chemical oxygen demands. Likewise, the water's heavy metal contents (i.e., Cr, Fe, Pb, Mn, and Ni) decreased compared to the control without the inoculation of bacterial strains (Shahid et al. 2019). In addition, the biomass of bacteria has varying capacities to treat wastewater through biosorption. Its treatment and experimental conditions determine the biosorption capacity of each bacterium cell. Bacterial cells must adapt to changing physical, chemical, and bioreactor conditions to enhance biosorption (Yangbenro and Babalola 2017). Bacteria are biosorbents because of their large size, ability to grow in controlled environments, and ability to respond to changing environmental conditions (Srivastava et al. 2015).

8 Bacteria–Plant Interaction in Bioremediation

In the agroecosystem, bioremediation occurs naturally, though at a very slow rate compared to when this process is manipulated by introducing the right plant-microbe pairs. Scientific investigations have shown that matching contaminants in a biodegrading bacteria–plant mixture can further increase the output from the bioremediation process. One of the major advantages of using microbe–plant interaction in bioremediation is that it ensures that a lot of contaminants, such as heavy metals, are prevented from contaminating groundwater and the entire ecosystem by being held up in the plants and microbial cells. Several studies have shown that without using microbes, plants alone (phytoremediation) may not be sufficient for the bioremediation of contaminated soils (Chaudhry et al. 2005).

Furthermore, the use of a consortium of bacteria in the biodegradation of contaminants has also been advocated as their combined action and diversity could have an important effect on the rapid and efficient bioremediation process (Ledezma-Villanueva et al. 2015). Through quorum sensing, consortia of different bacterial strains and species can communicate and create complex systems that can enhance the degradation of contaminants in a more efficient way (Pileggi et al. 2020). Different species of bacteria can be introduced to contaminated sites using mechanical injection, which requires minimal labor, is less expensive, and has a very high efficacy (Ajona and Vasanthi 2021). Plants can also be used to introduce selected bacteria strains deeply into contaminated sites through their extensive root systems (Saravanan et al. 2020). The plant host can also furnish the microbes with photosynthates and provide better redox conditions for bacterial proliferation within the rhizosphere (Bodelier et al. 2000). Root exudations, including flavonoids, phytohormones, and organic acid anions, create a conducive environment for bacterial proliferation to effect rapid degradation of contaminants. For example, a well-known contaminant-degrading bacteria, *Pseudomonas putida* PML2, has been reported to exhibit a high level of proliferation when grown on flavonoid-supplemented media (Narasimhan et al. 2003). Overall, providing conditions that enhance bacterial colonization of plants' roots in the rhizosphere of contaminated soils is one of the best options for promoting contaminants bioremediation.

9 Some Examples of Bacteria Used in Bioremediation

Bioremediation with bacteria entails the use of bacteria that can break down and lessen the toxicity of environmental pollutants (Vidali 2001; Table 2). Microbes and their interactions with pollution and the environment are significant in determining the effectiveness of bioremediation (Tiedje 1993). These bacteria can be found naturally in the contaminated site or can be isolated from other locations and artificially introduced (Vidali 2001). Biodegradation is a common element of microbial metabolism, and in some situations, microorganisms can capture carbon and energy directly by decomposing pollutants (Vidali 2001).

Table 2 The Major Bacterial Genera Involved in Bioremediation

Strains of bacteria used	Benefits (pollutants degraded)	References
Acinetobacter baumanii	Cysteine-rich peptides such as metallothioneins, phytochelatins, and glutathione	Paul et al. (2005)
Acinetobacter sp.	Chlorpyrifos and methyl parathion	Hussaini et al. (2013)
Achromobacter sp.	Petrol contains diesel, polycyclic aromatic hydrocarbons, toluene, pesticides, hydrocarbons, alkanes, and polyaromatic chemicals	Singh et al. (2014); Safiyanu et al. (2015); Sani et al. (2015)
Achromobacter sp. AO22	Hg	Ng et al. (2009)
Acidithiobacillus ferrooxidans	Hg	Sasaki et al. (2005)
Acinetobacter baumanii	*Dye effluents containing azo dyes*	Kumar et al. (2016a, b)
Acinetobacter sp.	Pesticides, hydrocarbons, alkanes, polyaromatic chemicals, chromium, aromatic hydrocarbons, diesel polycyclic aromatic hydrocarbons, toluene, petrol, and Fitoraz WP 76	Singh et al. (2014); De Jaysankar et al. (2008); Simarro et al. (2013); Safiyanu et al. (2015); Sani et al. (2015); Mónica et al. (2016); Hussaini et al. (2013); De et al. (2008)
Aerococcus sp.	Cd, Cr, Pb	Sinha and Paul (2014); Sinha and Biswas (2014)
Aeromonas sp.	Cr, U, Cu, Ni, and petroleum hydrocarbons	Sinha et al. (2011); Thapa et al. (2012)
Alcaligenes faecalis	Cadmium and lead	De Jaysankar et al. (2008)
Alcaligenes sp.	Pesticides, hydrocarbons, alkanes, and polyaromatic chemicals	Singh et al. (2014)

(continued)

Table 2 (continued)

Strains of bacteria used	Benefits (pollutants degraded)	References
Alcanivorax borkumensis	Oil	Hassanshahian et al. (2014)
Arthrobacter sp.	Reduction of chromium (Cr) and decomposition of petroleum hydrocarbons absorb toxic heavy metal ions, pesticides, hydrocarbons, alkanes, and polyaromatic chemicals Fitoraz WP 76 Diesel oil Petroleum hydrocarbons	Singh et al. (2014); Thapa et al. (2012); De Jaysankar et al. (2008); Roanne and Pepper (2001); Mónica et al. (2016); Hussaini et al. (2013); Sukumar and Nirmala (2016); De et al. (2008)
Bacilli sp.	Depetroleum hydrocarbons.	Thapa et al. (2012)
Bacillus sp.	Pesticides, hydrocarbons, alkanes, polyaromatic chemicals, and endosulfan Diesel oil Useful for hydrocarbon bioremediation applications such as PAHs and crude oil	Singh et al. (2014); Mosa et al. (2016); Gupta et al. (2000); Kim et al. (2007); Mohamed et al. (2011); Sukumar and Nirmala (2016); Feitkenhauer et al. (2003)
Bacillus sp. JF8	Useful for hydrocarbon bioremediation applications such as naphthalene and PCBs	Shimura et al. (1999)
Bacillus spp. ETL-2012	Sulfonated di-azo dye, textile dye, and RNB dye	Shah et al. (2013); Yogesh and Akshaya (2016); Das et al. (2015)
Bacillus brevis	Crude oil	El-Borai et al. (2016)
Bacillus cereus	Detoxify hexavalent chromium, such as Cr, azo dye effluents, diesel oil, and crude oil	Kanmani et al. (2012); Coelho et al. (2015); Singh et al. (2013b); Kumar et al. (2016a, b); Kehinde and Isaac (2016)
Bacillus cereus A	Diesel oil	Maliji et al. (2013)
Bacillus cereus strain XMCr-6	Cr	Dong et al. (2013)
Bacillus coagulans	Crude oil and diesel oil	Kehinde and Isaac (2016).
Bacillus firmus	Textile effluents and vat dyes	Chen and Wang (2007)
Bacillus idriensis strains	As	Liu et al. (2011)
Bacillus licheniformis	Crude oil Naphthalene	Sivakumar et al. (2012); Lin et al. (2010); El-Borai et al. (2016)
Bacillus macerans	Vat dyes and textile effluents	Chen and Wang (2007)
Bacillus megaterium	Lead, chromium, and cadmium	Abioye et al. (2018)
Bacillus pumilus	Lead cadmium	De Jaysankar et al. (2008)
Bacillus pumilus HKG212	RNB dye, textile dye, sulfonated di-azo dye, and reactive Red HE8B	Shah et al. (2013); Yogesh and Akshaya (2016); Das et al. (2015)

(continued)

Table 2 (continued)

Strains of bacteria used	Benefits (pollutants degraded)	References
Bacillus safensis (JX126862) strain (PB-5 and RSA-4)	Cd	Priyalaxmi et al. (2014)
Bacillus sphaericus	Crude oil	El-Borai et al. (2016)
Bacillus stearothermophilus	Crude oil	Sorkhoh et al. (1993)
Bacillus subtilis	Lead, chromium, and cadmium Oil Phenol	Balamurugan et al. (2014); Bhakta et al. (2014); Abioye et al. (2018); Singh et al. (2013a)
Bacillus subtilis BR151	Cadmium, zinc, mercury, and lead	Bondarenko et al. (2008)
Bacillus subtilis strains NAP1, NAP2, NAP4	Oil-based paints	Phulpoto et al. (2016)
Bacillus sphaericus	Crude oil	El-Borai et al. (2016)
Bacillus thermoleovorans	Long-chain alkanes and naphthalene	Kato et al. (2001)
Bacillus thermoleovorans DSM 10561	Aromatic	Markl et al. (1999)
Beijerinckia sp.	Petroleum hydrocarbons	Thapa et al. (2012)
Brevibacterium iodinum	Cadmium and lead	De Jaysankar et al. (2008)
Burkholderia sp.	Absorb toxic heavy metal ions Petroleum hydrocarbons	Jiang et al. (2008); Adebusoye et al. (2007)
Burkholderia xenovorans LB400 (ohb)	Arochlor 1242	Rodrigues et al. (2006)
Caulobacter crescentus JS4022/p723-6H	Cd	Patel et al. (2010)
Chrobacteria sp.	Petroleum hydrocarbons	Thapa et al. (2012)
Citrobacter sp.	Fe (III)	Renninger et al. (2001); Venkidusamy et al. (2018)
Citrobacter koseri	Diesel oil and crude oil	Kehinde and Isaac (2016)
Corynebacterium sp.	Pesticides, hydrocarbons, alkanes, and polyaromatic chemicals	Singh et al. (2014)
Corynebacterium propinquum	Phenol Oil	Singh et al. (2013a)
Cupriavidus metallidurans	Mercury	Roanne and Pepper (2001); Bravo et al. (2020)
Cupriavidus necator RW112	Arochlor 1221 and 1232	Wittich and Wolff (2007)
Cyanobacteria	Decompose petroleum hydrocarbons and naphthalene	Thapa et al. (2012); Gupta et al. (2000); Sivakumar et al. (2012); Lin et al. (2010)
Cycloclasticus sp.	Petroleum aromatics	Teramoto et al. (2010)
Dechloromonas aromatica	Benzoate, chlorobenzoate, and toluene	JGI Genome Portal (2016)

(continued)

Table 2 (continued)

Strains of bacteria used	Benefits (pollutants degraded)	References
Deinococcus radiodurans	Mercury and toluene	Hassan et al. (2008)
Deinococcus geothermalis	Glutathione, phytochelatins, and metallothioneins	Paul et al. (2005)
Deinococcus radiodurans	Mercury	Brim et al. (2000)
Desulfovibrio desulfuricans	Chromium, copper, and nickel	Joo et al. (2015)
Enterobacter	Teichoic acid, chlorpyrifos, and methyl parathion	Mosa et al. (2016); Hussaini et al. (2013)
Enterobacter cloacae	Lead, cadmium, and nickel	Gupta et al. (2000); Banerjee et al. (2015)
Enterobacter cloacae B2-DHA	Chromium	Rahman et al. (2015)
Enterococcus faecium	Remove heavy metals	Bhakta et al. (2014)
Escherichia coli	Hg and Cd	Bhakta et al. (2014); Bae et al. (2001); Murtaza et al. (2002)
Escherichia coli AtzA	Atrazine	Strong et al. (2000)
Escherichia coli JM109	Hg	Zhao et al. (2005)
Escherichia coli MC1061	Zn, Pb, Hg, and Cd	Bondarenko et al. (2008)
Escherichia coli SE5000	Nickel	Deng et al. (2005)
Exiguobacterium indicum	Azo dye effluents	Kumar et al. (2016a, b)
Exiguobacterium aurantiacum	Azo dye effluents	Kumar et al. (2016a, b)
Flavobacteria sp.	Petroleum hydrocarbons	Thapa et al. (2012)
Flavobacterium sp.	Pesticides, hydrocarbons, alkanes, and polyaromatic chemicals Toluene and other polycyclic aromatic Hydrocarbons are found in gasoline and diesel	Singh et al. (2014); Mosa et al. (2016); Safiyanu et al. (2015); Sani et al. (2015)
Geobacter metallireducens	Uranium	Kumar et al. (2011)
Geobacter thermoleovorans	Long-chain alkanes	Marchant et al. (2017)
Klebsiella oxytoca	Remove heavy metals from water bodies, vat dyes, textile effluents, endosulfan, and vat dyes	Chen and Wang (2007)
Kocuria flava	Cu	Achal et al. (2011); Coelho et al. (2015)
Marinobacter sp.	Crude oil	Chernikova et al. (2020); Zenati and Badis (2019)
Mesorhizobium huakuii B3	Cd2	Sriprang et al. (2003)
Methylibium petroleiphilum	Methyl tert-butyl ether	Jessica et al. (1999)
Methylococcus capsulatus (Bath)	Cr	Hasin et al. (2010)

(continued)

Table 2 (continued)

Strains of bacteria used	Benefits (pollutants degraded)	References
Microbacterium profundi strain Shh49T	Iron	Wu et al. (2006)
Micrococcus sp.	Teichoic acid	Mosa et al. (2016)
Microbacterium sp.	Aromatic hydrocarbons	Simarro et al. (2013)
Micrococcus luteus	Remove a huge quantity of Pb or textile azo dyes	Puyen et al. (2012); Hassan et al. (2013)
Modococci sp.	Decompose petroleum hydrocarbons	Thapa et al. (2012)
Moraxella sp.	Decompose petroleum hydrocarbons Cd and Hg	Thapa et al. (2012); Bae et al. (2001)
Mycobacterium sp.	Pesticides, hydrocarbons, alkanes, polyaromatic chemicals, and aromatic hydrocarbons	Singh et al. (2014); Simarro et al. (2013)
Nitrosomonas sp.	Pesticides, hydrocarbons, alkanes, and polyaromatic chemicals	Singh et al. (2014)
Nitrosomonas europaea	Mineral nitrogen compounds (i.e., ammonia).	
Nitrobacter hamburgensis	Mineral nitrogen compounds (i.e., nitrate).	
Nocardia sp.	Petroleum hydrocarbons, pesticides, hydrocarbons, alkanes, and polyaromatic chemicals	Thapa et al. (2012); Singh et al. (2014)
Oleispira sp.	Crude oil	Yakimov et al. (2007)
Paracoccus denitrificans	Nitrate	Nitrification and Denitrification Wastewater Treatment (1996)
Photobacterium sp.	Methyl parathion and chlorpyrifos	Hussaini et al. (2013)
Pseudomonas sp.	Teichoic acid, petroleum hydrocarbons, pesticides, hydrocarbons, alkanes, polyaromatic chemicals, and aromatic hydrocarbons Chlorpyrifos and methyl parathion Useful for hydrocarbon bioremediation 3 phenoxybenzoic acid Petroleum hydrocarbons	Singh et al. (2014); Mosa et al. (2016); Thapa et al. (2012); Simarro et al. (2013); Hussaini et al. (2013); Halden et al. (1999); Adebusoye et al. (2007)
Pseudomonas K-62	*Mercury*	Kiyono and Pan-Hou (2006)

(continued)

Table 2 (continued)

Strains of bacteria used	Benefits (pollutants degraded)	References
Pseudomonas aeruginosa	Cadmium, U, Cu, Ni, Cr, iron, zinc, lead, Mn^{2+}, Cu^2, lead (Pb), naphthalene, salicylate, camphor degradation, textile dye, sulfonated di-azo dye reactive red HE8B, and RNB dye Phenol Diesel oil	Wasilkowski et al. (2012); De Jaysankar et al. (2008); Dias et al. (2002); Zhang et al. (2005); Shah et al. (2013); Yogesh and Akshaya (2016); Das et al. (2015); Singh et al. (2013a); Sinha et al. (2011); Paranthaman and Karthikeyan (2015); Sukumar and Nirmala (2016)
Pseudomonas aeruginosa KH6	Crude oil	El-Borai et al. (2016)
Pseudomonas aeruginosa AP02-1	Crude and diesel oil	Perfumo et al. (2006)
Pseudomonas cepacia	Diesel oil and crude oil	Kehinde and Isaac (2016)
Pseudomonas fluorescens	Iron, zinc, lead, Mn^{2+}, and Cu^2 Naphthalene 4F39 Ni	Paranthaman and Karthikeyan (2015); Wasilkowski et al. (2012); Lopez et al. (2002)
Pseudomonas fluorescens CS2	Biphenyl ethylbenzene	Parameswarappa et al. (2008)
Pseudomonas fluorescens F113rifPCB	Biphenyl and polychlorinated	Brazil et al. (1995)
Pseudomonas fluorescens F113rifpcbrmBP1:gfpmut3	Biphenyl and polychlorinated biphenyl	Boldt et al. (2004)
Pseudomonas fluorescens MP	2,4-dinitrotoluene	Monti et al. (2005)
Pseudomonas fluorescens RE	2,4-dinitrotoluene	Monti et al. (2005)
Pseudomonas fluorescens OS8	Zinc, cadmium, lead, and *mercury*	Bondarenko et al. (2008)
Pseudomonas mendocina	Petrol, diesel polycyclic aromatic, and hydrocarbons toluene	Safiyanu et al. (2015); Sani et al. (2015)
Pseudomonas putida	Salicylate and camphor Naphthalene Ridomil MZ 68 MG, Fitoraz WP 76, Decis 2.5 EC, malathion Diesel oil 4 chlorobenzoic acid, chlorobenzoates Cr Benzene, xylene, toluene, and oselene (BTX)	Balamurugan et al. (2014); Wasilkowski et al. (2012); Mónica et al. (2016); Hussaini et al. (2013); Sukumar and Nirmala (2016); Wittich and Wolff (2007); Ackerley et al. (2004); Sukumar and Nirmala (2016); Safiyanu et al. (2015); Sani et al. (2015); Sarang et al. (2013)
Pseudomonas putida 06909	Cadmium	Wu et al. (2006)

(continued)

Table 2 (continued)

Strains of bacteria used	Benefits (pollutants degraded)	References
Pseudomonas putida B13ST1(pPOB)	3 phenoxybenzoic acid	Halden et al. (1999)
Pseudomonas putida BCRc14349	Phenol and trichloroethane	Chen et al. (2007)
Pseudomonas putida KT2442	Naphthalene	Filonov et al. (2005)
Pseudomonas veronii	Cadmium, zinc, copper, petrol, and diesel Polycyclic aromatic hydrocarbons like toluene	Vullo et al. (2008); Coelho et al. (2015)
Ralstonia sp.	Aromatic hydrocarbons	Simarro et al. (2013)
Ralstonia eutropha	CH34 Cd2	Valls et al. (2000)
Rhodococcus sp.	Pesticides, hydrocarbons, alkanes, and polyaromatic chemicals Petroleum hydrocarbons	Singh et al. (2014); Adebusoye et al. (2007)
Rhodococcus sp. F92	Petroleum	Quek et al. (2006)
Rhodococcus sp. strain RHA1	4-chlorobenzoate	Rodrigues et al. (2001)
Rhodopseudomonas palustris	Lead, cadmium, and chromium	Sinha and Paul (2014); Sinha and Biswas (2014)
Serratia ficaria	Diesel oil and crude oil	Kehinde and Isaac (2016)
Sphingomonas sp.	Pesticides, hydrocarbons, alkanes, and polyaromatic chemicals	Singh et al. (2014); Thapa et al. (2012)
Sphingomonas desiccabilis	As	Liu et al. (2011)
Sporosarcina ginsengisoli	Absorb toxic heavy metal ions such as	Achal et al. (2011); Coelho et al. (2015)
Staphylococcus sp.	Endosulfan, vat dyes, and textile effluents	Mohamed et al. (2011); Chen and Wang (2007)
Staphylococcus aureus	Remove heavy metals from water bodies, vat dyes, and textile effluents	Bhakta et al. (2014); Chen and Wang (2007)
Staphylococcus aureus RN4220	Cadmium, mercury, zinc, and lead	Bondarenko et al. (2008)
Stenotrophomonas spp.	Au	Song et al. (2008)
Streptomyces sp.	Petroleum hydrocarbons	Thapa et al. (2012); Dias et al. (2002)
Thallassolituus sp.	Petroleum hydrocarbons	Mahjoubi et al. (2018)
Thermus sp.	Useful for hydrocarbon bioremediation applications as PAHs	Feitkenhauer et al. (2003)
Xanthobacter sp.	Pesticides, hydrocarbons, alkanes, and polyaromatic chemicals	Singh et al. (2014)
Zoogloea ramigera	*Iron*, cobalt, *nickel*, copper, zinc, and nickel ions	Gupta et al. (2000); Sağ and Kutsal (2000)

Bacteria must come in contact with the contaminant(s) in order to degrade it. Bacteria are divided into two categories for use in bioremediation: aerobic and anaerobic bacteria. Aerobic bacteria can break down complex compounds in the presence of oxygen. Bacteria can degrade hydrocarbons, pesticides, polyaromatic compounds, and alkanes. Aerobic bacteria are found in a variety of environments. Anaerobic bacteria do not require oxygen to function and are not commonly utilized in bioremediation. Some anaerobic bacteria are used in the bioremediation of chlorinated aromatic compounds, the dechlorination of some solvents such as trichloroethylene, polychlorinated biphenyls (PCBs), and chloroform.

10 Future Perspectives on the Use of Bacteria in Polluted Environment Bioremediation

To be used in the future, genetic engineering techniques have been used to develop new and exceptional bacterial species for deployment in the biological treatment of contaminated environments to be more efficient. The strategy of using genetic modification is fundamental to modifying the routes of metabolism of genetically engineered bacteria (GEB) and inhibiting the toxic effects of heavy metals. It is now extremely simple to create genetically edited bacteria by combining genes and promoters, as well as improve their performance in situ and improve the capabilities of these bacteria using DNA techniques to analyze components harmful to the wastewater by discovering and converting genes into suitable bioprocessing agents. In recombinant DNA technology, diverse techniques such as polymerase chain reaction (PCR), antisense RNA, and site-directed mutagenesis are being investigated. Bacteria can be modified through genetic engineering to achieve unique properties such as rapid growth of bacteria and pH changes, unfavorable environmental conditions, and low-cost agriculture.

With recent developments in genetic engineering, bacteria that have had their genetic material altered through artificial genetic exchange between them and other bacteria are known as genetically engineered bacteria. This is the term for recombinant DNA technology. To eliminate hazardous waste from wastewater, recombinant DNA techniques, or the transfer of natural genetic material across bacteria, were used to create genetically engineered bacteria (Jain et al. 2011). It is now possible to put the proper gene into a cell to develop a specialized enzyme that can break down various wastewater contaminants. It has been found that genetically engineered bacteria can improve some bacteria's ability to digest certain substances, such as pesticides and hydrocarbons (Perpetuo et al. 2011; Singh et al. 2014). Inorganic-form metals are converted into organic-form metals through a few processes, such as oxidation and reduction, enhancing the solubility of heavy metals, bacteria interactions, skin reactions, changes in hydrogen, absorption, bioaccumulation, sedimentation, and encapsulation, thus improving the solubility of heavy metals or rendering them inert in the environment (Nicolaou et al. 2010). It also breaks down some pollutants, allowing the creation of a synthetic mixture of genes not found together in nature.

The most common ways are genetic engineering using single genes or components, pathway development, and gene sequence alteration (Wasilkowski et al. 2012). We can learn more about the possible actions of bacteria in contaminated environments using genetic and biochemical techniques such as polymerase chain reaction, in situ hybridization, and antibody use. DNA probes that target specific genetic sequences, such as genes that affect a microorganism's ability to break down, can be used to analyze contaminated locations and determine general community structure and catabolic activity during the bioremediation process (Stapleton et al. 1998). Genetically engineered bacteria (GEB) have exhibited several bioremediation uses, such as groundwater, sewage effluent, and soil. For example, increased decomposition rates in wastewater using genetically modified bacteria can be achieved by the following:

1. Determination of metabolic pathways that are genetically known.
2. The introduction of completely new metabolic pathways into bacterial strains to break down previously insoluble compounds.
3. Enzyme specificity and affinity can be altered.
4. Create and organize a path.
5. Vital process development, monitoring, and control. This bio-reporter sensor can be used for bio-applications such as affinity, toxicity reduction, chemical sensing, and end-point analysis. DNA testing can detect the presence of specific bacteria that can participate in biodegradation, and bacteria-specific antibodies can detect changes in the composition of bacterial populations due to the stimulation of catabolic enzymes during wastewater treatment. During wastewater treatment operations, alterations in bacterial population composition can also be noticed, as well as differences when compared to uncontaminated sites.

Approaches based on molecules such as genomics, proteomics, transcriptomics, metabolomics, and fluxomics have provided better knowledge of microbial communities that live in a specific niche in the environment (Gupta et al. 2020; De Sousa et al. 2018). Researchers have used next-generation sequencing technologies and in silico analysis to help solve challenges linked to uncultivated microorganisms (Malla et al. 2018; Maphosa et al. 2010). These innovations have aided in the acquisition of extensive biological data about bacteria, their biological systems, and biomolecules involved in bioremediation, such as enzymes (Misra et al. 2018; Zhang et al. 2010; Pandey et al. 2019). The first stage is to better understand bacteria, their interactions with the environment, and the factors that influence their genotypic expression (Callaghan 2013; Chakraborty et al. 2012).

Genomic technologies are becoming more widely available, which has accelerated the long-term treatment of polluted areas. Bacterial culture-independent analyses from locations that are contaminated have increased awareness of community dynamics and the diversity of bacteria participating in bioremediation (Desai et al. 2010; Kapley et al. 2007). The use of advanced genomic technologies to identify microbial populations has permitted the discovery of previously unreachable microbes by culture-based methods. In examining the microbial population, techniques for gene amplification (PCR) and sequencing (Gołebiewski and Tretyn 2020), it was identified that 16S rRNA has the highest quality level for describing phylogenetic similarities between bacteria in microbial communities. The gene sequence of all bacteria is highly conserved and diverse (Lovley 2003). Sequence analysis of the 16S rRNA gene can be used to thoroughly examine bacterial diversity because the hyper-variable portions of the 16S rRNA gene were specifically amplified and sequenced. It is a cost-effective and highly efficient technology that is easily available. It has become a widely used technique for profiling complex microbial communities using a variety of bioinformatics tools (Han et al. 2020). It can be used to identify either novel or uncultivable bacteria that have not yet been phenotypically identified (Clarridge 2004).

The phylogenetic order of bacteria involved in bioremediation processes can be figured out using 16S rRNA sequences taken from contaminated settings (Lovley 2003). The genetics of PAH breakdown have been studied using a coal tar mixing plant (Sakshi and Haritash 2020). In Shanghai, Kou et al. (2018) used 16S rRNA gene amplicon sequencing to investigate the diversity and richness of the microbial community in soil contaminated with heavy metals such as zinc, copper, and lead. Pseudomonas species, a bacteria discovered in soil, capable of digesting polyurethane from a location where fragile plastic debris abounds, was identified using sequencing of the 16S rRNA gene and the fatty acid content of the membrane (Cárdenas Espinosa et al. 2020).

Genes encoding enzymes required for the catabolism of some of these unusual substrates may be carried on plasmids, although most of the bacteria's essential genes are found on a single chromosome. The catabolism has been linked to plasmids. As a result, GEB can be exploited efficiently for biodegradation, indicating a future research frontier with far-reaching ramifications (Kulshreshtha 2013).

11 Conclusions

Globally, bioremediation is recognized as a suitable technique to decontaminate the ecosystem. The deployment of bacteria in the bioremediation of contaminated sites is of high ecological significance as they are ubiquitous in the environment. In the last few years, industrial wastewater treatment has been done by reducing concentrations of toxic metals derived from human activities as well as developing cost-effective techniques. It has been proven that bacteria have a significant role in removing metals from solutions by absorbing toxic metals. The synergistic use of bacteria and plants in bioremediation is also an optimal way of achieving effective and rapid decontamination of our agricultural lands and the environment at large. Moreover, it is necessary to encourage a preventive approach to the problems of mineral pollution by using genetic engineering of bacteria. Some bacteria can naturally break down pollutants, but this is not enough on a global scale. Identified naturally occurring contaminant-degrading bacteria can be isolated, cultivated in large quantities, and studied for gene transformation and transfer to other strains of bacteria. Therefore, genetically engineered bacteria developed with the help of different genomic approaches and bacterial strains must be used at a high degree of efficiency in bioremediation, considering their availability and cost. It should also be easy to obtain and cultivate identified contaminant-degrading bacteria. Some important points still need to be studied to decontaminate polluted water with genetically engineered bacteria by developing new clean, environmentally acceptable, and commercially feasible techniques.

References

Abioye OP, Oyewole OA, Oyeleke SB, Adeyemi MO, Orukotan AA (2018) Biosorption of lead, chromium and cadmium in tannery effluent using indigenous microorganisms. Braz J Biol Sci 5(9):25–32. https://doi.org/10.21472/bjbs.050903

Achal V, Pan X, Zhang D (2011) Remediation of copper-contaminated soil by *Kocuria flava* CR1, based on microbially induced calcite precipitation. Ecol Eng 37(10):1601–1605

Ackerley DF, Donzalez CF, Keyhan M et al (2004) Mechanism of chromate reduction by the Escherichia coli protein, NfsA, and the role of different chromate reductases in minimizing oxidative stress during chromate reduction. Environ Microbiol 6:851–860

Adebusoye SA, Illori MA, Amund OO et al (2007) Microbial degradation of petroleum hydrocarbons in a polluted tropical strain. J Microbiol Biotechnol 23(8):1149–1159

Ahmad Khan MS, Singh B, Singh Cameotra S (2015) Biological applications of biosurfactants and strategies to potentiate commercial production. In: CRC Press (ed) Biosurfactants: production and utilization: processes, technologies and economics, vol 159. Taylor & Francis Group, Boca Raton

Ajona M, Vasanthi P (2021) Bio-remediation of crude oil contaminated soil using recombinant native microbial strain. Environ Technol Innov 23:101635. https://doi.org/10.1016/j.eti.2021.101635

Albers P, Weytjens B, De Mot R, Marchal K, Springael D (2018) Molecular processes underlying synergistic linuron mineralization in a triple-species bacterial consortium biofilm revealed by differential transcriptomics. Microbiology 7:e00559

Antizar-Ladislao B, Beck AJ, Spanova K, Lopez-Real J, Russell NJ (2007) The influence of different temperature programmes on the bioremediation of polycyclic aromatic hydrocarbons (PAHs) in a coal-tar contaminated soil by in-vessel composting. J Hazard Mater 14:340–347

Antizar-Ladislao B, Spanova K, Beck AJ, Russell NJ (2008) Microbial community structure changes during bioremediation of PAHs in an aged coal-tar contaminated soil by in-vessel composting. Int Biodeterior Biodegradation 61:357–364

Asira EE (2013) Factors that determine bioremediation of organic compounds in the soil. Acad J Interdiscip Stud 2:125–128. https://goo.gl/aZ2ntP

Bae W, Chen W, Mulchandani A et al (2001) Genetic engineering of Escherichia coli for enhance uptake and bioaccumulation of mercury. Appl Environ Microbiol 67:5335–5338

Bains J, Capalash N, Sharma P (2003) Laccase from a non-melanogenic, alkalotolerant gammaproteobacterium JB isolated from industrial wastewater drained soil. Biotechnol Lett 25(14):1155–1159. https://doi.org/10.1023/a:1024569722413

Balamurugan D, Udayasooriyan C, Kamaladevi B (2014) Chromium (VI) reduction by *Pseudomonas putida* and *Bacillus subtilis* isolated from contaminated soils. Int J Environ Sci 5:522. [Google Scholar]

Banerjee G, Pandey S, Ray AK, Kuma R (2015) Bioremediation of heavy metals by a novel bacterial strain *Enterobacter cloacae* and its antioxidant enzyme activity, flocculant production, and protein expression in presence of lead, cadmium, and nickel. Water Air Soil Pollut 226:91. https://doi.org/10.1007/s11270-015-2359-9

Bansal N, Kanwar SS (2013) Peroxidase(s) in environment protection. Sci World J 2013:714639

Barman S, Das AC (2015) Residual effect of pre-emergence herbicides on microbial activities in relation to mineralization of C, N and P in the Gangetic alluvial soil of West Bengal, India. Environ Monit Assess 187:465

Beffa R, Menne H, Köcher H (2019) Herbicide resistance action committee (HRAC): herbicide classification, resistance evolution, survey, and resistance mitigation activities. In: Jeschke P, Witschel M, Krämer W, Schirmer U (eds) Modern crop protection compounds. Wiley-VCH Verlag GmbH & Co. KGaA, Weinheim, pp 5–32

Beiyuan J, Fang L, Chen H, Li M, Liu D, Wang Y (2020) Nitrogen of EDDS enhanced removal of potentially toxic elements and attenuated their oxidative stress in a phytoextraction process. Environ Pollut 268:115719

Bhakta JN, Munekage Y, Ohnishi K, Jana BB, Balcazar JL (2014) Isolation and characterization of cadmium and arsenic-absorbing bacteria for bioremediation. Water Air Soil Pollut 225:2150–2159

Bodelier PLE, Roslev P, Henckel T (2000) Stimulation by ammonium-based fertilizers of methane oxidation in soil around rice roots. Nature 403:421–424

Boldt TS, Sorensen J, Karlson U et al (2004) Combined use of different Gfp reporters for monitoring single-cell activities of a genetically modified PCB degrader in the rhizosphere of alfalfa. FEMS Microbiol Ecol 48(2):139–148

Bondarenko O, Rolova T, Kahru A et al (2008) Bioavailability of Cd, Zn and Hg in soil to nine recombinant luminescent metal sensor bacteria. Sensors 8:6899–6923

Boopathy R (2000) Factors limiting bioremediation technologies. Bioresour Technol 74:63–67. https://goo.gl/eQhPh7

Bravo G, Vega-Celedón P, Gentina JC, Seeger M (2020) Bioremediation by *Cupriavidus metallidurans* strain MSR33 of mercury-polluted agricultural soil in a rotary drum bioreactor and its effects on nitrogen cycle microorganisms. Microorganisms 8:1–20

Brazil GM, Kenefick L, Callanan M et al (1995) Construction of a rhizosphere *pseudomonad* with potential to degrade polychlorinated biphenyls and detection of bph gene expression in the rhizosphere. Appl Environ Microbiol 61(5):1946–1952

Brim H, McFarlan SC, Fredrickson JK et al (2000) Engineering *Deinococcus radiodurans* for metal remediation in radioactive mixed waste environments. Nat Biotechnol 18:85–90

Callaghan AV (2013) Metabolomic investigations of anaerobic hydrocarbon-impacted environments. Curr Opin Biotechnol 24:506–515. https://doi.org/10.1016/j.copbio.2012.08.012

Cárdenas Espinosa MJ, Colina Blanco A, Schmidgall T, Atanasoff-Kardjalieff AK, Kappelmeyer U, Tischler D et al (2020) Toward biorecycling: isolation of a soil bacterium that grows on a polyurethane oligomer and monomer. Front Microbiol 11:404. https://doi.org/10.3389/fmicb.2020.00404

Čertík M, Dercová K, Sejáková Z, Finďová M, Jakubík T (2003) Effect of polyaromatic hydrocarbons (PAHs) on the membrane lipids of bacterial cell. Biology 58:1111–1117. https://doi.org/10.2478/s11756-014-0337-0

Chakraborty R, Wu CH, Hazen TC (2012) Systems biology approach to bioremediation. Curr Opin Biotechnol 23:483–490. https://doi.org/10.1016/j.copbio.2012.01.015

Chandra R, Chowdhary P (2015) Properties of bacterial laccases and their application for bioremediation of industrial wastes. Environ Sci 17(2):326–342

Chaudhry Q, Blom-Zandstra M, Gupta S (2005) Utilising the synergy between plants and rhizosphere microorganisms to enhance breakdown of organic pollutants in the environment. Environ Sci Pollut Res 12:34–48

Chen C, Wang JL (2007) Characteristics of Zn2+ biosorption by *Saccharomyces cerevisiae*. Biomed Environ Sci 20:478–482. https://goo.gl/a3DH9q

Chen YM, Lin TF, Huang C et al (2007) Degradation of phenol and TCE using suspended and chitosan-bead immobilized *Pseudomonas putida*. J Hazard Mater 148(3):660–670

Chernikova TN, Bargiela R, Toshchakov SV, Shivaraman V et al (2020) Hydrocarbon-degrading bacteria Alcanivorax and Marinobacter associated with microalgae *Pavlova lutheri* and *Nannochloropsis oculata*. Front Microbiol 11:572931

Chu C-W, Liu B, Li N, Yao S-G, Cheng D, Zhao J-D et al (2017) A novel aerobic degradation pathway for thiobencarb is initiated by the TmoAB two-component flavin mononucleotide-dependent monooxygenase system in *Acidovorax* sp. strain T1. Appl Environ Microbiol 83:e01490-17

Clarridge JE (2004) Impact of 16S rRNA gene sequence analysis for identification of bacteria on clinical microbiology and infectious diseases. Clin Microbiol Rev 17:840–862. https://doi.org/10.1128/CMR.17.4.840-862.2004

Coelho LM, Rezende HC, Coelho LM, de Sousa PA, Melo DF, Coelho NM (2015) Bioremediation of polluted waters using microorganisms. In: Shiomi N (ed) Advances in bioremediation of wastewater and polluted soil. IntechOpen, Shanghai. [Google Scholar]

Couto N, Fritt-Rasmussen J, Jensen PE, Højrup M, Rodrigo AP et al (2014) Suitability of oil bioremediation in an Artic soil using surplus heating from an incineration facility. Environ Sci Pollut Res 21:6221–6227. https://goo.gl/eXKYxm

Das N, Chandran P (2011) Microbial degradation of petroleum hydrocarbon contaminants: an overview. Biotechnol Res Int 2011:1–13. https://goo.gl/jt4THZ

Das A, Mishra S, Verma VK (2015) Enhanced biodecolorization of textile dye Remazol navy blue using an isolated bacterial strain Bacillus pumilus HKG212 under improved culture conditions. J Biochem Technol 6:962–969. https://goo.gl/pwCbPy

De Jaysankar, Ramaiah N, Vardanyan L (2008) Detoxification of toxic heavy metals by marine bacteria highly resistant to mercury. Mar Biotechnol 10(4):471–477

De Sousa CS, Hassan SS, Pinto AC, Silva WM, De Almeida SS, Soares SDC et al (2018) Microbial omics: applications in biotechnology. In: Barh D, Azevedo V (eds) Omics technologies and bio-engineering. Academic Press, Cambridge, MA, pp 3–20. https://doi.org/10.1016/B978-0-12-815870-8.00001-2

De J, Ramaiah N, Vardanyan L (2008) Detoxification of toxic heavy metals by marine bacteria highly resistant to mercury. Mar Biotechnol 10(4):471–477

Deng X, Li QB, Lu YH et al (2005) Genetic engineering of Escherichia coli SE5000 and its potential for Ni2+ bioremediation. Process Biochem 40:425–430

Denich TJ, Beaudette LA, Lee H, Trevor SJT (2003) Effect of selected environmental and physicochemical factors on bacterial cytoplasmic membranes. J Microbiol Methods 52:149–182

Desai C, Pathak H, Madamwar D (2010) Advances in molecular and "-omics" technologies to gauge microbial communities and bioremediation at xenobiotic/anthropogen contaminated sites. Bioresour Technol 101:1558–1569. https://doi.org/10.1016/j.biortech.2009.10.080

Dias MA, Lacerda ICA, Pimentel PF, Castro HF, Rosa CA (2002) Removal of heavy metals by an Aspergillus terreus strain immobilized in a polyurethane matrix. Lett Appl Microbiol 34:46–50

Dixit R, Wasiullah, Malaviya D, Pandiyan K, Singh UB, Sahu A, Shukla R, Singh BP, Rai JP, Sharma PK, Lade H, Paul D (2015) Bioremediation of heavy metals from soil and aquatic environment: an overview of principles and criteria of fundamental processes. Environ Sustain Appl 7(2):2189–2212. https://doi.org/10.3390/su7022189

Dombrowski N, Donaho JA, Gutierrez T, Seitz KW, Teske AP, Baker BJ (2016) Reconstructing metabolic pathways of hydrocarbon-degrading bacteria from the Deepwater Horizon oil spill. Nat Microbiol 1:16057. https://doi.org/10.1038/nmicrobiol.2016.57

Dong W, Chen Q, Hou Y, Li S, Zhuang K, Huang F et al (2015) Metabolic pathway involved in 2-methyl-6-ethylaniline degradation by *Sphingobium* sp. strain MEA3-1 and cloning of the novel flavin-dependent monooxygenase system meaBA. Appl Environ Microbiol 81:8254–8264

Dong et al. (2013) Dong G, Wang Y, Gong L, Wang M, Wang H, He N, Zheng Y, Li Q (2013) Formation of soluble Cr (III) end-products and nanoparticles during Cr (VI) reduction by *Bacillus cereus* strain XMCr-6. Biochem Eng J 15;70:166–72

Duldhardt I, Gaebel J, Chrzanowski L, Nijenhuis I, Härtig C, Schauer F, Heipieper HJ (2010) Adaptation of anaerobically grown *Thauera aromatica*, *Geobacter sulfurreducens* and

Desulfococcus multivorans to organic solvents on the level of membrane fatty acid composition. Microb Biotechnol 3(2):201–209

Duran N, Esposito E (2000) Potential applications of oxidative enzymes and phenoloxidase-like compounds in wastewater and soil treatment: a review. Appl Catal B 28(2):83–99

Efeovbokhan VE, Hymore FK, Ayoola AA, Adeeyo OA (2014) Comparison of aerobic and anaerobic bioremediation of polluted water samples. Am Int J Contemp Res 4(4):120–126

El-Borai AM, Eltayeb KM, Mostafa AR, El-Assar SA (2016) Biodegradation of industrial oil-polluted wastewater in Egypt by bacterial consortium immobilized in different types of carriers. Pol J Environ Stud 25:1901–1909. https://goo.gl/JtCFdd

El-Sheekh MM, Farghl AA, Galal HR, Bayoumi HS (2016) Bioremediation of different types of polluted water using microalgae. Rend Fis Acc Lincei 27:401–410. https://doi.org/10.1007/s12210-015-0495-1

Fantroussi ELS, Agathos SN (2005) Is bioaugmentation a feasible strategy for pollutant removal and site remediation? Curr Opin Microbiol 8:268–275. https://goo.gl/y6kLsc, https://goo.gl/S0MQ3

Feitkenhauer H, Muller R, Markl H (2003) Degradation of polycyclic. FEMS Microbiol Lett 178:87–93

Fernandes P, Marques PCM, Carvalho F, de Carvalho CCRC (2011) Organic-solvent tolerant gram-positive bacteria: applications and mechanisms of tolerance. In: Carter RE (ed) Organic solvents: properties, toxicity, and industrial effects. Nova Science, New York, pp 89–104

Filonov AE, Akhmetov LI, Puntus IF et al (2005) The construction and monitoring of genetically tagged, plasmid-containing, naphthalene-degrading strains in soil. Microbiology 74(4):453–458

Fomina M, Gadd GM (2014) Biosorption: current perspectives on concept, definition and application. Bioresour Technol 160:3–14

Fuentes S, Barra B, Caporaso JG, Seeger M (2015) From rare to dominant: a fine-tuned soil bacterial bloom during petroleum hydrocarbon bioremediation. Appl Environ Microbiol 82:888–896. https://doi.org/10.1128/AEM.02625-15

Gadd GM, White C (1993) Microbial treatment of metal pollution—a working biotechnology? Trends Biotechnol 11:353–359

Gianfreda L, Rao AM (2004) Potential of extra cellular enzymes in remediation of polluted soils: a review. Enzym Microb Technol 35(4):339–354. https://doi.org/10.1016/j.enzmictec.2004.05.006

Gołebiewski M, Tretyn A (2020) Generating amplicon reads for microbial community assessment with next generation sequencing. J Appl Microbiol 128:330–354. https://doi.org/10.1111/jam.14380

Gravina F, Dobrzanski T, Olchanheski LR, Galvão CW, Reche PM, Pileggi SAV, Azevedo RA, Sadowsky MJ, Pileggi M (2017) Metabolic Interference of sod gene mutations on catalase activity in *Escherichia coli* exposed to Gramoxone® (paraquat) herbicide. Ecotoxicol Environ Saf 139:89–96

Guerra AB, Oliveira JS, Silva-Portela RC, Araujo W, Carlos AC, Vasconcelos ATR et al (2018) Metagenome enrichment approach used for selection of oil-degrading bacteria consortia for drill cutting residue bioremediation. Environ Pollut 235:869–880. https://doi.org/10.1016/j.envpol.2018.01.014

Gupta P, Diwan B (2017) Bacterial exopolysaccharide mediated heavy metal removal: a review on biosynthesis, mechanism and remediation strategies. Biotechnol Rep 13:58–71

Gupta R, Ahuja P, Khan S, Saxena RK, Mohapatra H (2000) Microbial biosorbents: meeting challenges of heavy metal pollution in aqueous solutions. Curr Sci 78(8):967–973

Gupta K, Biswas R, Sarkar A (2020) Advancement of omics: prospects for bioremediation of contaminated soils. In: Shah M (ed) Microbial bioremediation & biodegradation. Springer, Singapore, pp 113–142

Halden RU, Tepp SM, Halden BG et al (1999) Degradation of 3-phenoxybenzoic acid in soil by *Pseudomonas pseudoalcaligenes* POB310 (pPOB) and two modified *Pseudomonas*. Appl Environ Microbiol 65(8):3354–3359

Han D, Gao P, Li R, Tan P, Xie J, Zhang R et al (2020) Multicenter assessment of microbial community profiling using 16S rRNA gene sequencing and shotgun metagenomic sequencing. J Adv Res 26:111–121. https://doi.org/10.1016/j.jare.2020.07.010

Hasin AA, Gurman SJ, Murphy LM et al (2010) Remediation of chromium (VI) by a methane-oxidizing bacterium. Environ Sci Technol 44:400–405

Hassan Saad SM, Nasser S Awwad, Awaad HA. Aboterika (2008) Removal of mercury(II) from wastewater using camel bone charcoal. Journal of Hazardous Materials, 154:1–3, 992–997 https://doi.org/10.1016/j.jhazmat.2007.11.003

Hassan M, Alam MZ, Anwar MN (2013) Biodegradation of textile azo dyes by bacteria isolated from dyeing industry effluent. Int Res J Biol Sci 2:27–31. https://goo.gl/uwRa6r

Hassanshahian M, Emtiazi G, Caruso G, Cappello S (2014) Bioremediation (bioaugmentation/ biostimulation) trials of oil polluted seawater: a mesocosm simulation study. Mar Environ Res 95:28–38

Heipieper HJ, de Bont JAM (1994) Adaptation of *Pseudomonas putida* S 12 to ethanol and toluene. at the level of fatty acid composition of membranes. Appl Environ Microbiol 60:4440–4444

Heipieper HJ, Diefenbach R, Keweloh H (1992) Conversion of cis unsaturated fatty acids to trans, a possible mechanism for the protection of phenol-degrading *Pseudomonas putida* P8 from substrate toxicity. Appl Environ Microbiol 58:847–1852

Horel A, Schiewer S (2020) Microbial degradation of different hydrocarbon fuels with mycoremediation of volatiles. Microorganisms 8(2):163. https://doi.org/10.3390/microorganisms8020163

Hussaini S, Shaker M, Asef M (2013) Isolation of bacterial for degradation of selected pesticides. Bull Environ Pharmacol Life Sci 2:50–53. https://goo.gl/F8H5jz

Irazusta V, Bernal RA, Estévez CM, de Figueroa ICL (2018) Proteomic and enzymatic response under Cr (VI) overload in yeast isolated from textile-dye industry effluent. Ecotoxicol Environ Saf 148:490–500

Jain PK, Gupta VK, Bajpai V, Lowry M, Jaroli DP (2011) GMO's: perspective of bioremediation. In: Jain PK, Gupta VK, Bajpai V (eds) Recent advances in environmental biotechnology. LAP Lambert Academic Publishing AG and Co. KG, Saarbrücken, pp 6–23

Jessica R, Ackerman CE, Scow KM (1999) Biodegradation of methyl tert-butyl ether by a bacterial pure culture. Appl Environ Microbiol 11:4788–4792

JGI Genome Portal (2016) Dechloromonas aromatica RCB. 16 Feb 2016. http://genome.jgi.doe.gov/decar/decar.home.html

Jiang CY, Sheng XF, Qian M, Wang QY (2008) Isolation and characterization of heavy metal resistant Burkholderia species from heavy metal contaminated paddy field soil and its potential in promoting plant growth and heavy metal accumulation in metal polluted soil. Chemosphere 72:157–164

Joo JO, Choi JH, Kim IH et al (2015) Effective bioremediation of Cadmium (II), nickel (II), and chromium (VI) in a marine environment by using *Desulfovibrio desulfuricans*. Biotechnol Bioproc Eng 20:937–941. https://doi.org/10.1007/s12257-015-0287-6

Jurkiewicz P, Olżyńska A, Cwiklik L (2012) Biophysics of lipid bilayers containing oxidatively modified phospholipids: insights from fluorescence and EPR experiments and from MD simulations. Biochim Biophys Acta 1818(10):2388–2402

Kadirvelu K, Senthilkumar P, Thamaraiselvi K, Subburam V (2002) Activated carbon prepared from biomass as adsorbent: elimination of Ni (II) from aqueous solution. Bioresour Technol 81:87–90

Kang JW (2014) Removing environmental organic pollutants with bioremediation and phytoremediation. Biotechnol Lett 36(6):1129–1139. https://doi.org/10.1007/s10529-014-1466-9

Kanissery RG, Sims GK (2011) Biostimulation for the enhanced degradation of herbicides in soil. Appl Environ Soil Sci 843450:10. https://doi.org/10.1155/2011/843450

Kanmani P, Aravind J, Preston D (2012) Remediation of chromium contaminants using bacteria. Int J Environ Sci Technol 9:183–193. https://doi.org/10.1007/s13762-011-0013-7. [Google Scholar]

Kapley A, Prasad S, Purohit HJ (2007) Changes in microbial diversity in fed-batch reactor operation with wastewater containing nitroaromatic residues. Bioresour Technol 98:2479–2484. https://doi.org/10.1016/j.biortech.2006.09.012

Karigar SC, Rao SS (2011) Role of microbial enzymes in the bioremediation of pollutants. Enzyme Res 2011:805187. https://doi.org/10.4061/2011/80518

Katarína D, Murínová S, Dudášová H, Lászlová K, Horváthová H (2018) The adaptation mechanisms of bacteria applied in bioremediation of hydrophobic toxic environmental pollutants: how indigenous and introduced bacteria can respond to persistent organic pollutants-induced stress? In: Donyinah SK (ed) Persistent organic pollutants. IntechOpen. https://doi.org/10.5772/intechopen.79646

Kato T, Haruki M, Imanaka T, Morikawa M, Kanaya S (2001) Isolation and characterization of long-chain-alkane degrading *Bacillus thermoleovorans* from deep subterranean petroleum reservoirs. J Biosci Bioeng 91(1):64–70

Kehinde FO, Isaac SA (2016) Effectiveness of augmented consortia of Bacillus coagulans, Citrobacter koseri and Serratia ficaria in the degradation of diesel polluted soil supplemented with pig dung. Afr J Microbiol Res 10:1637–1644. https://goo.gl/g8Sh3U

Kim SU, Cheong YH, Seo DC, Hu JS, Heo JS, Cho JS (2007) Characterisation of heavy metal tolerance and biosorption capacity of bacterium strain CPB4 (Bacillus spp.). Water Sci Technol 55(1–2):105–111

Kiyono M, Pan-Hou H (2006) Genetic engineering of bacteria for environmental remediation of mercury. J Health Sci 52:199–204

Kleindienst S, Paul JH, Joye SB (2015) Using dispersants after oil spills: impacts on the composition and activity of microbial communities. Nat Rev Microbiol 13:388–396. https://doi.org/10.1038/nrmicro3452

Kou S, Vincent G, Gonzalez E, Pitre FE, Labrecque M, Brereton NJB (2018) The response of a 16S ribosomal RNA gene fragment amplified community to lead, zinc, and copper pollution in a Shanghai field trial. Front Microbiol 9:366. https://doi.org/10.3389/fmicb.2018.00366

Krasowska A, Sigler K (2014) How microorganisms use hydrophobicity and what does this mean for human needs? Front Cell Infect Microbiol 4:112. https://doi.org/10.3389/fcimb.2014.00112

Kulshreshtha S (2013) Genetically engineered microorganisms: a problem solving approach for bioremediation. J Bioremed Biodegr 4:1–2

Kumar JIN, Oommen C (2012) Removal of heavy metals by biosorption using freshwater alga Spirogyra hyaline. J Environ Biol 33:27–31

Kumar A, Bisht BS, Joshi VD, Dhewa T (2011) Review on bioremediation of polluted environment: a management tool. Int J Environ Sci 1(6):1079–1093

Kumar A, Trefault N, Olaniran AO (2016a) Microbial degradation of 2,4-dichlorophenoxyacetic acid: insight into the enzymes and catabolic genes involved, their regulation and biotechnological implications. Crit Rev Microbiol 42(2):194–208

Kumar S, Chaurasia P, Kumar A (2016b) Isolation and characterization of microbial strains from textile industry effluents of Bhilwara, India: analysis with bioremediation. J Chem Pharm Res 8:143–150. https://goo.gl/nrYoz6

Ledezma-Villanueva A, Adame-Rodríhuez JM, O'Connor-Sánchez IA, Villarreal-Chiu JF, Aréchiga-Carvajal ET (2015) Biodegradation kinetic rates of diesel-contaminated sandy soil samples by two different microbial consortia. Ann Microbiol 66:197–206

Lee TH, Byun IG, Kim YO, Hwang IS, Park TJ (2006) Monitoring biodegradation of diesel fuel in bioventing processes using in situ respiration rate. Water Sci Technol 53(4–5):263–272

Li H, Feng Y, Li X, Zeng D (2018a) Analytical confirmation of various herbicides in drinking water resources in sugarcane production regions of Guangxi, China. Bull Environ Contam Toxicol 100:815–820

Li T, Horton RM, Bader DA, Liu F, Sun Q, Kinney PL (2018b) Long-term projections of temperature-related mortality risks for ischemic stroke, hemorrhagic stroke, and acute ischemic heart disease under changing climate in Beijing, China. Environ Int 112:1–9. https://doi.org/10.1016/j.envint.2017.12.006

Lin C, Gan L, Chen ZL (2010) Biodegradation of naphthalene by strain Bacillus fusiformis (BFN). J Hazard Mater 182:771–777. https://goo.gl/L828ta

Liu S, Zhang F, Chen J et al (2011) Arsenic removal from contaminated soil via biovolatilization by genetically engineered bacteria under laboratory conditions. J Environ Sci 23:1544–1550

Lopez A, Lazaro N, Morales S et al (2002) Nickel biosorption by free and immobilized cells of Pseudomonas fluorescens 4F39: a comparative study. Water Air Soil Pollut 135:157–172

Lovley DR (2003) Cleaning up with genomics: applying molecular biology to bioremediation. Nat Rev Microbiol 1:35–44. https://doi.org/10.1038/nrmicro731

Ma YL, Lu W, Wan LL, Luo N (2015) Elucidation of fluoranthene degradative characteristics in a newly isolated *Achromobacter xylosoxidans* DN002. Appl Biochem Biotechnol 175:1294–1305. https://doi.org/10.1007/s12010-014-1347-7

Macaulay BM (2014) Understanding the behavior of oil-degrading microorganisms to enhance the microbial remediation of spilled petroleum. Appl Ecol Environ Res 13:247–262. https://goo.gl/JfFVWd

Madhavi GN, Mohini DD (2012) Review paper on – Parameters affecting bioremediation. International journal of life science and pharma research 2:77–80. https://goo.gl/tBP2C6

Maggi F, Tang FHM, la Cecilia D, McBratney A (2019) PEST-CHEMGRIDS, global gridded maps of the top 20 crop-specific pesticide application rates from 2015 to 2025. Sci Data 6:170

Mahjoubi M, Cappello S, Souissi Y, Jaouani A, Cherif A (2018) Microbial bioremediation of petroleum hydrocarbon contaminated marine environments. In: Recent insights in petroleum science and engineering. IntechOpen

Maliji D, Olama Z, Holail H (2013) Environmental studies on the microbial degradation of oil hydrocarbons and its application in Lebanese oil polluted coastal and marine ecosystem. Int J Curr Microbiol App Sci 2:1–18. https://goo.gl/qsW9tE

Malik A (2004) Metal bioremediation through growing cells. Environ Int 30(2):261–278

Malla MA, Dubey A, Yadav S, Kumar A, Hashem A, Abd Allah EF (2018) Understanding and designing the strategies for the microbe-mediated remediation of environmental contaminants using omics approaches. Front Microbiol 9:1132. https://doi.org/10.3389/fmicb.2018.01132

Maphosa F, de Vos WM, Smidt H (2010) Exploiting the ecogenomics toolbox for environmental diagnostics of organohalide-respiring bacteria. Trends Biotechnol 28:308–316. https://doi.org/10.1016/j.tibtech.2010.03.005

Marchant R, Sharkey FH, Banat IM, Rahman TJ, Perfumo A (2017) Marine environments. In: Recent insights in petroleum science and engineering. IntechOpen, pp 326–350. [CrossRef]

Markl H, Antranikian G, Becker P, Markossian S (1999) Aerobic biodegradation of aromatic compounds having low water solubility using *Bacillus thermoleovorans* strain DSM 10561. Patent US5965431, October 12

Mazhar SH, Herzberg M, Fekih IB, Zhang C, Bello SK, Li YP, Su J, Xu J, Feng R, Zhou S, Rensing C (2020) Comparative insights into the complete genome sequence of highly metal resistant *Cupriavidus metallidurans* strain BS1 isolated from a gold–copper mine. Front Microbiol 11:47. https://doi.org/10.3389/fmicb.2020.00047

Mejáre M, Bülow L (2001) Metal-binding proteins and peptides in bioremediation and phytoremediation of heavy metals. Trends Biotechnol 19(2):67–73

Misra BB, Langefeld CD, Olivier M, Cox LA (2018) Integrated omics: tools, advances, and future approaches. J Mol Endocrinol 62:R21–R45. https://doi.org/10.1530/JME-18-0055

Mohamed AT, El Hussein AA, El Siddig MA, Osman AG (2011) Degradation of oxyfluorfen herbicide by soil microorganisms: biodegradation of herbicides. Biotechnology 10:274–279. https://goo.gl/94wToV

Mohammed AS, Kapri A, Goel R (2011) Heavy metal pollution: source, impact, and remedies. In: Khan MS, Zaid A, Goel R, Musarrat J (eds) Biomanagement of metal contaminated soils, Environmental pollution, vol 20. Springer Science. https://doi.org/10.1007/978-94-007-1914-9_1

Mondal M, Halder G, Oinam G, Indrama T, Tiwari NO (2019) Bioremediation of organic and inorganic pollutants using microalgae. In: Gupta VK, Pandey A (eds) New and future developments in microbial biotechnology and bioengineering. Elsevier, Amsterdam, pp 223–235

Mónica P, Darwin RO, Manjunatha B, Zúñiga JJ, Diego R et al (2016) Evaluation of various pesticides-degrading pure bacterial cultures isolated from pesticide-contaminated soils in Ecuador. Afr J Biotechnol 15:2224–2233. https://goo.gl/Fs8VS7

Monti MR, Smania AM, Fabro G et al (2005) Engineering *Pseudomonas fluorescens* for biodegradation of 2, 4-dinitrotoluene. Appl Environ Microbiol 71(12):8864–8872

Mosa KA, Saadoun I, Kumar K, Helmy M, Dhankher OP (2016) Potential biotechnological strategies for the cleanup of heavy metals and metalloids. Front Plant Sci 7:303. View at: Google Scholar

Mrozik A, Cycoń M, Piotrowska-Seget Z (2010) Changes of FAME profiles as a marker of phenol degradation in different soils inoculated with *Pseudomonas sp.* CF600. Int Biodeterior Biodegradation 64:86–96

Mukherjee AK, Bhagowati P, Biswa BB, Chanda A, Kalita B (2017) A comparative intracellular proteomic profiling of Pseudomonas aeruginosa strain ASP-53 grown on pyrene or glucose as sole source of carbon and identification of some key enzymes of pyrene biodegradation pathway. J Proteome 167:25–35. https://doi.org/10.1016/j.jprot.2017.07.020

Mulligan CN, Yong R, Gibbs BF (2001) Remediation technologies for metal contaminated soils and groundwater: an evaluation. Eng Geol 60(1–4):193–207

Murínová S, Dercová K (2014) Response mechanisms of bacterial degraders to environmental contaminants on the level of cell walls and cytoplasmic membrane. Int J Microbiol 2014:873081. https://doi.org/10.1155/2014/873081

Murtaza I, Dutt A, Ali A (2002) Biomolecular engineering of Escherichia coli organomercurial lyase gene and its expression. Indian J Biotechnol 1:117–120

Naik MM, Dubey SK (2013) Lead resistant bacteria: lead resistance mechanisms, their applications in lead bioremediation and biomonitoring. Ecotoxicol Environ Saf 98:1–7

Narasimhan K, Basheer C, Bajic VB (2003) Enhancement of plant microbe interactions using rhizosphere metabolomics driven approach and its applications in the removal of poly-chlorinated biphenyls. Plant Physiol 132:146–153

Ng SP, Davis B, Polombo EA et al (2009) Tn5051 like mer containing transposon identified in a heavy metal tolerant strain Achromobacter sp. AO22. BMC Res Notes 7:2–38

Nicolaou SA, Gaida SM, Papoutsakis ETA (2010) Comparative view of metabolite and substrate stress and tolerance in microbial bioprocessing: from biofuels and chemicals, to biocatalysis and bioremediation. Metab Eng 12:307–331

Nitrification and Denitrification Wastewater Treatment (1996) Patent No 5536407, 16

Niu G, Zhang J, Zhao S, Liu H, Boon N, Zhou N (2009) Bioaugmentation of a 4-chloronitrobenzene contaminated soil with *Pseudomonas putida* ZWL73. Environ Pollut 57:763–771

Orji FA, Ibiene AA, Dike EN (2012) Laboratory scale bioremediation of petroleum hydrocarbon—polluted mangrove swamps in the Niger Delta using cow dung. Malays J Microbiol 8:21–228

Pacwa-Płociniczak M, Płaza GA, Piotrowska-Seget Z, Cameotra SS (2011) Environmental applications of biosurfactants: recent advances. Int J Mol Sci 12(1):633–654

Pandey A, Tripathi PH, Tripathi AH, Pandey SC, Gangola S (2019) Omics technology to study bioremediation and respective enzymes. In: Bhatt P (ed) Smart bioremediation technologies: microbial enzymes. Academic Press, New Delhi, pp 23–43. https://doi.org/10.1016/B978-0-12-818307-6.00002-0

Parameswarappa S, Karigar C, Nagenahalli M (2008) Degradation of ethylbenzene by free and immobilized *Pseudomonas fluorescens*-CS2. Biodegradation 19(1):137–144

Paranthaman SR, Karthikeyan B (2015) Bioremediation of heavy metal in paper mill effluent using *Pseudomonas* spp. Int J Microbiol 1:1–5. https://goo.gl/fU79iK

Patel J, Zhang Q, Michael R et al (2010) Genetic engineering of Caulobacter crescentus for removal of cadmium from water. Appl Biochem Biotechnol 160:232–243

Paul D, Pandey G, Jain RK (2005) Suicidal genetically engineered microorganisms for bioremediation: need and perspectives. BioEssays 27(5):563–573

Pepi M, Heipieper JH, Fischer J, Ruta M, Volterrani M, Focardi ES (2008) Membrane fatty acids adaptive profile in the simultaneous presence of arsenic and toluene in *Bacillus* sp. ORAs2 and *Pseudomonas* sp. ORAs5 strains. Extremophiles 12(3):343–349

Perpetuo EA, Souza CB, Nascimento CAO (2011) Engineering bacteria for bioremediation. In: Carpi A (ed) Progress in molecular and environmental bioengineering—from analysis and modeling to technology applications. InTech, Rijeka, pp 605–632

Perfumo A, Banat IM, Canganella F et al. (2006) Rhamnolipid production by a novel thermophilic hydrocarbondegrading Pseudomonas aeruginosa AP02-1. Appl Microbiol Biotechnol 72:132–138. https://doi.org/10.1007/s00253-005-0234-0

Phulpoto H, Qazi MA, Mangi S, Ahmed S, Kanhar NA (2016) Biodegradation of oil-based paint by Bacillus species monocultures isolated from the paint warehouses. Int J Environ Sci Technol 13:125–134. https://goo.gl/j4oESf

Pileggi M, Pileggi SAV, Olchanheski LR, da Silva PAG, Munoz Gonzalez AM, Koskinen WC, Barber B, Sadowsky MJ (2012) Isolation of mesotrione-degrading bacteria from aquatic environments in Brazil. Chemosphere 86:1127–1132

Pileggi M, Pileggi SAV, Sadowsky MJ (2020) Herbicide bioremediation: from strains to bacterial communities. Heliyon 6(12):e05767. https://doi.org/10.1016/j.heliyon.2020.e05767

Prione LP, Olchanheski LR, Tullio LD, Santo BCE, Reche PM, Martins PF, Carvalho G, Demiate IM, Pileggi SAV, Dourado MN, Prestes RA, Sadowsky MJ, Azevedo RA, Pileggi M (2016) GST activity and membrane lipid saturation prevents mesotrione-induced cellular damage in *Pantoea ananatis*. AMB Express 6:70

Priyalaxmi R, Murugan A, Raja P, Raj KD (2014) Bioremediation of cadmium by *Bacillus safensis* (JX126862), a marine bacterium isolated from mangrove sediments. Int J Curr Microbiol App Sci 3:326–335. https://goo.gl/qmHJgu

Puyen ZM, Villagrasa E, Maldonado J, Diestra E, Esteve I, Solé A (2012) Biosorption of lead and copper by heavy-metal tolerant Micrococcus luteus DE2008. Bioresour Technol 126:233–237. https://doi.org/10.1016/j.biortech.2012.09.036. Google Scholar

Qu Y, Zhang X, Xu J, Zhang W, Guo Y (2014) Removal of hexavalent chromium from wastewater using magnetotactic bacteria. Sep Purif Technol 136:10–17

Quek E, Ting YP, Tan HM (2006) *Rhodococcus* F92 immobilized on polyurethane foam shows ability to degrade various petroleum products. Bioresour Technol 97(1):32–38

Rahman A, Nahar N, Nawani NN, Jass J, Hossain K, Saud ZA, Saha AK, Ghosh S, Olsson B, Mandal A (2015) Bioremediation of hexavalent chromium (VI) by a soil-borne bacterium, Enterobacter cloacae b2-dha. J Environ Sci Health A 50:1136–1147. https://doi.org/10.1080/10934529.2015.1047670

Rangarajan V, Narayanan M (2018) Biosurfactants in soil bioremediation. In: Advances in soil microbiology: recent trends and future prospects. Springer, Singapore, pp 193–204

Rani A, Shouche YS, Goel R (2008) Declination of copper toxicity in pigeon pea and soil system by growth promoting *Proteus vulgaris* KNP3 strain. Curr Microbiol 57:78–82

Rani A, Shouche YS, Goel R (2009) Comparative assessment of in situ bioremediation potential of cadmium resistant acidophilic *Pseudomonas putida* 62BN and alkalophilic *Pseudomonas monteilli* 97AN strains on soybean. Int Biodeterior Biodegrad 63:62–66

Rendón-von Osten J, Dzul-Caamal R (2017) Glyphosate residues in groundwater, drinking water and urine of subsistence farmers from intensive agriculture localities: a survey in Hopelchén, Campeche, Mexico. Int J Environ Res Public Health 14:595

Renninger N, McMahon KD, Knop R, Nitsche H, Clark DS, Keasling JD (2001) Uranyl precipitation by biomass from an enhanced biological phosphorus removal reactor. Biodegradation 12:401–410

Riggle PJ, Kumamoto CA (2000) Role of a *Candida albicans* P1-type ATPase in resistance to copper and silver ion toxicity. J Bacteriol 182(17):4899–4905
Roanne TM, Pepper IL (2001) Environmental microbiolology. In: Roane TM, Pepper IL (eds) Microorganisms and metal pollutants, vol 17. Academic Press, pp 403–423
Rodrigues JL, Maltseva OV, Tsoi TV et al (2001) Development of a *Rhodococcus* recombinant strain for degradation of products from anaerobic dechlorination of PCBs. Environ Sci Technol 35(4):663–668
Rodrigues JL, Kachel CA, Aiello MR et al (2006) Degradation of aroclor 1242 dechlorination products in sediments by *Burkholderia xenovorans* LB400 (ohb) and *Rhodococcus* strain RHA1 (fcb). Appl Environ Microbiol 72(4):2476–2482
Ron EZ, Rosenberg E (2014) Enhanced bioremediation of oil spills in the sea. Curr Opin Biotechnol 27:191–194. https://doi.org/10.1016/j.copbio.2014.02.004
Safiyanu I, Isah AA, Abubakar US, Rita Singh M (2015) Review on comparative study on bioremediation for oil spills using microbes. Res J Pharm, Biol Chem Sci 6:783–790. https://goo.gl/LSj5vR
Sağ Y, Kutsal T (2000) Determination of the biosorption heats of heavy metal ions on Zoogloea ramigera and Rhizopus arrhizus. Biochem Eng J 6(2):145–151
Šajbidor J (1997) Effect of some environmental factors on the content and composition of microbial membrane lipids. Crit Rev Biotechnol 17(2):87–103. https://doi.org/10.3109/07388559709146608
Sakshi, Haritash AK (2020) A comprehensive review of metabolic and genomic aspects of PAH-degradation. Arch Microbiol 202:2033–2058. https://doi.org/10.1007/s00203-020-01929-5
Sánchez-Riego AM, López-Maury L, Florencio FJ (2014) Genomic responses to arsenic in the cyanobacterium *Synechocystis* sp. PCC 6803. PLoS One 9:e96826
Sani I, Safiyanu I, Rita SM (2015) Review on bioremediation of oil spills using microbial approach. IJESR 3:41–45
Sarang B, Richa K, Ram C (2013) Comparative study of bioremediation of hydrocarbon fuel. Int J Biotechnol Bioeng Res 4:677–686. https://goo.gl/CCnV57
Saravanan A, Jeevanantham S, Anantha Narayanan V, Senthil Kumar P, Yaashikaa PR, Mathan Muthu CM (2020) Rhizoremediation – a promising tool for the removal of soil contaminants: a review. J Environ Chem Eng 8(2):103543. https://doi.org/10.1016/j.jece.2019.103543
Sarkar P, Roy A, Pal S, Mohapatra B, Kazy SK, Maiti MK, Sara P (2017) Enrichment and characterization of hydrocarbon-degrading bacteria from petroleum refinery waste as potent bioaugmentation agent for in situ bioremediation. Bioresour Technol 242:15–27. https://doi.org/10.1016/j.biortech.2017.05.010
Sasaki Y, Minakawa T, Miyazaki A et al (2005) Functional dissection of a mercuric ion transporter Mer C from Acidithiobacillus ferrooxidans. Biosci Biotechnol Biochem 69:1394–1402
Satlewal A, Goel R, Garg GK (2010) Industrially useful microbial bioresources. In: Maheshwari DK, Dubey RC, Saravanamuthu R (eds) Industrial exploitation of microorganisms. IK International Publication Ltd, New Delhi, pp 390–405
Sei K, Nakao M, Mori KM, Kohno IT, Fujita M (2001) Design of PCR primers and a gene probe for extensive detection of poly (3-hydroxybutyrate) (PHB)-degrading bacteria possessing fibronectin type III linker type-PHB depolymerases. Appl Microbiol Biotechnol 55:801–806
Shah MP, Patel KA, Nair SS, Darji AM (2013) Microbial degradation of textile dye (Remazol black B) by *Bacillus spp*. ETL-2012. J Bioremed Biodegr 4:1–5. https://goo.gl/hk89fD
Shahid MJ, Tahseen R, Siddique M, Ali S, Iqbal S, Afzal M (2019) Remediation of polluted river water by floating treatment wetlands. Water Supply 19(3):967–977
Shimura M, Mukerjee-Dhar G, Kimbara K, Nagato H, Kiyohara H, Hatta T (1999) Isolation and characterization of a thermophilic *Bacillus* sp. JF8 capable of degrading polychlorinated biphenyls and naphthalene. FEMS Microbiol Letters 178:87–93
Simarro R, Gonzalez N, Bautista LF, Molina MC (2013) Assessment of the efficiency of in situ bioremediation techniques in a creosote polluted soil: change in bacterial community. J Hazard Mater 262:158–167. https://goo.gl/RceNkr

Singh A, Van Hamme JD, Ward OP (2007) Surfactants in microbiology and biotechnology: part 2. Application aspects. Biotechnol Adv 25(1):99–121
Singh A, Kumar V, Srivastava JN (2013a) Assessment of bioremediation of oil and phenol contents in refinery waste water via bacterial consortium. J Pet Environ Biotechnol 4:1–4. https://goo.gl/yavRNm
Singh N, Tuhina V, Rajeeva G (2013b) Detoxification of hexavalent chromium by an indigenous facultative anaerobic Bacillus cereus strain isolated from tannery effluent. Afr J Biotechnol 12(10):1091–1103
Singh R, Singh P, Sharma R (2014) Microorganism as a tool of bioremediation technology for cleaning environment: a review. Int Acad Ecol Environ Sci 4(1):1–6
Sinha SN, Biswas K (2014) Bioremediation of lead from river water through lead-resistant purple-nonsulfur bacteria. Glob J Microbiol Biotechnol 2:11–18. https://goo.gl/B7mKGH
Sinha SN, Paul D (2014) Heavy metal tolerance and accumulation by bacterial strains isolated from waste water. J Chem Biol Phys Sci 4:812–817. https://goo.gl/HMLnQQ
Sinha SN, Biswas M, Paul D, Rahaman S (2011) Biodegradation potential of bacterial isolates from tannery effluent with special reference to hexavalent chromium. Biotechnol Bioinformatics Bioeng 1:381–386. https://goo.gl/Hwz87L
Sivakumar G, Xu J, Thompson RW, Yang Y, Randol-Smith P et al (2012) Integrated green algal technology for bioremediation and biofuel. Bioresour Technol 107:1–9. https://goo.gl/U6UYUw
Song HP, Li XG, Sun JS, Xu SM, Han X (2008) Application of a magnetotatic bacterium, Stenotrophomonas sp. to the removal of Au(III) from contaminated wastewater with a magnetic separator. Chemosphere 72:616–621
Song M, Yang Y, Jiang L, Hong Q, Zhang D, Shen Z, Yin H, Luo C (2017) Characterization of the phenanthrene degradation-related genes and degrading ability of a newly isolated copper-tolerant bacterium. Environ Pollut 220:1059–1067. https://doi.org/10.1016/j.envpol.2016.11.037
Sorkhoh NA, Ibrahim AS, Ghannoum MA, Radwan SS (1993) High-temperature hydrocarbon degradation by Bacillus stearothermophilus from oil-polluted Kuwaiti desert. Appl Microbiol Biotechnol 39:123–126
Sriprang R, Hayashi M, Ono H, Takagi M (2003) Enhanced accumulation of Cd2+ by a Mesorhizobium sp. transformed with a gene from *Arabidopsis thaliana* coding for phytochelatin synthase. Appl Environ Microbiol 69(3):1791–1796. https://doi.org/10.1128/AEM.69.3.1791-1796.2003
Srivastava S, Agrawal SB, Mondal MK (2015) A review on progress of heavy metal removal using adsorbents of microbial and plant origin. Environ Sci Pollut Res 22(20):15386–15415. https://doi.org/10.1007/s11356-015-5278-9. [CrossRef], [PubMed], [Web of Science ®], [Google Scholar]
Stapleton RD, Ripp S, Jimenez L, Cheol-Koh S, Fleming JT, Gregory IR, Sayler GS (1998) Nucleic acid analytical approaches in bioremediation: site assessment and characterization. J Microbiol Methods 32:165–178
Strong LC, McTavish H, Sadowsky MJ et al (2000) Field-scale remediation of atrazine-contaminated soil using recombinant *Escherichia coli* expressing atrazine chlorohydrolase. Environ Microbiol 2(1):91–98
Sukumar S, Nirmala P (2016) Screening of diesel oil degrading bacteria from petroleum hydrocarbon contaminated soil. Int J Adv Res Biol Sci 3:18–22. https://goo.gl/PAoc9z
Svenningsen NB, Damgaard M, Rasmussen M, Pérez-Pantoja D, Nybroe O, Nicolaisen MH (2017) *Cupriavidus pinatubonensis* AEO106 deals with copper-induced oxidative stress before engaging in biodegradation of the herbicide 4-chloro-2-methylphenoxyacetic acid. BMC Microbiol 17:211
Sviridov AV, Shushkova TV, Ermakova IT, Ivanova EV, Epiktetov DO, Leontievsky AA (2015) Microbial degradation of glyphosate herbicides. Appl Biochem Microb 51:188–95. https://goo.gl/tBP2C6
Székács A, Mörtl M, Darvas B (2015) Monitoring pesticide residues in surface and ground water in Hungary: surveys in 1990–2015. J Chem 2015:1–15

Teramoto M, Suzuki M, Hatmanti A, Harayama S (2010) The potential of Cycloclasticus and Altererythrobacter strains for use in bioremediation of petroleum-aromatic-contaminated tropical marine environments. J Biosci Bioeng 110(1):48–52. https://doi.org/10.1016/j.jbiosc.2009.12.008

Terra Nova's Environmental Remediation Resources (2009) Terranovabiosystems.com. 31 Aug 2009. http://www.terranovbiosystems.com/science/remidiation-resourceshtml. Retrieved 22 Mar 2011

Texier AC, Andres Y, le Cloirec P (1999) Selective biosorption of lanthanide (La, Eu, Yb) ions by *Pseudomonas aeruginosa*. Environ Sci Technol 33:489–495

Thapa B, Kumar A, Ghimire A (2012) A review on bioremediation of petroleum hydrocarbon contaminants in soil. Kathmandu Univ J Sci Eng Technol 8(1):164–170

Thatoi H, Das S, Mishra J, Rath BP, Das N (2014) Bacterial chromate reductase, a potential enzyme for bioremediation of hexavalent chromium: a review. J Environ Manag 146:383–399

Tiedje JM (1993) Bioremediation from an ecological perspective. In: In situ bioremediation: when does it work. The National Academies Press, Washington, DC, pp 110–120

Torres S, Pandey A, Castro RG (2011) Organic solvent adaptation of Gram-positive bacteria: applications and biotechnological potentials. Biotechnol Adv 29(4):442–452

Tremblay J, Yergeau E, Fortin N, Cobanli S, Elias M, King LT, Lee K, Greer WC (2017) Chemical dispersants enhance the activity of oil-and gas condensate-degrading marine bacteria. ISME J 11:2793–2808. https://doi.org/10.1038/ismej.2017.129

Tsezos M, Volesky B (1981) Biosorption of uranium and thorium. Biotechnol Bioeng 23:583–604

Valle A, Boschin G, Negri M, Abbruscato P, Sorlini C, D'Agostina A, Zanardini E (2006) The microbial degradation of azimsulfuron and its effect on the soil bacterial community. J Appl Microbiol 101:443–452

Varjani SJ, Upasani VN (2016) Biodegradation of petroleum hydrocarbons by oleophilic strain of Pseudomonas aeruginosa NCIM 5514. Bioresour Technol 222:195–201. https://doi.org/10.1016/j.biortech.2016.10.006

Varjani SJ, Rana DP, Jain AK, Bateja S, Upasani VN (2015) Synergistic ex-situ biodegradation of crude oil by halotolerant bacterial consortium of indigenous strains isolated from on shore sites of Gujarat, India. Int Biodeterior Biodegrad 103:116–124. https://doi.org/10.1016/j.ibiod.2015.03.030

Valls M, Atrian S, de Lorenzo V. et al. (2000) Engineering a mouse metallothionein on the cell surface of Ralstonia eutropha CH34 for immobilization of heavy metals in soil. Nat Biotechnol 18, 661–665 https://doi.org/10.1038/76516

Venkidusamy K, Hari AR, Megharaj M (2018) Petrophilic, Fe (III) reducing exoelectrogen *Citrobacter* sp. KVM11, isolated from hydrocarbon fed microbial electrochemical remediation systems. Front Microbiol 9:349

Vidali M (2001) Bioremediation. An overview. Pure Appl Chem 73(7):1163–1172

Villaverde J, Rubio-Bellido M, Merchan F, Morillo E (2017) Bioremediation of diuron contaminated soils by a novel degrading microbial consortium. J Environ Manag 188:379–386

Vullo DL, Ceretti HM, Daniel MA, Ramírez SA, Zalts A (2008) Cadmium, zinc and copper biosorption mediated by *Pseudomonas veronii* 2E. Bioresour Technol 99(13):5574–5581

Wang XB, Chi CQ, Nie Y, Tang YQ, Tan Y, Wu G, Wu XL (2011) Degradation of petroleum hydrocarbons (C6–C40) and crude oil by a novel Dietzia strain. Bioresour Technol 102:7755–7761. https://doi.org/10.1016/j.biortech.2011.06.009

Wasilkowski D, Swedziol Ż, Mrozik A (2012) The applicability of genetically modified microorganisms in bioremediation of contaminated environments. Chemik 66(8):822–826

Weber FJ, Isken S, De Bont AMJ (1994) Cis/trans isomerization of fatty adds as a defence mechanism of *Pseudomonas putida* strains to toxic concentrations of toluene. Microbiology 140(8):2013–2017

Wittich RM, Wolff P (2007) Growth of the genetically engineered strain *Cupriavidus necator* RW112 with chlorobenzoates and technical chlorobiphenyls. Microbiology 153(Pt 1):186–195

Wu CH, Wood TK, Mulchandani A et al (2006) Engineering plant-microbe symbiosis for rhizoremediation of heavy metal. Appl Environ Microbiol 72:1129–1134

Xu X, Zhai Z, Li H, Wang Q, Han X, Yu H (2017) Synergetic effect of bio-photocatalytic hybrid system: g-C3N4, and Acinetobacter, sp. JLS1 for enhanced degradation of C16 alkane. Chem Eng J 323:520–529. https://doi.org/10.1016/j.cej.2017.04.138

Xu X, Liu W, Tian S, Wang W, Qi Q, Jiang P, Gao X, Li F, Li H, Yu H (2018) Petroleum hydrocarbon-degrading bacteria for the remediation of oil pollution under aerobic conditions: a perspective analysis. Front Microbiol 9:2885. https://doi.org/10.3389/fmicb.2018.02885

Xu M, Cui Y, Beiyuan J, Wang X, Duan C, Fang L (2021) Heavy metal pollution increases soil microbial carbon limitation: evidence from ecological enzyme stoichiometry. Soil Ecol Lett 3(3):230–241. https://doi.org/10.1007/s42832-021-0094-2

Yakimov MM, Timmis KN, Golyshin PN (2007) Obligate oil-degrading marine bacteria. Curr Opin Biotechnol 18:257–266. https://doi.org/10.1016/j.copbio.2007.04.006

Yang SZ, Jin HJ, Wei Z, He RX, Ji YJ et al (2009) Bioremediation of oil spills in cold environments: a review. Pedosphere 19:371–381

Yangbenro AS, Babalola OO (2017) A new strategy for heavy metal polluted environments: a review of microbial biosorbents. Int J Environ Res Public Health 14(1):94. https://doi.org/10.3390/ijerph14010094. [CrossRef], [Web of Science ®], [Google Scholar]

Yogesh P, Akshaya G (2016) Evaluation of bioremediation potential of isolated bacterial culture YPAG-9 (Pseudomonas aeruginosa) for decolorization of sulfonated di-azodye reactive red HE8B under optimized culture conditions. Int J Curr Microbiol App Sci 5:258–272. https://goo.gl/NuA1va

Zenati B, Badis A (2019) A non-toxic microbial surfactant from *Marinobacter hydrocarbonoclasticus* SdK644 for petroleum contaminated seawater bioremediation ation1

Zhang GL, Wu YZ, Qian XP, Meng Q (2005) Biodegradation of crude oil by Pseudomonas aeruginosa in the presence of rhamnolopids. J Zhejiang Univ Sci 6(8):725–730

Zhang W, Li F, Nie L (2010) Integrating multiple "omics" analysis for microbial biology: application and methodologies. Microbiology 156:287–301. https://doi.org/10.1099/mic.0.034793-0

Zhang X, Zhang Q, Yan T, Jiang Z, Zhang X, Zuo YY (2015) Quantitatively predicting bacterial adhesion using surface free energy determined with a spectrophotometric method. Environ Sci Technol 49:6164–6171. https://doi.org/10.1021/es5050425

Zhao XW, Zhou MH, Li QB et al (2005) Simultaneous mercury bioaccumulation and cell propagation by genetically engineered Escherichia coli. Process Biochem 40:1611–1616

Zorádová S, Dudášová H, Lukáčová L, Dercová K, Čertík M (2011) The effect of poly-chlorinated biphenyls (PCBs) on the membrane lipids of *Pseudomonas stutzeri*. Int Biodeterior Biodegrad 65:1019–1023

Viral Remediation in a Healthy Environment

Poonam Sharma

Abstract Viral remediation is of utmost importance in the environmental domain due to the substantial risks it poses to ecosystems through means such as disease transmission to host organisms, disruption of food chains, and impact on biodiversity. Viral contamination of water bodies can hurt human health and compromise the water's quality and purity. Various physical methods, such as filtration, ultraviolet (UV) irradiation, and heat treatment, are employed to eliminate or reduce viruses. Disinfectants such as ozone and chlorine are used to inactivate viruses via chemical processes. Biological methods employ microorganisms and plants that selectively target and eliminate specific viruses. Currently, researchers are exploring biological remediation as a potential remedy for viral contamination. The effectiveness of each method depends on various factors, including the specific characteristics of the virus, the environmental conditions, and the level of contamination, among others. Integrating a variety of methodologies has the potential to produce the most efficient solutions for the elimination of viruses.

Keywords Virus · Remediation · Zoonotic · COVID-19 · Disinfectants · Filtration · Nanotechnology

1 Introduction

Viruses are tiny infectious agents comprising genetic material (DNA or RNA) enveloped within a proteinaceous coat referred to as a capsid (Louten 2016). Viruses are characterized by their inability to carry out metabolic processes and replicate independently, necessitating their need for host cells for reproduction. Certain viruses also possess an external envelope that originates from the host cell's

P. Sharma (✉)
Department of Zoology, Gargi College, University of Delhi, Delhi, India
e-mail: poonam.sharma@gargi.du.ac.in

A. A. H. Abdel Latef et al. (eds.), *Sustainable Remediation for Pollution and Climate Resilience*, https://doi.org/10.1007/978-981-96-5674-5_15

Table 1 Viruses and human disease

Virus	Family	Disease	Genome type
Adenovirus	Adenoviridae	Respiratory infections	dsDNA
Rabies virus	Rhabdoviridae	Rabies	(−)ssRNA
Rotavirus	Reoviridae	Enteritis	dsRNA
Ebola virus	Filoviridae	Hemorrhagic fever	(−)ssRNA
Herpes simplex Type I or type II	Herpesviridae	Cold sores or genital warts	dsDNA
Smallpox virus	Poxviridae	Smallpox	dsDNA
Polio virus	Picornaviridae	Poliomyelitis	(−)ssRNA
Yellow fever virus	Flaviviridae	Yellow fever	(+)ssRNA
Human papillomavirus	Papovaviridae	Warts	dsDNA
SARS-CoV-2 virus	Coronaviridae	Covid-19	(+)ssRNA
Influenza virus	Orthomyxoviridae	Influenza	(−)ssRNA
Rhinovirus	Picornaviridae	Common cold	(+)ssRNA
HIV	Retroviridae	Acquired immunodeficiency syndrome (AIDS)	(+)ssRNA
Hepatovirus A	Picornaviridae	Hepatitis A	(+)ssRNA
Paramyxovirus	Paramyxoviridae	Mumps	(−)ssRNA
Epstein–Barr virus	Herpesviridae	Infectious mononucleosis	dsDNA
Morbillivirus	Paramyxoviridae	Measles	(−)ssRNA
Varicella-zoster	Herpesviridae	Chicken pox	dsDNA

membrane. A wide variety of viruses are responsible for a plethora of diseases (Chappell and Dermody 2015). The impact of viruses on human health and welfare is immense (Table 1). Viruses are the etiological factors responsible for several human diseases, such as the common cold and more serious illnesses like acquired immunodeficiency syndrome (AIDS), COVID-19, and *Ebola* virus disease (EVD *or Ebola*). These pathogens invade host cells, control their biological processes for reproduction, and then spread throughout the organism. Other viruses, like *human immunodeficiency virus* (HIV), specifically attack the immune system, leading to the development of acquired immunodeficiency syndrome (AIDS). On the other hand, viruses like the herpes simplex virus induce recurrent outbreaks of painful sores. Hepatitis viruses can cause hepatic inflammation and long-term damage, whereas the human papillomavirus (HPV) is the primary driver in cervical cancer development. Moreover, viruses such as measles and mumps have the potential to induce systemic infections that can lead to numerous issues. The recent COVID-19 pandemic, which the SARS-CoV-2 virus has triggered, serves as a prominent example of how viruses can greatly impact worldwide public health (Hu et al. 2021). These viral infections can lead to various health complications, including respiratory impairments and neurological manifestations (Sharma et al. 2020; Cavallieri et al. 2022).

Various mechanisms facilitate the transmission of viruses in the environment. Air pollution has the potential to facilitate the transmission of respiratory viruses such as influenza and COVID-19. This is because particulate matter can serve as carriers for these viruses, allowing them to spread further (Domingo and Rovira 2020). Outbreaks of rotavirus (RV), enterovirus, adenovirus, noroviruses, and caliciviruses can occur due to contaminated water and soil, particularly in regions with inadequate sanitation and water treatment (Lanrewaju et al. 2022a). It has been observed that the presence of pollutants can lead to an increase in virus mutation rates, which in turn can give rise to strains that are more virulent or resistant to drugs (Sloan et al. 2011). In addition, respiratory viruses spread through respiratory droplets released when an infected person coughs or sneezes (Wang et al. 2021). Vector-borne diseases are caused by arthropod-transmitted infections. The main vectors are mosquitoes and ticks. Climate change indirectly affects vector-borne diseases by altering vector dispersion and behavior. Thus, vector type determines influence levels. Global warming has been reported to impact mosquito-borne infectious diseases such as malaria, dengue, Japanese encephalitis (JE), and tick-borne encephalitis (Uwishema et al. 2023; Socha et al. 2022).

Free viral vectors used in healthcare, such as retrovirus, adenovirus, and herpes simplex, can enter the environment and cause immunological reactions or insertional mutagenesis in host genomes if accidentally released (Brown et al. 2020). These vectors are replication-deficient and can recombine with wild-type viruses to restore replication capabilities. Similarly, some live-attenuated virus vaccines induce an immune response without producing disease. These live, attenuated viruses can contaminate the environment and return to virulence, infecting immunocompromised individuals. Medical waste from virus-related procedures, such as diagnostic testing and treatments containing viral particles, if improperly disposed of, can contaminate the water bodies and soil. Even research facilities can discharge viruses through spills, leaks, or insufficient disinfection. Thus, employing viruses in healthcare and research is beneficial, but biosafety and waste management must be followed to prevent environmental viral propagation (Baldo et al. 2013). An in-depth understanding of virus transmission in the environment is crucial for effectively managing outbreaks and implementing preventive measures. Effective viral remediation strategies can significantly reduce the risk of human exposure to these infectious pathogens (Merenda and Dumée 2021). Consequently, the pressure on healthcare systems can be reduced, and community health can be preserved by preventing disease outbreaks (Sellaoui et al. 2021). In addition to human health, viral remediation is essential for ensuring clean water supplies. Adenoviruses, caliciviruses, enteroviruses, and hepatitis viruses, according to the US Environmental Protection Agency (USEPA 2018), are some of the most prevalent microbiological contaminants that are frequently found in drinking water. Among the most dangerous infectious diseases, according to 2011 WHO guidelines, are astroviruses, hepatitis E, sapoviruses, and rotaviruses. High virus levels in wastewater from households and hospitals pose a significant health and environmental challenge, potentially impacting wastewater treatment and freshwater supply (Sodhi and Singh 2023). Several studies have indicated that various viruses, such as rotavirus, norovirus, and enterovirus, have the potential to be transmitted through water (Soltane and Allayeh 2023; Jain et al. 2016; Gall et al. 2015; Scarcella et al. 2009). As a result, it is critical to inactivate or remove waterborne viruses present in municipal wastewater discharges to prevent their spread and mitigate their adverse impact on human health.

Agricultural ecosystems are other ecosystems affected by viral contamination. Plant viruses can destroy crops, compromising food security and disturbing global trade (Hamim et al. 2023; Roossinck and García-Arenal 2015). Hence, agricultural systems must employ efficient viral cleanup techniques to safeguard food supplies, preserve yields, and keep food production systems stable. Another essential component of viral remediation is maintaining the health of our wider ecosystems. Viruses can disrupt ecosystems and put biodiversity at risk in aquatic and wildlife habitats, causing ecological disturbances and endangering biodiversity (Keesing and Ostfeld 2021; Kim et al. 2005). Virus remediation is critical for preventing the spread of disease among animals as well as the overall health of the ecosystem.

The COVID-19 pandemic has highlighted our world's interdependence and the importance of effectively controlling zoonotic virus transmission through viral remediation (Fig. 1). In recent times, there has been a rising incidence of epidemics caused by zoonotic viruses such as the Nipah virus, monkeypox, and Langya virus, especially in their endemic regions (Bardhan et al. 2023; Aggarwal and Sharma 2023). The lack of vaccinations and medications specifically designed to combat these viral infections restricts patients' treatment options to supportive care and preventative measures. Although environmental management strategies strengthen the resilience of world health by reducing the likelihood of spreading zoonotic diseases and limiting the emergence of novel infectious pathogens, it is crucial to thoroughly prepare for any outbreaks of viral infections and reduce their frequency through appropriate viral remediation measures.

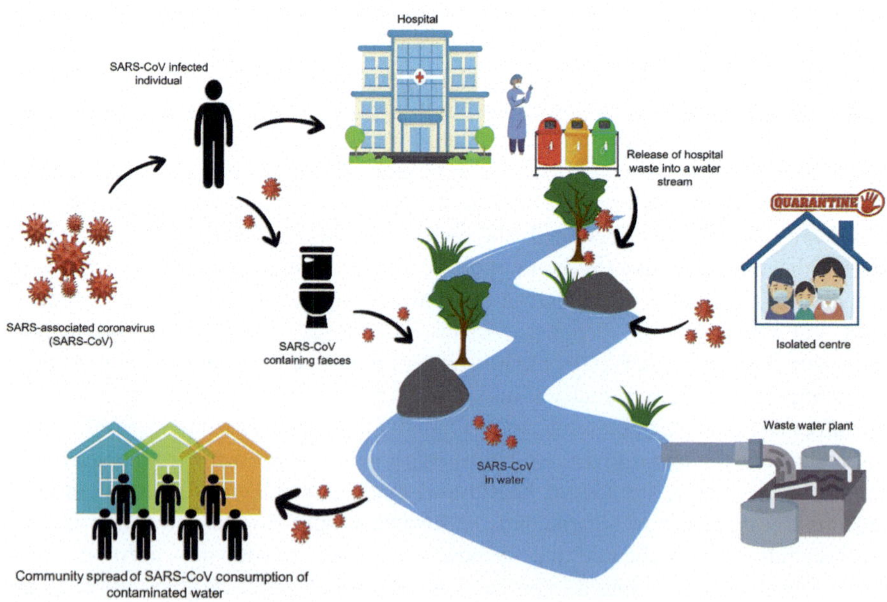

Fig. 1 Transmission of SARS-CoV-2 through wastewater

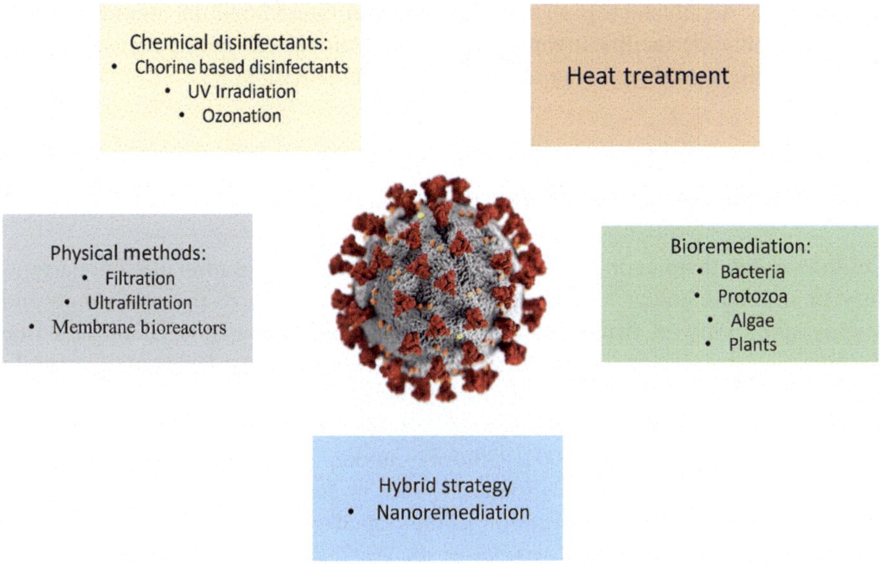

Fig. 2 Various strategies involved in viral remediation

Hence, the prerequisite for viral remediation in the environment is multifaceted and is associated with maintaining ecological balance, human health, and international stability. Communities and governments may collaborate to design and execute effective viral remediation plans, acknowledging viral remediation's significant role in these areas. This collaboration enables the successful management of viral contamination risks for the planet's and humankind's benefit. In order to address the issues related to viral remediation mentioned above, a field of study known as environmental virology has emerged. This multidisciplinary field specifically investigates viruses' presence, fate, and impacts in various environmental systems, such as water, soil, and air. In this chapter, the wide range of methods used in environmental virology, which can be classified into several techniques, each of which serves a specific purpose in assessing, detecting, and addressing viral contaminants, will be discussed (Fig. 2). Healthcare facilities, agriculture, and regular environments can employ these approaches in diverse scenarios.

2 Disinfectants

Disinfection is an essential step in the potable water treatment processes aimed at destroying contagious viruses, as primary treatment is not as efficient in eliminating viruses (Lanrewaju et al. 2022a, b). The efficacy of disinfectants differs based on the type of virus being targeted. It is imperative to choose a suitable disinfectant depending on a specific virus and the environment in which it will be used. The consistent application of disinfectants is an essential aspect of viral mitigation measures, especially in healthcare facilities, water treatment facilities, and public areas. The following are often employed disinfectants in viral remediation:

2.1 Chlorine-Based Disinfection

Chlorine-based disinfection is a widely used method in viral remediation, particularly in water treatment and sanitation. Chlorine is effective in inactivating and destroying a variety of viruses, bacteria, and other microorganisms. Employing free chlorine as a disinfectant in water has existed since the early 1900s and is widely adopted in various water systems globally (Fair et al. 1948). The predominant disinfectants utilized in the United States are chlorine and monochloramine (American Water Works Association 2000). Chlorine undergoes chemical reactions with organic and inorganic chemicals found in water or surfaces, resulting in oxidation. This mechanism induces a disruption in the organization and functioning of viral particles (Kong et al. 2021; Szczuka et al. 2022). Chlorine also interacts with viruses' functional groups, damaging their proteins and nucleic acids, impairing the viruses' capacity to multiply and induce infection (Na and Olson 2007). Studies have demonstrated the exceptional efficiency of chlorine in combating enveloped viruses. Chlorine disrupts the lipid membrane that covers enveloped viruses, rendering them inactive and incapable of infecting host cells (Ye et al. 2018).

Various applications commonly use chlorine-based chemicals, such as free chlorine, chlorine dioxide, and monochloramine, for viral remediation. These chlorine-based compounds have unique properties and applications in disinfection and water treatment. Water treatment frequently employs free chlorine, an effective disinfectant, to inactivate and eliminate a range of pathogens, including viruses. It is frequently used as a secondary disinfectant after filtering in distribution plants. It inactivates the virus's enzymes required for its crucial biological functions, such as replication. This impedes the virus's capacity to replicate and spread, which is a vital aspect of viral remediation as it governs the spread of diseases. In water, free chlorine exists in two main forms: hypochlorous acid (HOCl) and hypochlorite ion (OCl^-). These two types' proportions depend on the pH level of the water. When not dissociated, hypochlorous acid (HOCl) effectively disinfects by penetrating the lipid envelope surrounding enveloped viruses such as many influenza viruses and coronaviruses, thereby killing them (Onodera and Kirisawa 2023). Furthermore, hypochlorous acid is an oxidizing reagent that may react with numerous virus components, including proteins and nucleic acids, causing significant damage to viral structures and activities (Dianty et al. 2023).

Most of the early studies were centered around the chlorination process for the removal of polioviruses (Floyd et al. 1976; Hejkal et al. 1979; O'Brien and Newman 1979; Scarpino et al. 1972; Varma et al. 1974; Weidenkopf 1958). Initial investigations have suggested several factors that require regulation while disinfecting viruses. These factors include the duration of contact, temperature, concentration of ions, pH level, chlorine concentration, and aggregation of viruses (Sharp and Leong 1980; Sharp et al. 1980). These investigations have demonstrated that we can only accurately assess the effectiveness of disinfection against viruses if we employ identical disinfecting conditions. Several studies have assessed the effectiveness of chlorine disinfection against viruses over the years (Saguti et al. 2023; Totaro et al. 2021; Girones et al. 2014). Studies on coxsackievirus, adenovirus, and adenovirus 41 have shown that chlorine kills these viruses (Ballester and Malley 2004; Baxter et al. 2007; Sobsey et al. 1988). Similarly, another study found that hepatitis A virus (HAV), coxsackievirus B5, and coliphages MS2 and ϕX174 were rendered inactive using 0.5 mg/l of free chlorine at a pH range of 6–10, and 10 mg/l of monochloramine at a pH of 8. These treatments were conducted in 0.01 M phosphate buffer at a temperature of 5 °C. But only ϕX174 was rendered inactive by 10 mg/l of monochloramine, indicating that viruses were less susceptible to monochloramine. Studies have shown that inactivation of human adenoviruses, coxsackieviruses, echoviruses, and murine norovirus (MNV) can be effectively achieved by using 0.2 mg of free chlorine or 1 mg of monochloramine per liter of water (Cromeans et al. 2010; Sobsey et al. 1988). This effectiveness has been observed at pH levels of 7 and 8, in buffered reagent-grade water, and at a temperature of 5 °C. Based on these findings, it appears that free chlorine is more effective at inactivating various viruses compared to monochloramine. Nevertheless, there was a noticeable difference in the inactivation rates among viruses when exposed to monochloramine disinfection as opposed to chlorine treatment (Cromeans et al. 2010). Similarly, the SARS-CoV-2 virus, which can be found in respiratory secretions and feces, poses a high risk of transmission through swimming pools. Studies have indicated that within 30 s of exposure to 2.05 mg/L free chlorine, SARS-CoV-2 can be significantly reduced by 3.5 log, which is equivalent to a reduction of over 99.9%. Furthermore, within 2 min, the virus was reduced by more than 4.17 log (limit of detection), which is greater than 99.99%. Chlorine dioxide (ClO_2), another chlorine-based chemical, has emerged as a versatile and potent alternative disinfectant. ClO_2 is a highly efficient oxidizing agent with a low tendency to produce harmful organochlorinated byproducts (Ge et al. 2021; Zhu et al. 2019). Vincenti et al. 2019 utilize ClO_2 for viral remediation, chlorine-resistant bacteria elimination, and algae eradication. During the process of water treatment, ClO_2 undergoes a redox reaction, resulting in the formation of a significant number of chlorite ions (ClO_2^-) and chlorate ions (ClO_3^-). These byproducts have the potential to pose a threat to human health. In the United States, the maximum contaminant level (MCL) for ClO_2^- is set at 1.0 mg L^{-1}, while in China it is set at 0.7 mg L^{-1} (Xu et al. 2022). Since ClO_2^- and ClO_3^- are present during the ClO_2 oxidation process, they limit the amount of ClO_2 that can be used unless technologies are used later to get rid of them. Chlorine dioxide effectively inactivates bacteriophage and rotaviruses at 0.06–1.45 mg L^{-1} min, removing 99.99% water of viruses (4 logs) without disinfectants at varying pH and temperature (Berman and Hoff 1984; Hauchman et al. 1986; Sanekata et al. 2010).

According to Sanekata et al. (2010), canine parvovirus does not show any inactivation within 2 min at 1.0 mg L^{-1} chlorine dioxide exposure, and the cell-associated virus of simian rotavirus SA11 is less sensitive to chlorine dioxide killing than freely suspended viruses. Furthermore, many closely related viruses exhibit varying levels of ClO_2 susceptibility. For example, the disinfection period necessary to achieve 99% inactivation of Coxsackievirus B5 and A9 varies, implying that even minor genetic changes can impact viral inactivation (Zoni et al. 2007). Furthermore, ClO_2 susceptibility differs for enveloped and non-enveloped viruses. 1.0 mg L^{-1} ClO_2 is more harmful to viruses that are not enveloped, like human adenovirus, canine parvovirus, and feline calicivirus (FCV) (Aquino de Carvalho et al. 2017; Chen et al. 2016; Das et al. 2022; Sanekata et al. 2010). These viruses include Ebola, measles, Zika, avian influenza, SARS, MERS, and COVID-19. A possible reason is that ClO_2 can interact with proteins on the virus's membrane, specifically the spike glycoprotein. This interaction can cause damage to the spike glycoprotein, inhibiting host cell attachment. Consequently, the virus is unable to invade and infect the cell successfully (Guan et al. 2003; Yang et al. 2004). The inactivation rates of viruses during ClO_2 disinfection do not show an obvious correlation with their genome types, such as single-stranded DNA, double-stranded DNA, single-stranded RNA, and double-stranded RNA, unlike UV disinfection (Ge et al. 2021). During the processing phase, it is important to be aware that the presence of free chlorine may cause disinfection byproducts (DBPs) to be made, such as haloacetic acids and trihalomethanes. Effective management of these waste materials is crucial to protect water quality and adhere to regulatory standards.

2.2 *Ultraviolet (UV) Irradiation*

UV radiation is another efficient method for inactivating viruses. This method is cost-efficient, eco-friendly, simple to install, and consumes far less energy. UV irradiation is a widely employed technique for viral cleanup, particularly for disinfecting air, surfaces, and water. UV light possesses germicidal capabilities, enabling it to inactivate microorganisms, including viruses. UV light is classified into various categories according to its wavelength: UVA, UVB, and UVC. UVC light, which has a wavelength ranging from 200 to 280 nm, is highly effective in killing microbes and is mostly employed for treating viral infections.

According to a study by Araud et al. (2020), the proliferation of rotavirus (RV) and Tulane virus (TV) was effectively inhibited by UV irradiation at a wavelength of 220 nanometers. In contrast to RV, UV irradiation with a wavelength of 254 nm resulted in a greater degree of inactivation in TV. Additional experiments found that UV irradiation negatively impacted many parts of the RV and TV life cycles. Exposure to UV radiation decreased TV's affinity for its host receptor and caused mutations in its genome. While UV radiation at a wavelength of 220 nm appears to increase the interaction between the RV virus and its host receptor, it inhibits RV genome multiplication. These findings provide insights into disinfecting waterborne viruses, enabling the safe reuse or discharge of treated effluent. Reports also indicate that UV irradiation successfully inactivates SARS-CoV-2. Combining UVA and UVC radiation for 9 min completely inactivated SARS-CoV-2. Nonlinear regression analysis showed that 50% of the virus remained dormant after 1.4 min of UV exposure. After 9 min of radiation at 292 mJ/cm^2, the virus load decreased by 1 log. However, the virus was completely killed after 9 min of UVC at 1048 mJ/cm^2. These results confirm that UVC destroys viruses better and highlight UVC irradiation's effectiveness in neutralizing SARS-CoV-2 (Heilingloh et al. 2020). Another study examined the inactivation of herpes simplex virus type 1 (HSV-1) using UV-LEDs and found UV-LEDs emitting light at wavelengths of 260 and 280 nm had a virucidal effect through the production of photoproducts in the viral DNA and disintegrating viral proteins and phosphoglycerolipid molecules (Bui et al. 2022).

The susceptibility of various viruses and bacteriophages, such as hepatitis A virus AV strain HM-175, coxsackievirus type B-5, rotavirus strain SA-11, and bacteriophages MS2 and øX174, as well as poliovirus 1 [PV1] and F-specific RNA bacteriophages (MS2, GA, and Qbeta), to ultraviolet radiation with a wavelength of 254 nm in phosphate buffered water was determined. The findings showed that the following radiation doses were needed to lower viral activity by 4 log10 (99.99%): 16, 29, 42, and 9 mW sec/cm^2 for HAV, CB5, SA-11, and ΔX174, in that order (Eischeid et al. 2011). UV exposures of 30, 23.6, 14, 8.7, and 4.1 mW s/cm2 were necessary to inactivate coliphages MS-2 and PRD-1, poliovirus type 1, adenovirus 40, and adenovirus 41 at a 90% level, respectively (Simonet and Gantzer 2006; Sommer et al. 2001; Meng and Gerba 1996; Battigelli et al. 1993). The findings showed that the following radiation doses were needed to lower viral activity by 4 log10 (99.99%): 16, 29, 42, and 9 mW sec/cm2 for HAV, CB5, SA-11, and ΔX174, in that order (Eischeid et al. 2011). UV exposures of 30, 23.6, 14, 8.7, and 4.1 mW s/cm^2 were necessary to inactivate coliphages MS-2 and PRD-1, poliovirus type 1, adenovirus 40, and adenovirus 41 at a 90% level, respectively (Simonet and Gantzer 2006; Sommer et al. 2001; Meng and Gerba 1996; Battigelli et al. 1993).

UVC radiation, in contrast to chemical oxidants, directly degrades photolabile virus components, irrespective of the solvent's accessibility. The susceptibility of virus genomes to UVC radiation can be predicted using genome size and the prevalence of pyrimidine doublets (Lytle and Sagripanti 2005; Kowalski et al. 2009). Nevertheless, numerous reports highlight discrepancies suggesting virus components, such as genome and protein sequences, are not reliable indicators of susceptibility (Simonet and Gantzer 2006; Sommer et al. 2001). Hence, a deep understanding of the capsid and genome composition of viruses is essential to accurately predict their susceptibility to UV radiation and identify vulnerable regions. The capsid, made up of specific proteins, can have different levels of protection against UV radiation, based on specific arrangements and protein types being more susceptible to breakage and cross-linking when exposed to UV. In the same way, the composition of the viral genome, such as if it comprises DNA or RNA, single-stranded or double-stranded, and its nucleic acid sequence, plays a role in how UV radiation causes dimer formation, especially in regions that are rich in pyrimidines. This valuable information allows for the improvement of UVC radiation disinfection protocols, ensuring that maximum viral inactivation is achieved in the most effective way possible.

2.3 Ozonation

Ozonation is a process that employs ozone (O_3), a strong oxidizing agent, to disinfect different substances, such as air and water. Viral remediation uses ozonation to reduce or remove the concentration of viruses in a particular environment. O_3 has been extensively used as a disinfectant for over a century for the treatment of wastewater and the recycling of water in recent years (Morrison et al. 2022; Wang et al. 2018; von Gunten 2018; Gerrity et al. 2012). O_3 in its gaseous state can rapidly disperse across the entire area being disinfected. The concentration of O_3 directly influences the rate and efficacy of the sterilization it induces. The application of gaseous O_3 for the purpose of disinfection is prevalent in medical fields and is recognized as a highly effective strategy for clearing microorganisms by the World Health Organization (Blanco et al. 2021). Ozone disrupts the stability of viral proteins, lipoproteins, lipids, glycolipids, and glycoproteins, making them susceptible to oxidation and engaging in redox reactions (Fig. 3). In addition, it inflicts damage to the viral envelope glycoproteins and peplomers, which are essential for the virus's capacity to attach to host receptors (Ataei-Pirkooh et al. 2021; Manjunath et al. 2021).

Thurston-Enriquez et al. (2005) found that ozonation can inactivate feline calicivirus (FCV), which has a comparable nucleic acid structure and capsid to human caliciviruses, as well as adenovirus type 40 (AD40). Studies were conducted using pH -7-free water at a temperature of 5 °C, which was buffered with disinfectant. The Ct values for a 4-log (99.99%) reduction in ozone concentration at a temperature of 5 °C and a pH of 7 were found to be in the range of 0.07–0.60 mg/l min for AD40

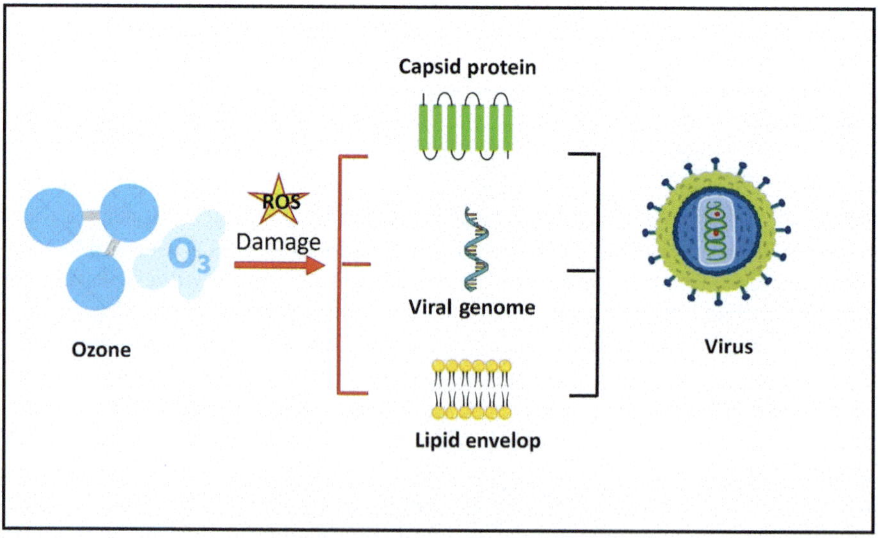

Fig. 3 Ozone generation ROS play a crucial role in inactivating viruses by disrupting their nucleic acid, lipid, and protein components

and less than 0.01–0.03 mg/l min for FCV. The study found that ozone residuals below 0.01 mg/l effectively deactivated FCV and AD40. In another study by Wang et al. (2018), it was found that the ozone treatment applied to traditionally treated wastewater significantly reduced the presence of viruses to undetectable levels. These results were observed for pecovirus, picobirnavirus, parvovirus-like virus, and gokushovirus over 3 weeks. Ozonation resulted in a decrease in adenovirus concentration ranging from 55 to 91%, a reduction in norovirus GII ranging from 85 to 100%, and a decrease in astrovirus 4 and parvovirus ranging from 99 to 100%. Although undetectable after ozone treatment, a number of these viruses were found in the outlet water following the dams, including parvovirus in weeks 49 and 50, as well as parvovirus-like virus, astrovirus 4, pecovirus, and adenovirus 41 in week 50. In the outlet water, the level of virus detected ranged from 4 to 100% of the amount that was observed in the traditionally treated sewage samples.

Despite being a promising technique for virus removal in water treatment, ozonation has several challenges. The need for customized treatments stems from the virus's susceptibility to ozone variability, making it difficult to establish a broadly effective treatment. The basic instability of ozone and its capacity to decompose are influenced by several parameters, such as temperature and pH (Epelle et al. 2023). Therefore, precise control is necessary to maintain appropriate concentrations. The presence of organic matter presents a challenge for ozone on viruses, thereby necessitating pretreatment measures (Farooq et al. 1977). Moreover, the production of disinfection byproducts and the necessity to eliminate residual ozone raise health and regulatory concerns. The expenses associated with the installation and maintenance of ozonation systems, together with the use of energy, pose financial challenges. The challenges of expanding operations and ensuring adherence to regulatory requirements hinder the widespread implementation of ozonation methods in large-scale water treatment facilities. To effectively use ozonation for viral cleanup in water, it is crucial to tackle these problems.

3 Heat Treatment

Heat treatment is a commonly used disinfection technique in the water, sanitation, and food industries, particularly in areas with limited resources. Both temperature and duration of exposure determine the effectiveness of heat treatment in inactivating pathogens (Rogers 2012). In addition, heat is employed to sterilize diverse materials, including personal safety gear, diagnostic and surgical equipment, cultures, and transportation media, as well as biological material (Gamble et al. 2021; Fischer et al. 2020; Darnell and Taylor 2006). The Centers for Disease Control and Prevention, United States, recommends moist heat as a viable method for eliminating SARS-CoV-2 (2020). Heat treatment is a powerful technique that successfully kills or deactivates viruses by the denaturation of proteins, the degradation of nucleic acids, and the breakdown of membranes. Increased temperatures cause the structural and enzymatic proteins of the virus to lose their activity, which in turn affects the virus's capacity to infect (Gamble et al. 2021).

Heat can have different moisture levels, and the ability of heating devices to transfer heat can differ. For example, heat blocks have the potential to be more efficient at transferring heat than ovens. As a result, samples placed in heat blocks can reach and maintain the desired temperature more efficiently and rapidly. Different evaporation levels can be permitted, like samples placed on flat surfaces or on open plates, which will undergo greater evaporation compared to samples stored in sealed vials. These types of containers are commonly used. The surrounding humidity and temperature can influence the rates at which viruses inactivate. This is due to the alterations in the interactions between molecules and the concentrations of solutes involved (Morris et al. 2021). Therefore, we can deduce that elements like heat transmission and evaporation possess the capacity to influence solute concentrations and modify the ambient temperature via evaporative cooling. Consequently, these factors can potentially control the rate at which viruses are inactivated, much like ambient temperature does.

Heat-based disinfection treatments have been employed to eliminate several viruses, such as hepatitis viruses (Abu-Aisha et al. 1995), influenza viruses (Jeong et al. 2010), parvoviruses (Eterpi et al. 2009), and human immunodeficiency viruses (WHO 1989). In a study conducted by Gamble et al. in 2021, researchers examined the effects of different heating techniques on a cell culture medium solution containing the SARS-CoV-2 virus. The study explored four specific techniques used to heat the solution to a temperature of 70 °C: (i) 24-well plates without a cover, (ii) 24-well plates with a cover made of unsealed plastic, (iii) 2-mL vials sealed in a dry oven preheated to 70 °C, and (iv) 24-well vials sealed in a water-filled heat block preheated to 70 °C. Before applying heat, 1 mL of solution was carefully added to the wells or vials of 24-well plates. These were removed from the uncovered 24-well plates at 10, 20, 30, and 60 min, or at 30, 60, and 90 min for the remaining three methods. They observed significant differences in the duration required for SARS-CoV-2 to become undetectable among the four heat treatment methods, resulting in significant variations in the inactivation rate despite identical initial numbers. When the vials were sealed and put in a heat block, the initial titer averaged between 4.7 and 5.5 log10 50% tissue culture infective dose (TCID50)/ml. The plate was left uncovered in the oven between 4.5 and 5.0 [4.1 to 5.0].

When heated, sealed vials were in a heat block or dry oven for 30 min or 90 min on covered plates, no active virus was found. After 60 min, uncovered plates had a 1 log10 TCID50/ml viral reduction. At collection, all samples were supplied with 1 ml of deionized water to account for open plate evaporation. Evaporation was practically complete after 60 min. Due to delayed inactivation and not evaporation, uncovered plates and some covered plates decline viral concentration slower than sealed vials. The findings indicate that evaporation plays a crucial role in the inactivation of viruses by dry heat. An additional study documented the resistance of parvoviruses to dry heat, moist heat at 70 °C for 10 min, and moist heat at 80 or 90 °C for 10 min. Adenovirus, which exhibited resistance to dry heat for 10 min at 80 °C and 90 °C for 1 min, was completely inactivated when exposed to moist conditions. On the contrary, under all heat conditions, poliovirus and vaccinia virus turned inactive (Eterpi et al. 2009). Furthermore, when exposed to a temperature of 60 °C, influenza viruses were observed to undergo a reduction of at least 4 orders of magnitude within around 30 min in virtually all types of media (Hessling et al. 2022). Another study found that thermal incubation at 80 °C can inactivate the hepatitis C virus (HCV). This process disrupts the HCV envelope, making the viral capsid vulnerable to proteolytic digestion (Espinosa et al. 2020).

The process of heat deactivation of viruses presents numerous challenges that impact its overall effectiveness and practical application. The variation in temperature sensitivity among different viruses poses a significant challenge. Achieving the optimal equilibrium between ensuring complete deactivation and preventing damage to heat-sensitive materials is complex. Extreme heat can have a negative impact on the stability and functionality of biological components in samples, putting their structural integrity at risk. Ensuring uniform heating in larger quantities or complex structures is quite a technological hurdle, as temperature fluctuations can lead to insufficient viral inactivation. The time and energy demand for efficient heat treatment might be significant, giving rise to questions regarding feasibility and resource utilization. In addition, it is crucial to have a clear understanding of the specific heat resistance of certain viruses to create effective methods for deactivating them. In order to tackle these concerns, it is crucial to develop an effective strategy that considers the different characteristics of viruses and materials. This will guarantee that heat inactivation methods are efficient and safe.

4 Filtration

The chemical disinfection methods mentioned above typically only eliminate 50–90% of waterborne viruses. This happens because viruses are protected by sludge and biosolids, enabling them to remain active and be released into the environment through solid waste discharged from wastewater treatment facilities (Dey et al. 2022; Gholipour et al. 2022; O'Brien et al. 2017; Saba et al. 2021). Drinking water treatment may not always effectively remove viruses and could potentially spread viral infections if disinfection is insufficient. This is because viruses are typically challenging to completely inactivate due to their small size and resistance against conventional disinfection processes. Additionally, viruses can persist in the environment for a specific duration, posing a risk to public health by contaminating drinking water sources. Therefore, to overcome the challenges linked to traditional unit operations for eliminating viruses, filtration techniques emerged as an appropriate replacement (Zeng et al. 2023; Kheirieh et al. 2018). Filtration is a time-efficient, energy-saving, and cost-effective method for removing viruses from water using fewer compounds.

During the filtration process, virus elimination is based on the interplay between viral particles and membranes. For example, the ability of membranes to capture viruses can be improved by incorporating complex porous structures. Sinclair et al.

(2019) found that hydrophobic membranes demonstrated higher virus retention rates in comparison to hydrophilic membranes. Thus, physical or chemical properties can be altered via membrane modification to improve pathogen removal. Both membrane filtration and ultrafiltration (UF) are widely recognized as reliable and efficient methods for eliminating viruses in aqueous environments (Nasir et al. 2022). The application's specific requirements, the size and type of viruses present, and the intended level of filtration efficiency all influence the selection between the two. The specifications of the usage, the characteristics features, and the size of the viruses that are present, and the expected level of filtration effectiveness are factors that influence the selection between the two methods (Zeng et al. 2023).

Effective elimination of waterborne viruses and bacteria from wastewater and drinking water can be accomplished by employing ultrafiltration (UF) membranes. Ultrafiltration membranes, with pore sizes ranging from 1 to 100 nm, are increasingly being used in water treatment to remove virus particles (Gentile et al. 2018). They are considered to be a highly effective barrier at the nanoscale level. Viral particles can be eliminated through various methods, such as electrostatic repulsion, adhesion, and size exclusion (Gentile et al. 2018). The surface and charge of viruses play a vital role in how they interact with other surfaces and the surrounding aqueous environment. Therefore, it is crucial to comprehensively understand the conditions in which viruses are generated, purified, and stored on a laboratory scale before evaluating their interactions in ultrafiltration processes (Gentile et al. 2018). In their study, Jacquet et al. (2021) estimated virus removal efficiency using ultrafiltration membrane in drinking water. They found that the virus removal rate was up to 3.0 log for coxsackievirus-B5 (CV-B5) and MS2 bacteriophage, and 3.5 log for adenovirus type 41 (ADV 41).

Membrane bioreactors (MBRs) are another advanced and effective technology for eliminating viruses and other contaminants from wastewater (O'Brien and Xagoraraki 2020). MBR is a highly effective process that combines together a microfiltration or ultrafiltration membrane with a suspended growth bioreactor. This innovative approach has gained significant popularity in both municipal and industrial wastewater treatment plants (WWTPs). MBR differs from traditional biological wastewater treatment methods because it uses a membrane that allows for effective separation of sludge. This membrane also acts as a protective barrier against waterborne pathogens, enhancing water safety (Iorhemen et al. 2016). Furthermore, the inclusion of biosolids not only allows for the successful elimination of pathogens larger than the average membrane pore size, such as pathogenic aggregated virus particles, but it also facilitates effective virus removal. Initially, MBR systems were implemented to clean municipal wastewater, with a particular focus on water recycling and reuse. The MBR's compact size, ability to generate reusable water, and flawless operation rendered it an optimal method for recycling municipal wastewater in environments with restricted space and water (Le-Clech 2010).

In 1997, Madaeni reported poliovirus elimination using a 0.2 μm MF membrane. The findings indicate that the adsorption of viruses onto a membrane is determined by the physiochemical properties of both the viral particle and the membrane used, as well as the ratio of the virus particle diameter to the pore diameter. Adsorption is reported to be most effective when the charges of these two components are opposite or negligible, and typically a low pH level can promote this interaction (Zhu et al. 2021; Kosiol et al. 2019). MBRs have been reported to remove viruses at rates ranging from 3.0 log to more than 6.0 log for domestic wastewater treatment. Simmons et al. (2011) found that MBR without disinfection effectively removed adenoviruses, enteroviruses, and noroviruses with efficiencies of 6.3, 6.8, and 4.8 log. In contrast, observed that MBR effectively eradicated noroviruses belonging to genogroup I and genogroup II by up to 5.5 and 5.2 log, respectively. In addition, Katayama et al. (2008) observed that the viral clearance rate via MBR varied between 3.4 and 6.3 log. This results in a decrease in viral concentrations ranging from 99.9% to 99.9999%, indicating that out of every 1000 to 1000,000 virus particles that enter the system, only one is able to get through. The high rates at which MBR membranes remove contaminants make them very successful in purifying water to prevent virus contamination.

But Zhou et al. (2015) reported that even after undergoing final disinfection, enteroviruses and noroviruses remained detectable in the MBR filtrate. An increase in the number of viruses eliminated by MBRs with an extended hydraulic retention time (HRT) and a decreased solids retention time (SRT) was found by Wu et al. (2010), suggesting that the operation of MBRs can also affect virus clearance. In contrast, Shang et al. (2005) proposed that bacteriophage elimination could be eased by an extended SRT or reduced mixed liquor suspended solids (MLSS). Serra-Compte et al. 2021 conducted a comparative analysis to determine the efficacy of three secondary treatments: activated sludge, nutrient removal, and membrane bioreactor (MBR), in clearing SARS-CoV-2 RNA. The study found that samples of MBR effluent lacked SARS-CoV-2 RNA. There was a discernible trend in the log removals: samples subjected to activated sludge and clarification exhibited an average value of 1.03 ± 0.59 log, followed by samples treated with active sludge plus nutrient removal and clarification at 1.37 ± 0.72 log. MBR treatment yielded the maximum log removal, exceeding 1.97 ± 0.93 log. Hence, the range of virus elimination efficiency for achieving a maximum log reduction value is around 91.87 to 99.89%. These results showed that MBR was the best secondary treatment for wastewater line SARS-CoV-2 RNA removal. MBR is far more efficient than traditional sludge, reducing SARS-CoV-2 logs in a range similar to other viruses. Membrane bioreactor (MBR) technology is highly regarded in the field of wastewater treatment for its ability to eliminate viruses effectively. As we move forward into the future, membrane bioreactors (MBRs) provide a viable approach for addressing the issue of viral contamination in wastewater.

Nevertheless, some obstacles need to be managed, despite the positive outlook. A key disadvantage of UF and MBR membranes is fouling, as it considerably affects the membrane's efficiency and longevity, resulting in increased expenses for operation and maintenance. Membrane fouling arises due to suspended matter like microbes and cell waste, colloids, and sewage flocs. Such substances accumulate on the membrane's surface and through the pores, blocking them and reducing membrane permeability (Iorhemen et al. 2016). Membrane fouling is an unavoidable challenge that is hard to control in long-term UF membranes and MBR applications. Membrane fouling reduction in MBRs has been a major focus of substantial studies to improve the wider applicability of MBR technology in wastewater engineering (Iorhemen et al. 2016).

Membrane fouling necessitates innovative membrane design and operation techniques to maintain long-term efficiency. Furthermore, the appearance of novel virus strains necessitates that MBR systems be adaptable and sensitive to maintain their effectiveness. Addressing the need to balance energy consumption and cost-effectiveness raises an additional challenge, prompting investigations into incorporating energy from renewable sources and optimizing processes. Ensuring compliance with regulations and gaining public acceptability are significant challenges that require clear interaction and collaboration to build conviction and maintain adherence to standards. However, within these challenges, there are possibilities for progress. Partnerships across disciplines, technological advancement, and community involvement may achieve the full potential of MBRs in protecting public health and the natural environment against viral threats in wastewater.

5 Nanoremediation

Applications of nanotechnology have led to significant advancements in the field of water and wastewater treatment. These advancements have demonstrated exponential development and enormous promise. Nanoremediation, a subfield within the broader field of environmental remediation, employs nanoparticles (NPs) to eliminate or reduce the number of pollutants present in the soil, air, and water. Nanoparticles are extremely small particles that have sizes that relate to the nanometer scale and typically range from one to one hundred nanometers. The unique properties exhibited by these particles result from their small size and high surface area-to-volume ratio (Sharma et al. 2022). We can either engineer these particles or produce them naturally. Recent advances in nanotechnology have developed numerous advantageous nanoscale materials (containing nanoparticles) to facilitate wastewater remediation. These materials consist of nanometals, nanoadsorbents, photocatalysts, and nanomembranes (Adeel et al. 2021; Gehrke et al. 2015). Notwithstanding the potential of nanoremediation to eliminate diverse contaminants, its implementation in addressing environmental challenges associated with viruses remains a comparatively emerging field of study (Sharma et al. 2022).

Nanoparticles with customized attributes can be developed to efficiently identify and eliminate viruses in water bodies using filtration mechanisms (Adeel et al. 2021; Zodrow et al. 2009). In nanofiltration, the membranes are designed to have pores at the nanometer scale, which allows them to separate particles based on size and charge selectively. These membranes can be made from various materials, including polymers, ceramics, or composites, and they are engineered to have specific properties suitable for the separation process. Nanofiltration operates on the principle of size exclusion, where particles larger than the membrane's pore size are retained while smaller ones pass through. Nanofilters have a high surface area and a small pore size, allowing them to trap and eliminate viruses and other contaminants in water effectively. Incorporating nanoparticles into complex composite membranes impregnated with various nanomaterials enables the elimination or degradation of a wide variety of viruses (Table 2).

While nanoparticle filtration and nanomembranes have been reported to be quite successful in laboratory settings, the process of implementing these systems for real-world applications poses challenges related to costs, stability, and scalability

Table 2 The use of various nanofilters reported for the elimination of various viruses

S. No.	Viruses	Nanofilters	References
1.	SARS-CoV-2, H1N1, and enterovirus 71	The in situ modification of a BLG AF–Fe nanofilter, which was developed using amyloid nanofibrils derived from β-lactoglobulin (BLG) and iron oxyhydroxide nanoparticles (NPs), effectively decreased the infectious virus concentration from around 106 PFU ml-1 to an undetectable level	Palika et al. (2021)
2.	Parvoviruses, picornaviruses, caliciviruses, papovaviruses, flaviviruses, togaviruses, reovirus, retroviruses, rhabdoviruses, and herpesviruses	15–20 nm pore-size nanofilters made by Asahi Kasei Corp., Pall Corp., Merck Millipore Corp., and Sartorius were capable of eliminating picornaviruses, caliciviruses, and SV40 viruses that are 25–40 nm in size	Roth et al. (2020)
3.	Semliki Forest virus	A mixture of tetraethoxysilane and ammonium tetrathiomolybdate was used in the electrospinning process of nanofiber membranes. This method yielded structures that were devoid of defects and exhibited extensive interconnections among pores, measuring between 0.8 and 3.1 μm in length, with an approximate specific surface area of 45.6 m^2/g. the aforementioned filters exhibited virus elimination effectiveness of as high as 98.9% per pass while permitting exceedingly high water permeation rates of 26,000 Lm − 2 h^{-1} bar^{-1}	Al-Attabi et al. (2019)
4.	Qβ bacteriophage virus	On a polysulfone support layer, photopolymerization is done to produce thin nanostructured bicontinuous cubic liquid-crystalline (LC) films. It has been shown that these membranes effectively remove the Qβ bacteriophage virus, characterized by a diameter of around 20 nm, achieving a rejection rate exceeding 99.9999%	Marets et al. (2017)

(continued)

Table 2 (continued)

S. No.	Viruses	Nanofilters	References
5..	Hepatitis C virus	Cellulose nanocomposite membranes are composed of two distinct layers: a barrier layer composed of regenerated cellulose gel that possesses minute interconnected pores and a layer of electrospun cellulose nanofabric subjected to hot-pressing to augment its mechanical strength. By their electronegative characteristics, this composite membrane effectively eliminates all positively charged 50 nm latex particles and 98.68 ± 0.71% hepatitis C virus	Huang et al. (2016)
6.	Poliovirus, norovirus, and coxsackievirus	Nanofilter developed from a carbon nanotube-silver composite material demonstrated the efficacy of eliminating waterborne viruses	Kim et al. (2016)
7.	Xenotropic murine leukemia virus (xMuLV)	Cellulose nanofibers made from Cladophora sp. algae were shown to efficiently remove xMuLV from contaminated solutions (LRV ≥5.25)	Asper et al. (2015)
8.	Pseudorabies virus, human immunode Ficiency virus, bovine viral diarrhea Virus, West Nile virus, hepatitis A virus, Murine encephalomyocarditis virus, And porcine parvovirus	A variety of process conditions were applied to nanofilters ranging in size from 18 nm to 200 nm as part of Grifol's plasma-derived manufacturing processes. The reduction factors (RFs) of these nanofilters exceed 4 log10, leading to a virus elimination rate of over 99.99%. Considering the RF values within the examined range, the process conditions examined (pH, ionic strength, temperature, filtration ratio, and protein concentration) had no significant effect	Caballero et al. (2014)
9.	MS2 bacteriophage	By employing a carbon nanotube filter, the model virus MS2 was removed. It was found that the efficacy of electric current in facilitating the elimination of viral particles ranged from 5.8 to 7.4 log removal. Fouling of the membrane was minimal after four cycles	Rahaman et al. (2012)
10.	MS2 bacteriophage	Nanofibrous membrane made of extremely thin cellulose fibers from ecologically friendly, economical, and sustainable wood pulp was mixed with 5–10 nm polyacrylonitrile electrospun nanofiber membranes (PAN ENMs). Due to its electrostatic positive charge, the membrane trapped viruses with a small negative surface charge	Sato et al. (2011)
11.	Bovine parvovirus, HIV, bovine diarrhea virus, porcine pseudorabies virus, reovirus type 3, and simian virus 40.	The cellulose-based nanofilter membranes with pore sizes of 15 and 35 nm are capable of achieving approximately 8 log elimination viruses. Hence, can be a viable option to remove viruses from blood plasma	Burnouf-Radosevich et al. (1994)

(Sharma et al. 2022). Further research can be conducted to address these challenges and develop economically viable nanoparticle-based filtration devices for practical use in areas such as wastewater treatment and drinking water purification. Additionally, studies have reported the effectiveness of metal nanoparticles in combination with natural membranes in viral remediation. For example, TiO_2 nanoparticles possess photocatalytic qualities that can eliminate many microorganisms, including gram-positive and gram-negative bacteria, viruses, algae, fungi, and protozoa (Sharma et al. 2022). Based on these studies, it is highly possible that these methodologies may prove efficient in combating SARS-CoV-2. For instance, untreated wastewater containing SARS-CoV-2 could be cleaned up using nano-filters/membranes composed of a variety of affordable and deployable nanomaterials.

In order to effectively address the specific challenges presented by waterborne viruses, it is imperative to enhance and optimize nanoremediation techniques. It is crucial to recognize that, despite the promising potential of nanoremediation, there are concerns regarding nanomaterials' possible ecological and health implications. In order to ensure the safe and effective integration of nanotechnology into water treatment processes, it is essential to conduct extensive studies and careful risk assessments (Sharma et al. 2022).

6 Bioremediation

In the aforementioned parts, we learned about different chemical and physical strategies for the elimination of viruses in the ecosystem. Even though these treatments lower the risk of dangerous viruses, there is still a possibility that they might present a greater health risk for humans due to their byproducts, and some viruses may persist in the treated wastewater. Treated waters harboring potentially hazardous viruses pose a high risk to public health. One way to deal with this problem is to implement wastewater treatment techniques that can eliminate an array of pathogens while also removing carbon and nutrients. An extensive array of abiotic factors predominantly shapes the fate of pathogens such as viruses in the aquatic environment. These factors play a crucial role in the inactivation and elimination of these infectious agents. However, the environment contains intricate networks of microbial communities that engage in many types of interactions, such as commensalism, antagonism, and parasitism. These biotic interactions significantly impact the composition of natural microbial communities. Interactions among species can either be mutually beneficial or antagonistic. Hence, using biological agents such as bacteria, algae, protozoa, and plants can be an eco-friendly, efficient, cost-effective, and sustainable alternative method for virus remediation (Feichtmayer et al. 2017).

Bacteria can play a vital role in eliminating viruses in wastewater treatment processes through antagonistic biotic interactions. Bacteria play an active role in degrading organic substances and eliminating contaminants, such as viruses, in biological wastewater treatment systems. Certain bacteria can break down organic compounds, thereby indirectly decreasing the number of viruses by targeting the compounds that viruses may be related to (Hibbing et al. 2010). Moreover, specific bacteriophages, which are viral agents that invade bacteria, can aid in diminishing bacterial populations in wastewater, consequently decreasing viral concentration. Additionally, the development of biofilms by bacteria can establish conditions in which viruses can be trapped or targeted by other microorganisms. Although bacteria play a crucial role in eliminating viruses during wastewater treatment, it is essential to recognize that additional treatment measures, such as filtration and disinfection processes, may be required to ensure thorough virus removal (Feichtmayer et al. 2017). In addition, many bacteria produce antimicrobial peptides or proteins known as bacteriocins. These are involved in the inhibition of the growth of closely related competing bacteria (Sharma and Yadav 2023). Although bacteriocins are most widely known for their antibacterial properties, many studies have also shown their antiviral activity. It has been reported that bacteriocins such as pediocin and nisin, which are produced by lactic acid bacteria (LAB), exhibit antiviral properties against both HSV and HIV (Al Kassaa et al. 2014). Likewise, plantaricins, a kind of bacteriocin made by *Lactobacillus plantarum*, have been established to impede the growth of enteric viruses such as rotavirus (Chamoun-Emanuelli et al. 2014). Furthermore, Domínguez-Díaz et al. (2019) have reported the production of pro-inflammatory cytokines (IL-4, TNFα, and IFN-γ) by bacteria. These cytokines promote viral inhibition and function as a protective mechanism against viruses through their inactivation. Nevertheless, additional research is necessary to assess bacteriocins' effectiveness, safety, and compatibility with current treatment processes before their use in effluent treatment for virus elimination.

Protozoan grazing is another significant cause of prokaryote mortality in natural aquatic environments (Feichtmayer et al. 2017; Menon et al. 2003). Although bacteria and other small protozoa primarily consume nonattached viral particles, research indicates that Tetrahymena and other ciliates, heterotrophic nanoflagellates, and amoeba also partake in this practice. A study conducted by Deng et al. (2014) revealed that implementing three heterotrophic flagellates resulted in a substantial decline in the population of the bacteriophage model MS2. This study utilized three distinct species of flagellates: *Salpingoeca* sp., which was a filter-feeding flagellate; *Thaumatomonas coloniensis*, which was a benthivorous grazer; and *Goniomonas truncate*, an aggressive raptorial feeder. The bacteriophage MS2 titer was observed to be reduced by six orders of magnitude over 90 days when it was grazed by *Salpingoeca* sp. or T. *coloniensis*. It was found that the antagonistic activities of the naturally occurring bacterial community in groundwater were responsible for the limited reduction of only two log units that were observed when protozoa were not present. According to Deng et al. (2014), while consumed, viruses only make an insignificant contribution to the carbon content of the protozoan diet. Indeed, it has been demonstrated that protozoa can directly utilize viral amino acids for their protein synthesis (Feichtmayer et al. 2017).

Algae and plants have also been reported to eliminate viruses through adsorption and capture (Sellaoui et al. 2021). They can serve as natural filters, capturing the viruses and preventing their spread in the water column. These biological agents can compete with viruses for nutrition and sunlight, potentially limiting virus proliferation (Al-Hazmi et al. 2022; Umamaheswari and Shanthakumar 2016). Also, polysaccharides derived from plants, marine vegetables, algae, and lichens are being examined for their capacity to scavenge free radicals, prevent lipid peroxidation, and hinder viral replication (Al-Hazmi et al. 2022; Sellaoui et al. 2021,). Viruses that are responsible for severe infectious diseases, including dengue virus type 1 (Hidari et al. 2008), yellow fever virus (Ono et al. 2003), herpes simplex viruses (Malagoli et al. 2014), and HIV (Besednova et al. 2019), have been reported to be inhibited by sulfated polysaccharides derived from algae and plants, including fucoidans, carrageenans, heparin, neem, laminarans, and ulvans.

According to Kim and Li (2011), marine sulfated polysaccharides have several benefits, including low production costs, a wide range of antiviral activities, minimal toxicity, safety, widespread acceptance, and unique mechanisms of action compared to other types of antivirals. Therefore, they proposed taking advantage of these sulfated polysaccharides as very effective decontamination solutions in the future period. Zhang et al. (2021) have reported the development of a highly efficient biohybrid microrobotic system for actively eliminating the SARS-CoV-2 spike protein and pseudovirus from different water sources by utilizing microalgae that have been engineered with ACE2 receptors. The ACE2-algae-robot's ability to bind and remove substances was tested using the SARS-CoV-2 spike protein and SARS-CoV-2 pseudovirus. The results showed an elimination effectiveness of 95% and 89%, respectively, under the given experimental conditions. These results demonstrate a high efficacy when compared to conventional approaches to eliminating viral contamination in wastewater. The improved elimination of the viral spike protein and pseudovirus was reported due to the continuous self-propulsion of the ACE2-algae-robot, which leads to increased mixing and interaction with the target protein and virus. This study also examined the potential for reusing and treating the microrobot, showing significant potential in effectively eliminating waterborne harmful viruses from contaminated water. The current approach can eliminate 89% of viruses; however, it is hard to link wastewater viral concentration to host illness, but any viral load drop will affect viral infection management and patient wellbeing. Algal microrobots in wastewater reservoirs have the potential to have better movement and lifetime than in small test tubes, making them promising for large-scale applications. Each approach specifically targets viral functions to eliminate viruses from natural ecosystems and the environment. Considering these findings and understanding the pros and cons of each approach will help develop more efficient and improved strategies for preventing viral infections transmitted by water and food. Furthermore, it can improve our understanding of how viruses adapt to various stresses, whether they arise naturally or are induced by human activities. We must also understand that human actions contribute to the rise in the virus population in the environment through many different routes. Poor wastewater and sewage from residential, industrial, or agricultural sources can transfer viruses into waterways, jeopardizing human and environmental health. Using untreated animal manure as fertilizer and field runoff can spread viruses into soil and water. Industrial effluent can include viruses. Pollution, climate change, urbanization, and land use changes disrupt natural ecosystems, making virus propagation and mutation easier. Moreover, engaging in leisure activities in polluted water sources and the spread of viruses through the air in densely populated interior areas worsen the issue. Additionally, pollution, climate change, urbanization, and changes in land use can disturb natural ecosystems, making it easier for viruses to propagate. Insufficient waste management, specifically of medical waste, can also contribute to environmental pollution. To tackle these difficulties, it is necessary to implement comprehensive strategies that encourage sustainable practices, strengthen sanitary infrastructure, and raise public knowledge about the significance of environmental stewardship in preventing virus infection.

7 Conclusion

The public and authorities are becoming more concerned about the potential health risks and the need for a proper sanitation strategy in water reclamation. To prevent the spread of waterborne diseases and ensure efficient garbage disposal, modern sewage systems have been under development since the middle of the nineteenth century to effectively separate and treat wastewater. Human activities produce wastewater from both municipalities and industries. This wastewater can contain pathogenic microbes like bacteria, viruses, and other pollutants. Unsafe treatment and distribution of these substances could lead to acute and chronic illnesses in humans. The recent global breakout of the COVID-19 pandemic has heightened awareness of the risks of water- and airborne viral infections and their potential consequences. The growing number of deadly disease outbreaks caused by unknown infections and the discovery of new pandemic strains necessitate the immediate and pressing development of more effective technologies for cleaning viral contamination.

We can develop a powerful and broad strategy for eliminating viruses in various environments by integrating physical, chemical, and biological methodologies. Physical methods, including ultraviolet irradiation and filtration, effectively eliminate and inactivate viruses based on their structure and size. On the contrary, chemical disinfection agents cause additional damage to viral membranes and genetic material. Biological processes like bioremediation and microbial degradation, which harness the activity of microorganisms to eliminate contaminants and decompose organic matter, further enable virus elimination. By integrating these methodologies, it is possible to attain synergistic outcomes that will improve overall efficiency and ensure the thorough elimination of viruses from different environments, including water bodies, wastewater, surfaces, and other places.

References

Abu-Aisha H, Mitwalli A, Huraib SO, Al-Wakeel J, Abid J, Yousif KI, Algayyar F, Ramia S (1995) The effect of chemical and heat disinfection of the hemodialysis machines on the spread of hepatitis C virus infection: a prospective study. Saudi J Kidney Dis Transpl 6(2):174–178

Adeel M, Farooq T, Shakoor N, Ahmar S, Fiaz S, White JC, Gardea-Torresdey JL, Mora-Poblete F, Rui Y (2021) COVID-19 and nanoscience in the developing world: rapid detection and remediation in wastewater. Nanomaterials (Basel, Switzerland) 11(4):991. https://doi.org/10.3390/nano11040991

Aggarwal J, Sharma P (2023) *In silico* analysis of Tecovirimat a repurposed drug against THE monkeypox virus, ITS off-target human proteins, and impact on human health. Int J Appl Pharm 15(6):206–210. https://doi.org/10.22159/ijap.2023v15i6.49248

Al Kassaa I, Hober D, Hamze M, Chihib NE, Drider D (2014) Antiviral potential of lactic acid bacteria and their bacteriocins. Probiotics Antimicrob Proteins 6(3–4):177–185. https://doi.org/10.1007/s12602-014-9162-6

Al-Attabi R, Rodriguez-Andres J, Schütz JA, Bechelany M, Des Ligneris E, Chen X et al (2019) Catalytic electrospun nano-composite membranes for virus capture and remediation. Sep Purif Technol 229:115806

Al-Hazmi HE, Shokrani H, Shokrani A, Jabbour K, Abida O, Mousavi Khadem SS, Habibzadeh S, Sonawane SH, Saeb MR, Bonilla-Petriciolet A, Badawi M (2022) Recent advances in aqueous virus removal technologies. Chemosphere 305:135441. https://doi.org/10.1016/j.chemosphere.2022.135441

American Water Works Association (2000) Committee report: disinfection at large and medium-sized systems. J Am Water Works Assoc 92:32–43

Aquino de Carvalho N, Stachler EN, Cimabue N, Bibby K (2017) Evaluation of Phi6 persistence and suitability as an enveloped virus surrogate. Environ Sci Technol 51(15):8692–8700. https://doi.org/10.1021/acs.est.7b01296

Araud E, Fuzawa M, Shisler JL, Li J, Nguyen TH (2020) UV inactivation of rotavirus and Tulane virus targets different components of the Virions. Appl Environ Microbiol 86(4):e02436–e02419. https://doi.org/10.1128/AEM.02436-19

Asper M, Hanrieder T, Quellmalz A, Mihranyan A (2015) Removal of xenotropic murine leukemia virus by nanocellulose based filter paper. Biologicals 43(6):452–456. https://doi.org/10.1016/j.biologicals.2015.08.001

Ataei-Pirkooh A, Alavi A, Kianirad M, Bagherzadeh K, Ghasempour A, Pourdakan O, Adl R, Kiani SJ, Mirzaei M, Mehravi B (2021) Destruction mechanisms of ozone over SARS-CoV-2. Sci Rep 11(1):18851. https://doi.org/10.1038/s41598-021-97860-w

Baldo A, van den Akker E, Bergmans HE, Lim F, Pauwels K (2013) General considerations on the biosafety of virus-derived vectors used in gene therapy and vaccination. Curr Gene Ther 13(6):385–394. https://doi.org/10.2174/15665232113136660005

Ballester NA, Malley JP (2004) Sequential disinfection of adenovirus type 2 with UV-chlorine-chloramine. J Am Water Works Assoc 96:97–103

Bardhan M, Ray I, Roy S, Bhatt P, Patel S, Asri S, Shariff S, Shree A, Mitra S, Roy P, Anand A (2023) Emerging zoonotic diseases and COVID-19 pandemic: global perspective and Indian scenario. Ann Med Surg 85(8):3997–4004. https://doi.org/10.1097/MS9.0000000000001057

Battigelli DA, Sobsey MD, Lobe DC (1993) The inactivation of hepatitis a virus and other model viruses by UV irradiation. Water Sci Technol 27(3–4):339–342

Baxter CS, Hofmann R, Templeton MR, Brown M, Andrews RC (2007) Inactivation of adenovirus types 2, 5, and 41 in drinking water by UV light, free chlorine, and monochloramine. J Environ Eng ASCE 133:95–103

Berman D, Hoff JC (1984) Inactivation of simian rotavirus SA11 by chlorine, chlorine dioxide, and monochloramine. Appl Environ Microbiol 48(2):317–323. https://doi.org/10.1128/aem.48.2.317-323.1984

Besednova NN, Zvyagintseva TN, Kuznetsova TA, Makarenkova ID, Smolina TP, Fedyanina LN, Kryzhanovsky SP, Zaporozhets TS (2019) Marine algae metabolites as promising therapeutics for the prevention and treatment of HIV/AIDS. Meta 9(5):87. https://doi.org/10.3390/metabo9050087

Blanco A, Ojembarrena FB, Clavo B, Negro C (2021) Ozone potential to fight against SAR-COV-2 pandemic: facts and research needs. Environ Sci Pollut Res Int 28(13):16517–16531. https://doi.org/10.1007/s11356-020-12036-9

Brown AM, Blind J, Campbell K, Ghosh S (2020) Safeguards for using viral vector Systems in Human Gene Therapy: a resource for biosafety professionals mitigating risks in health care settings. Appl Biosaf 25(4):184–193. https://doi.org/10.1177/1535676020934917

Bui TKN, Mawatari K, Emoto T, Fukushima S, Shimohata T, Uebanso T, Akutagawa M, Kinouchi Y, Takahashi A (2022) UV-LED irradiation reduces the infectivity of herpes simplex virus type 1 by targeting different viral components depending on the peak wavelength. J Photochem Photobiol B 228:112410. https://doi.org/10.1016/j.jphotobiol.2022.112410

Burnouf-Radosevich M, Appourchaux P, Huart JJ, Burnouf T (1994) Nanofiltration, a new specific virus elimination method applied to high-purity factor IX and factor XI concentrates. Vox Sang 67(2):132–138. https://doi.org/10.1111/j.1423-0410.1994.tb01647.x

Caballero S, Diez JM, Belda FJ, Otegui M, Herring S, Roth NJ, Lee D, Gajardo R, Jorquera JI (2014) Robustness of nanofiltration for increasing the viral safety margin of biological products. Biologicals 42(2):79–85. https://doi.org/10.1016/j.biologicals.2013.10.003

Cavallieri F, Sellner J, Zedde M, Moro E (2022) Neurologic complications of coronavirus and other respiratory viral infections. Handb Clin Neurol 189:331–358. https://doi.org/10.1016/B978-0-323-91532-8.00004-5

Centers for Disease Control and Prevention (2020) Implementing filtering facepiece respirator (FFR) reuse, including reuse after decontamination, when there are known shortages of N95 respirators. https://www.cdc.gov/coronavirus/2019-ncov/hcp/ppe-strategy/decontamination-reuse-respirators.html

Chamoun-Emanuelli AM, Bobardt M, Moncla B, Mankowski MK, Ptak RG, Gallay P, Chen Z (2014) Evaluation of PD 404,182 as an anti-HIV and anti-herpes simplex virus microbicide. Antimicrob Agents Chemother 58(2):687–697. https://doi.org/10.1128/AAC.02000-13

Chappell JD, Dermody TS (2015) Biology of viruses and viral diseases. Mandell, Douglas, and Bennett's Principles and Practice of Infectious Diseases. 31:1681–1693.e4. https://doi.org/10.1016/B978-1-4557-4801-3.00134-X

Chen Y, Liu F, Wang C, Zhao M, Deng L, Zhong J, Zhang Y, Ye J, Jing S, Cheng Z, Guan Y, Ma Y, Sun Y, Zhu B, Zhang Q (2016) Molecular identification and epidemiological features of human adenoviruses associated with acute respiratory infections in hospitalized children in southern China, 2012-2013. PLoS One 11(5):e0155412. https://doi.org/10.1371/journal.pone.0155412

Cromeans TL, Kahler AM, Hill VR (2010) Inactivation of adenoviruses, enteroviruses, and murine norovirus in water by free chlorine and monochloramine. Appl Environ Microbiol 76(4):1028–1033. https://doi.org/10.1128/AEM.01342-09

Darnell ME, Taylor DR (2006) Evaluation of inactivation methods for severe acute respiratory syndrome coronavirus in noncellular blood products. Transfusion 46(10):1770–1777. https://doi.org/10.1111/j.1537-2995.2006.00976.x

Das A, Mohapatra M, Basu S (2022) Recent developments focusing disinfectant systems for effective inactivation of coronaviruses. In: Mohapatra M, Choudhury B, Basu S (eds) COVID-19 pandemic: research and development activities from modeling to realization. Springer, Singapore, pp 111–128

Deng L, Krauss S, Feichtmayer J, Hofmann R, Arndt H, Griebler C (2014) Grazing of heterotrophic flagellates on viruses is driven by feeding behaviour. Environ Microbiol Rep 6(4):325–330. https://doi.org/10.1111/1758-2229.12119

Dey P, Haldar D, Rangarajan V, Suggala VS, Saji G, Dilip KJ (2022) Paradigm shift from conventional processes to advanced membrane adsorption-mediated inactivation processes towards holistic management of virus– a critical review. J Environ Chem Eng 10(6):108568

Dianty R, Hirano J, Anzai I, Kanai Y, Hayashi T, Morimoto M, Kataoka-Nakamura C, Kobayashi S, Uemura K, Ono C, Watanabe T, Kobayashi T, Murakami K, Kikuchi K, Hotta K, Yoshikawa T, Taguwa S, Matsuura Y (2023) Electrolyzed hypochlorous acid water exhibits potent disinfectant activity against various viruses through irreversible protein aggregation. Front Microbiol 14:1284274. https://doi.org/10.3389/fmicb.2023.1284274

Domingo JL, Rovira J (2020) Effects of air pollutants on the transmission and severity of respiratory viral infections. Environ Res 187:109650. https://doi.org/10.1016/j.envres.2020.109650

Domínguez-Díaz C, García-Orozco A, Riera-Leal A, Padilla-Arellano JR, Fafutis-Morris M (2019) Microbiota and its role on viral evasion: is it with us or against us? Front Cell Infect Microbiol 9:256. https://doi.org/10.3389/fcimb.2019.00256

Eischeid AC, Thurston JA, Linden KG (2011) UV disinfection of adenovirus: present state of the research and future directions. Crit Rev Environ Sci Technol 41(15):1375–1396

Epelle EI, Macfarlane A, Cusack M, Burns A, Okolie JA, Mackay W, Rateb M, Yaseen M (2023) Ozone application in different industries: a review of recent developments. Chem Eng J (Lausanne, Switzerland: 1996) 454:140188. https://doi.org/10.1016/j.cej.2022.140188

Espinosa MF, Sancho AN, Mendoza LM, Mota CR, Verbyla ME (2020) Systematic review and meta-analysis of time-temperature pathogen inactivation. Int J Hyg Environ Health 230:113595. https://doi.org/10.1016/j.ijheh.2020.113595

Eterpi M, McDonnell G, Thomas V (2009) Disinfection efficacy against parvoviruses compared with reference viruses. J Hosp Infect 73(1):64–70. https://doi.org/10.1016/j.jhin.2009.05.016

Fair GM, Morris JC, Lu Chang S, Weil I, Burden RP (1948) The behavior of chlorine as a water disinfectant. J (Am Water Works Assoc) 40(10):1051–1061. http://www.jstor.org/stable/41234959

Farooq S, Chian ES, Engelbrecht RS (1977) Basic concepts in disinfection with ozone. J Water Pollut Control Fed 49(8):1818–1831

Feichtmayer J, Deng L, Griebler C (2017) Antagonistic microbial interactions: contributions and potential applications for controlling pathogens in the aquatic systems. Front Microbiol 8:2192. https://doi.org/10.3389/fmicb.2017.02192

Fischer RJ, Morris DH, van Doremalen N, Sarchette S, Matson MJ, Bushmaker T, Yinda CK, Seifert SN, Gamble A, Williamson BN, Judson SD, de Wit E, Lloyd-Smith JO, Munster VJ (2020) Effectiveness of N95 respirator decontamination and reuse against SARS-CoV-2 virus. Emerg Infect Dis 26(9):2253–2255. https://doi.org/10.3201/eid2609.201524

Floyd R, Johnson JD, Sharp DG (1976) Inactivation by bromine of single poliovirus particles in water. Appl Environ Microbiol 31(2):298–303. https://doi.org/10.1128/aem.31.2.298-303.1976

Gall AM, Mariñas BJ, Lu Y, Shisler JL (2015) Waterborne viruses: a barrier to safe drinking water. PLoS Pathog 11(6):e1004867. https://doi.org/10.1371/journal.ppat.1004867

Gamble A, Fischer RJ, Morris DH, Yinda CK, Munster VJ, Lloyd-Smith JO (2021) Heat-treated virus inactivation rate depends strongly on treatment procedure: illustration with SARS-CoV-2. Appl Environ Microbiol 87(19):e0031421. https://doi.org/10.1128/AEM.00314-21

Ge Y, Zhang X, Shu L, Yang X (2021) Kinetics and mechanisms of virus inactivation by chlorine dioxide in water treatment: a review. Bull Environ Contam Toxicol 106(4):560–567. https://doi.org/10.1007/s00128-021-03137-3

Gehrke I, Geiser A, Somborn-Schulz A (2015) Innovations in nanotechnology for water treatment. Nanotechnol Sci Appl 8:1–17. https://doi.org/10.2147/NSA.S43773

Gentile GJ, Cruz MC, Rajal VB, de Cortalezzi MMF (2018) Electrostatic interactions in virus removal by ultrafiltration membranes. J Environ Chem Eng 6(1):1314–1321

Gerrity D, Gamage S, Jones D, Korshin GV, Lee Y, Pisarenko A, Trenholm RA, von Gunten U, Wert EC, Snyder SA (2012) Development of surrogate correlation models to predict trace organic contaminant oxidation and microbial inactivation during ozonation. Water Res 46(19):6257–6272. https://doi.org/10.1016/j.watres.2012.08.037

Gholipour S, Ghalhari MR, Nikaeen M, Rabbani D, Pakzad P, Miranzadeh MB (2022) Occurrence of viruses in sewage sludge: a systematic review. Sci Total Environ 824:153886. https://doi.org/10.1016/j.scitotenv.2022.153886

Girones R, Carratalà A, Calgua B, Calvo M, Rodriguez-Manzano J, Emerson S (2014) Chlorine inactivation of hepatitis E virus and human adenovirus 2 in water. J Water Health 12(3):436–442. https://doi.org/10.2166/wh.2014.027

Guan Y, Zheng BJ, He YQ, Liu XL, Zhuang ZX, Cheung CL, Luo SW, Li PH, Zhang LJ, Guan YJ, Butt KM, Wong KL, Chan KW, Lim W, Shortridge KF, Yuen KY, Peiris JS, Poon LL (2003) Isolation and characterization of viruses related to the SARS coronavirus from animals in southern China. Science New York, N.Y. 302(5643):276–278. https://doi.org/10.1126/science.1087139

Hamim I, Komatsu K, Prager SM, Wu B (2023) Editorial: viruses in agricultural systems: interactions with plants, insect pollinators and fungi. Front Microbiol 14:1170402. https://doi.org/10.3389/fmicb.2023.1170402

Hauchman FS, Noss CI, Olivieri VP (1986) Chlorine dioxide reactivity with nucleic acids. Water Res 20(3):357–361

Heilingloh CS, Aufderhorst UW, Schipper L, Dittmer U, Witzke O, Yang D, Zheng X, Sutter K, Trilling M, Alt M, Steinmann E, Krawczyk A (2020) Susceptibility of SARS-CoV-2 to UV irradiation. Am J Infect Control 48(10):1273–1275. https://doi.org/10.1016/j.ajic.2020.07.031

Hejkal TW, Wellings FM, LaRock PA, Lewis AL (1979) Survival of poliovirus within organic solids during chlorination. Appl Environ Microbiol 38(1):114–118. https://doi.org/10.1128/aem.38.1.114-118.1979

Hessling M, Gierke AM, Sicks B, Fehler N, Vatter P (2022) Sensitivity of influenza virus to ultraviolet irradiation. GMS Hyg Infect Control 17:Doc20. https://doi.org/10.3205/dgkh000423

Hibbing ME, Fuqua C, Parsek MR, Peterson SB (2010) Bacterial competition: surviving and thriving in the microbial jungle. Nat Rev Microbiol 8(1):15–25. https://doi.org/10.1038/nrmicro2259

Hidari KI, Takahashi N, Arihara M, Nagaoka M, Morita K, Suzuki T (2008) Structure and anti-dengue virus activity of sulfated polysaccharide from a marine alga. Biochem Biophys Res Commun 376(1):91–95

Hu B, Guo H, Zhou P, Shi ZL (2021) Characteristics of SARS-CoV-2 and COVID-19. Nat Rev Microbiol 19(3):141–154. https://doi.org/10.1038/s41579-020-00459-7

Huang W, Wang Y, Chen C, Law JL, Houghton M, Chen L (2016) Fabrication of flexible self-standing all-cellulose nanofibrous composite membranes for virus removal. Carbohydr Polym 143:9–17. https://doi.org/10.1016/j.carbpol.2016.02.011

Iorhemen OT, Hamza RA, Tay JH (2016) Membrane bioreactor (MBR) technology for wastewater treatment and reclamation: membrane fouling. Membranes 6(2):33. https://doi.org/10.3390/membranes6020033

Jacquet N, Wurtzer S, Darracq G, Wyart Y, Moulin L, Moulin P (2021) Effect of concentration on virus removal for ultrafiltration membrane in drinking water production. J Membr Sci 634:119417. https://doi.org/10.1016/j.memsci.2021.119417

Jain S, Thakur N, Grover N, Vashistt J, Changotra H (2016) Prevalence of rotavirus, norovirus and enterovirus in diarrheal diseases in Himachal Pradesh, India. Virusdisease 27(1):77–83. https://doi.org/10.1007/s13337-016-0303-2

Jeong EK, Bae JE, Kim IS (2010) Inactivation of influenza a virus H1N1 by disinfection process. Am J Infect Control 38(5):354–360. https://doi.org/10.1016/j.ajic.2010.03.003

Katayama H, Haramoto E, Oguma K, Yamashita H, Tajima A, Nakajima H, Ohgaki S (2008) One-year monthly quantitative survey of noroviruses, enteroviruses, and adenoviruses in wastewater collected from six plants in Japan. Water Res 42(6–7):1441–1448. https://doi.org/10.1016/j.watres.2007.10.029

Keesing F, Ostfeld RS (2021) Impacts of biodiversity and biodiversity loss on zoonotic diseases. Proc Natl Acad Sci 118(17):e2023540118. https://doi.org/10.1073/pnas.2023540118

Kheirieh S, Asghari M, Afsari M (2018) Application and modification of polysulfone membranes. Rev Chem Eng 34(5):657–693

Kim SK, Li YX (2011) Medicinal benefits of sulfated polysaccharides from sea vegetables. Adv Food Nutr Res 64:391–402

Kim K, Dobson AP, Gulland FM, Harvell CD (2005) Diseases and the conservation of marine biodiversity. In: Marine conservation biology. Island Press, pp 149–168
Kim JP, Kim JH, Kim J, Lee SN, Park HO (2016) A nanofilter composed of carbon nanotube-silver composites for virus removal and antibacterial activity improvement. J Environ Sci (China) 42:275–283. https://doi.org/10.1016/j.jes.2014.11.017
Kong J, Lu Y, Ren Y, Chen Z, Chen M (2021) The virus removal in UV irradiation, ozonation and chlorination. Water Cycle 2:23–31. https://doi.org/10.1016/j.watcyc.2021.05.001
Kosiol P, Kahrs C, Thom V, Ulbricht M, Hansmann B (2019) Investigation of virus retention by size exclusion membranes under different flow regimes. Biotechnol Prog 35(2):e2747. https://doi.org/10.1002/btpr.2747
Kowalski WJ, Bahnfleth W, Hernandez MT (2009) A genomic model for predicting the ultraviolet susceptibility of viruses. IUVA News 11(2):15–28
Lanrewaju AA, Enitan-Folami AM, Sabiu S, Edokpayi JN, Swalaha FM (2022a) Global public health implications of human exposure to viral contaminated water. Front Microbiol 13:981896. https://doi.org/10.3389/fmicb.2022.981896
Lanrewaju AA, Enitan-Folami AM, Sabiu S, Swalaha FM (2022b) A review on disinfection methods for inactivation of waterborne viruses. Front Microbiol 13:991856. https://doi.org/10.3389/fmicb.2022.991856
Le-Clech P (2010) Membrane bioreactors and their uses in wastewater treatments. Appl Microbiol Biotechnol 88(6):1253–1260. https://doi.org/10.1007/s00253-010-2885-8
Louten J (2016) Virus structure and classification. In: Essential human virology, pp 19–29. https://doi.org/10.1016/B978-0-12-800947-5.00002-8
Lytle CD, Sagripanti JL (2005) Predicted inactivation of viruses of relevance to biodefense by solar radiation. J Virol 79(22):14244–14252
Madaeni SS (1997) Mechanism of virus removal using membranes. Filtr Sep 34(1):61–65. https://doi.org/10.1016/S0015-1882(97)84823-9
Malagoli BG, Cardozo FT, Gomes JH, Ferraz VP, Simões CM, Braga FC (2014) Chemical characterization and antiherpes activity of sulfated polysaccharides from Lithothamnion muelleri. Int J Biol Macromol 66:332–337. https://doi.org/10.1016/j.ijbiomac.2014.02.053
Manjunath SN, Sakar M, Katapadi M, Geetha Balakrishna R (2021) Recent case studies on the use of ozone to combat coronavirus: problems and perspectives. Environ Technol Innov 21:101313. https://doi.org/10.1016/j.eti.2020.101313
Marets N, Kuo D, Torrey JR, Sakamoto T, Henmi M, Katayama H, Kato T (2017) Highly efficient virus rejection with self-organized membranes based on a Crosslinked Bicontinuous cubic liquid crystal. Adv Healthc Mater 6(14). https://doi.org/10.1002/adhm.201700252
Meng QS, Gerba CP (1996) Comparative inactivation of enteric adenoviruses, poliovirus and coliphages by ultraviolet irradiation. Water Res 30(11):2665–2668
Menon P, Billen G, Servais P (2003) Mortality rates of autochthonous and fecal bacteria in natural aquatic ecosystems. Water Res 37(17):4151–4158. https://doi.org/10.1016/S0043-1354(03)00349-X
Merenda A, Dumée LF (2021) Virus remediation in water engineering: are our current technologies up to the challenge? J Water Process Eng 44:102370. https://doi.org/10.1038/s41545-020-00096-w
Morris DH, Yinda KC, Gamble A, Rossine FW, Huang Q, Bushmaker T, Fischer RJ, Matson MJ, Van Doremalen N, Vikesland PJ, Marr LC, Munster VJ, Lloyd-Smith JO (2021) Mechanistic theory predicts the effects of temperature and humidity on inactivation of SARS-CoV-2 and other enveloped viruses. eLife 10:e65902. https://doi.org/10.7554/eLife.65902
Morrison CM, Hogard S, Pearce R, Gerrity D, von Gunten U, Wert EC (2022) Ozone disinfection of waterborne pathogens and their surrogates: a critical review. Water Res 214:118206. https://doi.org/10.1016/j.watres.2022.118206
Na C, Olson TM (2007) Relative reactivity of amino acids with chlorine in mixtures. Environ Sci Technol 41(9):3220–3225. https://doi.org/10.1021/es061999e

Nasir AM, Adam MR, Mohamad Kamal SNEA, Jaafar J, Othman MHD, Ismail AF, Aziz F, Yusof N, Bilad MR, Mohamud R, Rahman MA, Wan Salleh WN (2022) A review of the potential of conventional and advanced membrane technology in the removal of pathogens from wastewater. Sep Purif Technol 286:120454. https://doi.org/10.1016/j.seppur.2022.120454

O'Brien E, Xagoraraki I (2020) Removal of viruses in membrane bioreactors. J Environ Eng 146(7):03120007

O'Brien RT, Newman J (1979) Structural and compositional changes associated with chlorine inactivation of polioviruses. Appl Environ Microbiol 38(6):1034–1039. https://doi.org/10.1128/aem.38.6.1034-1039.1979

O'Brien E, Nakyazze J, Wu H, Kiwanuka N, Cunningham W, Kaneene JB, Xagoraraki I (2017) Viral diversity and abundance in polluted waters in Kampala, Uganda. Water Res 127:41–49. https://doi.org/10.1016/j.watres.2017.09.063

Ono L, Wollinger W, Rocco IM, Coimbra TL, Gorin PA, Sierakowski MR (2003) In vitro and in vivo antiviral properties of sulfated galactomannans against yellow fever virus (BeH111 strain) and dengue 1 virus (Hawaii strain). Antivir Res 60(3):201–208

Onodera T, Kirisawa R (2023) New advances in viral and microorganism disinfectants. Microorganisms 11(10):2530. https://doi.org/10.3390/microorganisms11102530

Palika A, Armanious A, Rahimi A, Medaglia C, Gasbarri M, Handschin S, Rossi A, Pohl MO, Busnadiego I, Gübeli C, Anjanappa RB, Bolisetty S, Peydayesh M, Stertz S, Hale BG, Tapparel C, Stellacci F, Mezzenga R (2021) An antiviral trap made of protein nanofibrils and iron oxyhydroxide nanoparticles. Nat Nanotechnol 16(8):918–925. https://doi.org/10.1038/s41565-021-00920-5

Rahaman MS, Vecitis CD, Elimelech M (2012) Electrochemical carbon-nanotube filter performance toward virus removal and inactivation in the presence of natural organic matter. Environ Sci Technol 46(3):1556–1564

Rogers W (2012) Sterilisation of biomaterials and medical devices. Elsevier, pp 20–55

Roossinck MJ, García-Arenal F (2015) Ecosystem simplification, biodiversity loss and plant virus emergence. Curr Opin Virol 10:56–62. https://doi.org/10.1016/j.scitotenv.2021.145172

Roth NJ, Dichtelmüller HO, Fabbrizzi F, Flechsig E, Gröner A, Gustafson M, Jorquera JI, Kreil TR, Misztela D, Moretti E, Moscardini M, Poelsler G, More J, Roberts P, Wieser A, Gajardo R (2020) Nanofiltration as a robust method contributing to viral safety of plasma-derived therapeutics: 20 years' experience of the plasma protein manufacturers. Transfusion 60(11):2661–2674. https://doi.org/10.1111/trf.16022

Saba B, Hasan SW, Kjellerup BV, Christy AD (2021) Capacity of existing wastewater treatment plants to treat SARS-CoV-2. A review. Bioresour Technol Rep 15:100737. https://doi.org/10.1016/j.biteb.2021.100737

Saguti F, Kjellberg I, Churqui MP, Wang H, Tunovic T, Ottoson J, Bergstedt O, Norder H, Nyström K (2023) The Virucidal effect of the chlorination of water at the initial phase of disinfection may be underestimated if contact time calculations are used. Pathogens (Basel, Switzerland) 12(10):1216. https://doi.org/10.3390/pathogens12101216

Sanekata T, Fukuda T, Miura T, Morino H, Lee C, Maeda K, Araki K, Otake T, Kawahata T, Shibata T (2010) Evaluation of the antiviral activity of chlorine dioxide and sodium hypochlorite against feline calicivirus, human influenza virus, measles virus, canine distemper virus, human herpesvirus, human adenovirus, canine adenovirus and canine parvovirus. Biocontrol Sci 15(2):45–49. https://doi.org/10.4265/bio.15.45

Sato A, Wang R, Ma H, Hsiao BS, Chu B (2011) Novel nanofibrous scaffolds for water filtration with bacteria and virus removal capability. J Electron Microsc 60(3):201–209

Scarcella C, Carasi S, Cadoria F, Macchi L, Pavan A, Salamana M, Alborali GL, Losio MM, Boni P, Lavazza A, Seyler T (2009) An outbreak of viral gastroenteritis linked to municipal water supply, Lombardy, Italy, June 2009. Euro Surveill 14(29):19274. https://doi.org/10.2807/ese.14.29.19274-en

Scarpino PV, Berg G, Chang SL, Dahling D, Lucas M (1972) A comparative study of the inactivation of viruses in water by chlorine. Water Res 6:959–965

Sellaoui L, Badawi M, Monari A, Tatarchuk T, Jemli S, Luiz Dotto G, Bonilla-Petriciolet A, Chen Z (2021) Make it clean, make it safe: a review on virus elimination via adsorption. Chem Eng J (Lausanne, Switzerland: 1996) 412:128682. https://doi.org/10.1016/j.cej.2021.128682

Serra-Compte A, González S, Arnaldos M, Berlendis S, Courtois S, Loret JF, Schlosser O, Yáñez AM, Soria-Soria E, Fittipaldi M, Saucedo G, Pinar-Méndez A, Paraira M, Galofré B, Lema JM, Balboa S, Mauricio-Iglesias M, Bosch A, Pintó RM, Bertrand I et al (2021) Elimination of SARS-CoV-2 along wastewater and sludge treatment processes. Water Res 202:117435. https://doi.org/10.1016/j.watres.2021.117435

Shang C, Wong HM, Chen G (2005) Bacteriophage MS-2 removal by submerged membrane bioreactor. Water Res 39(17):4211–4219. https://doi.org/10.1016/j.watres.2005.08.003

Sharma P, Yadav M (2023) Enhancing antibacterial properties of bacteriocins using combination therapy. J App Biol Biotech 11(01):232–243. https://doi.org/10.7324/JABB.2023.110206

Sharma A, Tiwari S, Deb MK, Marty JL (2020) Severe acute respiratory syndrome coronavirus-2 (SARS-CoV-2): a global pandemic and treatment strategies. Int J Antimicrob Agents 56(2):106054. https://doi.org/10.1016/j.ijantimicag.2020.106054

Sharma P, Kumari R, Yadav M, Lal R (2022) Evaluation of TiO_2 nanoparticles physicochemical parameters associated with their antimicrobial applications. Indian J Microbiol 62(3):338–350. https://doi.org/10.1007/s12088-022-01018-9

Sharp DG, Leong J (1980) Inactivation of poliovirus I (Brunhilde) single particles by chlorine in water. Appl Environ Microbiol 40(2):381–385. https://doi.org/10.1128/aem.40.2.381-385.1980

Sharp DG, Young DC, Floyd R, Johnson JD (1980) Effect of ionic environment on the inactivation of poliovirus in water by chlorine. Appl Environ Microbiol 39(3):530–534. https://doi.org/10.1128/aem.39.3.530-534.1980

Simmons FJ, Kuo DH, Xagoraraki I (2011) Removal of human enteric viruses by a full-scale membrane bioreactor during municipal wastewater processing. Water Res 45(9):2739–2750. https://doi.org/10.1016/j.watres.2011.02.001

Simonet J, Gantzer C (2006) Inactivation of poliovirus 1 and F-specific RNA phages and degradation of their genomes by UV irradiation at 254 nanometers. Appl Environ Microbiol 72(12):7671–7677. https://doi.org/10.1128/AEM.01106-06

Sinclair TR, Patil A, Raza BG, Reurink D, Van den Hengel SK, Rutjes SA et al (2019) Cationically modified membranes using covalent layer-by-layer assembly for antiviral applications in drinking water. J Membr Sci 570:494–503

Sloan C, Moore ML, Hartert T (2011) Impact of pollution, climate, and sociodemographic factors on spatiotemporal dynamics of seasonal respiratory viruses. Clin Transl Sci 4(1):48–54. https://doi.org/10.1111/j.1752-8062.2010.00257.x

Sobsey MD, Fuji T, Shields PA (1988) Inactivation of hepatitis a virus and model viruses in water by free chlorine and monochloramine. Water Sci Technol 20:385–391

Socha W, Kwasnik M, Larska M, Rola J, Rozek W (2022) Vector-borne viral diseases as a current threat for human and animal health-one health perspective. J Clin Med 11(11):3026. https://doi.org/10.3390/jcm11113026

Sodhi KK, Singh CK (2023) A systematic review on the occurrence, fate, and remediation of SARS-CoV-2 in wastewater. Int J Environ Sci Technol 20(7):8073–8086. https://doi.org/10.1007/s13762-022-04326-1

Soltane R, Allayeh AK (2023) Occurrence of enteroviruses, noroviruses, rotaviruses, and adenoviruses in a wastewater treatment plant. J Umm Al-Qura Univ Appl Sci 9(4):449–454. https://doi.org/10.1007/s43994-023-00053-6

Sommer R, Pribil W, Appelt S, Gehringer P, Eschweiler H, Leth H, Cabaj A, Haider T (2001) Inactivation of bacteriophages in water by means of non-ionizing (UV-253.7 nm) and ionizing (gamma) radiation: a comparative approach. Water Res 35(13):3109–3116. https://doi.org/10.1016/s0043-1354(01)00030-6

Szczuka A, Horton J, Evans KJ, DiPietri VT, Sivey JD, Wigginton KR (2022) Chloride enhances DNA reactivity with chlorine under conditions relevant to water treatment. Environ Sci Technol 56(18):13347–13356. https://doi.org/10.1021/acs.est.2c03267

Thurston-Enriquez JA, Haas CN, Jacangelo J, Gerba CP (2005) Inactivation of enteric adenovirus and feline calicivirus by ozone. Water Res 39(15):3650–3656. https://doi.org/10.1016/j.watres.2005.06.006

Totaro M, Badalucco F, Costa AL, Tuvo B, Casini B, Privitera G, Menchini Fabris GB, Baggiani A (2021) Effectiveness of disinfection with chlorine dioxide on respiratory transmitted, enteric, and Bloodborne viruses: a narrative synthesis. Pathogens (Basel, Switzerland) 10(8):1017. https://doi.org/10.3390/pathogens10081017

Umamaheswari J, Shanthakumar S (2016) Efficacy of microalgae for industrial wastewater treatment: a review on operating conditions, treatment efficiency and biomass productivity. Rev Environ Sci Biotechnol 15:265–284

US Environmental Protection Agency (USEPA) (2018). https://www.epa.gov/ccl/types-drinking-water-contaminants

Uwishema O, Masunga DS, Naisikye KM, Bhanji FG, Rapheal AJ, Mbwana R, Nazir A, Wellington J (2023) Impacts of environmental and climatic changes on future infectious diseases. Int J Surg (London, England) 109(2):167–170. https://doi.org/10.1097/JS9.0000000000000160

Varma MM, Christian BA, McKinstry DW (1974) Inactivation of Sabin oral poliomyelitis type I virus. J Water Pollut Cont Fed 46:987–992

Vincenti S, de Waure C, Raponi M, Teleman AA, Boninti F, Bruno S, Boccia S, Damiani G, Laurenti P (2019) Environmental surveillance of legionella spp. colonization in the water system of a large academic hospital: analysis of the four-year results on the effectiveness of the chlorine dioxide disinfection method. Sci Total Environ 657:248–253. https://doi.org/10.1016/j.scitotenv.2018.12.036

von Gunten U (2018) Oxidation processes in water treatment: are we on track? Environ Sci Technol 52(9):5062–5075. https://doi.org/10.1021/acs.est.8b00586

Wang H, Sikora P, Rutgersson C, Lindh M, Brodin T, Björlenius B, Larsson DGJ, Norder H (2018) Differential removal of human pathogenic viruses from sewage by conventional and ozone treatments. Int J Hyg Environ Health 221(3):479–488. https://doi.org/10.1016/j.ijheh.2018.01.012

Wang CC, Prather KA, Sznitman J, Jimenez JL, Lakdawala SS, Tufekci Z, Marr LC (2021) Airborne transmission of respiratory viruses. Science New York, N.Y. 373(6558):eabd9149. https://doi.org/10.1126/science.abd9149

Weidenkopf SJ (1958) Inactivation of type 1, poliomyelitis virus with chlorine. Virology 5(1):56–67. https://doi.org/10.1016/0042-6822(58)90005-9

WHO (2011) Guidelines for drinking-water quality, 4th edn. World Health Organization, Geneva. http://www.who.int/water_sanitation_health/publications/drinking-water-quality-guidelines-4-including-1st-addendum/en/index.html

World Health Organization (1989) Guidelines on sterilization and disinfection methods effective against human immunodeficiency virus (HIV), 2nd edn. World Health Organization, Geneva. https://apps.who.int/iris/handle/10665/41538

Wu J, Li H, Huang X (2010) Indigenous somatic coliphage removal from a real municipal wastewater by a submerged membrane bioreactor. Water Res 44(6):1853–1862. https://doi.org/10.1016/j.watres.2009.12.013

Xu MY, Lin YL, Zhang TY, Hu CY, Tang YL, Deng J, Xu B (2022) Chlorine dioxide-based oxidation processes for water purification: a review. J Hazard Mater 436:129195. https://doi.org/10.1016/j.jhazmat.2022.129195

Yang ZY, Huang Y, Ganesh L, Leung K, Kong WP, Schwartz O, Subbarao K, Nabel GJ (2004) pH-dependent entry of severe acute respiratory syndrome coronavirus is mediated by the spike glycoprotein and enhanced by dendritic cell transfer through DC-SIGN. J Virol 78(11):5642–5650. https://doi.org/10.1128/JVI.78.11.5642-5650.2004

Ye Y, Chang PH, Hartert J, Wigginton KR (2018) Reactivity of enveloped virus genome, proteins, and lipids with free chlorine and UV_{254}. Environ Sci Technol 52(14):7698–7708. https://doi.org/10.1021/acs.est.8b00824

Zeng J, Desmond P, Ngo HH, Lin W, Liu X, Liu B et al (2023) Membrane modification in enhancement of virus removal: a critical review. J Environ Sci 146:198

Zhang F, Li Z, Yin L, Zhang Q, Askarinam N, Mundaca-Uribe R et al (2021) ACE2 receptor-modified algae-based microrobot for removal of SARS-CoV-2 in wastewater. J Am Chem Soc 143(31):12194–12201

Zhou J, Wang XC, Ji Z, Xu L, Yu Z (2015) Source identification of bacterial and viral pathogens and their survival/fading in the process of wastewater treatment, reclamation, and environmental reuse. World J Microbiol Biotechnol 31(1):109–120. https://doi.org/10.1007/s11274-014-1770-5

Zhu Z, Guo Y, Yu P, Wang X, Zhang X, Dong W, Liu X, Guo C (2019) Chlorine dioxide inhibits the replication of porcine reproductive and respiratory syndrome virus by blocking viral attachment. Infect Genet Evol 67:78–87. https://doi.org/10.1016/j.meegid.2018.11.002

Zhu Y, Chen R, Li YY, Sano D (2021) Virus removal by membrane bioreactors: a review of mechanism investigation and modeling efforts. Water Res 188:116522. https://doi.org/10.1016/j.watres.2020.116522

Zodrow K, Brunet L, Mahendra S, Li D, Zhang A, Li Q, Alvarez PJ (2009) Polysulfone ultrafiltration membranes impregnated with silver nanoparticles show improved biofouling resistance and virus removal. Water Res 43(3):715–723. https://doi.org/10.1016/j.watres.2008.11.014

Zoni R, Zanelli R, Riboldi E, Bigliardi L, Sansebastiano G (2007) Investigation on virucidal activity of chlorine dioxide. Experimental data on feline calicivirus, HAV and Coxsackie B5. J Prev Med Hyg 48(3):91

The Role of Mycoremediation in the Circular Economy

Vinay Kumar Pandey, Sarvesh Rustagi, and Gurmeet Singh

Abstract The circular economy is a transformative model based on the decoupling of economic growth from the utilization of finite resources and the design of waste outside the system. Consequently, mycoremediation offers an essential strategy for sustainable waste disposal and environmental management. First, due to fungi's numerous enzymatic capabilities, they can break down intricate organic chemicals, including intransigent environmental toxins like heavy metals, pesticides, and synthetic dyes. These organisms can transform dangerous chemicals into less harmful forms, promoting the healthy recycling of nutrients and restoring polluted properties. With these characteristics, current applications employ mycoremediation to clean soil, purify water, and clean air. Water purification methods utilize fungi to purify industrial effluents and wastewater, eliminating harmful chemicals like pharmaceuticals and hazardous chemicals. In addition, air purification utilizes fungal bio-filters to absorb and digest contaminant waste, thus contributing to pure air quality. Fungi are employed in agriculture for composting and pesticide residue mycoremediation, which uses *mycorrhizae* to extract and degrade chemical residues. This chapter explores the pivotal role of fungi in bioremediation processes, focusing on their ability to degrade and detoxify various pollutants, thereby contributing to the circular economy.

Keywords Fungi · Remediation · Circular economy · Waste utilization

V. K. Pandey
Research and Development Cell, Biotechnology Department, Manav Rachna International Institute of Research and Studies (Deemed to Be University), Faridabad, Haryana, India

S. Rustagi (✉)
Department of Food Technology, School of Agriculture, Maya Devi University, Dehradun, Uttarakhand, India

G. Singh
R&D Department, Guru Nanak College of Pharmaceutical Sciences, Dehradun, Uttarakhand, India

A. A. H. Abdel Latef et al. (eds.), *Sustainable Remediation for Pollution and Climate Resilience*, https://doi.org/10.1007/978-981-96-5674-5_16

1 Introduction

A circular economy model seeks to change traditional economic principles toward greater reuse, refurbishment, and recycling of materials to reduce waste and the consumption of limited resources. Thus, mycoremediation, or the usage of a fungus's unique biological abilities to degrade and detoxicate pollutants, is actively pursued to develop this new economic framework. Integrating mycoremediation into the circular economy also has substantial economic advantages. As aforementioned, the traditional remediation methods, such as chemical treatment and physical removal of contaminants, can be costly and resource-consuming. In contrast, mycoremediation is often more affordable, and the cultivation of fungi can be performed using low-cost, readily available substrates, such as agricultural and industrial waste products. At the same time, the by-products of fungal processes of degradation can be repurposed for other industrial applications, such as biofuels, bioplastics, and pharmaceuticals, creating additional revenue streams and enhancing resource efficiency (Nandal et al. 2021). Fungi bioremediation can reduce reliance on nonrenewable resources and decrease the impact of waste disposal. For example, fungal composting of organic waste can produce high-quality compost, enriching soil fertility and supporting sustainable agriculture. It aligns with the principles of the circular economy, as waste becomes a valuable resource (Serbent et al. 2024).

For the circular economy to harness the full potential of mycoremediation, scientists and policymakers must establish new avenues of research and development and incentivize innovation. Genetic and metabolic engineering of fungi can further improve the productivity and specificity of degradation. The need for interdepartmental collaboration in these directions and close interaction with industrial representatives is paramount. This will develop a framework of rights and obligations that promotes the treatment, especially through fungi-based systems. Fungi contain incredibly diverse and robust enzymatic machineries, which has a broad range of complex organic pollutants such as hydrocarbons, pesticides, dyes, and pharmaceuticals (Meyer et al. 2020). Fungi are beneficial in bioremediation, as most, if not all, of their enzymatic activities convert otherwise toxic and persistent substances to harmless by-products. This process is critical in ensuring the degeneration and elimination of persistent pollutants, which would otherwise be impossible. Hence, bioremediation is critical in ensuring the general cleaning of highly contaminated environments, thus ensuring high ecological health. Mycorrhizal fungi establish symbiotic relationships with the roots of plants. The fungi aid in nutrient uptake, strengthen soil structure, and stimulate plant development. Using fungi to treat contaminated soils can help improve soil health and revive its fertility. This is particularly important for agricultural soils because contamination leads to low crop yields and increased food scarcity (Jeffries et al. 2003).

Fungi are used to purify contaminated water, whereby their mycelium acts as a filter. When exposed to waste, they can remove and digest pollutants such as heavy metals and endocrine-disrupting compounds. Including a fungal purification stage in water purification may result in cleaner water sources vital to humans and aquatic species. Mycoremediation can turn waste materials into salvageable products (Zhu et al. 2019). For instance, agricultural waste products and food waste can be broken down by fungi to produce compost used as soil manure. This role not only reduces waste in landfills but also sets up a sustainable scenario in which such wastes are either recycled or reused, contributing to the sustainable resource. Fungi can bioconvert lignocellulosic biomass to biofuels and biochemicals used in renewable energy production, thus reducing dependency on fossil fuels. This role in bioeconomy contributes to sustainability in energy generation, reducing carbon fixation. This chapter discusses the various types of fungi used in bioremediation and how they can be further explored at the industrial level (Mujtaba et al. 2023).

2 Types of Fungi Used in Bioremediation

Bioremediation, the use of organisms to detoxify and restore polluted environments, has seen significant advances with the incorporation of fungi. Fungi, with their remarkable enzymatic capabilities and adaptability, play a crucial role in breaking down various environmental pollutants (Deshmukh et al. 2016).

2.1 White-Rot Fungi (WRF)

White-rot fungi are key contributors to bioremediation because of their unique and powerful enzymatic abilities. These basidiomycetes are well-known for their ability to decompose lignin, a highly complex and recalcitrant polymer in plant cell walls, by producing various extracellular enzymes, such as lignin peroxidases (LiPs), manganese peroxidases (MnPs), and laccases. The ligninolytic system allows WRF to decompose a wide range of persistent biotechnological organic contaminants, making them essential tools in bioremediation to reclaim and clean up polluted environments (Zhuo and Fan 2021). One of WRF's most important contributions to bioremediation is that they decay polycyclic aromatic hydrocarbons (PAHs). Stable and toxic forms are common in oil spills, industrial discharges, and soils damaged by pollution. They pose a challenge to the environment because they resist the action of decomposers and do not decompose naturally. However, the oxidative enzymes produced by WRF break down and eventually metabolize these compounds, reducing their toxicity. This allows the recovery of ill-occupied properties and prevents long-term ecological damage (Honda and Suzuki 2020). Apart from polycyclic aromatic hydrocarbons (PAHs), white-rot fungi (WRF) have proven effective in degrading various other pollutants, including pesticides, industrial dyes, and pharmaceutical residues. Ligninolytic enzymes degrade various pesticides, including several aromatic compounds, and can deactivate them before they enter the ecosystem and food chains, causing numerous ecological and human health impacts. In the same context, industrial dyes, such as azo and triphenylmethane dyes, used while dyeing textiles, are highly recalcitrant, making them difficult to remove from dye effluents in conventional wastewater treatment plants. However, WRF, which can mineralize the industrial dyes, withdrawing their toxic effects on the environment, can contribute to eliminating industrial effluents. These further extend to soil and wastewater treatment; WRF might be spread onto the soil to degrade complex organic pollutants for the soil to recuperate (Torres-Farradá et al. 2017).

In soil remediation, WRF is used in soils as a candidate propagule, where the minimum number of unique genotypes can be introduced into contaminated soil. In the presence of appropriate nutrients, organisms are expected to grow, spread, build their genotypes, and degrade pollutants. Composting and solid waste management, such as WRF, are impressive at breaking down nonwoody polymers of the woods. They will decrease composting time or lower the amount of waste left in a landfill (Zhang et al. 2020). Moreover, the sustainability and eco-friendliness of bioremediation with WRF are other crucial advantages. Unlike chemical pathways of pollutant decomposition, which may generate numerous toxic substances themselves, WRF presents an opportunity to detoxify the environment by applying a natural mechanism. Therefore, no additional harm is imposed on the environment, and its natural tendency to be restored is triggered. While the potential of WRF looks promising, the future perspectives for its utilization in bioremediation are incredibly broad (Lakhani et al. 2022). Scientists suggest that ongoing research and development will provide numerous new opportunities for managing these fungi and for their widespread application in various environment-oriented activities. More so, the genetic engineering area may offer ways of improving WRF to make them even more effective in pollution control. Finally, further recognition of WRF directions may identify their potential for being mixed with other biological, chemical, and physical methods, making the offerings more complete (Vaksmaa et al. 2023).

Fungi play a crucial role in bioremediation by contributing to the detoxification of various environmental pollutants found in water, air, and soil, thereby helping to restore ecological balance and promote a healthier, more sustainable environment on a global scale. White-rot fungi (WRF) are able to effectively degrade a variety of pollutants, including pharmaceutical residues, industrial dyes, and polycyclic aromatic hydrocarbons (PAHs) present in water. Fungi secrete extracellular enzymes that can degrade complex organic pollutants, including lignin peroxidases (LiPs) and manganese peroxidases (MnPs), so fungi have been regarded as significant participants in wastewater treatment (Leng et al. 2021). Fungi, particularly in soil, also play a crucial role in mycoremediation, the use of fungi to degrade toxic substances like pesticides, metals, and hydrocarbons. They decompose pollutants via their mycelial networks and improve soil structure and nutrient cycling to support plant growth, fostering sustainable ecosystems.

Certain molds and other fungi in the air can collect volatile organic compounds (VOCs) as well as airborne chemicals. Mycofiltration is the utilization of fungal mycelium for filtering both air and water, having gained popularity among effluent treatment methods in recent years due to it being an eco-friendly method for pollutant uptake. Fungi have become celebrated for their ecological possibilities from North America and Europe to emerging economies worldwide. But further research and investment are, of course, necessary to have widespread uptake in industrial, agricultural, and municipal waste management facilities around the world (Noman et al. 2019).

2.2 *Brown-Rot Fungi (BRF)*

Brown-rot fungi are an important group in bioremediation due to their novel wood decay mechanism, which largely differs from that used by their bioremediation counterpart, white-rot fungi. Whereas white-rot fungi break down lignin to access cellulose and hemicellulose of wood, BRF first degrade cellulose and hemicellulose, leaving depleted residual lignin. BRF achieve this selectivity by secreting a combination of oxidant enzymes and reactive oxygen species (ROS), with the hydroxyl radical being a critical factor in wood polysaccharide degradation. The biochemistry of BRF, as discussed above, forms the basis for their contributions to bioremediation, especially their role in breaking down complex organic pollutants and stabilizing derivate contaminated soils (Goodell 2003). One of the most significant advantages of BRF in bioremediation is the amount of ROS they produce, predominantly because BRF use the Fenton reaction to create ROS. In this waste reduction procedure, iron ions are reduced to assume the ferrous state, supporting the production of hydroxyl free radicals. These free radicals are then released into the cell, capable of nonspecifically breaking down a wide range of organic pollutants. In wood, the structural polysaccharides and other complex organic compounds, such as PAH, phenols, and hydrocarbons, are broken down by the radicals, making them beneficial. Consequently, BRF are efficient in decomposing pollutants, and this efficiency in the decomposition of pollutants is one of their most significant benefits. For example, BRF are used in industrial site cleanups as they partially and harmlessly reduce the toxicity of pollutants like wood preservatives or creosote (Abdal Dayem et al. 2017).

Aside from oxidative abilities, BRF also serve the soil as a stabilizing factor. Cellulose and hemicellulose degrade due to the activity of BRF, and it leads to the production of humic substances that shape various qualities of soil, particularly its fertility and structure. The ability to transform barren, polluted soil into arable soil that supports plant life is instrumental in reclaiming lands devastated by industry. By facilitating soil formation and fertility, BRF help rebuild ecosystems that have suffered from human activity and careless pollution (Weimer 2022). Additionally, BRF degrade synthetic polymers with the same straightforward process. Due to their ability to produce oxidative enzymes, they break the long-chained polymer in the plastics, known as polyethylene, and the resulting molecules become less endurable. Innovations in bioremediation that offer this approach may soon provide a remedy to the global issue of corrupted plastic waste (Mohanan et al. 2020). In addition to the above, the employment of BRF in conjunction with other microbes and chemical treatments can also improve bioremediation's productivity. Symbiotic solutions that incorporate BRF with bacteria or other fungi can generate an extensive degradation system capable of breaking down a wide range of pollutants. For example, the BRF may be used to start the disassembling of complicated organic substances for reprocessing, during which the resulting simpler substances can be completely disassembled to the extent that they can only be found in nature (Bala et al. 2022).

2.3 *Mycorrhizal Fungi*

Mycorrhizal fungi are critical facilitators in bioremediation, given their mutually beneficial associations with plant roots. The fungi are mutualistic with the roots of almost all terrestrial plant species. The fungi assist in nutrient exchange, benefiting both the fungi and their hosts. Especially in bioremediation, mycorrhizal fungi facilitate the ability of plants to withstand and accumulate pollutants. The fungi are, therefore, part of the phytoremediation process due to these and other roles. One of the primary mechanisms by which mycorrhizal fungi enable bioremediation is by boosting plant nutrient acquisition (Wahab et al. 2023). Mycorrhizal fungi's enormous hyphal networks enhance the roots' surface area for nutrient absorption, providing plants with access to nutrients they would otherwise not get from the soil. Therefore, the fungal function benefits the plants by causing them to grow stronger and more resistant to toxic pollutants. Instead, by supporting plants' good health and vitality, mycorrhizal fungi enhance the plants' potential to soak up and store pollutants like heavy metals, radionuclides, and organics. Mycorrhizal fungi also contribute directly to the detoxification of contaminants due to their ability to produce a range of extracellular enzymes and organic acids capable of degrading complex organic pollutants, including polycyclic aromatic hydrocarbons, pesticides, and industrial solvents (Begum et al. 2019). These enzymatic activities cause the degradation of hazardous compounds to relatively nontoxic forms that can be further destroyed or stabilized by other soil microorganisms. Therefore, the biochemical transformation is essential for efficiently detoxifying and removing hazardous pollutants from the ecosystem (Miglani et al. 2022).

Mycorrhizal fungi biosorb heavy metals and other inorganic pollutants on their cell walls or intracellular structures, thus immobilizing the toxicants and preventing their leaching into groundwater. As a result, the bioavailability and mobility of the toxic metals are mitigated since the immobilized metals are taken up by the plants and sequestered in their biomass in a process known as phytoextraction when the mycorrhizal colonization enhances the capability of the host to take up and translocate the toxic metals. Another process enhanced by the symbiotic relationship between mycorrhizal fungi and plants is the improvement of soil structure and health, which supports the long-term recovery and stability of ecosystems (Boorboori and Zhang 2022). The hyphal network of mycorrhizal fungi binds soil particles together, thus improving soil aggregation and soil porosity. In this case, soil

Table 1 Various fungi, their applications in environmental bioremediation, and when their use began

Name of fungi	Extent of use	Beginning of use
White-rot fungi (WRF)	Widely used for degrading pollutants like PAHs, pesticides, dyes, and pharmaceutical residues. Common in soil and water bioremediation projects	1980s
Brown-rot fungi (BRF)	Primarily used for selective degradation of lignin and wood-based pollutants. Emerging applications in waste treatment and soil remediation	1990s
Aspergillus spp.	Applied in the treatment of industrial waste and mycoremediation of heavy metals and organic pollutants	1970s
Trametes versicolor	Known for breaking down various toxic pollutants in water, including synthetic dyes and phenolic compounds	1980s
Pleurotus ostreatus	Used in soil remediation, breaking down complex organic compounds and heavy metals. Common in agricultural waste recycling	1990s
Phanerochaete chrysosporium	One of the earliest fungi studied for its lignin-degrading enzymes, used for degrading a wide range of organic pollutants	1970s
Aspergillus niger	Applied in industrial waste treatment, including heavy metal biosorption and enzyme production	1970s
Penicillium spp.	Effective in mycoremediation of metals, organics, and even oil spills. Found in soil and water remediation	1980s
Lentinula edodes (Shiitake)	Explored for breaking down lignin and other organic materials, used in waste management	1990s

aggregation enhances soil infiltration and water retention, which helps minimize soil erosion. Additionally, networking enhances the establishment and growth of vegetation. The robust plant covers help improve soil health and structure, which are necessary for minimal environmental effects. In this case, the healthy soil and well-established vegetation reduce the effect of stressor factors necessary for continuous bioremediation. Mycorrhizal fungi also facilitate the establishment of plant species within a contaminated site where the condition is too severe for plants to grow. The fungi help improve plant growth in harsh conditions, which helps stabilize the soil while initiating the ecological process necessary for recovery. This case is more relevant in metal-contaminated areas, where plant establishment minimizes soil erosion and contamination spread (Lehmann et al. 2017) (Table 1).

2.4 *Endophytic Fungi*

Endophytic fungi, living under the tissues of plants in a parasitic but unobtrusive way, have demonstrated potential for application in bioremediation. The mutualistic relationship between plants and endophytic fungi gained much interest after it was found that these fungi could degrade pollutants and increase the plants' fitness in polluted environments. Endophytic fungi have been tested in various bioremediation mechanisms, including uptake, intercellular transformation, and detoxification of pollutants (Wen et al. 2022). From this perspective, endophytic fungi represent an attractive option for a range of environmental restoration strategies. One of the direct aspects where endophytic fungi can contribute to bioremediation is their ability to absorb and sequester pollutants from the local environment. Endophytic fungi can easily absorb and sequester many pollutants, such as heavy metals, hydrocarbons, pesticides, and pharmaceuticals, through their hyphae and mycelium (Prabha et al. 2024). The subsequent uptake of pollutants into fungal cells allows their transformation via biologics like different enzymatic activities and creation in immobilized states. After these processes, the bioavailability of these pollutants is significantly reduced, ensuring avoidance of future environmental risks (Wen et al. 2022).

In addition, due to their tremendous metabolic versatility, endophytic fungi have also been able to metabolize complex organic pollutants. Due to their ability, fungi secrete a variety of enzymes, including ligninases and peroxidases, laccases, and cytochrome P450s. These enzymes break down the parent compound and produce several metabolites with mineralizable structures. For example, various endophytic fungi have the competence to degrade polycyclic aromatic hydrocarbons (PAHs) and genistein. Enzymatic degradation enables contaminating sites, and thus, excess incomplete pollutants into simple compounds are based on micelles that are more variable to microbial destruction. Furthermore, the fungi have helped in phytoremediation (Liu et al. 2021). Phytoremediation is the elocution of a plant along with filamentous menteries for remediation. They lead to plant growth and designs, nutrient and elongation, and stress endurance. Endophytic populations have mutualistic relations with floras, providing them with fertile nutrients, enlargement of anchorage, and deterrence from pathogens and surrounding forces. Plants also provide a method for the fungus's liberal germination and growth. The phenomenon has increased fungi's phytoremediation, including the cooperative remediation potential. Moreover, endophytic fungi's detoxification capacities play a vital role in bioremediation. The fungi can detoxify pollutants through enzymatic transformation, bioaccumulation, or encapsulation (Ma et al. 2016). For instance, some endophytic fungi convert toxic levels of heavy metals such as cadmium and lead into less toxic states by producing methylated metallothionein or predation by complexing the metals. The detoxification reduces the pollutants' deleterious impact on plants and ecosystems. It also speeds up the recovery process of polluted areas (Su et al. 2021).

2.5 *Yeasts*

Yeasts are a broad group of single-celled fungi that are essential components of the bioremediation process due to their wide metabolic capacities and adjustability to various environmental factors. Yeasts can degrade a variety of pollutants, transform toxic substances, and bioaccumulate them. In this regard, yeasts are indispensable for the bioremediation of natural environments. Hence, yeasts are involved in several basic mechanisms that promote the elimination and decontamination of pollutants in polluted habitats. First, yeasts can metabolize a broad spectrum of organic pollutants (Deshmukh et al. 2016). Various yeasts produce several enzymes, including oxidases, dehydrogenases, and esterases, among others, that effectively degrade diverse organic substances, such as hydrocarbons, pesticides, and aromatic compounds. For instance, some yeast species demonstrated high efficiency in degrading petroleum hydrocarbons, particularly the components referred to as benzene, toluene, ethylbenzene, and xylene, or BTEX in short, which are the common pollutants in oil spills and industrial areas. Thus, the metabolic breakdown of pollutants by yeasts provides for end-product contaminants of lesser complexity and hazard. Yeasts are highly tolerant and able to detoxify such contaminants as heavy metals in soil and water systems. Thanks to a mechanism called biosorption, where the metal ions are taken up by the surface of a cell or internalized within its organelles, these yeasts can chelate heavy metal ions via accumulation (Okal et al. 2023). Some types of yeasts develop certain proteins, such as phytochelatins and metallothioneins, which help to keep and remove heavy metals from their bodies. As a result of reducing the potential of heavy metals to accumulate in organisms, yeasts' capacity to detoxify heavy metals significantly reduces their bioavailability in the environment and bioaccumulation in the system, thus reducing their harmful influence. Yeasts are also capable of helping in biodegradation and detoxification of xenobiotics, artificially created substances. Yeasts engage in the processes of bioaccumulation and biosorption, wherein they gather toxins on their surfaces or inside their cells. Pollutant concentrations in the surrounding environment may be decreased due to this buildup, which may cause the concentration and sequestration of pollutants. By releasing chelating agents or surfactants that solubilize and mobilize hydrophobic chemicals and facilitate their breakdown by other microbes or chemical processes, yeasts can also increase the bioavailability of certain contaminants (Balzano et al. 2020).

3 Applications of Mycoremediation

3.1 Soil Decontamination

Another common application of fungi in soil decontamination is their ability to degrade organic pollutants. Some fungi, for instance, white-rot fungi, e.g., *Phanerochaete chrysosporium*, are characterized by the enzymatic machinery capable of degrading complex organic pollutants such as polycyclic aromatic hydrocarbons or PAHs, petroleum hydrocarbons, and pesticides. These fungi produce ligninolytic enzymes, including lignin peroxidase, manganese peroxidase, and laccase, which can oxidize and degrade recalcitrant pollutants. In the process of enzymatic degradation, the fungi reduce hazardous pollutants into less toxic or nontoxic

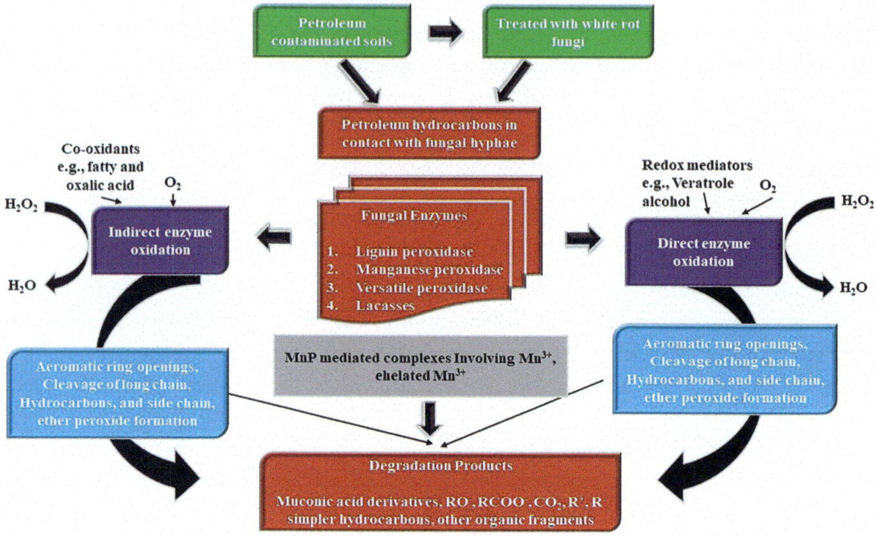

Fig. 1 Mycoremediation of soil contaminated with petroleum compounds

metabolites, thereby minimizing the environmental impact of pollutants while restoring the soil's quality (Peralta et al. 2017). Fungi facilitate soil decontamination through other mechanisms such as biosorption and bioremediation. Certain fungi species have a strong affinity for heavy metals, through which they accumulate heavy metals inside their cells or on their cell walls. The fungi cell wall components such as chitin, glucans, and melanin-like pigments facilitate this mechanism, which can bind and chelate heavy metal ions. The process reduces the bioavailability of heavy metals in the soil and minimizes their uptake by plants to reduce their toxic effects on soil organisms and ecosystems. Fungi also participate in the biotransformation of xenobiotic compounds in the soil (Dwivedi 2023). Xenobiotics are chemical compounds synthesized by man and include synthetic herbicides, insecticides, and industrial pollutants that resist degradation by natural processes. Fungi such as *Fusarium* spp., *Penicillium* spp., and *Trichoderma* spp. produce enzyme systems capable of degrading xenobiotics to simpler, less toxic molecules. The mechanisms lower soil detoxification and reduce the persistence of pollutants, enhancing their degradation by other soil microbial communities (Miglani et al. 2022) (Fig. 1).

3.2 *Water Treatment*

Fungi have wide applications in water treatment areas. Fungi enable the cleanup of water bodies from contaminants and the purification of wastewater. Their different metabolic abilities and adaptive mechanisms make fungi powerful tools for addressing water degradation challenges. Various processes involved in the use of fungi in

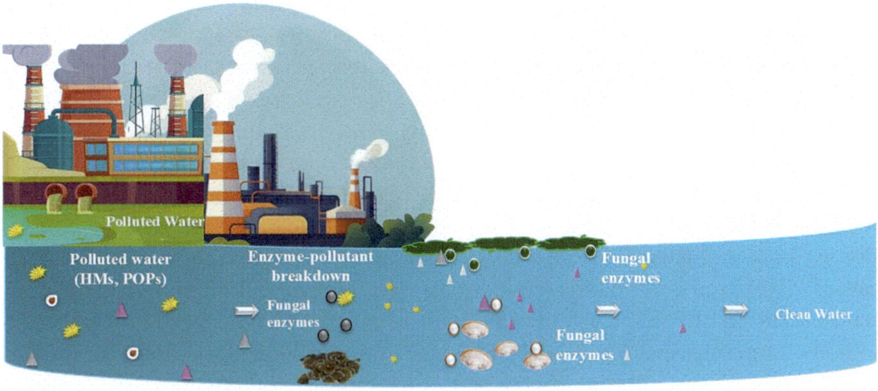

Fig. 2 Diagrammatic representation of the path showing the treatment procedure of polluted water to clean water

water treatment include enzymatic degradation, biosorption, bioaccumulation, and nutrient recycling. Fungi are essential in the removal of organic pollutants, heavy metals, and other pollutants from water bodies. One of the uses of fungi in water treatment is enzyme degradation of organic pollutants (Amobonye et al. 2023). Some fungi, particularly white-rot fungi, such as *Phanerochaete chrysosporium* and *Pleurotus ostreatus*, produce ligninolytic enzymes, such as lignin peroxidase, manganese peroxidase, and laccase. These enzymes decompose various organic compounds, including dyes, pesticides, polycyclic aromatic hydrocarbons, and pharmaceutical residues. The pollutants broken down by the fungi into more straightforward and less harmful by-products are easily removed from water bodies and excreted; thereby, the fungi facilitate the degradation of industrial effluents. Fungi have utilized biosorption and bioaccumulation mechanisms in water treatment (Kumar and Chandra 2020). Many fungal species include fungal cell wall chitin, glucans, and melanin-like pigments. The cell wall components have a high affinity for heavy metals and other pollutants. The fungi absorb heavy metals and other pollutants onto their surfaces or accumulate them in the cells. The two processes primarily reduce the concentration of heavy metals and metalloids in water (Garcia-Rubio et al. 2020) (Fig. 2).

3.3 *Air Purification*

Fungi play a vital role in air purification as they utilize different mechanisms and applications to exploit their metabolic capacities and symbiotic relationships. There are four common applications and methods of fungal air purification: biofiltration, bioremediation, and biodegradation (Cantele, 364). They are associated with biotic interactions and are used to remove airborne pollutants, volatile organic compounds, and odors from the air in indoor and public spaces. Thus, one of the primary applications of fungi in air purification is accomplished through biofiltration, in which fungal biofilms or mats are utilized for air filtration and degradation of pollutants. Biofiltration, such as mold, is a feasible process with such as fungi *Aspergillus* spp., *Trichoderma* spp., and *Cladosporium* spp. They are usually used in filtration systems because of their high metabolic rates and ability to metabolize numerous organic chemicals and volatile organic compounds. Fungi form mats on the filter's solid media (Deshmukh et al. 2016). This media absorbs, adsorbs, and biodegrades pollutants via which air is passed. The same materials passively absorb odorous contaminants from crevices, while alcohol helps remove polyurethane foam (PUF)-based compounds from the air in buildings and other indoor environments. Additionally, fungi also play a role in air purification through biodegradation methods, targeting specific toxic pollutants. Certain filamentous fungi, or sternebrae, act on certain filtrates and wood smoke these compounds. For instance, *Penicillium* spp., *Aspergillus niger*, and *Pleurotus* spp. can all biodegrade mycotoxins, a class of toxic substances secreted by molds and fungi. Enzymes inactivate mycotoxins by breaking their chemical bonds (Khaling et al. 2020).

4 Mycoremediation in Industrial Processes

4.1 Waste Management in Agriculture

Fungi are also instrumental in agricultural waste management, providing sustainable approaches to the recycling, decomposition, and utilization of organic waste materials. Due to their metabolic capacities, symbiotic relationships with plants, and the ability to decompose complex organic compounds, fungi can be implemented in numerous agricultural waste management practices, becoming an essential asset in the process. Fungi are employed in composting, biodegradation, nutrient cycling, soil enrichment, and plant health promotion, allowing agricultural professionals to convert waste into valuable resources that maintain agricultural ecosystems (Tomer et al. 2021). Composting, which utilizes fungi to convert the material into a valuable soil amendment, is one of the fungi's fundamental applications in waste management. Fungi such as *Aspergillus* spp., *Penicillium* spp., and *Trichoderma* spp. are responsible for the initiation of composting, where they decompose high molecular weight organic compounds and lignocellulosic materials (Abdullah et al. 2017). Enzymes, like cellulases, hemicellulases, and ligninases, break the cellulose, hemicellulose, and polyphenolic critical components of plant residue to facilitate their decay and conversion to humus. Composting by fungi reduces waste volume and converts material into humus, improving soil physical characteristics, such as large size and aggregation, and promoting microbial activity in agricultural fields (Andlar et al. 2018).

Fungi also play a role in biodegradation. As part of agricultural waste management, fungi assist in the mineralization and breakdown of specific organic

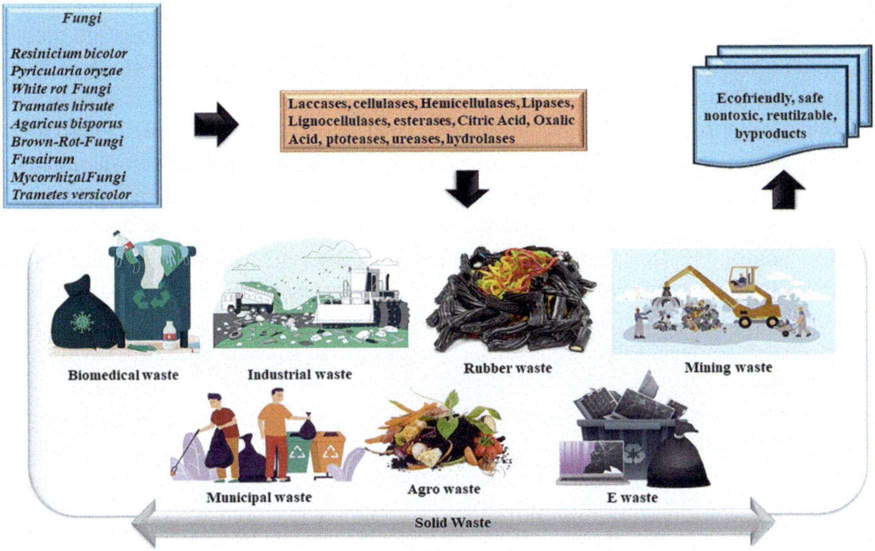

Fig. 3 Schematic representation of the bioremediation of solid waste

pollutants and contaminants. For instance, certain fungi such as *Pleurotus ostreatus* and *Agaricus bisporus* are helpful in the biodegradation of lignin-rich materials such as crop residues, straw, and sawdust. The fungi release ligninolytic enzymes, such as peroxidases and laccases, which degrade the lignocellulosic substrates and provide nutrient-enriched substrates that are proper for mushroom formation or incorporation into the soil. Therefore, fungi biodegradation prevents the accumulation of organic waste in agricultural fields, producing useful biomass and organic matter for soil enrichment (Gupta et al. 2018). Additionally, fungi help in nutrient cycling and soil enrichment in agricultural waste management. Mycorrhizal fungi create root symbiotic affiliations with plants, allowing them to absorb and utilize nutrients. For example, these fungi assist in the transfer of compost and organic matter nutrients to plants, reducing the requirement for artificial fertilizers. Fungi also increase nutrient availability and soil fertility as a result, supporting eco-friendly agricultural activities that rely on reducing ecological harm and promoting long-term soil health (Wahab et al. 2023) (Fig. 3).

4.2 *Bioremediation in Mining and Metallurgy*

In the field of mining and metallurgy, fungi bioremediation presents an advanced approach that provides sustainable solutions for mining waste, mine tailings, and mining-impacted site restoration and remediation. Fungi's exceptional metabolic properties, like tolerance to extreme conditions, diverse enzymatic nature, and metal-binding characteristics, make them ideal candidates that can be applied in mining and metallurgy. The prospects of fungi bioremediation in mining and metallurgy include metal bioaccumulation, biosorption, biomineralization, and soil recycling and reclamation (Newsome and Falagán 2021). These applications contribute to reducing environmental implications and risks associated with mining activities. Primarily, fungi can be utilized to accumulate heavy metals from contaminated soils and mine tailings and detoxify the metals in the process. Biometal-tolerant fungi such as *Aspergillus niger*, *Penicillium* spp., and *Saccharomyces cerevisiae* have adaptive techniques that enable them to thrive in the presence of heavy metals by accumulating large amounts of metals such as copper, zinc, cadmium, and lead through bioaccumulation. The accumulated metals are then removed, sequestered, or converted to less toxic substances, reducing their bioavailability (Vaksmaa et al. 2023).

Fungi have biosorption abilities, leading to the adsorption of heavy metals on the surface of their cells or the secreted extracellular matrix. Fungi contain cell wall components like melanin-like pigments, chitin, and glucans with a higher chelating capacity to metal ions and immobilize heavy metals in polluted soils and sediments. This biosorption activity helps remove and stabilize heavy metals and reduces the risk of metal leaching into groundwater or surface water bodies for subsequent ecological contamination (Dhankhar and Hooda 2011). Finally, fungi also promote biomineralization resulting from heavy metals' weathering and dissolution and their combination into newly formed metallic minerals and complexes. Some fungi species, such as *Aspergillus niger* and *Penicillium* spp., produce organic acids that promote the precipitation of metals in sequence insoluble mineral forms like oxalates, phosphates, and sulfides. This innovation enables soluble metal species to convert into less mobile and ecologically available biominerals, promoting long-term containment and immobilization (Šebesta et al. 2022). Table 2 explains the significance and mechanisms of different mycoremediation in various industries.

5 Economic and Environmental Benefits

The method of fungus remediation, also referred to as mycoremediation, is one of the elements that influence its cost-effectiveness. The principle behind mycoremediation is the use of fungus to break down, purify, or absorb contaminants from a range of environmental media, including soil, water, and air. The low-cost generation and maintenance of fungal biomass, multifactorial bioremediation, and the viability of mycoremediation as a crucial component of sustainable remediation and phytoremediation ideas all contribute to the cost-effectiveness of fungal remediation. Firstly, the cheap cost of producing and maintaining fungal biomass provides the foundation for the cost-effectiveness of fungal remediation (Akhtar and Mannan 2020). Many fungi, especially those that cause white rot (*Phanerochaete chrysosporium* and *Pleurotus ostreatus*), proliferate quickly on inexpensive growing media, including sawdust, agricultural waste, and industrial wastes. Renewing biocatalysts

Table 2 Role of different fungi and their mechanism in remediation in various industries

Industry	Type of fungi	Mechanism in bioremediation	Significance	References
Water treatment	White-rot fungi	Enzymatic degradation of organic pollutants	Break down complex pollutants like dyes, pesticides, and pharmaceuticals in wastewater	Torres-Farradá et al. (2024)
	Yeasts	Biosorption of heavy metals, biodegradation of xenobiotics	Remove heavy metals and detoxify synthetic chemicals from wastewater	Tripathi et al. (2023)
Agriculture	Arbuscular Mycorrhizal Fungi	Nutrient uptake, soil structure improvement	Enhance plant growth, improve soil fertility and structure	Wahab et al. (2023)
	Trichoderma spp.	Biocontrol of plant pathogens, nutrient cycling	Suppress plant diseases, promote nutrient availability in soil	Yao et al. (2023)
Soil management	Endophytic fungi	Pollutant uptake, transformation, detoxification	Enhance plant tolerance to pollutants, detoxify soil contaminants	He et al. (2020)
	Mycorrhizal fungi	Nutrient cycling, soil aggregation, water retention	Improve soil structure, increase water and nutrient uptake by plants	Dierks et al. (2021)
Textile industries	*Aspergillus* spp.	Enzymatic degradation of textile dyes, effluent treatment	Treat textile wastewater, reduce the environmental impact of dye discharge	Baghel et al. (2020)
	Phanerochaete chrysosporium	Lignin degradation, bioremediation of industrial waste	Break down lignin-rich compounds, assist in industrial waste cleanup	Singh et al. (2021)
Pharma industries	*Penicillium* spp.	Enzymatic degradation of pharmaceutical compounds	Degrade and detoxify pharmaceutical residues in wastewater	Olicón-Hernández et al. (2017)
	Fusarium spp.	Mycoremediation of pesticide residues	Assist in the degradation of pesticide contaminants in soil	Pandey et al. (2018)
Food industries	*Saccharomyces cerevisiae*	Fermentation, detoxification of food by-products	Aid in fermentation processes, detoxify food processing waste	Londoño-Hernandez et al. (2020)
	Rhizopus spp.	Enzymatic breakdown of food wastes, biodegradation in composting	Assist in composting organic food waste and reduce the environmental impact of food waste disposal	Palaniveloo et al. (2020)

that may be produced from carbon sources from sustainable agriculture is inexpensive and simple to get in the form of fungi. Fungal biomass culture is frequently far less expensive than the manufacture of chemicals and the environmental risks associated with their usage, as it does not require complex machinery or costly gear (Dhiman et al. 2024).

Fungi have a wide target spectrum of pollutants that includes a large variety of organic compounds, heavy metals, pesticides, hydrocarbons, and radioactive contaminants. This enables fungi to be used in diverse environmental remediation cases and applies the same remediation technology to various pollutant types. The simultaneous impact on several pollutants can even make it possible to replace several remediation technologies or treatments, saving costs and ensuring an efficient operation in remediation projects (Deshmukh et al. 2016). Mycoremediation is considered a promising approach for integrated and sustainable environmental cleanup, as it aligns with green and eco-friendly remediation principles by minimizing physical and chemical disturbances to the affected area. Along with natural processes, mycoremediation can be supported with other methods, such as phytoremediation or utilization of specific microbial consortia, to create a synergistic effect. For example, mycorrhizal fungi increase the growth of plants and their nutrient uptake, and white-rot fungi can degrade organic pollutants in the soil (Ghosh et al. 2023).

Mycoremediation is cost-effective due to its compatibility with various environmental conditions, including rigorous pH, temperature, and moisture levels. As a result, fungi can function in various environmental settings ranging from stringent homeostatic conditions to openly polluted sites, unlike numerous remediation technologies that are only feasible in a particular niche (Deshmukh et al. 2016). Acclimation to different conditions eliminates other site-specific costly changes because fungi function independently despite the environmental conditions. Hence, fungi are cost-effective due to the subsidization of expensive alterations to operate in specific conditions. Also, fungi can remediate contaminants noninvasively and nondestructively, unlike traditional excavation and dumping. The widespread use of fungi reduces the need for on-site modifications typically required by conventional remediation methods, making it a more cost-effective solution (Tomer et al. 2021).

Mycoremediation has a crucial role in reducing environmental pollution, as it provides sustainable and effective remedial solutions for polluted environments. Fungi reduce environmental pollution via numerous mechanisms and applications, including biodegradation, biosorption, bioaccumulation, nutrient cycles, and soil restoration. Fungi target several pollutants, including organic compounds, heavy metals, pesticides, hydrocarbons, and pharmaceuticals. The numerous processes involved in fungi to target pollutants help lead to cleaner air, water, and soil ecosystems (Varjani and Patel 2017). Biodegradation of organic pollutants is one of fungi's most significant targets designed to reduce environmental pollution. Some fungi, such as white-rot fungi, release the ligninolytic enzyme, which breaks down complex organic pollutants like PAHs, dyes, pesticides, and petroleum hydrocarbons. This degradation process involves producing simple and less hazardous PAH forms by fungi. It purifies soil, water, and air by breaking down organic pollutants, thus reducing environmental pollution (Ghosh et al. 2023).

In addition, fungi are involved in biosorption and bioaccumulation processes through which they adsorb and accumulate various pollutants, including heavy metals, metalloids, and organic contaminants. The cell wall components of fungi, such as chitin, glucans, and melanin-like pigments, have a high affinity for pollutants and can efficiently bind and sequester them away from the soil and water bodies. This reduces the bioavailable and mobile forms of pollutants, thus reducing their dispersal and minimizing their impacts on the environment and human and animal health (Joshi et al. 2011). Over time, the sequestration reduces the pollution levels in the environment, allowing for the natural uptake of the pollutants or their permanent sequestration away from natural resources. Similarly, fungi are involved in nutrient cycling and soil restoration, which are essential for maintaining healthy ecosystems. Mycorrhizal fungi form associations with plant roots to facilitate nutrient pickup, improve soil structure, and increase soil fertility and microbial activity. The fungi aid in nutrient recycling, thereby reducing loss through leaching, soil erosion, and nutrient imbalances that may lead to pollution in water bodies. Mycoremediation supports healthy soils, which are important for life, and promotes sustainable livelihoods (Boorboori and Zhang 2022).

6 Future Prospects and Innovations

Overall, the future prospects and innovations in the role of mycoremediation in the circular economy present tremendous potential to solve environmental challenges, promote sustainability, and promote resource efficiency. As the circular economy takes root around the world, fungi in remediation play a critical role in closing the loop in waste management, reducing the environmental impact, and creating value from waste streams. There is a need to develop tailored fungal strains and biotechnologies that are optimized for pollutant degradation and resource recovery for mycoremediation in the future. Further, advances in genetic engineering, synthetic biology, and bioprocess optimization are likely to increase the degradation efficiency of fungi rupturing complex pollutants such as plastics, pharmaceuticals, and emerging contaminants. Finally, engineered fungi with constructed metabolic pathways and pollutant-degrading enzymes may drive mycoremediation and the resource recovery field into newer avenues under this circular economy paradigm. In addition, combining mycoremediation with other circular economy principles, such as bioenergy generation, bio-based materials production, and waste valorization, provides novel prospects for sustainable resourcefulness. Fungi-mediated procedures, such as anaerobic digestion, fermentation, and biotransformation, can transform organic waste into biofuels, bioplastics, biochemicals, and other high-value products. Closed-loop systems can be developed using fungi's metabolic and biocatalytic potential to convert waste into resources, reducing reliance on scarce resources while simultaneously reducing environmental pollution. Furthermore, the use of mycoremediation to rehabilitate soil health and agricultural durability contributes to the circular economy's goals of regenerative farming and ecosystem stability. Mycorrhizal fungi improve nutrient recycling, soil carbon storage, and plant–microbe connections while also boosting crop production and soil health. Groundbreaking technologies in fungal biostimulants, biofertilizers, and soil inoculants can boost agriculture and sustainability by maximizing land usage, lowering chemical utilization, and increasing the chances of circularity principles.

7 Conclusion

Mycoremediation plays a crucial role in the circular economy by transforming waste management practices, promoting sustainability, and supporting the regenerative use of resources. Mycoremediation offers numerous benefits that are closely associated with the core principles of the circular economy, such as waste reduction, resource recovery, environmental conservation, and the generation of economic value. First, it reduces waste through the process of using fungi to convert organic waste streams into valuable resources. This type of remediation combines enzymatic degradation, biosorption, and bioaccumulation processes that help fungi convert a wide range of pollutants—from organic compounds to heavy metals—into less harmful or recoverable forms. As a result, mycoremediation removes waste from storage in landfills and mitigates the negative effects of pollution while also enabling the recovery of valuable materials. Second, mycoremediation offers remediation of affected sites and revitalization of healthy soil, ensuring that contaminated sites can be repurposed for industrial activities or residences. As mentioned earlier, fungi also convert other types of waste into bio-based products, renewable energy, and sustainable materials. Many fungi-mediated processes, including biodegradation, fermentation, and biotransformation, can convert organic waste sources into biofuels, biopolymers, biochemicals, and bioactive compounds. It reduces the reliance on fossil fuels and virgin feedstock and drives the transition to a bioeconomy that encompasses renewable resources and circular supply complementarities. Finally, mycoremediation contributes to protecting the environment by ensuring that the affected sites are remedied and healthy soils are revitalized.

References

Abdal Dayem A, Hossain MK, Lee SB, Kim K, Saha SK, Yang GM, Choi HY, Cho SG (2017) The role of reactive oxygen species (ROS) in the biological activities of metallic nanoparticles. Int J Mol Sci 18(1):120. https://doi.org/10.3390/ijms18010120

Abdullah JJ, El-Imam AA, Greetham D, Du C, Tucker GA (2017) The application of fungi for bioleaching of municipal solid wastes for the production of environmental acceptable compost production. J Environ Sci Public Health 1(3):167–194

Akhtar N, Mannan MAU (2020) Mycoremediation: expunging environmental pollutants. Biotechnol Rep 26:e00452

Amobonye A, Aruwa CE, Aransiola S, Omame J, Alabi TD, Lalung J (2023) The potential of fungi in the bioremediation of pharmaceutically active compounds: a comprehensive review. Front Microbiol 14:1207792. https://doi.org/10.3389/fmicb.2023.1207792

Andlar M, Rezić T, Marđetko N, Kracher D, Ludwig R, Šantek B (2018) Lignocellulose degradation: an overview of fungi and fungal enzymes involved in lignocellulose degradation. Eng Life Sci 18(11):768–778. https://doi.org/10.1002/elsc.201800039

Baghel S, Sahariah BP, Anandkumar J (2020) Bioremediation of lignin-rich pulp and paper industry effluent. In: Combined application of physico-chemical and microbiological processes for industrial effluent treatment plant. Springer, Singapore, pp 261–278

Bala S, Garg D, Thirumalesh BV, Sharma M, Sridhar K, Inbaraj BS, Tripathi M (2022) Recent strategies for bioremediation of emerging pollutants: a review for a green and sustainable environment. Toxics 10(8):484

Balzano S, Sardo A, Blasio M, Chahine TB, Dell'Anno F, Sansone C, Brunet C (2020) Microalgal metallothioneins and phytochelatins and their potential use in bioremediation. Front Microbiol 11:517. https://doi.org/10.3389/fmicb.2020.00517

Begum N, Qin C, Ahanger MA, Raza S, Khan MI, Ashraf M et al (2019) Role of arbuscular mycorrhizal fungi in plant growth regulation: implications in abiotic stress tolerance. Front Plant Sci 10:1068

Boorboori MR, Zhang HY (2022) Arbuscular mycorrhizal fungi are an influential factor in improving the phytoremediation of arsenic, cadmium, lead, and chromium. J Fungi 8(2):176. https://doi.org/10.3390/jof8020176

Deshmukh R, Khardenavis AA, Purohit HJ (2016) Diverse metabolic capacities of fungi for bioremediation. Indian J Microbiol 56(3):247–264. https://doi.org/10.1007/s12088-016-0584-6

Dhankhar R, Hooda A (2011) Fungal biosorption–an alternative to meet the challenges of heavy metal pollution in aqueous solutions. Environ Technol 32(5):467–491

Dhiman S, Kaur P, Narang J, Mukherjee G, Thakur B, Kaur S, Tripathi M (2024) Fungal bioprocessing for circular bioeconomy: exploring lignocellulosic waste valorization. Mycology 15:1–26

Dierks J, Blaser-Hart WJ, Gamper HA, Nyoka IB, Barrios E, Six J (2021) Trees enhance abundance of arbuscular mycorrhizal fungi, soil structure, and nutrient retention in low-input maize cropping systems. Agric Ecosyst Environ 318:107487

Dwivedi SK (2023) Fungi mediated detoxification of heavy metals: insights on mechanisms, influencing factors and recent developments. J Water Process Eng 53:103800

Garcia-Rubio R, de Oliveira HC, Rivera J, Trevijano-Contador N (2020) The fungal cell wall: *Candida*, *Cryptococcus*, and *Aspergillus* species. Front Microbiol 10:2993. https://doi.org/10.3389/fmicb.2019.02993

Ghosh S, Rusyn I, Dmytruk OV, Dmytruk KV, Onyeaka H, Gryzenhout M, Gafforov Y (2023) Filamentous fungi for sustainable remediation of pharmaceutical compounds, heavy metal and oil hydrocarbons. Front Bioeng Biotechnol 11:1106973. https://doi.org/10.3389/fbioe.2023.1106973

Goodell B (2003) Brown-rot fungal degradation of wood: our evolving view. ACS Symp Ser 845:97–118

Gupta S, Annepu SK, Summuna B, Gupta M, Nair SA (2018) Role of mushroom fungi in decolourization of industrial dyes and degradation of agrochemicals. In: Biology of macrofungi. Springer, Cham, pp 177–190

He W, Megharaj M, Wu CY, Subashchandrabose SR, Dai CC (2020) Endophyte-assisted phytoremediation: mechanisms and current application strategies for soil mixed pollutants. Crit Rev Biotechnol 40(1):31–45

Honda M, Suzuki N (2020) Toxicities of polycyclic aromatic hydrocarbons for aquatic animals. Int J Environ Res Public Health 17(4):1363. https://doi.org/10.3390/ijerph17041363

Jeffries P, Gianinazzi S, Perotto S, Turnau K, Barea JM (2003) The contribution of arbuscular mycorrhizal fungi in sustainable maintenance of plant health and soil fertility. Biol Fertil Soils 37:1–16

Joshi PK, Swarup A, Maheshwari S, Kumar R, Singh N (2011) Bioremediation of heavy metals in liquid media through fungi isolated from contaminated sources. Indian J Microbiol 51(4):482–487. https://doi.org/10.1007/s12088-011-0110-9

Khaling E, Agyei T, Jokinen S, Holopainen JK, Blande JD (2020) The phytotoxic air-pollutant O_3 enhances the emission of herbivore-induced volatile organic compounds (VOCs) and affects the susceptibility of black mustard plants to pest attack. *Environ Pollut* 265(Pt A):115030. https://doi.org/10.1016/j.envpol.2020.115030

Kumar A, Chandra R (2020) Ligninolytic enzymes and its mechanisms for degradation of lignocellulosic waste in environment. Heliyon 6(2):e03170. https://doi.org/10.1016/j.heliyon.2020.e03170

Lakhani S, Acharya D, Sakariya R, Sharma D, Patel P, Shah M, Prajapati M (2022) A comprehensive study of bioremediation for pharmaceutical wastewater treatment. Clean Chem Eng 4:100073

Lehmann A, Leifheit EF, Rillig MC (2017) Mycorrhizas and soil aggregation. In: Mycorrhizal mediation of soil. Elsevier, pp 241–262

Leng L, Li W, Chen J, Leng S, Chen J, Wei L et al (2021) Co-culture of fungi-microalgae consortium for wastewater treatment: a review. Bioresour Technol 330:125008

Liu X, Zhou ZY, Cui JL, Wang ML, Wang JH (2021) Biotransformation ability of endophytic fungi: from species evolution to industrial applications. Appl Microbiol Biotechnol 105:1–19

Londoño-Hernandez L, Chávez-González ML, Ascacio-Valdés JA, Ruiz HA, Toro CR, Aguilar CN (2020) Fermentative bioprocesses for detoxification of agri-food wastes for production of bioactive compounds. In: Food microbiology and biotechnology. Apple Academic Press, pp 287–318

Ma Y, Rajkumar M, Zhang C, Freitas H (2016) Beneficial role of bacterial endophytes in heavy metal phytoremediation. J Environ Manag 174:14–25. https://doi.org/10.1016/j.jenvman.2016.02.047

Meyer V, Basenko EY, Benz JP, Braus GH, Caddick MX, Csukai M et al (2020) Growing a circular economy with fungal biotechnology: a white paper. Fungal Biol Biotechnol 7(1):5

Miglani R, Parveen N, Kumar A, Ansari MA, Khanna S, Rawat G, Panda AK, Bisht SS, Upadhyay J, Ansari MN (2022) Degradation of xenobiotic pollutants: an environmentally sustainable approach. Metabolites 12(9):818. https://doi.org/10.3390/metabo12090818

Mohanan N, Montazer Z, Sharma PK, Levin DB (2020) Microbial and enzymatic degradation of synthetic plastics. Front Microbiol 11:580709

Mujtaba M, Fraceto LF, Fazeli M, Mukherjee S, Savassa SM, de Medeiros GA et al (2023) Lignocellulosic biomass from agricultural waste to the circular economy: a review with focus on biofuels, biocomposites and bioplastics. J Clean Prod 402:136815

Nandal P, Arora A, Virmani S (2021) An appraisal on valorization of lignin: a byproduct from biorefineries and paper industries. Biomass Bioenergy 155:106295

Newsome L, Falagán C (2021) The microbiology of metal mine waste: bioremediation applications and implications for planetary health. Geohealth 5(10):e2020GH000380. https://doi.org/10.1029/2020GH000380

Noman E, Al-Gheethi A, Mohamed RMSR, Talip BA (2019) Myco-remediation of xenobiotic organic compounds for a sustainable environment: a critical review. Top Curr Chem 377:1–41

Okal EJ, Heng G, Magige EA, Khan S, Wu S, Ge Z et al (2023) Insights into the mechanisms involved in the fungal degradation of plastics. Ecotoxicol Environ Saf 262:115202

Olicón-Hernández DR, González-López J, Aranda E (2017) Overview on the biochemical potential of filamentous fungi to degrade pharmaceutical compounds. Front Microbiol 8:256025

Palaniveloo K, Amran MA, Norhashim NA, Mohamad-Fauzi N, Peng-Hui F, Hui-Wen L et al (2020) Food waste composting and microbial community structure profiling. Processes 8(6):723

Pandey C, Prabha D, Negi YK (2018) Mycoremediation of common agricultural pesticides. In: Mycoremediation and environmental sustainability, vol 2. Springer, Cham, pp 155–179

Peralta RM, da Silva BP, Côrrea RCG, Kato CG, Seixas FAV, Bracht A (2017) Enzymes from basidiomycetes—peculiar and efficient tools for biotechnology. In: Biotechnology of microbial enzymes. Academic Press, pp 119–149

Prabha R, Singh DP, Kumar A, Kumar M, Ritika, Singh S et al (2024) Endophytic fungi for microbial phytoremediation: prospects for agricultural and environmental sustainability. In: Endophytic fungi: the hidden sustainable jewels for the pharmaceutical and agricultural industries. Springer International Publishing, Cham, pp 325–345

Šebesta M, Vojtková H, Cyprichová V, Ingle AP, Urík M, Kolenčík M (2022) Mycosynthesis of metal-containing nanoparticles-fungal metal resistance and mechanisms of synthesis. Int J Mol Sci 23(22):14084. https://doi.org/10.3390/ijms232214084

Serbent MP, Magario I, Saux C (2024) Immobilizing white-rot fungi laccase: toward bio-derived supports as a circular economy approach in organochlorine removal. Biotechnol Bioeng 121(2):434–455

Singh AK, Bilal M, Iqbal HM, Meyer AS, Raj A (2021) Bioremediation of lignin derivatives and phenolics in wastewater with lignin modifying enzymes: status, opportunities and challenges. Sci Total Environ 777:145988

Su Z, Zeng Y, Li X, Perumal AB, Zhu J, Lu X, Dai M, Liu X, Lin F (2021) The endophytic fungus Piriformospora indica-assisted alleviation of cadmium in tobacco. J Fungi 7(8):675. https://doi.org/10.3390/jof7080675

Tomer A, Singh R, Singh SK, Dwivedi SA, Reddy CU, Keloth MRA, Rachel R (2021) Role of fungi in bioremediation and environmental sustainability. In: Mycoremediation and environmental sustainability, vol 3. Springer, Cham, pp 187–200

Torres-Farradá G, Manzano León AM, Rineau F, Ledo Alonso LL, Sánchez-López MI, Thijs S, Colpaert J, Ramos-Leal M, Guerra G, Vangronsveld J (2017) Diversity of ligninolytic enzymes and their genes in strains of the genus *Ganoderma*: applicable for biodegradation of xenobiotic compounds? Front Microbiol 8:898. https://doi.org/10.3389/fmicb.2017.00898

Torres-Farradá G, Thijs S, Rineau F, Guerra G, Vangronsveld J (2024) White rot fungi as tools for the bioremediation of xenobiotics: a review. J Fungi 10(3):167. https://doi.org/10.3390/jof10030167

Tripathi M, Singh P, Singh R, Bala S, Pathak N, Singh S, Chauhan RS, Singh PK (2023) Microbial biosorbent for remediation of dyes and heavy metals pollution: a green strategy for sustainable environment. Front Microbiol 14:1168954. https://doi.org/10.3389/fmicb.2023.1168954

Vaksmaa A, Guerrero-Cruz S, Ghosh P, Zeghal E, Hernando-Morales V, Niemann H (2023) Role of fungi in bioremediation of emerging pollutants. Front Mar Sci 10:1070905

Varjani SJ, Patel RK (2017) Fungi: a remedy to eliminate environmental pollutants. In: Mycoremediation and environmental sustainability, vol 1. Springer, Cham, pp 53–67

Wahab A, Muhammad M, Munir A, Abdi G, Zaman W, Ayaz A, Khizar C, Reddy SPP (2023) Role of arbuscular mycorrhizal fungi in regulating growth, enhancing productivity, and potentially influencing ecosystems under abiotic and biotic stresses. Plants 12(17):3102. https://doi.org/10.3390/plants12173102

Weimer PJ (2022) Degradation of cellulose and hemicellulose by ruminal microorganisms. Microorganisms 10(12):2345

Wen J, Okyere SK, Wang S, Wang J, Xie L, Ran Y, Hu Y (2022) Endophytic fungi: an effective alternative source of plant-derived bioactive compounds for pharmacological studies. J Fungi 8(2):205. https://doi.org/10.3390/jof8020205

Yao X, Guo H, Zhang K, Zhao M, Ruan J, Chen J (2023) Trichoderma and its role in biological control of plant fungal and nematode disease. Front Microbiol 14:1160551

Zhang C, Wu D, Ren H (2020) Bioremediation of oil contaminated soil using agricultural wastes via microbial consortium. Sci Rep 10(1):9188. https://doi.org/10.1038/s41598-020-66169-5

Zhu W, Lei J, Li Y, Dai L, Chen T, Bai X et al (2019) Procedural growth of fungal hyphae/Fe3O4/graphene oxide as ordered-structure composites for water purification. Chem Eng J 355:777–783

Zhuo R, Fan F (2021) A comprehensive insight into the application of white rot fungi and their lignocellulolytic enzymes in the removal of organic pollutants. Sci Total Environ 778:146132

Rhizosphere and Remediation Soil, Water, and Air Pollution

Ehab M. Zayed, Muhammad Imran, Mohammed Saba, and Kamal A. M. Abo-Elyousr

Abstract Soil microbiology examines the diverse microorganisms in soil, such as bacteria, fungi, algae, and protozoa, highlighting their critical role in maintaining soil fertility and soil health. These organisms contribute to nitrogen fixation, nutrient cycling, and reshaping soil structure, thereby improving plant growth and soil ecosystem. A notable aspect is the abundance of bacteria, with about 10 billion bacterial cells per gram of soil, showcasing their responsiveness to environmental changes. Additionally, the interaction between plants and root-associated microorganisms, particularly plant growth-promoting Rhizobacteria (PGPR), improves the availability of essential nutrients and enhances plant resilience, indicating their importance in sustainable agriculture and ecological health. Nature-based solutions (NBSs) in aquatic sediments provide sustainable methods for treating agricultural runoff and wastewater, particularly for nitrogen removal. These solutions, including constructed wetlands, buffer strips, planted canals, and aquatic sediment control ponds, utilize physical, chemical, and biological processes to reduce nitrogen levels in polluted water effectively. While the effectiveness of these systems has previously been documented, there are still key gaps in understanding the specific

E. M. Zayed
Department of Cell Study Research, Field Crops Research Institute, Agricultural Research Center, Giza, Egypt

M. Imran
School of Biological Engineering, Henan University of Technology, Zhengzhou, China
e-mail: Imranpathologist@cau.edu.cn

M. Saba
Department of Agriculture, Faculty of Environmental Sciences, King AbdulAziz University, Jeddah, Saudi Arabia

K. A. M. Abo-Elyousr (✉)
Department of Agriculture, Faculty of Environmental Sciences, King AbdulAziz University, Jeddah, Saudi Arabia

Department of Plant Pathology, Faculty of Agriculture, University of Assiut, Assiut, Egypt
e-mail: kaaboelyousr@agr.au.edu.eg

A. A. H. Abdel Latef et al. (eds.), *Sustainable Remediation for Pollution and Climate Resilience*, https://doi.org/10.1007/978-981-96-5674-5_17

mechanisms and the factors involved that influenced their performance. Future research on enhancing the design and application of these systems can promote a circular economy by reusing treated wastewater for agricultural irrigation.

Keywords Rhizosphere · Remediation · Water · Soil

1 Introduction

Soil microbiology explores the microscopic organisms living in the soil and their crucial roles in maintaining soil health and ecosystem functions. These organisms, which first appeared billions of years ago, played a pivotal role in atmospheric nitrogen fixation and producing oxygen, leading to the development of more complex life forms (Sylvia et al. 2005; Reeder 2012; Wilson 2013). Nowadays, rhizosphere microorganisms, including bacteria, fungi, algae, and protozoa, offer numerous benefits, such as enhancing soil fertility and soil structure (Chauhan et al. 2023; Wang et al. 2024). In the rhizosphere, around 10 billion bacterial cells per gram of soil interact with plant roots (Berg and Smalla 2009). For instance, rhizobium bacteria are critical in nitrogen fixation, converting atmospheric nitrogen into a form that plants can easily absorb (Abdelfattah et al. 2021). Nitrogen-fixing bacteria like *Nitrobacter* are the key to the soil ecosystems (Mir et al. 2021), while actinomycetes, which share characteristics with fungi, produce antibiotics vital for medicine (Putri et al. 2020) and also assist in decomposing the organic matter by promoting the plant growth and producing hormones that strengthen the plant immune responses to various biotic and abiotic stresses and enhance their tolerance by enhancing the crop production (Singh et al. 2018). However, the excessive use of synthetic chemicals utterly harms the soil microbiome, thereby reducing the soil fertility and enhancing the dependency on agricultural inputs, while symbiotic relationships in crops such as rhizobia and legumes or mycorrhizae and plant roots underscore the importance of microorganisms in nutrient uptake and plant resilience (Barka et al. 2016; De Simeis and Serra 2021) and a healthy microbial ecosystem causes the reduction in the disease incidence, supporting the plant diversity and improving soil health (Naylor et al. 2022).

Human agricultural practices such as tillage of soil can influence microbial diversity and their functions, highlighting the need for sustainable soil management (Daunoras et al. 2024). Initiatives aimed to promote the awareness of soil ecosystems and the management of natural resources are vital for sustainable agriculture and environmental protection (Meenakshi et al. 2024; Wei et al. 2024), and these efforts align with global goals to protect ecosystems and fight emerging threats of climate change by encouraging green agricultural production and consumption.

2 Diverse Microbial Communities in the Rhizosphere

2.1 Bacterial and Fungal Communities and Their Functions

Bacteria and archaea are among the smallest living organisms in the soil, apart from viruses, and they are prokaryotes, meaning they have a simpler cell structure without internal organelles (van Wolferen et al. 2022). In contrast, other microorganisms are eukaryotic, featuring more complex cells with organelles and the ability to reproduce sexually. Despite their simplicity, these microorganisms have been found abundantly in soil, which plays an essential role in nitrogen fixation (Odelade and Babalola 2019). Among the functions of bacteria, one of the most prominent functions is their biochemical conversion, for example, *Pseudomonas* bacterial species cause the breakdown of various chemicals within the soil while *Nitrobacter* bacteria derive energy by converting nitrites into nitrates during the bio-oxidation (Ahmed et al. 2023; Diggle and Whiteley 2020; Grzyb et al. 2021). Further, *Clostridium* is another bacterial genus that thrives in oxygen-free environments through anaerobic respiration, and notably, some bacterial species like *Pseudomonas aeruginosa* have been reported to perform both aerobic and anaerobic respiration using nitrate as an electron acceptor (Ahmed et al. 2023). Interestingly, some bacterial species are considered a crucial component for nitrogen fixation, converting atmospheric nitrogen into plant-usable forms like ammonia (Gupta et al. 2016; Paśmionka et al. 2021; Mus et al. 2016), whereas autotrophic bacteria like *Nitrobacter* are less common than heterotrophs and they play a vital role in the growth and development of plant by enhancing the soil fertility.

Apart from this, some Actinomycetes, which are soil-dwelling microorganisms classified as bacteria, but exhibit several fungal-like characteristics, such as mycelium formation, spore production, and branching, likely due to evolutionary convergence (Farda et al. 2022; Nazari et al. 2022; Li et al. 2016; Ventura et al. 2007; Barka et al. 2016). These organisms are renowned for producing important antibiotics, including streptomycin, neomycin, erythromycin, and tetracycline, which are used to treat bacterial infections such as tuberculosis, pneumonia, and bronchitis (Krause et al. 2016; Yingyong et al. 2024; De Simeis and Serra 2021). Additionally, actinomycetes play a significant role in the decomposition of organic matter in the soil by enhancing nutrient cycling, thereby promoting the growth and development of plants (Kumar et al. 2022) and as promising biocontrol agents, they play a vital role in inhibiting various plant pathogens by producing volatile organic metabolites (Sharma et al. 2021; Imran et al. 2024).

Fungi, another group of microorganisms that are abundant in soil, thrive especially in acidic and dry environments where bacteria and actinomycetes struggle to survive (Bahram and Netherway 2022). They play a vital role in decomposing organic matter and serve as a food source for other organisms while also forming a symbiotic relationship that helps in improving soil health (Frąc Magdalena et al. 2018; Dick 2009). However, fungi can also cause plant diseases and are influenced by the same environmental factors that affect the growth of bacteria and actinomycetes (Dick 2009).

2.2 *Role and Functions of Algae in Rhizosphere*

Algae are autotrophic organisms that produce food through photosynthesis, relying on light, moderate humidity, and suitable temperature conditions for growth. They thrive beneath the soil surface and can fix atmospheric nitrogen, similar to bacteria (Ibrahim et al. 2024; Nesrine et al. 2024; Bold and Wynne 1978; Baweja and Sahoo 2015; Tilden 1933). Algae are divided into three main families: *Cyanophyceae* (blue-green algae), *Chlorophyceae* (green algae), and *Bacillariaceae* (brown algae). *Cyanophyceae* absorb red and blue light and fix nitrogen, with the extent of fixation influenced more by physiological and environmental factors such as sunlight intensity, nitrogen source concentration, ambient temperature, and environmental stability than by the inherent capabilities of organisms (Mazur et al. 1980). *Chlorophyceae* enhance the soil ecosystems by contributing organic matter, improving soil structure, and promoting nutrient cycling through photosynthesis, which releases oxygen and essential nutrients, while *Bacillariaceae* improve soil health by aiding nutrient cycling, enhancing aeration and water retention, and promoting soil structure with their silica remains, thereby supporting biodiversity and stability in soil ecosystems (Smol and Stoermer 2010).

2.3 *Overview of Key Microorganisms: Protozoa, Diatoms, Amoebas, and Paramecium*

Protozoa are eukaryotic microorganisms that reproduce sexually and are influenced by the same environmental conditions as many soil microorganisms. They are divided into three classes: flagellates, amoebae, and paramecium (Yaeger 1996; Shanan et al. 2015). Diatoms, the smallest protozoan phyla, can participate in photosynthesis depending on the genus (Kemp and Villareal 2013; Kuczynska et al. 2015). Amoebae are distinguished by their pseudopodia, which aid in movement and feeding (Salazar-Ardiles et al. 2022), while paramecium, which is the largest protozoan phylum, employs numerous cilia for rapid movement and maneuverability, allowing it to spread efficiently compared to flagellates or amoebae (Abello et al. 2020; Valentine and Van Houten 2021; Van Houten 2019, 2023).

3 Symbiotic Interactions in the Rhizosphere: The Role of Plant Hormones

Soil microorganisms play a crucial role in making nutrients from minerals available to plants, producing growth-stimulating hormones, and enhancing plant immune systems. They can either stimulate or weaken plant stress responses to drought or moisture (Haoye et al. 2024). Generally, the biological activity of soil microorganisms leads to greater plant diversity, reduced diseases, and increased yields. However, agricultural practices like chemical fertilizers and pesticides can harm soil rhizobia, disrupting this vital ecosystem (Jones et al. 2009; Wang et al. 2022a; Bag et al. 2022). Healthy soil enhances fertility by supplying nutrients such as nitrogen, protecting against pests and diseases, and reducing water needs. Rhizobia, a group of bacteria, fixes atmospheric nitrogen into useful compounds for legumes (Karnena et al. 2024; Shuo et al. 2024). Mycorrhizae, or root fungi, extend roots to facilitate water and nutrient uptake. Plants release up to 30% of their fixed carbon as root exudates, which include sugars, amino acids, and fatty acids, attracting beneficial microbes and repelling harmful ones (Sneha et al. 2023; Chiaranunt and White 2023; Jones et al. 2009; Wang et al. 2022a; Bag et al. 2022; Karnena et al. 2024; Shuo et al. 2024; Haoye et al. 2024).

Phytohormones, such as salicylic acid (SA), jasmonic acid (JA), and ethylene, as immune modulators in plants, play a crucial role in enhancing resistance. For instance, plants with mutations that hinder SA secretion become highly susceptible to microbial parasites, while those lacking JA and ethylene exhibit hypersensitivity to insect and microbial threats (Jiang et al. 2022; Yim et al. 2020; Pantigoso et al. 2022). Managing the symbiotic and parasitic relationships in plant roots involves more intricate immune mechanisms than simply removing the pathogens from leaves (Yim et al. 2020; Pantigoso et al. 2022). A study on arabidopsis revealed that mutations affecting hormone secretion increased the sensitivity of plants to SA, leading to enhanced bacterial growth within root tissues (Pantigoso et al. 2022; Wang et al. 2024), and this phenomenon indicates that SA may serve as a key regulator of plant immunity against harmful parasites (Taylor and Menge 2018; Han et al. 2020). In addition to mediating responses to biotic stress, plant immune hormones contribute to increased growth rates and metabolic activity, fine-tuning the immune regulation of harmful microbial growth (Laranjo et al. 2014; Lazcano et al. 2021). The selection of beneficial traits occurs during plant fertilization, but the symbiotic relationship with microorganisms often remains unconsidered. Notably, minor alterations in bacterial symbiosis can significantly influence the immunity of plants as well as their essential functions which ultimately mitigates the adverse effects of microbial interactions (Laranjo et al. 2014; Lazcano et al. 2021; Jiang et al. 2022; Yim et al. 2020; Pantigoso et al. 2022; Wang et al. 2024; Taylor and Menge 2018; Han et al. 2020).

Natural biocides developed from various microorganisms generate about $1 billion annually, representing less than 1% of the $110 billion chemical modification market (Degenkolb and Vilcinskas 2016). Key examples include *Trichoderma*, a fungus that inhibits pathogenic fungi, and *Bacillus thuringiensis*, a bacterium lethal to various microbes. Products like Serenade, containing *Bacillus*, are utilized in both conventional and organic agriculture for their antifungal and antibacterial properties, promoting plant growth while combating pathogens (Corbu et al. 2023; Demain and Sanchez 2009; Ayaz et al. 2023). Major fertilizer companies, such as Bayer, are investing in biotechnology; Bayer acquired AgraQuest in 2012 for $425 million to fund research on alternatives to chemical pesticides (Degenkolb and Vilcinskas 2016; Faria et al. 2018). Novozymes has also partnered with Monsanto to develop biofertilizers with *Penicillium bilaiae* and bioinsecticides from

Metarhizium anisopliae (Corbu et al. 2023). The trend is echoed by Syngenta and BASF's acquisitions of bioproduct firms, with DuPont following in 2015. A study showed that *Dichanthelium lanuginosum* thrives in the geothermal soils of Yellowstone National Park and was introduced to the U.S. market in 2014 to enhance heat stress resilience in corn and rice. In the U.S. and Europe, biofertilizer manufacturers must provide scientific evidence of product safety, leading many to rebrand as "biostimulants" instead of "biopesticides."

Beneficial microorganisms like *Bacillus subtilis* can face challenges that threaten their survival, particularly due to limited adaptability to environmental changes. While Serenade enhances the growth of these bacteria, factors such as soil salinity and heavy metal contamination from chemical fertilizers can impair their functionality. Moreover, chemical fertilizers can deplete soil organic matter and potentially convert beneficial microorganisms into harmful forms. Despite being a stimulant for *Bacillus subtilis*, the growth of these bacteria may decline when they cannot adapt to changing conditions, although Serenade remains a key method for promoting beneficial symbiotic strains (Hashem et al. 2019). The disadvantages of chemical fertilizers include not only the depletion of organic matter but also increased soil salinity and chlorine levels, which may adversely affect the health of beneficial symbiotic bacteria (Miljaković et al. 2020).

4 Pilot Project on Sustainable Agricultural Practices

In Europe, a pilot project using a flying plow aimed to reduce soil disturbance and pollution while enhancing microbial diversity. The 100-hectare site was divided into three sections: one treated with chemical fertilizers, another with pesticides, and a third with varying amounts of organic biofertilizers derived from fermented grape residues, along with diverse bacteria, fungi, and four types of mycorrhizal fungi (Congress of the European Society for Agronomy 2022, https://express.converia.de/custom/media/ESA2022/Book_of_Abstract_ESA_2022.pdf). Crops receiving the highest organic fertilizers grew nearly twice as tall as those treated with chemicals and surpassed those in the pesticide area by one inch. Yields in the organic treatment areas were significantly higher than in conventional methods. Mycorrhizae facilitated rock breakdown by secreting hydrolytic acids, allowing plant roots to penetrate up to 2 meters into rocky soil for groundwater access.

5 The Vital Role of Plants and Soil Microorganisms in Ecosystem Health

Plants can effectively remove organic pollutants from soil through natural absorption and metabolism, aided by interactions with soil microorganisms that enhance the breakdown of petroleum compounds. The success of phytoremediation depends on selecting appropriate plant species for the specific ecosystem. While nanomaterials and organic fertilizers can promote plant growth and pollutant absorption, excessive use can be harmful, necessitating a balanced approach for effective environmental restoration. The demand for petroleum fuels has led to widespread pollution, prompting traditional treatment methods like thermal, physical, chemical, and biological approaches, which are often expensive and have significant side effects. Plants play a crucial role in restoring ecological balance through their absorption capabilities and interactions with microorganisms. Therefore, careful species selection is vital for effective ecosystem repair. Current efforts using nanomaterials have successfully reduced suspended particulate dust pollution in Greater Cairo and the Delta by 25%, meeting the targets set for 2025 (He et al. 2023; Gkorezis et al. 2016; Daâssi and Qabil 2022; Gospodarek et al. 2021; Kabenge et al. 2017).

Soil microorganisms like bacteria and fungi play crucial roles in nutrient cycling, soil structure maintenance, and pathogen protection. These microbes mediate 80–90% of soil processes, including mineralization and element turnover, and enhance plant growth by utilizing root exudates and producing growth regulators. They also improve soil texture, water retention, and aeration through polymer production. However, factors like soil composition, moisture, pH, and agricultural practices, especially improper tillage, can significantly impact microbial diversity and activity. Remarkably, a single gram of soil can contain 4000–6000 distinct bacterial genomes, though traditional methods isolate only 0.1–10% of these populations, underscoring the need to understand their functional roles (Adomako et al. 2022; Wang et al. 2024).

6 Harnessing the Power of Plant Growth-Promoting Rhizobacteria for Sustainable Agriculture

Plant growth-promoting rhizobacteria (PGPR) significantly enhance the growth and productivity of crops and forest trees by colonizing the rhizosphere, where they improve nutrient availability, hormone production, resistance to pathogens, and soil quality. Researchers have documented that PGPR can enhance soil fertility, support nitrogen fixation, and positively influence plant health, making them valuable for sustainable agriculture (de Andrade et al. 2023). However, harmful practices like excessive chemical use can damage these beneficial communities, leading to reduced crop yields and soil degradation. Sustainable land management is vital to maintaining a healthy rhizosphere, essential for ecological balance and food security.

Farmers can ensure healthier and more productive crops by harnessing the power of beneficial bacteria within plants, leading to higher yields and better food security de Andrade et al. 2023. The diversity of bacterial species in the rhizosphere serves as a bioindicator for soil quality and fertility, contributing to nitrogen fixation, hormone production, and nutrient distribution. The rhizosphere is home to abundant bacterial communities, including betaproteobacteria (e.g., *Burkholderia*), bacteroidetes, alphaproteobacteria (e.g., *Rhizobia*), gammaproteobacteria (e.g., *Pseudomonas*), and firmicutes (e.g., *Bacillus*). These bacteria can be utilized as biocontrol agents, bioremediation agents, and biofertilizers, promoting soil health, fertility, and crop productivity while ensuring environmental sustainability, which emphasizes enhancing biological and physical systems without compromising ecosystem diversity and well-being.Global intervention is necessary to address inappropriate land use practices, and a shift towards nonchemical methods for pest control is encouraged. Climate change poses additional challenges, making it vital to understand and mitigate its impacts at the root-soil interface. We urge soil scientists, researchers, policymakers, and governments to prioritize the conservation of the soil root microbiome for future generations (Brunel et al. 2020; del Carmen Orozco-Mosqueda et al. 2022; Compant et al. 2021; Nuccio et al. 2020; Hassan et al. 2019; Vetterlein et al. 2020; de Andrade et al. 2023; Adedayo et al. 2022).

7 Roots of Resilience: Leveraging Belowground Traits for Climate Adaptation

The critical role of belowground traits, particularly roots and the rhizosphere, is highlighted in adapting to and mitigating climate change impacts on plant growth, crop productivity, soil health, and ecosystem services. Key discussions focus on the effects of combined stress on plants, the role of microbial communities in the rhizosphere, and the potential for monitoring these traits for breeding purposes. Additionally, rhizosphere soil and associated bacterial communities are explored for biotechnological applications, including biofuel production and bioremediation (Trenberth et al. 2014; Evelin et al. 2019; Lohani et al. 2020). Belowground traits will be affected by climate change, with plastic responses to stress emphasizing the importance of root microbiota in enhancing plant resilience. Modeling is suggested as a tool for understanding complex interactions and identifying optimal targets for climate adaptation, with reflections on research uncertainties and areas needing further investigation. While the rhizosphere is recognized as a hub for bacterial concentrations, little attention has been paid to bacterial inoculants and soil-like substrates (Fitzpatrick et al. 2018; Finkel et al. 2019; Nephali et al. 2020). The focus has largely been on chemical fertilizers, limiting the exploration of natural bacterial materials. Key areas include rhizosphere soil as a bacterial reservoir, cellulose biodegradation for biofuels, rhizosphere soil as a soil conditioner, and its role in bioremediation and biofiltration (Bilal et al. 2020; Cai et al. 2022). To maximize the potential of rhizosphere soil for biotechnological purposes, critical research is needed to explore the metabolic potential of its bacteria and optimize factors that enhance soil restoration (Fitzpatrick et al. 2020; Pantigoso et al. 2022). The implications of water microbiology are also discussed, emphasizing the importance of addressing pollution in drinking and irrigation sources, as well as the consequences of untreated sewage (Fourcroy et al. 2014). Overall, the need for targeted research to unlock the full potential of rhizosphere soil bacteria for sustainable agricultural practices and environmental remediation is underscored (Bilyera et al. 2022; George et al. 2024).

8 Bioremediation of Agricultural Soil: Addressing Pollution with Microbial Solutions

Sewage water presents significant environmental challenges due to its complex physical, chemical, and biological characteristics, including flow velocity, volume, pH levels, and contaminants. It consists mainly of microorganisms and organic matter, comprising about 99% of wastewater. Biological oxygen demand (BOD) is a crucial measure of the oxygen needed for microbial decomposition of organic materials, reflecting the organic load in sewage (Protonotariou et al. 2021a; Poulou et al. 2014). Effective sewage treatment involves solid waste removal, biological decomposition, and advanced methods like chlorination to eliminate pathogens (Driehuis et al. 1999; Ferreira et al. 2007). Soil pollution from agricultural effluents is concerning, necessitating bioremediation, a sustainable, cost-effective approach using microorganisms to detoxify contaminated soils. Bioremediation is vital in treating agricultural soils and enhancing microbial mechanisms for pollutant degradation. Pesticides, particularly herbicides, are essential for improving crop yield but contribute to soil and water pollution when overused (Danner et al. 2003; Ferreira 2000). Microbial metabolism effectively degrades these pollutants, offering a viable alternative to conventional treatment methods. However, the efficiency of these processes depends on the type of pollutants and soil conditions (Faust et al. 2015; Chen et al. 2021; Boli Zhang et al. 1992; Kim et al. 2021; Rhodes et al. 2017). Regulatory measures are crucial for mitigating pesticide use and ensuring safe agricultural practices, emphasizing the importance of bioremediation in managing pesticide-contaminated soils and relevant global legislative frameworks.

9 Microbial Bioremediation: Tackling Soil Pollution from Pesticides and Industrial Waste

Microbiological characteristics, including growth rates, environmental tolerances, and pathogenicity, significantly influence how microorganisms interact with their environment and affect the quality of biological products. Certain microbial polysaccharides are recognized for their nontoxic, biodegradable properties and stability across extreme conditions (Protonotariou et al. 2021a; Poulou et al. 2014; Protonotariou et al. 2021b; Risa et al. 2021; Chen et al. 2021; Boli Zhang et al. 1992; Kim et al. 2021; Rhodes et al. 2017). Soil pollution is a significant global challenge exacerbated by rapid industrialization and agricultural practices, leading to serious health impacts on humans and animals. Bioremediation offers an environmentally friendly strategy for removing toxic pollutants from agricultural fields, particularly due to the high biodegradability of agricultural waste. This review evaluates the feasibility of using microbes in bioremediation, assessing their performance in reducing hazardous pollutants and the potential health risks associated with contaminated soils (Faust et al. 2015; Sepulveda et al. 2020; He et al. 2022).

Soil bioremediation employs microorganisms like bacteria, fungi, and yeast to clean up contaminated soils, especially those polluted with organic chemicals. This environmentally friendly and cost-effective method relies on specific or indigenous microorganisms for the degradation of pollutants (Wang et al. 2024; Driehuis et al. 1999). Despite numerous laboratory studies, real-world applications are limited due to poor collaboration among research labs, local authorities, and bioremediation companies (Kalpana et al. 2022). Additionally, legislation gaps hinder effective remediation efforts (Ferreira 2000). In areas lacking sewage treatment, rural households often discharge untreated wastewater, posing environmental risks (Courtin and Spoelstra 1990).

Soil microbial activity, measured by total microbial biomass, showed a positive response to TOC and TN, with treated wastewater fostering greater bacterial abundance at the expense of fungi, reducing the fungi-to-bacteria ratio. Additionally, treated wastewater influenced microbial community structures, particularly increasing the abundance of proteobacteria, acidobacteria, and specific genera such as *Balneonas*, *Rubobacter*, and *Steroidobacter*. Despite the advantages of using treated wastewater, potential detrimental effects on soil microorganisms, especially when sodium hypochlorite is involved, were noted, highlighting the importance of soil pH. Overall, while reusing treated wastewater offers a solution to water scarcity exacerbated by urbanization and population growth, it may also pose challenges for plant growth and soil ecosystems, necessitating careful management.

Graphical abstract of constructed wetland mitigating N and P from agricultural runoff (Li et al. 2021). The rhizosphere microbial community plays a critical role in breaking down organic pollutants like polycyclic aromatic hydrocarbons (PAHs) and phthalates (PAEs). Soil contamination disrupts microbial diversity, depleting sensitive species and enriching tolerant ones. Plant species, root-associated environments, and soil properties influence microbial responses to pollutants and co-contaminants like heavy metals (Chan 2003; Behera et al. 2017). The rhizosphere enriches pollutant-degrading microbes and functional genes, enhancing soil remediation efficiency (Cheon et al. 2015). Advances in metagenomics show that the rhizosphere has higher adaptability than bulk soil (Ban et al. 2018). Future research is needed to better understand these microbial responses and adaptations (Rahmati et al. 2022).

Building on the previous insights, the rhizosphere outperforms bulk soil in breaking down organic pollutants like polycyclic aromatic hydrocarbons (PAHs) and polychlorinated biphenyls (PCBs) (Défago et al. 1990; Dell'Amico et al. 2005). Plant-microbe interactions in the rhizosphere promote the enrichment of pollutant-degrading microbes and genes, boosting the degradation process (Jing 2007). While some studies show a direct correlation between microbial diversity and pollutant removal, results are inconsistent (Guerra et al. 2021). The adaptation of microbial communities, influenced by root exudates, enhances the expression of degradation genes, expediting pollutant breakdown. Updating this knowledge remains essential (Mahajan and Kaushal 2018).

10 Advances in Phytoremediation: Plant-Microbe Interactions and Organic Pollutant Dissipation

The rhizosphere, a microbial activity hotspot surrounding plant roots, promotes pollutant dissipation more effectively than bulk soil, with plant species and microbial diversity playing key roles (Défago et al. 1990; Dell'Amico et al. 2005). Studies have shown that microbial communities in the rhizosphere respond to contaminants by enriching organisms and genes involved in degradation, influenced by factors like plant species, root niches, and co-contaminants such as heavy metals (Jing 2007). Root exudates selectively stimulate pollutant-degrading microbes, enhancing phytoremediation (Behera et al. 2017; Cheon et al. 2015). Phytoremediation involves mechanisms such as phytoextraction (uptake), phytotransformation (enzymatic degradation), and rhizosphere biodegradation (microbial breakdown) (Ban et al. 2018). Rhizosphere microbes play a crucial role in transforming persistent pollutants like PAHs and PCBs into less harmful forms, with genes responsible for degradation more abundant in the rhizosphere (Brandl et al. 2008; Jing 2007). Factors such as molecular size, hydrophobicity, pH, and dissolved organic carbon content influence phytoremediation efficiency (Aponte et al. 2018). Recent advances in genomic technologies have deepened our understanding of the rhizosphere microbiome's response to pollution, revealing greater microbial diversity and resilience in polluted environments (Guerra et al. 2021). Genetic engineering and gene editing hold promise for enhancing phytoremediation by modifying plants for better pollutant degradation (Mahajan and Kaushal 2018). Future research should address knowledge gaps, especially regarding microbial adaptation to pollutants and co-contamination, to improve large-scale environmental remediation.

Phytoremediation of organic pollutants in soil primarily relies on specific pollutant-degrading microbes rather than the entire microbial community. Recent advancements in rhizosphere microbial research have uncovered the enrichment of specific pollutant-degrading microbes and genes in response to contaminants (Li et al. 2011, 2013). These microbes play a vital role in the biodegradation and bioremediation of pollutants such as volatile organic compounds (VOCs) and particulate matter (PMs) (Rodrigo Mendes et al. 2013; Kumar et al. 2011). Although significant progress has been made, knowledge gaps persist regarding the adaptation, feedback processes, and interactions between plants and microbes in polluted environments (Kotoky and Pandey 2020). Plant-microbe interactions, facilitated by mutualistic exchanges of nutrients and chemicals, enhance the detoxification of air pollutants and pollutant uptake by roots and leaves (Azaizeh et al. 2011; Eze and Amuji 2024). Future research should focus on identifying optimal plant-microbe systems for efficient pollutant degradation and promoting sustainable environmental practices (Marinari et al. 2014; Maseko and Dakora 2013).

11 Initiative of Soil Microbiology in Egypt

11.1 The Role of the Egyptian Ministry of Environment

Launched on World Environment Day, June 5, 2023, a groundbreaking initiative seeks to raise awareness about the ecological footprint and its connections to water and carbon footprints, promoting sustainable production and consumption. Running until 2030, the initiative aims to educate the community and private sector on the importance of fair natural resource use and environmental accounting. Central to this effort is the measurement of the total ecosystem product (TEP), which quantifies the annual value of ecosystem goods and services essential for human well-being. By focusing on ecosystem assets as valuable natural resources, the initiative helps evaluate ecosystem health, contributions to human development, and the impact of conservation efforts (Ouyang et al. 2020; Adomako et al. 2022; Hejnowicz and Rudd 2017). The history of agricultural microbiology research in Egypt dates back to 1948 when the Ministry of Agriculture established the Soil Microbiology section within the Department of Agricultural Chemistry. In 1954, this section became the Department of Agricultural Microbiology, marking a pivotal shift in research focus. Between 1954 and 1963, researchers concentrated on nitrogen fixation, blue-green algae, and the microbiological activity involved in agricultural and industrial waste management. In 1963, the department created the Soil Microbiology and Agricultural Waste Divisions to monitor developments in agricultural microbiology. By 1973, the department had expanded into three primary units: atmospheric nitrogen fixation (biofertilizers), microbiological activity, and organic matter. By 2006, the department employed 100 researchers across multiple locations, making significant strides in agricultural microbiology. Today, as climate change and biodiversity loss increasingly intersect, Egypt's agricultural future hinges on a deep understanding of soil microorganisms' role in both agriculture and human health. Recent studies suggest that nanofertilizers, compared to chemical fertilizers, significantly enhance soil microbial populations, pointing to innovative microbiological methods for improving Egyptian soil and agriculture (Egyptian Journal of Soil Science). These findings are crucial, given that the global green economy relies on ecosystem services valued between \$120 and \$170 billion annually, with few reliable indicators to track growth or decline. The initiative underscores the need to go beyond GDP as a measure of prosperity, instead focusing on environmental sustainability and social well-being (Ouyang et al. 2020; Adomako et al. 2022; Hejnowicz and Rudd 2017).

11.2 Policy and Efforts on Climate Change Issues

Egypt plays a significant role in climate change negotiations, representing Arab and African interests as a member of groups like the Group of 77 and OAPEC. The country emphasizes that climate change results from emissions by industrialized nations since the Industrial Revolution and insists that these nations fulfill their obligations to reduce emissions, transfer technology, and finance adaptation efforts. Egypt's National Climate Change Strategy aims for sustainable growth, increased renewable energy use, improved energy efficiency, and reduced greenhouse gas emissions. It focuses on mitigating climate impacts, preserving natural resources, and enhancing resilience. To tackle climate change, Egypt has ratified the UNFCCC and Kyoto Protocol, established a National Committee for the Clean Development Mechanism, and launched projects to promote renewable energy, improve energy efficiency, and protect coastal areas. Under President Abdel Fattah El-Sisi, the country prioritizes environmental sustainability in its development projects.

In preparation for the COP27 Climate Conference in Sharm El-Sheikh, Egypt and England held a workshop on May 28, 2022, focusing on scientific cooperation and solar energy for smart agriculture and sustainable cities. Dr. Rania Al-Mashat, Egypt's Minister of International Cooperation, engaged with international officials at the World Economic Forum in Davos to advance climate action and green economy initiatives. Discussions included green transformation projects and emissions reduction efforts, particularly in renewable energy and sustainable infrastructure. Additionally, on May 19, 2022, a protocol was signed between the Arab Organization for Industrialization, the Egyptians Investment Group, and the Emirati Barq Electric Vehicles Company to localize the production of environmentally friendly electric vehicles in Egypt, aiming for integrated manufacturing to support the national strategy for green transformation and enhance transport solutions for Arab and African markets.

11.3 National Climate Change Strategy 2050

On May 19, 2022, the Egyptian government launched the "National Climate Change Strategy 2050" to ensure sustainable development and mitigate climate disasters. This strategy reflects Egypt's commitment to addressing the climate crisis, which threatens various sectors, including coasts, agriculture, water resources, health, and infrastructure. Developed collaboratively among state agencies, it aligns with Egypt's preparations for the COP 27 Climate Summit in Sharm El-Sheikh. The strategy aims to reduce emissions and adapt to potential climate changes while promoting economic development through a low-emission approach. It is a critical component of Egypt's Vision 2030, prioritizing an integrated and sustainable environmental system to enhance resilience against climate-related risks. The National Climate Change Strategy 2050 has five main objectives: achieving sustainable economic growth and low-emission development, building resilience and adaptive capacity to climate impacts, enhancing governance and management in climate action, improving financing infrastructure for climate initiatives (including green banking), and promoting scientific research, technology transfer, and public awareness regarding climate change. This consultative approach incorporates lessons from other countries facing similar challenges, establishing a roadmap for addressing climate change in Egypt.

12 Biopesticides: A Green Initiative of Egypt to Combat Climate Changes

The Egyptian government is championing green and sustainable agriculture as a cornerstone of its sustainable development strategy. By collaborating with businesses, civil society, academia, and international partners, Egypt is fostering innovative agricultural practices that enhance food security and environmental resilience. Central to this effort is the Environmental Sustainability Standards Guide, which guides green investments in agriculture. To support these initiatives, Egypt has proposed \$750 million in green bonds to fund sustainable projects, along with \$120–200 million in private sector bonds, making it the first MENA country to take such a step. These funds will promote sustainable farming practices, including the use of biopesticides like Magic Mix, Matco Gold, and Magican Cu, essential for protecting crops while minimizing chemical reliance. In response to climate challenges, Egypt is actively rehabilitating 1.5 million acres of agricultural land, modernizing irrigation systems, and protecting northern coasts from rising sea levels. Moreover, the country is investing in renewable energy, with the Benban solar power facility generating 2 gigawatts of electricity and the Gebel El Zeit wind farm contributing an additional 580 megawatts. These initiatives not only support agricultural operations but also align with Egypt's commitment to a sustainable, green economy.

Biopesticides are gaining prominence in Egypt as an effective means to combat harmful pests that threaten agricultural productivity. These environmentally friendly alternatives to chemical pesticides are essential in protecting crops, flowers, and various agricultural products from insects, bacteria, fungi, and viruses. The increasing awareness of the advantages of biopesticides is driving their adoption among farmers looking for sustainable solutions in agriculture.

12.1 Key Biopesticides in Egypt

Magic Mix is a unique blend of microelements that enhances plant health and productivity. Its composition includes iron at 5%, crucial for chlorophyll production; zinc at 7%, which supports plant growth and development; manganese at 4%, aiding in photosynthesis; boron at 0.5%, essential for cell wall formation; molybdenum at 0.05%, facilitating nitrogen fixation; copper at 0.5%, important for photosynthesis and respiration; and citric acid at 20%, enhancing nutrient absorption. This biopesticide is highly effective and easily dissolves in water, ensuring efficient nutrient delivery to plants. It significantly boosts the respiration rate in plants, promoting vitality and improving vital processes. Magic Mix is particularly effective in combating yellowing due to iron deficiency, enhancing overall plant health and growth.

Matco Gold is a versatile biopesticide designed to treat and prevent a variety of pest issues affecting agricultural crops. It contains metalaxyl, which adheres well to plant surfaces, allowing it to penetrate effectively and combat fungal infections. Additionally, it includes mancozeb, a temporary fungicide that forms a protective barrier against fungal spores, preventing their germination. This biopesticide is ideal for maintaining healthy crops by eliminating various fungal diseases, thereby safeguarding plant health and increasing yields.

Magican Cu is a potent fungicide effective against a broad range of plant diseases. Its water-soluble formulation is easy to apply, making it suitable for various crops, including citrus fruits, mango trees, potatoes, and tomatoes. It addresses common issues such as flower scent damage and anthracnose in mangoes, early and late blight in potatoes and tomatoes, curling of leaves in apricot, plum, and peach trees, and downy mildew in grapes. This biopesticide is particularly effective post-pruning and is essential for maintaining healthy plant growth by controlling bacterial and viral infections.

Protecto is a bacterial biopesticide containing *Bacillus thuringiensis kurstaki*, known for its efficacy against lepidopteran larvae. It effectively controls pests like the cotton leafworm, tomato miner (*Tuta absoluta*), and fall armyworm. Produced by the Biopesticide Production Unit at the Plant Protection Research Institute, Protecto is applied in a series of sprays starting from the seventh day after planting, ensuring continuous protection against these harmful pests. For optimal results, it's recommended to control the pupae in the soil and apply the biopesticide at specified intervals to manage the pest population effectively.

The registration and use of agricultural pesticides in Egypt are governed by several regulations to ensure safety and efficacy. Stakeholders must apply to the Agricultural Pesticides Committee, submitting necessary technical data and fees. Applications undergo thorough review, ensuring compliance with standards set by international bodies like the US EPA and the EU. Approved pesticides undergo a testing period of two consecutive agricultural seasons to assess their effectiveness against targeted pests. Once a pesticide passes all tests, a registration certificate is issued, valid for six years, with an option for renewal. These regulations aim to maintain high safety standards and effectiveness in agricultural practices while promoting the use of biopesticides.

13 Conclusion and Future Direction

The rhizosphere microbiome emerges as a crucial player in the fight against pollution in water, air, and soil. Its capacity to promote plant growth and enhance pollutant accumulation positions it as a key facilitator in phytoremediation, particularly under challenging conditions such as elevated cadmium levels. By remodeling microbial communities in the rhizosphere, we can significantly enhance the degradation and removal of various pollutants, paving the way for innovative bioremediation strategies. To harness the full potential of the rhizosphere microbiome, we recommend a multifaceted approach that focuses on deepening our understanding of plant-microbe interactions. This includes the development of targeted rhizoremediation techniques aimed at organic pollutants, especially polycyclic aromatic hydrocarbons (PAHs), which pose significant environmental risks. By exploring the diverse taxonomic and functional traits of the rhizosphere microbiome, we can tailor bioremediation processes to improve their efficacy. Future research should prioritize microbial community engineering to intentionally reshape rhizosphere microbiomes for optimized pollutant degradation, utilizing synthetic biology and advanced microbiome analysis. Cross-disciplinary collaboration among ecologists, microbiologists, and agricultural scientists is essential to develop integrated management strategies that address both pollution and agricultural productivity. Implementing large-scale field trials will validate laboratory findings and ensure that proposed strategies are effective in real-world conditions. Leveraging cutting-edge technologies, such as metagenomics and bioinformatics, can provide insights into analyzing and manipulating the rhizosphere microbiome, leading to more efficient bioremediation techniques. Additionally, promoting awareness of the rhizosphere microbiome's role in pollution treatment through educational programs and supportive policies will be crucial for sustainable agricultural practices. By establishing successful bioremediation strategies that harness the natural capabilities of the rhizosphere microbiome, we can significantly advance sustainable environmental management, addressing pollution challenges while enhancing agricultural resilience and contributing to a healthier planet for future generations.

References

Abdelfattah A, Wisniewski M, Schena L, Tack AJM (2021) Experimental evidence of microbial inheritance in plants and transmission routes from seed to phyllosphere and root. Environ Microbiol 23:2199–2214. https://doi.org/10.1111/1462-2920.15392

Abello A, Régnier V, Arnaiz O, Le Bars R, Bétermier M, Bischerour J (2020) Functional diversification of Paramecium Ku80 paralogs safeguards genome integrity during precise programmed DNA elimination. PLoS Genet 16:e1008723

Adedayo AA, Babalola OO, Prigent-Combaret C, Cruz C, Stefan M, Kutu F, Glick BR (2022) The application of plant growth-promoting rhizobacteria in Solanum lycopersicum production in the agricultural system: a review. PeerJ 10:e13405. https://doi.org/10.7717/peerj.13405

Adomako MO, Roiloa S, Fei-Hai Y (2022) Potential roles of soil microorganisms in regulating the effect of soil nutrient heterogeneity on plant performance. Microorganisms 10(12):2399. https://doi.org/10.3390/microorganisms10122399

Ahmed T, Noman M, Qi Y, Shahid M, Hussain S, Masood HA, Lihui X, Ali HM, Negm S, El-Kott AF et al (2023) Fertilization of microbial composts: a technology for improving stress resilience in plants. Plan Theory 12(20):3550. https://doi.org/10.3390/plants12203550

Aponte M, Anastasio A, Marrone R, Mercogliano R, Peruzy MF, Murru N (2018) Impact of gaseous ozone coupled to passive refrigeration system to maximize shelf-life and quality of four different fresh fish products. LWT 93:412–419

Ayaz M, Li CH, Ali Q, Zhao W, Chi YK, Shafiq M, Ali F, Yu XY, Yu Q, Zhao JT, Yu JW, Qi RD, Huang WK (2023) Bacterial and fungal biocontrol agents for plant disease protection: journey from lab to field, current status, challenges, and global perspectives. Molecules 28(18):6735. https://doi.org/10.3390/molecules28186735
Azaizeh H, Castro PML, Kidd P (2011) Biodegradation of organic xenobiotic pollutants in the rhizosphere. In: Schröder P, Collins C (eds) Organic xenobiotics and plants. Plant ecophysiology, vol 8. Springer, Dordrecht. https://doi.org/10.1007/978-90-481-9852-8_9
Bag S, Mondal A, Majumder A, Mondal SK, Banik A (2022) Flavonoid mediated selective cross-talk between plants and beneficial soil microbiome. Phytochem Rev 21(5):1739–1760. https://doi.org/10.1007/s11101-022-09806-3
Bahram M, Netherway T (2022) Fungi as mediators linking organisms and ecosystems. FEMS Microbiol Rev 46(2):fuab058. https://doi.org/10.1093/femsre/fuab058
Ban C, Lee DH, Jo Y, Bae H, Seong H, Kim SO, Lim S, Choi YJ (2018) Use of superheated steam to inactivate Salmonella enterica serovars Typhimurium and Enteritidis contamination on black peppercorns, pecans, and almonds. J Food Eng 222:284–291
Barka EA, Vatsa P, Sanchez L, Gaveau-Vaillant N, Jacquard C, Klenk HP et al (2016) Taxonomy, physiology, and natural products of Actinobacteria. Microbiol Mol Biol Rev 80:1–43. https://doi.org/10.1128/MMBR.00019-15
Baweja P, Sahoo D (2015) Classification of algae. In: Sahoo D, Seckbach J (eds) The algae world. Cellular origin, life in extreme habitats and astrobiology, vol 26. Springer, Dordrecht. https://doi.org/10.1007/978-94-017-7321-8_2
Behera G, Sutar P, Aditya S (2017) Development of novel high power-short time (HPST) microwave assisted commercial decontamination process for dried turmeric powder (Curcuma longa L.). J Food Sci Technol 54(12):4078–4091
Berg G, Smalla K (2009) Plant species and soil type cooperatively shape the sturcutre and function of microbial communities in the rhizosphere. FEMS Microbiol Ecol 68:1–13
Bilal S, Shahzad R, Imran M, Jan R, Kim KM, Lee IJ (2020) Synergistic association of endophytic fungi enhances Glycine max L. resilience to combined abiotic stresses: heavy metals, high temperature and drought stress. Ind Crops Prod 143:111931
Bilyera N, Hummel C, Daudin G, Santangeli M, Zhang X, Santner J, Lippold E, Schlüter S, Bertrand I, Wenzel W, Spielvogel S, Vetterlein D, Razavi BS, Oburger E (2022) Co-localised phosphorus mobilization processes in the rhizosphere of field-grown maize jointly contribute to plant nutrition. Soil Biol Biochem 165:108497
Bold HC, Wynne MJ (1978) Introduction to algae: structure and reproduction. Prentice Hall of India, New Delhi, p 720
Boli Zhang B, Zhao Y, Chen, J (1992) Processes of pollutant degradation in soil and their influencing factors. J Environ Sci,14(3), 205–212
Brandl MT, Pan Z, Huynh S, Zhu YI, Mchugh TH (2008) Reduction of Salmonella enteritidis population sizes on almond kernels with infrared heat. J Food Prot 71(5):897–902
Brunel C, Pouteau R, Dawson W, Pester M, Ramirez KS, van Kleunen M (2020) Towards unraveling macroecological patterns in rhizosphere microbiomes. Trends Plant Sci 25:1017–1029. https://doi.org/10.1016/j.tplants.2020.04.015
Cai G, Ahmed MA, Abdalla M, Carminati A (2022) Root hydraulic phenotypes impacting water uptake in drying soils. Plant Cell Environ 45:650–663
Chan K (2003) Some aspects of toxic contaminants in herbal medicines. Chemosphere 52(9):1361–1371
Chauhan P, Sharma N, Tapwal A, Kumar A, Verma GS, Meena M, Seth CS, Swapnil P (2023) Soil microbiome: diversity, benefits and interactions with plants. Sustain For 15(19):14643. https://doi.org/10.3390/su151914643
Chen H et al (2021) Microbial characteristics across different tongue coating types in a healthy population. J Oral Microbiol 13(1):1946316
Cheon H-L, Shin J-Y, Park K-H, Chung M-S, Kang D-H (2015) Inactivation of foodborne pathogens in powdered red pepper (Capsicum annuum L.) using combined UV-C irradiation and mild heat treatment. Food Control 50:441–445. https://doi.org/10.1002/fsn3.2707

Chen J, Wang Y, Liu X (2021) Impacts of biochar on soil microbial communities and pollutant degradation. Environmental Science Journal 45(3), 215–230
Chiaranunt P, White JF (2023) Plant beneficial bacteria and their potential applications in vertical farming systems. Plan Theory 12(2):400. https://doi.org/10.3390/plants12020400
Compant S, Cambon MC, Vacher C, Mitter B, Samad A, Sessitsch A (2021) The plant endosphere world–bacterial life within plants. Environ Microbiol 23:1812–1829. https://doi.org/10.1111/1462-2920.15240
Corbu VM, Gheorghe-Barbu I, Dumbravă AȘ, Vrâncianu CO, Șesan TE (2023) Current insights in fungal importance – a comprehensive review. Microorganisms 11(6):1384. https://doi.org/10.3390/microorganisms11061384
Courtin MG, Spoelstra SF (1990) A simulation model of the microbiological and chemical changes accompanying the initial stage of aerobic deterioration of silage. Grass Forage Sci 45:153–165
Daâssi D, Qabil AF (2022) Petroleum-contaminated soil: environmental occurrence and remediation strategies. 3 Biotech 12(6):139. https://doi.org/10.1007/s13205-022-03198-z
Danner H, Holzer M, Mayrhuber E et al (2003) Acetic acid increases stability of silage under aerobic conditions. Appl Environ Microbiol 69(1):562–567
Daunoras J, Kačergius A, Gudiukaitė R (2024) Role of soil microbiota enzymes in soil health and activity changes depending on climate change and the type of soil ecosystem. Biology 13, no. 2:85. https://doi.org/10.3390/biology13020085
de Andrade LA, Santos CHB, Frezarin ET, Sales LR, Rigobelo EC (2023) Plant growth-promoting Rhizobacteria for sustainable agricultural production. Microorganisms 11(4):1088. https://doi.org/10.3390/microorganisms11041088
De Simeis D, Serra S (2021) Actinomycetes: a never-ending source of bioactive compounds – an overview on antibiotics production. Antibiotics (Basel) 10(5):483. https://doi.org/10.3390/antibiotics10050483
Défago G, Berling CH, Burger U et al (1990) Suppression of black root rot of tobacco and other root diseases by strains of Pseudomonas fluorescens: potential applications and mechanisms. In: Hornby D (ed) Biological control of soilborne plant pathogens. CAB International, Oxon, pp 93–108
Degenkolb T, Vilcinskas A (2016) Metabolites from nematophagous fungi and nematicidal natural products from fungi as an alternative for biological control. Part I: Metabolites from nematophagous ascomycetes. Appl Microbiol Biotechnol 100:3799–3812. https://doi.org/10.1007/s00253-015-7233-6
del Carmen Orozco-Mosqueda M, Fadiji AE, Babalola OO, Glick BR, Santoyo G (2022) Rhizobiome engineering: unveiling complex rhizosphere interactions to enhance plant growth and health. Microbiol Res 263:127137. https://doi.org/10.1016/j.micres.2022.127137
Dell'Amico E, Cavalca L, Andreoni V (2005) Analysis of rhizobacterial communities in perennial Graminaceae from polluted water meadow soil, and screening of metal-resistant, potentially plant growth-promoting bacteria. FEMS Microbiol Ecol 52(2):153–162. https://doi.org/10.1016/j.femsec.2004.11.005
Demain AL, Sanchez S (2009) Microbial drug discovery: 80 years of progress. J Antibiot (Tokyo) 62(1):5–16. https://doi.org/10.1038/ja.2008.16
Dick R (2009) Lecture on soil fungus in soil microbiology, personal collection of R. Dick. The Ohio State University School of Environment and Natural Resources, Columbus
Diggle SP, Whiteley M (2020) Microbe profile: Pseudomonas aeruginosa: opportunistic pathogen and lab rat. Microbiology (Reading) 166(1):30–33. https://doi.org/10.1099/mic.0.000860. Erratum in: Microbiology (Reading). 2021;167(8). https://doi.org/10.1099/mic.0.001073
Driehuis F, Oude Elferink SJWH, Spoelstra SF (1999) Anaerobic lactic acid degradation during ensilage of whole crop maize inoculated with Lactobacillus buchneri inhibits yeast growth and improves aerobic stability. J Appl Microbiol 87:583–594
Evelin H, Devi TS, Gupta S, Kapoor R (2019) Mitigation of salinity stress in plants by arbuscular mycorrhizal symbiosis: current understanding and new challenges. Front Plant Sci 10:470. https://doi.org/10.3389/fpls.2019.00470

Eze MO, Amuji CF (2024) Elucidating the significant roles of root exudates in organic pollutant biotransformation within the rhizosphere. Sci Rep 14:2359. https://doi.org/10.1038/s41598-024-53027-x

Farda B, Djebaili R, Vaccarelli I, Del Gallo M, Pellegrini M (2022) Actinomycetes from caves: an overview of their diversity, biotechnological properties, and insights for their use in soil environments. Microorganisms 10(2):453. https://doi.org/10.3390/microorganisms10020453

Faria WM, Figueiredo CCD, Coser TR (2018) Sewage sludge biochar as a source of nutrients in plant-soil systems. Journal of Environmental Management, 215:1–10. https://doi.org/10.1016/j.jenvman.2018.03.017

Fadiji AE, Santoyo G, Babalola OO (2023) Understanding the impact of abiotic stressors on plantmicrobe interactions for sustainable agriculture. Frontiers in Plant Science, 14, 1482739. https://doi.org/10.3389/fpls.2024.1482739

Faust K et al (2015) Cross-biome comparison of microbial association networks. Front Microbiol 6:1200

Ferreira DF (2000) Análises estatísticas por meio do SISVAR para windows versão 4.0. In: Reunião anual da região brasileira da sociedade internacional de biometria, 45. UFSCar, São Carlos. Anais… São Carlos, pp 225–258

Ferreira DA, Gonçalves LC, Molina LR et al (2007) Características de fermentação da silagem de cana-de-açúcar tratada com uréia, zeólita, inoculante bacteriano e inoculante bacteriano/enzimático. Arquivo Brasileiro de Medicina Veterinária e Zootecnia 59(2):423–433

Finkel OM, Salas-González I, Castrillo G, Spaepen S, Law TF, Teixeira PJPL et al (2019) The effects of soil phosphorus content on plant microbiota are driven by the plant phosphate starvation response. PLoS Biol 17(11):e3000534

Fitzpatrick CR, Copeland J, Wang PW, Guttman DS, Kotanen PM, Johnson MT (2018) Assembly and ecological function of the root microbiome across angiosperm plant species. Proc Natl Acad Sci USA 115(6):E1157–E1165

Fitzpatrick CR, Salas-González I, Conway JM, Finkel OM, Gilbert S, Russ D et al (2020) The plant microbiome: from ecology to reductionism and beyond. Ann Rev Microbiol 74:81–100

Fourcroy P, Sisó-Terraza P, Sudre D, Savirón M, Reyt G, Gaymard F et al (2014) Involvement of the ABCG 37 transporter in secretion of scopoletin and derivatives by Arabidopsis roots in response to iron deficiency. New Phytol 201(1):155–167

George TS, Bulgarelli D, Carminati A et al (2024) Bottom-up perspective – the role of roots and rhizosphere in climate change adaptation and mitigation in agroecosystems. Plant Soil 500:297–323. https://doi.org/10.1007/s11104-024-06626-6

Gkorezis P, Daghio M, Franzetti A, Van Hamme JD, Sillen W, Vangronsveld J (2016) The interaction between plants and bacteria in the remediation of petroleum hydrocarbons: an environmental perspective. Front Microbiol 7:1836. https://doi.org/10.3389/fmicb.2016.01836

Gospodarek J, Rusin M, Kandziora-Ciupa M, Nadgórska-Socha A (2021) The subsequent effects of soil pollution by petroleum products and its bioremediation on the antioxidant response and content of elements in Vicia faba plants. Energies 14(22):7748. https://doi.org/10.3390/en14227748

Grzyb A, Wolna-Maruwka A, Niewiadomska A (2021) The significance of microbial transformation of nitrogen compounds in the light of integrated crop management. Agronomy 11(7):1415. https://doi.org/10.3390/agronomy11071415

Gupta A, Gupta R, Singh RL (2016) Microbes and environment. In: Principles and applications of environmental biotechnology for a sustainable future, pp 43–84. https://doi.org/10.1007/978-981-10-1866-4_3

Guerra CA, Delgado-Baquerizo M, Duarte E, Marigliano O, Görgen C, Maestre FT, Eisenhauer N (2021) Global projections of the soil microbiome in the Anthropocene. Global Ecology and Biogeography, 30(5):987–999. https://doi.org/10.1111/geb.13273

Han Q, Ma Q, Chen Y et al (2020) Variation in rhizosphere microbial communities and its association with the symbiotic efficiency of rhizobia in soybean. ISME J 14:1915–1928. https://doi.org/10.1038/s41396-020-0648-9

Haoye Li, Lei Chang, Huijia Liu, Yuefen Li (2024) Diverse factors influence the amounts of carbon input to soils via rhizodeposition in plants: A review, Science of The Total Environment, 948. https://doi.org/10.1016/j.scitotenv.2024.174858

Hashem A, Tabassum B, Allah FA, E. (2019) Bacillus subtilis: a plant-growth promoting rhizobacterium that also impacts biotic stress. Saudi J Biol Sci 26(6):1291–1297. https://doi.org/10.1016/j.sjbs.2019.05.004

Hassan MK, McInroy JA, Kloepper JW (2019) The interactions of rhizodeposits with plant growth-promoting rhizobacteria in the rhizosphere: a review. Agriculture 9:142. https://doi.org/10.3390/agriculture9070142

Hejnowicz AP, Rudd MA (2017) The value landscape in ecosystem services: value, value wherefore art thou value? Sustain For 9(5):850. https://doi.org/10.3390/su9050850

He Y, Li X, Zhang T (2023) Nanoparticles-induced potential toxicity on human health. PMC

Ibrahim Hamad I. Al-Rajab and Bashar Tareq Al-Shandah (2024) IOP Conf. Ser.: Earth Environ. Sci. 1371022005, https://doi.org/10.1088/1755-1315/1371/2/022005

Imran M, Sun Z, Abo-Elyousr KA, Ali H, Aldayel MF, Li C (2024) One stone two birds: endophytes alleviating trace elements accumulation and suppressing soilborne pathogen by stimulating plant growth, photosynthetic potential and defense related gene expression. J Hazard Mater 476:135084

Jiang G, Zhang Y, Gan G et al (2022) Exploring rhizo-microbiome transplants as a tool for protective plant-microbiome manipulation. ISME Commun 2:10. https://doi.org/10.1038/s43705-022-00094-8

Jing Q (2007) Improving resource use efficiency in rice-based cropping systems: experimentation and Modelling. Ph.D. thesis, Wageningen University, Wageningen, Netherlands

Jones DL, Nguyen C, Finlay RD (2009) Carbon flow in the rhizosphere: carbon trading at the soil–root interface. Plant Soil 321:5–33. https://doi.org/10.1007/s11104-009-9925-0

Kabenge I, Katimbo A, Kiggundu N, Banadda N (2017) Bioremediation technology potential for management of soil and water pollution from anticipated rapid industrialization and planned oil and gas sector in Uganda: a review. J Environ Prot 8:1393–1423. https://doi.org/10.4236/jep.2017.811085

Kemp AES, Villareal TA (2013) High diatom production and export in stratified waters – a potential negative feedback to global warming. Prog Oceanogr 119:4–23

Kim MJ et al (2021) The herbal formula CWBSD improves sleep quality dependent on oral microbial type and tongue diagnostic features in insomnia. J Pers Med 11(5):325

Kotoky R, Pandey P (2020) Rhizosphere assisted biodegradation of benzo(a)pyrene by cadmium resistant plant-probiotic Serratia marcescens S2I7, and its genomic traits. Sci Rep 10:5279. https://doi.org/10.1038/s41598-020-62285-4

Krause KM, Serio AW, Kane TR, Connolly LE (2016) Aminoglycosides: an overview. Cold Spring Harb Perspect Med 6(6):a027029. https://doi.org/10.1101/cshperspect.a027029

Kuczynska P, Jemiola-Rzeminska M, Strzalka K (2015) Photosynthetic pigments in diatoms. Mar Drugs 13(9):5847–5881. https://doi.org/10.3390/md13095847

Kumar S, Chaudhuri S, Maiti SK (2011) Phosphatase activity in natural and mined soil – a review. Indian J Environ Prot 31:955–962

Kumar R, Adhikari S, Driver EM, Smith T, Bhatnagar A, Lorkiewicz PK, Zhengzhi Xie J, Hoetker D, Halden RU (2022) Towards a novel application of wastewater-based epidemiology in population-wide assessment of exposure to volatile organic compounds. Sci Total Environ 845:157008., ISSN:0048-9697. https://doi.org/10.1016/j.scitotenv.2022.157008

Laranjo M, Alexandre A, Oliveira S (2014) Legume growth-promoting rhizobia: an overview on the Mesorhizobium genus. Microbiol Res 20:2–17

Lazcano C, Boyd E, Holmes G et al (2021) The rhizosphere microbiome plays a role in the resistance to soil-borne pathogens and nutrient uptake of strawberry cultivars under field conditions. Sci Rep 11:3188. https://doi.org/10.1038/s41598-021-82768-2

Li Y-S, Gao Y, Tian Q-Y, Shi F-L, Li L-H, Zhang W-H (2011) Stimulation of root acid phosphatase by phosphorus deficiency is regulated by ethylene in Medicago falcate. Environ Exp Bot 71:114–120

Li Y, Sun H, Li H, Yang L, Ye B, Wang W (2013) Dynamic changes of rhizosphere properties and antioxidant enzyme responses of wheat plants (Triticum aestivum L.). Chemosphere 93:972–977
Li Q, Chen X, Jiang Y, Jiang C (2016) Morphological identification of Actinobacteria. InTech, Actinobacteria – basics and biotechnological applications. https://doi.org/10.5772/61461
Li J, Zheng B, Chen X, Li Z, Xia Q, Wang H, Yang Y, Zhou Y, Yang H (2021) The use of constructed wetland for mitigating nitrogen and phosphorus from agricultural runoff: a review. Water 13(4):476. https://doi.org/10.3390/w13040476
Lohani N, Jain D, Singh MB, Bhalla PL (2020) Engineering multiple abiotic stress tolerance in canola, Brassica napus. Front Plant Sci 11:3. https://doi.org/10.3389/fpls.2020.00003
Magdalena F, Hannula SE, Marta B, Małgorzata J (2018) Fungal biodiversity and their role in soil health. Front Microbiol 9. https://doi.org/10.3389/fmicb.2018.00707
Mahajan P, Kaushal J (2018) Role of phytoremediation in reducing cadmium toxicity in soil and water. J Toxicol 2018:4864365. https://doi.org/10.1155/2018/4864365
Marinari S, Moscatelli C, Grego S (2014) Enzymes at plant-soil interface. In: Gianfreda L, Rao MA (eds) Enzymes in agricultural sciences. OMICS eBooks Group, pp 94–109
Maseko ST, Dakora FD (2013) Rhizosphere acid and alkaline phosphatase activity as a marker of Pb nutrition in nodulated Cyclopia and Aspalathus species in the Cape fynbos of South Africa. S Afr J Bot 89:289–295
Mazur BJ, Rice D, Haselkorn R (1980) Identification of blue-green algal nitrogen fixation genes by using heterologous DNA hybridization probes. Proc Natl Acad Sci USA 77(1):186–190. https://doi.org/10.1073/pnas.77.1.186
Meenakshi S, Hiremath J, Meenakshi MH, Shivaveerakumar S (2024) Actinomycetes: isolation, cultivation and its active biomolecules. J Pure Appl Microbiol 18(1):118–143. https://doi.org/10.22207/JPAM.18.1.48
Mendes R, Garbeva P, Raaijmakers JM (2013) The rhizosphere microbiome: significance of plant beneficial, plant pathogenic, and human pathogenic microorganisms. FEMS Microbiol Rev 37(5):634–663. https://doi.org/10.1111/1574-6976.12028
Miljaković D, Marinković J, Balešević-Tubić S (2020) The significance of Bacillus spp. in disease suppression and growth promotion of field and vegetable crops. Microorganisms 8(7):1037. https://doi.org/10.3390/microorganisms8071037
Mir MI, Kumar BK, Gopalakrishnan S, Vadlamudi S, Hameeda B (2021) Characterization of rhizobia isolated from leguminous plants and their impact on the growth of ICCV 2 variety of chickpea (Cicer arietinum L.). Heliyon 7(11):e08321. https://doi.org/10.1016/j.heliyon.2021.e08321
Mus F, Crook MB, Garcia K, Garcia Costas A, Geddes BA, Kouri ED, Paramasivan P, Ryu MH, Oldroyd GED, Poole PS, Udvardi MK, Voigt CA, Ané JM, Peters JW (2016) Symbiotic nitrogen fixation and the challenges to its extension to nonlegumes. Appl Environ Microbiol 82(13):3698–3710. https://doi.org/10.1128/AEM.01055-16
Naylor D, McClure R, Jansson J (2022) Trends in microbial community composition and function by soil depth. Microorganisms 10(3):540. https://doi.org/10.3390/microorganisms10030540
Nazari MT, Machado BS, Marchezi G, Crestani L, Ferrari V, Colla LM, Piccin JS (2022) Use of soil actinomycetes for pharmaceutical, food, agricultural, and environmental purposes. 3 Biotech 12(9):232. https://doi.org/10.1007/s13205-022-03307-y
Nephali L, Piater LA, Dubery IA, Patterson V, Huyser J, Burgess K, Tugizimana F (2020) Biostimulants for plant growth and mitigation of abiotic stresses: a metabolomics perspective. Meta 10:505. https://doi.org/10.3390/metabo10120505
Nesrine H, Khattab AR, Saad HH, Abib B, Farag MA (2024) A multiplex metabolomic approach for quality control of Spirulina supplement and its allied microalgae (Amphora & Chlorella) assisted by chemometrics and molecular networking. Sci Rep 14(1):2809. https://doi.org/10.1038/s41598-024-53219-5
Nuccio EE, Starr E, Karaoz U, Brodie EL, Zhou J, Tringe SG, Malmstrom RR, Woyke T, Banfield JF, Firestone MK (2020) Niche differentiation is spatially and temporally regulated in the rhizosphere. ISME J 14:999–1014. https://doi.org/10.1038/s41396-019-0582-x

Odelade KA, Babalola OO (2019) Bacteria, fungi and archaea domains in rhizospheric soil and their effects in enhancing agricultural productivity. Int J Environ Res Public Health 16(20):3873. https://doi.org/10.3390/ijerph16203873

Ouyang Z, Song C, Zheng H, Polasky S, Xiao Y, Bateman IJ, Liu J, Ruckelshaus M, Shi F, Xiao Y, Xu W, Zou Z, Daily GC (2020) Using gross ecosystem product (GEP) to value nature in decision making. Proc Natl Acad Sci USA 117(25):14593–14601. https://doi.org/10.1073/pnas.1911439117

Pantigoso HA, Newberger D, Vivanco JM (2022) The rhizosphere microbiome: plant-microbial interactions for resource acquisition. J Appl Microbiol 133(5):2864–2876. https://doi.org/10.1111/jam.15686

Paśmionka IB, Bulski K, Boligłowa E (2021) The participation of microbiota in the transformation of nitrogen compounds in the soil—A review. Agronomy, 11(5), 977

Poulou A, Grivakou E, Vrioni G et al (2014) Modified CLSI extended-spectrum β-lactamase (ESBL) confirmatory test for phenotypic detection of ESBLs among Enterobacteriaceae producing various β-lactamases. J Clin Microbiol 52:1483–1489

Protonotariou E, Meletis G, Papadopoulou D et al (2021a) Evaluation of the "AMR Direct Flow Chip Kit" DNA microarray for detecting antimicrobial resistance genes directly from rectal and nasopharyngeal clinical samples upon ICU admission. Enferm Infecc Microbiol Clin (Engl Ed) 39:276–278

Protonotariou E, Mantzana P, Meletis G, Tychala A, Kassomenaki A, Vasilaki O, Kagkalou G, Gkeka I, Archonti M, Kati S, Metallidis S, Skoura L (2021b) Microbiological characteristics of bacteremias among COVID-19 hospitalized patients in a tertiary referral hospital in Northern Greece during the second epidemic wave. FEMS Microbes. Dec 2;2:xtab021. https://doi.org/10.1093/femsmc/xtab021. PMID: 35311247; PMCID: PMC8847882

Putri DKT, Amirda F, Muzadi H, Carabelly AN, Erlita I (2020) The antibacterial activity of actinomycetes against the growth of Streptococcus mutans and Lactobacillus acidophilus. In: BIO web of conferences, vol 20. EDP Sciences, Banjarbaru, p 03006. https://doi.org/10.1051/bioconf/20202003006

Reeder RC (2012) Photograph of rhizobium bacteria and nodules infecting soybean roots. Personal collection of R. Reeder, food, agricultural and biological engineering. The Ohio State University, Columbus

Rhodes A, Evans L, Alhazzani W, Levy M, Antonelli M, Ferrer R et al (2017) Surviving sepsis campaign: international guidelines for management of sepsis and septic shock: 2016. Intens Care Med 43:304–377. https://doi.org/10.1007/s00134-017-4683-6

Risa E, Roach D, Budak JZ et al (2021) Characterization of secondary bacterial infections and antibiotic use in mechanically ventilated patients with COVID-19 induced acute respiratory distress syndrome. J Intensive Care Med; 36:1167–75

Salazar-Ardiles C, Boyero L, Cornejo A. (2022). Amoebae: Beyond pathogens—Exploring their benefits and future perspectives. PMC

Sepulveda J, Westblade LF, Whittier S et al (2020) Bacteremia and blood culture utilization during COVID-19 surge in New York City. J Clin Microbiol 58:e00875–e00820

Shanan S, Abd H, Bayoumi M, Saeed A, Sandström G (2015) Prevalence of protozoa species in drinking and environmental water sources in Sudan. Biomed Res Int 2015:345619. https://doi.org/10.1155/2015/345619

Sharma M et al. (2021) Actinomycetes as biocontrol agents against fungal phytopathogens. J Appl Microbiol 130(2), 1–15

Shuo Liang, Nan Sun, Jeroen Meersmans, Bernard Longdoz, Gilles Colinet, Minggang Xu, Lianhai Wu (2024) Impacts of climate change on crop production and soil carbon stock in a continuous wheat cropping system in southeast England, Agriculture, Ecosystems and Environment, 365 https://doi.org/10.1016/j.agee.2024.108909

Singh DP, Patil HJ, Prabha R, Yandigeri MS, Prasad SR (2018) Actinomycetes as potential plant growth-promoting microbial communities. In: Prasad R, Gill SS, Tuteja N (eds) Crop improvement through microbial biotechnology. Elsevier, Amsterdam, pp 27–38. https://doi.org/10.1016/B978-0-444-63987-5.00002-5

Smol JP, Stoermer EF (eds) (2010) The diatoms: applications for the environmental and earth sciences. Cambridge University Press

Sneha Gupta, Sibel Yildirim, Benjamin Andrikopoulos, Uta Wille, Ute Roessner (2023) Deciphering the Interactions in the Root–Soil Nexus Caused by Urease and Nitrification Inhibitors: A Review, Agronomy, 13 (6) https://doi.org/10.3390/agronomy13061603

Sylvia DM, Hartel PG, Fuhrmann JJ, Zuberer DA (2005) In: Sylva DM (ed) Principles and applications of soil microbiology, 2nd edn. Pearson Prentice Hall, Upper Saddle River

Taylor BN, Menge DNL (2018) Light regulates tropical symbiotic nitrogen fixation more strongly than soil nitrogen. Nat Plants 4:655–661

Tilden JE (1933) The algae and their life relations. Fundamentals of phycology, 1st edn. Hofner Publishing Company, New York

Total Coliform Rule: A Handbook for Small Noncommunity Water Systems Serving less than 3,300 Persons (2006) Office of Water (4606M). EPA, Washington, DC. EPA 816-B-06–001

Trenberth KE, Dai A, Van Der Schrier G, Jones PD, Barichivich J, Briffa KR, Sheffield J (2014) Global warming and changes in drought. Nat Clim Chang 4:17. https://doi.org/10.1038/nclimate2067

Valentine M, Van Houten J (2021) Using paramecium as a model for ciliopathies. Genes (Basel) 12(10):1493. https://doi.org/10.3390/genes12101493

Van Houten J (2019) Paramecium biology. In: Tworzydlo W, Bilinski S (eds) Evo-Devo: non-model species in cell and developmental biology. Results and problems in cell differentiation, vol 68. Springer, Cham. https://doi.org/10.1007/978-3-030-23459-1_13

Van Houten J (2023) A review for the special issue on paramecium as a modern model organism. Microorganisms 11(4):937. https://doi.org/10.3390/microorganisms11040937

van Wolferen M, Pulschen AA, Baum B, Gribaldo S, Albers SV (2022) The cell biology of archaea. Nat Microbiol 7(11):1744–1755. https://doi.org/10.1038/s41564-022-01215-8

Ventura M, Canchaya C, Tauch A, Chandra G, Fitzgerald GF, Chater KF, van Sinderen D (2007) Genomics of Actinobacteria: tracing the evolutionary history of an ancient phylum. Microbiol Mol Biol Rev 71(3):495–548. https://doi.org/10.1128/MMBR.00005-07

Vetterlein D, Carminati A, Kögel-Knabner I, Bienert GP, Smalla K, Oburger E, Schnepf A, Banitz T, Tarkka MT, Schlüter S (2020) Rhizosphere spatiotemporal organization – a key to rhizosphere functions. Front Agron 2:8. https://doi.org/10.3389/fagro.2020.00008

Wang L, Chen M, Lam PY, Dini-Andreote F, Dai L, Wei Z (2022) Multifaceted roles of flavonoids mediating plant-microbe interactions. Microbiome 10(1):233. https://doi.org/10.1186/s40168-022-01420-x

Wang X, Chi Y, Song S (2024) Important soil microbiota's effects on plants and soils: a comprehensive 30-year systematic literature review. Front Microbiol 15:1347745. https://doi.org/10.3389/fmicb.2024.1347745

Wei X, Xie B, Wan C, Song R, Zhong W, Xin S, Song K (2024) Enhancing soil health and plant growth through microbial fertilizers: mechanisms, benefits, and sustainable agricultural practices. Agronomy 14(3):609. https://doi.org/10.3390/agronomy14030609

Wilson TJ (2013) Photographs of ciliates, flagellates, and various bacteria. From Compost, Soil & Compost Tea Microorganisms

Yaeger RG (1996) Chapter 77: Protozoa: structure, classification, growth, and development. In: Baron S (ed) Medical microbiology, 4th edn. University of Texas Medical Branch at Galveston, Galveston. Available from: https://www.ncbi.nlm.nih.gov/books/NBK8325/

Yim B, Baumann A, Grunewaldt-Stöcker G et al (2020) Rhizosphere microbial communities associated to rose replant disease: links to plant growth and root metabolites. Hortic Res 7:144. https://doi.org/10.1038/s41438-020-00365-2

Yingyong Luo, Yunfeng Zhang, Zhuang Xiong, Xiaodie Chen, Ajia Sha, Wenqi Xiao, Lianxin Peng, Liang Zou, Jialiang Han, Qiang Li, (2024) Peptides Used for Heavy Metal Remediation: A Promising Approach, International J Mol Sci, 25(12)https://doi.org/10.3390/ijms25126717

Biocontrol and Remediation Help in a Safe Environment Production

Ayat M. Ali and Nashwa A. H. Fetyan

Abstract The ever-increasing population and the depletion of land and other resources provide the agriculture sector with enormous hurdles in satisfying the expanding demand for food. In an attempt to overcome these challenges, some essential inputs, such as fertilizers, pesticides, and other agrochemicals, are abused, resulting in environmental damage. Neglecting ecological sustainability and favoring massive agricultural production has led to declining soil health, degraded land, and severe ecological problems. Increasing agricultural output under sustainable conditions should be the main goal of future efforts to feed the expanding population. In modern agriculture, one major concern is that agricultural pests are evolving resistance to pesticides. Future crop production must be assured by applying innovative strategies that increase crop yields while preserving the environment. Among all the biota in soil, soil microbes are the most prevalent and are responsible for various aspects of ecosystem functioning, such as soil fertility, plant health, primary production, controlling plant pests, and bioremediation of both organic and inorganic. A considerable number of these organisms are naturally found in the soil, but in certain circumstances, it could be advantageous to boost their numbers by inoculation or the use of certain agricultural management strategies that improve their activity and abundance. This chapter provides a comprehensive overview of the mechanisms of soil microorganisms involved in pests' biocontrol and polluted soil remediation. Furthermore, the advantages and disadvantages of the microbial rhizosphere for agricultural sustainability are discussed.

Keywords Bio-control agent · Competition · Hyperparasitism · Antibiosis · Secondary metabolites · Bioremediation

A. M. Ali
Central Lab of Organic Agriculture, Agricultural Research Center, Giza, Egypt

N. A. H. Fetyan (✉)
Department of Microbiology, Soils, Water & Environment Research Institute, Agricultural Research Center, Giza, Egypt
e-mail: nashwa.fetyan@arc.sci.eg

A. A. H. Abdel Latef et al. (eds.), *Sustainable Remediation for Pollution and Climate Resilience*, https://doi.org/10.1007/978-981-96-5674-5_18

1 Introduction

Since the discovery of some synthetic organic pesticides in the 1940s, numerous highly effective chemical pesticides have been developed. They are widely used to control plant pathogens, agricultural pests, and plant weeds (Sharma et al. 2021). Pesticides are necessary to safeguard crop harvests, but excessive use and repeated application have negatively impacted the environment and endangered human health. Moreover, certain pesticide residues might cause serious phytotoxicity to crops that are susceptible to rotation (Huang et al. 2023).

Pesticide use has increased significantly, leading to environmental pollution, particularly in water, soil, and food. Synthetic pesticides come in various forms, but the most often used pesticides are organophosphates, organochlorines, carbamates, and pyrethroids, all of which pose health and environmental risks.

Pesticides are commonly used to protect crops but can also pollute our soil and food. Therefore, the elimination of pesticide residue from polluted soils has been the exclusive focus of researchers and the general public. Biological control agents (BCAs) and bioremediation techniques are instances of environmentally conscious pesticide residue management strategies and examples of nontoxic green alternatives. When comparing the use of pesticides and BCAs, it became clear that to obtain food free of contaminants, BCAs must be applied (Fig. 1). Bioremediation and biocontrol techniques rely on beneficial microorganisms and their enzymes to break down synthetic pesticides. The global sustainability mission is driving up demand for pesticide reduction in agricultural cultivation while maintaining a profitable yield and high-quality products. Biological control is regarded as one of the most essential uses of organic agriculture since synthetic chemical pesticides are prohibited.

2 Biological Control

One of the biological control definitions is using micro- or macroorganisms (except man and plant itself) to reduce the number or effect of certain pathogens or pests. Abd El-Moity (1976) reported that in Egypt, *Trichoderma harzianum*, *Bacillus subtilis*, and *Streptomyces* spp. have antagonistic effects against various pathogenic fungi and bacteria. Many investigators who followed this work confirmed these studies (Morsy et al. 2005; Singh et al. 2008; Montero et al. 2011; Anis et al. 2013; Ali 2021).

Many studies have shown that BCAs effectively protect plants against various soil or airborne pathogens. Some researchers explain this protective effect as antibiosis action occurring in the court of infection. Other researchers discovered plant physiology and chemical composition variations in plants treated with these bioagents. This chapter will look at recent advancements in the exciting fields of biocontrol work.

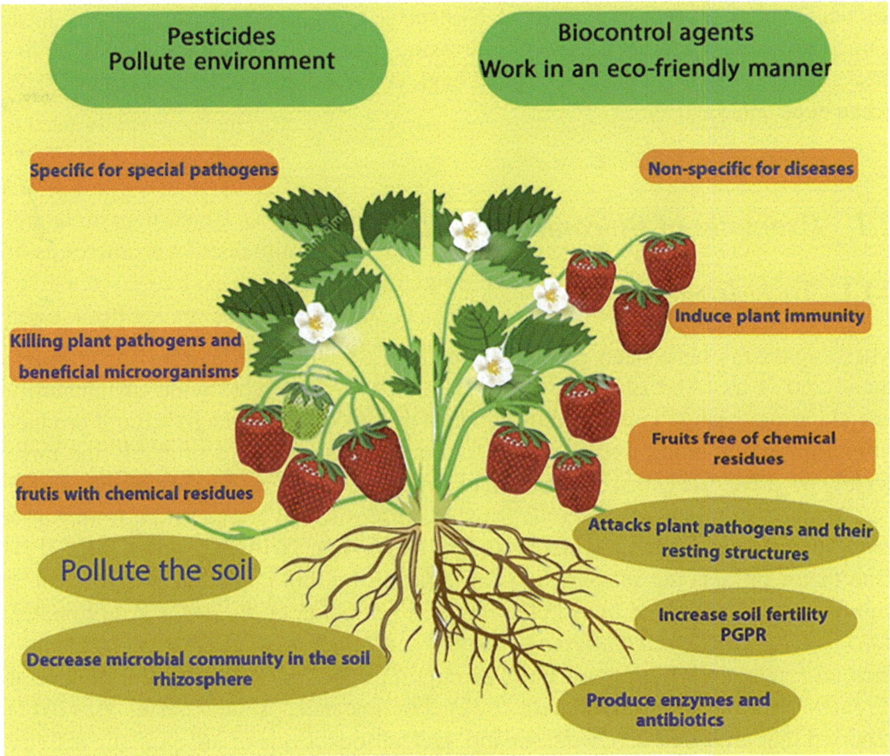

Fig. 1 A comparison between biocontrol agents and synthetic pesticides on plants

Importance of Biological Control Agents (BCAs)

1. BCAs are effective against many soil or airborne pathogens when chemical fungicides are not an option (especially in organic agriculture).
2. BCAs avoid problems of pest resistance to chemical pesticides.
3. BCA biosynthesis targets metabolites like plant growth regulators, enzymes, siderophores, and antibiotics. BCAs enhance plant defense against pathogenic microbes like viruses, bacteria, and fungi by promoting resistance mechanisms.
4. BCAs are safe for both the environment and those implementing them, and avoiding pollution.
5. BCAs are nontoxic to plants and improve agricultural yields through enhancing root and plant growth.

Traditionally, microorganism-based biocontrol relied on a single strain of bacteria or fungi. However, current trends in crop protection research support the use of microorganisms with many mechanisms of action. Multiple bioactive BCAs are thought to increase the application's effectiveness and broaden the spectrum of the product's efficacy against various pathogen species (Sarma et al. 2015; Minchev et al. 2021). Following the development of the first commercial biological agent, several microorganisms, including fungi, bacteria, and actinomycetes, became known for plant disease management. Consortia, or mixes of microorganisms, have gained greater interest as a way of addressing multiple diseases and effective disease control (Mazaro et al. 2008; Hafez et al. 2012; Ali 2021; Mergawy et al. 2022; Khan et al. 2023).

2.1 Examples of Biocontrol Agents (BCAs)

2.1.1 *Trichoderma* spp.

Microorganisms have been recognized to interact with one another for approximately 50 years (Abd El-Moity 1976; Baker 1987). The worldwide sustainability agenda is increasing the demand for reduced pesticide inputs in agricultural production while maintaining excellent yields. *Trichoderma* spp. are probably the most widely used BCAs against various phytopathogenic fungi worldwide. *Trichoderma* acts against pathogens by mycoparasitism, antibiosis, and competition; it may also promote resistance in treated plants (Prasad and Kumar 2013; Ali and Ayoub 2017). One factor influencing its success is an efficient coiling procedure followed by a significant generation of hydrolytic enzymes (Almeida et al. 2007). Additionally, *Trichoderma viride*, *Trichoderma hamatum*, and *T. harzianum* have successful commercial products in Egypt and India as important BCAs for the control of several diseases (Kumar et al. 2014; Ali 2021). For instance, Abou-Zeid et al. (2011) reported that all tested *Trichoderma* spp. had various levels of antagonistic activity against *Rhizoctonia solani* and *Pythium ultimum*. *Trichoderma* isolates overlapped against *Rhizoctonia solani* and *Pythium ultimum* within 5 days. Substantial evidence suggests that *T. harzianum's* extraordinary ability to attack phytopathogenic fungi is related to the synergistic action of competition for key nutrients, antagonism by antibiotics, production of cell wall-degrading enzymes, stimulating of plant defense systems, and a combination of these opportunities (Mayo et al. 2015; Marques et al. 2018; Kamel et al. 2020). *Trichoderma's* mycoparasitic properties are essential for destroying pathogen cell walls, it was studied between *T. harzianum* and *R. solani* under light and scanning microscopy by Ali (2021) (Fig. 2). Chitinases, glucanases, proteases, and cellulases are *Trichoderma* enzymes that dissolve the cell wall and have antifungal action against *S. sclerotiorum* (Lopez-Mondejar et al. 2011). Furthermore, utilizing more than one isolate increases the likelihood of a synergistic effect between different defenses against the pathogen. In this regard, Cook (1995) suggested that various *Trichoderma* strains may be included in a formulation to broaden the scope of control.

2.1.2 Actinomycetes (*Actinobacteria*)

Actinobacteria are Gram-positive bacteria, and they are mostly aerobic. They are a widely distributed group of soil microbes, constituting about 10–50% of the soil microflora community (Waksman 1959; Passari et al. 2016; Mitra 2018). The

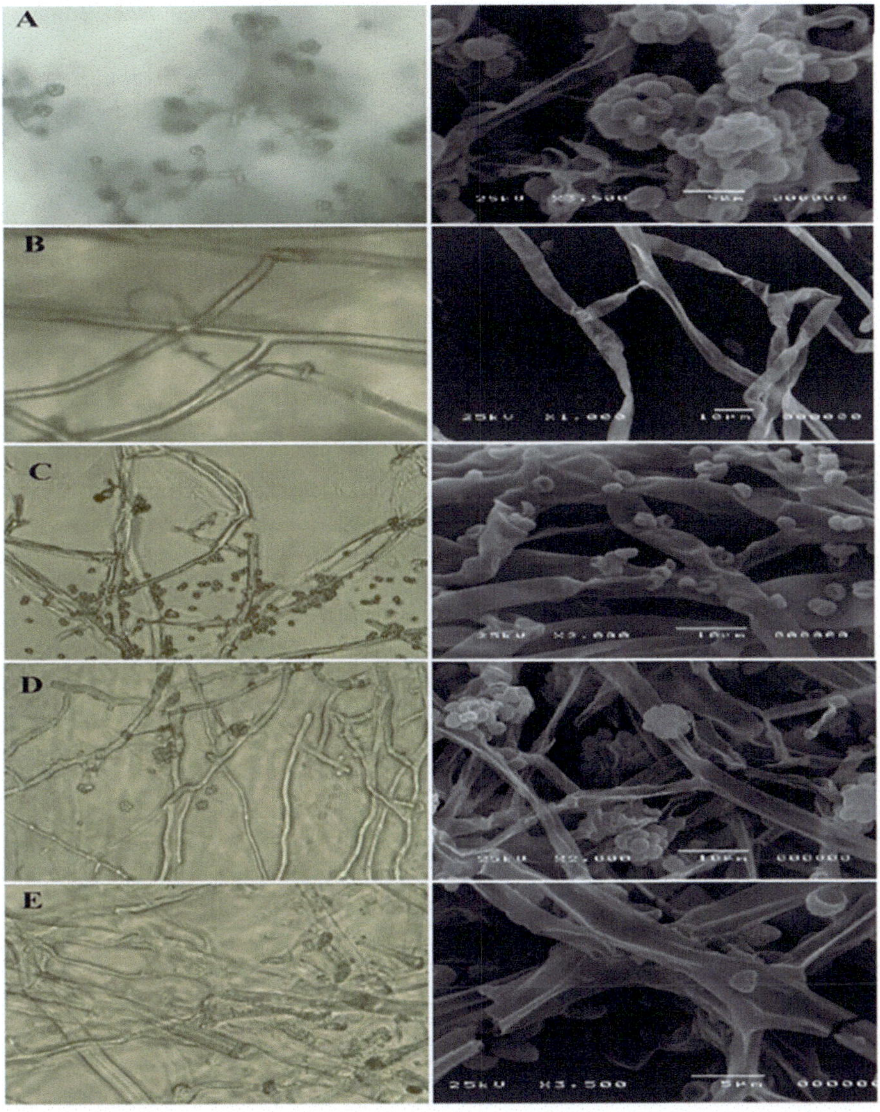

Fig. 2 Light and scanning electron photomicrographs of hyphal interactions between *Rhizoctonia solani* and *T. harzianum*. (**a**) culture of *T. harzianum* (**b**) normal *R. solani* mycelium (**c**) coiling and spore formation without penetration of *T. harzianum* in the mycelium over *R. solani* (**d** and **e**) Spore formation of *T. harzianum* over *R. solani* mycelia and degrading the cell walls of *R. solani* (Ali 2021)

important secondary metabolites of actinobacteria are antibiotics, pigments, enzymes, odors, organic volatile substances, and other factors. The metabolites produced are diversified in their biological activities and functions such as antifungal (e.g., antibiotics and/or low molecular weight antifungal compounds or lytic enzymes like chitinase), antibacterial, antiviral, insecticidal, and other activities

(Lyu et al. 2017; Gomes et al. 2018; Gebily et al. 2021). Most examined actinomycetes plant growth-promoting species possess antibacterial or antifungal activity, evident during their biocontrol agents screening (Nanjappan et al. 2018; Oluwaseyi and Olubukola 2019). For example, Chen et al. (2018) found that after treatment with a crude extract of *Streptomyces* sp. CB-75, the pathogenic fungus *Colletotrichum gloeosporioides* deformed, shrank, and collapsed. The crude extract of this isolate comprises 18 chemical components, with (Z)-13-docosenamide being the most abundant.

2.1.3 *Bacillus* spp.

Bacillus subtilis, the model system for Gram-positive organisms, is capable of producing over two dozen antibiotics with amazing structural diversity that can control fungal pathogens through competition, direct antibiosis, inducing systemic resistance (ISR) in plants, and competing for ecological niches with plant pathogens (Stein 2005; Fira et al. 2018). *B. subtilis* mode of action includes the formation of several different kinds of antibiotics, the production of extracellular enzymes, competition for nutrients and habitats, and the induction of systemic resistance in plants toward infections. For example, *B. velezensis* inhibits hyphal growth in *R. solani* and produces fungal cell wall-degrading enzymes such as protease, cellulase, and β-1,3-glucanase, as well as growth-promoting compounds like indole-3-acetic acid (IAA) and siderophore (Huang et al. 2023). *B. subtilis* XZ18-3 sterile filtrates showed strong antifungal activity against *R. cerealis*. *Bacillus* spp. produce antimicrobial peptides, which have been related to the biocontrol of various plant diseases. Also, *B. subtilis* SG6 effectively suppress *Fusarium. graminearum* development and sporulation. SG6 can break down the cell wall of *F. graminearum* causing cell contents to spill. The lipopeptide profile of strain SG6 was analyzed; fengycins and surfactins are the most significant products of strain SG6. (Stein 2005; Zhao et al. 2014; Yanjie et al. 2022).

2.2 Mechanisms of Action Contributing to Biocontrol

2.2.1 Competition

In the rhizosphere, competitive colonization and successful establishment in the root zone are required for effective biocontrol, regardless of the mechanism(s) involved. Direct antagonism includes the synthesis of several microbial metabolites, such as exudates, consumption, siderophore scavenging, and lipopeptides. Utilization of resources (carbon, nitrogen, phosphorus, and other essential resources). Active colonization of the rhizosphere in the root zone is required for effective biocontrol, regardless of which mechanism is utilized. Direct antagonism is characterized by the creation of many microbial metabolites, the most important of which play a critical role. Recently, there has been an increase in rhizosphere competence-based competition. It is significant because a biocontrol agent cannot compete for space or resources if it is unable to establish itself in the rhizosphere. *Trichoderma* species, whether introduced to the soil or supplied as seed treatments, grow rapidly alongside the developing root system of the treated plant (Alizadeh et al. 2013; Samaras et al. 2018).

2.2.2 Hyperparasitism

Hyperparasitism occurs when a particular BCA attacks the pathogen directly, killing it or its propagules (direct antagonism). In this scenario, hyperparasitism would be seen as the most direct form of antagonism. There are various fungal infections of plant diseases, some of which attack sclerotia (e.g., *Trichoderma* spp.), while others attack hyphae (e.g., *Pythium oligandrum*). Additionally, numerous hyperparasites can attack the same fungal pathogen. For example, Ampelomyces quisqualis grows on mildew pathogens, and the fungus *Gliocladium virens* can also parasitize powdery mildew pathogens and can serve as a biocontrol agent (Heydari and Pessarakli 2010).

A scanning electron microscopic investigation showed that Trichoderma hyphae developed in parallel and were attached to *F. solani* hyphae. Additionally, found that *Trichoderma* hyphae coil and go through *F. solani* hyphae. Then *Trichoderma* can eliminate nutrients from the cell, resulting in *F. solani* cell death. It causes morphological deformations and collapses in the structure of the cell wall (Prasad and Kumar 2013). Also, the parasitic activity of *Streptomyces* spp. on *S. sclerotiorum* and interaction caused deformation, tightening, and breakdown of the pathogen's mycelium. According to the author, mycoparasitism in *Streptomyces* could be called Actinoparasitism (Gebily 2021).

2.2.3 Antibiosis

Antibiotics are organic chemicals produced by microorganisms that, at low doses, can limit their growth or kill other microbes. The BCA suppresses the pathogen by its release of toxic secondary (specialized) metabolites.

For instance, *Agrobacterium radiobacter* K-84 was first identified as a promising crown gall control agent in 1973. It is highly efficient against *Agrobacterium tumefaciens*, which attacks stone fruits (such as plums and peaches). The majority of microorganisms secrete one or more antibiotic compounds. In every instance, the antibiotics were found to be extremely effective at inhibiting the growth of the pathogen that was targeted in vitro. To be profitable, antibiotics must be generated in sufficient quantities. Formica (1990) cites Agrocin 84 as an example of an antibiotic produced by BCAs. Gliotoxin, an epithiodiketopiperazine produced by *Trichoderma virens*, is essential for curtailing the growth and multiplication of phytopathogens (Anitha and Murugesan 2005), Iturin, *B. subtilis* (Kloepper et al. 2004), and 2, 4-diacetyl-phloroglucinol, *Pseudomonas fluorescens* (Samavat et al. 2011). *Trichoderma* spp. the most common chemicals discovered by GC-MS analysis included hexadecanol, nonadecene, 1,2-benzenedicarboxylic acid, and octadecenoic acid (Ali 2021). The application of mixture strains performs better than individual strains for controlling pests and diseases. This can be explained in light of the fact that different isolates produce different antibacterial and antifungal substances; by combining two complementary isolates, the resultant mixture showed a high protection effect. *T. harzianum* works through different modes of action, i.e., mycoparasitism, and antifungal substance production such as gliotoxin and trichodermin, these modes of action of *T. harzianum* when combined with modes of action of *B. subtilis* lead to an increase in the scope of antagonism and give the highest effect of plant protection. Also, a synergistic effect was noticed when two isolates of *B. subtilis* were combined. The combined use of the bio-agents *T. asperellum* and *S. rochei* was highly effective in both reducing the disease and improving plant growth parameters. *T. harzianum* + *B. subtilis* significantly increased the activities of defense-related enzymes i.e., catalase, polyphenol oxidase, peroxidase, and the total content of phenols (Singh et al. 2008; Ali 2013; Mergawy et al. 2022).

2.2.4 Lytic Enzyme Production

Numerous BCAs can produce and excrete lytic enzymes capable of hydrolyzing various polymeric materials, including hemicellulose, cellulose, and chitin. Actinomycetes may synthesize a wide range of extracellular enzymes, allowing them to break down diverse biopolymers in soil. Actinomycetes' capability to synthesize extracellular enzymes has received additional attention because of their use in biocontrol. Numerous correlations have been documented between fungal antagonism and bacterial production of chitinases or glucanases (Fayad et al. 2001; Kumar et al. 2012). This rhizosphere-colonizing actinomycete chitinase, most likely, significantly improves its antifungal biocontrol efficiency. *Streptomyces* spp. produce a variety of extracellular hydrolytic enzymes degrading cell walls (Gebily et al. 2021). Many recent studies have focused on enzymes produced by *Trichoderma* spp., which attack the cell walls of pathogenic fungi. These mainly consist of ß-1,3-glucanase, chitinase, and cellulose. Enzymatic activity increased when *T. harzianum* became connected to *S. rolfsii* or *R. solani*. (Prasad and Kumar 2013; Kamel et al. 2020). Biocontrol competes strongly with other microbes due to its capacity to rapidly make use of seed and root exudates, colonize and reproduce in the rhizosphere, and create a diverse range of bioactive metabolites (e.g., antibiotics, siderophores, volatiles, chitinase, and growth-promoting compounds). Samavat et al. (2011) found that *P. fluorescens* isolates produced siderophore, hydrocyanic acid (HCN), IAA, and exopolysaccharide. Resistance is induced by an increase in both defensive metabolites and enzymes. Similarly, *T. harzianum*-derived chitinase led to the development of extremely tolerant or completely resistant transgenic lines against soil-borne pathogens like *R. solani*, as well as foliar pathogens like *Alternaria alternata*, *A. solani*, and *Botrytis cinerea* (Howell 2003).

2.2.5 Volatile Secondary Metabolites

Most biocontrol agents are known to produce volatile chemicals as pathogen inhibitors. Common volatile chemicals include HCN, alcohols, ketones, aldehydes, and sulfides (Bouizgarne 2013). Wei et al. (1991) reported that HCN production plays an important role in disease suppression. For example, strains of *P. fluorescens* can produce HCN, which aids in the inhibition of *Macrophomina phaseolina* (Reetha et al. 2014). According to Chen et al. (2014), *B. subtilis* EDR4 initiated unusual swelling of mycelia in *S. sclerotiorum*. The antibiosis was caused by the release of antibiotics and volatile compounds (ammonia, cyanide, alcohols, esters, ketones, enzymes, etc.). Volatile organic compounds (VOCs) components were identified as antibacterial and antifungal to control bacterial and fungal diseases in vitro by destroying the cell membrane. Also, different *Streptomyces* spp. were screened for the production of volatile compounds. The VOCs' mixture had antifungal activity and included alkanes, alkenes, alcohols, esters, ketones, sulfur compounds, and isoprene (Scholler et al. 2002; Cordovez et al. 2015; Wu et al. 2019 and Le et al. 2022). Volatile organic compounds produced by *Streptomyces philanthi* RM1-138 reduced the growth of various plant pathogenic fungi (*R. solani* and *Fusarium fujikuroi*). They are classified chemically as alcohols, alkenes, sulfides, ketones, and esters (Boukaew et al. 2013).

2.2.6 Plant Growth-Promoting Rhizobacteria

Plant growth-promoting rhizobacteria (PGPR) enhance plant root development, stimulating plant growth and/or reducing the incidence of plant diseases. The stimulation of plant growth is the dominant aspect of some PGPR, which are known as biofertilizers. Nitrogen fixation, phosphate solubilization, and the synthesis of phytohormones (such as auxin and cytokinin) and volatile growth stimulants (such as ethylene and 2,3-butanediol). PGPR plays a significant role in plant immunity to diseases by inducing systemic resistance (ISR), antibiotics, hydrogen cyanide, lytic enzyme, toxin breakdown, resource competition, and parasitic activity. Phosphate-solubilizing bacteria are PGPR that convert organic and inorganic insoluble P molecules into soluble forms that plants can use. They help to create hormones, including auxin and gibberellins, siderophores, ammonia, and aminocyclopropane-1-carboxylate (Chen et al. 2018; Vurukonda et al. 2018; Kumar et al. 2021). Actinobacteria can enhance nitrogen-fixing symbiosis between legumes and nonlegume plants by working as assistant bacteria (Arumugam et al. 2017). Yobo et al. (2009) studied the efficiency of single and dual inoculations using selected *Trichoderma* and Bacillus isolates on the performance of dry bean plants. The *Trichoderma* sp. and *Bacillus* sp. treated plants showed an increase in dry shoot biomass and the greatest degree of nodulation. These results confirm that these organisms can play a role in root-plant growth promoting rhizobacteria or fungi. Sreevidya et al. (2016) also investigated four actinomycetes for their plant growth-promoting activities on chickpeas in the field. They found that they all increased nodule number, shoot weight, and yield. These isolates produced siderophore, cellulase, lipase, protease, chitinase, and hydrocyanic acid. In field trials, Ghanem et al. (2022) stated that foliar application of *Streptomyces* spp. and their metabolites was found to be a promising biocontrol agent for suppressing *S. sclerotiorum* while also increasing plant growth and production. *S. rochei* and *S. griseus* were good plant growth promoters and ecologically friendly fungicides. Similarly, Vurukonda et al. (2018) discovered that *S. rochei* may increase plant growth by releasing secondary metabolites such as auxin, gibberellin, and cytokinin.

2.2.7 Induction of Plant Systemic Resistance (ISR)

Stimulation of plant defense mechanisms by BCAs is the most indirect form of antagonism. Application of BCAs triggers an enhanced resistance state, also termed induced systemic resistance (ISR), in the host against a broad range of pathogens. Upon the activation of ISR, plants employ long-distance systemic signaling to induce rapid and strong immune responses against pathogen invasions. The transmission of ISR signals was assumed to be dependent on jasmonic acid- and ethylene-dependent. *Bacillus amyloliquefaciens*, *B. atrophaeus*, *B. cereus*, *P. fluorescens*, etc., were demonstrated to be effective against fungal, bacterial, and viral invasion by inducing early plant ISR (Guo et al. 2019; Yiyang et al. 2022). Several stimulants released by *Trichoderma* hyphae can stimulate various forms of signals interacting inside the plant, such as salicylic acid (SA), jasmonic acid, or reactive oxygen species, resulting in the formation of defense proteins. For example, *T. asperellum* colonizing roots activated ISR via the SA signaling pathway. Furthermore, plants treated with *Trichoderma* filtrates activated the SA and jasmonate/ethylene pathways. In the present concept of the plant immune system, pattern-recognition receptors have developed to recognize common microbial components, such as bacterial flagellin or fungal chitin, known as pathogen-related molecular patterns (Yoshioka et al. 2011; Alizadeh et al. 2013; Nawrocka and Małolepsza 2013; Mayo et al. 2015).

Gas chromatography-mass analysis of *Streptomyces actuosus* (SCF20) demonstrated the majority of fatty acids (FAs) and their derivatives, Citroflex A, hexanoic acid, hexadecanoic acid, and ethyl ester, which proposed that they act as elicitor molecules for the induction of systemic acquired resistance (SAR) against potato virus Y. Furthermore, JCK-6131 shown antagonistic activity in vitro against (*Erwinia amylovora*) apple fire blight and *Fusarium oxysporum* f. sp. *Cucumerinum* caused cucumber wilt disease. Levels of salicylate and jasmonate in the treated plants were increased when JCK-6131 was applied as a foliar application (Nasr-Eldin et al. 2019). *Streptomyces* sp. AcH 505 suppressed oak powdery mildew infection by inducing some systemic defensive responses in oak. The RNA-Seq results show that the salicylic acid-dependent pathway is also activated in addition to triggering the jasmonic acid/ethylene-dependent pathway. Transcripts involving tryptophan and phenylalanine production were increased (Kurth et al. 2014).

Zhang et al. (2020) proposed that antifungal mycin derived from *Streptomyces* sp. N_2 could promote resistance to *R. solani* in rice seedlings by enhancing defense response activation. The defense-related physiological and biochemical parameters, such as the production of reactive oxygen species, enzymatic activities involving antioxidant systems (superoxide dismutase, catalase, peroxidase), phenolics metabolism (phenylalanine ammonia-lyase, polyphenol oxidase), and pathogenesis-related proteins (β-1,3-glucanase and chitinase).

3 Soil Remediation

Owing to the growing global population, the land is now needed for more than just forestry and agriculture—it is also required for housing, transportation, industry, large-scale tourism development programs, and other urban infrastructure requirements. Here, pressures on the soil have been intense, and most have been destroyed or severely degraded. Over the last few decades, environmental contamination has increased globally due to population growth, rapid industrialization, anthropogenic activities, and unregulated agricultural practices. The accumulation of hazardous waste in the environment, including waste from plastic, industry, rubber, metallic sources, and agriculture, is another effect of these issues. This hazardous waste could contain heavy metals, petroleum oil, pesticides, hydrocarbons (such as aliphatic, aromatic, and polycyclic aromatic hydrocarbons (PAHs)), nitroaromatic compounds, chlorinated hydrocarbons (like perchloroethylene and polychlorinated biphenyls (PCBs), organophosphorus compounds, organic solvents (phenolic compounds), and phthalates.

Land restoration is a key component of the sustainable development concept. Most definitions agree that equity is important not only for current generations but also for those to come. Thus, it is imperative to take action beyond simply stopping or slowing down land degradation; as the world's population increases, land cleanup will be acknowledged as a critical need. Urban areas are currently the focus of cleanup efforts in developed nations for the reasons outlined above. There is a brief assessment of the scope of land restoration, a brief definition of bioremediation, and an outline of the approach.

The polluted environment or ecosystem harms the biogeochemical cycle of the elements present in the environment, plants, aquatic organisms, soil microbial growth and metabolic activities, and soil composition (structure and fertility). These factors further affect the ecosystem, its functioning, and human health. Since environmental and living quality are inextricably intertwined, there is a global urgency to protect and improve environmental quality (Oladoye et al. 2021).

Remedial treatments have advanced somewhat in terms of their cost, environmentally benign nature, and efficacy in cleaning up contaminated areas (Priya et al. 2023). However, traditional waste remediation methods like soil washing, flocculation, adsorption, thermal desorption, chemical oxidation, soil vapor extraction, landfilling, pyrolysis, and incineration are expensive, time- and energy-consuming, ineffective, and unfriendly to the environment when it comes to eliminating the pollutants from the contaminated site. Due to their affordability, sustainability, and naturalness, biological remediation techniques are becoming increasingly popular for the detoxification and biodegradation of pollutants in waste management and cleanup programs to return contaminated sites to their original state (Kumar et al. 2018).

3.1 Types of Remediation

Environmental contamination is one of the world's most pressing challenges today. Significant measures are taken to prevent the release of dangerous substances into the environment, and overtime, numerous techniques for cleaning up soil, surface water, and groundwater have been established. There are various kinds of remediation, including physical, chemical, and biological remediation techniques. The features of the polluted site, the process's cost, and the amount of time available will all influence the method selection. Skimmers, booms, and sorbent materials are employed for the physical cleanup. Boom is a material-based physical barrier that gathers and traps oil pollutants until further treatment techniques are applied, according to Vocciante et al. (2019). Strategies for physical cleanup include skimmers and sorbent materials, which are used after booms to absorb and adsorb contaminants. The reliance of boom remediation technology on roll response and buoyancy is its primary impediment to application. Booms float on the water's surface for extended periods while remaining buoyant.. The term "roll response" describes the torque required to rotate the bloom away from its vertical position (Kumari et al. 2019). In other words, a higher remediation process follows an enhanced roll reaction (Dhaka and Chattopadhyay 2021).

Chemical remediation solutions are particularly intriguing since they may remove and degrade pollutants in settings that have been damaged. Among the many chemical remediation methods are chemical oxidation, flocculation, adsorption, ion exchange, precipitation, washing and flushing of the soil, and electrokinetic remediation. Despite the widespread usage of certain technologies, research ought to concentrate on improving the current technologies and creating new remediation technologies. (Ong and Kolz 2007; Khan et al. 2004; Jovanović et al. 2021). Another approach to treating pollution is bioremediation, which is a safe, inexpensive, and sustainable remediation process (Kumar et al. 2022; Patel et al. 2022). Microbes and plants are among the organic materials used in the technology. The pollution's type, extent, and location determine whether this strategy is feasible (Patel et al. 2022). However, it has been shown that microbes effectively clean up environmental contaminants. They are chosen over plants for remediation because they are easier to grow, develop quickly, and are more straightforward to work with. Therefore, in order to support a sustainable ecosystem, it is imperative to enhance the employment of microorganisms as agents of bioremediation.

3.2 Bioremediation

Agriculture pesticides, heavy metals, and petroleum hydrocarbons all pose a major risk to human health and the environment because they alter soil physicochemical and microbiological properties and have mutagenic, carcinogenic, immune toxic, and teratogenic effects. To reduce the amount of harm caused by soil contamination, several physicochemical and biological procedures and treatments must be applied immediately. Bioremediation is one of them; it has been demonstrated to be an alternative that can provide an economically feasible means of restoring contaminated areas. Bioremediation is "the elimination, attenuation or transformation of polluting or contaminating substances by the use of biological processes" (BBSRC 1999). The technique known as "bioremediation" uses living organisms such as plants, algae, and microorganisms to clean up, remove, or eliminate environmental pollution (Saxena and Bharagava 2020).

Optimizing operating conditions is necessary to enable microbial cells to biodegrade pollutants fast, as bioremediation is facilitated by the action of microbial enzymes that transform pollutants into nonhazardous substances (Sergio et al. 2016). This bioprocess was initially used in 1989 to clean up the environmental devastation caused by the Exxon Valdez and is one of the most promising novel therapies for eliminating oil and derivatives (Mandy et al. 2018). PGPR-assisted bioremediation is a viable, sustainable, and environmentally benign way to remove hazardous from the environment. To clean up heavy metal-contaminated environments, PGPR employs a variety of techniques, including efflux systems, siderophores and chelation, biotransformation, biosorption, bioaccumulation, precipitation, 1-aminocyclopropane-1-carboxylate (ACC) deaminase activity, biodegradation, and biomineralization strategies (Gupta et al. 2023).

3.3 The Role of Microorganisms in Soil Bioremediation

Soil pollution is a global issue arising from both natural and anthropogenic sources. Pollutants have accumulated over time in the environment due to the demand for chemicals and substances driven by urbanization, industrialization, and increased food consumption. Polycyclic aromatic hydrocarbons (PAHs), heavy metals, and pesticides are the three types of pollutants that are most frequently found in soil (Mirsal 2008). Pesticides are chemical substances used to eliminate pests. For the following purposes, they are referred to: larvicides, bactericides, fungicides, herbicides, insecticides, and rodenticides (Raffa and Chiampo 2021).

The accumulation of pesticide residue in the soil impacts the microorganisms that inhabit it. Human exposure can happen when pesticides are used directly in the household, business, or agricultural sector or when contaminated food, water, or air are ingested. Pesticides can enter the human body through the mouth, skin, eyes, and respiratory systems (Kim et al. 2017). The toxicity of pesticides depends on their electrical properties, molecular makeup, dosage, and length of exposure (Hamadache et al. 2016; Heard et al. 2017). Pesticides increase a plant's resistance to insects and change its morphology, physiology, biochemistry, and molecular composition, all of which adversely affect production and growth (Varjani et al. 2019).

Heavy metals (HMs) and metalloids are environmental pollutants. They are also pollutants in agricultural soil because HMs can harm crop health and productivity when they are present in the soil at high concentrations (Maksymiec 2007; Shahid et al. 2015). In the absence of leaching or plant uptake, HMs have the potential to accumulate in the soil and persist for an extended period. They can withstand deterioration (Ghori et al. 2019; Ali et al. 2019). The elements Cd, Pb, Cr, As, Hg, Ni, Cu, and Zn are often reported to pollute agricultural soils and have harmful effects on plants at elevated levels (Tóth et al. 2016). Almost all levels of contamination threaten plant health, with Cd, Pb, As, Hg, and Cr being the most hazardous (Tóth et al. 2016; Rai et al. 2019). Soil contamination by HMs is a worldwide issue. The main causes of the current HM-induced environmental contamination appear to be rising industrial, municipal, agricultural, residential, medical, and technical applications, fast urbanization, and different geological and climatic conditions. But the issue is more pervasive in many underdeveloped nations, maybe because people are less aware of how harmful these compounds are to human and agricultural health (Sharma et al. 2008; Kumar et al. 2016; Ahmed et al. 2019). Microorganisms play an important role in heavy metal remediation (Say et al. 2003; Hasan and Srivastava 2009; Ajaz et al. 2010; Wang et al. 2019; Kumar and Chandra 2020; Atuchin et al. 2023).

It is expected to find persistent organic pollutants (POPs) as contaminants originating from dispersed sources (Tremolada et al. 2008). POPs are exceedingly dangerous materials that are resistant to breaking down; some of them can lead to mutations or cancer. These include total petroleum hydrocarbons (TPHs) from the use and exploration of fossil fuels, polychlorinated terphenyls (PCTs), polycyclic aromatic hydrocarbons (PAHs), polychlorinated biphenyls (PCBs), and halogenated chemicals (such as trichloroethylene and perchloroethylene, chlorpyrifos) (Lu et al. 2019; Ali et al. 2020; Ejaz et al. 2021; Kumar et al. 2022). POPs are primarily deposited into the soil through atmospheric dry and wet deposition, and they can travel great distances in the atmosphere in both gaseous and particle forms (Cousins et al. 1999). Compounds may be deposited indirectly through vegetation (Wania and Mclachlan 2001). As waste falls to the ground and decomposes, chemicals absorbed by plants may find their way into the soil. POPs might then gather for years in horizons rich in organic matter (OM) (Masih and Taneja 2006).

Bioremediation and remediation technologies, which use naturally occurring or genetically modified microbes, are continuously improved to eliminate environmental pollution (Mahjoubi et al. 2017). Plant growth-promoting rhizobacteria (PGPR) are increasingly being used in bioremediation, according to a recent study, due to their significant effects on plant growth promotion and various capacities to detoxify and break down toxins (Ali et al. 2020; Haider et al. 2021). Soil bacteria of the *Entero cloacae* species can break down trinitrotoluene (TNT) into relatively less hazardous byproducts (Zhu et al. 2012). Deuteromycetes and white-rot fungi can degrade PAHs having more than one ring (Liu et al. 2017). Several scientists proposed using the association of microorganisms, i.e., phosphate-solubilizing bacteria found in phytoremediation, to remediate contaminated soils effectively (Jia et al. 2016).

An overview of some research on the microorganisms used in bioremediation is shown in Table 1.

3.4 Mechanisms of Microbial Bioremediation

3.4.1 Bioaugmentation

In this process, a group of indigenous microbes, or microbes created through genetic engineering, are added to the contaminated site. Bioaugmentation finds extensive application in soil bioremediation (Sharma 2012). Adding microbial biomass can significantly improve the biodegradation performance of contaminated areas. It might be argued that no bioaugmentation process works for all environmental circumstances, pollutants, and technological concepts. However, additional investigation of bioaugmentation outcomes in the laboratory and practice will strengthen the theoretical basis for better prediction of bioremediation processes in certain scenarios (Muter 2023). The majority of accessible cultures are a consortium of microbial cultures based on research, comprising all essential microorganisms. It has occasionally been observed that non-native microorganisms compete with native ones for population expansion; as a result, appropriate land treatment unit management is necessary (Sharma 2012).

Effective soil augmentation requires the selection of the appropriate microbial strains or consortia. Microorganisms have quick growth, are easily cultured, can endure high contamination concentrations, and thrive in many environments. There are numerous techniques for selecting a suitable bacterial strain (Niu et al. 2009). One method involves isolating a bacterial strain from contaminated soil, growing it in a laboratory setting to aid in preadaptation, and then reintroducing the strain to the contaminated soil. The dosage ratio and administration period have a major impact on the functional flora of the bioaugmentation system and its ability to coordinate and adapt (Saeed et al. 2022). Bioaugmentation in the field in under-regulated settings is challenging due to the system's overall biodiversity, competition between native microorganisms and microbial agents, ambient variables, substrate competition, remediation cycles, and other considerations. To select the optimal bioaugmentation method, more investigation is needed into the relationships between different functional bacteria and their resistance to different stressors, enzyme activity, and system robustness (Chen et al. 2022). Comprehensive research, a critical assessment of the databases that are now in place, and the application of systems biology, genetic engineering, and nanotechnology can all help to promote bioremediation (Bokade et al. 2023).

Table 1 Microbes used in soil bioremediation

Microorganisms	Species	Contaminants	References
Bacteria			
	Arthrobacter sp., *Bacillus* sp.	Copper	Hasan and Srivastava (2009) and Tunali et al. (2005)
	Alcaligenes eutrophus	Cadmium, lead	Diels et al. (1999)
	B. subtilis, *B. thuringiensis*	Copper, cadmium, zinc, nickel	Ajaz et al. (2010)
	Klebsiella oxytoca, *Rhizobium radiobacter*	Lead	Atuchin et al. (2023)
	Achromobacter denitrificans, *Klebsiella oxytoca*, *Rhizobium radiobacter*	Nickel	Atuchin et al. (2023)
	Bacillus coagulans, *citrobacter koseri*, *serratia ficaria*	Diesel	Kehinde and Isaac (2016)
	Bacillus subtilis	Petroleum products	Montagnolli et al. (2015)
	Azospirillum lipoferum L. and *Azospirillum brasilense* L.	Crude oil	Muratova et al. (2005)
	Actinobacteria sp., *Caulobacterales* sp., *Rhizobiales* sp., *Rhodococcus* sp., *Xanthomonadales* sp.	Phenanthrene	Thomas and Cébron (2016)
	Archaea sp., *Acinetobacter* sp., *Achromobacter* sp., *Alcaligenes* sp., *Arthrobacter* sp., *Azotobacter* sp., *Bacillus cereus* sp., *Flavobacterium* sp., *Neurospora* sp., *Leptolyngbya* sp.	Phenolic compounds	Baraldi et al. (2008)
	Pseudomonas kilonensis MB490	Dimethoate	Yasmin et al. (2022)
	Pseudomonas aeruginosa L.	Dichlorodiphenyltrichloroethane (DDT)	Verma et al. (2014)
	B. thuringiensis MOS-5	Malathion	Kamal et al. 2008
	Acinetobacter baumannii and *Pseudomonas putida*	Malathion Chlorpyrifos	Ewida et al. (2016)
	Bacillus thuringiensis	Pyrethroid Cypermethrin	Birolli et al. (2021)
	Burkholderia sp.	Aromatic compounds	Morya et al. (2020)
	Entero cloacae	TNT	Zhu et al. (2012)
Fungi			
	Aspergillus niger, *Rhizopus arrhizus*	Copper	Mukhopadhyay et al. (2007) Subudhi and Kar (2008)

Microorganisms	Species	Contaminants	References
	Penicillium canescens	Arsenic	Say et al. (2003)
	Mucor rouxii	Lead	Yan and Viraraghavan (2003)
	Funalia trogii	Mercury, cadmium, zinc	Yakup et al. (2004)
	Aspergillus niger	Nickel	Ahmad et al. (2006)
	Arbuscular mycorrhizal (AM)	Cadmium, Lead, copper	Gonzalez-Chavez et al. (2004)
	Debaryomyces subglobosus L.	Phenolic compounds	Akardere et al. (2011)
	Aspergillus fumigatus, *Aspergillus niger*, *Penicillium chrysogenum* and *Aspergillus flavus*	Malathion Chlorpyrifos	Ewida et al. (2016)
	Phanerochaete chrysosporium	2, 4, 2, 4-tetrabrominated biphenyl ether	Liu and Wu (2018)
	Aspergillus sp., *Cladosporium* sp., *Fusarium* sp., *Monicillium* sp., *Penicillium* sp., and *Phanerochaete* sp.	Phenol, o-cresol, m-cresol, and p-cresol	Atagana (2004)
	Penicillium sp.	Petrol	Vanishree et al. (2014)
	Coriolus versicolor	Chloronitrofen	Hiratsuka et al. (2001)
	Phlebia radiata	2,4,6-trinitrotoluene (TNT)	Van Aken et al. (1999)
	Phanerochaete chrysosporium	Polycyclic aromatic hydrocarbons (PAHs)	Liu et al. (2017)
Algae			

(continued)

Table 1 (continued)

Microorganisms	Species	Contaminants	References
	Microcystis aeruginosa, *Stichococcus bacillaris*	Arsenic	Wang et al. (2019), Pawlik-Skowrońska et al. (2004)
	Chaetoceros calcitrans, *Scenedesmus abundans*	Cadmium	Kumar et al. (2015)
	Spirogyra sp.	Mercury	Rezaee et al. (2006)
	Chlorella vulgaris, *Coelastrella* sp.	Copper	Akhtar et al. (2003), Wu et al. (2022)
	Spirulina sp.	Mercury, cadmium, lead	Khan et al. (2022)
	Ankistrodesmus braunii and *Sc. Quadricauda*	Phenols	Pinto et al. (2003)
	Sc. Obliquus-bacterial consortium	Oil wastes	Verasoundarapandian et al. (2022)
	Scenedesmus obliquus and *Chlorella vulgaris*	Crude oil	El-Sheekh et al. (2013)
	Micractinium pusillum, *Chlorella singularis*	Malathion	Nanda et al. (2019)
	Chlamydomonas reinhardtii	Isoproturon	Lynn et al. (2007)
	Chlorella vulgaris	Diazinon	Kurade et al. (2016)
	Spirulina platensis	Chlorpyrifos	Thengodkar and Sivakami (2010)
	Scytonema hofmanni	Methyl parathion	Tiwari et al. (2017a, b)

Bioaugmentation has primarily been used in soils with fewer pollutant-degrading microorganisms and contaminants that require multi-process treatment (Mrozik and Piotrowska-Seget 2010). Aside from these, a few more factors influence the rate of bioaugmentation in soil. The pH, temperature, moisture, organic matter concentration, aeration, nutrient content, and soil type are some of the parameters or elements that have been seen to influence the bioaugmentation process (Hong et al. 2007). The lack of any of these soil properties in natural conditions renders the remediation procedure inefficient (Jimenez et al. 2006).

3.4.2 Biostimulation

Biostimulation is a remediation technology that is very efficient, cost-effective, and environmentally friendly. Biostimulation is the technique of modifying the environment to stimulate preexisting microorganisms capable of bioremediation (Dehnavi and Ebrahimipour 2022). Biostimulation is the supply of rate-limiting nutrients such as phosphorus, nitrogen, oxygen, and electron donors to heavily contaminated areas to stimulate the existing bacteria to digest hazardous and toxic pollutants. (Rhykerd et al. 1999; Tyagi et al. 2010).

The addition of rate-limiting nutrients effectively boosts the inhabitant bacteria's degrading capacity while dramatically increasing the decontamination rate (Adams et al. 2015). Biostimulation is regarded as the most efficient method of hydrocarbon remediation (Kanissery and Sims 2011; Delille et al. 2004).

This is primarily due to the high carbon (C) content available from petroleum contamination, a rate-limiting nutrient required by the native bacteria for metabolic activities. As a result, except for carbon, adding a modest amount of other rate-limiting nutrients to the soil accelerates the breakdown of petroleum. In addition to the aforementioned rate-limiting nutrients, the utilization of nutrient-rich organic materials can greatly speed up the restoration process (Goswami et al. 2018).

Several environmental conditions, such as pH, moisture content, temperature, and so on, influence the bioremediation of a polluted site via biostimulation (Abdulsalam and Omale 2009). In addition to these considerations, the current ambient physiology influences the biostimulation rate (Goswami et al. 2018).

3.4.3 Bioleaching

Bioleaching is an effective technology, also known as microbial leaching or bio-hydrometallurgy, which employs naturally occurring bacteria such as mesophiles, moderately thermophilic bacteria, and extremophiles to dissolve metals from refractory ores (Mishra et al. 2005). This method transforms metal sulfide in ore/minerals into metal cations in the leaching media (Rohwerder et al. 2003). Bioleaching, or bio-oxidation, is the process of solubilizing nonrelevant metal sulfides from precious metal ore. For example, iron and arsenic sulfides are solubilized in arsenopyrite bioleaching, and the residual gold is recovered via cyanidation

(Rohwerder et al. 2003). This bioleaching procedure was employed 3000 years ago to recover copper from mine waters in the Mediterranean. Since then, it has been used seldom. Approximately 40 factories have recently used this process to extract copper, gold, zinc, cobalt, and uranium. The organisms used in this method are single-celled and have chemosynthetic metabolism. They derive carbon dioxide and oxygen from the atmosphere and require an acidic pH (Rendón-Castrillón et al. 2023).

Acidophiles, or bacteria and archaea, produce sulfuric acid and ferric ions during bioleaching, which solubilizes the metals from the ore or mineral by thiosulphate or polysulfide routes. Metals are soluble in heterotrophic fungi by a variety of processes, including (i) complexolysis and acidolysis via organic acids, (ii) redoxolysis, and (iii) bioaccumulation (Dusengemungu et al. 2021).

Acidithiobacillus ferrooxidans (A. ferrooxidans) are the most extensively researched microorganisms involved in sulfide ore bioleaching (Johnson 2014). This chemolithotrophic bacterium produces electrons and energy by oxidizing reduced forms of iron (Fe^{2+}) and sulfur, with CO_2 serving as a carbon source (Watling 2006). Bioleaching is a method that solubilizes metals from ore or minerals via thiosulphate or polysulfide routes. Acidophiles, also known as bacteria and archaea, create sulfuric acid and ferric ions during this process. Among the many ways metals can become soluble in heterotrophic fungi are complexolysis and acidolysis through organic acids, redoxolysis, and bioaccumulation. Many hazardous wastes, such as electronic trash, spent catalysts, sewage sludge, slag, and so forth, are being studied for their potential metal components, and the bioleaching process is one such waste method. One can extract metal in batch, continuous, one-step, two-step, and wasteful medium-step modes from bioleaching. Depending on the ore preprocessing activities, biohydrometallurgy processes have varying degrees of complexity (Mahmoud et al. 2017). There are several, varyingly advanced biohydrometallurgy techniques, depending on the level of ore preprocessing. Stirred reactor bioleaching (crushed ore/concentrate), heap (crushed/agglomerated ores), "in situ" (underground leaching), and dump (no ore size reduction) are some examples (Natarajan 2018). Heap bioleaching is the most frequently used commercial technology for treating ores with nickel, zinc, cobalt, and copper (Natarajan 2018). Heap bioleaching can also be used to detoxicate tailings containing notable levels of heavy metals, such as arsenic (Borja et al. 2015). To remove metal components from a broader range of hazardous wastes, such as electronic waste, spent catalysts, sewage sludge, slag, etc., the bioleaching technique is currently the subject of substantial research. One-step, two-step, and wasted medium-step batch and continuous modes are available for the bioleaching process to extract metals. By using different pretreatment techniques (such as acetone washing, decoking, etc.), adding catalysts, adapting the microorganisms beforehand, and optimizing the process parameters (such as pH, pulp density, temperature, etc.), metal solubilization efficiency could be further increased (Rawlings 2005; Watling 2006; Nascimento et al. 2019).

3.4.4 Bioventing

The process of "bioventing" involves stimulating indigenous microorganisms through the addition of a gas, this promotes the organic pollutants' breakdown that has been absorbed into the soil. By creating specially designed wells in contaminated soils, nutrients and oxygen or air flow are incorporated into the soil's unsaturated zone, increasing microbial activity (Raffa and Chiampo 2021). Because of the light ventilation, only the minimal amount of oxygen crucial to encourage microbial activity and avoid pollutants volatilization may be provided. Regarding aeration, bioventing can be implemented in either an active or passive manner.

The bioventing remediation process might take anywhere from six months to five years, contingent upon the kind and level of contamination, rates of biodegradation, and soil properties including permeability and water content (Raffa and Chiampo 2021). The following factors may limit the process's applicability and effectiveness: (1) low permeability soils, which reduce the effectiveness of bioventing; (2) the air nearest to the structure of concern must be extracted to prevent vapor buildup in basements within the air injection wells' influence radius; (3) soil surface off-gas monitoring might be necessary; (4)) Low soil moisture content, which may be brought on by bioventing, restricts biodegradation; and (5) It might not be efficient for many chlorinated compounds to degrade aerobically (Hinchee 1993).

3.4.5 Biosorption

The biodegradation process occurs in the biosparging strategy by infusing air into groundwater to increase the oxygen concentration and promote the naturally occurring bacteria (Gadd 2009). Except for introducing air into the saturated zone during biosparging, the procedure is similar to that of bioventing. Unlike bioventing, air is injected into the saturated region, which causes volatile organic molecules to ascend into the unsaturated zone and facilitate biodegradation (Michalak et al. 2013). The process's effectiveness depends on two major factors: pollutant biodegradability and soil permeability, which control a pollutant's bioavailability to microorganisms (Raffa and Chiampo 2021). Biosorption can occur in a wide range of pH (3–9) and temperature (4–90 °C). The optimal biosorbent particle size is between 1 and 2 mm, thus equilibrium states for both adsorption and desorption are attained rather quickly. This method does not require a large upfront investment, thus its running costs are low. Furthermore, biological materials that are frequently affordable can be produced from industrial or agricultural waste (Kuyucak 1990).

3.4.6 Bioaccumulation

This process can occur during the biogeochemical cycling to produce microfossils, mineralized manganese and silver, and the deposition of metals (iron and manganese) due to enzyme activity and secondary metabolite galactosis (Sharma et al.

2022). Ion pumps, protein channels, and carrier-mediated transport are just a few of the methods that bacteria employ to move heavy metals across their membranes (Mir-Tutusaus et al. 2018). Many reports exist of very active organisms that take up heavy metals through bioaccumulation. As per Sharma et al. (2022), *Pseudomonas putida* of bioremediating cadmium, *Rhizopus arrhizus* is capable of bioremediating mercury and *Aspergillus niger* of bioremediating thorium. Microorganisms that can bioaccumulate and sequester metal ions within their intracellular spaces may be further employed for cellular functions such as catalysis, enzyme signaling, stabilizing charges on proteins, and so on. Soil pollution management has new opportunities thanks to applying microbiological techniques for heavy metal cleanup and treatment.

Monitoring chemical concentrations in biota can be used to identify chemical sources, estimate the environmental risk associated with events, and verify compliance with regulatory mandates (ECHA 2012; Schäfer et al. 2015). Chemical concentrations in biota can be monitored to determine chemical sources, quantify environmental risk related to particular events, and confirm that regulations are being followed. Field bioaccumulation evaluation is challenging because so many factors need to be considered to determine the amount of a chemical that accumulates in an organism. Passive sampling can be used in combination with biota monitoring since samplers with certain partition properties can be used over a wide temporal and geographical range (Schäfer et al. 2015).

3.4.7 Precipitation

This process can occur during the biogeochemical cycling to create mineralized manganese and silver, microfossils, and the deposition of metals (manganese and iron) due to secondary metabolite galactosis and enzyme activity. Turning contaminants or heavy metals into precipitates or crystals, lowers their toxicity level (Sharma et al. 2022). For instance, sulfate-reducing bacteria can change organophosphate into orthophosphate in an alkaline pH environment greater than seven (Pratush et al. 2018). Heavy metal pollution in soil can be recovered, but the techniques to be used will depend on the specific metals present, their speciation, and the level of contamination (Maity et al. 2019; Wang et al. 2019).

According to Wang et al. (2023), biomineralization is the process by which organisms produce minerals. Microorganisms can mediate the creation of heavy metal precipitation through nonliving organic substrates, hence mediating the occurrence of mineralization and helping to achieve the goal of heavy metals remediation (Dhami et al. 2017). Accumulation of silicate in algae and diatoms (Gorlich, et al. 2019), carbonate in invertebrates (Aizenberg et al. 2002), calcium phosphate, and carbonate in vertebrates' hard tissues (Vuola et al. 1996; Zhu and Dittrich 2016). Microbial mineralization is the category that currently includes the majority of biomineralization applications in engineering (Jiang et al. 2020). Utilizing urea hydrolysis as one type of microbial mineralization to remediate heavy metals with microbial-induced carbonate precipitation (MICP) has garnered significant attention (Wang et al. 2023).

3.4.8 Siderophores and Heavy Metal Remediation

Plant growth-promoting rhizobacteria produce siderophores. Siderophores have an impact on iron phosphate solubility in soil (Rajkumar et al. 2010). In soil contaminated with trace metals, plant development can be enhanced by bacterial activity that generates siderophores, which are defined as low-molecular-weight proteins (Gaonkar and Bhosle 2013). Bacteria, plants, and fungi in low-iron environments develop secondary metabolites that chelate iron (Chandra et al. 2015).

Plant development can be stimulated directly by siderophore-producing PGPR through the improvement of plant iron nutrition, or indirectly through the reduction of plant pathogen activity in the soil root zone by limiting iron supply. (Chandra et al. 2015; Marques et al. 2013). Remarkably, phytosiderophores need less iron for optimum growth than bacterial siderophores, despite having a lower affinity for iron than the latter (Ma et al. 2011).

Phytoextraction is increased by enhancing plant nutrition and mobilizing metals (Ma et al. 2015, Kumar et al. 2021). Siderophores are produced by several microorganisms including *Penicillium aculeatum*, *Agromyces* AR33; *Enterobacter* sp., and *Klebsiella* spp. (Demjanová et al. 2021; Babu et al. 2014; Belimov et al. 2005). *Pseudomonas aeruginosa* produces pseudobactin and pyoverdine, which are capable of chelating all of 16 different metals including Ag^{+}, Al^{3+}, Cd^{2+}, Co^{2+}, Cr^{2+}, Cu^{2+}, Eu^{3+}, Ga^{3+}, Hg^{2+}, Mn^{2+}, Ni^{2+}, Pb^{2+}, $Sn2^{+}$, Tb^{3+}, Tl^{+}, and Zn^{2+} (Braud et al. 2009a, b); mixed siderophores, such as azotobactin and yersiniabactin, are produced by Yersinia pestis and *Azotobacter vinelandii*, respectively (Knosp et al. 1984; Haag et al. 1993). Siderophore-producing bacteria can improve crop development, chlorophyll content, and biomass in trace metal-contaminated soils by selectively increasing iron uptake from competing cations (Rajkumar et al. 2010). Moreover, rhizobacterial siderophores' complexation of trace metals in the rhizosphere could prevent oxidative stress and free radicals from developing (Chouychai et al. 2009).

3.4.9 Biosurfactants in Organic and Inorganic Remediation

Biosurfactants, amphiphilic molecules with large hydrophobic and hydrophilic groups, play a crucial role in removing pollutants from the rhizosphere (Kumar and Chandra 2020). A variety of microorganisms, including *Sphingomonas* sp., *Pseudomonas* sp., *Candida tropicalis, Achromobacter* sp., *Brevibacillus* sp., *Pusillimonas* sp., *Dietzia* sp., and *Sphingopyxis* sp. (Tambekar and Gadakh 2013; Zeng et al. 2011; Cui et al. 2008), produce biosurfactants as secondary metabolites that are released extracellularly or cling to cell surfaces. Biosurfactants can be used in place of chemical surfactants because they foam better, are less dangerous, eco-friendly, and have lower critical micelle concentration (CMC) values. They are advantageous in many applications, including bioremediation, healthcare, cosmetics, food, and oil, because of their potential benefits (Jahan et al. 2020). Biosurfactants, microbial surface-active metabolites, are metal-complexing agents that can help remediate heavy metal-contaminated environments (Das et al. 2016; Kramer et al.

2020). *Mucor hiemalis* UCP 0039; *Penicillium chrysogenum*, *Fusarium*, *Penicillium* and *Trichoderma* SNP5, *Achromobacter* sp., *Brevibacillus* sp., *Pusillimonas* sp., *Dietzia* sp., and *Sphingopyxis* sp. have all been identified as emulsifiers (Ferreira et al. 2020; Gautam et al. 2014; Castillo-Mendez et al. 2017; Tambekar and Gadakh 2013). Recent studies have shown that in metal-contaminated soils, Cd, Zn, and Pb have higher affinities for biosurfactants like rhamnolipid produced by various *Pseudomonas aeruginosa* (Juwarkar et al. 2007; Cui et al. 2008; Chakraborty and Das 2014).

Much emphasis has been paid to the complex formation and subsequent mobilization process used in the biosurfactant-assisted removal of heavy metal ions (Lawniczak et al. 2013; Tambekar and Gadakh 2013). Since biosurfactants are tiny, biodegradable, low toxicity, economical, and highly specific, they are widely used in organic and trace metal-contaminated soil (Sharma et al. 2021).

3.4.10 Enzymatic Bioremediation

Enzymes are biological catalysts, that provide favorable conditions for reducing a reaction's activation energy, hence enhancing substrate conversion into products. Enzymes are glycoproteins or proteins. They include at least one polypeptide moiety. Microbial oxidoreductases, such as monooxygenases, dioxygenases, peroxidases, and laccases, play a key role in the bioremediation of diverse xenobiotics. Microbes obtain energy through energy-yielding biochemical reactions catalyzed by enzymes that cleave chemical bonds and facilitate the transfer of electrons from a reduced organic substrate (donor) to another chemical component (acceptor). Oxidation and reduction reactions turn contaminants into harmless molecules (Chandrakant and Shwetha 2011).

Through their oxi-reduction capacity, microbial oxidoreductases can reduce heavy metals, reverse xenobiotic toxicity, and destroy both natural and manmade pollutants. (Okino-Delgado et al. 2019). Monooxygenases and dioxygenases can play important roles in the breakdown and detoxification of aromatic substances by hydroxylation and ring cleavage (Xu et al. 2021; Mohamed et al. 2023). PAHs and PCBs are two main aromatic contaminants. The aerobic metabolism of polycyclic aromatic hydrocarbons (PAHs), and polychlorinated biphenyls (PCBs) begins with the oxidation of the benzene ring by dioxygenase enzymes, which introduces two hydroxyl groups into aromatic molecules (Xu et al. 2021). Consider naphthalene as an example; there are a variety of bacteria that can oxidize naphthalene using naphthalene dioxygenase including those from the genera *Pseudomonas*, *Rhodococcus*, and *Mycobacterium* (Yang et al. 2014; Kumari et al. 2021). Peroxidases are used in bioremediation methods due to their thermostability and ability to oxidize a wide range of pollutants (Fetyan et al. 2016; Bhattacharya and Gupta 2022). Using oxidative processes, decarboxylation, and demethylation, laccases have the ability to act on a wide range of pollutants, such as petroleum derivatives (PHAs), phenols, polyphenols, dyes, and plastics (Martínková et al. 2016; Santo et al. 2013; Fetyan et al. 2017; Copete-Pertuz et al. 2018). Chrome reductase converts hexavalent chromium to trivalent chromium, while azoreductase breaks down azo compounds into azo bonds (Saxena et al. 2020).

Hydrolytic enzymes also break down critical chemical bonds in organic contaminants, which reduces their toxicity. This technique is usually effective in breaking down oil spills, organophosphate, and carbamate pesticides (Kamal et al. 2008; Bhatt et al. 2021).

Amylases, proteases, lipases, DNases, pullulanases, and xylanases hydrolytic enzymes have enormous applications in soil remediation (Mesbah 2022). Hemicellulase, cellulase, and glycosidase are key enzymes aiding biomass decomposition (Chandrakant and Shwetha 2011; Abdel-Sater et al. 2019). A type of soil bacterium called *Entero cloacae* produces enzymes called nitroreductase and pentaerythritol tetranitrate reductase (PETN reductase) which are able to break down TNT into much less dangerous byproducts.

Challenges in Soil Bioremediation

The following are the primary challenges to soil bioremediation which have to be avoided:

1. The pollutants' accessibility or bioavailability: Pollutants' adsorption, toxicity, and utilization affect the soil bacteria's capacity to absorb them. The soil contaminated with organic compounds contains a wide variety of extremely volatile chemical substances, such as pesticides and polycyclic aromatic hydrocarbons (PAHs). As a result, reduced bioavailability, rather than low density of PAH-degrading bacteria, has a stronger influence on pesticide and PAH-contaminated soil bioremediation. Overtime, pollutants' bioavailability varies. This phenomenon is referred to as "aging." One of the primary problems in remediating soil pollution is the aging of contaminants. Determining how to increase the bioavailability of pollutants in soil.
2. Reversing the stability of pollutants: Composting the organic matter that remains after waste treatment is a practical and long-term solution to the issue of an accumulation of organic products. On the other hand, organic matter that has become mineralized in the environment changes the pH, which makes the pollutant less stable. Microbes' adaptation:
3. Exogenous microorganisms that do not comply well with contaminated soils find adapting to a new environment more difficult. As a result, bioaugmentation loses its ability to accelerate the breakdown of contaminants. Appropriate techniques could be required to investigate and resolve this issue.
4. Metabolic processes: Little is known about the microbial metabolic pathways involved in the bioremediation of organic contaminants and heavy metals. Consequently, in order to alter the genetic composition of external microbes, metabolic pathways and the dynamics of microbial communities must be thoroughly studied. High-throughput sequencing and other molecular biology techniques should be used to understand the genetic structure of natural bacteria better. Synthetic biology techniques and technology must overcome the aforementioned limits to effectively clean up environmental toxins (Banerjee et al. 2016; Bordbar et al. 2014; Rawls et al. 2019).

5. Enzymatic studies: Because the basic catalytic action of distinct enzymes is responsible for the breakdown of pollutants, which is essentially concealed, several enzymatic properties, such as kinetics, molecular structure, activity, and inhibitory mechanism, require significant research.
6. Pollutant interactions: It is still unknown how contaminants from the same soil/compost mixture would interact with one another in a synergistic or antagonistic way to cause deterioration. To overcome this obstacle, more research is necessary.
7. Quality of end product: It is essential to make sure that the final product discharged during bioremediation does not include any harmful organic compounds or metals beyond a particular level.

4 Conclusion

Recently, crop yield has been seriously threatened by various biotic and abiotic stressors. Increasing agricultural production on limited land and feeding the growing population are the two main concerns facing scientists and policymakers. Moreover, the ecosystem is seriously endangered by contemporary farming practices that degrade soil, such as the overuse of pesticides, chemicals, fertilizers, and irrigation with untreated wastewater. Furthermore, urbanization and industrialization have resulted in a significant decline in agricultural land, prompting scientists to develop sustainable ways to increase food yields from the already decreasing planted area. The application of biocontrol and bioremediation techniques using microbial inoculants, biofertilizers, and bio(chemicals) has long been recognized as a means of improving soil biology, fertility, agricultural productivity, and soil remediation. Biological control employs several strategies, both direct and indirect, to enhance plant development in a range of agricultural environments. Microbes that inhabit the rhizosphere hire a range of strategies, including competition, hyperparasitism, the synthesis of volatile secondary metabolites, antibiosis, colonization of roots, and enhanced nutrition availability (Samaras et al. 2018; Gebily 2021; Ghanem et al. 2022). These microbes indirectly support plant growth by fostering plant resilience to a range of biotic and abiotic stressors, including pathogen invasions and heavy metal contamination. Furthermore, bioaccumulation, biosorption, enzyme activity, biosurfactants, and siderophores are among the methods employed by rhizobacteria for the bioremediation of contaminated soils (Schäfer et al. 2015; Goswami et al. 2018; Kumari et al. 2021; Jahan et al. 2020; Kumar et al. 2021). The effectiveness and sustainability of integrated control strategies will likely be improved shortly by the synergistic combination of microbial control agents with other technologies (pesticides, bioaccumulation, biosorption, enzymatic bioremediation, biosurfactants, other natural enemies, and resistant plants).

References

Abd El-Moity TH (1976) Studies on the biological control of white rot disease of onion. M.Sc. thesis, Fac Agric, Minufiya Univ, p 121

Abdel-Sater M, Hussein N, Fetyan NF, Gad S (2019) Immobilization of cellulases produced by *Penicillium brevicompactum* AUMC 10987, using cross-linkage, chitosan-coating and encapsulation. Catrina Inter J of Enviro Sci 18(1):139–149

Abdulsalam S, Omale AB (2009) Comparison of biostimulation and bioaugmentation techniques for the remediation of used motor oil contaminated soil. Braz Arch Biol Technol 52(3):747–754

Abou-Zeid NM, Gado EAM, Mahmoud NA, Mosa AA, Ahmed AY (2011) Trichoderma species in Egypt and their biocontrol potential against some plant pathogenic fungi. Egypt J Biol Pest Cont 21(2):233–244

Adams GO, Fufeyin PT, Okoro SE (2015) Bioremediation, biostimulation and bioaugmentation: a review. Int J Environ Bioremed Biodegrad 3(1):28–39

Ahmad I, Ansari MI, Aqil F (2006) Biosorption of Ni, Cr and Cd by metal tolerant *Aspergillus niger* and *Penicillium* sp. using single and multi-metal solution. Indian J Exp Biol 44:73–76

Ahmed M, Matsumoto M, Ozaki A, Thinh NV, Kurosawa K (2019) Heavy metal contamination of irrigation water, soil, and vegetables and the difference between dry and wet seasons near a multi-industry zone in Bangladesh. Water 11:583

Aizenberg J, Lambert G, Weiner S, Addadi L (2002) Factors involved in the formation of amorphous and crystalline calcium carbonate: a study of an ascidian skeleton. J Am Chem Soc 124:32–39

Ajaz HM, Arasuc RT, Narayananb VKR, Zahir HMI (2010) Bioremediation of heavy metal contaminated soil by the exigobacterium and accumulation of Cd, Ni, Zn and Cu from soil environment. Int J Biol Technol 1(2):94–101

Akardere E, Karakaya A, Laleli Y, Takac S (2011) Variations in phenolic compounds during biodegradation of olive mill wastewater by *Debaryomyces hansenii*. Curr Opin Biotechnol 1:22–34

Akhtar N, Saeed A, Iqbal M (2003) Chlorella sorokiniana immobilized on the biomatrix of vegetable sponge of *Luffa cylindrica*: a new system to remove cadmium from contaminated aqueous medium. Bioresour Technol 88:163–165

Ali AM (2013) Control of strawberry fungal diseases under organic agriculture system. Ph.D. thesis, Depart Plant Pathol Fac of Agric, Cairo Univ, p 123

Ali AM (2021) The competitive potential of different *Trichoderma* spp. to control Rhizoctonia root rot disease of pepper (*Capsicum annuum* L.). Egypt J of Phytopathol 49(1):136–150. https://doi.org/10.21608/EJP.2021.73456.1030

Ali AM, Ayoub FH (2017) Effect of some bio-control agents, natural salts and planting densities on controlling sweet pepper powdery mildew and some horticultural characteristics under greenhouse conditions. Egypt J Phytopathol 45(1):117–134

Ali H, Khan E, Ilahi I (2019) Environmental chemistry and ecotoxicology of hazardous heavy metals: environmental persistence, toxicity, and bioaccumulation. J Chem 6730305

Ali MH, Sattar MT, Khan MI, Naveed M, Rafique M, Alamri S, Siddiqui MH (2020) Enhanced growth of Mungbean and remediation of petroleum hydrocarbons by Enterobacter sp. MN17 and biochar addition in diesel contaminated soil. Appl Sci 10:8548

Alizadeh H, Behboudi K, Ahmadzadeh M, Javan-Nikkhah M, Zamioudis C, Pieterse CMJ, Bakker PAHM (2013) Induced systemic resistance in cucumber and Arabidopsis thaliana by the combination of *Trichoderma harzianum* Tr6 and *Pseudomonas* sp. Ps14. Biol Control 65:14–23

Almeida FBR, Cerqueira FM, Silva RN, Ulhoa C (2007) Mycoparasitism studies of *Trichoderma harzianum* strains against *Rhizoctonia solani*: evaluation of coiling and hydrolytic enzyme production. Biotechnol Lett 29:1189–1193

Anis M, Zaki MJ, Abbasi MW (2013) Effects of seed coating with *Trichoderma* species on colonization of *Macrophomina phaseolina* and the growth of sunflower under field conditions. Int J Biol Biotechnol 10(2):203–208

Anitha R, Murugesan K (2005) Production of gliotoxin on natural substrates by *Trichoderma virens*. J Basic Microbiol 45(1):12–19

Arumugam S, Rajendran V, Subramaniam G (2017) Plant growth-promoting actinobacteria: a new strategy for enhancing sustainable production and protection of grain legumes. Biotech 7(102):1–10

Atagana HI (2004) Biodegradation of phenol, o-cresol, m-cresol and p-cresol by indigenous soil fungi in soil contaminated with creosote. World J Microbiol Biotechnol 20:851–858

Atuchin VV, Asyakina LK, Serazetdinova YR, Frolova AS, Velichkovich NS, Prosekov AY (2023) Microorganisms for bioremediation of soils contaminated with heavy metals. Microorganisms 11(4):864. https://doi.org/10.3390/microorganisms11040864

Babu AG, Shim J, Shea PJ, OHBT (2014) *Penicillium aculeatum* PDR-4 and *Trichoderma* sp. PDR-16 promote phytoremediation of mine tailing soil and bioenergy production with sorghum-Sudan grass. Ecol Eng 69:186–191

Baker KF (1987) Evolving concepts of biological control of plant pathogens. Annu Rev Phytopathol 25:67

Banerjee C, Singh PK, Shukla P (2016) Microalgal bioengineering for sustainable energy development: recent transgenesis and metabolic engineering strategies. Biotechnol J 11:303–314. https://doi.org/10.1002/biot.201500284

Baraldi EA, Damianovic M, Manfio GR, Foresti E, Vazoller RF (2008) Performance of a horizontal-flow anaerobic immobilized biomass (HAIB) reactor and dynamics of the microbial community during degradation of pentachlorophenol (PCP). Anaerobe 14:268–274. [CrossRef]

BBSRC (1999) A joint research council review of bioremediation research in the United Kingdom. BBRRC, EPSRC and NERC, Swindon

Belimov AA, Hontzeas N, Safronova VI, Demchinskaya SV, Piluzza G, Bullitta S, Glick BR (2005) Cadmium-tolerant plant growth promoting rhizobacteria associated with the roots of Indian mustard (Brassica juncea L. Czern.). Soil Biol Biochem 37:241–250

Bhatt P, Zhou X, Huang Y, Zhang W, Chen S (2021) Characterization of the role of esterases in the biodegradation of organophosphate, carbamate, and pyrethroid pesticides. J Hazard Mater 411:125026

Bhattacharya A, Gupta A (2022) Current trends in applicability of thermophiles and thermozymes in bioremediation of environmental pollutants. In: Microbial extremozymes. Academic, pp 161–176

Birolli WG, Dos Santos A, Pilau E, Rodrigues-Filho E (2021) New role for a commercially available bioinsecticide: *Bacillus thuringiensis* Berliner biodegrades the *Pyrethroid Cypermethrin*. Environ Sci Technol 55:4792–4803. https://doi.org/10.1021/acs.est.0c06907

Bokade P, Gaur VK, Tripathi V, Bobate S, Manickam N, Bajaj A (2023) Bacterial remediation of pesticide polluted soils: exploring the feasibility of site restoration. J Hazard Mater 441:129906. https://doi.org/10.1016/j.jhazmat.2022.129906

Bordbar A, Monk JM, King ZA, Palsson BO (2014) Constraint-based models predict metabolic and associated cellular functions. Nat Rev Genet 15:107. https://doi.org/10.1038/nrg3643

Borja D, Lee E, Silva RA, Kim H, Park JH, Kim H (2015) Column bioleaching of arsenic from mine tailings using a mixed acidophilic culture: a technical feasibility assessment. J Korean Inst Resour Recycl 24:69–77

Bouizgarne B (2013) Bacteria for plant growth promotion and disease management. In: Maheshwari DK (ed) Bacteria in agrobiology: disease management. Springer, Berlin, pp 15–47

Boukaew S, Plubrukam A, Prasertsan P (2013) Effect of volatile substances from *Streptomyces philanthi* RM-1-138 on growth of *Rhizoctonia solani* on rice leaf. BioControl 4:471–482

Braud A, Hannauer M, Milsin GLA, Schalk IJ (2009a) The Pseudomonas aeruginosa pyochelin-iron uptake pathway and its metal specificity. J Bacteriol 191:5317–5325

Braud A, Hoegy F, Jezequel K, Lebeau T, Schalk IJ (2009b) New insights into the metal specificity of the Pseudomonas aeruginosa pyoverdine-iron uptake pathway. Environ Microbiol 11:1079–1091

Castillo-Mendez L, Prieto-Ccorrea E, Jimenez-Junca C (2017) Identification of fungi isolated from banana rachis and characterization of their surface activity. Appl Microbiol 64(3):246–251. https://doi.org/10.1111/lam.12712

Chakraborty J, Das S (2014) Biosurfactant-based bioremediation of toxic metals. In: Microbial biodegradation and bioremediation. Elsevier, Amsterdam, pp 167–201

Chandra R, Saxena G, Kumar V (2015) Phytoremediation of environmental pollutants: an eco-sustainable green technology to environmental management. In: Chandra R (ed) Advances in biodegradation and bioremediation of industrial waste. CRC Press, Boca Raton, pp 1–29

Chandrakant SK, Shwetha SR (2011) Role of microbial enzymes in the bioremediation of pollutants: a review. Enzyme Res:1–11. https://doi.org/10.4061/2011/80518

Chen Y, Gao X, Chen Y, Qin H, Huang L, Han Q (2014) Inhibitory efficacy of endophytic *Bacillus subtilis* EDR4 against *Sclerotinia sclerotiorum* on rapeseed. Biol Control 78:67–76

Chen Y, Zhou D, Qi D, Gao Z, Xie J, Luo Y (2018) Growth promotion and disease suppression ability of a Streptomyces sp. CB-75 from banana rhizosphere soil. Front Microbiol 8:2704

Chen Y, Wang S, Geng N, Wu Z, Xiong W, Su H (2022) Artificially constructing mixed bacteria system for bioaugmentation of nitrogen removal from saline wastewater at low temperature. J Environ Manag 324:116351

Chouychai W, Thongkukiatkul A, Upatham S, Lee H, Pokethitiyook P, Kruatrachue M (2009) Plant-enhanced phenanthrene and pyrene biodegradation in acidic soil. J Environ Biol 30:139–144

Cook RJ, Thomasow LS, Weller DM, Fujimoto D, Mazolla M, Bangera G, Kim DS (1995) Molecular mechanisms of defense by rhizobacteria against root disease. Proc Natl Acad Sci USA 92:4197–4201

Copete-Pertuz LS, Plácido J, Serna-Galvis EA, Torres-Palma RA, Mora A (2018) Elimination of isoxazolyl-penicillins antibiotics in waters by the ligninolytic native Colombian strain leptosphaerulina sp. considerations on biodegradation process and antimicrobial activity removal. Sci Total Environ 630:1195–1204. https://doi.org/10.1016/j.scitotenv.2018.02.244

Cordovez VVJ, Carrion DW, Etalo R, Mumm HZ, Wezel GP, Rajmakers JM, Raaijmakers JM (2015) Diversity and functions of volatile organic compounds produced by Streptomyces from a disease-suppressive soil. Front Microbiol 6(1081):1–13

Cousins IT, Beck AJ, Jones KC (1999) A review of the processes involved in the exchange of semi-volatile organic compounds (Svoc) across the air-soil interface. Sci Total Environ 228:5–24

Cui CZ, Zeng C, Wan X, Chen D, Zhang JY, Shen P (2008) Effect of Rhamnolipids in degradation of anthracene by two newly isolated strains, Sphingomonas sp. 12A and Pseudomonas sp. 12B. World J Microb Biot 18:63–66

Das AJ, Lal S, Kumar R, Verma C (2016) Bacterial biosurfactants can be an ecofriendly and advanced technology for remediation of heavy metals and co-contaminated soil. Int J Environ Sci Technol 14:1343–1354

Dehnavi SM, Ebrahimipour G (2022) Comparative remediation rate of biostimulation, bioaugmentation, and phytoremediation in hydrocarbon contaminants. Int J Environ Sci Technol 19:11561–11586. https://doi.org/10.1007/s13762-022-04343-0

Delille D, Coulon F, Pelletier E (2004) Effects of temperature warming during a bioremediation study of natural and nutrient-amended hydrocarbon-contaminated sub-Antarctic soils. Cold Reg Sci Technol 40(1–2):61–70

Demjanová S, Jevinová P, Pipová M, Regecová I (2021) Identification of Penicillium verrucosum, Penicillium commune, and Penicillium crustosum isolated from chicken eggs. PRO 9:53. https://doi.org/10.3390/pr9010053

Dhaka A, Chattopadhyay P (2021) A review on physical remediation techniques for treatment of marine oil spills. J Environ Manag 288:112428. https://doi.org/10.1016/j.jenvman.2021.112428

Dhami NK, Quirin MEC, Mukherjee A (2017) Carbonate biomineralization and heavy metal remediation by calcifying fungi isolated from karstic caves. Ecol Eng 103:106–117

Diels L, De Smet M, Hooyberghs L, Corbisier P (1999) Heavy metals bioremediation of soil. In: Molecular biotechnology. Humana Press

Dusengemungu L, Kasali G, Gwanama C, Mubemba B (2021) Overview of fungal bioleaching of metals. Environ Adv 5:100083

ECHA (2012) Guidance on information requirements and chemical safety assessment. Chapter R.11: PBT assessment. Version 1.1. European Chemicals Agency, Helsinki

Ejaz M, Zhao B, Wang X, Bashir S, Haider FU, Aslam Z, Khan MI, Shabaan M, Naveed M, Mustafa A (2021) Isolation and characterization of oil degrading *Enterobacter* sp. from naturally hydrocarbon contaminated soils and their potential against bioremediation of crude oil. Appl Sci 11:3504

El-Sheekh MM, Hamouda RA, Nizam AA (2013) Biodegradation of crude oil by Scenedesmus obliquus and Chlorella vulgaris growing under heterotrophic conditions. Int Biodeterior Biodegrad 82:67–72. https://doi.org/10.1016/j.ibiod.2012.12.015

Ewida AY, El-Sesy ME, El-Aswar EI, ElSayed EE (2016) Biodegradation of malathion and chlorpyrifos by some microorganisms isolated from agricultural drainage water in Egypt. Tripathi A; Dixit S. Bioremediation of phenolic compounds by higher fungi – a review. Int J Curr Adv Res 4:14–35

Fayad KPAM, Simao-Beaunoir A, Gautheir C, Leclerc H, Mamady C, Beaulieu M, Brzezinski R (2001) Purification and properties of a β-1,6-glucanase from Streptomyces sp. EF-14, an actinomycete antagonistic to Phytophthora spp. Appl Microbiol Biotechnol 57:117–123

Ferreira IN, Rodríguez DM, Campos-Takaki GM, da Silva Andrade RF (2020) Biosurfactant and bioemulsifier as promising molecules produced by *Mucor hiemalis* isolated from Caatinga soil. Electron J Biotechnol 1:47–51

Fetyan NAH, Ahmed ZA, Ismail MI, Shaban SA (2016) Oxidative decolorization of direct blue 71 azo dye by Saccharomyces cerevisiae catalyzed by nano zero-valent iron. Annu Res Rev Biol 11(2):1–12

Fetyan NAH, Azeiz AI, Salem T (2017) Biodegradation of Cibacron Redazo dye and industrial textile effluent by *Pseudomonas aeruginosa* immobilized on chitosan-Fe2O3 composite. J Adv Biol Biotechnol 12:1–15

Fira DI, Dimkić T, Berić J, Lozo S, Stankovic C (2018) Biological control of plant pathogens by Bacillus species. J Biotechnol 285:44–55

Formica JV (1990) Conditions for the production of Agrocin 84 by *Agrobacterium radiobacter* K84. J Indust Microbiol 5:33–44

Gadd GM (2009) Biotechnology. J Chem Technol Biotechnol 84:13–28. https://doi.org/10.1002/jctb.1999

Gaonkar T, Bhosle S (2013) Effect of metals on a siderophore producing isolate and its implications on microbial assisted bioremediation of metal contaminated soils. Chemosphere 93:9

Gautam G, Mishra V, Verma P, Pandey AK, Negi SA (2014) Cost effective strategy for production of bio-surfactant from locally isolated *Penicillium chrysogenum* SNP5 and its applications. J Bioprocess Biotechnol 4(6):1–7. https://doi.org/10.4172/2155-9821.1000177

Gebily DAS (2021) Use of some Streptomyces species as bio-agents to control bean white rot disease caused by *Sclerotinia sclerotiorum*. PhD thesis, Fac Agric, Cairo Univ, 202 pp.

Gebily DAS, Ghanem GAM, Ragab MM, Ali AM, Soliman NEK, Abd El-Moity TH (2021) Characterization and potential antifungal activities of three Streptomyces spp. as biocontrol agents against *Sclerotinia sclerotiorum* (Lib.) de Bary infecting green bean. Egypt J Biol Pest Control 31(33):1–15

Ghanem GAM, Gebily DAS, Ragab MM, Ali AM, Soliman NEK, Abd El-Moity TH (2022) Efficacy of antifungal substances of three *Streptomyces* spp. against different plant pathogenic fungi. Egypt J Biol Pest Control 32(112):1–13

Ghori NH, Ghori T, Hayat MQ, Imadi SR, Gul A, Altay V, Ozturk M (2019) Heavy metal stress and responses in plants. Int J Environ Sci Technol 16:1807–1828

Gomes EDB, Léo-Ruben LD, Ritade C, Miranda M (2018) Actinomycetes bioactive compounds: biological control of fungi and phytopathogenic insect. Afr J of Biotechnol 17(17):552–559

Gonzalez-Chavez MC, Carrillo-Gonzalez R, Wright SF, Nichols KA (2004) The role of glomalin, a protein produced by arbuscular mycorrhizal fungi, in sequestering potentially toxic elements. Environ Pollut 130:317–323

Gorlich S, Pawolski D, Zlotnikov I, Kroger N (2019) Control of biosilica morphology and mechanical performance by the conserved diatom gene Silicanin-1. Commun Biol 2:245

Goswami M, Chakraborty P, Mukherjee K (2018) Bioaugmentation and biostimulation: a potential strategy for environmental remediation. J Microbiol Exp 6(5):223–231. https://doi.org/10.15406/jmen.2018.06.00219

Guo Q, Li Y, Lou Y, Shi M, Jiang Y, Zhou J, Sun Y, Xue Q, Lai H (2019) Bacillus amyloliquefaciens Ba13 induces plant systemic resistance and improves rhizosphere microecology against tomato yellow leaf curl virus disease. Appl Soil Ecol 137:154–166

Gupta R, Khan F, Alqahtani FM (2023) Plant Growth–Promoting Rhizobacteria (PGPR) assisted bioremediation of heavy metal toxicity. Appl Biochem Biotechnol. https://doi.org/10.1007/s12010-023-04545-3

Haag H, Hantke K, Drechsel H, Stojiljkovic I, Jung G, Zahner H (1993) Purification of yersiniabactin: a siderophore and possible virulence factor of Yersinia enterocolitica. J Gen Microbiol 139:2159–2165

Hafez E, Balbaa MM, Kabeil SS, El-Saadani MA, Ahmed SA (2012) Molecular studies on the biocontrol effect of *Trichoderma viride* and *Bacillus subtilis* on *Fusarium oxysporum* and *Rhizoctonia solani* infected tomato plants. World Appl Sci J 19(1):89–99

Haider FU, Liqun C, Coulter JA, Cheema AS, Wu J, Zhang R, Wenjun M, Farooq M (2021) Cadmium toxicity in plants: impacts and remediation strategies. Ecotoxicol Environ Saf 211:111887

Hamadache M, Benkortbi O, Hanini S, Amrane A, Khaouane L, Si Moussa CA (2016) Quantitative structure activity relationship for acute oral toxicity of pesticides on rats: validation, domain of application and prediction. J Hazard Mater 303:28–40. https://doi.org/10.1016/j.jhazmat.2015.09.021

Hasan SH, Srivastava P (2009) Batch and continuous biosorption of Cu (2+) by immobilized biomass of *Arthrobacter* sp. J Environ Manage [Internet] 90(11):3313–3321. https://doi.org/10.1016/jjenvman

Heard MS, Baas J, Dorne JL, Lahive E, Robinson AG, Rortais A, Spurgeon DJ, Svendsen C, Hesketh H (2017) Comparative toxicity of pesticides and environmental contaminants in bees: are honey bees a useful proxy for wild bee species? Sci Total Environ 578:357–365. https://doi.org/10.1016/j.scitotenv.2016.10.180

Heydari A, Pessarakli M (2010) A review on biological control of fungal plant pathogens using microbial antagonists. J Biol Sci 10(4):273–290

Hinchee RE (1993) Bioventing of petroleum hydrocarbons, in handbook of bioremediation. CRC Press, Boca Raton

Hiratsuka N, Wariishi H, Tanaka H (2001) Degradation of diphenyl ether herbicides by the lignin-degrading basidiomycete Coriolus versicolor. Appl Microbiol Biotechnol 57:563–571

Hong Q, Zhang Z, Hong Y (2007) A microcosm study on bioremediation of fenitrothion-contaminated soil using *Burkholderia* sp. FDS-1. Int Biodeterior Biodegrad 59(1):55–61

Howell CR (2003) Mechanisms employed by *Trichoderma* species in the biological control of plant diseases: the history and evolution of current concepts. Plant Dis 87:4–10

Huang TY, Zhang Z, Zhuang YW (2023) *Bacillus velezensis* BV01 has broad spectrum biocontrol potential and the ability to promote plant growth. Microorganisms 11(11):2627. https://doi.org/10.3390/microorganisms11112627

Jahan R, Bodratti AM, Tsianou M, Alexandridis P (2020) Biosurfactants, natural alternatives to synthetic surfactants: physicochemical properties and applications. Adv Colloid Interf Sci 275:102061. https://doi.org/10.1016/j.cis.2019.102061

Jia X, Liu C, Song H, Ding M, Du J, Ma Q, Yuan Y (2016) Design, analysis and application of synthetic microbial consortia. Synth Syst Biotechnol 1(2):109–117. https://doi.org/10.1016/j.synbio.2016.02.001

Jiang L, Liu X, Yin H, Liang Y, Liu H, Miao B, Peng Q, Meng D, Wang S, Yang J (2020) The utilization of biomineralization technique based on microbial induced phosphate precipitation in remediation of potentially toxic ions contaminated soil: a mini review. Ecotoxicol Environ Saf 191:110009

Jimenez N, Vinas M, Sabate J (2006) The prestige oil spill. 2. Enhanced biodegradation of a heavy fuel oil under field conditions by the use of an oleophilic fertilizer. Environ Sci Technol 40(8):2578–2585

Johnson DB (2014) Biomining – biotechnologies for extracting and recovering metals from ores and waste materials. Curr Opin Biotechnol 30:24–31

Jovanović T, Petrović M, Kostić M, Bojić D, Bojić A (2021) Chemical remediation technologies. Facta Universitatis Ser Phys Chem Technol 19(1):1–15. https://doi.org/10.2298/FUPCT2101001J

Juwarkar AA, Nair A, Dubey KV, Singh SK, Devotta S (2007) Biosurfactant technology for remediation of cadmium and lead contaminated soils. Chemosphere 68:1996–2002

Kamal MZ, Fetyan NAH, Ibrahim MA, El-Nagdy S (2008) Biodegradation and detoxification of malathion by of *Bacillus thuringiensis* MOS-5. Aust J Basic Appl Sci 2(3):724–732

Kamel SM, Farag FM, Arafa RA, Essa TA (2020) Bio-control potentials of Trichoderma spp. against *Sclerotium rolfsii* the causative of root and crown rot in tomato, common bean and cabbage. Egypt J Phytopathol 48:122–136

Kanissery RG, Sims GK (2011) Biostimulation for the enhanced degradation of herbicides in soil. Appl Environ Soil Sci 2011:Article ID 843450. https://doi.org/10.1155/2011/843450. http://www.hindawi.com/journals/aess/2011/843450/

Kehinde FO, Isaac SA (2016) Effectiveness of augmented consortia of *Bacillus coagulans, citrobacter koseri* and *serratia ficaria* in the degradation of diesel polluted soil supplemented with pig dung. Afr J Microbiol Res 10:1637–1644. https://doi.org/10.5897/AJMR2016.8249

Khan FI, Husain T, Hejazi R (2004) An overview and analysis of site remediation technologies. J Environ Manag 71:95–122. https://doi.org/10.1016/j.jenvman

Khan AA, Gul J, Naqvi SR, Ali I, Farooq W, Liaqat R, AlMohamadi H, Štěpanec L, Juchelková D (2022) Recent progress in microalgae-derived biochar for the treatment of textile industry wastewater. Chemosphere 306:135565

Khan S, Srivastava S, Karnwal A, Malik T (2023) Streptomyces as a promising biological control agents for plant pathogens. Front Microbiol 14:1–22

Kim KH, Kabir E, Jahan SA (2017) Exposure to pesticides and the associated human health effects. Sci Total Environ 575:525–535. https://doi.org/10.1016/j.scitotenv.2016.09.009

Kloepper JW, Ryu CM, Zhang S (2004) Induced systemic resistance and promotion of plant growth by *Bacillus* spp. Phytopathology 11(94):1259–1266

Knosp O, von Tigerstrom M, Page WJ (1984) Siderophore-mediated uptake of iron in *Azotobacter vinelandii*. J Bacteriol 159:341–347

Kramer J, Özkaya Ö, Kümmerli R (2020) Bacterial siderophores in community and host interactions. Nat Rev Microbiol 18:152–163

Kumar V, Chandra R (2020) Bacterial-assisted phytoextraction mechanism of heavy metals by native hyperaccumulator plants from distillery waste–contaminated site for eco-restoration. In: Microbes for sustainable development and bioremediation. CRC Press, Boca Rato

Kumar V, Bharti A, Negi YK, Gusain O, Pandey P, Bisht GS (2012) Screening of actinomycetes from earthworm castings for their antimicrobial activity and industrial enzymes. Braz J Microbiol 43:205–214

Kumar S, Thakur M, Rani A (2014) Trichoderma: mass production, formulation, quality control, delivery and its scope in commercialization in India for the management of plant diseases. Afr J Agric Res 9(53):3838–3852

Kumar KS, Dahms HU, Won EJ, Lee JS, Shin KH (2015) Microalgae/cyanobacteria – a promising tool for heavy metal remediation. Ecotoxicol Environ Saf 113:329–352

Kumar K, Singh DP, Barman SC, Kumar N (2016) Chapter 2: Heavy metal and their regulation in plant system: an overview. In: Singh A, Prasad SM, Singh RP (eds) Plant responses to xenobiotics. Springer, Singapore, pp 19–38

Kumar V, Shahi SK, Singh S (2018) Bioremediation: an eco-sustainable approach for restoration of contaminated sites. In: Microbial bioprospecting for sustainable development. Springer, Singapore, pp 115–136

Kumar M, Giri P, Pandey S, Gupta A, Patel MK, Bajpai AB, Jenkinsand S, Siddique KHM (2021) Plant-growth-promoting rhizobacteria emerging as an effective bioinoculant to improve the growth, production, and stress tolerance of vegetable crops. Int J Mol Sci 22(22):12245

Kumar G, Lal S, Soni SK, Maurya SK, Shukla PK, Chaudhary P (2022) Mechanism and kinetics of chlorpyrifos co-metabolism by using environment restoring microbes isolated from rhizosphere of horticultural crops under subtropics. Front Microbiol 13:2796. https://doi.org/10.3389/fmicb.2022.891870

Kumari A, Kaur R, Kaur R (2019) A review on fate and remediation techniques of oil spills. Int J Res Pharm Sci 10:111

Kumari S, Mangwani N, Das S (2021) Naphthalene catabolism by biofilm forming marine bacterium *Pseudomonas aeruginosa* n6p6 and the role of quorum sensing in regulation of dioxygenase gene. J Appl Microbiol 130:1217–1231. https://doi.org/10.1111/jam.14867

Kurade MB, Kim JR, Govindwar SP, Jeon BH (2016) Insights into microalgae mediated biodegradation of Diazinon by Chlorella vulgaris: microalgal tolerance to xenobiotic pollutants and metabolism. Algal Res 20:126–134

Kurth F, Mailänder S, Bönn M, Feldhahn L, Herrmann S, Große I, Buscot F, Schrey SD, Tarkka MT (2014) Streptomyces-induced resistance against oak powdery mildew involves host plant responses in defense, photosynthesis, and secondary metabolism pathways. Mol Plant-Microbe Interact 27(9):891–900

Kuyucak N (1990) Feasibility of biosorbents application. In: Volesky B (ed) Biosorption of heavy metals. CRC Press, Boca Raton, pp 371–378

Lawniczak L, Marecik R, Chrzanowski L (2013) Contributions of biosurfactants to natural or induced bioremediation. Appl Microbiol Biotechnol 97(6):2327–2339. https://doi.org/10.1007/s00253-013-4740-1

Le KD, Yu NH, Park AR, Park DJ, Kim CJ, Kim JC (2022) Streptomyces sp. AN090126 as a biocontrol agent against bacterial and fungal plant diseases. Microorganisms 10(4):791. https://doi.org/10.3390/microorganisms10040791

Liu N, Wu Z (2018) Toxic effects of linear alkylbenzenesulfonate on Chara vulgaris L. Environ Sci Pollut Res 25:4934–4941

Liu SH, Zeng GM, Niu QY, Liu Y, Zhou L, Jiang LH, Cheng M (2017) Bioremediation mechanisms of combined pollution of PAHs and heavy metals by bacteria and fungi: a mini review. Bioresour Technol 224:25–33. https://doi.org/10.1016/j.biortech.2016.11.095

Lopez-Mondejar R, Ros M, Pascual JA (2011) Mycoparasitism-related genes expression of *Trichoderma harzianum* isolates to evaluate their efficiency as biological control agents. Biol Control 56:59–66

Lu L, Zhang J, Peng C (2019) Shift of Soil Polycyclic Aromatic Hydrocarbons (PAHs) dissipation pattern and microbial community composition due to rhamnolipid supplementation. Water Air Soil Pollut 230:107

Lynn SG, Price DJ, Birge WJ, Kilham SS (2007) Effect of nutrient availability on the uptake of PCB congener 2,2′,6,6′-tetrachlorobiphenyl by a diatom (*Stephanodiscus minutulus*) and transfer to a zooplankton (*Daphnia pulicaria*). Aquat Toxicol 83:24–32. https://doi.org/10.1016/j.aquatox.2007.03.007

Lyu A, Liu H, Che H, Yang L, Zhang J, Wu M (2017) Reveromycins A and B from Streptomyces sp. 3–10: antifungal activity against plant pathogenic fungi in vitro and in a strawberry food model system. Front Microbiol 8:550

Ma Y, Prasad MNV, Rajkumar M, Freitas H (2011) Plant growth promoting rhizobacteria and endophytes accelerate phytoremediation of metalliferous soils. Biotechnol Adv 29:248–258

Ma Y, Rajkumar M, Oliveira R, Freitas H (2015) Serpentine bacteria influence metal translocation and bioconcentration of Brassica juncea and Ricinus communis growing in multi-pollute soils. Front Plant Sci 5:1–10

Mahjoubi M, Cappello S, Souissi Y, Jaouani A, Cherif A (2017) Microbial bioremediation of petroleum hydrocarbon–contaminated marine environments. In: Zoveidavianpoor M (ed) Recent insights in petroleum science and engineering. IntechOpen, London

Mahmoud A, Cézac P, Hoadley AFA, Contamine F, D'Hugues P (2017) A review of sulfide minerals microbially assisted leaching in stirred tank reactors. Int Biodeterior Biodegrad 119:118–146

Maity JP, Chen GS, Huang YH, Sun AC, Chen CY (2019) Ecofriendly heavy metal stabilization: microbial induced mineral precipitation (MIMP) and biomineralization for heavy metals within the contaminated soil by indigenous bacteria. Geomicrobiol J 36(7):612–623. https://doi.org/10.1080/01490451.2019.15

Maksymiec W (2007) Signaling responses in plants to heavy metal stress. Acta Physiol Plant 29:177–187

Mandy RL, Maselko J, Ron AH, Corey JF, Holland L (2018) Conditions of persistent oil on beaches in Prince William Sound 26 years after the Exxon Valdez spill. Deep-Sea Res II Top Stud Oceanogr 147:9–19. https://doi.org/10.1016/j.dsr2.2017.07.011

Marques GC, Moreira H, Franco AR, Rangel AOSS, Castro PML (2013) Inoculating Helianthus annuus (sunflower) grown in zinc and cadmium contaminated soils with plant growth promoting bacteria – effects on phytoremediation strategies. Chemosphere 92:74–83

Marques E, Martins I, de Mell CM (2018) Antifungal potential of crude extracts of *Trichoderma* spp. Biota Neotrop 18(1):1–5

Martínková L, Kotik M, Marková E, Homolka L (2016) Biodegradation of phenolic compounds by Basidiomycota and its phenol oxidases: a review. Chemosphere 149:373–382

Masih A, Taneja A (2006) Polycyclic aromatic hydrocarbons (PAHs) concentrations and related carcinogenic potencies in soil at a semi-arid region of India. Chemosphere 65:449–456

Mayo S, Gutiérrez S, Malmierca MG, Lorenzana A, Campelo MP, Hermosa R, Casquero A (2015) Influence of *Rhizoctonia solani* and *Trichoderma* spp. in growth of bean (*Phaseolus vulgaris* L.) and in the induction of plant defense-related genes. Front Plant Sci 6(685):1–11

Mazaro SM, Deschamps C, Mio LM, Biasi LA, Gouvea A (2008) Post harvest behavior of strawberry fruits after pre harvest treatment with chitosan and acibenzolar-S-methyl. Rev Brasil de Fruticul 30(1):185–190

Mergawy MM, Metwaly HA, Shoeip AMO (2022) Evaluation of the efficacy of some bioagents accompanied with bio- and mineral fertilizers in controlling early blight of tomato and improvement yield. Egypt J Phytopathol 50(1):31–50

Mesbah NM (2022) Industrial biotechnology based on enzymes from extreme environments. Front Bioeng Biotechnol 2022(10):870083. https://doi.org/10.3389/fbioe.2022.870083

Michalak I, Chojnacka K, Witek-Krowiak A (2013) State of the art for the biosorption process--a review. Appl Biochem Biotechnol 170(6):1389–1416. https://doi.org/10.1007/s12010-013-0269-0

Minchev Z, Kostenko O, Soler R, Pozo MJ (2021) Microbial consortia for effective biocontrol of root and foliar diseases in tomato. Front Plant Sci 12:756368

Mirsal IA (2008) Soil pollution origin, monitoring & remediation, 2nd edn. Springer, Berlin/Heidelberg

Mir-Tutusaus JA, Baccar R, Caminal G, Sarrà M (2018) Can white-rot fungi be a real wastewater treatment alternative for organic micropollutants removal? A review. Water Res 138:137–151. https://doi.org/10.1016/j.watres.2018.02.056

Mishra D, Kim DJ, Ahn JG, Rhee YH (2005) Bioleaching: a microbial process of metal recovery; a review. Met Mater Int 11:249

Mitra A (2018) A review on the biological control of plant diseases using various microorganisms. J Res Med Dent Sci 6(4):30–35

Mohamed MS, Asair AA, Fetyan NA, Elnagdy SM (2023) Complete biodegradation of diclofenac by new bacterial strains: postulated pathways and degrading enzymes. Microorganisms 11(6):1445

Montagnolli RN, Lopes PRM, Bidoia ED (2015) Assessing *Bacillus subtilis* biosurfactant effects on the biodegradation of petroleum products. Environ Monit Assess 187:1–17. https://doi.org/10.1007/s10661-014-4116-8

Montero SL, Grondona M, Llobell I, Monte A (2011) In vitro antifungal activity of *Trichoderma harzianum, T. longibrachiatum, T. asperellum and T. atroviride* against *Botrytis cinerea* to strawberry. Bull OILB/SROP 25(10):253–256

Morsy EM, Abdel-Kawi KA, Khalil MN (2005) Efficiency of *Trichoderma viride* and *Bacillus subtilis* as biocontrol agents against *Fusarium solani* on tomato plants. Egypt J Phytopathol 37(1):47–57

Morya R, Salvachúa D, Thakur IS (2020) Burkholderia: an untapped but promising bacterial genus for the conversion of aromatic compounds. Trends Biotechnol 38(9):963–975

Mrozik A, Piotrowska-Seget Z (2010) Bioaugmentation as a strategy for cleaning up of soils contaminated with aromatic compounds. Microbiol Res 165(5):363–375

Mukhopadhyay M, Noronha SB, Suraishkumar GK (2007) Kinetic modeling for the biosorption of copper by pretreated *Aspergillus niger* biomass. Bioresour Technol 98(9):1781–1787

Muratova AY, Turkovskaya OV, Antonyuk LP, Makarov OE, Pozdnyakova LI, Ignatov VV (2005) Oil-oxidizing potential of associative rhizobacteria of the genus Azospirillum. Microbiology 74:210–215

Muter O (2023) Current trends in bioaugmentation tools for bioremediation: a critical review of advances and knowledge gaps. Microorganisms 11:710. https://doi.org/10.3390/microorganisms11030710

Nanda M, Kumar V, Fatima N, Pruthi V, Verma M, Chauhan PK, Vlaskin MS, Grigorenko AV (2019) Detoxification mechanism of organophosphorus pesticide via carboxylestrase pathway that triggers de novo TAG biosynthesis in oleaginous microalgae. Aquat Toxicol 209:49–55

Nanjappan K, Kuppusamy P, Pramod KS, Ramakrishnan S, Udai BS (2018) Actinomycetes: a promising tool for plant growth promotion and disease control. Int J Curr Microbiol App Sci 7(7):2418–2429

Nascimento DNOD, Lucheta AR, Palmieri MC, Carmo ALVD, Silva PMP, Ferreira RVDP, Junca E, Grillo FF, Alves JO (2019) Bioleaching for copper extraction of marginal ores from the Brazilian Amazon Region. Metals 9:81. https://doi.org/10.3390/met9010081

Nasr-Eldin M, Messiha N, Othman B, Megahed A, Elhalag K (2019) Induction of potato systemic resistance against the potato virus Y (PVYNTN), using crude filtrates of Streptomyces spp. under greenhouse conditions. Egypt Biol Pest Control 29(62):1–12

Natarajan KA (2018) Methods in biohydrometallurgy and developments. In: Natarajan KA (ed) Biotechnology of metals principles, recovery methods and environmental concerns. Elsevier, Amsterdam, p 502

Nawrocka JD, Małolepsza U (2013) Diversity in plant systemic resistance induced by Trichoderma. Biol Control 67(2):149–156

Niu G, Zhang J, Zhao S, Liu H, Boon N, Zhou N (2009) Bioaugmentation of a 4-chloronitrobenzene contaminated soil with Pseudomonas putida ZWL73. Environ Pollut (1987) 157(3):763–771. https://doi.org/10.1016/j.envpol.2008.11.024

Okino-Delgado CH, Zanutto-Elgui MR, do Prado DZ, Pereira MS, Fleuri LF (2019) Enzymatic bioremediation: current status, challenges of obtaining process, and applications. Microb Metab Xenobiot Compd:79–101

Oladoye PO, Olowe OM, Asemoloye MD (2021) Phytoremediation technology and food security impacts of heavy metal contaminated soils: a review of literature. Chemosphere 288:132555. https://doi.org/10.1016/j.chemosphere:132555

Oluwaseyi SO, Olubukola OB (2019) Streptomyces: implications and interactions in plant growth promotion. Appl Microbiol Biotechnol 103:1179–1188

Ong SK, Kolz A (2007) Chemical treatment technologies. In: Bhandari A, Surampalli RY, Champagne P, Ong SK, Tyagi RD, Lo IMC (eds) Remediation technologies for soils and ground water. American Society of Civil Engineering, Reston, pp 79–132

Passari A, Vineet KM, Vijaik KG, Ratuk S, Bhim PS (2016) Distribution and identification of endophytic *Streptomyces* species from *Schima wallichii* as potential biocontrol agents against fungal plant pathogens. Polish J Microbiol 65(3):319–329

Patel AK, Singhania RR, Albarico FPJB, Pandey A, Chen CW, Dong CD (2022) Organic wastes bioremediation and its changing prospects. Sci Total Environ 824:153889. https://doi.org/10.1016/j.scitotenv.2022.153889

Pawlik-Skowrońska B, Pirszel J, Kalinowska R, Skowroński T (2004) Arsenic availability, toxicity and direct role of GSH and phytochelatins in as detoxification in the green alga Stichococcus bacillaris. Aquat Toxicol 70(3):201–212

Pinto G, Pollio A, Previtera L, Stanzione M, Temussi F (2003) Removal of low molecular weight phenols from olive oil mill wastewater using microalgae. Biotechnol Lett 25:1657–1659. https://doi.org/10.1023/A:1025667429222

Prasad BN, Kumar MR (2013) Scanning electron microscopic studies on mycoparasitic activity of *Trichoderma* spp. against *Rhizoctonia solani*, incitant of sheath blight of rice. Res J Pharmaceut Biol Chem Sci 4(3):88–96

Pratush A, Kumar A, Hu Z (2018) Adverse effect of heavy metals (As, Pb, Hg, and Cr) on health and their bioremediation strategies: a review. Int Microbiol 21(3):97–106. https://doi.org/10.1007/s10123-018-0012-3

Priya AK, Muruganandam M, Ali SS, Kornaros M (2023) Clean-up of heavy metals from contaminated soil by phytoremediation: a multidisciplinary and eco-friendly approach. Toxics 11(5):422. https://doi.org/10.3390/toxics11050422

Raffa CM, Chiampo F (2021) Bioremediation of agricultural soils polluted with pesticides: a review. Bioengineering (Basel) 8(7):92. https://doi.org/10.3390/bioengineering8070092

Rai PK, Leeb SS, Zhangc Tsangd UF, Kime KH (2019) Heavy metals in food crops: health risks, fate, mechanisms, and management. Environ Int 125:365–385

Rajkumar M, Ae N, Prasad MN, Freitas H (2010) Potential of siderophore-producing bacteria for improving heavy metal phytoextraction. Trends Biotechnol 28:142–149

Rawlings DE (2005) Characteristics and adaptability of iron- and sulfur-oxidizing microorganisms used for the recovery of metals from minerals and their concentrates. Microb Cell Factories 4:13

Rawls KD, Dougherty BV, Blais EM, Stancliffe E, Kolling GL, Vinnakota K (2019) A simplified metabolic network reconstruction to promote understanding and development of flux balance analysis tools. Comput Biol Med 105:64–71. https://doi.org/10.1016/j.compbiomed.2018.12.010

Reetha AK, Pavani SL, Mohan S (2014) Hydrogen cyanide production ability by bacterial antagonist and their antibiotics inhibition potential on *Macrophomina phaseolina* (Tassi.) Goid. Int J Curr Microbiol Appl Sci 3(5):172–178

Rendón-Castrillón L, Ramírez-Carmona M, Ocampo-López C, Gómez-Arroyave L (2023) Bioleaching techniques for sustainable recovery of metals from solid matrices. Sustain For 15(13):10222. https://doi.org/10.3390/su151310222

Rezaee A, Ramavandi B, Ganati F, Ansari M, Solimanian A (2006) Biosorption of mercury by biomass of filamentous algae Spirogyra species. J Biol Sci 6(4):695–700

Rhykerd RL, Crews B, McInnes KJ (1999) Impact of bulking agents, forced aeration and tillage on remediation of oil-contaminated soil. Bioresour Technol 67(3):279–285

Rohwerder T, Gehrke T, Kinzler K, Sand W (2003) Bioleaching review part A. Appl Microbiol Biotechnol 63:239

Saeed M, Ilyas N, Jayachandran K, Shabir S, Akhtar N, Shahzad A, Sayyed RZ, Bano A (2022) Advances in biochar and PGPR engineering system for hydrocarbon degradation: a promising strategy for environmental remediation. Environ Pollut 305:119282. https://doi.org/10.1016/j.envpol.2022.119282

Samaras A, Efthimiou K, Roumeliotis E, Karaoglanidis G (2018) Biocontrol potential and plant-growth-promoting effects of *Bacillus amyloliquefaciens* MBI 600 against *Fusarium oxysporum* f. sp. *radicis-lycopersici* on Tomato. Acta Hortic 1207:139–145. https://doi.org/10.17660/ActaHortic.2018.1207.18

Samavat S, Samavat S, Besharati H, Behboudi K (2011) Interactions of rhizobia cultural filtrates with *Pseudomonas fluorescens* on bean damping-off control. J Agric Sci Tech 13(6):965–976

Santo M, Weitsman R, Sivan A (2013) The role of the copper-binding enzyme–laccase–in the biodegradation of polyethylene by the actinomycete *Rhodococcus ruber*. Int Biodeterior Biodegrad 84:204–210

Sarma BK, Yadav SK, Singh S (2015) Microbial consortium-mediated plant defence against phytopathogens: readdressing for enhancing efficacy. Soil Biol Biochem 87:25–33

Saxena G, Bharagava RN (2020) Bioremediation of industrial waste for environmental safety. In: Saxena G, Bharagava RN (eds) Volume I: Industrial waste and its management. Springer, Singapore. ISBN:978-981-13-1891-7

Saxena G, Kishor R, Bharagava RN (2020) Application of microbial enzymes in degradation and detoxification of organic and inorganic pollutants. In: Bioremediation of industrial waste for environmental safety. Springer, New York, pp 41–51

Say R, Yilmaz N, Denizli A (2003) Biosorption of cadmium, lead, mercury, and arsenic ions by the fungus *Penicillium purpurogenum*. Sep Sci Technol 38:2039–2053

Schäfer S, Buchmeier G, Claus E, Duester L, Heininger P, Körner A, Mayer P, Paschke A, Rauert C, Reifferscheid G, Rüdel H, Schlechtriem C, Schröter-Kermani C, Schudoma D, Smedes F, Steffen D, Vietoris F (2015) Bioaccumulation in aquatic systems: methodological approaches, monitoring and assessment. Environ Sci Eur 27(1):5. https://doi.org/10.1186/s12302-014-0036-z

Scholler CG, Pedersen R, Molin S, Wilkins K (2002) Volatile metabolites from Actinomycetes. J Agric Food Chem 50:2615–2621

Sergio CLC, Hernández-Rodríguez C, Soto-Cruz NÓ, Contreras JAR, Lopez-Miranda J (2016) Changes in bacterial populations during bioremediation of soil contaminated with petroleum hydrocarbons. Water Air Soil Pollut 227:e91. https://doi.org/10.1007/s11270-016-2789-z

Shahid M, Khalid S, Abbas G, Shahid N, Nadeem M, Sabir M, Aslam M, Dumat C (2015) Heavy metal stress and crop productivity. In: Hakeem HR (ed) Crop production and global environmental issues. Springer, Cham

Sharma S (2012) Bioremediation: features, strategies and applications. Asian J Pharm Life Sci 2231:4423. https://api.semanticscholar.org/CorpusID:17190307

Sharma RK, Agrawal M, Marshall FM (2008) Atmospheric deposition of heavy metals (Cu, Zn, Cd and Pb) in Varanasi City, India. Environ Monit Assess 142:269–278

Sharma J, Sundar D, Srivastava P (2021) Biosurfactants: potential agents for controlling cellular communication, motility, and antagonism. Front Mol Biosci 8. https://doi.org/10.3389/fmolb.2021.727070

Sharma P, Dutta D, Udayan A, Nadda AK, Lam SS, Kumar S (2022) Role of microbes in bioaccumulation of heavy metals in municipal solid waste: impacts on plant and human being. Environ Pollut 305:119248. https://doi.org/10.1016/j.envpol.2022.119248

Singh N, Pandey P, Dubey RC, Maheshwari DK (2008) Biological control of root rot fungus *Macrophomina phaseolina* and growth enhancement of *Pinus roxburghii* (Sarg.) by rhizosphere competent *Bacillus subtilis* BN1. World J Microbiol Biotechnol 24(9):1669–1679

Sreevidya M, Gopalakrishnanb S, Kudapab H, Varshney RK (2016) Exploring plant growth-promotion actinomycetes from vermicompost and rhizosphere soil for yield enhancement in chickpea. Braz J Microbiol 47:85–95

Stein T (2005) Bacillus subtilis antibiotics: structures, syntheses and specific functions. Mol Microbiol 56(4):845–857

Subudhi E, Kar RN (2008) *Rhizopus arrhizus* – an efficient fungus for copper effluent treatment. Int J Integr Biol 2(3):166–171

Tambekar DH, Gadakh PV (2013) Biochemical and molecular detection of biosurfactant producing bacteria from soil. Int J Life Sci Biotechnol Pharma Res 2:204–211

Thengodkar RRM, Sivakami B (2010) Degradation of chlorpyrifos by alkaline phosphatase from the cyanobacterium spirulina platensis. Biodegradation:637–644. https://doi.org/10.1007/s10532-010-9331-6

Thomas F, Cébron A (2016) Short-term rhizosphere effect on available carbon sources, phenanthrene degradation, and active microbiome in an aged contaminated industrial soil. Front Microbiol 7:92

Tiwari B, Singh S, Chakraborty S, Verma E, Mishra AK (2017a) Sequential role of biosorption and biodegradation in rapid removal degradation and utilization of methyl parathion as a phosphate source by a new cyanobacterial isolate Scytonema sp. BHUS-5. Int J Phytoremediat 19(10):884–893

Tiwari S, Tripathi A, Gaur R (2017b) Bioremediation of plant refuges and xenobiotics. In: Principles and applications of environmental biotechnology for a sustainable future, vol 2017, pp 85–142

Tóth G, Hermann T, Silva MR, Montanarella C (2016) Heavy metals in agricultural soils of the European Union with implications for food safety. Environ Int 88:299–309

Tremolada P, Villa S, Bazzarin P, Bizzotto E, Comolli R, Vighi M (2008) Pops in mountain soils from the Alps and Andes: suggestions for a 'precipitation effect' on altitudinal gradients. Water Air Soil Pollut 188:93–109

Tunali S, Cabuk A, Akar T (2005) Removal of lead and copper ions from aqueous solutions by bacterial strain isolated from soil. Chem Eng J [Internet] 115(3):203–211. https://doi.org/10.1016/j.cej.2005.09.023

Tyagi M, da Fronseca MM, de Carvalho CC (2010) Bioaugmentation and biostimulation strategies to improve the effectiveness of bioremediation processes. Biodegradation 22(2):231–241

Van Aken B, Hofrichter M, Scheibner K, Hatakka AI, Naveau H, Agathos SN (1999) Transformation and mineralization of 2, 4, 6-trinitrotoluene (TNT) by manganese peroxidase from the white-rot basidiomycete *Phlebia radiata*. Biodegradation 10(2):83–91

Vanishree M, Thatheyus AJ, Ramya D (2014) Biodegradation of petrol using the fungus *Penicillium* sp. Sci Int 2(1):26–31

Varjani S, Kumar G, Rene ER (2019) Developments in biochar application for pesticide remediation: current knowledge and future research directions. J Environ Manag 232:505–513

Verasoundarapandian G, Lim ZS, Radziff SBM, Taufik SH, Puasa NA, Shaharuddin NA, Merican F, Wong CY, Lalung J, Ahmad SA (2022) Remediation of pesticides by microalgae as feasible approach in agriculture: bibliometric strategies. Agronomy 12:117. https://doi.org/10.3390/agronomy12010117

Verma JP, Jaiswal DK, Sagar R (2014) Pesticide relevance and their microbial degradation: a-state-of-art. Rev Environ Sci Biotechnol 13:429–466

Vocciante M, Finocchi A, Folly D, D'Auris A, Conte A, Tonziello J, Pola A (2019) Enhanced oil spill remediation by adsorption with interlinked multilayered graphene. Materials 12(14):2231. https://doi.org/10.3390/ma12142231

Vuola J, Göransson H, Böhling T, Asko-Seljavaara S (1996) Bone marrow induced osteogenesis in hydroxyapatite and calcium carbonate implants. Biomaterials 17:1761–1766

Vurukonda SKP, Giovanardi D, Stefani E (2018) Plant growth -promoting and biocontrol activity of *Streptomyces* spp. as endophytes. Int J Mol Sci 19:952

Waksman SA (1959) The actinomycetes V.1. Nature, occurrence and activities. The Williams & Wilkins Co, Baltimore, p 327

Wang P, Sun Z, Hu Y, Cheng H (2019) Leaching of heavy metals from abandoned mine tailings brought by precipitation and the associated environmental impact. Sci Total Environ 695:133893

Wang Y, Liu Z, Pan H, Tang R (2023) Biomineralization inspired crystal growth for biomimetic materials preparation. J Cryst Growth 603:127029

Wania F, Mclachlan MS (2001) Estimating the influence of forests on the overall fate of semi-volatile organic compounds using a multimedia fate model. Environ Sci Technol 35:582–590

Watling HR (2006) The bioleaching of sulfide minerals with emphasis on copper sulfides – a review. Hydrometallurgy 84:81–108

Wei G, Kloepper JW, TuZun S (1991) Induction of systemic resistance of cucumber to *Colletotrichum orbiculare* by select strains of plant growth-promoting rhizobacteria. Phytopathology 81:1508–1512

Wu Y, Zhou J, Li C, Ma Y (2019) Antifungal and plant growth promotion activity of volatile organic compounds produced by *Bacillus amyloliquefaciens*. Microbiol:1–14. https://doi.org/10.1002/mbo3.813

Wu W, Ke T, Zhou X, Li Q, Tao Y, Zhang Y, Zeng Y, Cao J, Chen L (2022) Synergistic remediation of copper mine tailing sand by microalgae and fungi. Appl Soil Ecol 175:104453

Xu A, Zhang X, Wu S, Xu N, Huang Y, Yan X, Zhou J, Cui Z, Dong W (2021) Pollutant degrading enzyme: catalytic mechanisms and their expanded applications. Molecules 26(16):4751. https://doi.org/10.3390/molecules26164751

Yakup AM, Bayramog LG, Yilmaz M, Bektas S, Genc O (2004) Biosorption of Hg (2+), Cd (2+), and Zn (2+) by Ca-alginate and immobilized wood rotting fungus *Funalia trogii*. J Hazard Mater 109(1–3):191–199

Yan G, Viraraghavan T (2003) Mechanism of biosorption of heavy metals by Mucor rouxii. Eng Life Sci 8(4):363–371

Yang Y, Wang J, Liao J, Xie S, Huang Y (2014) Distribution of naphthalene dioxygenase genes in crude oil-contaminated soils. Microb Ecol 68:785–793. https://doi.org/10.1007/s00248-014-0457-7

Yanjie Y, Luan P, Liu S, Shan Y, Hou Z, Zhao S, Jia S, Li R (2022) Efficacy of Bacillus subtilis XZ18-3 as a biocontrol agent against *Rhizoctonia cerealis* on wheat. Agriculture 12(2):258

Yasmin A, Ambreen S, Shabir S (2022) Biotransformation of dimethoate into novel metabolites by bacterial isolate *Pseudomonas kilonensis* MB490. J Environ Sci Health Part B 57:13–22. https://doi.org/10.1080/03601234.2021.2017723

Yiyang Y, Gui Y, Li Z, Jiang C, Guo J, Niu D (2022) Induced systemic resistance for improving plant immunity by beneficial microbes. Plan Theory 11(3):386. https://doi.org/10.3390/plants11030386

Yobo KS, Laing MD, Hunter CH (2009) Effects of single and dual applications of selected Trichoderma and Bacillus isolates on performance of dry bean seedlings grown in composted pine bark growth medium under shadehouse conditions. J Plant Nutr 32:1271–1289

Yoshioka Y, Ichikawa H, Naznin HA, Kogure A, Hyakumachi M (2011) Systemic resistance induced in Arabidopsis thaliana by *Trichoderma asperellum* SKT-1, a microbial pesticide of seed borne diseases of rice. Pest Manag Sci 68:60–66

Zeng G, Liu Z, Zhong H, Li J, Yuan X, Fu H, Ding Y, Wang J, Zhou M (2011) Effect of mono rhamnolipid on the degradation of n-hexadecane by *Candida tropicalis* and the association with cell surface properties. Appl Microbiol Biotechnol 90:1155–1161. https://doi.org/10.1007/s00253-011-3125-6

Zhang S, Yang Y, Wu Z, Li K (2020) Induced defense responses against *Rhizoctonia solani* in rice seedling by a novel antifungalmycin N2 from Streptomyces sp. N2. Australasian Plant Pathol 49:267–276

Zhao Y, Selvaraj JN, Xing F, Zhou L, Wang Y (2014) Antagonistic action of *Bacillus subtilis* strain SG6 on *Fusarium graminearum*. PLoS One 9(3):1–11. https://doi.org/10.1371/journal

Zhu T, Dittrich M (2016) Carbonate precipitation through microbial activities in natural environment, and their potential in biotechnology: a review. Front Bioeng Biotechnol 4:4

Zhu B, Peng RH, Fu XY, Jin XF, Zhao W, Xu J, Yao QH (2012) Enhanced transformation of TNT by Arabidopsis plants expressing an old yellow enzyme. PLoS One 7(7):e39861

The Role of Insect and Insect Enzymes in Environment with References to Plastic Waste Management

Raj Narayan Roy

Abstract Over time, the exploitation of natural resources has accelerated to meet the demand due to the global explosion of the human population. To corroborate with such demand, revolution is necessitated in various industrial sectors. Such a revolution is associated with the production of various organic and inorganic chemicals that pollute our habitats. Pollution associated with plastic, heavy metals, petroleum, pesticides, etc., is alarming for our blue planet. Efforts are being made to use insects and their enzymes to mitigate these hazardous pollutants. Insects, being the most varied group of organisms, are able to colonize almost all ecological niches on Earth. To adapt to such diverse and occasionally extreme habitats, insects have evolved varied biological and chemical systems. Enzymes are pivotal in these systems, enabling insects to digest food from different sources. To overcome metabolic limitations, insects have established various adaptations, including microbial symbiotic associations as an evolutionary approach. These symbionts particularly the gut symbionts contribute significantly to overcoming the metabolic limitation of their host. It is possible through increasing and diverging their food digestibility, and synthesis of various essential metabolites. The insect-associated enzymes are getting more and more attention in various biotechnology fields, including bioremediation. Such technology for treating pollutants through biological systems is an important technology from both an economical and environmental point of view. The present effort of searching effective fields, particularly plastic-degrading insects and their enzymes including the scope of the use of native or genetically modified microorganisms for the degradation of plastics, is a potent technology for bioremediation.

Keywords Plastic waste · Biodegradation · Insect · Insects enzymes

R. N. Roy (✉)
Microbiology Research Laboratory, Department of Botany, Dr. Bhupendra Nath Dutta Smriti Mahavidyalaya, Burdwan, West Bengal, India

A. A. H. Abdel Latef et al. (eds.), *Sustainable Remediation for Pollution and Climate Resilience*, https://doi.org/10.1007/978-981-96-5674-5_19

1 Introduction

Exploitation of natural resources is as old as history itself. Overtime, the global explosion of the human population has accelerated such exploitation to meet the gradually increasing demands of the population for food, energy, and all other necessities. The revolution of various industrial sectors was a response to these necessities. Such a revolution is linked to the production of various organic and inorganic chemicals. These chemicals have directly and indirectly led to a prolonged negative impact on habitats. Environmental contamination by any hazardous material poses a strong threat to the life of all organisms on land and in water. The inclination towards environmental pollution is so quick and massive that such contamination becomes detectable in the farthest ocean waters at a steady rate. The exercise of using natural agents to eradicate harmful wastes from the environment is called bioremediation. The term bioremediation consists of two words, "bios," which means the life that denotes living entities, and "remediate," which means to solve an issue. It is a technology for eliminating environmental pollutants to restore the original natural atmosphere and prohibit further pollution. However, bioremediation may be defined comprehensively as the environmentally friendly process of eliminating pollutants from environments using the metabolic capabilities of diverged organisms through converting contaminants to harmless products by mineralization, forming carbon (IV) oxide, water, or by transforming into biomass. The biodegradability of different compounds varies considerably. However, through bioremediation, it is possible to overcome such issues related to such an environment.

In the past decades, the habitually used wood, leather, and metal have been substituted by plastic materials due to their flexibility, toughness, and various other properties (Sivan et al. 2006; Singh and Sharma 2008). The permanence and undesirable dumping of synthetic polymers are major environmental threats. At present, next to climate change, plastic pollution is the most vital among the fastest emerging environmental issues. Such pollution may result in multiple adverse consequences to the health of ecosystems. Thus, resolving these consequences is being sought at different levels of socio-ecological organization (Rochman 2018; Przemieniecki et al. 2020). The reutilization of plastic waste may be a way to minimize the extent of the crisis related to plastic pollution. Simultaneously, it is also possible to render a massive potential reservoir of carbon (Ellis et al. 2021). Recycling through physical and chemical processes is another procedure for plastic upcycling, but massive energy and environmental issues might still obstruct it (Zhang et al. 2022). However, the biodegradation process is emerging widely as a favorable approach for the disposal of plastic residues.

Insects evolved nearly 385 million years ago (Engel 2015). Through stable diversification, they are the richest group comprising almost one million species. They are acclimatized to nearly all ecological conditions with various mechanisms including holding a highly diverse microbial reservoir (Banerjee et al. 2022). The ubiquitous prokaryotic microorganisms have developed diverse associates with a number of eukaryotic taxa, including insects (Banerjee et al. 2017). These symbionts can withstand harsh environmental conditions. They may tolerate various properties useful for a number of insects to tide over the difficulties generated by different pollutants (Engel and Moran 2013). Thus, the prevalent ones are encouraged to explore certain unusual nutritive factors (Engel and Moran 2013). These insect gut symbionts contribute significantly to overcoming their host's metabolic limitation by increasing and diverging their food digestibility (Banerjee et al. 2016, 2020).

Microorganisms are natural decomposers of ecosystems possessing high generation rates, higher chemotactism, complex enzymatic response, and rich secretion systems. These features make the microbes more valuable for alternative bioremediation agents than chemical and physical ones. Symbiotic microbes are promising but mostly unexploited sources for potential applications in bioremediation. The diverse efficacy of these symbiotic microbes has been proven for millions of years in a eukaryotic host through natural selection (Banerjee et al. 2017). Thus, the products of these symbionts possibly reduce the threat of harmful side effects. The treatment of pollutants through biological systems is an important technology from both an economic and environmental aspect. The present effort for the use of native or genetically modified organisms for the degradation or elimination of plastics is a potent technology for in situ remediation.

2 Plastic as a Pollutant

Plastic is one of the most commonly used materials on earth. It is extensively used in packaging food, pharmaceuticals, cosmetics, chemicals, and detergents, as well as in construction, transportation, textile, electrical and electronics, and industrial machinery. The commercial manufacturing of plastic started during the 1950s, and then the production rate accelerated incredibly. Plastic production was nearly 270 million tons in 2010, but as of 2021, it rose to 390.7 million tons. It is estimated that by 2025, the global annual production of plastics will be 445.25 million metric tons, and by 2050 the amount will rise to around 590 million metric tons (Global plastics production forecast 2025-2050 | Statista12). Succeeding climate change, plastic waste is becoming one of the most problems of our time. Less than 50% of the plastic waste has either been disposed of in landfills or recycled. Thus, a huge amount of the leftover plastic waste litter, the white pollution, creates a "Plastic world," scattering through continents, oceans, and every corner of our blue planet (Rochman et al. 2013). In 2016, the amount of mismanaged plastic was 91 million tons which is likely to be 239 million tons by 2040 (Tang 2023).

Plastic is one of the toughest polymers of polycarbonate, acrylonitrile butadiene styrene (ABS), and polydicyclopentadiene. This versatile polymer is used nearly in every industry for its synthesis, flexibility, and consumer-satisfying properties. The widest-used forms of plastic are polyethylene (PE) plastic bags, polyvinyl chloride (PVC) pipes, polypropylene-based bottles, and polystyrene ((PS) cups. Its low conductivity of electricity, availability in varied forms, durability, transparency, and low density make it suitable for varied applications (Rodriguez et al. 2014) including as wrapping material, and storage containers for several commodities (Hedenqvist 2018). In the healthcare industry, the use of plastics is also favored in the device for drug delivery and packaging of drugs (Rosato and Rosato 2003). Even high degradation-resistant plastics (e.g., polycarbonates and acrylonitrile butadiene styrene) are also used for industrial and household purposes. Conductive plastics are also getting their way in the electronic industry (Shrivastava 2018; Gulrez et al. 2014).

The indiscriminate dumping of plastics on land and burning in the open air consequences in the release of harmful chemicals into the air, impacting all life forms, natural habitats, and public health (Alabi et al. 2012). Buried plastic waste under the soil may decrease agricultural productivity by disrupting drainage systems, decreasing soil quality, influencing soil fauna, and decreasing soil quality. It is estimated that annually 0.48–1.27 million tons of plastic pollutants move into the sea (Crompton 2007). Microplastic particles eluted from plastic waste affect important ecological species such as salt marsh grasses, mussels, and corals, which are more likely to enter food chains (Browne et al. 2008; Uhrin and Schellinger 2011). The contamination of plastic particles in the ecosystem, particularly in food materials anticipated for human intake, is very alarming (Lusher et al. 2017). In the course of the incineration of plastic wastes, a number of carcinogenic polycyclic aromatic hydrocarbons are released as airborne pollutants (Al-Salem et al. 2009).

At present, plastic degradation, i.e., turning them into tiny particles is a challenging environmental issue. In the course of such processing, small toxic particles may release toxicity. The phthalate esters, commonly used as plasticizers, introduce plasticity, flexibility, and durability in plastic products (Giuliani et al. 2020; Roy 2020). Humans are exposed to phthalates mainly through food and contaminated water. Such exposure may lead to various health issues including disruption of endocrine glands, disorders in metabolic processes, and reproductive toxicity (Roy 2020; Zhou et al. 2020). Bisphenol A, used for manufacturing several polymer materials, is a potent human health hazard (Konieczna et al. 2015). The association between microplastic and plastic additives augments their toxicity (Wrigth and Kelly 2017). The absorption of these additives on microplastic surfaces is responsible for the intensification of toxicity. Human exposure to microplastics and additives through inhalation, ingestion, and skin contact may be responsible for disrupting the balance of natural gut microbiota.

The problem connected with plastic pollution has drastically gripped the entire planet. These plastics may be degraded in nonhostile and in harsh environments of temperature, pH, radiation, etc. Naturally, it is difficult to build such harsh environments in order to degrade plastic. A single remediation approach is too difficult to overcome this crisis. Thus, multidisciplinary physicochemical and biological strategies are gaining interest to nullify the emergent problem of the aftermath of plastic pollution. It is noteworthy that amidst the various approaches, biological methods are gaining momentum to combat issues related to pollution problems.

3 The Challenges in Biodegradation of Plastics

Plastic polymers are long-chain macromolecules having a single functional group. These are not enriched with functional groups and hydrolyzing bonds, and thus, are unsuitable for attachment of microbes and enzymatic reactions (Al Hosni et al. 2019). Plastics' intrinsic configuration and properties make them much more durable and degradation-resistant. The existing approaches for the disposal of plastic waste include burning, abandoning into marine water, and landfills. However, all of these methods may lead to secondary pollution (e.g., chloride and dioxins) and greenhouse gases (Barnes et al. 2009; Restrepo-Florez et al. 2014). Thus, the development of a new potent method for biodegradation is an ongoing process (Silva et al. 2018; Tang and Chen 2019). Through biodegradation, organic polymers are disintegrated into smaller compounds e.g., CO_2 and H_2O (Lucas et al. 2008; Shah et al. 2008). The recalcitrant polymers are not entirely degraded thus becoming a challenge for environmental assimilation (Geyer et al. 2017). The microbial degradability is also significantly reduced due to their higher molecular weight, crystallinity, and hydrophobicity. These less flexible long-chain recalcitrant polymers effectively degrade into tiny monomers, under prolonged enzymatic activities. Such treatment leads to the formation of specific delicate bonds having chemical groups on the polymer or its side chain to increase their fragmentation (Maity et al. 2021).

Conventional petroleum-derived plastics are not entirely decomposed and assimilated by microbes under abiotic factors, involving UV radiation, temperature, and prolonged physical stress (Urbanek et al. 2018). Plastics devoid of pro-oxidants lack functional groups/enzyme-sensitive ester bonds and may not be subjected to enzymatic attacks. Plastics with proper functional groups or enzyme-susceptible ester linkages are attacked by thermo-, photo- and enzymatic oxidation (Narancic and O'Connor 2019). Long-chain plastic polymers are less inclined to degradation than others (e.g., emerging biodegradable plastics) owing to their small size, flexibility, low molecular weights and possibly having numerous functional groups (Narancic and O'Connor 2019).

The complications of plastic degradation are observable in traditional plastics having distinct chemical configurations. Technological advancement and capital investment are not in line with social progression. Thus, the available microorganisms are not capable of degrading plastics efficiently. For plastic biodegradation, introducing oxygen within the plastic polymeric chain is a prerequisite (Albertsson and Karlsson 1990; Gryn'ova et al. 2011). This introduction of oxygen within the plastic polymeric structure is much more complicated. This oxidation resistance is deliberately planned to enhance the durability of plastic. Such introduction may form carbonyl groups leading to the production of smaller molecules from the long-chain of hydrocarbons. These smaller molecules may be digested by microorganisms (Roy et al. 2011). Nevertheless, for polymer oxidation, the critical initial step for the degradation of plastics is chain fragmentation. We must depend on abiotic factors including UV light, high temperature, etc. Such abiotic treatments are neither cost-effective nor eco-friendly. Through rigorous investigation, such hurdles have been attempted to overcome under laboratory conditions (Sanluis-Verdes et al. 2022).

4 Insect for Bioremediation of Plastic

The amount of literature in the last couple of decades related to the involvement of insects in bioremediation, including plastics is gradually increasing. Reports related to such capability concerning plastic polymers' degradation are restricted to members of the four orders (Lepidoptera, *Coleopteran Orthoptera, and Isoptera) only* (Table 1). The desire to eat plastic may be linked to the digestive activity of intestinal microbes, which aids in the digestion process. The masticatory mechanism of lepidopteran larvae, which consists of a pair of jaws, is atrophied in adults. Furthermore, it is supported by the chewing mouthpiece of the larvae of these animals (Pivato et al. 2022). However, reports about the ability of some insect larvae to chew and eat plastic as the sole carbon source is a new approach to mitigate white (plastics) pollution. These reports are very encouraging for scientists toward the

Table 1 A list of plastivore insects

Insect order	Common name	Scientific name	References
Lepidoptera	*Greater wax moth*	*Galleria mellonella*	Jiang et al. (2021) and Wang et al. (2022)
	Indianmeal moth	*Plodia interpunctella*	Yang et al. (2014) and Mahmoud et al. (2020)
	The rice moth	*Corcyra cephalonica*	Kesti and Sharana (2019)
	Full armyworm	*Spodoptera frugiperda*	Zhu et al. (2022)
	Lesser wax moth	*Achroia grisella*	Kundungal et al. (2019)
	Greater wax moth	*Galleria mellonella*	Brandon et al. (2018)
Coleoptera	Yellow mealworm	*Tenebrio molitor*	Yang et al. (2018) and Peng et al. (2020a)
	Dark mealworm	*Tenebrio obscurus*	Peng et al. (2019)
	Red flour beetle	*Tribolium castaneum*	Wang et al. (2020)
	Confused flour beetle	*Tribolium confusum*	Abdulhay (2020)
	Super worm	*Zophobas atratus*	Luo et al. (2021),
	Darkling beetles	*Plasiophthamus davidis*	Woo et al. (2020)
	Lesser mealworm	*Alphitobius diaperinus*	Cucini et al. (2020) and (2022)
	Uloma darkling beetle	*Uloma* spp.	Kundungal et al. (2021)
Orthoptera	Two-spotted cricket	*Gryllus bimaculatus*	Khan et al. (2021)
Isoptera	*Acacia termite*	*Coptotermes acinaciformis*	Lenz et al. (2011)
	Super termite	*Coptotermes formosanus*	Terzi et al. (2009)
	West Indian drywood termites	*Cryptotermes brevis*	Lopez et al. (2020)
	Darwin termite	*Mastotermes darwiniensis*	Lenz et al. (2011)
	Conehead termites	*Nasutitermes corniger*	Lopez et al. (2020)
	Black-headed nasute termite	*Nasutitermes nigriceps*	Gonçalves et al. (2014)

insect as a powerful bio-resource for plastic biodegradation (Kannan et al. 2019). However, insect larvae and their intestinal microflora have a mutually beneficial symbiotic relation. The individual involvement of the cells of the digestive tract and gut symbiotic microflora for the degradation of plastic polymers has not been distinctly established so far. Przemieniecki et al. (2020) observed much more activity of the lysates of the digestive tract (with symbionts) in comparison to that of the microorganisms alone. Thus, both the intestinal microflora of the insect larvae and the enzyme system of the insect itself play a significant role in the biodegradation of the ingested plastic. However, further observation is required to determine the interaction between these insects and their gut microorganisms to reveal the degradation mechanism of plastic by plastivore insects (Jiang et al. 2021).

4.1 Role of Insect Mandible and Saliva in Degradation of Plastics

For many years, the experimental approach to the gut microbiota has been primarily confined to the field of biodegradation of plastic by insect larvae. The mandibulate insects can masticate plastic materials to break them into smaller pieces before consumption. The ability of a number of insect larvae to pierce and chew plastics, particularly of the members of the orders coleopteran and lepidopteran, has been detected dating back a decade (Pivato et al. 2022). In contrast, experimental observation of the larvae of *Galleria. mellonella* revealed a notable prompt ability to oxidize unprocessed PE within an hour (Sanluis-Verdes et al. 2022; Bombelli et al. 2017). The larva is capable of breaking PE using its saliva. In an aqueous solution with a neutral pH, the saliva can oxidize the polymer at room temperature. Tiny oxidized molecules are created through such degradation of plastic polymers (Sanluis-Verdes et al. 2022). The active enzyme of the larval saliva of *G. mellonella* belongs to the family of phenol oxidases known as PEases (Sanluis-Verdes et al. 2022). A gas chromatography-mass spectrometry study confirmed the existence of tiny oxidized aliphatic chains as polymer degradation products (Fig. 1). The existence of a few enzymes of the prophenoloxidase/hexamerin family in the saliva of *G. mellonella* has also been marked (Sanluis-Verdes et al. 2022). The salivary enzyme is active on plastic at room temperature and at neutral pH. It is a promising tactic for plastic polymer biodegradation. This agenda defies the existing demarcation for plastic biodegradation.

4.2 Role of Insect Gut on the Degradation of Plastics

In insects, the process of degradation of plastic polymers takes place mainly in their gut. The insect and gut microbiota cells are responsible for such degradation processes.

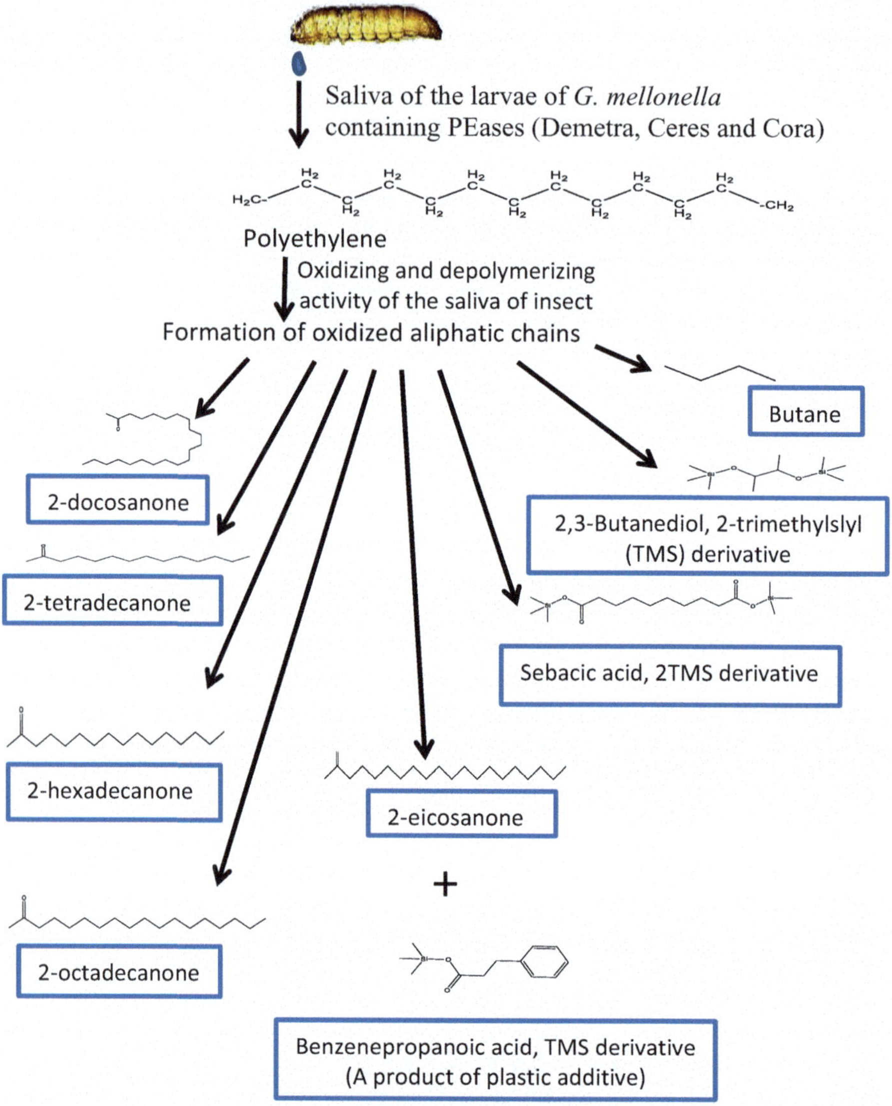

Fig. 1 Oxidizing and depolymerizing activity of the saliva of G. *mellonella* larvae on the polyethylene and plastic additive. (Ref: Sanluis-Verdes et al. 2022, Spínola-Amilibia et al. 2023)

4.2.1 Role of the Cells of the Intestinal Tract of Insects on the Degradation of Plastics

Presently, the exclusive involvement of gut microbiota in the degradation of plastic is facing some reservations. The perception is based on the fact that the insect may have an advanced mechanism to some extent for the degradation of plastic (Kong et al. 2019; Yang et al. 2021). Przemieniecki et al. (2020) reported the fact that the lysates of the digestive tract (with symbionts) are much more active than the microorganisms alone. The qualitative analysis shows that some enzymes are produced from the cells of the digestive tract and symbiotic microbes. Yang et al. (2015a, b) reported more efficacy of the symbiotic bacteria inside the gut (47.7% gut inside and 7.4% external medium, in 60 days) than in the medium in terms of plastics depolymerization. It is an indication of the presence of certain favorable factors inside the gut. The synergistic effect and the other gut enzyme secreted by the larva itself may be probable reasons behind such augmentation (Yang et al. 2015a, b). However, larvae of some lepidopteran and coleopteran insects are capable of degrading untreated plastic polymers (Yang et al. 2014; Yang et al. 2015a, b; Bombelli et al. 2017; Brandon et al. 2018; Yang et al. 2020).

4.2.1.1 Variability in Insect Gut Structure

The insect gut harbors and maintains a well-established microflora. The structure of insect guts varies considerably. This variation enables insects to harbor microbes to meet their requirements. Insect symbiotic gut microflora secretes various enzymes which are organized as auspicious sources for prospective applications in bioremediation. The pretreatments of plastic material reduced the hydrophobicity of the polymer to make it more compatible with microorganisms. It also introduces groups like C=O or – OH, which are more prone to microbial degradation (Arutchelvi et al. 2008). The adaptation of insects at diverse feeding habitats is mainly responsible for alterations in their digestive tracts.

Generally, the insect digestive tract is divided into foregut, midgut, and hindgut. The foregut is segmented into the pharynx and esophagus with the crop/diverticula for transitory food storage and also additional proventriculus for food crushing. The hindgut is segmented into several regions including fermentation compartments and a discrete rectum that retains feces before discharge from the body. However, the midgut of several insects works as a key organ for digestion as well as absorption of ingested food material (Terra and Ferreira 2020). The peritrophic matrix works as a protective lining for the tender epithelial cell layer of the midgut of several insects. This matrix compartmentalize the midgut into ecto- and endotrophic regions that protect the midgut epithelium from direct contact with coarse food particles and microorganisms. Thus, this prevents the epithelium from injury and pathogenic infection, etc. This compartmentalization also accelerates the digestibility of the insect through selective transporting of solutes and enzymes in between ecto- and endoperitrophic spaces. The peritrophic matrix also neutralizes the toxic substances consumed, particularly pesticides and metallic elements (Hegedus et al. 2009). This matrix also generates a current flow between ecto- and endoperitrophic spaces that favor nutrient absorption, minimizing the loss of digestive enzymes by frass excretion and increasing digestion efficiency (Lehane 1997; Terra and Ferreira 2020).

4.2.2 Insect Gut Symbionts Involved in the Digestion of Plastic Polymer

The insects' gut isolates are getting attention due to their capability to break down various uncommon foods including plastics. The involvement of such microbes in the degradation and digestion of food can be recognized through the analysis of the degrading ability of gut microbes (Peng et al. 2020a). However, a concise list of insects along with the activities of active gut microbe (s) is presented in Table 2. *Tenebrio. obscurus* chews plastic foam into small fragments and increases the surface area of the polymers, which becomes susceptible to depolymerization by extracellular enzymes (Singh and Sharma 2008). Feeding on varied plastics may change the biodegrading intestinal microbial diversity of the larval gut of the insect. The depolymerization of plastic was dramatically slowed down due to the suppression of gut microbiota with mixed antibiotics (Peng et al. 2020a). Depression of gut microbes by the antibiotic treatment resulted in significant inhibition of depolymerization of plastic polymers (Yang et al. 2020). Thus, insect gut bacteria are drawing extensive attention to the field of bioremediation. Yang et al. (2015a, b) reported on the formation of biofilm on plastic film by the gut isolates after 1 month of incubation. Numerous pits and holes were detected on the surfaces of the plastic film. The hydrophobicity of the plastic film was also decreased. C–O polar groups' formation was also detected (Yang et al. 2015a, b). Zhang et al. 2022 found that the gut isolate EMBL-1 can use PVC film as a sole energy source. After 90 days of incubation, the isolate formed a compact biofilm on the plastic surface. The film was damaged due to such biofilm formation. The formation of additional pits and cracks below the biofilm advocates the degradation activities of the isolate.

The capability of *Tenebrio molitor* larvae for depolymerization of polyester-Polyurethane (PU) foam was recorded as high as 67% (Liu et al. 2022). A broad range of plastic degradation capabilities was observed in *Tenebrio molitor*, including PE (Brandon et al. 2018), PS (Brandon et al. 2021), PVC (Peng et al. 2020a), and polyester—PU (Liu et al. 2022). The larvae of *Tenebrio molitor* were found capable of degrading a wide range of PE, including low-density polyethylene (LDPE), linear low-density polyethylene (LLDPE), and high-density polyethylene (HDPE). The depolymerization competency depends on the types of plastic, molecular weight, and the number of branches (Yang et al. 2022). The larvae of *Tenebrio molitor* were able to degrade PS completely and the resultant CO_2 was transformed into their biomass (Yang et al. 2015a, b). The degradation capability of *Zophobas atratus* as well as *Tenebrio obscurus* was found to have higher PS than *Tenebrio molitor* (Peng et al. 2019; Yang et al. 2020). It was found that larvae of *Zophobas atratus* could degrade PS as well as PE; however, they were unable to generate NPs in their frass (Peng et al. 2020b, 2022). This indicates that plastic biodegradation by insects is a more environmentally friendly option. The PS degradation capability of the *Tribolium castaneum* larvae was also observed. The gut symbiont, *Acinetobacter* was found connected with the degradation of *T. castaneum* (Wang et al. 2020). The PE plastics are chewed by *Plodia interpunctella* and the gut microbe, *Enterobacter asburiae,* and *Bacillus* sp. were linked with the degradation process of PE (Bowditch 1997). Plastic polymers' degradation has also been established in several insects including *Achroia grisella* (Kundungal et al. 2019), and *Tribolium confusum* (Kundungal et al. 2019). The termites (under order Isoptera), the most important soil insects, are also reported to have plastic degradation capability with their gut microbiota (López-Naranjo et al. 2013; Kumar et al. 2022).

The chemical characteristics of plastics vary considerably. The degradability of the plastic polymers by insects also varies accordingly. *Galleria mellonella* was able to degrade polyethylene (PE) faster than PS (Lou et al. 2020). The preference to ingest the mixtures of plastic polymers by the larvae of Tenebrio molitor for better degradability of plastics was observed (Brandon et al. 2018; Kundungal et al. 2019). The PS degradation rate doubled when *Tenebrio molitor*, the larvae, were served bran (Brandon et al. 2018). PE-fed-larvae supplementation with beeswax increased the species abundance and consistency of the intestinal microflora (Lou et al. 2020). Like the larvae of *Achroia grisella*, the larvae of beetle also showed enhanced degradation of PS leading to better growth due to a continual PS intake with supplemental nourishment. The PS degradation capability of the larvae of *Uloma* sp. advocates the omnipresent plastic degradation phenomena among the bee-wax ingesting species (Kundungal et al. 2021). The degradation products through such mechanisms probably meet the demand for high nutrients, resulting in the upsurging of intestinal microfloral diversity in the larvae. It necessitates the search for potent organisms for the degradation of specific plastic polymers through the formulation of the best diet for blending plastics and required nutrients. Various parameters (insect species, variation of polymers, nature of gut symbiont, effect of nutrient supplementation, and fermentation parameters) are interlinked with the efficiency of plastic polymer degradation of insects. A concise list of such plastic degradation by insects is represented in Table 2.

5 Mechanism of Bioremediation of Plastic Polymer by Insects and their Enzymes

The mandibulate insects can masticate plastic materials to break them into smaller pieces before consumption. Due to such fragmentation, the surface area of plastic has increased. Insect such as *G. mellonella* (larvae) mix their saliva with these fragments and promptly oxidize unprocessed plastic polymers within an hour at room temperature (Sanluis-Verdes et al. 2022). Such physicochemical "management" of chewing, ingesting, and blending with intestinal contents, as well as the enzymes secreted by larvae may be crucial for the prompt in vivo plastic polymers degradation (Yang et al. 2015a, b). The enzymes of gut bacteria are also capable of changes in surface topography, minimizing the hydrophobicity as well as depolymerization of untreated plastic polymers (Yang et al. 2015a, b). The larvae and gut microbiota of *T. molitor* secreted one or more factor(s) to emulsify the ingested plastic polymers to increase their bioavailability (Brandon et al. 2018; Zhang et al. 2020). In the course of plastic polymer degradation, the upregulation of two genes, Afla_053930

Table 2 Insects and thier gut symbionts in plastic waste management

Plastic material	Name of the insect	Symbiont microbes	Major findings	Reference
Polyethylene (PE)	*Achroia grisella*	–	Over 8 days, the rate of PE ingestion was 1.83 mg/day/larva. The formation of new carbonyl and alcoholic groups with an increase in unsaturated hydrocarbon in the post depredated frass of the larvae indicated the formation of biodegraded intermediates.	Kundungal et al. (2019)
	Galleria mellonella	*Enterobacter* sp. D1	The D1 strain resulted in roughness, depressions, and cracks on the PE film. The bacteria increased certain alcohols, esters, and acids due to oxidation of the surface of the PE film.	Ren et al. (2019)
	Galleria mellonella	*Aspergillus flavus*	Two laccase-like multicopper oxidases genes, (AFLA_ 006190 and AFLA_053930) displayed upregulated expression during the degradation process, expected to be the PE - degrading enzymes.	Zhang et al. (2019)
	Galleria mellonella	*Pseudomonas* sp., *Pseudomonas citronellolis, Lactobacillus* sp., *Fibrobacter*	Feeding with polyethylene changed the bacterial gut microbiome of the insect larvae at the phyla and genus levels. Both bacterial and fungal communities were associated with a plastivore insect.	Barrionuevo et al. (2022)
	Galleria mellonella	*Acinetobacter*	The intestinal microbiome is related to the in vivo polyethylene biodegradation. An intricate relationship between an intact microbiome, low-density polyethylene (LDPE) biodegradation, and the production of glycol as a metabolic by-product was observed. The excretions were reduced by antibiotic treatment.	Cassone et al. (2020)
	Galleria mellonella	*Bacillus* sp., *Serratia* sp.	Plastic consumption after 21 days was 13 mg/larvae. Depolymerization and biodegradation form of long-chain fatty acids along with C=O and C-O-containing functional group compounds was observed.	Luo et al. (2020)
	Plodia interpunctella	*Enterobacter tabaci YIM Hb-3(B1), Bacillus subtilis NBRC 101239(B2)*	Changes in the surface topography of plastic samples (damage, including cavities and pits) increased with the time of treatment.	Mahmoud et al. (2020)

	Plodia interpunctella	*Enterobacter asburiae* YT1, *Bacillus* sp. YP1	The hydrophobicity of PE films was decreased, and obvious damage, together with pits and cavities (0.3–0.4 μm in depth), was observed on the surfaces of the PE films over 28 days of incubation. Suspension culture of the YT1 and YP1 strain (10^8cells/ml) degraded PE nearly 6.1 and 10.7% respectively the strain of the PE on 60th day incubation.	Yang et al. (2014)
	Tenebrio molitor	*Citrobacter* sp., *Kosakonia* sp.	Nearly 49.0% of ingested PE degraded into CO_2. The molecular weights of egested polymer residues decreased by 40.1% in PE-fed mealworms.	Brandon et al. (2018)
	Zophobas atratus	*Pseudomonas aeruginosa* PDB-1	Over 36 days, 24.04% low-density polyethylene (LDPE) and 15.12% linear low-density polyethylene (LLDPE) was consumed on average by the larvae, with survival rates of 85% and 87%, respectively. LDPE- and LLDPE-degrading enzymes e.g., P450 monooxygenase, alkane monooxygenase, alcohol dehydrogenase, etc. were identified in PDB-1.	Zaman et al. (2024)
	Tenebrio molitor	*Bacillus* sp. PELW2042	After 42 days of incubation, a large number of pits and cracks were observed on the polyethylene film surface accompanied by carbonyl groups (-C=O) and ether groups (–C–O) formation.	Zhang et al. (2023)
	Tenebrio molitor	*Pantoea* sp., *Lactococcus* sp., *Elizabethkingia* sp.	The structure of the gut microbiome was changed depending on the diet. The weight of the polyethylene substrate decreased by 16.6%, a clear indication of the link of the nutrition of larvae with the tested substrate.	Przemieniecki et al. (2020)
	Tenebrio molitor	*Citrobacter* sp., *Kosakonia* sp.	The microbes were able to degrade 49% of ingested PE into a putative gas fraction (CO_2). The molecular weights of the egested polymer decreased by 40.1%. Chemical modifications were consistent with the degradation and partial oxidation of the polymer. The gut microbes were able to degrade chemically dissimilar plastics.	Brandon et al. (2018)
	Tenebrio molitor	*Dyella, Lysobacter, Leptothrix*	The consumption rate was 0.0333–0.0345 mg/larvae/day. Suppression of gut microbes through antibiotic treatment significantly inhibits Expanded Polystyrene (EPS) depolymerization but does not stop LDPE depolymerization. Thus PS biodegradation was gut microbe-dependent but LDPE biodegradation was less dependent or independent of gut microbes. Formation of new oxygen-containing functional groups, change in physical property, chemical modification, indicated biodegradation of EPS and LDPE occurred.	Yang et al. (2020)

(continued)

Table 2 (continued)

Plastic material	Name of the insect	Symbiont microbes	Major findings	Reference
	Tribolium confusum	–	The larval survival rate was found 50% after 30 days.	Abdulhay (2020)
	Zophobas atratus	*Enterococcus* sp., *Citrobacter* sp.	The close relation between the diversity of gut microflora and plastics diets. A sole plastic diet significantly changes the activities of digestive enzymes.	Luo et al. (2021)
	Zophoba atratus	*Pseudomona aeruginosa*	The insertion of oxygen on the plastic surface was an indication of oxidation during the biodegradation process. PE had a weight loss of 0.64% per day.	Lee et al. (2020)
Polystyrene (PS)	*Alphitobius diaperinus*	*Pseudomonas* sp., *Kocuria* sp., *Cronobacter* sp., *Pseudogracilibacillus* *Virgibacillus* *Aspergillus, Hypdermella, Trichohoderma*	A loss of 89% weight of PS after 20 days of incubation. 28 days-old liquid cultures showed that thin fibrillar structures attached the concerned bacteria on the surfaces of the PS.	Cucini et al. (2020)
	Alphitobius diaperinus	*Klebsiella, pseudomonas, Stenotrophomonas*	Isolated bacteria were able to attach to PS surfaces and SEM observations showed the presence of thin fibrillar structures connecting the bacterial cells to the abiotic surface.	Cucini et al. (2022)
	Galleria mellonella	*Pseudomonas* sp., *Pseudomonas citronellolis, Lactobacillus* sp., *Schumannela* sp.	Feeding with polystyrene changes the bacterial gut microbiome of the insect larvae at the phyla and genus levels. Both bacterial and fungal communities are associated with the larvae of the plastivore insect.	Barrionuevo et al. (2022)
	Galleria mellonella	*Massilia* sp. FS1903	The PS film turns from initial hydrophobic to hydrophobic leading to structure and morphology change. More oxygen-containing functional groups were generated during PS degradation. After 30 days of incubation, the weight of the PS film significantly decreased (12.97 ± 1.05%).	Jiang et al. (2021)
	Galleria mellonella	*Acinetobacter* sp.		Zhang et al. (2020)

	Galleria mellonella	*Massilia* sp. FS1903	Degradation of PS was conducted using *Massilia* sp. FS1903 isolated from *galleria mellonella* larvae. Water contact angle measurements indicated that the hydrophobic surface of PS was changed to hydrophilic due to the appearance of more oxygen-containing functional groups upon larvae degradation.	Jiang et al. (2021)
	Galleria mellonella		Depolymerization and biodegradation resulted in the formation of long-chain fatty acids along with C=O and C-O-containing functional group compounds.	Lou et al. (2020)
	Galleria mellonella	*Bacillus* sp., *Serratia* sp.	Plastic consumption after 21 days was 5.87 mg /larvae, depolymerization and biodegradation form of long-chain fatty acids along with C=O and C-O-containing functional group compounds was observed.	Luo et al. (2020)
	Plesiophthalmus davidis	*Serratia* sp. WSW	Introduction of C-O and C=O bonds into the biodegraded PS film. Analysis of frass revealed a decrease in the average molecular weight of the residual PS in frass than that of feed styrofoam.	Woo et al. (2020)
	Tenebrio molitor	*Lactococcus and Elizabethkingia Agrobacterium, Nitrosomonas, Nitrospira*	The weight of the polystyrene substrate decreased by 12.14%. The enzyme activities were found higher in phosphatases, esterases, leucine arylamidase, β-galactosidase, β-glucuronidase, α-glucosidase, β-glucosidase, chitinase, α-mannosidase and α-fucosidase.	Przemieniecki et al. (2020)
	Tenebrio molitor	*Citrobacter* sp. *Kosakonia* sp.	Nearly 49.0% of ingested PE degraded into CO_2. The molecular weights of egested polymer residues decreased by 12.8% in PS-fed mealworms.	Brandon et al. (2018)
	Tenebrio molitor	*Exiguobacterium* sp. *YT2*	Within a 16-day 13-test period, 47.7% of the ingested styrofoam carbon was converted into CO_2.	Yang et al. (2015a)
	Tenebrio molitor	*Exiguobacterium* sp. *YT2*	Gentamicin-feeding mealworms lost their ability to depolymerize PS and mineralize PS into CO_2, over a 28-day incubation period making obvious pits and cavities (0.2–0.3 mm in width). Suspension culture of strain YT2 (10^8 cells/ml) was able to degrade 7.4 ± 0.4% of the PS pieces over a 60-day incubation period.	Yang et al. (2015b)
	Tenebrio molitor	*Aeromonas* sp. *TM1*	Yeast extract will be a very important co-factor for the TM1 with more efficient PS degrading ability.	Tang et al. (2017)

(continued)

Table 2 (continued)

Plastic material	Name of the insect	Symbiont microbes	Major findings	Reference
	Tenebrio molitor	*Klebsiella* sp., *Pseudomonas* sp., *Serratia* sp.	Gut symbionts were able to use diverse types of bioplastic as a sole source of carbon.	Urbanek et al. (2020)
	Tenebrio molitor	*Coprococcus*	The consumption rate of PS was 0.04.27 mg/ larvae/ day. The formation of new oxygen-containing functional groups, change in physical properties, and chemical modification, indicated biodegradation of EPS and LDPE occurred.	Yang et al. (2020)
	Tenebrio obscures *Tenebrio molitor*	*Spiroplasmataceae*, *Enterococcaceae*, *Enterobacteriaceae*	Treatment with antibiotics inhibited the depolymerization of PS and significant shifting in the gut microbial community in both species to three predominate families (Enterobacteriaceae, Spiroplasmataceae, and Enterococcaceae).PS consumption rate for *T. Obscurus* and *T. Molitor* was 0.3244 and 0.2430 mg/ larvae/day, respectively.	Peng et al. (2019)
	Tenebrio molitor	*Citrobacter* sp., *Kosakonia* sp.	The molecular weights of egested polymer residues decreased by 12.8 ± 3.1% in PS-fed mealworms. Chemical modifications were consistent with the degradation and partial oxidation of the polymer. The gut microbes were able to degrade chemically dissimilar plastics.	Brandon et al. (2018)
	Tribolium confusum		The larval survival rate was found in the PS diet (70%), after 30 days.	Abdulhay (2020)
	Tribolium castanium	*Acinetobacter* sp. *AnTc-1*	After 60 days, the mass weight (12.14%) PS was significantly reduced.	Wang et al. (2020)
	Zophobas atratus	*Pseudomonas* sp. DSM 50071	Formation of carbonyl group (C=O) in the oxidation pathway during PS biodegradation. The level of gene expression of serine hydrolase (SH) in DSM 50071 was highly increased during the degradation of PS. The enzyme-mediated biodegradation of PS was further confirmed by the treatment of an inhibitor.	Kim et al. (2020)
	Zophobas atratus	*Klebsiella pneumonia ZM1*	Yeast extract very important co-factor for the ZM1 with a more efficient PS degrading ability.	Tang et al. (2017)

	Zophobas atratus	*Pseudomonas aeruginosa* PDB-2	Over 36 days, 20.01% expanded polystyrene was consumed, the average survival rate of larvae was 90%., Phenylacetaldehyde dehydrogenase and other enzymes involved in EPS degradation.	Zaman et al. (2024)
	Uloma spp.	Not identified	After 28-day feeding consumption rate of PS was found as 0.3714 mg/ day/ larva. Supplementing waxcomb with the PS feed facilitated enhanced PS degradation. The percentage of residual PS in the frass reduced from 63 ± 1.0% on day 7 to 38.1 ± 1.0% by day 28.	Kundungal et al. (2021)
	Zophobas atratus Larvae	*Enterococcus, Dysgonomonas, Sphingobacterium*	Plastic diet-dependent diversity, characteristics of gut microbiomes as well as digestive enzyme activities.	Luo et al. (2021)
	Zophobas atratus	*Pseudomona aeruginosa*	Oxygen insertion on the plastic surface during the biodegradation process. The weight reduction of PS was 0.098%.	Lee et al. (2020)
Polyphenylene sulfide (PPS)	*Zophobas atratus*	*Pseudomona aeruginosa*	Oxygen insertion on the plastic surface indicates oxidation during the biodegradation process. The weight reduction of PPS was 0.53% per day	Lee et al. (2020)
Ethylene-vinyl acetate (EVA)	*Tribolium confusum*	Not identified	The highest larval survival rate was found (30%) after 30 days.	Abdulhay (2020)
Polyvinyl chloride (PVC)	*Tenebrio molitor*	*Lactococcus* sp.	The ingestion rate of plastic was 0.36.62 mg /larva/ during 16 days. Biodegradation and oxidation of the PVC plastics were supported by the formation of O-C and O=C functional groups.	Peng et al. 2020a
	Spodoptera frugiperda	*Enterococcus Klebsiella* sp. EMBL-1	The strain EMBL-1 forms a biofilm on the PVC film causing initial destruction of its hydrophobic surface through the secretion of extracellular enzymes. The genes and proteins (e.g., catalase-peroxidase, dehalogenases, enolase, aldehyde dehydrogenase, and oxygenase) are potentially involved in PVC degradation and utilization as a sole source of energy. The purified catalase-peroxidase takes part in PVC depolymerization.	Zhang et al. (2022)
Polyester polyurethane (PUF)	*Gryllus bimaculatus*	*Aspergillus flavus* G10	In the first report of direct aiding degradation of PUF fungi as insect gut microflora, the degrade rate of PUF was found 0.28 mg/g/d.	Khan et al. (2021)
Polyurethane (PU)	*Zophobas atratus*	*Enterococcus* sp. *Mangrovibacter* sp.	Sole plastic feeding affects the activities of digestive enzymes of the gut, strong association between gut microbiome diversity and plastic diets and the plastic-dependent characteristics of gut microbiomes.	Luo et al. (2021)

and Afla_006190, linked to laccase-like multicopper oxidase was also reported (Zhang et al. 2020). Several degrading enzymes, including PETase and MHETase secreted by the concerned gut microbes, degrade the skeleton of C-C groups of plastic into the -C=O- (carbonyl) resulting in depolymerization of the plastic polymers into monomers, dimers, or oligomers ((Yang et al. 2015a, b; Yoshida et al. 2016; Palm et al. 2019). A single bacteria strain can depolymerize diverse plastic polymers, having different efficiencies (Li et al. 2020).

A good number of extracellular enzymes such as lipases, depolymerase, cutinase, esterase, proteinase dehydrase K, and urease are secreted by gut symbionts for the degradation of plastic polymers (Shahnawaz et al. 2019; Taniguchi et al. 2019). These enzymes mainly depolymerize the plastic polymers into short fragments or small molecules. These fragments are then transported to cells to complete the degradation process (Amobonye et al. 2021). The PET hydrolase is responsible for the degradation of PET by changing the polymer chain or ring structure to make it susceptible to enzymatic hydrolysis (Kawai et al. 2019). These products of depolymerization are mineralized into CO_2. The increased gene expression level for the concerned hydrolase (e.g., serine hydrolase) evident from the analysis of the reverse transcription-quantitative polymerase chain reaction has also been reported (Kim et al. 2020). The leftover fragments, along with some gut microorganisms are excreted as fecula. Even after excretion, the process of degradation continues. However, extensive study at the molecular level on the physiological process of plastic degradation in the insect gut is required for an effective fight against the growing tide of plastic pollution. The overall role of insects and their enzymes in plastic management is presented in Fig. 2.

6 Enzymes and Challenges in Microbial Plastic Degradation

Insect-based degradation of plastic polymer is mainly confined to the enzymatic activities of insect gut microflora. The microbial degradation of plastics points toward longer incubation time to give considerable results. The key bottleneck of such difficulty is the inadequate amount of degradation enzymes such as hydrolases, depolymerases, and peroxidases, affecting the ability of such plastic bioremediating microbes (Amobonye et al. 2021). Several microbial enzymes such as hydrolases, laccases, esterases, cutinases, lipases, etc are linked to the degradation of plastic (Bhardwaj et al. 2013). Global research is still at the stage of preparing the list of enzyme cocktails presuming their promising activity in connection with plastic degradation. However, further research in the direction of increased yield of cocktails with such plastic degrading enzymes should be attempted. For such yield, fermentation technology-based scaling up, or even molecular tools are required. These answers are directed toward further uninterrupted research in the direction of plastic bioremediation. From the viewpoint of the upcoming days, insect-based prognoses of bioremediation, the advancement of scientific acquaintance with microbial populations, and their interactions with the gut interior environment and contaminants

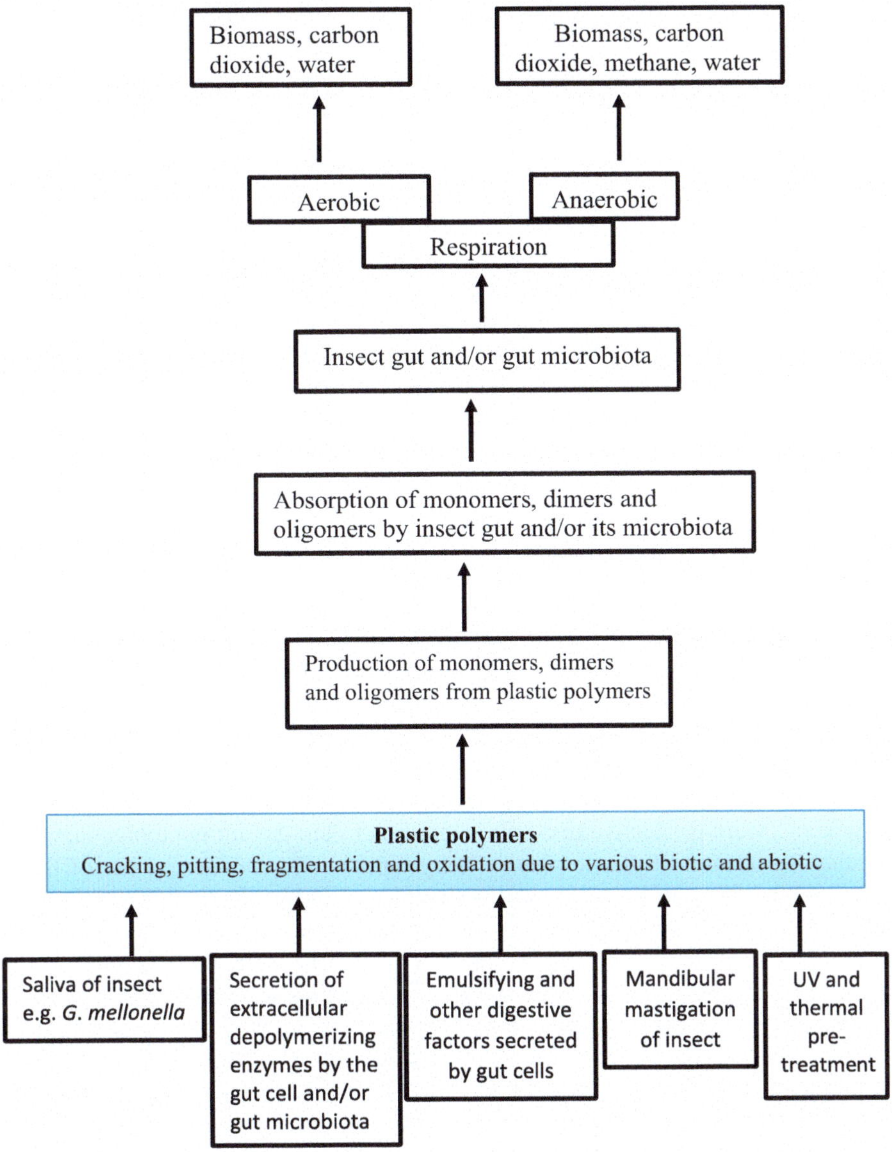

Fig. 2 Schematic representation of the role of insect and their gut microbiota in plastic waste management

should be thoroughly studied. Looking through their increased genetic capabilities to degrade plastics, the long-term field trial for new economical bioremediation techniques can increase the potential for significant advances. There is no doubt that bioremediation is a necessity of the current scientific research. It can lead to the shielding and maintenance of natural wealth that we have browsed from the next generations.

7 Conclusions and Future Perspective

From the literature, it is found that the use of plastic materials is unavoidable in meeting our day-to-day life essentials. Necessity and application of plastics is an ever-growing practice. This practice should be corroborated with plastic waste management. Negligence in plastic product disposal has resulted in enormous pollution, and there is no optimal solution for dealing with discarded plastics. Traditional techniques of decomposing plastic waste, such as landfilling, and physical and chemical treatment, have been shown to have limitations. These are neither economically viable nor eco-friendly. On the other hand, the utilization of insect larvae for plastic degradation is cost-effective and devoid of secondary pollution. Moreover, the larvae of insects may be used as valued sources of animal nourishment. Waste plastics can be recycled by feeding these to insects. To avoid toxicological dangers or the accumulative effects of dangerous materials in organisms, toxicological analysis must be conducted on insects that consume plastic in order to implement this approach. The outcome of current research on plastic degradation by potent plastivore insect larvae and their gut microbiota has consequences for real-world applications to some extent. Thus, efforts have been focused on identifying new plastivore insects and increasing the efficacy of insect larvae's plastic digestibility. The literature about the ability of plastic degradation is confined to the larvae of coleopteran and lepidopteran insects. Thus, it also necessitates continued investigation for the documentation of new insect species for plastic biodegradation from other insect orders. Toxicological analysis of these plastivore insects must be carried out to avoid any issue linked with toxicity.

Bioremediation of hazardous material through microbial activity is yet at a stage of infancy. More productive research is required to tap the gut microbial enzymes for the degradation of plastic polymers. Systematic and commercial use of these microbes should be well-organized. Combined application of genomic and fermentation techniques could direct positive results in this direction. The proper consortium of physical and chemical methods will also be helpful in getting rid of hazardous materials in the near future. Adequate learning and counseling have to be provided to our succeeding generation about the essentiality of eliminating hazardous material from the environment and its suitable disposal. The collective approach of mechanical methods to produce microplastics with further microbial methods of bioremediation of the microplastic is favorable. A continuous search for new species and strains for bioremediation processes is highly recommended.

The salivary enzyme is active on plastic at room temperature and at neutral pH. It is a promising tactic for plastic polymer biodegradation. This agenda defies the existing demarcation for plastic biodegradation. The knowledge is completely centered on the transformation of plastic to CO_2 through microbial metabolism. The pragmatic oxidation and plastic degradation depend on any microbial activity. The hassle-free environment for decay and degradation products (e.g., ketones and additives) advocates in favor of plastic decay and recycling/ upcycling potentialities of these enzymes. Thus, the saliva's enzymatic activities may be used as a substitute process for transforming the plastic to CO_2 through metabolic activities. Alternatively, this may be employed as the primary oxidative phase for the pathway of microbial degradation. This saliva of insects might be of consequence as a stockpile of degrading enzymes particularly for plastic polymers may revolutionize the field of bioremediation. For a deeper consideration, further research is necessary for the stepwise evolution related to the degradation of plastic with the salivary enzymes of insect larvae. Such a discovery is acquainted with another prospective line for dealing with plastic degradation. Considering the economic setup, this knowledge could unlock a prospective field for upcycling and forthcoming plastic manufacturing with ad hoc preparations liable to degradation by certain enzymes. Plastic degradation through biological systems using reutilization of the degraded products could be an approach to overcome the global threat of plastic waste management (Yang et al. 2015a, b). Bioremediation of plastic by insects, particularly utilizing the enzymes of their gut symbionts, would play a crucial role in the upcoming days.

References

Abdulhay H (2020) Biodegradation of plastic wastes by confused flour beetle *Tribolium confusum* Jacquelin du Val larvae. Asian J Agric Biol 8:201–206. https://doi.org/10.35495/ajab.2019.11.515

Al Hosni AS, Pittman JK, Robson GD (2019) Microbial degradation of four biodegradable polymers in soil and compost demonstrating polycaprolactone as an ideal compostable plastic. Waste Manag 97:105–114. https://doi.org/10.1016/j.wasman.2019.07.042

Alabi OA, Bakare AA, Xu X, Li B, Zhang Y, Huo X (2012) Comparative evaluation of environmental contamination and DNA damage induced by electronic-waste in Nigeria and China. Sci Total Environ 423:62–72. https://doi.org/10.1016/j.scitotenv.2012.01.056

Albertsson AC, Karlsson S (1990) The influence of biotic and abiotic environments on the degradation of polyethylene. Prog Polym Sci 15:177–192. https://doi.org/10.1016/0079-6700(90)90027-X

Al-Salem SM, Lettieri P, Baeyens J (2009) Recycling and recovery routes of plastic solid waste (PSW): a review. Waste Manag 29(10):2625–2643. https://doi.org/10.1016/j.wasman.2009.06.004

Amobonye A, Bhagwat P, Singh S et al (2021) Plastic biodegradation: frontline microbes and their enzymes. Sci Total Environ 759:143536

Arutchelvi J, Sudhakar M, Arkatkar A, Doble M, Bhaduri S, Uppara PV (2008) Biodegradation of polyethylene and polypropylene. Indian J Biotechnol 7:9–22

Banerjee S, Maity TK, Roy RN (2016) Identification and product optimization of amylolytic *Rhodococcus opacus* GAA 31.1 isolated from gut of Gryllotalpa africana. J Genet Eng Biotechnol 10:490–496. https://doi.org/10.1016/j.jgeb.2016.05.005

Banerjee S, Maity TK, Roy RN (2017) Protease production by thermo-alkaliphilic novel gut isolate *Kitasatospora cheerisanensis* GAP 12.4 from *Gryllotalpa Africana*. Biocatal Biotransformation 35(3):168–176. https://doi.org/10.1080/10242422.2017.1306739

Banerjee S, Maity TK, Roy RN (2020) Production, purification, characterization of cellulase from *Acinetobacter junii* GAC 16.2, a novel cellulolytic gut isolate of *Gryllotalpa africana* and its effects on cotton fiber and saw dust. Ann Microbiol 70:28. https://doi.org/10.1186/s13213-020-01569-6

Banerjee S, Maiti TK, Roy RN (2022) Enzyme producing insect gut microbes: an unexplored biotechnological aspect. Crit Rev Biotechnol 42(3):384–402. https://doi.org/10.1080/07388551.2021.1942777

Barnes DK, Galgani F, Thompson RC, Barlaz M (2009) Accumulation and fragmentation of plastic debris in global environments. Phil Trans R Soc 364(1526):1985–1998. https://doi.org/10.1098/rstb.2008.0205

Barrionuevo JMR, Vilanova-Cuevas B, Alvarez A, Martín E, Malizia A, Galindo-Cardona A, de Cristóbal RE, Occhionero MA, Chalup A, Monmany-Garzia AC, Godoy-Vitorino F (2022) The bacterial and fungal gut microbiota of the greater wax moth, Galleria mellonella L. consuming polyethylene and polystyrene. Front Microbiol 13:918861

Bhardwaj H, Gupta R, Tiwari A (2013) Communities of microbial enzymes associated with biodegradation of plastics. J Polym Environ 21(2):575–579

Bombelli P, Howe CJ, Bertocchini F (2017) Polyethylene bio-degradation by caterpillars of the wax moth Galleria mellonella. Curr Biol 27(8):R292–R293

Bowditch GT (1997) Penetration of Polyvinyl Chloride and Polypropylene packaging films by Ephestia cautella (Lepidoptera: Pyralidae) and Plodia interpunctella (Lepidoptera: Pyralidae) Larvae, and Tribolium confusum (Coleoptera: Tenebrionidae) adults. J Econ Entomol 90:1028–1031

Brandon AM, Gao SH, Tian R, Ning D, Yang SS, Zhao J et al (2018) Biodegradation of polyethylene and plastic mixtures in mealworms (larvae of Tenebrio molitor) and effects on the gut microbiome. Environ Sci Technol 52(11):6526–6533

Brandon AM, Garcia AM, Khlystov NA, Wu WM, Criddle CS (2021) Enhanced bioavailability and microbial biodegradation of polystyrene in an enrichment derived from the gut microbiome of *Tenebrio molitor* (mealworm larvae). Environ Sci Technol 55(3):2027–2036

Browne MA, Dissanayake A, Galloway TS, Lowe DM, Thompson RC (2008) Ingested microscopic plastic translocates to the circulatory system of the mussel, Mytilus edulis (L.). Environ Sci Technol 42:5026–5031

Cassone BJ, Grove HC, Elebute O, Villanueva SM, LeMoine CM (2020) Role of the intestinal microbiome in low-density polyethylene degradation by caterpillar larvae of the greater wax moth, Galleria mellonella. Proc R Soc B.287(1922):20200112

Crompton T (2007) Additive migration from plastics into foods. A guide for analytical chemistry. Smithers Rapra Publishing, Shrewsbury. ISBN: 1847350569

Cucini C, Leo C, Vitale M, Frati F, Carapelli A, Nardi F (2020) Bacterial and fungal diversity in the gut of polystyrene-fed Alphitobius diaperinus (Insecta: Coleoptera). Anim Gene 17:200109. https://doi.org/10.1016/j.angen.2020.200109

Cucini C, Funari R, Mercati D, Nardi F, Carapelli A, Marri L (2022) Polystyrene shaping effect on the enriched bacterial community from the plastic-eating *Alphitobius diaperinus* (Insecta: Coleoptera). Symbiosis 86:305–313. https://doi.org/10.1007/s13199-022-00847-y

Ellis LD et al (2021) Chemical and biological catalysis for plastics recycling and upcycling. Nat Catal 4:539–556

Engel MS (2015) Insect evolution. Curr Biol 25(19):868–872

Engel P, Moran NA (2013) The gut microbiota of insects – diversity in structure and function. FEMS Microbiol Rev 37(5):699–735

Geyer R, Jambeck JR, Law KL (2017) Production, use, and fate of all plastics ever made. Sci Adv 3:e1700782. https://doi.org/10.1126/sciadv.1700782

Giuliani A, Zuccarini M, Cichelli A, Khan H, Reale M (2020) Critical review on the presence of phthalates in food and evidence of their biological impact. Int J Environ Res Public Health 17(16):5655. https://doi.org/10.3390/ijerph17165655

Gonçalves FG, Brocco VF, Paes JB, Loiola PL, Lelis RCC (2014) Resistência de Painéisaglomerados de Acacia mangiumWilld. colados com ureia-formaldeído e taninos a organism osxilófagos. Floresta e Ambiente 21(3):409–415. https://doi.org/10.1590/2179-8087.059113

Gryn'ova G, Hodgson JL, Coote ML (2011) Revising the mechanism of polymer autooxidation. Org Biomol Chem 9:480–490

Gulrez SK, Ali Mohsin ME, Shaikh H et al (2014) A review on electrically conductive polypropylene and polyethylene. Polym Compos 35(5):900–914

Hedenqvist MS (2018) Chapter 26 – Barrier packaging materials. In: Kutz M (ed) Handbook of environmental degradation of materials, 3rd edn. William Andrew Publishing, New York, pp 559–581

Hegedus D, Erlandson M, Gillott C, Toprak U (2009) New insights into peritrophic matrix synthesis, architecture, and function. Annu Rev Entomol 54(1):285–302

Jiang S, Su T, Zhao J, Wang Z (2021) Isolation, identification, and characterization of polystyrene degrading bacteria from the gut of *Galleria mellonella* (Lepidoptera: Pyralidae) larvae. Front Bioeng Biotechnol 9:736062. https://doi.org/10.3389/fbioe.2021.736062

Kannan M, Mubarakali D, Thiyonila B, Krishnan M, Padmanaban B, Shantkriti S (2019) Insect gut as a bioresource for potential enzymes – an unexploited area for industrial biotechnology. Biocatal Agr Biotechnol 18:101010. https://doi.org/10.1016/j.bcab.2019.01.048

Kawai F, Kawabata T, Oda M (2019) Current knowledge on enzymatic PET degradation and its possible application to waste stream management and other fields. Appl Microbiol Biotechnol 103(11):4253–4268. https://doi.org/10.1007/s00253-019-09717-y

Kesti S, Sharana S (2019) First report on biodegradation of low density polyethylene by rice moth larvae, Corcyra cephalonica (Stainton). Holist Appr Environ 9:79–83. https://doi.org/10.33765/thate.9.4.2

Khan S, Dong Y, Nadir S, Schaefer DA, Mortimer PE, Xu J, Ye L, Gui H, Wanasinghe DN, Dossa GGO, Yu M, Sheng J (2021) Valorizing plastic waste by insect consumption. Circ Agric Syst 1(1):1–9. https://doi.org/10.48130/CAS-2021-0007

Kim HR, Lee HM, Yu HC, Jeon E, Lee S, Li J et al (2020) Biodegradation of polystyrene by Pseudomonas sp. isolated from the gut of superworms (larvae of Zophobas atratus). Environ Sci Technol 54(11):6987–6996

Kong HG, Kim HH, Chung JH, Jun J, Lee S, Kim HM et al (2019) The *Galleria mellonella* hologenome supports microbiota-independent metabolism of long-chain hydrocarbon beeswax. Cell Rep 26(9):2451–2464. https://doi.org/10.1016/j.celrep.2019.02.018

Konieczna A, Rutkowska A, Rachoń D (2015) Health risk of exposure to Bisphenol A (BPA). Rocz Panstw Zakl Hig 66:5–11

Kumar A, Kalleshwaraswamy CM, Sharma R, Sharma P, Poonia A (2022) Biodegradation of plastic using termites and their gut microbiota: a mini review IOP Conf Ser: earth. Environ Sci 1057:012016. https://doi.org/10.1088/1755-1315/1057/1/012016

Kundungal H, Gangarapu M, Sarangapani S (2019) Efficient biodegradation of polyethylene (HDPE) waste by the plastic-eating lesser wax worm (*Achroia grisella*). Environ Sci Pollut Res 26:18509–18519. https://doi.org/10.1007/s11356-019-05038-9

Kundungal H, Sarangapani S, Devipriya SP (2021) Biodegradation of polystyrene wastes by a newly reported honey bee pest *Uloma* sp. larvae: an insight to the ability of polystyrene-fed larvae to complete its life cycle. Environ Challenges 4:100083. https://doi.org/10.1016/j.envc.2021.100083

Lee H, Kim H, Jeon E, Yu HC, Lee S, Li J, Kim DH (2020) Evaluation of the biodegradation efficiency of four various types of plastics by *Pseudomonas aeruginosa* isolated from the Gut Extract of Superworms. Micro Org 8:1341

Lehane MJ (1997) Peritrophic matrix structure and function. Annu Rev Entomol 42(1):525–550

Lenz M, Lee CY, Lacey MJ, Yoshimura T, Tsunoda K (2011) The potential and limits of termites (Isoptera) as decomposers of waste paper products. J Econ Entomol 104(1):232–242. https://doi.org/10.1603/ec10155

Li J, Kim HR, Lee HM (2020) Rapid biodegradation of polyphenylene sulfide plastic beads by Pseudomonas sp. Sci Total Environ 720:137616

Liu J, Liu J, Xu B, Xu A, Cao S, Wei R (2022) Biodegradation of polyether-polyurethane foam in yellow mealworms (*Tenebrio molitor*) and effects on the gut microbiome. Chemosphere 304:135263. https://doi.org/10.1016/j.chemosphere.2022.135263

Lopez YM, Goncalves FG, Paes JB, Gustave D, Nantet ACT, Sales TJ (2020) Resistance of wood plastic composite produced by compression to termites *Nasutitermes corniger* (Motsch.)

and *Cryptotermes brevis* (Walker). Inte Biodeterior Biodegrad 152:104998. https://doi.org/10.1016/j.ibiod.2020.104998

López-Naranjo EJ, Alzate-Gaviria LM, Hernández-Zárate G, Reyes-Trujeque J, Cupul-Manzano CV, Cruz-Estrada RH (2013) Effect of biological degradation by termites on the flexural properties of pinewood residue/recycled high-density polyethylene composites. J Appl Polym Sci 128(5):2595–2603. https://doi.org/10.1002/app.38212

Lou Y, Ekaterina P, Yang SS, Lu B, Liu B, Ren N et al (2020) Biodegradation of polyethylene and polystyrene by greater wax moth larvae (*Galleria mellonella* L.) and the effect of co-diet supplementation on the core gut microbiome. Environ Sci Technol 54(5):2821–2831

Lucas N, Bienaime C, Belloy C, Queneudec M, Silvestre F, Nava-Saucedo JE (2008) Polymer biodegradation: mechanisms and estimation techniques–a review. Chemosphere 73:429–442. https://doi.org/10.1016/j.chemosphere.2008.06.064

Luo Y, Ekaterina P, Yang SS, Lu B, Liu B, Ren N, Corvini PF-X, Xing D (2020) Biodegradation of polyethylene and polystyrene by greater wax moth larvae (*Galleria mellonella* L.) and the effect of co-diet supplementation on the core gut microbiome. Environ Sci Technol 54:2821–2831. https://doi.org/10.1021/acs.est.9b07044

Luo L, Wang Y, Guo H, Yang Y, Qi N, Zhao X et al (2021) Biodegradation of foam plastics by Zophobas atratus larvae (Coleoptera: Tenebrionidae) associated with changes of gut digestive enzymes activities and microbiome. Chemosphere 282:131006

Lusher AL et al (2017) Sampling, isolating and identifying microplastics ingested by fish and invertebrates. Anal Methods 9:1346–1136. https://doi.org/10.1039/c6ay02415g

Mahmoud E, Al-Hagar O, Abdel-aziz M (2020) Gamma radiation effect on the midgut bacteria of *Plodia interpunctella* and its role in organic wastes biodegradation. Int J Trop Insect Sci 41:261–272

Maity S, Banerjee S, Biswas C, Guchhait R, Chatterjee A, Pramanick K (2021) Functional interplay between plastic polymers and microbes: a comprehensive review. Biodegradation 32:487–510. https://doi.org/10.1007/s10532-021-09954-x

Narancic T, O'Connor KE (2019) Plastic waste as a global challenge: are biodegradable plastics the answer to the plastic waste problem? Microbiology 165:129–137. https://doi.org/10.1099/mic.0.000749

Palm GJ, Reisky L, Böttcher D, Müller H, Michels EAP, Walczak MC, Berndt L, Weiss MS, Bornscheuer UT, Weber G (2019) Structure of the plastic-degrading Ideonella sakaiensis MHETase bound to a substrate. Nat Commun 10:1717

Peng BY, Su Y, Chen Z, Chen J, Zhou X, Benbow ME et al (2019) Biodegradation of polystyrene by dark (Tenebrio obscurus) and yellow (Tenebrio molitor) mealworms (Coleoptera: Tenebrionidae). Environ Sci Technol 53(9):5256–5265

Peng B-Y, Chen Z, Chen J, Yu H, Zhou X, Criddle CS, Zhang Y (2020a) Biodegradation of Polyvinyl Chloride (PVC) in *Tenebrio molitor* (Coleoptera: Tenebrionidae) larvae. Environ Int 145:106106

Peng BY, Li Y, Fan R, Chen Z, Chen J, Brandon AM et al (2020b) Biodegradation of low-density polyethylene and polystyrene in superworms, larvae of Zophobas atratus (Coleoptera: Tenebrionidae): broad and limited extent depolymerization. Environ Pollut 266:115206. https://doi.org/10.1016/j.envpol.2020.115206

Peng BY, Sun Y, Wu Z, Chen J, Shen Z, Zhou X, Wu WM, Zhang Y (2022) Biodegradation of polystyrene and low-density polyethylene by Zophobas atratus larvae: fragmentation into microplastics, gut microbiota shift, and microbial functional enzymes. J Cleaner Prod 367:132987. https://doi.org/10.1016/j.jclepro.2022.132987

Pivato AF, Miranda GM, Prichula J, Lima JEA, Ligabue RA, Seixas A, Trentin DS (2022) Hydrocarbon-based plastics: progress and perspectives on consumption and biodegradation by insect larvae. Chemosphere 293:133600. https://doi.org/10.1016/j.chemosphere.2022.133600

Przemieniecki SW, Kosewska A, Ciesielski S, Kosewska O (2020) Changes in the gut microbiome 1707 and enzymatic profile of Tenebrio molitor larvae biodegrading cellulose, polyethylene, and 1708 polystyrene waste. Environ Pollut 256:113265. https://doi.org/10.1016/j.envpol.2019.113265

Ren L, Men L, Zhang Z, Guan F, Tian J, Wang B, Wang J, Zhang Y, Zhang W (2019) Biodegradation of Polyethylene by *Enterobacter* sp D1 from the Guts of Wax Moth *Galleria mellonella*. Int J Environ Res Public Health 16:1941

Restrepo-Florez J-M, Bassi A, Thompson MR (2014) Microbial degradation and deterioration of polyethylene. Int Biodeterior Biodegrad 88:83–90

Rochman CM (2018) Microplastics research –from sink to source. Science 360:28–29

Rochman CM, Browne MA, Halpern BS, Hentschel BT, Hoh E, Karapanagioti H, Rios-Mendoza LM, Takada H, Tehe S, Thompson RC (2013) Classify plastic waste as hazardous. Nature 494:169–171. https://doi.org/10.1038/494169a

Rodriguez F, Cohen C, Ober CK et al (2014) Principles of polymer systems. CRC Press

Rosato D, Rosato D (2003) 4 – Product design, in plastics engineered product design. In: Rosato D, Rosato D (eds) In plastics engineered product design. Elsevier Science p, New York, pp 198–343

Roy RN (2020) Bioactive natural derivatives of phthalate ester. Crit Rev Biotechnol 40:913–929. https://doi.org/10.1080/07388551.2020.1789838

Roy PK, Hakkarainen M, Varma IK, Albertsson AC (2011) Degradable polyethylene: fantasy or reality. Environ Sci Technol 45(10):4217–4227

Sanluis-Verdes A, Colomer-Vidal P, Rodriguez-Ventura F, Bello-Villarino M, Spinola-Amilibia M, Ruiz-Lopez E, Illanes-Vicioso R, Castroviejo P, Aiese Cigliano R, Montoya M, Falabella P, Pesquera C, Gonzalez-Legarreta L, AriasPalomo E, Solà M, Torroba T, Arias CF, Bertocchini F (2022) Wax worm saliva and the enzymes therein are the key to polyethylene degradation by *Galleria mellonella*. Nat Commun 13:5568. https://doi.org/10.1038/s41467-022-33127-w

Shah AA, Hasan F, Hameed A, Ahmed S (2008) Biological degradation of plastics: a comprehensive review. Biotechnol Adv 26:246–265. https://doi.org/10.1016/j.biotechadv.2007.12.005

Shahnawaz M, Sangale MK, Ade AB (2019) In: Shahnawaz M, Sangale MK, Ade AB (eds) Bacteria as key players of plastic bioremediation, in bioremediation technology for plastic waste. Springer, Singapore, pp 45–69. https://doi.org/10.1007/978-981-13-7492-0

Shrivastava A (2018) Introduction to plastics engineering. In: Shrivastava A (ed) Introduction to plastics engineering. William Andrew Publishing, Cambridge, MA, pp 1–16

Silva AB, Costa MF, Duarte AC (2018) Biotechnology advances for dealing with environmental pollution by micro(nano)plastics: lessons on theory and practices. Curr Opin Environ Sci Health 1:30–35. https://doi.org/10.1016/j.coesh.2017.10.005

Singh B, Sharma N (2008) Mechanistic implications of plastic degradation. Polym Degrad Stab 93(3):561–584

Sivan A, Szanto M, Pavlov V (2006) Biofilm development of the polyethylene-degrading bacterium *Rhododcoccus ruber*. Appl Microbiol Biotechnol 72(20):346–352. https://doi.org/10.1007/s00253-005-0259-4

Spínola-Amilibia M, Illanes-Vicioso R, Ruiz-Lopez E, Colomer-Vidal P, Rodriguez-Ventura F, Perez RP, Arias CF, Torroba T, Sola M, Arias-Palomo E, Bertocchini F (2023) Plastic degradation by insect hexam-erins: near-atomic resolution structures of the polyethylene-degrading proteins from the wax worm saliva. Sci Adv 9(38):eadi6813. https://doi.org/10.1126/sciadv.adi6813

Tang KHD (2023) Enhanced plastic economy: a perspective and a call for international action. Environ Sci Adv 2:1011–1018. https://doi.org/10.1039/D3VA00057E

Tang X, Chen EYX (2019) Toward infinitely recyclable plastics derived from renewable cyclic esters. Chem 5:284–312. https://doi.org/10.1016/j.chempr.2018.10.011

Tang ZL, Kuo TA, Liu HH (2017) The study of the microbes degraded polystyrene. Adv Technol Innov 2(1):13–17

Taniguchi I, Yoshida S, Hiraga K, Miyamoto K, Kimura Y, Oda K (2019) Biodegradation of PET: current status and application aspects. ACS Catal 9:4089–4105. https://doi.org/10.1021/acscatal.8b05171

Terra WR, Ferreira C (2020) Evolutionary trends of digestion and absorption in the major insect orders. Arthropod Struct Dev 56:100931

Terzi E, Köse C, Büyüksarı Ü, Avcı E, Ayrılmış N, Kartal SN (2009) Evaluation of possible decay and termite resistance of particleboard containing waste tire rubber. Int Biodeterior Biodegradation 63(6):806–809

Uhrin AV, Schellinger (2011) Marine debris impacts to a tidal fringing-marsh in North Carolina Mar Pollut Bull 62(12):2605–2610. https://doi.org/10.1016/j.marpolbul.2011.10.006

Urbanek AK, Rymowicz W, Mirończuk AM (2018) Degradation of plastics and plastic-degrading bacteria in cold marine habitats. Appl Microbiol Biotechnol 102:7669–7678. https://doi.org/10.1007/s00253-018-9195-y

Urbanek A, Rybak J, Wróbel M, Leluk K, Mirończuk A (2020) A comprehensive assessment of microbiome diversity in *Tenebrio molitor* fed with polystyrene waste. Environ Pollut 262:114281

Wang Z, Xin X, Shi X, Zhang YL (2020) A polystyrene-degrading Acinetobacter bacterium isolated from the larvae of Tribolium castaneum. Sci Total Environ 726:138564

Wang S, Shi W, Huang Z, Zhou N, Xie Y, Tang Y et al (2022) Complete digestion/biodegradation of polystyrene microplastics by greater wax moth (*Galleria mellonella*) larvae: direct in vivo evidence, gut microbiota independence, and potential metabolic pathways. J Hazardous Mater 423:127213. https://doi.org/10.1016/j.jhazmat.2021.127213

Woo S, Song I, Cha HJ (2020) Fast and facile biodegradation of polystyrene by the gut microbial flora of *Plesiophthalmus davidis* Larvae. Appl Environ Microbiol 86:e01361–e01320. https://doi.org/10.1128/AEM.01361-20

Wrigth SL, Kelly FJ (2017) Plastic and human health: a micro issue? Environ Sci Technol 51(12):6634–6647. https://doi.org/10.1021/acs.est.7b00423

Yang J, Yang Y, Wu W-M, Zhao J, Jiang L (2014) Evidence of polyethylene biodegradation by bacterial strains from the guts of plastic-eating wax worms. Environ Sci Technol 48(23):13776–13784

Yang Y, Yang J, Wu W, Zhao J, Song Y, Gao L, Yang R, Jiang L (2015a) Biodegradation and mineralization of polystyrene by plastic-eating mealworms: part 1. Chemical and physical characterization and isotopic tests. Environ Sci Technol 49:12080–12086

Yang Y, Yang J, Wu W, Zhao J, Song Y, Gao L, Yang R, Jiang L (2015b) Biodegradation and mineralization of polystyrene by plastic-eating mealworms part 2. Role of gut microorganisms tests. Environ Sci Technol 49:12087–12093

Yang SS, Brandon AM, Flanagan AJC, Yang J, Ning D, Cai SY et al (2018) Biodegradation of polystyrene wastes in yellow mealworms (larvae of Tenebrio molitor Linnaeus): factors affecting biodegradation rates and the ability of polystyrene-fed larvae to complete their life cycle. Chemosphere 191:979–989. https://doi.org/10.1016/j.chemosphere.2017.10.117

Yang L, Gao J, Liu Y, Zhuang G, Peng X, Wu W-M, Zhuang X (2020) Biodegradation of expanded polystyrene and low-density polyethylene foams in larvae of *Tenebrio molitor* Linnaeus (Coleoptera: Tenebrionidae): broad versus limited extent depolymerization and microbe-dependence versus independence. Chemosphere 262(12):127818. https://doi.org/10.1016/j.chemosphere.2020.127818

Yang XL, Wang SH, Gong Y, Yang ZG (2021) Effect of biological degradation by termites on the abnormal leakage of buried HDPE pipes. Eng Fail Anal 124:105367. https://doi.org/10.1016/j.engfailanal.2021.105367

Yang S-S, Ding M-Q, Ren X-R, Zhang Z-R, Li M-X, Zhang L-L, Pang J-W, Chen C-X, Zhao L, Xing D-F, Ren N-Q, Ding J, Wu W-M (2022) Impacts of physical-chemical property of polyethylene on depolymerization and biodegradation in yellow and dark mealworms with high purity microplastics. Sci Total Environ 828:154458. https://doi.org/10.1016/j.scitotenv.2022.154458

Yoshida S, Hiraga K, Takehana T, Taniguchi I, Yamaji H, Maeda Y, Toyohara K, Miyamoto K, Kimura Y, Oda K (2016) A bacterium that degrades and assimilates poly (ethylene terephthalate). Science 351:1196–1199. https://doi.org/10.1126/science.aad635

Zaman I, Turjya RR, Shakil MS, Shahariar AL, Emu MRRH, Ahmed A, Hossain MM (2024) Biodegradation of polyethylene and polystyrene by *Zophobas atratus* larvae from Bangladeshi source and isolation of two plastic-degrading gut bacteria. Environ Pollut 345:123446

Zhang X, Tang H, Chen G, Qiao L, Li J, Liu B et al (2019) Growth performance and nutritional profile of mealworm reared on corn stover, soybean meal, and distiller's grains. Eur Food Res Technol 245(12):2631–2640

Zhang H, Liu Q, Wu H, Sun W, Yang F, Ma Y, Qi Y (2023) Biodegradation of polyethylene film by the Bacillus sp. PELW2042 from the guts of Tenebrio molitor (Mealworm Larvae) Process Biochem 130:236–244. https://doi.org/10.1016/j.procbio.2023.04.006

Zhang J, Gao D, Li Q, Zhao Y, Li L, Lin H, Bi Q, Zhao Y (2020) Biodegradation of polyethylene microplastic particles by fungi *Aspergillus flavus* from the guts of moth *Galleria mellonella*. Sci Total Environ 704:135931. https://doi.org/10.1016/j.scitotenv.2019.135931

Zhang Z, Peng H, Yang D, Zhang G, Zhang J, Ju F (2022) Polyvinyl chloride degradation by a bacterium isolated from the gut of insect larvae. Nat Commun 13:5360

Zhou B, Zhao L, Wang Y, Sun Y, Li X, Xu H, Weng L, Pan Z, Yang S, Chang X et al (2020) Spatial distribution of phthalate esters and the associated response of enzyme activities and microbial community composition in typical plastic-shed vegetable soils in China. Ecotoxicol Environ Saf 195:110495. https://doi.org/10.1016/j.ecoenv.2020.110495

Zhu P, Shen Y, Li X, Liu X, Qian G, Zhou J (2022) Feeding preference of insect larvae to waste electrical and electronic equipment plastics. Sci Total Environ 807:151037. https://doi.org/10.1016/j.scitotenv.2021.151037

Biofertilizer and Remediation of Soil and Other Environments

Saikat Dey, Anannya Dhar, Soujanya Jana, Sudipta Tripathi, Shilajit Barua, Mahadev Bera, Sukamal Sarkar, Sourav Garai, and Debojyoti Moulick

Abstract Hiking in food demand to feed the mammoth increasing population, intensive agricultural practices have been the prerequisite for modern agriculture in the recent few years. Intensive agriculture and profuse crop production often rely on optimum cultivable land and excessive use of synthetic fertilizers. Despite high nutrient content, ease of availability to plants, and ability to promote faster growth, excessive application of agrochemicals has been found to pose risks to both soil health and the environment while also potentially polluting groundwater and the atmosphere in the long-term. Therefore, to overcome these challenges, biofertilizers, derived from beneficial microorganisms, offer a sustainable alternative to chemical fertilizers, contributing to soil fertility enhancement and environmental restoration. The utilization of biofertilizers promotes soil health by fostering microbial activity, improving nutrient availability, and enhancing soil structure. Moreover, biofertilizers have shown promising results in mitigating soil pollution and remediating contaminated environments through biodegradation, bioremediation, biostimulation, bioaugmentation, and phytoremediation. This chapter explores the potential applications of biofertilizers in addressing various environmental challenges, highlighting their efficacy, benefits, and limitations. Future research directions and

S. Dey · A. Dhar · S. Jana · S. Sarkar · S. Garai
Division of Agronomy, School of Agriculture and Rural Development, Ramakrishna Mission Vivekananda Educational and Research Institute, Narendrapur, Kolkata, West Bengal, India

S. Tripathi · M. Bera
School of Environment and Disaster Management, Ramakrishna Mission Vivekananda Educational and Research Institute, Narendrapur, Kolkata, West Bengal, India

S. Barua
Department of Microbiology, Vijaygarh Jyotish Ray College, Jadavpur, Kolkata, India

D. Moulick (✉)
School of Environment and Disaster Management, , Ramakrishna Mission Vivekananda Educational and Research Institute, Narendrapur, Kolkata, West Bengal, India

Department of Environmental Science, University of Kalyani, Nadia, West Bengal, India

A. A. H. Abdel Latef et al. (eds.), *Sustainable Remediation for Pollution and Climate Resilience*, https://doi.org/10.1007/978-981-96-5674-5_20

potential advancements in biofertilizer technology for sustainable soil and environmental management are also discussed.

Keywords Soil health · Soil pollution · Microbial activity · Sustainable soil management

1 Introduction

Soil health improvements in soil fertility status, organic matter content, soil quality indices, and belowground ecosystems are the major factors for crop production anywhere in the world. As the world's population expands, ensuring food security and adequate nutrition becomes increasingly imperative (Majumdar et al. 2024a, b; Santosh et al. 2024; Moulick et al. 2024a, b, c, d). In 2022, the world's population reached 8 billion and is projected to grow to around 9 billion by the end of 2040 (Choudhury and Moulick 2022; Roy and Hossain 2024). Despite the present demand for food production, the available land remains static in quantity; therefore, farmers often rely on the excessive application of chemical fertilizers to augment crop production with limited available land resources. However, most of the cultivable soils in the world lose their capacity to produce optimum yield due to several factors that accelerated the negative impacts on soil health, including intensified cropping systems, imbalance fertilization, overgrazing, excessive tillage, heavy metals contamination (Deb et al. 2023; Ghosh et al. 2023; Rizvi et al. 2020), and traditional farming practices, making the soil inappropriate for crop production (Pattnaik et al. 2023). Even under anaerobic conditions, nitrogenous fertilizers are denitrified through several microorganisms and promote dangerous Green House Gas (GHGs) like nitrous oxide into the atmosphere which in turn causes ozone layer depletion.

Furthermore, heavy metals and metalloids generated from geogenic as well as several anthropogenic means. These HMs generally accumulated in the soil, are not readily taken up by plants and thus persist for longer duration in the soil due to their nondegradable characteristics (Saif and Khan 2017; Moulick et al. 2016a, 2018a, 2019a, b; Saha et al. 2019; Sharma et al. 2020; Moulick et al. 2021); therefore, they may deteriorate the soil physicochemical functions and contaminate groundwater table. When the concentration of heavy metals exceeds their threshold boundary/limits, disbalance in the microbial population, stunted plant growth, and consequently reduces crop yield are noticed (Goyal et al. 2020; Majumdar et al. 2023a, b; Rizvi and Khan 2017b, 2018). The harmful effects of toxic metals often involve damaging genetic material, disturbing essential physiological functions like lipid peroxidation and photosynthesis, inhibiting respiration, interfering with protein synthesis, and impacting carbohydrate metabolism (Moulick et al. 2016b, 2017, 2018c; Rizvi and Khan 2017a; Jaishankar et al. 2014; Rizvi et al. 2022), leading to the yield reduction and plant death (Moulick et al. 2018a, b; Waqas et al. 2019; Amari et al. 2017). Apart from this, fertilizers accumulated in the ponds, sea, and nearby lakes along with runoff, promote nutrient concentrations like nitrogen and phosphates resulting in algal blooms and biological oxygen demand (BOD) and subsequent death of aquatic species.

Therefore, efforts have been made to remediate the polluted soil by implementing sustainable approaches to safeguard the crop rhizosphere and restore soil fertility. In recent years, the world's focus on sustainable agriculture and environmental remediation has encouraged interest in biofertilizers as effective tools for improving soil fertility and mitigating environmental degradation. Biofertilizers act as sustainable agricultural components, encompassing a diverse range of microbial inoculants and offering promising solutions for enhancing soil health while reducing reliance on synthetic fertilizers and agrochemicals (Bhardwaj et al. 2014). Biofertilizers offer handy application methods, including foliar spray, seed treatment, and soil application. When microbial inoculants are applied to the soil, they show profuse multiplication in the plant rhizosphere or endophytes inside the plant roots, building an association by providing nutrients to the host plant in terms of food from the plant (Malusá and Vassilev 2014). Bioremediation involves the application of biofertilizers used as nitrogen fixers, phosphate mobilizers/solubilizers, potassic solubilizers, etc. (Afzal et al. 2017). These biofertilizers offer an inexpensive and environmentally feasible solution that detoxifies metals and possesses distinctive characteristics to enhance plant growth (Ouertani et al. 2020). Restoring polluted soil through plant-microbe interactions is a widely adopted and important microbial technology (Hansda et al. 2014). Bioremediation is becoming popular for managing polluted soil, near about 35% of people rely on it whereas, 16% of people choose plant-based remediation technology to remove hazardous metals and metalloids from the contaminated soil (Rizvi et al. 2022).

The application of biofertilizers in soil offers numerous advantages, including the provision of nutrients through processes such as fixation, mineralization, mobilization, and the recycling of nutrient pools (Zaidi et al. 2017). Moreover, biofertilizers contribute to soil organic carbon accumulation, thereby improving soil fertility and resilience. Khandare et al. (2020) in their experiment showed that 25% chemical fertilizer can be replaced by incorporating *Azotobacter* and PSB in wheat cultivation. The impact of biofertilizers extends beyond nutrient provision, encompassing both direct and indirect effects on soil's physical and chemical properties. This, in turn, improves soil porosity and reduces bulk density, facilitating better water infiltration and root growth. Indirectly, biofertilizers enhance soil chemical properties by increasing cation exchange capacity, improving nutrient retention and plant uptake availability. Additionally, biofertilizers contribute to the neutralization of soil pH, creating a more favorable environment for plant growth and microbial activity. Biofertilizers have immense potential to remediate heavy metals and promote plant growth by establishing a hormonal balance in plant physiology (Zerrouk et al. 2020). *Rhizobium* inoculated *Medicago sativa* plant play a significant role in mitigating toxicity hazards likewise, regulation of MT-gene and phytochelatin synthesis to reduce the copper stress in plants (Kaur et al. 2019).

Therefore, applying various types of biofertilizers comprising more than one advantage, including soil and other environmental remediation alongside enhancing crop performance and overall productivity, becomes essential to address global food requirements.

2 Biofertilizers: Definition and Types

2.1 *Definition of Biofertilizers*

Biofertilizers are microbial inoculants containing live or dormant microbial cells, which when applied to seed, soil, and plant surfaces promote plant growth and development by enhancing the bioavailability and bioaccessibility of essential plant nutrients and improving nutrient uptake efficiency, as well as providing other beneficial effects such as disease resistance and stress tolerance (Sarkar et al. 2023; Chaudhary et al. 2022; Santra et al. 2015; Moulick et al. 2020a, b, c; Mahmud et al. 2021). As the name suggests, they are bio-based organic fertilizers derived from bacteria, fungi, and other microorganisms that have the potential to transform the nutrients from unavailable to available forms for plants through various microbial processes like fixation of atmospheric nitrogen in the soil, solubilization of phosphorous, and mobilization of potassium, making them readily available for the crops (Majumdar et al. 2023; Itelima et al. 2018; Rokhzadi et al. 2008). They can also break down organic matter, enrich compost, solubilize zinc and other micronutrients, and enhance soil health and fertility (Hazra et al. 2024).

They can rightly be claimed to be one of the most advanced biotechnological tools that promote sustainable agriculture and environmental sustainability. Being an integral component of integrated nutrient management, they cutoff the excessive application of inorganic chemical fertilizers (Santra et al. 2015).

2.2 *Categories of Biofertilizers*

Based on the activity associated with the biofertilizer, it can be classified into numerous types. Nitrogen-fixing biofertilizers, phosphate-solubilizing/ mobilizing biofertilizers, and potassium mobilizing/ solubilizing biofertilizers are the prominent and often used ones. Plant growth-promoting biofertilizers, zinc solubilizing biofertilizers, sulfur-oxidizing biofertilizers, and compost accelerators are also being used in sustainable agricultural practices (Mahmud et al. 2021).

2.2.1 Nitrogen-Fixing Biofertilizers

Though an enormous amount of nitrogen (dinitrogen) is present in the atmosphere but is not available to the plants. Hence, it needs to be transformed into a usable form. Some of the nitrogen-fixing biofertilizers help plants use atmospheric nitrogen by converting it into ammonia (NH_4^+) in the root nodules via biological nitrogen fixation (Tairo and Ndakidemi 2013). This ammonia is further converted to nitrite and nitrate. The nitrogen-fixing biofertilizers can be further classified as symbiotic (*Rhizobium, Frankia, Cassuarina*, Blue-green algae), non-symbiotic/free-living (*Azotobacter, Clostridium*), and associative symbiotic (*Azosprillum*) based on the association of the microbe present in the biofertilizer and roots of the plants (Chaudhary et al. 2022). *Rhizobium* is a nodule-forming symbiont, while blue-green algae (*Anabaena azollae*) are nonnodule-forming symbionts (Santra et al. 2015).

2.2.2 Phosphorous-Solubilizing/Mobilizing Biofertilizers

Despite the presence of a good quantity of phosphorous in the soil, a very small portion is usable for crop nutrition, thus making it the second most limiting plant macronutrient after nitrogen (Bechtaoui et al. 2021). After fertilization with chemical fertilizers, soluble inorganic phosphorous gets immobilized in the soil (Mahanty et al. 2017). Moreover, phosphorous is immobile in the soil (Santra et al. 2015). Phosphorous-solubilizing bacteria (PSB) are the group of bacteria that have the potential to solubilize the fixed phosphorous in the soil and make it available for uptake by the plant roots (Santra et al. 2015). The phosphorous-mobilizing biofertilizers help mobilize the soluble phosphorous and mineralize the organic phosphorous compounds, which are the unavailable soil pools of phosphorous (Kirui et al. 2022). They even transfer the available phosphorous from the soil to the cortex of feeder roots (Asoegwu et al. 2020). *Bacillus* and *Pseudomonas* are PSB, while *Penicillium* and *Aspergillus* are phosphorous-solubilizing fungi (PSF). Among bacteria and fungi, the efficacy of solubilizing phosphorous is seen higher in the former (Alam et al. 2002). Arbuscular mycorrhiza (Vesicular-arbuscular mycorrhiza (VAM)) falls under the phosphorous mobilizers/ absorbers (Asoegwu et al. 2020).

2.2.3 Potassium Solubilizing/Mobilizing Biofertilizers

Potassium is an essential nutrient for plant growth and is crucial in various physiological processes. However, potassium availability in the soil can be limited, especially in soils with low potassium content or poor potassium release characteristics (Zorb et al. 2014). Potassium solubilizing biofertilizers are a promising solution to address this issue. These biofertilizers contain live microorganisms that can solubilize potassium from insoluble forms in the soil, making it readily available for plant uptake. They are either bacteria or fungi that have the potential to solubilize the usable potassium from its sources, like potassium-bearing minerals, by the process of releasing organic acids (Varga et al. 2020). They are also considered potassium-solubilizing/ dissolving bacteria (Ghaffari et al. 2018). *Frateuria aurantia, Acidithiobacillus ferrooxidans, Bacillus mucilaginous, and Herbaspirillum* spp. are some of the prominent potassium-solubilizing bacteria. Aspergillus and arbuscular mycorrhizal fungi (AMF) can also act as potassium solubilizers (Chaudhary et al. 2022).

2.2.4 Sulfur-Solubilizing Biofertilizers

Sulfur is emerging as the most important plant nutrient after N, P, and K due to its essential role in various crops' metabolic processes, especially in oilseeds and pulses. Sulfur deficiency is widespread in Indian soils and can severely impact the growth and yield of crops, especially in nutrient-deficient soils. To address this issue, researchers have focused on exploring the potential of sulfur-solubilizing bacteria to enhance sulfur availability in the soil (Kaur et al. 2020). Sulfur-solubilizing biofertilizers are sulfur-oxidizing bacteria (SOB) that have the potential to yield sulfate (available form) from hydrogen sulfide (unavailable form) (Wang et al. 2019). The common sulfur-oxidizing bacteria include *Azotobacter pasturianam, Wolinella succinogenes, Thiobacillus, Bacillus, Paracoccus, Rhodopseudomonas*, etc. (Kusale et al. 2021).

2.2.5 Plant Growth-Promoting Biofertilizers

These mainly include plant growth-promoting rhizobacteria, which dwell in the soil's rhizosphere. The rhizosphere is defined as the area surrounding the roots of the plants and is the zone of interaction between roots and the soil microflora (Kumar et al. 2015). They are known to affect crops' growth and development and make them tolerant of adverse environmental stresses (Ghosh et al. 2019). Plant growth-promoting rhizobacteria (PGPR) are brilliant sustainable agricultural tools in this era of climate change as they perform a variety of roles, starting with nitrogen fixation, the release of phytohormones, reducing heavy metals from the soil system, and even preventing disease and pest attacks (Mohanty et al. 2021). *Pseudomonas taiwanensis, Pantoea agglomerans* in maize, and *Rhizobium lupini* in alfalfa are some of the experimented with PGPR to enhance the overall performance of the mentioned crops (Chaudhary et al. 2022).

2.2.6 Organic Matter Decomposing Biofertilizers

Organic matter-decomposing biofertilizers are those microorganisms that break down the complex substances present in organic matter into simpler and more readily usable forms by plants (Chaudhary et al. 2022). They are also added to various composts to aid the decomposition process and hence are also regarded as compost accelerators. Cellulose, hemicellulose, and lignin are organic matter components generally broken down by bacteria, fungi, and actinomycetes into inorganic forms (Lou et al. 2022). They are further divided into cellulose-breaking organisms (cellulolytic) like *Trichoderma*, and *Cellulomonas* and lignin-breaking organisms (lingolytic) like *Humicola, Arthobacter,* and *Agaricus* (Phukongchai et al. 2022).

2.2.7 Iron-Mobilizing Biofertilizers

Iron is an essential micronutrient for plants having crucial roles in chlorophyll formation and photosynthesis. Plant-utilizable iron is generally scarce in aerobic field conditions as it is present in ferric conditions and thus forms insoluble complexes (Lurthy et al. 2020). Iron-mobilizing biofertilizers are also referred to as siderophore-producing biofertilizers, which help solubilize iron and make it usable for plants. Siderophores are the iron-binding proteins of lower molecular weights that capture the ferric iron cation and transform it into a ferrous cation, which the roots can uptake easily (Mahmud et al. 2021). Thus, they help in chelating the ferric cation of iron in the environment (Saha et al. 2016). *Pseudomonas aer*uginosa RSP8, *Pseudomonas aeruginosa* RSP5, and *Pseudomonas fluorescens* C7, are the siderophore-producing strains of bacteria (Sah et al. 2017).

2.2.8 Plant Hormone-Producing Biofertilizers

Biofertilizers that produce phytohormones are microbial formulations containing helpful microorganisms capable of generating plant hormones like auxins (IAA), cytokinins, gibberellins, and others (Eichmann et al. 2021). These biofertilizers are utilized in agriculture to improve plant growth, development, and overall productivity. For instance, certain strains of bacteria like *Azospirillum* and *Azotobacter* produce auxins, while some fungi like *Trichoderma* spp. and Mycorrhizal fungi produce auxins, cytokinins, and gibberellins (Vejan et al. 2016). These plant hormones are generally produced to balance the hormone level inside the host plant during environmental stress (Lopes et al. 2021).

2.2.9 Zinc Solubilizing Biofertilizers

Zinc is considered an essential micronutrient for plant growth and metabolism, but its deficiency in soil is increasing at an alarming rate (Ali et al. 2021). Within a week of zinc application in soil, an enormous amount of it gets transformed into unavailable forms such as zinc oxide, zinc phosphate, and zinc carbonate (Saravanan et al. 2004). Zinc-solubilizing bacteria (ZSB) help the plant by solubilizing the unavailable zinc to a readily available form promoting easy uptake by roots (Kumawat et al. 2019). Some of the zinc solubilizers are of the genera *Thiobacillus, Pseudomonas, Bacillus*, and *Acinetobacter* (Saravanan et al. 2007).

2.2.10 Silicon-Solubilizing Biofertilizers

Silicon is released in the soil as a result of the weathering of silicate-containing minerals, which is quite a slow process. Silicon-solubilizing bacteria are the group of microbes dwelling in soil that have the potential to solubilize silicon from silicate minerals, especially aluminum silicate, which are insoluble in soil solution by decomposing them and thus enhancing the uptake by feeder roots (Kumawat et al. 2019). *Bacillus caldolyticus, Bacillus mucilaginosus, Proteus mirabilis*, and *Pseudomonas* are some of the common silicate-solubilizing bacteria (SSB) (Meena et al. 2014a, b).

3 Mechanisms of Biofertilizers

3.1 Mechanism of Nitrogen-Fixing Biofertilizers

Nitrogen fixation is generally categorized as a free-living, associative symbiotic, and symbiotic association. The symbiotic association is again classified into nodule forming and nonnodule forming, of which nodule forming is the most commercially exploited (Santra et al. 2015). Different strains of *Rhizobium* form symbiotic associations with specific leguminous plants i.e., cross-inoculation group. The *Rhizobium* is generally present in the rhizosphere or is added via seed and soil treatment before sowing. The root hairs release various chemicals containing flavonoids and amino acids that attract the rhizobia toward itself. After assembling around the root hair, the nodulation factor (nod factor) is released by the *Rhizobia*, leading to the curling of the root hair. *Rhizobia* proliferates inside the root hair and causes the infection thread, which helps the bacteria get inside the root and form nodule primordia. The nodule primordia divides to form outgrowth called nodules, and the *Rhizobium* dwelling inside the nodule transforms into branched, swollen, nondividing structures called *Bacteroides*. The nitrogenase enzyme and the leghemoglobin pigment are produced inside the *Bacteroid*, both of which are essential for nitrogen fixation (Tairo and Ndakidemi 2013).

Nonnodule forming fixation is done by *Azolla*, an aquatic fern for symbiotic association with cyanobacteria named *Anabaena azollae*. Cyanobacteria reside in the leaf cavities of Azolla, and they perform nitrogen fixation (Gupta 2004).

3.2 Mechanism of Phosphorous-Solubilizing/Mobilizing Biofertilizers

The availability of phosphorous in the soil is made by solubilization and mineralization of nonavailable phosphorous, which co-occurs with the same bacteria (McComb et al. 2013). Phosphate solubilization is generally performed by phosphate-solubilizing bacteria (PSB) dwelling in higher amounts at the plant rhizosphere

(Youssef and Eissa 2014). These bacteria synthesize and secrete organic acids with low molecular weights (LMWs), such as citric and gluconic acids, oxalic acids, maleic acids, formic acids, lactic acids, fumaric acids, etc. (Mahmud et al. 2021). These acids lower the soil system's pH and help dissolve the bound phosphates (Asoegwu et al. 2020). The conversion of insoluble phosphorous to soluble organic phosphorous in the soil system is created by the chelation of the cations (iron and aluminum) bound with the phosphate by the carboxyl (COOH) and hydroxyl (OH) group present in the secreted organic acids (Mahanty et al. 2017). Phosphorous mineralization from immobile form is carried out by forming phosphatase enzymes which fasten the hydrolysis of phosphoric esters (McComb et al. 2013). The VAM helps mobilize and absorb phosphorous by root by transferring it from the soil to the cortex of the feeder roots. They get into the cortex of the roots by penetration, leading to an increment in root surface area, followed by displacing off the equilibrium of absorption of the phosphate and thus assisting the P ion transfer. The arbuscules uptake the phosphate into the roots (Asoegwu et al. 2020).

3.3 *Mechanism of Potassium-Solubilizing/Mobilizing Biofertilizers*

Potassium is present in the soil in the form of K-containing minerals like biotite, feldspar, muscovite, mica, orthoclase, and illite, but the feeder roots cannot directly uptake them in this form (Meena et al. 2014). For uptake by plants, they needed to be transformed into a simpler form. Potassium availability is enhanced by mobilizing and solubilizing the insoluble K-bearing minerals. Acidification of the rhizosphere owing to the release of various organic acids and protons like citric acid, oxalic acid, α-ketogluconic acid, succinic acid, and gluconic acid, leading to the solubilization of K from minerals is one of the prominent mechanisms of potassium mobilizing biofertilizers (Ahmad et al. 2016). These organic acids released by microorganisms either enable direct solubilization of mineral K or dissociate the lattice structure of the potassium-bearing minerals by chelating the iron, calcium, silicon, and aluminum ions leading to the release of potassium in the soil solution for better uptake by the roots (Olaniyan et al. 2022). Thus, the mechanism of potassium availability can be summarized as the lowering of soil pH, followed by chelation of the cations bound with K and acidification and protonation of the rhizospheric area, resulting in the release of available potassium for phyto- uptake (Zarjani et al. 2013; Parmar and Sindhu 2013; Uroz et al. 2009).

3.4 *Mechanism of Zinc-, Silicon-, Iron-, and Sulfur-Mobilizing Biofertilizers*

Zinc solubilization occurs mainly by two mechanisms i.e. lowering of rhizospheric pH due to organic acid production and zinc chelation (Sindhu et al. 2019). Organic acids and protons reduce the soil pH, resulting in the solubilization of zinc oxide, zinc carbonate, and zinc sulfide (Saravanan et al. 2004). A unit decrease in soil pH is reported to increase availability by 100 times (Havlin et al. 2005). The plant roots and the microbes release various zinc-chelating compounds that bind zinc and enhance their bioavailability in the rhizosphere for better uptake by feeder roots of the plants (Sindhu et al. 2019).

Silicon solubilization by silicon-solubilizing bacteria is achieved through several mechanisms. One mechanism is the production of organic acids by the bacteria, such as citric acid or oxalic acid. These organic acids help to solubilize silicon by breaking down insoluble silicon compounds into soluble forms (Kumawat et al. 2019). Another mechanism is the secretion of enzymes by the bacteria, such as phosphatases and glycosidases, that can hydrolyze silicon-containing compounds. These enzymes break down the bonds within silicon compounds, converting them into soluble forms that plants can readily take up (Guerriero et al. 2016).

Iron-mobilizing biofertilizers work by utilizing specific microorganisms that can solubilize and mobilize iron in the soil. These microorganisms produce siderophores, compounds that bind to iron and make it more soluble and available for plant uptake (Dimkpa et al. 2009). As a result, plants can access and absorb the iron they need for optimal growth and development. Furthermore, these microorganisms may also enhance iron bioavailability by releasing organic acids and enzymes that break down insoluble iron compounds in the soil.

Sulfur-mobilizing biofertilizers function by enhancing sulfur availability in the soil through microbial activity. These biofertilizers contain specialized strains of bacteria capable of solubilizing sulfur from insoluble forms such as elemental sulfur or sulfide minerals, making them accessible to plants (Chaudhary et al. 2022). They oxidize the hydrogen sulfide to sulfate, which is available to the feeder roots of the plants (Wang et al. 2019). Figure 1 depicts the mechanisms of various biofertilizers at the rhizosphere of the plant.

4 Role of Biofertilizers in Soil Health

4.1 Improvement of Soil Physicochemical Properties

Soil health and soil fertility are the two most important terms in the case of crop production and productivity, though these terms are significantly interrelated. Soil health encompasses the physicochemical and microbial ecosystems that enrich the soil's inherent capacity to supply vital plant nutrients appropriately, thereby justifying the term soil fertility. Therefore, soil structure and texture play an indispensable role in maintaining soil health and fertility. Biofertilizers contain several beneficial microorganisms, including nitrogen-fixing, phosphate, and potassium-solubilizing soil bacteria, mycorrhizal fungi, and other nutrient-mobilizing microbes. Soil microbial ecosystems play a crucial role in regulating the biophysical properties of soil by secreting organic compounds like exopolysaccharides, and organic acids

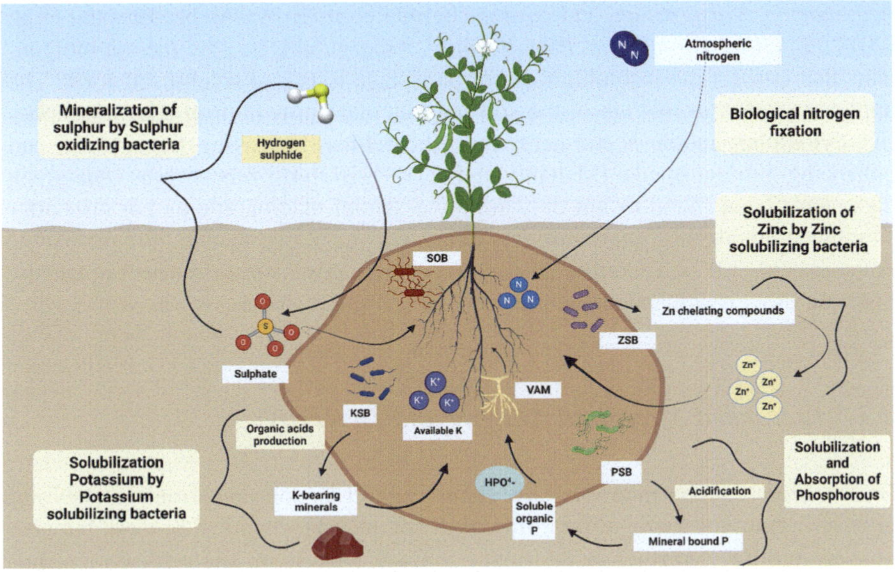

Fig. 1 The mechanisms of various biofertilizers in mobilizing and solubilizing essential plant nutrients at the root rhizosphere of the plant. (Image created with BioRender.com; License Number: MK26RAFOR7)

that aid in soil particle aggregation, contributing to the formation of a well-defined soil structure (Wani et al. 2015). Microbial decomposition and extracellular debris of microorganisms help in addition to organic matter in the soil, which enhances the nutrient availability and exchange capacity within the soil. Biofertilizers like phosphate-solubilizing microbes and *Azotobacter* act as buffering agents by reducing soil pH (Mohammadi and Sohrabi 2012). They can improve soil porosity by binding the particles together and promoting root expansion, moisture holding capacity, infiltration rate, and lowering bulk density (Kalhapure et al. 2013). Ultimately, biofertilizer application improves soil texture and structure, prevents soil erosion, and contributes to soil health and sustainable agricultural practices.

4.2 *Enhancement of Nutrient Availability*

Enrichment of microbial inoculants in the soil through the application of biofertilizers includes nitrogen-fixing, phosphate-solubilizing or mobilizing, potassium mobilizing, and plant growth-promoting bacteria, one of the key elements in integrated nutrient management that help in reducing the dependency on chemical fertilizers along with improving soil quality and sustainable crop production (Hossain et al. 2021c). Biofertilizers containing microbes enhance the soil nitrogen status by fixing atmospheric nitrogen into plant-available forms, such as *Rhizobium*,

Azotobacter, *Azospirillum*, and *Nostoc* (Itelima et al. 2018). As per Etesami et al. (2017), *Bacillus, Penicillium, Pseudomonas, Aspergillum*, etc., are the microorganisms that convert the insoluble form of phosphate ions into solubilized form. The application of *Azospirillum* in the soil induces plant growth-promoting hormones like cytokinins, indole acetic acid, and gibberellins, improving root growth and enhancing nutrient uptake (Mahanty et al. 2017). Biofertilizers such as blue-green algae have been found to not only supply essential macronutrients but also offer significant amounts of micronutrients like iron, zinc, molybdenum, and others (Al Abboud et al. 2013). Therefore, biofertilizers indirectly increase nutrient uptake by boosting root growth and exudation.

4.3 Increased Microbial Activity and Diversity

Biofertilizers contain microbial inoculants that act as soil conditioners. They not only have nutrient-fixing ability through a complex biochemical process but also possess multifaceted roles in improving the soil ecosystem. Biofertilizers encompass microbial formulation that can be applied through seed, soil, or foliar application, colonized in the root rhizosphere, and make nutrients insoluble to soluble forms (Rokhzadi et al. 2008). Several nitrogen-fixing bacteria including *Rhizobium*, *Azotobacter*, and *Azospirillum* have a significant role in capturing atmospheric nitrogen (N_2) and converting it into ammonia (NH_3). They secrete several plant growth-promoting hormones that induce complex enzyme structures, solubilizing and mobilizing insoluble plant nutrients, fastening organic matter decomposition, and providing a favorable niche for a diversified microbial population (Youssef and Eissa 2014). The application of diverse groups of microorganisms has been reported to be involved in mobilizing and solubilizing the fixed and insoluble phosphate and potassium in the soil that may be readily available to the plant. Mycorrhiza forms a mutualistic association between the roots of higher plants and soil-borne fungi, and they help extend the rhizosphere through the hyphal network (Frey-Klett et al. 2011). They have been involved in producing several enzymes that break down complex phosphorous, and nitrogen (Salvioli et al. 2012), secreting organic acid, and antioxidant enzymes; in return, plants provide carbohydrates to fungal organisms (Aggarwal et al. 2011). Therefore, they can form a symbiotic relationship, which could be attributed to increased microbial population in the soil, which means improved soil ecosystems, especially biological communities.

4.4 Reduction of Soil-Borne Diseases

Microorganisms present in the formulation of biofertilizers are an essential component in agricultural science, as they provide a prevention mechanism by inducing several antibiotics and antifungal agents to inhibit the growth of phytopathogens in plant rhizosphere (Mazurier et al. 2009). Several researchers reported that bacteria like *Azotobacter* perform against various soil-borne pathogens by secreting substances like gibberellins, naphthalene acetic acid, and other phytohormones, leading to profuse root growth and efficient mineral uptake (Mathivanan et al. 2015). According to Hermosa et al. (2012) rhizobacterial formulations available to the market for commercial purposes, most of these bioinoculants act as plant disease-resisting products rather than supplements of mineral nutrients. Several *Rhizobium* species, including *R. leguminosarum* (Afzal and Bano 2008), *R. japonicum* (Shaharoona et al. 2006), and *R. meliloti* (Prabha et al. 2013), play a significant role against phytopathogenic fungi and bacteria by producing antibiotics and antifungal components in the soil.

4.5 Reduction of Chemical Fertilizer Dependence

After the Green Revolution, the farming community in the agricultural sector entirely depends on synthetic fertilizers, weedicides, insecticides, and other pesticides to increase crop yield and productivity in terms of quality and quantity. Chemical fertilizers and pesticides have the potential to supply nutrients in available form and provide the efficient utilization of essential minerals to the plant, promote rapid growth, and increase crop yield (Ghosh et al. 2020a, b, 2021, 2022a, b). But in recent years, the outmost challenges in the agricultural sector, especially in developing countries are determined by the unavailability of vital crop nutrients in both sufficient quantity and types of chemical fertilizers (Itelima et al. 2018). While overexploitation of chemicals in agricultural land not only deteriorated the physico-chemical properties of the soil (Ye et al. 2020) but also posed an enormous threat to the soil's beneficial microbial population, which led to diminished crop production and quality of produce (Khandare et al. 2020). Excessive utilization of chemical fertilizers not only degrades the soil quality but also negatively impacts the soil-plant-environment continuum. Biofertilizers encompass beneficial microbial inoculants that hold living cell cultures of the effective strains of nitrogen fixer, phosphorus solubilizers or mobilizers, potassium solubilizers, and others (Itelima et al. 2018; Fasusi et al. 2021). Biofertilizers are a major component of integrated nutrient management approaches, in terms of both improving soil health and sustainable crop production while, in the meantime, safeguarding the environmental ecosystems, being economically viable, and providing eco-friendly, renewable crop nutrient sources to reduce the dependency on chemically induced fertilizers in sustainable agriculture practices (Yadav and Sarkar 2019). Biofertilizers not only provide essential macronutrients but also supply the major micronutrients to the plant likewise, *Azolla,* when applied as a biofertilizer, can significantly provide a certain amount of zinc, iron, molybdenum, and other micronutrients to the plant (Al Abboud et al. 2013).

4.6 *Increased Crop Yield and Quality*

The application of biofertilizers in the soil has gained special attention in recent years for their cost-effectiveness, environment-friendliness, and sustainability in nature, consequently increasing the yield and quality of the crop (Kumar et al. 2017). Plants rely on obtaining essential macro and micronutrients in adequate quantities, which they can acquire through the decomposition of organic matter and the supplementation of chemical fertilizers to enhance their growth and maximize yield (Mahmud et al. 2021). On the other hand, biofertilizers, which incorporate microbial cultures, accelerate the breakdown of organic matter, facilitating a steady release of plant nutrients in readily available forms within the soil. This prompts a rapid response from the plants and enhances the overall yield and productivity of the crop (Pii et al. 2015). The positive potential impact of biofertilizers on crop yield and quality largely depends on the continuity and consistency of their application on the soil. According to Panda (2011), his experiment estimated that the application of biofertilizers may increase the crop yield by up to 65%. *Azotobacter* as biofertilizer application in maize increases the yield by almost 63.11%, cob length (cm) by 362.2%, and grain weight (g) by 48.7% respectively over control (Mahato and Kafle 2018; Gao et al. 2020).

5 Role of Biofertilizers in Soil Remediation

Rapid urbanization and population expansion in recent years have generated enormous amounts of hazardous waste, disrupted socioeconomic activity, and jeopardized the sustainability of the environment (Awasthi et al. 2022). To ensure food security throughout the world, crop production is required to be optimized using fertile agroecosystems. Anthropogenic activities, such as excessive and prolonged use of chemical pesticides and fertilizers, have irreversibly harmed the soil to increase crop yield (Ji and Sang 2022). Environmental contaminants include organic compounds such as pesticides, plastics and their additives, perfluoroalkyl, and polyfluoroalkyl substances (PFAS), polyaromatic hydrocarbons (PAHs), mercury (Hg), and inorganic nutrients like phosphates and nitrates (Awasthi et al. 2022). Of these, potentially toxic elements (PTEs) like heavy metals like chromium (Cr), arsenic (As), cadmium (Cd), lead (Pb), and mercury (Hg), detrimentally affect the sustainability of crop production systems (Awasthi et al. 2022).

The biological non-degradative nature of heavy metals generally causes them to linger in the environment, and their subsequent absorption hurts the physiological processes linked to the uptake of biofertilizers that are beneficial to agriculture (Sharma et al. 2020). Because of these, it is essential to reduce soil pollution to restore the cultivability of stressed soils that have been rendered unfit for use. To preserve rhizospheres and restore soil fertility, efforts have been made to clean up the contaminated soils. The physical, chemical, and biological (bioremediation) processes are the main choices that have been used thus far. Techniques like adsorption (Kumar et al. 2022b); photodegradation; advanced oxidation process (AOP) (Kumar et al. 2022a); bioremediation; electrokinetic remediation, electro-bioremediation (Li et al. 2020); bioleaching (Xu et al. 2020); and phytoremediation have been identified to remove these pollutants from environmental matrices. Although the physical and chemical approaches are quick, they are complicated because of factors like cost, facilities for operations, and how differently they affect the biological qualities of soils, which leads to secondary contamination (Ali et al. 2013). Bioremediation is a sustainable and eco-friendly way to clean up pollutants with various benefits over conventional remediation techniques like excavation and burning (Sharma and Shukla 2023).

Bioremediation is a process that uses microorganisms, plants, or their enzymes to detoxify or remove pollutants from contaminated environments (Sharma et al. 2021; Yaashikaa et al. 2022a, b). This process harnesses the natural abilities of microorganisms such as bacteria, fungi, and plants to metabolize or transform pollutants into less harmful forms. These organisms can break down or detoxify a wide range of contaminants, including petroleum hydrocarbons, heavy metals, pesticides, and industrial chemicals (Sharma et al. 2021). Several factors influence bioremediation's effectiveness, including environmental conditions (such as temperature, pH, and oxygen levels), the availability of nutrients essential for microbial growth, and the type and concentration of contaminants present (Alori et al. 2022). Optimizing these factors can enhance the efficiency of bioremediation processes.

The presence of beneficial soil microbiota, which improves the nutrient pool through the biogeochemical cycling of the soil constituents, is one of the key components of healthy soil. Biofertilizers are substances containing living microorganisms, primarily bacteria, fungi, or algae, that enhance soil fertility and plant growth by facilitating nutrient uptake, improving soil structure, and promoting beneficial interactions within the rhizosphere (the soil zone influenced by plant roots) (Murugaragavan Ramasamy and Yuvaraj 2020). Using biofertilizers is one of the best modern strategies for agricultural sustainability. Integrated nutrient management in agricultural systems includes biofertilizers as a key component because they are affordable and sustainable alternatives to synthetic fertilizers that provide plant nutrients (Mącik et al. 2020; Kumar et al. 2022a, b). While their primary function is to improve agricultural productivity, biofertilizers indirectly contribute to bioremediation by fostering microbial activity in the soil. Biofertilizers play a vital role in soil remediation by enhancing soil fertility and promoting the growth of plants, which can aid in the restoration of degraded ecosystems. Unlike chemical fertilizers, which primarily provide essential nutrients to plants, biofertilizers contain beneficial microorganisms that contribute to soil health and productivity through various mechanisms.

One of the key roles of biofertilizers in soil remediation is their ability to fix atmospheric nitrogen into a form that plants can utilize (Singh et al. 2021a, b, c). Nitrogen-fixing bacteria, such as *Rhizobium* and *Azotobacter*, form symbiotic relationships with certain plant species or live freely in the soil, converting atmospheric nitrogen into ammonia or nitrate compounds that plants can absorb. This process reduces the reliance on synthetic nitrogen fertilizers, which can contribute to environmental pollution when overused (Abd-Alla et al. 2023).

Additionally, biofertilizers contain other beneficial microorganisms, such as phosphate-solubilizing bacteria and mycorrhizal fungi, which help solubilize and mobilize nutrients like phosphorus, potassium, and micronutrients in the soil, making them more available to plants (Teotia et al. 2017; Chakraborty and Akhtar 2021). These microorganisms also contribute to soil aggregation and structure, improving water retention and reducing erosion (Usharani et al. 2019).

Furthermore, biofertilizers can play a role in bioremediation by enhancing the degradation of organic pollutants in the soil (Thakur 2023). Certain microorganisms present in biofertilizers possess the metabolic capabilities to degrade contaminants like hydrocarbons (Girigiri et al. 2019), pesticides (Prodhan et al. 2023), and detoxify heavy metals (Rizvi et al. 2022) through enzymatic processes. By inoculating contaminated soils with specific microbial consortia, biofertilizers can accelerate the breakdown of pollutants, leading to their detoxification and eventual removal from the environment.

6 Bioremediation of Heavy Metals

Heavy metal pollution in soil poses significant environmental and health risks due to its persistence and toxicity (Ali et al. 2019). These metals, including cadmium (Cd), lead (Pb), mercury (Hg), arsenic (As), chromium (Cr), and nickel (Ni), originate from various natural and anthropogenic sources (Kumari and Mishra 2021) (Table 1). Such pollution in soil has multifaceted impacts. It disrupts soil health and fertility, hindering plant growth and reducing agricultural productivity (Hossain et al. 2021a; Chowardhara et al. 2019; Hossain et al. 2023; Moulick et al. 2020b; Rashid et al. 2023). Heavy metal pollution inhibits plant growth, disrupts nutrient uptake, and damages vital processes like photosynthesis (Moulick et al. 2023a, b; Hazra et al. 2023; Goyal et al. 2020). Toxic metals impair root development, induce leaf chlorosis and necrosis, and reduce overall biomass production (Sahli and Belhiouani 2022). This interference jeopardizes crop yields, ecosystem stability, and food security. Contaminated soil poses risks to ecosystems as metals leach into water sources, affecting aquatic life and biodiversity (Ogidi and Akpan 2022). Human health suffers as crops uptake metals, leading to food contamination and health ailments (Chowardhara et al. 2024; Moulick et al. 2024a, b, c, d; Zakaria et al. 2021) like organ damage and cancer (Table 2). Toxic metals like cadmium, lead, and arsenic disrupt soil structure and fertility (Khatun et al. 2022), inhibiting microbial activity crucial for nutrient cycling and soil ecosystem functioning (Aponte et al. 2020). Metal toxicity reduces microbial diversity and alters microbial composition, compromising soil resilience and ecosystem stability (Pan et al. 2020). This disruption impairs the soil's ability to support plant growth and hinders bioremediation processes. Moreover, some metal-resistant microbes proliferate, altering soil microbial dynamics (Hao et al. 2021).

Table 1 Sources of heavy metal pollution in soil

Heavy metal	Natural sources	Anthropogenic sources	References
Cadmium (Cd)	Volcanic eruptions, weathering of cadmium-containing rocks and minerals	Mining activities, industrial emissions, use of phosphate fertilizers	Hayat et al. (2019)
Lead (Pb)	Geological weathering of lead-bearing minerals, erosion of lead-rich soils	Industrial emissions, mining and smelting, lead-based paints, and gasoline	Raj and Das (2023)
Mercury (Hg)	Volcanic activity and degassing, natural weathering of mercury-containing rocks	Coal combustion, artisanal gold mining, industrial processes (e.g., chlor-alkali production)	Veeraswamy et al. (2023), Chalkidis et al. (2020)
Arsenic (As)	Geological processes releasing arsenic from minerals and ores	Mining and smelting, pesticides and herbicides, contaminated groundwater	Patel et al. (2023), Khan and Flora (2023)
Chromium (Cr)	Weathering of chromium-bearing rocks and soils	Industrial discharges, tanneries, steel manufacturing	Prasad et al. (2021), Dongre (2021)
Nickel (Ni)	Natural weathering and erosion of nickel-rich rocks and minerals	Stainless steel production, mining and smelting, coal combustion	Harasim (2018)

Agricultural practices also contribute to heavy metal pollution in soil. The use of chemical fertilizers, pesticides, and sewage sludge containing heavy metals can accumulate these contaminants in agricultural soils (Kumari and Mishra 2021). Additionally, irrigation with contaminated water sources can introduce heavy metals into the soil, affecting crop quality and posing risks to human health through food consumption (Balkhair and Ashraf 2016).

Urbanization and improper waste management exacerbate heavy metal pollution in soil. Urban areas often have high concentrations of heavy metals due to vehicular emissions, industrial activities, and the disposal of electronic waste containing metals like lead, mercury, and cadmium. Runoff from urban areas can transport these contaminants into nearby soil and water bodies, contaminating surrounding environments (Silva et al. 2021).

Mitigating heavy metal pollution in soil involves a multifaceted approach. Phytoremediation, using plants to absorb and accumulate contaminants, helps remediate polluted areas (Ali et al. 2013). Soil washing and electrokinetic remediation techniques can remove metals from soil (Yang et al. 2018). Implementing proper waste management practices reduces the input of heavy metals into the environment. Promoting sustainable agricultural practices like organic farming minimizes metal accumulation in soil (Alam et al. 2020). Regulatory measures to control industrial emissions (Khanam et al. 2023) and monitor agricultural runoff are crucial. Public awareness campaigns raise consciousness about the dangers of heavy metal pollution, fostering community engagement and support for pollution prevention and remediation efforts.

Table 2 Effect of heavy metal pollution in soil on plants and animals

Heavy metal	Effects on plants	Effects on animal systems	References
Cadmium (Cd)	Inhibits seed germination and root growth, disrupts photosynthesis and nutrient uptake, causes leaf chlorosis and necrosis	Kidney damage, bone weakening, reproductive issues, and increased cancer risks	Shaari et al. (2022), Reyes-Hinojosa et al. (2019)
Lead (Pb)	Reduces plant growth and development, damages root structure and inhibits water uptake, interferes with enzyme activity and metabolic processes	Damages the nervous system and cognitive function and impairs growth and development in children	Zeng et al. (2020), Ortega et al. (2021)
Mercury (Hg)	Inhibits seedling growth and root elongation, disrupts water balance and nutrient transport, causes leaf chlorosis and stunted growth	Neurotoxic effects, leading to cognitive and developmental impairments, bioaccumulates in aquatic food chains	Kalinhoff and Calderón (2022), Zafar et al. (2024)
Arsenic (As)	Impairs root growth and development, disrupts carbohydrate metabolism and enzyme function, leads to leaf wilting and necrosis	Carcinogenic properties, increasing cancer risks, disrupts cellular functions and DNA integrity	Chandrakar et al. (2016), Nabi et al. (2021), Partridge et al. (2007)
Chromium (Cr)	Reduces plant growth and biomass production, alters mineral nutrient uptake and distribution, causes leaf yellowing and necrosis	Allergic reactions and skin irritation, respiratory issues upon inhalation of chromium compounds	López-Bucio et al. (2022), Long et al. (2024)
Nickel (Ni)	Inhibits seed germination and root elongation, disrupts water and nutrient uptake, causes leaf chlorosis and reduced photosynthesis	Dermatitis and skin allergies upon skin contact, respiratory problems upon inhalation of nickel-containing particles	Hassan et al. (2019), Banerjee and Roychoudhury (2020)

Bioremediation offers a sustainable and cost-effective approach to mitigate heavy metal pollution in soil and water ecosystems. This process utilizes living organisms such as bacteria, fungi, algae, and plants to detoxify or remove heavy metals from contaminated environments (Ojuederie and Babalola 2017). Bioremediation offers several advantages over traditional remediation methods, including low environmental impact, minimal ecosystem disturbance, and the potential for long-term sustainable solutions. However, its effectiveness depends on various factors, such as the type and concentration of heavy metals in environmental conditions and the presence of suitable microbial or plant species (Pande et al. 2020).

Phytoremediation is a bioremediation method that uses plants to extract, stabilize, or accumulate heavy metals from soil or water (Bhat et al. 2022). Hyperaccumulating plants like certain species of willow, sunflower, and Indian mustard can absorb and concentrate metals in their tissues, which can be harvested and disposed of safely (Singh et al. 2022).

Microorganisms play a crucial role in bioremediation by enzymatically transforming or immobilizing heavy metals, rendering them less toxic or more easily. This heavy metal remediation approach involves harnessing microorganisms' capabilities to detoxify or remove metal contaminants from soil, water, and sediments. Microbes like bacteria and fungi employ various mechanisms such as biosorption, biotransformation, and bioaccumulation to sequester or transform heavy metals into less toxic forms (Chug et al. 2021). Certain bacterial species, like *Pseudomonas* and *Bacillus*, possess metal-resistant genes and enzymes that facilitate metal detoxification through processes such as biosorption, bioaccumulation, and biotransformation (Fakhar et al. 2022). These processes can occur naturally in the environment or be enhanced through bioaugmentation, where metal-resistant microbes are introduced to contaminated sites (Hassan et al. 2019; Purwanti et al. 2020). Microbial remediation offers a sustainable and cost-effective approach to mitigate heavy metal pollution, contributing to environmental restoration and safeguarding human health. Several bacterial and fungal species employ different mechanisms for detoxifying heavy metals in soil. Such instances are as follows.

6.1 Cadmium (Cd)

Cadmium remediation by microorganisms involves several species known for mitigating cadmium contamination in soil and water. Here are some notable examples:

- *Pseudomonas putida*: This bacterium is renowned for its cadmium detoxification capabilities through the production of metal-chelating compounds and biosorption onto its cell surface components (Halim et al. 2020).
- *Bacillus* spp.: Certain species of *Bacillus* like *B. subtilis*, *B. cereus*, and *B. megaterium* are effective in cadmium bioremediation due to their biosorption capacity, which involves binding cadmium to cell surface structures (Huang et al. 2014; Halim et al. 2020; Chi et al. 2020).
- *Rhizobium* spp.: Some strains of *Rhizobium* are employed in cadmium bioremediation through rhizofiltration, wherein the plant-associated bacteria assist in the uptake and accumulation of cadmium by plant roots (Lebrazi and Fikri-Benbrahim 2018).
- *Saccharomyces cerevisiae*: This yeast species is utilized for cadmium remediation via biosorption, where cadmium ions are bound to the cell wall polysaccharides of *Saccharomyces cerevisiae* (Hadi and El-Naas 2019).
- *Aspergillus niger*: This fungus is known for its biosorption and bioaccumulation capabilities, which involve the extracellular binding of cadmium to fungal biomass (Yang et al. 2005).

6.2 Lead (Pb)

Lead (Pb) remediation by microorganisms involves several strategies employed by various bacteria and fungi to mitigate lead contamination in soil and water. Some common approaches include:

- *Pseudomonas sp:* The cell surface structures like lipopolysaccharides in *Pseudomonas putida* and structures containing binding sites like Carbonyl (C=O), phosphate (PO_4^{3-}), hydroxyl (-OH), and amino (-NH) groups favor lead biosorption in *Pseudomonas aeruginosa* ASU 6a (Gabr et al. 2008). Pseudomonas fluorescence produces enzymes capable of detoxifying lead by converting it to less toxic forms (Naik and Dubey 2013). *Pseudomonas aeruginosa* can produce metallothioneins, which are small, cysteine-rich proteins capable of binding heavy metals such as lead. Metallothioneins sequester lead ions within the bacterial cell, reducing their toxicity and bioavailability (Naik et al. 2012). *Pseudomonas fluorescens* ATCC 13525 can actively transport lead ions across the cell membrane and sequester them within intracellular compartments. This process helps detoxify the cells and reduce lead concentration in the surrounding environment. *Pseudomonas putida* produce metal-chelating compounds (siderophores) that sequester lead ions, reducing their bioavailability in the environment (Tripathi et al. 2005). *Pseudomonas aeruginosa* and *Pseudomonas putida* produce extracellular polymeric substances (EPS), which contain functional groups such as carboxyl, hydroxyl, and amino groups. These groups have a high affinity for lead ions (Pb^{2+}), allowing them to adsorb onto the surface of the bacterial cells or the EPS matrix (Vimalnath and Subramanian 2018a, b).
- *Bacillus sp:* Certain species of *Bacillus* (*B. subtilis* and *B. licheniformis*) have cell wall components such as teichoic acids that contain functional groups (e.g., carboxyl, amino) capable of binding lead ions (Pb^{2+}), facilitating their adsorption onto the cell surface (Beveridge et al. 1982; Mirouze et al. 2018). *Bacillus sp* can also accumulate lead ions within their cells. Once adsorbed onto the cell surface, lead ions may be transported across the cell membrane and sequestered within intracellular compartments, reducing their concentration in the environment (Arifiyanto et al. 2017). Some *Bacillus* and *Paenibacillus* species produce EPS, which serves as a matrix for the adsorption of lead ions (Fella-Temzi et al. 2018). The EPS contains binding sites that interact with lead ions, enhancing their retention on the cell surface or in the surrounding environment.
- *Desulfovibrio sp:* Sulfate-reducing bacteria like *Desulfovibrio desulfuricans*, an anaerobic bacterium possess enzymes, such as hydrogenases and sulfite reductases, that can reduce toxic lead ions (Pb^{2+}) to less harmful forms, such as elemental lead (Pb^0) or insoluble lead sulfide (PbS). Sulfate is an electron acceptor in the reduction process, allowing the bacteria to transfer electrons to lead ions and facilitating their reduction. This reduction process immobilizes lead, making it less soluble and less mobile in the environment (Song 2003).
- *Pseudomonas* species could indirectly contribute to lead bioreduction by creating reduced environmental conditions. By metabolizing organic compounds,

Pseudomonas consumes oxygen, which can lead to the depletion of oxygen and the creation of anaerobic microenvironments (Schobert and Tielen 2010). In these anaerobic conditions, other microorganisms, such as sulfate-reducing bacteria like *Desulfovibrio desulfuricans*, may thrive and carry out the bioreduction of lead. Certain aerobic bacteria produce enzymes that can catalyze the reduction of lead ions to less toxic forms. Although this process is less common in aerobic bacteria compared to anaerobic bacteria, it can still contribute to lead bioreduction in oxygen-rich environments.

Fungi Fungi of the *Aspergillus* genus possess cell wall components containing functional groups (e.g., hydroxyl, amino) that bind lead ions. These binding sites facilitate the adsorption of lead onto the fungal biomass, including mycelium and hyphae (Tian et al. 2019). *Aspergillus* species can accumulate lead within their cellular structures. Lead ions may be adsorbed onto the cell wall (Bairagi et al. 2011) or taken up by the fungal cells (Qiu et al. 2021) and sequestered within intracellular compartments, reducing their concentration in the surrounding environment. Some *Aspergillus* species produce EPS capable of binding or sequestering lead ions (Paria et al. 2022), reducing their bioavailability in the environment.

6.3 Mercury (Hg)

Several bacteria and fungi are involved in mercury (Hg) bioremediation due to their unique abilities to transform or sequester mercury in contaminated environments. Some of the notable examples in this regard are:

Bacteria

- *Desulfovibrio desulfuricans*: This bacterium can reduce mercury ions (Hg^{2+}) to less toxic forms like elemental mercury (Hg^0) or methylmercury (CH_3Hg^+), particularly under anaerobic conditions (Ma et al. 2019).
- *Bacillus* spp.: Certain species of *Bacillus* are known for their ability to bioaccumulate mercury ions through biosorption (Sinha et al. 2012). Strain *Bacillus cereus* RC1 was reported to carry out intracellular accumulation, effectively removing mercury from the environment (Huang et al. 2014).
- *Pseudomonas* spp.: Strains of genus *Pseudomonas* like *Pseudomonas* sp. strain, AN-B15, *Pseudomonas aeruginosa, Pseudomonas putida, Pseudomonas stutzeri, Pseudomonas fluorescens* have been shown to possess, the physiological system responsible for detoxifying mercury by converting it to less toxic forms (Rani et al. 2021; Chang et al. 2022).
- Other bacteria: Several other bacterial strains like *Shigella flexneri, Klebsiella pneumoniae, Morganella morganii, Xanthomonas, Achromobacter, Acinetobacter calcoaceticus, Serratia marcescens, Mycobacterium marinum, Staphylococcus aureus, Bacillus* sp., *Enterobacter, Sphingobium* sp., *Sphingopyxis* sp., *Luteimonas* sp., *and Psychrobacter* sp. has been reported to harbor the mer operon system in its genome, responsible for detoxification of mercury (Mahbub et al. 2017).

Fungi

- *Aspergillus* spp.: Fungi of the *Aspergillus* genus can bioaccumulate mercury through biosorption onto their cell walls or intracellular accumulation, reducing mercury levels in contaminated environments (Anuar et al. 2020; Vacar et al. 2021).
- *Trichoderma* spp.: Certain species of *Trichoderma* have been studied for their ability to overcome mercury stress in plants growing in mercury-contaminated soils. Seed coating treatment with *Trichoderma harzianum* helped in overcoming mercury stress by triggering the production of ROS-scavenging proteins and preventing oxidative damage (Umber et al. 2019). In an in vitro analysis, a strain of *Trichoderma virens*, when experimentally inoculated along with *Arabidopsis thaliana* in Murashige and Skoog agar-solidified, was found to volatilize mercury resulting in reduced uptake by the plant (Hernández-Flores et al. 2018).
- *Other fungi:* Other fungal strains like *Aspergillus* sp. A31 and *Curvularia geniculate* P1, *Pseudomonodictys pantanalensis* sp. nov. A73 and *Westerdykella aquatica* P71, *Lindgomycetaceae* P87, and *Westerdykella* sp. exhibited efficiency in mercury bioremediation (de Siqueira et al. 2021; Senabio et al. 2023; Eltarahony et al. 2023).

6.4 Arsenic (As)

Bacteria and fungi play significant roles in arsenic (As) bioremediation through various mechanisms such as biosorption, biotransformation, and bioaccumulation. Specific bacteria and fungi known for their involvement in arsenic bioremediation:

Bacteria

- *Bacillus* spp.*:* Some species of *Bacillus*, such as *Bacillus subtilis Bacillus flexus, Bacillus licheniformis,* and *Bacillus cereus* carry out arsenic immobilization through precipitation, adsorption onto its cell surface structures (Titah et al. 2018; Atikpo et al. 2019; Marwa et al. 2019) and enzymatic transformation of arsenic species (Mujawar et al. 2021; Mandal et al. 2022), thus reducing the concentration of arsenic in the environment.
- *Pseudomonas* spp.*:* Certain strains of *Pseudomonas* bacteria possess arsenic resistance genes and can oxidize arsenite (As(III)) to arsenate (As(V)), making it more amenable to removal from the environment. *Pseudomonas alcaligenes, Pseudomonas vancouverensis strain m318 Pseudomonas stutzeri*, and other strains of *Pseudomonas* (Wang et al. 2012; Zhang et al. 2016; Yang et al. 2020; Thongnok et al. 2022) and plant growth-promoting *Pseudomonas fluorescence* (Pinter et al. 2017) has been reported to possess arsenic resistance genes and enzymes that oxidize arsenite to arsenate, reducing the mobilization of arsenic thus contributing to the lesser flow into biotic systems.

- Other mechanisms of arsenic remediation and bacteria involved: Extracellular polymeric substances (EPS) produced by several arsenic-resistant bacteria result in biofilm formation. This physiological trait has been reported to induce arsenite oxidation and immobilization by *Pseudomonas vancouverensis* (Zeng et al. 2018) and *Pseudomonas migulae* VC-19 (Herrera et al. 2021).
- *Desulfovibrio* spp.: Some strains of *Desulfovibrio* bacteria, like *Desulfovibrio desulfuricans*, are capable of reducing arsenate (As(V)) to arsenite (As(III)) under anaerobic conditions, facilitating arsenic immobilization (Bryukhanov and Khijniak 2022). *Desulfovibrio* bacteria are primary sulfate reducers and their metabolic activity results in the production of sulfide ions (S^{2-}) under anaerobic conditions. Sulfide can react with arsenic species to form insoluble precipitates, such as arsenic sulfides, reducing the mobility and bioavailability of arsenic in the environment (Neria-González and Aguilar-López 2021). These bacterial species can also form biofilms that promote arsenic bioremediation processes through sorption, reduction, and precipitation (Wu et al. 2023).

Other Bacterial Species Arsenic-tolerant iron-oxidizing bacteria, such as *Gallionella* spp. and *Leptothrix* spp., oxidize ferrous iron (Fe^{2+}) to ferric iron (Fe^{3+}) under aerobic conditions. This process results in the formation of insoluble ferric hydroxide [Fe(OH)3], which acts as a primary sorbent for arsenic (Purakayastha et al. 2011).

Fungi

Fungi, notably strains of *Aspergillus* and *Penicillium*, are integral to arsenic bioremediation. These species possess enzymatic pathways capable of metabolizing arsenic compounds, converting toxic inorganic arsenic into less harmful organic forms or insoluble precipitates. Fungi detoxify arsenic-contaminated environments through oxidation, reduction, and methylation. *Aspergillus nidulans, A. niger, Aspergillus* sp. isolate HKK4, and an isolate of the genus *Penicillium* exhibited arsenic tolerance and were found capable of removing arsenic from media (Pandey et al. 2022). Several fungi like *Penicillium* sp., *Aspergillus* sp., *Fusarium oxysporum, Saccharomyces cerevisiae, Zygosaccharomyces* sp., *Trichoderma viride, Scopulariopsis brevicaulis, Ulocladium* sp., *Sphaerotilus* sp., *Pichia* sp., *Ralstonia eutropha, Mucor rouxii, Candida humicola, Gliocladium roseum, and Rhizopus arrhizus* uses different mechanisms like biosorption, methylation, Fe-As coprecipitation, etc., to remediate arsenic from soil (Rai et al. 2020). The unique physiology of fungi allows them to thrive in harsh conditions, making them suitable for bioremediation in diverse ecosystems. Moreover, their ability to form symbiotic relationships with plants enhances arsenic removal from soil and water. Harnessing the natural capabilities of these fungi offers a sustainable and efficient solution for mitigating arsenic pollution.

6.5 Chromium (Cr)

Microbial remediation of chromium (Cr) bioremediation relies on several mechanisms like, chromate reductase to convert toxic hexavalent chromium (Cr(VI)) into less harmful trivalent chromium (Cr(III)), which precipitates and becomes immobilized. Others utilize uptake systems to transport chromium into their cells, where it undergoes detoxification or sequestration. Additionally, biofilm formation enhances bacterial chromium remediation by providing a matrix for microbial activity and chromium binding. Some of the notable examples in this regard are:

Bacteria Amongst the plethora of reports regarding bacterial reduction of toxic hexavalent chromium to less toxic trivalent form, some of the recent findings are as follows: *Acinetobacter* sp. (ADHR1) isolated from electroplating wastewater was reported capable of reducing toxic hexavalent chromium (Cr(VI)) into less harmful trivalent chromium (Cr(III)) (Montes-Robledo et al. 2024). *Arthrobacter chlorophenolicus* an indigenous bacteria of Mt. Merapi, Indonesia was also reported to be capable of reducing Cr(VI) to less toxic Cr(III) (Lathifah et al. 2021). *Serratia marcescens* AJRR-22 was reported to have mitigated Cr(VI) from textile washout–enriched agricultural soil and tannery effluent before discharge (Jasu et al. 2023). *Lysinibacillus macroides* isolated from the Chapala Lake, Mexico sediments were also capable of reducing chromium from hexavalent to trivalent form (Hernandez-Pena et al. 2021). In a recent extensive review, Bhunia et al. (2022) discussed the role of different bacterial species in the biosorption, accumulation, and transformation of Cr(VI) to Cr(III) from diverse niches and the author carried out extensive discussion regarding Cr(VI) removal by biofilm formation. In a study carried out by Maurya et al. (2022a), biofilm formed by *Bacillus vallismortis* (MT027009), *Bacillus haynesii* (MT027008), and *Alcaligenes aquatilis* (MT027005) isolated from tannery sludge exhibited higher degree of Cr(VI) removal (60–99%) when compared with that by the planktonic cells (43–94%). In another study exposure to chromium increased extracellular polymeric substances (EPS) production (a major component of biofilm) was increased by 10–35% by promoting the synthesis of protein (94–119%) and polysaccharide (2–33%) components in EPS (Maurya et al. 2022b).

Fungi Several fungal species possess enzymatic systems capable of reducing toxic hexavalent chromium (Cr(VI)) to less harmful trivalent chromium (Cr(III)). This reduction process often occurs intracellularly, where Cr(III) can then be sequestered or immobilized within fungal biomass or extracellular matrices. Additionally, fungi can form symbiotic relationships with plants, facilitating chromium uptake and accumulation in plant tissues. Recent reports revealed *Aspergillus tamarii, Trichoderma atroviride, and Aspergillus niger* produce good biomass growth, which is up to 0.23 g, and a high percentage of Cr(VI) reduction (up to 57.3%) from media (Hock et al. 2020). *Penicillium oxalicum* SL2 exhibited indirect Cr(VI) reduction by acid metabolites and form transformation in mycelium, thus reported as an effective mediator for chromium bioremediation (Long et al. 2020). The reduction of hexavalent chromium by *Fusarium proliferatum* S4 isolated from chromium-contaminated soil (Shan et al. 2022) and detoxification of Cr(VI) and extracellular formation of nanoparticles Cr_2O_3 by *Fusarium solani* SWUZF-1 (Tuo et al. 2023) has also been recently reported. The mycelia of *Gloeophyllum sepiarium*, a brown rot fungus, was reported to absorb chromium from tannery effluents (Ejechi et al. 2016).

7 Degradation of Organic Pollutants

Organic pollutants in soil encompass a spectrum of compounds, including pesticides, herbicides, petroleum hydrocarbons, and industrial chemicals (Saha et al. 2017). Pesticides and herbicides originate from agricultural activities, where their excessive use leads to runoff and soil leaching (Pérez-Lucas et al. 2019). Pesticides like petroleum hydrocarbons stem from spills and leaks in oil storage facilities or transportation networks. Industrial chemicals, such as solvents and polycyclic aromatic hydrocarbons (PAHs), result from manufacturing processes and improper disposal practices (Kanungo et al. 2024). These pollutants infiltrate soil through various pathways like atmospheric deposition, wastewater irrigation, and direct application, posing threats to ecosystems, human health, and agricultural productivity. Effective management strategies are crucial to mitigate their adverse impacts.

Firstly, these chemicals can disrupt the delicate balance of soil microorganisms essential for nutrient cycling, decomposition, and soil fertility. Pesticides and herbicides may kill beneficial bacteria, fungi, and other soil organisms, reducing soil biodiversity and microbial activity (Kalia and Gosal 2011). This can impair soil structure, decrease water infiltration rates, and increase erosion risk and nutrient runoff (Zhu et al. 2020).

Moreover, pesticides and herbicides can persist in soil for extended periods, depending on their chemical properties and environmental conditions (Tudi et al. 2021). The accumulation of these compounds in soil can have long-term toxic effects on soil biota, including earthworms, beneficial insects, and soil-dwelling organisms, disrupting soil food webs and ecosystem dynamics (Alengebawy et al. 2021).

Furthermore, pesticide and herbicide contamination in soil can have far-reaching ecological implications, including the loss of biodiversity, decline in pollinator populations, and disruption of natural ecosystems (Naz et al. 2023). These effects can have cascading impacts on ecosystem services such as pollination, soil fertility, and biological pest control, ultimately compromising agricultural productivity and ecosystem resilience (Brühl and Zaller 2021).

Petroleum hydrocarbons, which include a wide range of organic compounds found in crude oil, have significant detrimental effects on soil and ecosystems when released into the environment. When petroleum products such as gasoline, diesel, or crude oil spill or leak into the soil, they can persist for long periods due to their complex chemical composition and slow degradation rates (Jabbar et al. 2022). These pollutants can directly impact soil quality and fertility. They can alter soil pH, reduce soil permeability, and disrupt soil structure, decreasing water retention and nutrient availability for plants (Ossai et al. 2020; Gao et al. 2022). This can result in stunted plant growth, reduced crop yields, and diminished biodiversity in affected areas (Ismail et al. 2021). Microbial communities in soil play crucial roles in nutrient cycling, organic matter decomposition, and soil health maintenance. Petroleum hydrocarbons can be toxic to soil microorganisms, inhibiting their growth and activity. This disruption of microbial communities can further impair soil fertility and ecosystem functioning, exacerbating the long-term environmental impacts of hydrocarbon contamination (Alrumman et al. 2015; Gao et al. 2022). Industrial chemicals such as solvents and polycyclic aromatic hydrocarbons (PAHs) can severely impact soil health, posing risks to ecosystems, human health, and agricultural sustainability (Yadav et al. 2021). These pollutants often originate from industrial activities like manufacturing, mining, transportation, and improper disposal practices (Lawal 2017). When released into soil, they can persist for long periods due to their slow degradation rates, leading to chronic contamination. Various industrial solvents, including trichloroethylene (TCE), perchloroethylene (PCE), benzene, and toluene, can contaminate soil and negatively impact its health. TCE and PCE, chlorinated solvents commonly used in manufacturing and degreasing, persist in soil, disrupting microbial communities and inhibiting nutrient cycling (Anderson and Walton 1992). Their toxicity to soil organisms and potential to leach into groundwater pose risks to ecosystems and human health. Benzene, a volatile organic compound (VOC) found in petroleum products and industrial processes, can accumulate in soil, inhibiting plant growth and degrading soil fertility. Toluene and TCE, also a VOC, harmed the soil nematode population, thereby detrimentally affecting soil health (Fuller et al. 1997). Furthermore, solvents can be toxic to plants, inhibiting growth and compromising agricultural yields.

Polycyclic aromatic hydrocarbons (PAHs) are organic compounds with multiple fused aromatic rings. They are often formed during the incomplete combustion of organic matter, such as fossil fuels, wood, and tobacco, as well as industrial emissions (Lawal 2017). They are highly persistent in soil and can bioaccumulate in plants and organisms (Singh et al. 2021a, b, c). Low molecular weight (LMW) PAHs like naphthalene, acenaphthylene, acenaphthene, fluorene, and phenanthrene are generally more mobile and can penetrate deeper into the soil (Dai et al. 2022). They are often more readily biodegraded by microorganisms compared to high molecular weight (HMW) PAHs (Thavamani et al. 2012). However, their presence in soil can still have adverse effects on soil fertility, as they can inhibit the growth of plants and disrupt microbial activity essential for nutrient cycling and soil health. In contaminated soils, LMW PAHs can hinder seed germination, root development, and overall plant growth (Houshani and Salehi-Lisar 2020). High molecular weight (HMW) PAHs such as fluoranthene, pyrene, benzanthracene, chrysene, benzoapyrene, and Dibenzaanthracene are generally less mobile in soil but more persistent and resistant to biodegradation compared to LMW PAHs.

These compounds can alter soil microbial communities, disrupt soil food webs, and pose toxic risks to soil organisms. PAH contamination can inhibit plant growth (Maliszewska-Kordybach and Smreczak 2000; Jajoo 2017), reduce biodiversity, and impair ecosystem functioning. PAHs are classified into two main types based on their structure: low molecular weight (LMW) PAHs and high molecular weight (HMW) PAHs.

Bioremediation offers a versatile approach for mitigating various organic pollutants in soil, including pesticides, herbicides, industrial solvents, and polycyclic aromatic hydrocarbons (PAHs) (Megharaj et al. 2011; Haripriyan et al. 2022). Microorganisms, such as bacteria and fungi, play pivotal roles in breaking down these contaminants into less harmful byproducts through metabolic processes (Bala et al. 2022). For pesticides and herbicides, microbial degradation pathways can target specific chemical bonds, rendering them less toxic or completely inert (Yaashikaa et al. 2022a, b). Industrial solvents, like chlorinated compounds, are often degraded through anaerobic or aerobic microbial processes, converting them into harmless substances like carbon dioxide and water (Sharma et al. 2021). Polycyclic aromatic hydrocarbons (PAHs) undergo enzymatic degradation by specialized microorganisms (Gupta et al. 2015). Bioremediation strategies may involve bioaugmentation with engineered microbial strains (Gentry et al. 2004; Ferraro et al. 2021) or stimulating indigenous microbial communities through nutrient amendments and oxygenation (Sarkar et al. 2020; Pushkar and Sevak 2022).

Certain microbial species are adept at degrading pesticides and herbicides in soil. They employ several mechanisms like hydrolysis, oxidation, and biochemical conjugation (Gomathi et al. 2023; Thakkar and Vincent 2021), converting them into less toxic forms (Table 3). *Pseudomonas* spp., *Bacillus* spp., and *Rhodococcus* spp.

Table 3 Mechanism of degradation of pesticide and herbicide

Mechanism	Description	Enzymes
Hydrolysis	Enzymatic cleavage of ester or amide bonds in pesticides/herbicides, breaking them down into simpler compounds.	Meng et al. (2022)
Oxidation	The introduction of oxygen atoms into pesticide/herbicide molecules via oxygenase or peroxidase enzymes leads to their degradation into less toxic metabolites.	Mahmood et al. (2014), Mohapatra et al. (2022)
Conjugation	Attachment of small molecules (e.g., sugars, amino acids) to pesticides/herbicides, enhancing their solubility or facilitating excretion, rendering them less harmful.	Zhang and Yang (2021), Kaur et al. (2021)
Ring cleavage	Enzymatic cleavage of aromatic rings in pesticides/herbicides, breaking them into smaller fragments for further degradation.	Arya et al. (2017), Mbachu et al. (2020)
Co-metabolism	Degradation of pesticides/herbicides as secondary substrates during the metabolism of primary carbon sources is often facilitated by enzymes with broad substrate specificity.	Banwart et al. (2014), Mbachu et al. (2020)
Detoxification	Specialized pathways for neutralizing toxic effects of pesticides/herbicides, converting reactive metabolites into less harmful forms, or sequestering them within microbial cells.	Mahto et al. (2022)

are notable examples of their ability to metabolize various pesticides and herbicides (Table 4).

Microbial oxidation of pesticides and herbicides involves enzymes like oxygenases or peroxidases, which introduce oxygen atoms into the compounds, facilitating degradation (Mbachu et al. 2021). Examples include *Pseudomonas putida* Strain Ch2 oxidizing glyphosate (Esikova et al. 2023), *Bacillus thuringiensis* MB497 oxidizing organophosphate insecticides (Ambreen and Yasmin 2021), *Sphingomonas* spp. Oxidizing Atrazine (Chen et al. 2015), and *Rhodococcus* spp. metabolizing pyrethroid pesticides (Huang et al. 2023). These microorganisms play crucial roles in bioremediation by breaking down toxic compounds into harmless byproducts.

The conjugation of pesticides and herbicides by specific microorganisms involves attaching small molecules like sugars or amino acids to the compounds, enhancing solubility or facilitating excretion (Zulfiqar and Yasmin 2020). For instance, *Arthrobacter* spp. SC-JAK2 can conjugate atrazine, a widely used herbicide, with glutathione, rendering it more water-soluble and less toxic (Rajendran et al. 2021). Similarly, certain *Pseudomonas* spp. can conjugate organophosphate insecticides with sugars, aiding in their detoxification (Pailan et al. 2020).

Various microorganisms facilitate ring cleavage of pesticides and herbicides through enzymatic processes, aiding in their degradation. For instance, *Pseudomonas* sp. strains ADP and *Pseudomonas putida* can cleave the aromatic rings of herbicides like atrazine and glyphosate (Selvi and Manonmani, 2015; Pileggi et al. 2020), while *Arthrobacter* spp. are known for metabolizing pesticides such as simazine through ring cleavage pathways (Rajendran et al. 2021). Additionally, certain strains of *Alcaligenes* spp. possess enzymes capable of breaking down aromatic rings in pesticides like chlorpyrifos (Yadav et al. 2020).

Co-metabolism expands microbial metabolic capabilities, allowing for the simultaneous breakdown of pollutants and organic substrates (Espinosa-Ortiz et al. 2022). This process enhances bioremediation efficiency by utilizing diverse microbial communities and metabolic pathways, ultimately reducing pesticide and herbicide levels in the environment, and promoting ecosystem restoration. Co-metabolism of

Table 4 Hydrolytic Degradation of Pesticide and Herbicide by Different Microbes

Microbe	Pesticides/herbicides targeted	References
Pseudomonas spp.	Glyphosate, atrazine, chlorpyrifos, pyrethroids	Sehrawat et al. (2021), Abd Rani et al. (2022)
Bacillus spp.	Organophosphate insecticides, glyphosate	Elarabi et al. (2020)
Rhodococcus spp.	Atrazine, simazine, chlorpyrifos	Kharat et al. (2018), Kolekar et al. (2014)
Arthrobacter spp.	Atrazine, simazine, glyphosate	Fernández et al. (2013), Chen et al. (2022)
Alcaligenes spp.	Atrazine, simazine, chlorpyrifos	Geed et al. (2018), Yadav et al. (2020)
Burkholderia spp.	Atrazine, simazine, chlorpyrifos	Kim and Ahn (2009), Ahmad et al. (2023)

pesticides and herbicides involves their degradation as secondary substrates during various microorganisms' metabolism of primary carbon sources (Kumar et al. 2022a). For instance, *Bacillus* spp. can degrade organophosphate insecticides like malathion while utilizing organic carbon sources (Mahajan et al. 2023). Similarly, *Pseudomonas fluorescens* metabolize herbicides like *Trifluralin alongside* primary carbon sources (Faramarzi et al. 2023).

The detoxification of pesticides and herbicides by various microorganisms involves the conversion of toxic compounds into less harmful forms or sequestration within microbial cells (Sharma et al. 2020). For instance, *Rhodococcus phenolicus* strain MCP1 possesses enzymes that detoxify organophosphate insecticides like chlorpyrifos by breaking down their toxic metabolites (Singh et al. 2021a, b, c). Similarly, *Sphingomonas* spp. can detoxify herbicides, such as atrazine, by dechlorination through metabolic pathways (Luo et al. 2021).

Various microbes possess metabolic pathways capable of metabolizing different types of industrial solvents, including chlorinated solvents like trichloroethylene (TCE), perchloroethylene (PCE) (McCarty et al. 2020), and volatile organic compounds (VOCs) such as benzene, toluene, ethylbenzene, and xylene (BTEX) (Yoshikawa et al. 2017). Microorganisms employ different mechanisms, including reductive dichlorination (Xiao et al. 2020), aerobic degradation (Jindrova et al. 2002), oxidative metabolism (Mohapatra and Phale 2021), and anaerobic fermentation (Hatti-Kaul and Mattiasson 2016), to bioremediate industrial solvent contamination in the environment (Table 5).

Bioremediation of polycyclic aromatic hydrocarbons (PAHs) involves various microorganisms utilizing different mechanisms to degrade these compounds into less harmful forms (Ansari et al. 2023). Microbes employ mechanisms such as aerobic and anaerobic metabolism (Dhar et al. 2020), ring cleavage (Sakshi and Haritash 2020), and oxidative degradation (Dong et al. 2020) to bioremediate PAH-contaminated environments. Microorganisms (Table 6) utilize different mechanisms such as aerobic degradation, oxygenase-mediated oxidation, dioxygenase activity, and ligninolytic enzyme activity to bioremediate both low molecular weight (LMW)

Table 5 Microbial remediation of industrial solvents

Microbes	Industrial solvents	Mechanism	References
Dehalococcoides spp.	TCE, PCE	Reductive dechlorination: converts highly chlorinated solvents into less toxic end products.	Lu et al. (2021), Li et al. (2021)
Pseudomonas spp.	BTEX compounds	Aerobic degradation: utilizes aerobic metabolic pathways to break down solvents into CO_2 and H_2O.	Khodaei et al. (2017), Banerjee et al. (2022)
Rhodococcus spp.	Toluene, xylene	Aerobic degradation: metabolizes solvents as carbon and energy sources through aerobic pathways.	Di Canito et al. (2018), Feng et al. (2021)
Methylobacterium spp.	Methanol, ethanol, formaldehyde	Oxidative metabolism: utilizes specialized pathways to oxidize and detoxify VOCs.	Yanpirat et al. (2020)

Table 6 Microbial remediation of polycyclic aromatic hydrocarbons (PAHs)

Microbe	Low molecular weight (LMW) PAHs	High molecular weight (HMW) PAHs	Mechanism	References
Pseudomonas putida	Naphthalene, phenanthrene, anthracene	Fluoranthene, pyrene, benzapyrene	Aerobic degradation: utilizes oxygenase enzymes to oxidize PAHs, breaking down aromatic rings into simpler compounds.	Khan et al. (2014), Baig et al. (2022), Yao et al. (2017)
Mycobacterium spp.	Acenaphthene, fluorene, phenanthrene	Benzo[a]anthracene, chrysene, dibenzoanthracene	Oxygenase-mediated oxidation: initiates ring cleavage and subsequent degradation of PAHs under aerobic conditions.	Paterek et al. (2003), Zeng et al. (2010), Hennessee et al. (2016)
Sphingomonas spp.	Acenaphthylene, fluorene, anthracene	Benzofluoranthene, benzofluoranthene, benzoperylene	Dioxygenase activity: catalyzes oxidation reactions, leading to ring cleavage and degradation of PAHs.	Leys et al. (2004), Zhou et al. (2016), Song et al. (2021)
Burkholderia cepacia	Acenaphthene, fluorene, phenanthrene	Benzopyrene, dibenzoanthracene, indeno[1,2,3]pyrene	Aerobic and anaerobic degradation: utilizes dioxygenases under aerobic conditions and reductive dehalogenases under anaerobic conditions for PAH degradation.	Kim et al. (2003), Darmawan et al. (2015), Cauduro et al. (2021)
Phanerochaete chrysosporium	Acenaphthene, fluorene, phenanthrene	Benzopyrene, dibenzoanthracene, indeno[1,2,3]pyrene	Ligninolytic enzyme activity: utilizes laccase and peroxidase to oxidize and degrade PAHs.	Abo-State et al. (2021), Wang et al. (2009)

and high molecular weight (HMW) polycyclic aromatic hydrocarbons (PAHs) in contaminated environments (Perini et al. 2020).

Microbial remediation offers a promising solution for mitigating environmental organic pollutant contamination. By harnessing the metabolic capabilities of diverse microorganisms, bioremediation processes effectively degrade pollutants such as pesticides, herbicides, industrial solvents, and PAHs into less harmful forms. This eco-friendly approach promotes environmental sustainability by restoring soil fertility, protecting water quality, and safeguarding ecosystem health. Continuous research and development in microbial bioremediation strategies hold the potential to address organic pollutant contamination effectively, offering a cost-effective and sustainable means of environmental cleanup and restoration.

8 Application of Biofertilizers in Restoring Contaminated Sites

The use of biofertilizers for restoring contaminated sites is a sustainable and environmentally friendly approach to remediation. Beneficial microorganisms found in biofertilizers, such as mycorrhizal fungi, phosphate-solubilizing bacteria, and nitrogen-fixing bacteria, aid in the restoration of soil ecosystems that have been harmed by pollution (Rabani et al. 2023). In addition, biofertilizers play a crucial role in revitalizing degraded environments by enhancing nutrient availability, improving soil structure, and promoting plant growth (Khan et al. 2023). Some biofertilizers also have bioremediation capabilities, actively participating in the degradation or immobilization of contaminants like organic pollutants and heavy metals (Gupta et al. 2023) and this approach not only supports the restoration of microbial diversity but also reduces reliance on chemical fertilizers, thus fostering sustainable agriculture practices. The comprehensive use of biofertilizers contributes to the overall health of contaminated sites, promoting resilience and fostering the recovery of ecosystems for the benefit of both the environment and surrounding communities. The application of biofertilizers offers several benefits, including enhanced nutrient availability, improved soil structure, and the ability to remediate or mitigate the impact of contaminants (Osorio-Reyes et al. 2023). Using biofertilizers to restore contaminated sites is a global practice with different methods based on each country's unique agricultural practices and environmental challenges. Utilizing eco-friendly and sustainable solutions aligns with global initiatives to support soil health, biodiversity, and sustainable agriculture. Table 7 provides examples of how biofertilizers are applied in different countries to restore contaminated sites. Additionally, Fig. 2 illustrates the various approaches used in the biofertilizer application process to restore contaminated sites.

8.1 Based on Remediation Approaches

8.1.1 Bioaugmentation

It is a process of environmental remediation that involves using biofertilizers to restore contaminated sites (Rana and Gupta 2023; Tiwari et al. 2023). Biofertilizers are made up of microorganisms like bacteria, fungi, or algae that promote soil fertility and enhance plant growth (Gonçalves et al. 2023). By introducing these microorganisms into contaminated sites, biofertilizers serve two purposes. First, they can help break down certain pollutants present in the soil, leading to the natural

Table 7 Application of biofertilizers in the different countries of the World

Countries	Description	Statistics	References
India	Biofertilizers have been widely used in India, particularly in agricultural practices. Commonly used biofertilizers include *Rhizobium*, *Azotobacter*, and phosphate-solubilizing bacteria. These biofertilizers can help restore soil fertility, fix nitrogen, and encourage the growth of plants that can absorb or change pollutants in contaminated areas.	India's use of biofertilizers has been steadily increasing due to their environmental benefits and role in sustainable agriculture. In 2022, the country produced approximately 800,000 tons of biofertilizers, notably higher than in previous years. This growth is supported by government initiatives and subsidies promoting organic farming. Biofertilizers like *Rhizobium*, *Azotobacter*, and blue-green algae are extensively used in states like Maharashtra, Karnataka, and Tamil Nadu. Despite the increase, biofertilizers still represent a small fraction of India's total fertilizer market, indicating potential for further expansion and adoption.	Singh et al. (2021a, b, c), Soni et al. (2022)
China	Soil contamination in China caused by industrial activities is a significant issue. To make plants more resistant to pollutants and improve soil structure, biofertilizers containing plant growth-promoting bacteria (PGPB) are utilized. In order to combat soil pollution, the Chinese government has implemented policies that encourage the usage of biofertilizers as a crucial component of sustainable agricultural practices.	In recent years, China has seen a notable rise in the use of biofertilizers, driven by government support for sustainable agriculture. By 2022, production had exceeded 1.2 million tons, with provinces like Shandong, Jiangsu, and Sichuan leading in usage. Biofertilizers still represent a small segment of the overall market, indicating room for growth as China emphasizes ecological farming practices.	Ma et al. (2020), Zhao et al. (2024)
United States	Biofertilizers are becoming increasingly popular in ecological restoration projects in the United States to remediate contaminated sites, such as those affected by pesticides, heavy metals, and industrial pollutants. For instance, mycorrhizal fungi are deployed to enhance plant nutrient uptake and stimulate vegetation growth in polluted areas.	In the United States, the use of biofertilizers in sustainable agriculture is increasing. As of 2022, the US biofertilizer market was valued at $500 million, with a 12% annual growth rate. This growth is driven by rising demand for organic produce and government incentives promoting environmentally friendly farming practices. States like California, Iowa, and Texas are leading in biofertilizer adoption due to their significant agricultural activities. Biofertilizers still represent a small portion of the total fertilizer market, indicating potential for future expansion.	Raimi et al. (2021), Zhao et al. (2024)

Brazil	Brazil is known for its contribution toward environmental restoration and sustainable agriculture. In areas where the soil is contaminated, biofertilizers containing mycorrhizal fungi and nitrogen-fixing bacteria are used to enhance soil fertility and support the growth of native vegetation. The use of biofertilizers is in line with Brazil's efforts to promote ecologically sustainable farming practices.	In Brazil, the use of biofertilizers is rapidly increasing, driven by the country's large agricultural sector and a push toward sustainable farming practices. The market was valued at $200 million in 2022, with a 15% annual growth rate. Adoption is encouraged by government policies aimed at reducing the environmental impact of traditional fertilizers. Despite this progress, biofertilizers still make up a small fraction of the overall fertilizer market in Brazil, suggesting significant potential for further expansion.	Bomfim et al. (2021), Mrabet (2023)
European Union (EU) countries	Several EU countries are taking steps toward sustainable agriculture and tackling soil contamination through the use of biofertilizers. These biofertilizers contain beneficial microorganisms that can improve soil structure, enhance nutrient cycling, and assist in the restoration of contaminated areas.	In the EU, the use of biofertilizers is steadily growing as part of the region's commitment to sustainable agriculture. The biofertilizer market was valued at around €400 million in 2022, with a 10% annual growth rate. Leading countries in adoption include Germany, France, and Italy, driven by strict regulations and consumer demand for organic produce. The EU's Common Agricultural Policy and national initiatives further promote biofertilizer use, indicating room for growth in the green agriculture agenda.	Barros-Rodríguez et al. (2020), Pânzaru et al. (2023)
Australia	Australia is facing issues with salinity and soil degradation, which is causing a decline in soil fertility and health. To address this problem, Australian researchers and farmers are exploring innovative ways to incorporate biofertilizers, such as nitrogen-fixing bacteria and phosphate-solubilizing microorganisms, into land restoration strategies. These biofertilizers are proving to be effective in enhancing soil fertility and health in the affected areas.	In Australia, the use of biofertilizers is increasing due to a growing focus on sustainable farming and organic agriculture. As of 2022, the Australian biofertilizer market was valued at around AUD 50 million, with an annual growth rate of about 8%. Key agricultural regions such as New South Wales, Victoria, and Queensland are adopting biofertilizers, including products like nitrogen-fixing bacteria and mycorrhizal fungi. Government initiatives and increasing awareness among farmers about the environmental and soil health benefits of biofertilizers are driving this growth.	Dadzie et al. (2023), Ammar et al. (2023)

(continued)

Table 7 (continued)

Countries	Description	Statistics	References
World	The use of biofertilizers is increasing globally as countries prioritize sustainable agriculture. The global biofertilizer market was valued at around $2 billion in 2022, with a 13% annual growth rate. Major markets include North America, Europe, Asia–Pacific, and Latin America, with leading adoption in the United States, China, India, and Brazil. Biofertilizers like nitrogen-fixing bacteria and mycorrhizal fungi are being used to improve soil fertility and crop yields while reducing reliance on chemical fertilizers. Despite their benefits, biofertilizers still represent a small fraction of the overall fertilizer market, showing potential for future growth as sustainable farming practices become more prevalent worldwide.	Global use of biofertilizers is increasing due to the demand for sustainable agriculture. The market was valued at $2.3 billion in 2022 and is projected to reach $4.5 billion by 2028, with a compound annual growth rate (CAGR) of 11.2%. The surge is notable in Asia–Pacific and Latin America, with countries like India, Brazil, and China leading adoption due to government initiatives and environmental awareness.	Atieno et al. (2020)

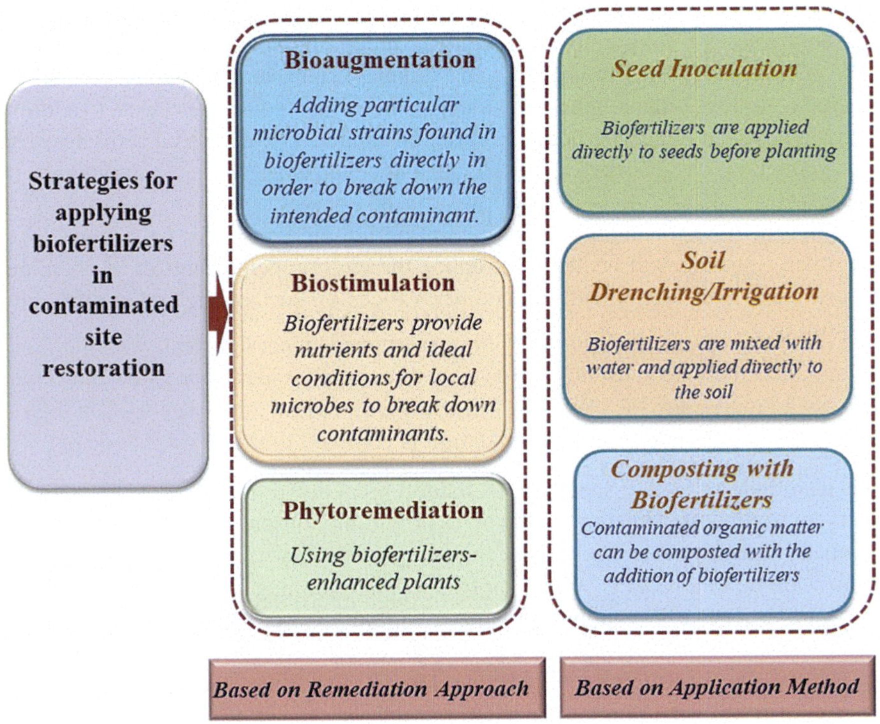

Fig. 2 Biofertilizer application strategies for the restoration of contaminated sites

remediation of the environment. Second, the improved soil fertility resulting from biofertilizer application creates a more favorable environment for plants to grow, aiding in the overall restoration of the ecosystem. This sustainable approach addresses contamination issues and supports ecological recovery by fostering a healthier and more balanced environment. The use of bioaugmentation with biofertilizers is an environmentally friendly practice that represents an innovative strategy for mitigating the impact of contaminants on soil and ecosystems (Ayilara et al. 2023). Bioaugmentation practices are being implemented across the globe to address environmental challenges, enhance industrial processes, and promote sustainable practices. Some examples of countries where bioaugmentation techniques are being used:

- In the United States, bioaugmentation is being used in environmental remediation projects such as the cleanup of oil spills. Microorganisms that are capable of breaking down oil are introduced to accelerate the natural degradation process.
- India has embraced bioaugmentation in agriculture by promoting the use of biofertilizers containing nitrogen-fixing bacteria and other beneficial microorganisms. This approach aims to improve soil fertility and reduce the reliance on chemical fertilizers.

- Canada uses bioaugmentation in wastewater treatment plants to enhance microbial processes for more efficient removal of contaminants. This approach aligns with the country's commitment to sustainable water management.
- China employs bioaugmentation strategies in soil remediation projects to address contamination issues, particularly in regions affected by industrial activities. The introduction of specialized microorganisms helps break down pollutants in the soil.
- In Brazil, bioaugmentation is applied in agricultural practices to enhance crop growth, improve soil health, and reduce the environmental impact of chemical fertilizers. This aligns with the country's focus on sustainable and eco-friendly farming.

8.1.2 Biostimulation

This technique is used to speed up the natural breakdown of pollutants in contaminated environments. This process involves providing nutrients or other amendments to enhance the growth and activity of microorganisms that can break down contaminants (Udume et al. 2023). While biofertilizers containing beneficial microorganisms can be part of biostimulation strategies, the term covers a broader range of approaches. In restoring contaminated sites, biostimulation may involve adding organic materials, nutrients, or electron acceptors to the contaminated area to create a more favorable environment for existing microorganisms to thrive and carry out biodegradation processes (Sharma 2020). It is important to note that biostimulation encourages the growth of indigenous microorganisms, while bioaugmentation involves the introduction of specific microorganisms, such as bacteria or fungi, to enhance the remediation process (Lopes et al. 2022). Both biostimulation and bioaugmentation are commonly used together to optimize the restoration of contaminated sites, providing a holistic and sustainable approach to environmental remediation. Biostimulation practices are being implemented worldwide to address environmental challenges and promote sustainable practices. Bioaugmentation techniques are used in many countries, including:

- Biostimulation has been used in several environmental remediation projects in the US, most notably in the cleanup of contaminated sites caused by industrial activities or oil spills.
- Germany has a history of employing biostimulation in soil and groundwater remediation projects that align with the country's commitment to sustainable environmental practices.
- Canada has used biostimulation techniques to remediate contaminated sites, especially in areas affected by industrial pollutants. The country emphasizes environmentally friendly approaches to address environmental challenges.
- Australia has explored biostimulation as part of its soil and groundwater remediation strategies. The country is committed to sustainable practices, and bioremediation techniques are considered in addressing environmental contamination.

- India has shown interest in biostimulation for agricultural and environmental purposes. The application of biofertilizers and other biostimulants is considered part of efforts to improve soil health and address pollution issues.
- China has been exploring biostimulation methods, particularly in the context of soil and water remediation. The country is investing in sustainable technologies to tackle environmental challenges arising from rapid industrialization.

8.1.3 Phytoremediation Using Biofertilizers-Enhanced Plants

This is a highly effective tool for environmental restoration. When combined with biofertilizer-enhanced plants, it becomes even more powerful. This approach takes advantage of the synergistic relationships between plants and beneficial microorganisms to optimize the remediation process (Sharma et al. 2023; Yadav et al. 2023c). By adding nitrogen-fixing bacteria, mycorrhizal fungi, phosphate-solubilizing bacteria, and other useful microbes as biofertilizers, the remediation potential of plants is greatly increased. These microorganisms contribute to improved nutrient uptake, stress tolerance, and the creation of a healthy rhizosphere, providing an ideal environment for plant growth and pollutant remediation (Anli et al. 2020). Furthermore, the use of organic amendments alongside biofertilizers enriches the soil, creating a nutrient-rich habitat that supports the development of robust plant-microbe interactions. This integrated approach addresses pollutant removal and aims to restore the ecological balance of contaminated sites, emphasizing a sustainable and holistic strategy for environmental rejuvenation (Sharma et al. 2023). Success in implementing biofertilizer-enhanced plants in phytoremediation relies on careful consideration of plant-microbe interactions, site-specific conditions, and the overall compatibility of these bio-enhanced systems with environmental restoration goals. Some examples of how phytoremediation can be applied with the help of biofertilizers in different countries:

- In the United States, a research initiative is being undertaken to enhance phytoremediation with biofertilizers in brownfield sites contaminated with heavy metals. The project aims to improve plant health and metal uptake by selecting hyperaccumulator plants and introducing nitrogen-fixing bacteria and mycorrhizal fungi as biofertilizers, contributing to the site's remediation.
- In China, biofertilizer-enhanced phytoremediation can be implemented in regions affected by agricultural runoff containing pesticides. Using specific plant species and introducing phosphate-solubilizing bacteria as biofertilizers could enhance the plants' ability to absorb and degrade these contaminants, addressing water quality concerns.
- In India, where soil salinity is a significant issue in some regions, a phytoremediation project can be designed to restore salt-affected soils. Biofertilizers containing salt-tolerant nitrogen-fixing bacteria and mycorrhizal fungi would be applied to support the growth of halophytic plants, aiding in soil improvement.

- In Brazil, biofertilizer-enhanced phytoremediation can be focused on areas impacted by oil spills. The project aims to accelerate the natural degradation of petroleum contaminants in the soil by introducing oil-degrading bacteria and mycorrhizal fungi to support the growth of hydrocarbon-tolerant plants.
- Phytoremediation, combined with biofertilizer-enhanced plants, is attracting global attention as an innovative method to address soil contamination and environmental degradation. By leveraging the natural abilities of plants and enhancing their remediation potential through biofertilizers, this approach offers a sustainable and cost-effective way to detoxify polluted sites while improving soil health. Initiatives to deploy phytoremediation technologies supported by biofertilizers are underway worldwide, from industrialized nations to developing regions. This method not only facilitates the removal of heavy metals, organic pollutants, and other contaminants from soil and water but also promotes the sustainable reuse of land for agriculture and other purposes (Misra et al. 2020).
- The United Nations and international organizations advocate for the use of phytoremediation with biofertilizer-enhanced plants as a sustainable solution for environmental remediation. The UN Food and Agriculture Organization (FAO) promotes its use to address soil degradation and contamination. They highlight its potential to restore ecosystem health and promote agricultural productivity (Osorio-Reyes et al. 2023). The United Nations Environment Programme (UNEP) and the International Union for Conservation of Nature (IUCN) recognize the benefits of phytoremediation, citing its cost-effectiveness, environmental friendliness, and its dual role in rehabilitating polluted sites and improving soil fertility through biofertilizers. These biofertilizers enhance the growth and efficiency of hyperaccumulator plants used to extract heavy metals and contaminants from the soil. International initiatives and partnerships, such as those under the Global Environment Facility (GEF) and various research consortia, actively promote research, development, and deployment of phytoremediation technologies (Odoh et al. 2019). These efforts include capacity-building programs, funding for pilot projects, and the dissemination of best practices to enhance the global adoption of phytoremediation. The UN and international organizations strongly support integrating phytoremediation and biofertilizers into environmental management for sustainable development goals related to land degradation, food security, and environmental health.

8.2 Based on the Application Method

8.2.1 Seed Inoculation

This agricultural practice involves applying biofertilizers directly to seeds before planting. This technique introduces beneficial microorganisms, such as *Rhizobium* bacteria, mycorrhizal fungi, *Azotobacter*, or *Azospirillum*, to seeds, which establish symbiotic relationships with plants (Gatabazi et al. 2023). *Rhizobium* bacteria, for instance, form associations with leguminous plants, fixing atmospheric nitrogen in root nodules and enhancing overall nitrogen availability. Mycorrhizal fungi, on the other hand, create mutualistic connections with various crops, promoting nutrient uptake, especially phosphorus. The seed inoculation process includes coating seeds with a biofertilizer solution, ensuring the microorganisms' viability and effectiveness (Sujatha et al. 2023). Once planted, the inoculated seeds facilitate establishing these beneficial relationships in the soil, leading to improved nutrient absorption, enhanced plant growth, and reduced reliance on chemical fertilizers. Seed inoculation is a sustainable agricultural approach that contributes to soil fertility and crop productivity while minimizing environmental impact. The most common types of biofertilizers used in the seed are *Rhizobium* inoculants, mycorrhizal inoculants, *Azotobacter*, and *Azospirillum* inoculants.

Seed inoculation is a promising strategy for restoring contaminated sites and it involves introducing beneficial microorganisms, such as rhizobacteria and mycorrhizal fungi, to seeds. This technique promotes phytoremediation, where plants help remove, absorb, or transform contaminants (Dawood et al. 2023). Microbes also help plants thrive in contaminated environments by enhancing nutrient uptake and helping them withstand stress. Some bacteria can even degrade or transform specific contaminants, contributing to soil detoxification. The symbiotic relationship between plants and microorganisms supports overall plant health and resilience. Seed inoculation also aids in erosion control by stabilizing the soil, and preventing further degradation. This sustainable and cost-effective approach leverages plants' natural abilities and microbial allies to restore the ecological balance of contaminated sites. However, site-specific factors and careful selection of appropriate microbial strains are crucial for the success of seed inoculation in contaminated site restoration.

8.2.2 Soil Drenching or Irrigation

This strategy using a blend of biofertilizers and water effectively improves soil fertility and promotes plant growth in agriculture (Cao et al. 2023; Meddich 2023). Biofertilizers, which contain helpful microorganisms like bacteria or fungi, form beneficial relationships with plants, aiding in nutrient absorption (Nosheen et al. 2021). To begin, carefully read and follow the manufacturer's instructions for the specific biofertilizer product, which typically includes guidance on application rates and methods, and calculate the appropriate application rate based on factors such as crop type and soil conditions. Measure the required amount of clean water and mix it with the recommended quantity of biofertilizer in a mixing container. Stir the mixture thoroughly to ensure an even distribution of the beneficial microorganisms. Apply the resulting solution directly to the soil, covering the entire root zone of the plants using watering cans, drip irrigation systems, or other appropriate methods. Timing is essential, usually done during early growth or when plants absorb nutrients. Regular monitoring of plant growth and soil health helps assess the effectiveness, allowing for adjustments to the application rate and frequency as needed for optimal results throughout the growing season.

Soil drenching or irrigation with a solution of biofertilizers and water is also a valuable strategy for restoring contaminated sites. Biofertilizers containing helpful microorganisms play a dual role in remediation (Parmar and Srivatsan 2023). Firstly, they engage in bioremediation, actively breaking down and metabolizing contaminants present in the soil, thereby reducing their concentration. At the same time, biofertilizers contribute to improved soil fertility by aiding in nutrient cycling, nitrogen fixation, and the solubilization of essential elements. This nutrient enrichment creates a conducive environment for vegetation to thrive, promoting plant growth and root development. The establishment of symbiotic relationships between the microorganisms and plant roots further enhances the process, as plants become more resilient and skilled at accumulating contaminants, contributing to phytoremediation. As a result, the overall improvement in soil structure, biodiversity, and ecosystem sustainability highlights the effectiveness of soil drenching with biofertilizers in the comprehensive restoration of contaminated sites.

8.2.3 Composting with Biofertilizers

This is an effective and sustainable way to manage contaminated organic matter (Al-Suhaibani et al. 2020). In situations where organic materials like agricultural residues or green waste are contaminated with pollutants, traditional composting may not be enough to address the issue. However, introducing biofertilizers into the composting process improves its remediation potential. Typically, biofertilizers contain beneficial microorganisms, including bacteria and fungi, that contribute to the breakdown of organic matter and the transformation of contaminants. These microorganisms play a crucial role in biodegradation by breaking down pollutants into less harmful forms (Ning et al. 2021). Additionally, their activity helps in nutrient cycling, promoting a healthier environment for composting. The resulting compost, enriched with beneficial microorganisms and free from the initial contaminants, can be used to improve soil fertility and structure. This eco-friendly approach addresses the issue of contaminated organic matter and transforms it into a valuable resource for sustainable agriculture and soil health. It is crucial to follow proper guidelines and recommendations for the application of biofertilizers to ensure effective and safe composting practices.

Composting with biofertilizers has emerged as a crucial strategy for restoring contaminated sites (Bhattacharjya et al. 2021). Biofertilizers play a dual role in remediating both organic matter and pollutants. They contain a blend of beneficial microorganisms that facilitate composting by breaking down contaminants within the organic material. This biodegradation transforms pollutants into less harmful forms, reducing their environmental impact. Biofertilizers also enhance the microbial activity that takes place during composting, resulting in an efficient composting environment. The decomposition of organic matter is accelerated, enriching nutrient cycling, and producing a nutrient-dense compost that supports plant growth and enhances soil fertility during restoration. Khandare et al. (2020) in their experiment showed that 25% chemical fertilizer can be replaced by incorporating *Azotobacter* and PSB in wheat cultivation. Ye et al. (2020) in their experiment at Yinchuan, Ningxia Province, China, reported that the application of *Trichoderma harzianum* T7 in Tomato plant yield can be maximized from 82.85 to 86.78 kg per plot. In conclusion, this eco-friendly approach transforms contaminated organic matter into a valuable resource, fostering sustainable soil management and contributing to the comprehensive restoration of degraded sites.

9 Challenges Associated with the Widespread Adoption of Biofertilizers

Despite being low-cost and environmentally friendly, biofertilizers encounter several challenges in their widespread adoption. These constraints include slow biofertilizer response, poor-quality bioinoculants, limited industrial production, and the requirement for enhanced formulations and novel biofertilizers. Surmounting these hurdles is crucial for successful biofertilizer commercialization.

9.1 Agroclimatic Constraints/Environmental Constraints

Several factors influence the effectiveness of biofertilizers, such as *Rhizobium* inoculants, including the crop genotype, the particular strain of microorganisms, environmental conditions like soil quality and weather patterns, and agricultural management practices (Hossain et al. 2021b; Woomer et al. 2014). Testing the effectiveness of these products under variable conditions, including abiotic stresses like low fertility of soil, soil acidity or drought, etc., is vital for developing apt recommendations for their use (Masso et al. 2015).

Soil pH plays a significant role in biofertilizer effectiveness as it affects the population of microorganisms, the strain's survival rate, and the availability of nutrients. Soils with low pH do not support a vigorous population of beneficial microorganisms for plant growth (Table 8).

The impact of soil pH on biofertilizer effectiveness can vary depending on the type of biofertilizer used. Using vesicular-arbuscular mycorrhiza (VAM) in less fertile and

Table 8 Population and survival of beneficial microorganisms affected by soil pH

Indicator	Acidic soil condition	Alkaline soil condition	References
Population of beneficial microorganisms (Rhizobia)	Low	High	Gupta et al. (2007)
Strain survival	Low	High	Howieson and Ballard (2004)

acidic soil has been proven beneficial in enhancing the uptake of nitrogen and phosphorus (Effiong and Ibia 2009). Field trials with cyanobacteria on varied soil types revealed that the application of biofertilizer could reduce urea nitrogen inputs by 25–35% in acidic and saline rice fields, but in calcareous and neutral soils, its effectiveness was lower (Hashem 2001). The efficacy of a biofertilizer hinges on its ability to survive and thrive in specific field conditions. Therefore, optimum pH quantification for each type of biofertilizer in different agro-climatic conditions is crucial.

9.2 Technological Constraints

One of the most crucial factors is the stability of microbial strains. It is essential for the effectiveness of biofertilizers both during storage and application. Genetic variability within microbial populations can make it challenging to maintain this stability. Different strains of biofertilizers can promote plant growth to different extents, affecting their performance in various soil conditions (Otieno et al. 2015). Therefore, for practical agricultural use, it is important to recognize and pick strains that consistently promote stable and effective plant growth.

Manufacturing and composition must be optimized to ensure constancy and efficiency. Various methods have been explored such as ultraviolet (UV) mutagenesis and immobilization technology through research to improve nitrogen fixation and enhance phosphate solubilization (Walpola et al. 2014). The cost of production is another challenge, requiring competitiveness with conventional fertilizers for market sustainability (Malusa et al. 2012).

Formulating a product requires careful consideration of its shelf life, especially when stored and handled under varying conditions. (Kumar 2014). The field performance of biofertilizers is affected by the shelf life of the product, carrier materials standard, storage specifications (e.g. temperature), handling (e.g. transportation) (Chen 2006; Tan et al. 2009), also the occurrence of impurities (Sivasakthivelan and Saranraj 2013) disturb the on-site performance of the biofertilizers and subsequently the adoption rate. Thus, the storage duration of locally formulated biofertilizers should be improved (Brahmaprakash and Sahu 2012; Kumar 2014) in different storage specifications to guarantee product integrity over an extended period. Moreover, the use of inappropriate, less effective strains for production, absence of qualified technical personnel in production units, short shelf life of inoculants, etc., come under various technological constraints.

9.3 Biosafety Concerns and Quality Control

Quality control and regulation are essential to ensure that biofertilizers conform to prescribed standards, are safe, and are effective (Banayo et al. 2012; Desyane and Wiyana 2012). If the quality control framework is not well-structured, poor-quality biofertilizers can enter the market and lead to subpar on-site efficiency (Catroux et al. 2001; Herridge et al. 2002).

Marketing and distribution of biofertilizers face several obstacles, including regulatory challenges and safety considerations. There are apprehensions regarding the possible influence of biofertilizers on ecosystems as well as human health, necessitating rigorous testing, safety evaluation, and efficacy before commercialization. However, traversing these protocols can be intricate and laborious, hampering commercialization. It is critical to assess the safety and effectiveness of biofertilizer products to address these challenges. For example, as a promoter of wheat plant growth phosphate-solubilizing *Pseudomonas fluorescens* PSM1 strain has been proved highly effective by the researchers, ensuring environmental safety as well (Yadav et al. 2016).

According to the World Health Organization (WHO 2015), biosafety levels (BSLs) and risk groups provide essential guidelines for handling microorganisms, including those used in biofertilizers. These guidelines classify microorganisms based on their pathogenicity to adult humans, modes of transmission, and the availability of preventive measures and treatments. There are four biosafety levels:

- *Biosafety Level 1 (BSL-1)*: This level includes microorganisms that pose low individual and community risk. These organisms are generally nonpathogenic and do not cause infections in healthy adult humans, laboratory personnel, or the environment. An example is *Azospirillum brasilense*, which, although it may cause skin irritation, can enhance the shoot and root fresh weight in tomato plants (Stets et al. 2015).
- *Biosafety Level 2 (BSL-2)*: Microorganisms at this level pose a moderate individual risk and low community risk.
- *Biosafety Level 3 (BSL-3)*: These microorganisms present a high individual risk and moderate to high community risk.
- *Biosafety Level 4 (BSL-4)*: This level includes microorganisms that pose a very high individual and community risk.

These classifications are crucial for ensuring safe practices in laboratories and agricultural settings, particularly when dealing with biofertilizers (Keswani et al. 2019).

Quality assurance issues encompass formulation and storage processes with many products missing strong quality control schemes, leading to unreliable formulations and restricted evidence on optimal storage conditions and application methods (Lauriano-Barajas et al. 2018). Furthermore, commercially available biofertilizers frequently consist of a limited variety of microbial species, which limits their efficacy and usability compared to the diverse microorganisms found in natural ecosystems.

9.4 Market Challenges

Consumer acceptance presents a huge challenge in the marketing of biofertilizers. To identify their effect on human health, the microorganisms used in these fertilizers must undergo rigorous evaluation, which is crucial not only for ultimate consumers but also for those engaged in the manufacturing process (Flores-Felix et al.

2015). Although for nonleguminous crops, specific strains of microbes from genera *Bacillus*, *Azospirillum*, *Azotobacter*, and *Gluconacetobacter* have been securely commercialized, further research is necessary to conclusively verify the use of biofertilizers safely across diverse crop varieties. One more obstacle, food technology neophobia, refers to the apprehension toward novel food technologies. This fear can affect consumer preferences for biofertilizers and is shaped by factors such as cultural background and socio-demographic characteristics (Yadav et al. 2023b).

The global biofertilizer market is substantial, with an estimated value of USD 3.27 billion in 2024. This market is anticipated to expand significantly, reaching USD 5.23 billion by 2029, driven by a compound annual growth rate (CAGR) of 9.85% over the forecast period. In 2022, North America emerged as the leading region in the biofertilizer market, commanding a 35.9% share, followed by Europe with a 31.9% share. Among various types of biofertilizers, vesicular-arbuscular mycorrhiza (VAM) ranks the most widely consumed globally, with *Rhizobium* as the second most utilized biofertilizer (***). In the context of India, the biofertilizer market demonstrated notable growth. It was valued at USD 99.59 million in 2021 and increased to USD 110.07 million in 2022. This upward trend is expected to continue, with projections indicating that the market will reach USD 243.61 million by 2029. This growth trajectory corresponds to a robust CAGR of 12% over the analysis period, highlighting the increasing adoption of biofertilizers in India's agricultural sector (***). Biofertilizers' elevated production and utilization costs persist as notable barriers, particularly for small-scale farmers (Haskett et al. 2021). Isolation and formulation of beneficial microorganisms are involved in the production process, which is labor-intensive and necessitates specialized facilities (Soumare et al. 2020). Additionally, expanding production and maintaining uniform quality can pose challenges and impact cost efficiency (Haskett et al. 2021).

In addition to these, there are other constraints to consider for the successful commercialization of biofertilizers:

- Absence of necessary equipment, power supply, and other essential resources
- Availability of space for laboratory, production, storage, etc.
- Inadequate facilities for cold storage of inoculant packets
- Financial limitations
- Insufficient funds and difficulties obtaining bank loans
- Lower returns from product sales in smaller production units
- Lack of awareness regarding the technology's benefits
- Challenges in farmer adoption of the technology due to varied inoculation methods
- Absence of immediate visual crop growth differences compared to inorganic fertilizers
- Lack of awareness regarding the ecosystem damage caused by continuous inorganic fertilizer application

10 Research and Development Needs for Improving Biofertilizer Efficiency

Combining biofertilizers with conventional fertilizers offers synergistic benefits like improved nutrient uptake and soil health. However, challenges include complex interactions, varying guidelines for farmers, and regulatory hurdles, which can delay commercialization and impact farmer trust. Understanding the rhizosphere dynamics is crucial for overcoming challenges in the commercialization of biofertilizers. The rhizosphere is a complex and ever-changing environment, impacted by factors like soil composition, plant varieties, and microbial interplay. Seasonal variations further complicate matters by affecting soil moisture, temperature fluctuations, and microbial engagement, impacting biofertilizers' effectiveness and hindering the development of a universal solution (Yadav et al. 2013). Although integrating biofertilizers with integrated nutrient management (INM) shows that numerous obstacles must be handled in advance to fulfill the potential for sustainable agriculture. These include the development of cost-effective manufacturing approaches, guaranteeing quality and effectiveness, and promoting adoption by farmers (Panazai et al. 2019). Furthermore, additional research is required to comprehend how biofertilizers relate with other components of INM like organic and inorganic fertilizers, to refine nutrient management approaches (Yadav et al. 2023a). Microbial ecology and biotechnology advancements could lead to novel biofertilizer production with enhanced nutrient mobilization and plant-growth-enhancing attributes (Aasfar et al. 2021). Incorporating biofertilizers into precision agriculture methods like data analytics, remote sensing, etc., could facilitate precise nutrient management and improve resource utilization efficiency (Arjjumend et al. 2021).

Biofertilizers offer a promising eco-friendly alternative, foreshadowing a potential second green revolution in grain production. However, certain challenges persist despite their significance in soil conservation and agricultural management. For instance, the elusive nature of AMF, present in 90% of symbiotic plant associations, hinders laboratory culture (Chakraborty et al. 2021). Extensive research is imperative to develop viable culturing techniques, with initial successes observed in culturing AMF in carrot root organs (Declerck et al. 2005).

Moreover, *Azospirillum*, known for nitrogen fixation, has demonstrated additional benefits in root development and architecture. Mutants, such as glutamine synthetase (GS) mutants, exhibit enhanced growth compared to parental types, suggesting avenues for further enhancement through genetic manipulation (Van-Dommelen et al. 2009). Exploration of tripartite associations and identification of enzymes like 1-aminocyclopropane-1-carboxylate (ACC) deaminase, crucial for ethylene elimination and plant growth promotion, underline the possibility of genetic engineering in creating plant growth-promoting microbes (Glick et al. 2007). In the quest for sustainable agriculture, ongoing research endeavors aim to unlock the potential of various microorganisms as biofertilizers. Harnessing biotechnological tools and genetic engineering holds promise in sculpting a pathway toward enduring agricultural growth.

11 Potential Innovations and Emerging Technologies in the Field

In recent years, significant progress has been made in improving soil and plant nutrient levels and growth dynamics to enhance crop production. This includes using specialized microorganisms for plant inoculation, selecting optimal technologies, and developing suitable carrier materials for inoculum delivery (Mahmud et al. 2021). To ensure cost-effectiveness and sustainability in biofertilizer production, current strategies such as artificial selection of microbiomes, employment of mixed inoculants, utilization of PGPR as biofertilizers, incorporation of biofilms, Nano-biofertilizers, and bio-organo-chemical fertilizers are gaining popularity (Saheewala et al. 2023).

11.1 PGPR Based Biofertilizer

PGPB contains symbiotically associated or free-living bacteria that colonize in root rhizosphere to enhance the uptake of soil-available nutrients and accelerate plant growth and productivity through some regulators and transporters (Gouda et al. 2018). PGPB biofertilizers mobilize the soil nutrients and protect the plant from some pathogenic elements (Kumar et al. 2010). *Bradyrhizobium, Bacillus megaterium, Anabaena, Azolla, and Bacillus polymyxa* are some bacterial species that improve plants' growth and performance (Olanrewaju et al. 2017).

11.2 Encapsulation and Lyophilization of Biofertilizer

Encapsulation and lyophilization are techniques used in the production of biofertilizers to improve their shelf life, stability, and efficacy (Singh et al. 2017). Here's an overview of these techniques:

1. Encapsulation: Encapsulation involves enclosing the biofertilizer (such as bacteria, fungi, or algae) within a protective matrix or coating (Balla et al. 2022). This matrix can be made from various materials such as polymers, proteins, or lipids. Encapsulation helps protect the biofertilizer from harsh environmental conditions (such as UV radiation, pH extremes, and desiccation) and from interactions with soil components that may reduce its effectiveness (Chaudhary et al. 2020). Encapsulation can also control the release of the biofertilizer, ensuring a gradual and sustained release overtime, which can improve its efficiency (Vassilev et al. 2020).
2. Lyophilization (freeze-drying): Lyophilization is a process that involves freezing the biofertilizer and then removing the frozen water by sublimation under a vac-

uum (Chaudhary et al. 2020). This process helps preserve the biofertilizer by removing water, a major factor in degradation reactions. Lyophilization can improve the stability of the biofertilizer at room temperature, prolonging its shelf life (Kumar et al. 2022a, b). It also helps maintain the viability and activity of the biofertilizer during storage and transportation.

Combining encapsulation with lyophilization can further enhance the stability and efficacy of biofertilizers, making them more practical and effective for agricultural applications. These techniques are particularly useful for preserving sensitive microorganisms and ensuring their survival and functionality when applied to the soil.

11.3 Nano-biofertilizer

Nano-biofertilizer are a type of fertilizer that consists of nanoparticles and bioinoculants that improve nutrient uptake and utilization by plants (Kumari and Singh 2019). These nanoparticles can be engineered to release nutrients slowly, protect them from leaching or volatilization, and enhance their availability to plants (Garg et al. 2023). Nano-biofertilizer have shown promising results in improving crop yields and nutrient efficiency, and they are considered a potential solution for sustainable agriculture (Moulick et al. 2020a, b, c). However, more research is needed to fully understand their long-term effects on soil health and the environment. Nano-biofertilizer improve crop quality by increasing the efficiency of indigenous microbial populations, and enhancing the utilization of essential nutrients like potassium, nitrogen, and phosphorus (Mala et al. 2017). Additionally, nano-biofertilizers can stimulate microbial enzyme activity, which further sustainably enhances soil fertility.

11.4 Biofilm Biofertilizer

Biofilm biofertilizer is a type of biofertilizer where beneficial microorganisms are enclosed within a protective biofilm matrix (Swarnalakshmi et al. 2013). This matrix shields the microorganisms and helps them adhere to plant roots, improving their effectiveness (Parween et al. 2017). The production process involves selecting suitable microorganisms and cultivating them under conditions that promote biofilm formation. The formed biofilm is then encapsulated in a matrix, often made from biodegradable materials. After drying and stabilization, the biofilm biofertilizer is ready for use (Swarnalakshmi et al. 2013). This technology offers a sustainable approach to agriculture, enhancing nutrient uptake and plant growth while reducing the need for chemical fertilizers (Setyawan 2021).

12 Conclusion

There is no doubt that biofertilizers have shown enormous potential in achieving sustainability in agriculture. However, issues like the easy availability of biofertilizers, and rapid commercialization remain challenging. The utilization of biofertilizers not only fosters a more sustainable and eco-friendly approach to agriculture but also supports soil remediation efforts. This dual functionality highlights the importance of integrating biofertilizers into agricultural practices to achieve long-term environmental sustainability and food security. As global agricultural demands increase, the adoption of biofertilizers will be crucial in maintaining soil health, reducing pollution, and ensuring the availability of essential nutrients for crops. Future research and technological advancements should focus on optimizing biofertilizer formulations and applications, promoting widespread adoption, and maximizing their benefits for both agriculture and environmental remediation. Moreover, prolonged studies on the impact of biofertilizer applications in different agroecosystems should be encouraged, with the specific goal of addressing multiple stress resilience issues.

References

Aasfar A, Bargaz A, Yaakoubi K, Hilali A, Meftah Kadmiri I (2021) Nitrogen-fixing *Azotobacter* species as potential soil biological enhancers for crop nutrition and yield stability. Front Microbiol 12:628379

Abd Rani NF, Ahmad Kamil K, Aris F, Mohamed Yunus N, Zakaria NA (2022) Atrazine-degrading bacteria for bioremediation strategy: a review. Biocatal Biotransformation 40(4):233–247

Abd-Alla MH, Al-Amri SM, El-Enany AWE (2023) Enhancing *Rhizobium*–Legume symbiosis and reducing nitrogen fertilizer use are potential options for mitigating climate change. Agriculture 13(11):2092

Abo-State MAM, Osman ME, Khattab OH, El-Kelani TA, Abdel-Rahman ZM (2021) Degradative pathways of polycyclic aromatic hydrocarbons (PAHs) by *Phanerochaete chrysosporium* under optimum conditions. J Radiat Res Appl Sci 14(1):507–520

Afzal AM, Rasool MH, Waseem M, Aslam B (2017) Assessment of heavy metal tolerance and biosorptive potential of Klebsiella variicola isolated from industrial effluents. Amb Express, 7: 1–9

Afzal A, Bano A (2008) Rhizobium and phosphate solubilizing bacteria improve the yield and phosphorus uptake in wheat (Triticum aestivum). Int J Agric Biol, 10(1): 85–88

Aggarwal A, Kadian N, Tanwar A, Yadav A, Gupta KK (2011) Role of arbuscular mycorrhizal fungi (AMF) in global sustainable development. J Appl Natl Sci 3(2):340–351

Ahmad M, Nadeem SM, Naveed M, Zahir ZA (2016) Potassium-solubilizing bacteria and their application in agriculture. In: Meena VS, Maurya BR, Verma JP, Meena RS (eds) Potassium solubilizing microorganisms for sustainable agriculture. Springer India, pp 293–313

Ahmad S, Chandrasekaran M, Ahmad HW (2023) Investigation of the persistence, toxicological effects, and ecological issues of S-Triazine Herbicides and their biodegradation using emerging technologies: a review. Microorganisms 11(10):2558

Al Abboud MA, Ghany TMA, Alawlaqi MM (2013) Role of biofertilizers in agriculture: a brief review. Mycopathologia 11(2):95

Alam S, Khalil S, Ayub N, Rashid M (2002) In vitro solubilization of inorganic phosphate by phosphate-solubilizing microorganism (PSM) from maize rhizosphere. Int J Agric Biol 4:454–458
Alam M, Hussain Z, Khan A, Khan MA, Rab A, Asif M et al (2020) The effects of organic amendments on heavy metals bioavailability in mine impacted soil and associated human health risk. Sci Hortic 262:109067
Alengebawy A, Abdelkhalek ST, Qureshi SR, Wang MQ (2021) Heavy metals and pesticides toxicity in agricultural soil and plants: ecological risks and human health implications. Toxics 9(3):42
Ali H, Khan E, Sajad MA (2013) Phytoremediation of heavy metals—concepts and applications. Chemosphere 91(7):869–881
Ali H, Khan E, Ilahi I (2019) Environmental chemistry and ecotoxicology of hazardous heavy metals: environmental persistence, toxicity, and bioaccumulation. Journal of chemistry, 2019(1): 6730305
Ali MF, Ammar A, Bilal S, Ali U, Huma N, Adnan M (2021) Mitigating zinc deficiency in plants and soils through agronomic techniques: a review. J Environ Agric Sci 23(1&2):1–10
Alori ET, Gabasawa AI, Elenwo CE, Agbeyegbe OO (2022) Bioremediation techniques as affected by limiting factors in soil environment. Front Soil Sci 2:937186
Alrumman SA, Standing DB, Paton GI (2015) Effects of hydrocarbon contamination on soil microbial community and enzyme activity. J King Saud Univ Sci 27(1):31–41
Al-Suhaibani N, Selim M, Alderfasi A, El-Hendawy S (2020) Comparative performance of integrated nutrient management between composted agricultural wastes, chemical fertilizers, and biofertilizers in improving soil quantitative and qualitative properties and crop yields under arid conditions. Agronomy 10(10):1503
Ambreen S, Yasmin A (2021) Novel degradation pathways for Chlorpyrifos and 3, 5, 6-Trichloro-2-pyridinol degradation by bacterial strain *Bacillus thuringiensis* MB497 isolated from agricultural fields of Mianwali, Pakistan. Pestic Biochem Physiol 172:104750
Ammar EE, Rady HA, Khattab AM, Amer MH, Mohamed SA, Elodamy NI et al (2023) A comprehensive overview of eco-friendly bio-fertilizers extracted from living organisms. Environ Sci Pollut Res 30(53):113119–113137
Amari T, Ghnaya T, Abdelly C (2017) Nickel, cadmium and lead phytotoxicity and potential of halophytic plants in heavy metal extraction. South African Journal of Botany, 111: 99–110
Anderson TA, Walton BT (1992) Comparative plant uptake and microbial degradation of trichloroethylene in the rhizospheres of five plant species—implications for bioremediation of contaminated surface soils (No. ORNL/TM-12017). Oak Ridge National Lab.(ORNL), Oak Ridge
Anli M, Baslam M, Tahiri A, Raklami A, Symanczik S, Boutasknit A et al (2020) Biofertilizers as strategies to improve photosynthetic apparatus, growth, and drought stress tolerance in the date palm. Front Plant Sci 11:516818
Ansari F, Ahmad A, Rafatullah M (2023) Review on bioremediation technologies of polycyclic aromatic hydrocarbons (PAHs) from soil: mechanisms and future perspective. Int Biodeterior Biodegradation 179:105582
Anuar SNIS, Othman F, Chay TC, Nasir NAHA (2020) Biosorption of mercury ion (Hg^{2+}) using live and dead cells of *Rhizopus oryzae* and *Aspergillus Niger*: characterization, kinetic and isotherm studies. J Pure Appl Microbiol 14(3):1749–1760
Aponte H, Meli P, Butler B, Paolini J, Matus F, Merino C et al (2020) Meta-analysis of heavy metal effects on soil enzyme activities. Sci Total Environ 737:139744
Arifiyanto A, Apriyanti FD, Purwaningsih P, Kalqutny SH, Agustina D, Surtiningsih T et al (2017) Lead (Pb) bioaccumulation; Genera *Bacillus* isolate S1 and SS19 as a case study. In: AIP conference proceedings, vol 1854, No. 1. AIP Publishing
Arjjumend H, Koutouki K, Neufeld S (2021) Comparative advantage of using biofertilizers in Indian agroecosystems: an analysis from the perspectives of stakeholders. Eur J Agric Food Sci 3(2):26–36

Arya R, Kumar R, Mishra NK, Sharma AK (2017) Microbial flora and biodegradation of pesticides: trends, scope, and relevance. In: Advances in environmental biotechnology. Springer, Singapore, pp 243–263

Asoegwu CR, Awuchi CG, Nelson KCT, Orji CG, Nwosu OU, Egbufor UC, Awuchi CG (2020) A review on the role of biofertilizers in reducing soil pollution and increasing soil nutrients. Himalayan J Agric 1(1):34–38

Atieno M, Herrmann L, Nguyen HT, Phan HT, Nguyen NK, Srean P et al (2020) Assessment of biofertilizer use for sustainable agriculture in the Great Mekong Region. J Environ Manag 275:111300

Atikpo E, Agori JE, Peretomode MT, Micheal A, Ofonedu EC (2019) Kinetic study of biosorption of arsenic from soil using microorganisms. Niger J Technol 38(3):777–783

Awasthi SK, Kumar M, Sarsaiya S, Ahluwalia V, Chen H, Kaur G et al (2022) Multi-criteria research lines on livestock manure biorefinery development towards a circular economy: from the perspective of a life cycle assessment and business models strategies. J Clean Prod 341:130862

Ayilara MS, Adeleke BS, Adebajo MT, Akinola SA, Fayose CA, Adeyemi UT et al (2023) Remediation by enhanced natural attenuation; an environment-friendly remediation approach. Front Environ Sci 11:1182586

Baig ZT, Abbasi SA, Memon AG, Naz A, Soomro AF (2022) Assessment of degradation potential of *Pseudomonas* species in bioremediating soils contaminated with petroleum hydrocarbons. J Chem Technol Biotechnol 97(2):455–465

Bairagi H, Khan MMR, Ray L, Guha AK (2011) Adsorption profile of lead on *Aspergillus versicolor*: a mechanistic probing. J Hazard Mater 186(1):756–764

Bala S, Garg D, Thirumalesh BV, Sharma M, Sridhar K, Inbaraj BS, Tripathi M (2022) Recent strategies for bioremediation of emerging pollutants: a review for a green and sustainable environment. Toxics 10(8):484

Balkhair KS, Ashraf MA (2016) Field accumulation risks of heavy metals in soil and vegetable crop irrigated with sewage water in western region of Saudi Arabia. Saudi J Biol Sci 23(1):S32–S44

Balla A, Silini A, Cherif-Silini H, Chenari Bouket A, Alenezi FN, Belbahri L (2022) Recent advances in encapsulation techniques of plant growth-promoting microorganisms and their prospects in the sustainable agriculture. Appl Sci 12:9020. https://doi.org/10.3390/app12189020

Banayo NPM, Cruz PCS, Aguilar EA, Badayos RB, Haefele SM (2012) Evaluation of biofertilizers in irrigated rice: effects on grain yield at different fertilizer rates. Agriculture 2:73–86

Banerjee A, Roychoudhury A (2020) Plant responses to environmental nickel toxicity. In: Plant micronutrients: deficiency and toxicity management. Springer, Cham, pp 101–111

Banerjee S, Bedics A, Tóth E, Kriszt B, Soares AR, Bóka K, Táncsics A (2022) Isolation of *Pseudomonas aromaticivorans* sp. nov from a hydrocarbon-contaminated groundwater capable of degrading benzene-, toluene-, m-and p-xylene under microaerobic conditions. Front Microbiol 13:929128

Banwart SA, Noellemeyer E, Milne E (eds) (2014) Soil carbon: science, management and policy for multiple benefits, vol 71. CABI

Barros-Rodríguez A, Rangseekaew P, Lasudee K, Pathom-aree W, Manzanera M (2020) Regulatory risks associated with bacteria as biostimulants and biofertilizers in the frame of the European Regulation (EU) 2019/1009. Sci Total Environ 740:140239

Bechtaoui N, Rabiu MK, Raklami A, Oufdou K, Hafidi M, Jemo M (2021) Phosphate-dependent regulation of growth and stresses management in plants. Front Plant Sci 12:679916

Beveridge TJ, Forsberg CW, Doyle RJ (1982) Major sites of metal binding in *Bacillus licheniformis* walls. J Bacteriol 150(3):1438–1448

Bhardwaj D, Ansari MW, Sahoo RK, Tuteja N (2014) Biofertilizers function as key player in sustainable agriculture by improving soil fertility, plant tolerance and crop productivity. Microb Cell Factories 13:1–10

Bhat SA, Bashir O, Haq SAU, Amin T, Rafiq A, Ali M et al (2022) Phytoremediation of heavy metals in soil and water: an eco-friendly, sustainable and multidisciplinary approach. Chemosphere 303:134788

Bhattacharjya S, Das S, Amat D (2021) Potential of microbial inoculants for organic waste decomposition and decontamination. In: Biofertilizers. Woodhead Publishing, pp 103–132

Bhunia A, Lahiri D, Nag M, Upadhye V, Pandit S (2022) Bacterial biofilm mediated bioremediation of hexavalent chromium: a review. Biocatal Agric Biotechnol 43:102397

Bomfim CA, Coelho LGF, do Vale HMM, de Carvalho Mendes I, Megías M, Ollero FJ, dos Reis Junior FB (2021) Brief history of biofertilizers in Brazil: from conventional approaches to new biotechnological solutions. Braz J Microbiol 52(4):2215–2232

Brühl CA, Zaller JG (2021) Indirect herbicide effects on biodiversity, ecosystem functions, and interactions with global changes. In: Herbicides. Elsevier, pp 231–272

Brahmaprakash GP, Sahu PK (2012) Biofertilizers for Sustainability, J Indian Inst Sci, 92(1):37–62

Bryukhanov AL, Khijniak TV (2022) The application of sulfate-reducing bacteria in the bioremediation of heavy metals and metalloids. Appl Biochem Microbiol 58(Suppl 1):S1–S15

Cao TND, Mukhtar H, Le LT, Tran DPH, Ngo MTT, Nguyen TB, Bui XT (2023) Roles of microalgae-based biofertilizer in sustainability of green agriculture and food-water-energy security nexus. Sci Total Environ 870:161927

Catroux G, Hartmann A, Revellin C (2001) Trends in Rhizobial Inoculant production and use. Plant Soil 230:21–30. Kluwer Academic Publishers

Cauduro GP, Leal AL, Marmitt M, de Ávila LG, Kern G, Quadros PD et al (2021) New benzo (a) pyrene-degrading strains of the *Burkholderia cepacia* complex prospected from activated sludge in a petrochemical wastewater treatment plant. Environ Monit Assess 193(4):163

Chandrakar V, Naithani SC, Keshavkant S (2016) Arsenic-induced metabolic disturbances and their mitigation mechanisms in crop plants: A review. Biologia, 71(4): 367–377

Chakraborty T, Akhtar N (2021) Biofertilizers: prospects and challenges for future. In: Biofertilizers: study and impact. Wiley, pp 575–590

Chalkidis A, Jampaiah D, Aryana A, Wood CD, Hartley PG, Sabri YM, Bhargava SK (2020) Mercury-bearing wastes: sources, policies and treatment technologies for mercury recovery and safe disposal. J Environ Manag 270:110945

Chang J, Yan Z, Dong J, Wu X, Meng Z, Shi Y, Chen J (2022) Mechanisms controlling the transformation of and resistance to mercury (II) for a plant-associated *Pseudomonas* sp. strain, AN-B15. J Hazard Mater 425:127948

Chaudhary T, Dixit M, Gera R, Shukla AK, Prakash A, Gupta G, Shukla P (2020) Techniques for improving formulations of bioinoculants. 3 Biotech 10:199

Chaudhary P, Singh S, Chaudhary A, Sharma A, Kumar G (2022) Overview of biofertilizers in crop production and stress management for sustainable agriculture. Front Plant Sci 13:930340

Chen J (2006) The combined use of chemical and organic fertilizers and/ or bio-fertilizer for crop growth and soil fertility. In: Proceedings of the international workshop on sustained management of the soil rhizosphere system for efficient crop production and fertilizer use, held 16–20 October, Bangkok, Thailand

Chen Q, Yang B, Wang H, He F, Gao Y, Scheel RA (2015) Soil microbial community toxic response to atrazine and its residues under atrazine and lead contamination. Environ Sci Pollut Res 22:996–1007

Chen Y, Chen WJ, Huang Y, Li J, Zhong J, Zhang W et al (2022) Insights into the microbial degradation and resistance mechanisms of glyphosate. Environ Res 215:114153

Chi Y, Huang Y, Wang J, Chen X, Chu S, Hayat K et al (2020) Two plant growth promoting bacterial *Bacillus* strains possess different mechanisms in adsorption and resistance to cadmium. Sci Total Environ 741:140422

Choudhury S, Moulick D (2022) Response of field crops to abiotic stress: current status and future prospects, 1st edn. CRC Press. https://doi.org/10.1201/9781003258063

Chowardhara B, Borgohain P, Saha B, Awasthi JP, Moulick D, Panda SK (2019) Phytotoxicity of Cd and Zn on three popular Indian mustard varieties during germination and early seedling growth. Biocatal Agric Biotechnol 21:101349

Chowardhara B, Saha B, Awasthi JP, Deori BB, Nath R, Roy S, Sarkar S, Santra SC, Hossain A, Moulick D (2024) An assessment of nanotechnology-based interventions for cleaning up toxic heavy metal/metalloid-contaminated agroecosystems: potentials and Issues. Chemosphere:142178. https://doi.org/10.1016/j.Chemosphere.2024.142178

Chug R, Mathur S, Kothari SL, Gour VS (2021) Maximizing EPS production from *Pseudomonas aeruginosa* and its application in Cr and Ni sequestration. Biochem Biophys Rep 26:100972

Dadzie FA, Egidi E, Stewart J, Eldridge DJ, Molesworth A, Singh BK, Muñoz-Rojas M (2023) Agricultural soil degradation in Australia. In: Impact of agriculture on soil degradation I: perspectives from Africa, Asia, America and Oceania. Springer International Publishing, Cham, pp 49–68

Dai C, Han Y, Duan Y, Lai X, Fu R, Liu S et al (2022) Review on the contamination and remediation of polycyclic aromatic hydrocarbons (PAHs) in coastal soil and sediments. Environ Res 205:112423

Darmawan R, Nakata H, Ohta H, Niidome T, Takikawa K, Morimura S (2015) Isolation and evaluation of PAH degrading bacteria. J Bioremed Biodegr 6(3):1

Dawood A, Majeed A, Ul-Allah S, Naveed M, Farooq S, Sarwar N, Hussain M (2023) Combined application of organic and inorganic nitrogen and seed inoculation with rhizobacteria (*Stenotrophomonas maltophilia* FA-9) improved productivity, nitrogen use efficiency, and economic returns of Pearl Millet. Sustain For 15(10):8248

de Siqueira KA, Senabio JA, Pietro-Souza W, de Oliveira Mendes TA, Soares MA (2021) *Aspergillus* sp. A31 and *Curvularia geniculata* P1 mitigate mercury toxicity to *Oryza sativa* L. Arch Microbiol 203(9):5345–5361

Deb P, Mandal A, Harendra CSR, Das A, Hazra S, Mukherjee A, Ghosh D, Choudhury S, Santra SC, Moulick D (2023) Improving plant nutrient use efficiency for climate-resilient agriculture. In: Climate-resilient agriculture, Vol 2: agro-biotechnological advancement for crop production. Springer International Publishing, Cham, pp 209–243

Declerck S, Strullu DG, Fortin A (2005) In vitro culture of mycorrhizas, vol 4. Springer Science & Business Media

Desyane HK, Wiyana AF (2012) Proposed Quality Improvement of Liquid Organic Fertilizers 'Herbafarm' to Meet National Standards in Indonesia. Indonesian J Busin Admin 1(6):343–352

Dhar K, Subashchandrabose SR, Venkateswarlu K, Krishnan K, Megharaj M (2020) Anaerobic microbial degradation of polycyclic aromatic hydrocarbons: a comprehensive review. Rev Environ Contam Toxicol 251:25–108

Di Canito A, Zampolli J, Orro A, D'Ursi P, Milanesi L, Sello G et al (2018) Genome-based analysis for the identification of genes involved in o-xylene degradation in *Rhodococcus opacus* R7. BMC Genomics 19:1–17

Dimkpa CO, Merten D, Svatoš A, Büchel G, Kothe E (2009) Metal-induced oxidative stress impacting plant growth in contaminated soil is alleviated by microbial siderophores. Soil Biol Biochem 41(1):154–162

Dong CD, Tsai ML, Wang TH, Chang JH, Chen CW, Hung CM (2020) Removal of polycyclic aromatic hydrocarbon (PAH)-contaminated sediments by persulfate oxidation and determination of degradation product cytotoxicity based on HepG2 and ZF4 cell lines. Environ Sci Pollut Res 27:34596–34605

Dongre RS (2021) Chromium & lead as soil pollutants: insights on toxicity profiles and their remediation. J Adv Biotechnol Bioeng 9:1–16

Effiong GS, Ibia TO (2009) Characteristics and constraints of some river flood plain soils to crop production in Southeastern Nigeria. Agric J 4(2):103–108

Eichmann R, Richards L, Schafer P (2021) Hormones as go-betweens in plant microbiome assembly. Plant J 105:518–541

Ejechi BO, Akpomie OO (2016) Removal of Cr (VI) from Tannery effluent and aqueous solution by sequential treatment with Microfungi and Basidiomycete-degraded Sawdust. J Environ Prot 7(6):771–777

Elarabi NI, Abdelhadi AA, Ahmed RH, Saleh I, Arif IA, Osman G, Ahmed DS (2020) Bacillus aryabhattai FACU: a promising bacterial strain capable of manipulate the glyphosate herbicide residues. Saudi J Biol Sci 27(9):2207–2214

Eltarahony M, Ibrahim E, Hegazy G, Sabry A (2023) Microbial remediation of mercury: an overview. In: Mercury toxicity: challenges and solutions. Springer, Singapore, pp 201–234

Esikova TZ, Anokhina TO, Suzina NE, Shushkova TV, Wu Y, Solyanikova IP (2023) Characterization of a new *Pseudomonas putida* Strain Ch2, a degrader of toxic anthropogenic compounds Epsilon-Caprolactam and Glyphosate. Microorganisms 11(3):650

Espinosa-Ortiz EJ, Rene ER, Gerlach R (2022) Potential use of fungal-bacterial co-cultures for the removal of organic pollutants. Crit Rev Biotechnol 42(3):361–383

Etesami H, Emami S, Alikhani HA (2017) Potassium solubilizing bacteria (KSB): mechanisms, promotion of plant growth, and future prospects A review. J Soil Sci Plant Nutr 17(4):897–911. https://doi.org/10.4067/S0718-95162017000400005

Fakhar A, Gul B, Gurmani AR, Khan SM, Ali S, Sultan T et al (2022) Heavy metal remediation and resistance mechanism of *Aeromonas, Bacillus, and Pseudomonas*: a review. Crit Rev Environ Sci Technol 52(11):1868–1914

Faramarzi M, Avarseji Z, GholamalipuorAlamdari E, Taliei F (2023) Biodegradation of the Trifluralin Herbicide by *Pseudomonas fluorescens*. Int J Environ Sci Technol 20(4):3591–3598

Fasusi OA, Cruz C, Babalola OO (2021) Agricultural sustainability: microbial biofertilizers in rhizosphere management. Agriculture 11:163. https://doi.org/10.3390/agriculture11020163

Fella-Temzi S, Yalaoui-Guellal D, Rodriguez-Carvajal MA, Belhadi D, Madani K, Kaci Y (2018) Removal of lead by exopolysaccharides from *Paenibacilluspeoriae* strainTS7 isolated from rhizosphere of durum wheat. Biocatal Agric Biotechnol 16:425–432

Feng S, Gong L, Zhang Y, Tong Y, Zhang H, Zhu D et al (2021) Bioaugmentation potential evaluation of a bacterial consortium composed of isolated *Pseudomonas* and *Rhodococcus* for degrading benzene, toluene and styrene in sludge and sewage. Bioresour Technol 320:124329

Fernández LA, Valverde C, Gómez MA (2013) Isolation and characterization of atrazine-degrading Arthrobacter sp. strains from Argentine agricultural soils. Ann Microbiol 63:207–214

Ferraro A, Massini G, Miritana VM, Panico A, Pontoni L, Race M et al (2021) Bioaugmentation strategy to enhance polycyclic aromatic hydrocarbons anaerobic biodegradation in contaminated soils. Chemosphere 275:130091

Flores-Félix JD, Silva LR, Rivera LP, Marcos-García M, García-Fraile P, Martínez-Molina E et al (2015) Plants probiotics as a tool to produce highly functional fruits: the case of *Phyllobacterium* and vitamin C in strawberries. PLoS One 10(4):e0122281

Frey-Klett P, Burlinson P, Deveau A, Barret M, Tarkka M, Sarniguet A (2011) Bacterial-fungal interactions: hyphens between agricultural, clinical, environmental, and food microbiologists. Microbiol Mol Biol Rev 75(4):583–609

Fuller ME, Scow KM, Lau S, Ferris H (1997) Trichloroethylene (TCE) and toluene effects on the structure and function of the soil community. Soil Biol Biochem 29(1):75–89

Gabr RM, Hassan SHA, Shoreit AAM (2008) Biosorption of lead and nickel by living and nonliving cells of Pseudomonas aeruginosa ASU 6a. Int Biodeterior Biodegradation 62(2):195–203

Gao C, El-Sawah AM, Ali DFI, Alhaj HY, Shaghaleh H, Sheteiwy MS (2020) The integration of bio and organic fertilizers improve plant growth, grain yield, quality and metabolism of hybrid maize (Zea mays L.). Agronomy 10 (3): 319

Gao H, Wu M, Liu H, Xu Y, Liu Z (2022) Effect of petroleum hydrocarbon pollution levels on the soil microecosystem and ecological function. Environ Pollut 293:118511

Garg D, Sridhar K, Stephen Inbaraj B, Chawla P, Tripathi M, Sharma M (2023) Nano-biofertilizer formulations for agriculture: a systematic review on recent advances and prospective applications. Bioengineering 10:1010. https://doi.org/10.3390/bioengineering10091010

Gatabazi A, Botha M, Mvondo-She MA (2023) Assessing liquid inoculant formulation of biofertilizer (Sinorhizobium meliloti) on growth, yield, and nitrogen uptake of lucerne (*Medicago sativa*). Nitrogen 4(1):125–134

Geed SR, Prasad S, Kureel MK, Singh RS, Rai BN (2018) Biodegradation of wastewater in alternating aerobic-anoxic lab scale pilot plant by Alcaligenes sp. S3 isolated from agricultural field. J Environ Manag 214:408–415

Gentry T, Rensing C, Pepper IAN (2004) New approaches for bioaugmentation as a remediation technology. Crit Rev Environ Sci Technol 34(5):447–494

Ghaffari H, Gholizadeh A, Biabani A, Fallah A, Mohammadian M (2018) Plant growth promoting rhizobacteria (PGPR) application with different nitrogen fertilizer levels in rice (*Oryza sativa* L.). Pertanika. J Trop Agric Sci 41:715–728

Ghosh D, Gupta A, Mohapatra S (2019) A comparative analysis of exopolysaccharide and phytohormone secretions by four drought-tolerant rhizobacterial strains and their impact on osmotic-stress mitigation in *Arabidopsis thaliana*. World J Microbiol Biotechnol 35:1–15

Ghosh D, Brahmachari K, Brestic M, Ondrisik P, Hossain A, Skalicky M, Sarkar S, Moulick D, Dinda NK, Das A, Pramanick B (2020a) Integrated weed and nutrient management improve yield, nutrient uptake and economics of maize in the rice-maize cropping system of Eastern India. Agronomy 10(12):1906

Ghosh D, Brahmachari K, Skalicky M, Hossain A, Sarkar S, Dinda NK, Das A, Pramanick B, Moulick D, Brestic M, Raza MA (2020b) Nutrients supplementation through organic manures influence the growth of weeds and maize productivity. Molecules 25(21):4924

Ghosh D, Brahmachari K, Das A, Hassan MM, Mukherjee PK, Sarkar S, Dinda NK, Pramanick B, Moulick D, Maitra S, Hossain A (2021) Assessment of energy budgeting and its indicator for sustainable nutrient and weed management in a rice-maize-green gram cropping system. Agronomy 11(1):166

Ghosh D, Brahmachari K, Skalický M, Roy D, Das A, Sarkar S, Moulick D, Brestič M, Hejnak V, Vachova P, Hassan MM (2022a) The combination of organic and inorganic fertilizers influence the weed growth, productivity and soil fertility of monsoon rice. PLoS One 17(1):e0262586

Ghosh D, Brahmachari K, Sarkar S, Dinda NK, Das A, Moulick D (2022b) Impact of nutrient management in rice-maize-greengram cropping system and integrated weed management treatments on summer green gram productivity. Indian J Weed Sci 54(1):25–30

Ghosh S, Sarkar S, Garai S, Roy A, Saha S, Dey S, Santra SC, Moulick D, Mondal M, Brahmachari K (2023) Climate change and global crop production: an inclusive insight. In: Climate-resilient agriculture, Vol 2: agro-biotechnological advancement for crop production. Springer International Publishing, Cham, pp 1–34

Girigiri B, Ariole CN, Stanley HO (2019) Bioremediation of crude oil polluted soil using biofertilizer from nitrogen-fixing and phosphate solubilizing bacteria. Am J NeuroSci 5:27

Glick BR, Cheng Z, Czarny J, Duan J (2007) Promotion of plant growth by ACC deaminase-producing soil bacteria. In: New perspectives and approaches in plant growth-promoting Rhizobacteria research. Springer, Dordrecht, pp 329–339

Gomathi S, Ambikapathy V, Panneerselvam A (2023) Biodegradation and bioaugmentation of pesticides using potential fungal species. In: Plant-microbe interaction-recent advances in molecular and biochemical approaches. Academic Press, pp 79–94

Gonçalves J, Freitas J, Fernandes I, Silva P (2023) Microalgae as biofertilizers: a sustainable way to improve soil fertility and plant growth. Sustain For 15(16):12413

Gouda S, Kerry RG, Das G, Paramithiotis S, Shin HS, Patra JK (2018) Revitalization of plant growth promoting rhizobacteria for sustainable development in agriculture. Microbiol Res 206:131–140. https://doi.org/10.1016/j.micres.2017.08.016

Goyal D, Yadav A, Prasad M, Singh TB, Shrivastav P, Ali A et al (2020) Effect of heavy metals on plant growth: an overview. In: Contaminants in agriculture: sources, impacts and management. Springer, Cham, pp 79–101

Guerriero G, Hausman JF, Legay S (2016) Silicon and the plant extracellular matrix. Front Plant Sci 7:188580

Gupta AK (2004) The complete technology book on biofertilizers and organic farming, vol 168. National Institute of Industrial Research Press India, p 242

Gupta RP, Anu K, Shammi K (2007) Bioinoculants: a step towards sustainable agriculture. New India Publishing

Gupta S, Pathak B, Fulekar MH (2015) Molecular approaches for biodegradation of polycyclic aromatic hydrocarbon compounds: a review. Rev Environ Sci Biotechnol 14:241–269

Gupta R, Khan F, Alqahtani FM, Hashem M, Ahmad F (2023) Plant growth–promoting Rhizobacteria (PGPR) assisted bioremediation of Heavy Metal Toxicity. Appl Biochem Biotechnol 196:1–29

Hadi B, El-Naas MH (2019) Biosorption of heavy metals: potential and applications of yeast cells for cadmium removal. In: Environmental contaminants: ecological implications and management. Springer, Singapore, pp 237–271

Halim MA, Rahman MM, Megharaj M, Naidu R (2020) Cadmium immobilization in the rhizosphere and plant cellular detoxification: role of plant-growth-promoting rhizobacteria as a sustainable solution. J Agric Food Chem 68(47):13497–13529

Hansda A, Kumar V, Anshumali A, Usmani Z (2014) Phytoremediation of heavy met als contaminated soil using plant growth promoting rhizobacteria (PGPR): a current perspective. Recent Res Sci Technol 6:131–134

Hao X, Zhu J, Rensing C, Liu Y, Gao S, Chen W et al (2021) Recent advances in exploring the heavy metal (loid) resistant microbiome. Comput Struct Biotechnol J 19:94–109

Harasim P (2018) Nickel resources and sources. In: Nickel in soils and plants. CRC Press, pp 87–104

Haripriyan U, Gopinath KP, Arun J, Govarthanan M (2022) Bioremediation of organic pollutants: a mini review on current and critical strategies for wastewater treatment. Arch Microbiol 204(5):286

Hashem MA (2001) Problems and prospects of cyanobacterial biofertilizer for rice cultivation. Aust J Plant Physiol 28(9):881–888

Haskett TL, Tkacz A, Poole PS (2021) Engineering rhizobacteria for sustainable agriculture. ISME J 15(4):949–964

Hassan A, Pariatamby A, Ahmed A, Auta HS, Hamid FS (2019) Enhanced bioremediation of heavy metal contaminated landfill soil using filamentous fungi consortia: a demonstration of bioaugmentation potential. Water Air Soil Pollut 230:1–20

Hatti-Kaul R, Mattiasson B (2016) Anaerobes in industrial-and environmental biotechnology. Springer International Publishing, pp 1–33

Havlin J, Beaton JD, Tisdale SL, Nelson WL (2005) Soil fertility and fertilizers: an introduction to nutrient management. Pearson Prentice Hall, Upper Saddle River

Hayat MT, Nauman M, Nazir N, Ali S, Bangash N (2019) Environmental hazards of cadmium: past, present, and future. In: Cadmium toxicity and tolerance in plants. Academic Press, pp 163–183

Hazra S, Moulick D, Mukherjee A, Sahib S, Chowardhara B, Majumdar A, Upadhyay MK, Yadav P, Roy P, Santra SC, Mandal S (2023) Evaluation of efficacy of non-coding RNA in abiotic stress management of field crops: current status and future prospective. Plant Physiol Biochem 203:107940

Hazra S, Das D, Moulick D, Hossain A (2024) Endophytes: the treasure house of bioactive compounds with potential applications in sustainable agriculture and other sectors. In: Biocontrol agents for improved agriculture. Academic Press, pp 477–506

Hermosa R, Viterbo A, Chet I, Monte E (2012) Plant-beneficial effects of Trichoderma and of its genes. Microbiology, 158(1): 17–25

Hennessee CT, Li QX (2016) Effects of polycyclic aromatic hydrocarbon mixtures on degradation, gene expression, and metabolite production in four Mycobacterium species. Appl Environ Microbiol 82(11):3357–3369

Hernández-Flores JL, Bernal Melo JG, Cruz Hernández A, Ramos López MA, Carlos SG, Romero Gómez SDJ et al (2018) Isolation and characterization of mercury resistant *Trichoderma* strains from soil with high levels of mercury and its effects on *Arabidopsis thaliana* mercury uptake. Adv Microbiol 8:600

Hernandez-Pena CC, Lares-Villa F, Santos-Villalobos SDL, Estrada-Alvarado MI, Cruz-Soto A, Flores-Tavizon E, Soto-Padilla MY (2021) Reduction in concentration of chromium (VI) by Lysinibacillus macroides isolated from sediments of the Chapala Lake, Mexico. An Acad Bras Cienc 93:e20190144

Herrera C, Moraga R, Bustamante B, Vilo C, Aguayo P, Valenzuela C et al (2021) Characterization of arsenite-oxidizing bacteria isolated from arsenic-rich sediments, atacama desert, Chile. Microorganisms 9(3):483

Herridge D, Gemell G, Hartley E (2002) Legume inoculants and quality control. In: Inoculants and nitrogen fixation of Legumes in Vietnam. ACIAR Proc 109e:105–115

Hock OG, Sing LS, Hua QC, Hee CW, Shing WL (2020) Study on the ability of fungi isolated from soil to bio-remediate chromium (VI). Toxicology 16:85

Hossain A, Ahmed S, Ahmad Z, Alam MJ, Moulick D, Saha B, Samanta S, Maitra S, Bhadra P, Bukhari MA, Aftab T (2021a) Next-generation genetic engineering tools for abiotic stress tolerance in plants. In: Frontiers in plant-soil interaction. Academic Press, pp 153–197

Hossain A, Ahmad Z, Moulik D, Maitra S, Bhadra P, Ahmad A, Garai S, Mondal M, Roy A, Sabagh AE, Aftab T (2021b) Jasmonates and salicylates: mechanisms, transport and signalling during abiotic stress in plants. In: Jasmonates and salicylates signaling in plants. Springer, Cham, pp 1–29

Hossain A, Pramanick B, Bhutia KL, Ahmad Z, Moulick D, Maitra S, Ahmad A, Aftab T (2021c) Emerging roles of osmoprotectant glycine betaine against salt-induced oxidative stress in plants: a major outlook of maize (Zea mays L.). In: Frontiers in plant-soil interaction. Academic Press, pp 567–587

Hossain A, Maitra S, Sarker S, Al Mahmud A, Ahmad Z, Emon RM, Vemuri H, Malek MA, Alam MA, Rahman MA, Alam MJ (2023) Aluminium stress in plants: consequences and mitigation mechanisms. In: Beneficial chemical elements of plants: recent developments and future prospects. Wiley, pp 123–168

Houshani M, Salehi-Lisar SY (2020) Agronomic crop responses and tolerance to polycyclic aromatic hydrocarbon toxicity. In: Agronomic crops: volume 3: stress responses and tolerance. Springer, Singapore, pp 265–283

Howieson J, Ballard R (2004) Optimizing the Legume Symbiosis in stressful and competitive environments within Southern Australia—some contemporary thoughts. Soil Biol Biochem 36(8):1261–1273

Huang F, Guo CL, Lu GN, Yi XY, Zhu LD, Dang Z (2014) Bioaccumulation characterization of cadmium by growing *Bacillus cereus* RC-1 and its mechanism. Chemosphere 109:134–142

Huang Y, Chen SF, Chen WJ, Zhu X, Mishra S, Bhatt P, Chen S (2023) Efficient biodegradation of multiple pyrethroid pesticides by *Rhodococcuspyridinivorans* strain Y6 and its degradation mechanism. Chem Eng J 469:143863

Ismail HY, Farouq AA, Rabah AB, Muhammad AB, Aliyu RU (2021) Effect of soil contamination with crude petroleum on cowpea: an insight into the prospects of crop production in Nigerian Frontier Basins. J Environ Agric Stud 2(2):50–62

Itelima JU, Bang WJ, Onyimba IA, Oj E (2018) A review: biofertilizer; a key player in enhancing soil fertility and crop productivity. J Microbiol Biotechnol Rep 2(1):22–28

Jabbar NM, Alardhi SM, Mohammed AK, Salih IK, Albayati TM (2022) Challenges in the implementation of bioremediation processes in petroleum-contaminated soils: a review. Environ Nanotechnol Monitor Manag 18:100694

Jajoo A (2017) Effects of environmental pollutants polycyclic aromatic hydrocarbons (PAH) on photosynthetic processes. In: Photosynthesis: structures, mechanisms, and applications. Springer, Cham, pp 249–259

Jaishankar M, Mathew BB, Shah MS, Gowda KRS (2014) Biosorption of few heavy metal ions using agricultural wastes. J Environ Pollut Hum Health, 2(1): 1–6

Jasu A, Dutta B, Das SC, Ray RR (2023) Application of *Serratia marcescens* AJRR-22 for effective Amelioration of Hexavalent chromium from industrial effluents contaminating bio-geosphere. Water Air Soil Pollut 234(12):752

Ji M, Sang W (2022) The remediation potential of biochar derived from different biomass for typical pollution in agricultural soil. In: Biochar in agriculture for achieving sustainable development goalS. Academic Press, pp 71–83

Jindrova E, Chocova M, Demnerova K, Brenner V (2002) Bacterial aerobic degradation of benzene, toluene, ethylbenzene and xylene. Folia Microbiol 47:83–93

Kalhapure AH, Shete BT, Dhonde MB (2013) Integrated nutrient management in maize (Zea mays l.) for increasing production with sustainability. Int J Agric Food Sci Technol 4:195–206

Kalia A, Gosal SK (2011) Effect of pesticide application on soil microorganisms. Arch Agron Soil Sci 57(6):569–596

Kalinhoff C, Calderón NT (2022) Mercury phytotoxicity and tolerance in three wild plants during germination and seedling development. Plan Theory 11(15):2046

Kanungo J, Sahoo T, Swain LP, Behera ID (2024) Toxicity of persistent hydrocarbon pollutants, sources and sustainable remediation process. In: Impact of petroleum waste on environmental pollution and its sustainable management through circular economy. Springer Nature Switzerland, Cham, pp 39–65

Kaur P, Purewal SS (2019) Biofertilizers and their role in sustainable agriculture. In: Biofertilizers for sustainable agriculture and environment. Springer, Cham, pp 285–300

Kaur H, Gosal SK, Walia SS, Kaur J (2020) Effect of sulphur oxidizing bacterial inoculants on microbial population and enzymatic activities in rhizosphere of *Brassica napus* (Var. GSC-7). Int J Curr Microbiol App Sci 9(8):753–763

Kaur R, Singh D, Kumari A, Sharma G, Rajput S, Arora S, Kaur R (2021) Pesticide residues degradation strategies in soil and water: a review. Int J Environ Sci Technol 20:1–24

Keswani C, Prakash O, Bharti N, Vílchez JI, Sansinenea E, Lally RD et al (2019) Re-addressing the biosafety issues of plant growth promoting rhizobacteria. Sci Total Environ 690:841–852

Khan SS, Flora SJS (2023) Arsenic: chemistry, occurrence, and exposure. In: Handbook of arsenic toxicology. Academic Press, pp 1–49

Khan Z, Roman D, Kintz T, delas Alas M, Yap R, Doty S (2014) Degradation, phytoprotection and phytoremediation of phenanthrene by endophyte *Pseudomonas putida*, PD1. Environ Sci Technol 48(20):12221–12228

Khan A, Panthari D, Sharma RS, Punetha A, Singh AV, Upadhayay VK (2023) Biofertilizers: a microbial-assisted strategy to improve plant growth and soil health. In: Advanced microbial techniques in agriculture, environment, and health management. Academic Press, pp 97–118

Khanam Z, Sultana FM, Mushtaq F (2023) Environmental pollution control measures and strategies: an overview of recent developments. In: Geospatial analytics for environmental pollution modeling: analysis, control and management. Springer, Cham, pp 385–414

Khandare RN, Chandra R, Pareek N, Raverkar KP (2020) Carrier-based and liquid bioinoculants of Azotobacter and PSB saved chemical fertilizers in wheat (Triticum aestivum L.) and enhanced soil biological properties in Mollisols. J Plant Nutr 43(1):36–50

Kharat AS, Parte SG, More NS (2018) MicrobesEcofriendly tools for bioremediation of pesticide-contaminated environments. In: Recent advances in environmental management. CRC Press, pp 391–416

Khatun J, Intekhab A, Dhak D (2022) Effect of uncontrolled fertilization and heavy metal toxicity associated with arsenic (As), lead (Pb) and cadmium (Cd), and possible remediation. Toxicology 477:153274

Khodaei K, Nassery HR, Asadi MM, Mohammadzadeh H, Mahmoodlu MG (2017) BTEX biodegradation in contaminated groundwater using a novel strain (*Pseudomonas* sp. BTEX-30). Int Biodeterior Biodegradation 116:234–242

Kim JR, Ahn YJ (2009) Identification and characterization of chlorpyrifos-methyl and 3, 5, 6-trichloro-2-pyridinol degrading *Burkholderia sp.* strain KR100. Biodegradation 20:487–497

Kim TJ, Lee EY, Kim YJ, Cho KS, Ryu HW (2003) Degradation of polyaromatic hydrocarbons by *Burkholderia cepacia* 2A-12. World J Microbiol Biotechnol 19:411–417

Kirui CK, Njeru EM, Runo S (2022) Diversity and phosphate solubilization efficiency of phosphate solubilizing bacteria isolated from semi-arid agroecosystems of Eastern Kenya. Microbiol Insights 17:15. https://doi.org/10.1177/11786361221088991

Kolekar PD, Phugare SS, Jadhav JP (2014) Biodegradation of atrazine by *Rhodococcus* sp. BCH2 to N-isopropylammelide with subsequent assessment of toxicity of biodegraded metabolites. Environ Sci Pollut Res 21:2334–2345

Kumar V (2014) Characterization, bio-formulation development and shelf-life studies of locally isolated bio-fertilizer strains. Octa J Environ Res 2(1):32–37

Kumar V, Chandra A, Singh G (2010) Efficacy of fly-ash based bio-fertilizers vs perfected chemical fertilizers in wheat (*Triticum aestivum*). Int J Sci Technol Res 2:31–35

Kumar A, Vandana RS, Singh M, Pandey KD (2015) Plant Growth Promoting Rhizobacteria (PGPR). In: A Promising approach for disease management. Microbes and environmental management. Studium Press, New Delhi, pp 195–209

Kumar R, Kumawat N, Kumar S, Singh AK, Bohra JS (2017) Effect of NPKS and Zn fertilization on, growth, yield and quality of baby corn-A review. Int J Curr Microbiol Appl Sci, 6(3): 1392–1428

Kumar G, Lal S, Soni SK, Chaudhary P (2022a) Mechanism and kinetics of chlorpyrifos co-metabolism by using environment restoring microbes isolated from rhizosphere of horticultural crops under subtropics. Front Microbiol 13:891870

Kumar S, Sindhu SS, Kumar R (2022b) Biofertilizers: an ecofriendly technology for nutrient recycling and environmental sustainability. Curr Res Microbial Sci 3:100094

Kumari S, Mishra A (2021) Heavy metal contamination. In: Soil contamination-threats and sustainable solutions. IntechOpen

Kumari R, Singh DP (2019) Nano-biofertilizer: an emerging eco-friendly approach for sustainable agriculture. Proc Natl Acad Sci India Sect B Biol Sci 90:1–9

Kumawat N, Kumar R, Khandkar UR, Yadav RK, Saurabh K, Mishra JS et al (2019) Silicon (Si)- and Zinc (Zn)-solubilizing microorganisms: role in sustainable agriculture. In: Biofertilizers for sustainable agriculture and environment. Springer, Cham, pp 109–135

Kusale SP, Attar YC, Sayyed RZ, Malek RA, Ilyas N, Suriani NL (2021) Production of plant beneficial and antioxidants metabolites by *Klebsiella variicola* under salinity stress. Molecules 26:1894

Lathifah AN, Asmara AA, Alfisyahr FAN, Maziya FB (2021) Potency of Indigenous Bacteria of Mt. Merapi, Arthrobacter chlorophenolicus for Chromium (VI) Bioremediation. In: IOP conference series: earth and environmental science, vol 933, No. 1. IOP Publishing, p 012011

Lauriano-Barajas J, Vega-Frutis R (2018) Infectivity and effectivity of commercial and native arbuscular mycorrhizal biofertilizers in seedlings of maize (*Zea mays*). Botan Sci 96(3):395–404

Lawal AT (2017) Polycyclic aromatic hydrocarbons. A review. Cogent Environ Sci 3(1):1339841

Lebrazi S, Fikri-Benbrahim K (2018) Rhizobium-Legume Symbioses: heavy metal effects and principal approaches for bioremediation of contaminated soil. In: Legumes for soil health and sustainable management. Springer, Singapore, pp 205–233

Leys NM, Ryngaert A, Bastiaens L, Verstraete W, Top EM, Springael D (2004) Occurrence and phylogenetic diversity of Sphingomonas strains in soils contaminated with polycyclic aromatic hydrocarbons. Appl Environ Microbiol 70(4):1944–1955

Li Q, Liu J, Gadd GM (2020) Fungal bioremediation of soil co-contaminated with petroleum hydrocarbons and toxic metals. Appl Microbiol Biotechnol,104(21): 8999–9008

Li T, Wen J, Li B, Ding S, Wang W (2021) Biological effects of tourmaline treatment on *Dehalococcoides* spp. during the reductive dechlorination of trichloroethylene. RSC Adv 11(20):12086–12094

Long B, Ye J, Ye Z, He J, Luo Y, Zhao Y, Shi J (2020) Cr (VI) removal by *Penicillium oxalicum* SL2: reduction with acidic metabolites and form transformation in the mycelium. Chemosphere 253:126731

Long C, Su Z, Hu G, Zhang Q, Zhang Y, Chen T et al (2024) Potential mechanisms of lung injury and repair after hexavalent chromium [Cr (VI)] aerosol whole-body dynamic exposure. Chemosphere 349:140918

Lopes MJS, Dias-Filho MB, Gurgel ESC (2021) Successful plant growth-promoting microbes: inoculation methods and abiotic factors. Front Sustain Food Syst 5:606454

Lopes PRM, Cruz VH, De Menezes AB, Gadanhoto BP, Moreira BRDA, Mendes CR et al (2022) Microbial bioremediation of pesticides in agricultural soils: an integrative review on natural attenuation, bioaugmentation and biostimulation. Rev Environ Sci Biotechnol 21(4):851–876

López-Bucio JS, Ravelo-Ortega G, López-Bucio J (2022) Chromium in plant growth and development: toxicity, tolerance and hormesis. Environ Pollut 312:120084

Lou X, Zhao J, Lou X, Xia X, Feng Y, Li H (2022) The biodegradation of soil organic matter in soil-dwelling humivorous fauna. Front Bioeng Biotechnol 9:808075

Lu Q, Liu J, He H, Liang Z, Qiu R, Wang S (2021) Waste activated sludge stimulates in situ microbial reductive dehalogenation of organohalide-contaminated soil. J Hazard Mater 411:125189

Luo S, Zhen Z, Zhu X, Ren L, Wu W, Zhang W et al (2021) Accelerated atrazine degradation and altered metabolic pathways in goat manure assisted soil bioremediation. Ecotoxicol Environ Saf 221:112432

Lurthy T, Cantat C, Jeudy C, Declerck P, Gallardo K, Barraud C (2020) Impact of bacterial siderophores on iron status and ionome in pea. Front Plant Sci 11:730

Ma M, Du H, Wang D (2019) Mercury methylation by anaerobic microorganisms: a review. Crit Rev Environ Sci Technol 49(20):1893–1936

Ma Y, Shen Y, Liu Y (2020) Food waste to biofertilizer: a potential game changer of global circular agricultural economy. J Agric Food Chem 68:5021

Mącik M, Gryta A, Frąc M (2020) Biofertilizers in agriculture: an overview on concepts, strategies and effects on soil microorganisms. Adv Agron 162:31–87

Mahajan R, Verma S, Chatterjee S (2023) Biodegradation of organophosphorus pesticide profenofos by the bacterium Bacillus sp. PF1 and elucidation of initial degradation pathway. Environ Technol 44(4):492–500

Mahato S, Kafle A (2018) Comparative study of Azotobacter with or without other fertilizers on growth and yield of wheat in Western hills of Nepal. Annals of Agrarian Science, 16(3): 250–256

Mahanty T, Bhattacharjee S, Goswami M, Bhattacharyya P, Das B, Ghosh A, Tribedi P (2017) Biofertilizers: a potential approach for sustainable agriculture development. Environ Sci Pollut Res 24:3315–3335

Mahbub KR, Bahar MM, Labbate M, Krishnan K, Andrews S, Naidu R, Megharaj M (2017) Bioremediation of mercury: not properly exploited in contaminated soils! Appl Microbiol Biotechnol 101(3):963–976. https://doi.org/10.1007/s00253-016-8079-2

Mahmood Q, Bilal M, Jan S (2014) Herbicides, pesticides, and plant tolerance: an overview. In: Emerging technologies and management of crop stress tolerance. Elsevier, pp 423–448

Mahmud AA, Upadhyay SK, Srivastava AK, Bhojiya AA (2021) Biofertilizers: a nexus between soil fertility and crop productivity under abiotic stress. Curr Res Environ Sustain 3:100063. https://doi.org/10.1016/j.crsust.2021.100063

Mahto KU, Priyadarshanee M, Samantaray DP, Das S (2022) Bacterial biofilm and extracellular polymeric substances in the treatment of environmental pollutants: Beyond the protective role in survivability. J Clean Prod 379:134759

Majumdar J, Moulik D, Santra SC, Hossain A (2023) Extremophile bacterial and archaebacterial population: metagenomics and novel enzyme reserve. In: Microbial symbionts and plant health: trends and applications for changing climate. Springer Nature Singapore, Singapore, pp 521–544

Majumdar A, Upadhyay MK, Giri B, Moulick D, Sarkar S, Thakur B, Yadav P, Sahu K, Srivastava AK, Buck M, Tibbett M (2023a) Sustainable irrigation reduces arsenic bioavailability in fluvio-alluvial soils promoting microbial responses, high rice productivity and economic profit. https://doi.org/10.31223/X50X0Q

Majumdar A, Dubey PK, Giri B, Moulick D, Srivastava AK, Roychowdhury T, Bose S, Jaiswal MK (2023b) Combined effects of dry-wet irrigation, redox changes and microbial diversity on soil nutrient bioavailability in the rice field. Soil Tillage Res 232:105752

Majumdar A, Moulick D, Srivastava S (2024a) Save Soil'by managing soil nutrient losses, agronomic practices and crop-microbial interaction: World Soil Day 2022. Front Sustain Food Syst 8:1360937

Majumdar A, Upadhyay MK, Giri B, Yadav P, Moulick D, Sarkar S, Thakur BK, Sahu K, Srivastava AK, Buck M, Tibbett M (2024b) Sustainable water management in rice cultivation reduces arsenic contamination, increases productivity, microbial molecular response, and profitability. J Hazard Mater 466:133610

Mala R, Selvaraj CA, Barathi Sundaram V, Siva B, Shanmuga Rajan R, Maheswari Gurusamy U (2017) Evaluation of nano structured slow release fertilizer on the soil fertility, yield and nutritional profile of vigna radiata. Recent Pat Nanotechnol 11(1):50–62

Maliszewska-Kordybach B, Smreczak B (2000) Ecotoxicological activity of soils polluted with polycyclic aromatic hydrocarbons (PAHs)-effect on plants. Environ Technol 21(10):1099–1110

Malusá E, Sas-Paszt L, Ciesielska JJTSWJ (2012) Technologies for beneficial microorganisms inocula used as biofertilizers. Sci World J 2012:491206

Malusá E, Vassilev N (2014) A contribution to set a legal framework for biofertilisers. Appl Microbiol Biotechnol, 98: 6599–6607

Mandal D, Sonar R, Saha I, Ahmed S, Basu A (2022) Isolation and identification of arsenic resistant bacteria: a tool for bioremediation of arsenic toxicity. Int J Environ Sci Technol 19(10):9883–9900

Marwa N, Singh N, Srivastava S, Saxena G, Pandey V, Singh N (2019) Characterizing the hypertolerance potential of two indigenous bacterial strains (*Bacillus flexus* and *Acinetobacter junii*) and their efficacy in arsenic bioremediation. J Appl Microbiol 126(4):1117–1127

Masso C, Ochieng JRA, Vanlauwe B (2015) Worldwide contrast in application of bio-fertilizers for sustainable agriculture: lessons for sub-Saharan Africa. J Biol Agric Healthc 5(12):34–50

Mathivanan R, Umavathi S, Ramasamy PK, Thangam Y (2015) Influence of vermicompost on the activity of the plant growth regulators in the leaves of the Indian butter bean plant, *Dolichos lab* L. Int J Adv Res Biol Sci 2(1):84–89

Maurya A, Kumar PS, Raj A (2022a) Characterization of biofilm formation and reduction of hexavalent chromium by bacteria isolated from tannery sludge. Chemosphere 286:131795

Maurya A, Kumar R, Yadav P, Singh A, Yadav A, Chowdhary P, Raj A (2022b) Biofilm formation and extracellular polymeric substance (EPS) production by *Bacillus haynesii* and influence of hexavalent chromium. Bioresour Technol 352:127109

Mazurier S, Corberand T, Lemanceau P, Raaijmakers JM (2009) Phenazine antibiotics produced by fluorescent pseudomonads contribute to natural soil suppressiveness to Fusarium wilt. The ISME journal, 3(8): 977–991

Mbachu AE, Chukwura EI, Mbachu NA (2020) Role of microorganisms in the degradation of organic pollutants: a review. Energy Environ Eng 7(1):1–11

Mbachu AE, Chukwura EI, Mbachu NA (2021) Microbial degradation and transformation of organic pollutants: an eco-friendly strategy for waste mitigation. Recent Prog Microbiol Biotechnol 6:19–34

McCarty PL, Criddle CS, Vogel TM (2020) Retrospective on microbial transformations of halogenated organics. Environ Sci: Processes Impacts 22(3):512–517

McComb RB, Bowers GN Jr, Posen S (2013) Alkaline phosphatase. Springer Science & Business Media, pp 79–436

Meddich A (2023) Biostimulants for resilient agriculture—Improving plant tolerance to abiotic stress: a concise review. Gesunde Pflanzen 75(4):709–727

Meena VD, Dotaniya ML, Coumar V, Rajendiran S, Ajay KS, Subba Rao A (2014a) A case for silicon fertilization to improve crop yields in tropical soils. Proc Natl Acad Sci India Sect B Biol Sci 84:505–518

Meena VS, Maurya BR, Verma JP (2014b) Does a rhizospheric microorganism enhance K+ availability in agricultural soils? Microbiol Res 169(5–6):337–347
Megharaj M, Ramakrishnan B, Venkateswarlu K, Sethunathan N, Naidu R (2011) Bioremediation approaches for organic pollutants: a critical perspective. Environ Int 37(8):1362–1375
Meng X, Guo Y, Wang Y, Fan S, Wang K, Han W (2022) A systematic review of photolysis and hydrolysis degradation modes, degradation mechanisms, and identification methods of pesticides. J Chem 2022:9552466. (Special issue: innovations in green plant protection and food storage technology)
Mirouze N, Ferret C, Cornilleau C, Carballido-López R (2018) Antibiotic sensitivity reveals that wall teichoic acids mediate DNA binding during competence in *Bacillus subtilis*. Nat Commun 9(1):5072
Misra M, Sachan A, Sachan SG (2020) Current aspects and applications of biofertilizers for sustainable agriculture. In: Plant microbiomes for sustainable agriculture. Springer, Cham, pp 445–473
Mohammadi K, Sohrabi Y (2012) Bacterial biofertilizers for sustainable crop production: a review. Arpn J Agric Biol Sci 7(5):307–316
Mohanty P, Singh PK, Chakraborty D, Mishra S, Pattnaik R (2021) Insight into the role of PGPR in sustainable agriculture and environment. Front Sustain Food Syst 5:667150
Mohapatra B, Phale PS (2021) Microbial degradation of naphthalene and substituted naphthalenes: metabolic diversity and genomic insight for bioremediation. Front Bioeng Biotechnol 9:602445
Mohapatra B, Dhamale T, Saha BK, Phale PS (2022) Microbial degradation of aromatic pollutants: metabolic routes, pathway diversity, and strategies for bioremediation. In: Microbial biodegradation and bioremediation. Elsevier, pp 365–394
Montes-Robledo A, Baena-Baldiris D, Baldiris-Avila R (2024) Reduction of Cr (VI) by planktonic cells and biofilm of *Acinetobacter* sp.(ADHR1) isolated from electroplating wastewater. Environ Technol Innov 33:103521
Moulick D, Ghosh D, Santra SC (2016a) An assessment of some physicochemical properties and cooking characteristics of milled rice and associated health risk in two rice varieties of arsenic contaminated areas of West Bengal, India. Int J Res Agric Food Sci 6(2):44–55
Moulick D, Ghosh D, Santra SC (2016b) Evaluation of effectiveness of seed priming with selenium in rice during germination under arsenic stress. Plant Physiol Biochem 109:571–578
Moulick D, Santra SC, Ghosh D (2017) Seed priming with Se alleviate As induced phytotoxicity during germination and seedling growth by restricting As translocation in rice (Oryza sativa L cv IET-4094). Ecotoxicol Environ Saf 145:449–456
Moulick D, Santra SC, Ghosh D (2018a) Consequences of paddy cultivation in arsenic-contaminated paddy fields of lower Indo-Gangetic plain on arsenic accumulation pattern and selected grain quality traits: a preliminary assessment. In: Mechanisms of arsenic toxicity and tolerance in plants. Springer Nature, Singapore, pp 49–78
Moulick D, Santra SC, Ghosh D (2018b) Effect of selenium induced seed priming on arsenic accumulation in rice plant and subsequent transmission in human food chain. Ecotoxicol Environ Saf 152:67–77
Moulick D, Santra SC, Ghosh D (2018c) Rice seed priming with se: a novel approach to mitigate as induced adverse consequences on growth, yield and as load in brown rice. J Hazard Mater 355:187–196
Moulick D, Santra SC, Ghosh D, Panda SK (2019a) An assessment of efficiency of zinc priming in rice (cv. MTU-7029) during germination and early seedling growth. In: Priming and pretreatment of seeds and seedlings: implication in plant stress tolerance and enhancing productivity in crop plants. Springer, Singapore, pp 495–507
Moulick D, Chowardhara B, Panda SK (2019b) Agroecotoxicological aspect of arsenic (As) and cadmium (Cd) on field crops and its mitigation: current status and future prospect. In: Plant-metal interactions. Springer, Cham, pp 217–246

Moulick D, Samanta S, Saha B, Mazumder MK, Dogra S, Panigrahi KC, Banerjee S, Ghosh D, Santra SC (2020a) Salinity stress responses in three popular field crops belonging to Fabaceae Family: current status and future prospect. In: The plant family Fabaceae: biology and physiological responses to environmental stresses. Springer, Singapore, pp 519–541

Moulick D, Sarkar S, Awasthi JP, Ghosh D, Choudhury S, Tata SK, Bramhachari K, Santra SC (2020b) Rice grain quality traits: neglected or less addressed? In: Rice research for quality improvement: genomics and genetic engineering: volume 1: breeding techniques and abiotic stress tolerance. Springer, Singapore, pp 729–745

Moulick RG, Das S, Debnath N, Bandyopadhyay K (2020c) Potential use of nanotechnology in sustainable and 'smart' agriculture: advancements made in the last decade. Plant Biotechnol Rep 14:505–513

Moulick D, Samanta S, Sarkar S, Mukherjee A, Pattnaik BK, Saha S, Awasthi JP, Bhowmick S, Ghosh D, Samal AC, Mahanta S (2021) Arsenic contamination, impact and mitigation strategies in rice agro-environment: an inclusive insight. Sci Total Environ 800:149477

Moulick D, Ghosh D, Mandal J, Bhowmick S, Mondal D, Choudhury S, Santra SC, Vithanage M, Biswas JK (2023a) A cumulative assessment of plant growth stages and selenium supplementation on arsenic and micronutrients accumulation in rice grains. J Clean Prod 386:135764

Moulick D, Bhutia KL, Sarkar S, Roy A, Mishra UN, Pramanick B, Maitra S, Shankar T, Hazra S, Skalicky M, Brestic M (2023b) The intertwining of Zn-finger motifs and abiotic stress tolerance in plants: current status and future prospects. Front Plant Sci 13:1083960

Moulick D, Mukherjee A, Das A, Roy A, Majumdar A, Dhar A, Pattanaik BK, Chowardhara B, Ghosh D, Upadhyay MK, Yadav P (2024a) Selenium–An environmentally friendly micronutrient in agroecosystem in the modern era: an overview of 50-year findings. Ecotoxicol Environ Saf 270:115832

Moulick D, Choudhury S, Brestic M, Hossain A (2024b) The impact of abiotic stresses on agriculture: mitigation through climate smart strategies. Front Plant Sci 15:1391051

Moulick D, Majumdar A, Choudhury A, Das A, Chowardhara B, Pattnaik BK, Dash GK, Murmu K, Bhutia KL, Upadhyay MK, Yadav P (2024c) Emerging concern of nano-pollution in agroecosystem: flip side of nanotechnology. Plant Physiol Biochem 211:108704. https://doi.org/10.1016/j.plaphy.2024.108704

Moulick D, Ghosh D, Gharde Y, Majumdar A, Upadhyay MK, Chakraborty D, Mahanta S, Das A, Choudhury S, Brestic M, Alahmadi TA (2024d) An assessment of the impact of traditional rice cooking practice and eating habits on arsenic and iron transfer into the food chain of smallholders of Indo-Gangetic plain of South-Asia: using AMMI and Monte-Carlo simulation model. Heliyon 10(7):e28296

Mrabet R (2023) Sustainable agriculture for food and nutritional security. In: Sustainable agriculture and the environment. Academic Press, pp 25–90

Mujawar SY, Vaigankar DC, Dubey SK (2021) Biological characterization of *Bacillus flexus* strain SSAI1 transforming highly toxic arsenite to less toxic arsenate mediated by periplasmic arsenite oxidase enzyme encoded by aioAB genes. Biometals 34(4):895–907

MurugaragavanRamasamy TG, Yuvaraj M (2020) Role of biofertilizers in plant growth and soil health. In: Nitrogen fixation, vol 1. Intechopen

Nabi A, Naeem M, Aftab T, Khan MMA, Ahmad P (2021) A comprehensive review of adaptations in plants under arsenic toxicity: physiological, metabolic and molecular interventions. Environ Pollut 290:118029

Naik MM, Dubey SK (2013) Lead resistant bacteria: lead resistance mechanisms, their applications in lead bioremediation and biomonitoring. Ecotoxicol Environ Saf 98:1–7

Naik MM, Pandey A, Dubey SK (2012) Pseudomonas aeruginosa strain WI-1 from Mandovi estuary possesses metallothionein to alleviate lead toxicity and promotes plant growth. Ecotoxicol Environ Saf 79:129–133

Naz S, Iqbal SS, Manan A, Chatha M, Zia M (2023) The web of life: role of pesticides in the biodiversity decline. Int J Forest Sci 3(2):72–94

Neria-González MI, Aguilar-López R (2021) Heavy metal removal processes by sulfate-reducing bacteria. In: Environmental pollution and remediation. Springer, Singapore, pp 367–394

Ning JY, Zhu XD, Liu HG, Yu GH (2021) Coupling thermophilic composting and vermicomposting processes to remove Cr from biogas residues and produce high value-added biofertilizers. Bioresour Technol 329:124869

Nosheen S, Ajmal I, Song Y (2021) Microbes as biofertilizers, a potential approach for sustainable crop production. Sustain For 13(4):1868

Odoh CK, Zabbey N, Sam K, Eze CN (2019) Status, progress and challenges of phytoremediation-An African scenario. J Environ Manag 237:365–378

Ogidi OI, Akpan UM (2022) Aquatic biodiversity loss: impacts of pollution and anthropogenic activities and strategies for conservation. In: Biodiversity in Africa: potentials, threats and conservation. Springer Nature Singapore, Singapore, pp 421–448

Ojuederie OB, Babalola OO (2017) Microbial and plant-assisted bioremediation of heavy metal polluted environments: a review. Int J Environ Res Public Health 14(12):1504

Olaniyan FT, Alori ET, Adekiya AO, Ayorinde BB, Daramola FY, Osemwegie OO, Babalola OO (2022) The use of soil microbial potassium solubilizers in potassium nutrient availability in soil and its dynamics. Ann Microbiol 72(1):45

Olanrewaju OS, Glick BR, Babalola OO (2017) Mechanisms of action of plant growth promoting bacteria. World J Microbiol Biotechnol 33(11):197. https://doi.org/10.1007/s11274-017-2364-9

Ortega DR, Esquivel DFG, Ayala TB, Pineda B, Manzo SG, Quino JM et al (2021) Cognitive impairment induced by lead exposure during lifespan: mechanisms of lead neurotoxicity. Toxics 9(2):23

Osorio-Reyes JG, Valenzuela-Amaro HM, Pizaña-Aranda JJP, Ramírez-Gamboa D, Meléndez-Sánchez ER, López-Arellanes ME et al (2023) Microalgae-based biotechnology as alternative biofertilizers for soil enhancement and carbon footprint reduction: advantages and implications. Mar Drugs 21(2):93

Ossai IC, Ahmed A, Hassan A, Hamid FS (2020) Remediation of soil and water contaminated with petroleum hydrocarbon: a review. Environ Technol Innov 17:100526

Otieno N, Lally RD, Germaine KJ, Dowling DN (2015) Plant growth promotion induced by phosphate solubilizing endophytic *Pseudomonas* isolates. Front Microbiol 6:148361

Ouertani, R., Ouertani, A., Mahjoubi, M., Bousselmi, Y., Najjari, A., Cherif, H., ... & Neifar, M. (2020) New plant growth-promoting, chromium-detoxifying microbacterium species isolated from a tannery wastewater: performance and genomic insights. Frontiers in Bioengineering and Biotechnology, 8:521

Pailan S, Sengupta K, Saha P (2020) Microbial metabolism of organophosphates: key. In: Microbial technology for health and environment, vol 22. Springer, Singapore, p 361

Pan X, Zhang S, Zhong Q, Gong G, Wang G, Guo X, Xu X (2020) Effects of soil chemical properties and fractions of Pb, Cd, and Zn on bacterial and fungal communities. Sci Total Environ 715:136904

Panazai GM, Khan A, Jan A, Kethran R, Agha SAH (2019) 95. Biozote-N Performance in peas production under the influence of compost application. Pure Appl Biol 8(1):931–941

Panda SC (2011) Organic Farming for Sustainable agriculture, third ed. Kalyani publisher, New Delhi

Pande V, Pandey SC, Sati D, Pande V, Samant M (2020) Bioremediation: an emerging effective approach towards environment restoration. Environ Sustain 3:91–103

Pandey D, Kehri HK, Zoomi I, Chaturvedi S, Chaudhary KL (2022) Seasonal effect on the diversity of soil fungi and screening for arsenic tolerance and their remediation. J Appl Biol Biotechnol 10(1):40–46

Pânzaru RL, Firoiu D, Ionescu GH, Ciobanu A, Medelete DM, Pîrvu R (2023) Organic Agriculture in the Context of 2030 Agenda Implementation in European Union Countries. Sustain For 15(13):10582

Paria K, Pyne S, Chakraborty SK (2022) Optimization of heavy metal (lead) remedial activities of fungi Aspergillus penicillioides (F12) through extra cellular polymeric substances. Chemosphere 286:131874

Parmar P, Sindhu SS (2013) (2013). Potassium solubilization by rhizosphere bacteria: influence of nutritional and environmental conditions. J Microbiol Res 3(1):25–31

Parmar P, Srivatsan V (2023) Microalgae as next generation plant growth additives: functions, applications, challenges and circular bioeconomy based solutions. Front Plant Sci 14:1073546

Partridge MA, Huang SX, Hernandez-Rosa E, Davidson MM, Hei TK (2007) Arsenic induced mitochondrial DNA damage and altered mitochondrial oxidative function: implications for genotoxic mechanisms in mammalian cells. Cancer Res 67(11):5239–5247

Parween T, Bhandari P, Siddiqui ZH, Jan S, Fatma T, Patanjali PK (2017) Biofilm: a next-generation biofertilizer. In: Mycoremediation and environmental sustainability, vol 1. Springer, Cham, pp 39–51

Patel KS, Pandey PK, Martín-Ramos P, Corns WT, Varol S, Bhattacharya P, Zhu Y (2023) A review on arsenic in the environment: contamination, mobility, sources, and exposure. RSC Adv 13(13):8803–8821

Paterek JR, Bogan WW, Trbovic V, Sullivan W (2003) Application of chemically accelerated biotreatment to reduce risk in oil-impacted soils. Gas Technology Institute (US)

Pattnaik BK, Sahu C, Choudhury S, Santra SC, Moulick D (2023) Importance of soil management in sustainable agriculture. In: Agriculture C-R (ed) Vol 1: Crop responses and agroecological perspectives. Springer International Publishing, Cham, pp 487–511

Pérez-Lucas G, Vela N, El Aatik A, Navarro S (2019) Environmental risk of groundwater pollution by pesticide leaching through the soil profile. In: Pesticides-use and misuse and their impact in the environment, vol 17. Intech Open, pp 1–28

Perini BLB, Bitencourt RL, Daronch NA, dos Santos Schneider AL, de Oliveira D (2020) Surfactant-enhanced in-situ enzymatic oxidation: a bioremediation strategy for oxidation of polycyclic aromatic hydrocarbons in contaminated soils and aquifers. J Environ Chem Eng 8(4):104013

Phukongchai W, Kaewpradit W, Rasche F (2022) Inoculation of cellulolytic and ligninolytic microorganisms accelerates decomposition of high C/N and cellulose rich sugarcane straw in tropical sandy soils. Appl Soil Ecol 172:104355

Pileggi M, Pileggi SA, Sadowsky MJ (2020) Herbicide bioremediation: from strains to bacterial communities. Heliyon 6(12):e05767

Pinter IF, Salomon MV, Berli F, Bottini R, Piccoli P (2017) Characterization of the As (III) tolerance conferred by plant growth promoting rhizobacteria to in vitro-grown grapevine. Appl Soil Ecol 109:60–68

Pii Y, Mimmo T, Tomasi N, Terzano R, Cesco S, Crecchio C (2015) Microbial interactions in the rhizosphere: beneficial influences of plant growth-promoting rhizobacteria on nutrient acquisition process. A review. Biol. Fertil. Soils 51 (4): 403–415

Prasad S, Yadav KK, Kumar S, Gupta N, Cabral-Pinto MM, Rezania S et al (2021) Chromium contamination and effect on environmental health and its remediation: a sustainable approaches. J Environ Manag 285:112174

Prabha C, Maheshwari DK, Bajpai VK (2013) Diverse role of fast growing rhizobia in growth promotion and enhancement of psoralen content in Psoralea corylifolia L. Pharmacognosy magazine, 9(1): S57

Prodhan MY, Rahman MB, Rahman A, Akbor MA, Ghosh S, Nahar MNEN et al (2023) Characterization of growth-promoting activities of Consortia of Chlorpyrifos mineralizing endophytic bacteria naturally harboring in rice plants—a potential bio-stimulant to develop a safe and sustainable agriculture. Microorganisms 11(7):1821

Purakayastha TJ (2011) Microbial remediation of arsenic contaminated soil. In: Detoxification of heavy metals. Springer, Berlin, Heidelberg, pp 221–260

Purwanti IF, Obenu A, Tangahu BV, Kurniawan SB, Imron MF, Abdullah SRS (2020) Bioaugmentation of Vibrio alginolyticus in phytoremediation of aluminium-contaminated soil using *Scirpusgrossus* and *Thypa angustifolia*. Heliyon 6(9):e05004

Pushkar B, Sevak P (2022) Role of indigenous microbial community in bioremediation: recent advances, challenges, and future outlook. In: Omics for environmental engineering and microbiology systems. Routledge, pp 221–255

Qiu J, Song X, Li S, Zhu B, Chen Y, Zhang L, Li Z (2021) Experimental and modeling studies of competitive Pb (II) and Cd (II) bioaccumulation by *Aspergillus Niger*. Appl Microbiol Biotechnol 105(16):6477–6488

Rabani MS, Hameed I, Gupta MK, Wani BA, Fayaz M, Hussain H et al (2023) Introduction of biofertilizers in agriculture with emphasis on nitrogen fixers and phosphate solubilizers. In: Microbiomes for the management of agricultural sustainability. Springer Nature Switzerland, Cham, pp 71–93

Rai S, Singh D, Singh AK (2020) Mycoremediation of arsenic from contaminated soil for sustainable agriculture. TTPP 575

Raimi A, Roopnarain A, Adeleke R (2021) Biofertilizer production in Africa: current status, factors impeding adoption and strategies for success. Sci African 11:e00694

Raj K, Das AP (2023) Lead pollution: impact on environment and human health and approach for a sustainable solution. Environ Chem Ecotoxicol 5:79–85

Rajendran K, Pujari L, Ethiraj K (2021) Biodegradation and bioremediation of s-triazine herbicides. Environ Biotechnol 3:31–54

Rana N, Gupta P (2023) Microbial assisted restoration of contaminated agricultural soil for sustainable environment. In: Integrative strategies for bioremediation of environmental contaminants, vol 2. Academic Press, pp 225–247

Rani L, Srivastav AL, Kaushal J (2021) Bioremediation: an effective approach of mercury removal from the aqueous solutions. Chemosphere 280:130654

Rashid A, Schutte BJ, Ulery A, Deyholos MK, Sanogo S, Lehnhoff EA, Beck L (2023) Heavy metal contamination in agricultural soil: environmental pollutants affecting crop health. Agronomy 13(6):1521

Reyes-Hinojosa D, Lozada-Pérez CA, Cuevas YZ, López-Reyes A, Martínez-Nava G, Fernández-Torres J et al (2019) Toxicity of cadmium in musculoskeletal diseases. Environ Toxicol Pharmacol 72:103219

Rizvi A, Khan MS (2017a) Biotoxic impact of heavy metals on growth, oxidative stress and morphological changes in root structure of wheat (Triticum aestivum L.) and stress alleviation by Pseudomonas aeruginosa strain CPSB1. Chemosphere, 185: 942–952

Rizvi A, Khan MS (2017b) Cellular damage, plant growth promoting activity and chromium reducing ability of metal tolerant Pseudomonas aeruginosa CPSB1 recovered from metal polluted chilli (Capsicum annuum) rhizosphere. Acta Sci Agric, 1: 36–46

Rizvi A, Khan MS (2018) Heavy metal induced oxidative damage and root morphology alterations of maize (Zea mays L.) plants and stress mitigation by metal tolerant nitrogen fixing Azotobacterchroococcum. Ecotoxicology and Environmental safety, 157: 9–20

Rizvi A, Zaidi A, Ahmad B, Ameen F, Al Kahtani MDF, Khan MS (2020) Heavy metal induced stress on wheat: phytotoxicity and microbiological management. RSC Adv 10:38379–38403

Rizvi A, Ahmed B, Umar S, Khan MS (2022) Bacterial biofertilizers for bioremediation: a priority for future research. In: Trends of applied microbiology for sustainable economy. Academic Press, pp 565–612

Rokhzadi A, Asgharzadeh A, Darvish F, Nourmohammadi G, Majidi E (2008) Infl uence of plant growth-promoting rhizobacteria on dry matter accumulation and yield of chickpea (*Cicer arietinum* L.) under field condition. Am-Euras J Agric Environ Sci 3(2):253–257

Roy S, Hossain A (eds) (2024) The nanotechnology driven agriculture: the future ahead, 1st edn. CRC Press. https://doi.org/10.1201/9781003376446

Saif S, Khan MS (2017) Assessment of heavy metals toxicity on plant growth promoting rhizobacteria and seedling characteristics of Pseudomonas putida SFB3 inoculated greengram. Acta Scientific Agriculture, 1: 47–56

Shaharoona B, Arshad M, Zahir ZA (2006) Effect of plant growth promoting rhizobacteria containing ACCdeaminase on maize (Zea mays L.) growth under axenic conditions and on nodulation in mung bean (Vigna radiate L.) Letters in applied microbiology, 42(2): 155–159

Sah S, Singh N, Singh R (2017) Iron acquisition in maize (*Zea mays* L.) using *Pseudomonas* siderophore. 3 Biotech 7(2):1–7

Saha M, Sarkar S, Sarkar B, Sharma BK, Bhattacharjee S, Tribedi P (2016) Microbial siderophores and their potential applications: a review. Environ Sci Pollut Res 23:3984–3999

Saha JK, Selladurai R, Coumar MV, Dotaniya ML, Kundu S, Patra AK et al (2017) Organic pollutants. In: Soil pollution-an emerging threat to agriculture. Springer, Singapore, pp 105–135

Saha B, Chowardhara B, Kar S, Devi SS, Awasthi JP, Moulick D, Tanti B, Panda SK (2019) Advances in heavy metal-induced stress alleviation with respect to exogenous amendments in crop plants. In: Priming and pretreatment of seeds and seedlings: implication in plant stress tolerance and enhancing productivity in crop plants. Springer, Singapore, pp 313–332

Saheewala H, Mahawar S, Sanadhaya S, Chauhan S, Kollah B, Mohanty SR, Chauhan PK, Upadhyay SK, Jain D (2023) Advances in biofertilizer production technologies: paradigm for improving agriculture through microbial resources. In: Nano-biofortification for human and environmental health. Springer International Publishing, Cham, pp 323–345

Sahli L, Belhiouani H (2022) Ficus retusa L. as possible indicator of air metallic pollution in urban environment. Int J Phytoremediation, 24(10): 1050–1059

Sakshi, Haritash AK (2020) A comprehensive review of metabolic and genomic aspects of PAH-degradation. Arch Microbiol 202(8):2033–2058

Salvioli A, Zouari I, Chalot M, Bonfante P (2012) The arbuscular mycorrhizal status has an impact on the transcriptome profile and amino acid composition of tomato fruit. BMC Plant Biol 12(1):1

Santosh DT, Debnath S, Maitra S, Sairam M, Sagar LL, Hossain A, Moulick D (2024) Alleviation of climate catastrophe in agriculture through adoption of climate-smart technologies. In: Climate crisis: adaptive approaches and sustainability. Cham, Springer Nature Switzerland, pp 307–332

Santra SC, Mallick A, Samal AC (2015) Biofertilizer for bioremediation. Recent trends in biofertilizers. IK International Publishing House Pvt. Ltd, pp 205–234

Saravanan VS, Subramoniam SR, Raj SA (2004) Assessing in vitro solubilization potential of different zinc solubilizing bacterial (ZSB) isolates. Braz J Microbiol 35(1–2):121–125

Saravanan VS, Kalaiarasan P, Madhaiyan M, Thangaraju M (2007) Solubilization of insoluble zinc compounds by *Gluconacetobacter diazotrophicus* and the detrimental action of zinc ion (Zn^{2+}) and zinc chelates on root knot nematode *Meloidogyne incognita*. Lett Appl Microbiol 44:235–224

Sarkar J, Roy A, Sar P, Kazy SK (2020) Accelerated bioremediation of petroleum refinery sludge through biostimulation and bioaugmentation of native microbiome. In: Emerging technologies in environmental bioremediation. Elsevier, pp 23–65

Sarkar S, Saha S, Ghosh S, Paul SK, Dey S, Moulick D, Santra SC, Brahmachari K (2023) Abiotic stress sensitivity and adaptation in field crops. In: Climate-resilient agriculture, Vol 2: agro-biotechnological advancement for crop production. Springer International Publishing, Cham, pp 319–362

Schobert M, Tielen P (2010) Contribution of oxygen-limiting conditions to persistent infection of *Pseudomonas aeruginosa*. Future Microbiol 5(4):603–621

Sehrawat A, Phour M, Kumar R, Sindhu SS (2021) Bioremediation of pesticides: an eco-friendly approach for environment sustainability. In: Microbial rejuvenation of polluted environment, vol 1. Springer, Singapore, pp 23–84

Selvi AA, Manonmani HK (2015) Purification and characterization of carbon–phosphorus bond-cleavage enzyme from glyphosate degrading *Pseudomonas putida* T5. Prep Biochem Biotechnol 45(4):380–397

Senabio JA, de Campos Pereira F, Pietro-Souza W, Sousa TF, Silva GF, Soares MA (2023) Enhanced mercury phytoremediation by *Pseudomonodictys pantanalensis* sp. nov. A73 and *Westerdykella aquatica* P71. Braz J Microbiol 54(2):949–964

Setyawan CY (2021) The effect of biofilm biofertilizer formula and organic fertilizer dosage to phosphorus uptake and yield of shallot on Vertisols. IOP Conf Ser Earth Environ Sci 905(1):012121

Shaari NEM, Tajudin MTFM, Khandaker MM, Majrashi A, Alenazi MM, Abdullahi UA, Mohd KS (2022) Cadmium toxicity symptoms and uptake mechanism in plants: a review. Braz J Biol 84:e252143

Shan B, Hao R, Xu H, Zhang J, Li J, Li Y, Ye Y (2022) Hexavalent chromium reduction and bioremediation potential of *Fusarium proliferatum* S4 isolated from chromium-contaminated soil. Environ Sci Pollut Res 29(52):78292–78302

Sharma I (2020) Bioremediation techniques for polluted environment: concept, advantages, limitations, and prospects. In: Trace metals in the environment-new approaches and recent advances. IntechOpen

Sharma A, Shukla S (2023) Bioremediation: remedy for emerging environmental pollutants. In: Current research trends and applications in waste management. Springer Nature Singapore, Singapore, pp 267–285

Sharma A, Kumar V, Kohli SK, Kaur R, Kaur T, Arora S et al (2020) Pesticide metabolism in plants, insects, soil microbes and fishes: an overview. In: Pesticides in crop production: physiological and biochemical action. Wiley, pp 35–53

Sharma J, Goutam J, Dhuriya YK, Sharma D (2021) Bioremediation of industrial pollutants. In: Microbial rejuvenation of polluted environment, vol 2. Springer, Singapore, pp 1–31

Sharma JK, Kumar N, Singh NP, Santal AR (2023) Phytoremediation technologies and their mechanism for removal of heavy metal from contaminated soil: an approach for a sustainable environment. Front Plant Sci 14:1076876

Silva HF, Silva NF, Oliveira CM, Matos MJ (2021) Heavy metals contamination of urban soils—A decade study in the city of Lisbon, Portugal. Soil Syst 5(2):27

Sindhu SS, Sharma R, Sindhu S, Phour M (2019) Plant nutrient management through inoculation of zinc-solubilizing bacteria for sustainable agriculture. In: Biofertilizers for sustainable agriculture and environment. Springer, Cham, pp 173–201

Singh R, Kumar M, Mittal A, Mehta PK (2017) Encapsulation technology: a promising technique for micro and nano carriers in delivering biofertilizers for agriculture. Biocatal Agric Biotechnol 11:264–271

Singh A, Kate P, Kumari K (2021a) Environmental fate, toxicity, and risk assessment of Persistent Organic Pollutants (POPs). In: Persistent organic pollutants. CRC Press, pp 59–90

Singh D, Thapa S, Geat N, Mehriya ML, Rajawat MVS (2021b) Biofertilizers: mechanisms and application. In: Biofertilizers. Woodhead Publishing, pp 151–166

Singh S, Kumar V, Kanwar R, Wani AB, Gill JPK, Garg VK et al (2021c) Toxicity and detoxification of monocrotophos from ecosystem using different approaches: a review. Chemosphere 275:130051

Singh D, Pandey H, Dhiman VK, Thakur K, Sharma I, Dhiman VK et al (2022) Assessment of environmental impacts of metal/metalloid pollution on plants. In: Metals metalloids soil plant water systems. Academic Press, pp 217–232

Sinha A, Pant KK, Khare SK (2012) Studies on mercury bioremediation by alginate immobilized mercury tolerant Bacillus cereus cells. Int Biodeterior Biodegradation 71:1–8

Sivasakthivelan P, Saranraj P (2013) *Azospirillum* and its formulations: a review. Int J Microbiol Res 4(3):275–287. IDOSI Publications

Song Y (2003) Mechanisms of lead and zinc removal from lead mine drainage in constructed wetland. University of Missouri-Rolla

Song L, Niu X, Zhang N, Li T (2021) Effect of biochar-immobilized *Sphingomonas* sp. PJ2 on bioremediation of PAHs and bacterial community composition in saline soil. Chemosphere 279:130427

Soni SK, Manhas R, Jakhar Y, Sharma A, Soni R (2022) Biofertilizers for sustainable agriculture: current trends and future perspective. In: Genomic, proteomics, and biotechnology. Routledge, pp 331–356

Soumare A, Boubekri K, Lyamlouli K, Hafidi M, Ouhdouch Y, Kouisni L (2020) From isolation of phosphate solubilizing microbes to their formulation and use as biofertilizers: status and needs. Front Bioeng Biotechnol, 7: 425

Stets MI, Alqueres SMC, Souza EM (2015) Quantification of Azospirillum brasilense FP2 bacteria in wheat roots by strain-specific quantitative PCR. Appl Environ Microbiol 81:6700–6709

Sujatha P, Madhavi M, Pallavi M, Bharathi Y, Rao PJM, Rajeswari B et al (2023) Biological seed coating innovations for sustainable healthy crop growth in tomato. In: Tomato cultivation and consumption-innovation and sustainability. IntechOpen

Swarnalakshmi K, Prasanna R, Saxena AK (2013) Biofilmed biofertilizers: a novel technology for green agriculture. Indian Farming 62:9–12

Tairo EV, Ndakidemi PA (2013) Possible benefits of Rhizobial inoculation and phosphorus supplementation on nutrition, growth and economic sustainability in grain legumes. Am J Res Commun 1(12):532–556

Tan GH, Nordin MS, Kert TL, Napsiah AB, Jeffrey LSH (2009) Isolation of Beneficial Microbes from Bio-fertilizer Products. J Trop Agric Food Sci 37(1):103–109

Teotia P, Kumar M, Prasad R, Kumar V, Tuteja N, Varma A (2017) Mobilization of micronutrients by mycorrhizal fungi. In: Mycorrhiza-function, diversity, state of the art. Springer, Cham, pp 9–26

Thakkar H, Vincent V (2021) Application of microorganisms in bioremediation. In: Bioprospecting of microorganism-based industrial molecules. Wiley, pp 77–103

Thakur R (2023) Organic contaminants are degraded by rhizospheric microorganisms: a review. Int J Plant Environ 9(01):1–6

Thavamani P, Megharaj M, Naidu R (2012) Bioremediation of high molecular weight polyaromatic hydrocarbons co-contaminated with metals in liquid and soil slurries by metal tolerant PAHs degrading bacterial consortium. Biodegradation 23:823–835

Thongnok S, Siripornadulsil W, Siripornadulsil S (2022) Responses to arsenic stress of rice varieties coinoculated with the heavy metal-resistant and rice growth-promoting bacteria *Pseudomonas stutzeri* and *Cupriavidus taiwanensis*. Plant Physiol Biochem 191:42–54

Tian D, Jiang Z, Jiang L, Su M, Feng Z, Zhang L et al (2019) A new insight into lead (II) tolerance of environmental fungi based on a study of *Aspergillus Niger* and *Penicilliumoxalicum*. Environ Microbiol 21(1):471–479

Titah HS, Abdullah SRS, Idris M, Anuar N, Basri H, Mukhlisin M et al (2018) Arsenic resistance and biosorption by isolated rhizobacteria from the roots of *Ludwigia octovalvis*. Int J Microbiol 2018:3101498

Tiwari S, Yadav D, Sarangi BK (2023) Bioaugmentation of metal phytoremediation through plant-microbe interaction. In: Metagenomics to bioremediation. Academic Press, pp 595–616

Tripathi M, Munot HP, Shouche Y, Meyer JM, Goel R (2005) Isolation and functional characterization of siderophore-producing lead-and cadmium-resistant *Pseudomonas putida* KNP9. Curr Microbiol 50:233–237

Tudi M, Daniel Ruan H, Wang L, Lyu J, Sadler R, Connell D et al (2021) Agriculture development, pesticide application and its impact on the environment. Int J Environ Res Public Health 18(3):1112

Tuo W, Zuo S, Dong J (2023) Detoxification of Cr (VI) and extracellular formation of nanoparticles Cr2O3 by a highly Cr (VI)-resistant fungus *Fusarium solani* SWUZF-1. Environ Technol Innov 32:103377

Udume OA, Abu GO, Stanley HO, Vincent-Akpu IF, Momoh Y, Eze MO (2023) Biostimulation of petroleum-contaminated soil using organic and inorganic amendments. Plan Theory 12(3):431

Umber M, Sultana R, Sehar R, Nasir F, Mubashir R (2019) Mitigation of hazardous effects of mercury pollution on wheat seedlings through *Trichoderma harzianum* seed coating treatment. Int J Chem Biochem Sci 15:1–4

Uroz S, Calvaruso C, Turpault MP, Frey-Klett P (2009) Mineral weathering by bacteria: ecology, actors and mechanisms. Trends Microbiol 17(8):378–387

Usharani KV, Roopashree KM, Naik D (2019) Role of soil physical, chemical and biological properties for soil health improvement and sustainable agriculture. J Pharmacogn Phytochem 8(5):1256–1267

Văcar CL, Covaci E, Chakraborty S, Li B, Weindorf DC, Frențiu T et al (2021) Heavy metal-resistant filamentous fungi as potential mercury bioremediators. J Fungi 7(5):386

Van Dommelen A, Croonenborghs A, Spaepen S, Vanderleyden J (2009) Wheat growth promotion through inoculation with an ammonium-excreting mutant of *Azospirillum brasilense*. Biol Fertil Soils 45:549–553

Varga T, Hixson KK, Ahkami AH, Sher AW, Barnes ME, Chu RK (2020) Endophyte-promoted phosphorus solubilization in populus. Front Plant Sci 11:567918

Vassilev N, Vassileva M, Martos V, Garcia del Moral LF, Kowalska J, Tylkowski B, Malusá E (2020) Formulation of microbial inoculants by encapsulation in natural polysaccharides: focus on beneficial properties of carrier additives and derivatives. Front Plant Sci 11:270

Veeraswamy D, Subramanian A, Mohan D, Ettiyagounder P, Selvaraj PS, Ramasamy SP, Veeramani V (2023) Exploring the origins and cleanup of mercury contamination: a comprehensive review. Environ Sci Pollut Res 31:1–30

Vejan P, Abdullah R, Khadiran T, Ismail S, Nasrulhaq Boyce A (2016) Role of plant growth promoting rhizobacteria in agricultural sustainability—a review. Molecules 21(5):573

Vimalnath S, Subramanian S (2018a) Studies on the biosorption of Pb (II) ions from aqueous solution using extracellular polymeric substances (EPS) of *Pseudomonas aeruginosa*. Colloids Surf B: Biointerfaces 172:60–67

Vimalnath S, Subramanian S (2018b) Studies on the biosorption of Pb (II) ions using *Pseudomonas putida*. Sep Sci Technol 53(16):2550–2562

Walpola BC, Kim AY, Jeon JH, Yoon MH (2014) Application of immobilization technology in solubilization of rock phosphate. Korean J Soil Sci Fertiliz 47(4):249–253

Waqas MA, Kaya C, Riaz A, Farooq M, Nawaz I, Wilkes A, Li Y (2019) Potential mechanisms of abiotic stress tolerance in crop plants induced by thiourea. Front Plant Sci, 10: 1336

Wang C, Sun H, Li J, Li Y, Zhang Q (2009) Enzyme activities during degradation of polycyclic aromatic hydrocarbons by white rot fungus *Phanerochaete chrysosporium* in soils. Chemosphere 77(6):733–738

Wang X, Rathinasabapathi B, Oliveira LMD, Guilherme LR, Ma LQ (2012) Bacteria-mediated arsenic oxidation and reduction in the growth media of arsenic hyperaccumulator *Pteris vittata*. Environ Sci Technol 46(20):11259–11266

Wang R, Lin J-Q, Liu X-M, Pang X, Zhang C-J, Yang C-L (2019) Sulfur oxidation in the acidophilic autotrophic *Acidithiobacillus* spp. Front Microbiol 9:3290

Wani FS, Ahmad L, Ali T, Mushtaq A (2015) Role of microorganisms in nutrient mobilization and soil health—a review. J Pure Appl Microbiol 9(2):1401–1410

Woomer PL, Huising J, Giller KE, Baijukya FP, Kantengwa S, Vanlauwe B, et al (2014) N2Africa final report of the first phase: 2009–2013

World Health Organization 2015 World Health Report (Geneva, Switzerland).

Wu Z, Chen Z, Wang H, Liu H, Wei Z (2023) Arsenic removal in flue gas through anaerobic denitrification and sulfate reduction cocoupled arsenic oxidation. Chemosphere 337:139350

Xiao Z, Jiang W, Chen D, Xu Y (2020) Bioremediation of typical chlorinated hydrocarbons by microbial reductive dechlorination and its key players: a review. Ecotoxicol Environ Saf 202:110925

Xu X, Du X, Wang F, Sha J, Chen Q, Tian G, Jiang Y (2020) Effects of potassium levels on plant growth, accumulation and distribution of carbon, and nitrate metabolism in apple dwarf rootstock seedlings. Front Plant Sci, 11: 904

Yaashikaa PR, Devi MK, Kumar PS (2022a) Advances in the application of immobilized enzyme for the remediation of hazardous pollutant: a review. Chemosphere 299:134390

Yaashikaa PR, Kumar PS, Jeevanantham S, Saravanan R (2022b) A review on bioremediation approach for heavy metal detoxification and accumulation in plants. Environ Pollut 301:119035

Yadav A, Yadav K (2013) Seasonal population dynamics of rhizosphere and non-rhizosphere soil microorganisms of chir pine seedlings (*Pinus roxburghii* Sarg.). Br Microbiol Res J 3(4):664–677

Yadav A, Yadav K, Vashistha A (2016) Phosphate solubilizing activity of Pseudomonas fluorescens PSM1 isolated from wheat rhizosphere. J Appl Nat Sci 8(1):93–96

Yadav KK, Sarkar S, (2019) Biofertilizers, impact on soil fertility and crop productivity under sustainable agriculture. Environ. Ecol. 37 (1): 89–93

Yadav U, Kushwaha S, Anand V, Kumar S, Prakash O, Singh PC (2020) Chlorpyrifos degradation by plant growth-promoting Alcaligenes faecalis bacteria isolated from oil-contaminated soil. Biorem J 25(1):80–88

Yadav R, Kumar A, Tokas D, Singh AN (2021) Soil contamination by polycyclic aromatic hydrocarbons in the agroecosystems. In: Sustainable agriculture reviews 50: emerging contaminants in agriculture. Springer, Cham, pp 211–234

Yadav AN, Kour D, Yadav N (2023a) Nano-biofertilizers for agricultural sustainability. J Appl Biol Biotechnol 11(5):i–iv

Yadav A, Yadav K, Ahmad R (2023b) Rhizosphere engineering for semiarid tropics: prospects and bottlenecks. Acad Biol 1:1–18

Yadav A, Yadav K, Ahmad R, Abd-Elsalam KA (2023c) Emerging frontiers in nanotechnology for precision agriculture: advancements, hurdles and prospects. Agrochemicals 2(2):220–256

Yang QI, Wang JL, Xing ZHI (2005) Biosorption of cadmium by fungal biomass of *Aspergillus Niger*. Biomed Environ Sci 18(3):141

Yang X, Mao X, Shao X, Han F, Chang T, Qin H, Li M (2018) Enhanced techniques of soil washing for the remediation of heavy metal-contaminated soils. Agric Res 7:99–104

Yang C, Ho YN, Makita R, Inoue C, Chien MF (2020) A multifunctional rhizobacterial strain with wide application in different ferns facilitates arsenic phytoremediation. Sci Total Environ 712:134504

Yanpirat P, Nakatsuji Y, Hiraga S, Fujitani Y, Izumi T, Masuda S et al (2020) Lanthanide-dependent methanol and formaldehyde oxidation in *Methylobacterium aquaticum* strain 22A. Microorganisms 8(6):822

Yao X, Tao F, Zhang K, Tang H, Xu P (2017) Multiple roles for two efflux pumps in the polycyclic aromatic hydrocarbon-degrading *Pseudomonas putida* strain B6-2 (DSM 28064). Appl Environ Microbiol 83(24):e01882–e01817

Ye L, Zhao X, Bao E, Li J, Zou Z, Cao K (2020) Bio-organic fertilizer with reduced rates of chemical fertilization improves soil fertility and enhances tomato yield and quality. *Sci Rep* 10:177

Yoshikawa M, Zhang M, Toyota K (2017) Biodegradation of volatile organic compounds and their effects on biodegradability under co-existing conditions. Microbes Environ 32(3):188–200

Youssef MM, Eissa MF (2014) Biofertilizers and their role in management of plant parasitic nematodes. A review. E3 J Biotechnol Pharm Res 5(1):1–6

Zafar A, Javed S, Akram N, Naqvi SAR (2024) Health risks of mercury. In: Mercury toxicity mitigation: sustainable nexus approach. Springer Nature Switzerland, Cham, pp 67–92

Zaidi A, Khan MS, Rizvi A, Saif S, Ahmad B, Shahid M (2017) Role of phosphate-solubilizing bacteria in legume improvement (pp. 175–197). Springer International Publishing.

Zakaria Z, Zulkafflee NS, MohdRedzuan NA, Selamat J, Ismail MR, Praveena SM et al (2021) Understanding potential heavy metal contamination, absorption, translocation and accumulation in rice and human health risks. Plan Theory 10(6):1070

Zarjani JK, Aliasgharzad N, Oustan S, Emadi M, Ahmadi A (2013) Isolation and characterization of potassium solubilizing bacteria in some Iranian soils. Arch Agro Soil Sci 2013(77):7569

Zeng J, Lin X, Zhang J, Li X (2010) Isolation of polycyclic aromatic hydrocarbons (PAHs)-degrading *Mycobacterium* spp. and the degradation in soil. J Hazard Mater 183(1–3):718–723

Zeng X-C, He Z, Chen X, Cao QAD, Li H, Wang Y (2018) Effects of arsenic on the biofilm formations of arsenite-oxidizing bacteria. Ecotoxicol Environ Saf 165:1–10
Zerrouk IZ, Rahmoune B, Auer S, Rößler S, Lin T, Baluska F, Ludwig-Müller J (2020) Growth and aluminum tolerance of maize roots mediated by auxin-and cytokinin-producing Bacillus toyonensis requires polar auxin transport. Environmental and Experimental Botany, 176: 104064
Zeng X, Huo X, Xu X, Liu D, Wu W (2020) E-waste lead exposure and children's health in China. Sci Total Environ 734:139286
Zhang JJ, Yang H (2021) Metabolism and detoxification of pesticides in plants. Sci Total Environ 790:148034
Zhang Z, Yin N, Cai X, Wang Z, Cui Y (2016) Arsenic redox transformation by *Pseudomonas* sp. HN-2 isolated from arsenic-contaminated soil in Hunan, China. J Environ Sci 47:165–173
Zhao G, Zhu X, Zheng G, Meng G, Dong Z, Baek JH et al (2024) Development of biofertilizers for sustainable agriculture over four decades (1980–2022). Geograph Sustain 5(1):19–28
Zhou L, Li H, Zhang Y, Han S, Xu H (2016) Sphingomonas from petroleum-contaminated soils in Shenfu, China and their PAHs degradation abilities. Braz J Microbiol 47:271–278
Zhu X, Liu W, Chen J, Bruijnzeel LA, Mao Z, Yang X et al (2020) Reductions in water, soil and nutrient losses and pesticide pollution in agroforestry practices: a review of evidence and processes. Plant Soil 453:45–86
Zorb C, Senbayram M, Peiter E (2014) Potassium in agriculture–status and perspectives. J Plant Physiol 171(9):656–669
Zulfiqar A, Yasmin A (2020) Microbiological aspects of pesticide remediation. Bioremediation and biotechnology. In: Persistent and recalcitrant toxic substances, vol 3. Springer, Cham, pp 139–171

Cement Factories Serve the Environment and Reduce Pollution

Mohamed Essam Sayed

Abstract The cement sector is one of the largest contributors to carbon dioxide (CO_2) emissions globally, generating an estimated 7% of CO_2 emissions. Easily seen as the second most-used material on Earth by humankind after fresh water, due to the perpetuated demand for cement through this unending urbanization process. The cement sector emits more than 800 kg of CO_2 per ton of cement, ranked among the highest in the world, reaching 14% of the carbon produced in Egypt, compared to the global average of 600 kg per ton of cement. In 2016, Egypt ratified the Paris Agreement, submitted the National Determined Contributions (NDC), and determined to minimize CO_2 emissions. However, headway toward implementing carbon capture for the cement industry has been slow-paced and uncoordinated. That is the question posed by this analysis of the case of the rollout of a green transition in the cement industry in Egypt to understand the obstacles and opportunities and develop policy actions for empowering the reduction of carbon emissions as soon as possessing a desire to bring technical assistance services elsewhere. The study adopts a quantitative methodology incorporating a content analysis of literature and industry reports to understand sustainability practices in Egypt's cement industry and the extent to which the industry participates in decarbonization. Alternatives can be found in using alternative fuels and raw materials (AFR), as well as more efficient production processes and renewable energy sources. This chapter can help achieve an important reduction of CO_2 and, at the same time, create demand for waste and renewable energy.

Keywords Egypt · Cement industry · Sustainable development · Decarbonization · Carbon footprint

M. E. Sayed (✉)
Basic Science Department, Modern Academy for Engineering and Technology in Maadi, Cairo, Egypt

A. A. H. Abdel Latef et al. (eds.), *Sustainable Remediation for Pollution and Climate Resilience*, https://doi.org/10.1007/978-981-96-5674-5_21

1 Introduction

The cement business in Egypt is of vital importance to the economy, but it has a substantial influence on the environment due to its high levels of greenhouse gas emissions and extensive energy consumption. It is imperative to adopt more sustainable and environmentally friendly techniques to tackle these problems. The sector represents 14% of Egypt's total emissions and has significant energy consumption obstacles. The export potential is restricted by logistical limitations, which are further exacerbated by broader economic challenges such as foreign currency shortages and a decline in local economic activity. The intricacies involved in the green transition of Egypt's cement sector pose a greater challenge than just adopting carbon reduction methods.

The emissions from the cement industry have a significant impact on air pollution. These emissions include gaseous pollutants such as nitric oxides, carbon monoxide, sulfur dioxide, and hydrocarbons, as well as fine and coarse particulate matter like diesel soot. Additionally, there are trace metals that are present in the airborne particles (Albeanu et al. 2004; Laschober et al. 2004). Cement factories are responsible for around 5–7% of CO_2 emissions worldwide, with the production of one ton of cement resulting in the release of 900 kg of CO_2 into the environment (Marland et al. 1989). Although other nations have implemented green transition models, and the European Bank for Reconstruction and Development in Egypt (EBRD 2016) has provided a study detailing a low-carbon roadmap for Egypt's cement industry, the progress has been sluggish. The lack of legislation and policies promoting emissions reduction, along with a low understanding of climate change mitigation, economic situations that impede investment in contemporary technology, and general resistance within the heavy industry sector, are the main factors contributing to this issue.

This study aims to gain insights from industry professionals regarding the specific requirements and implications of a green transition in Egypt. Additionally, the study attempts to identify the regulatory obstacles and potential advantages related to this shift. The aim is to fill the gaps in the existing literature concerning barriers to a sustainable transition and contribute to the ongoing discourse on minimizing the environmental footprint of the cement industry. In addition, the study examines the interdependencies between the cement industry and other economic sectors, such as solid waste management, to utilize these links for a successful transition toward sustainability.

2 The Egyptian Cement Industry's Carbon Footprint

The cement industry is one of the two largest producers of carbon dioxide (CO_2), creating up to 5% of worldwide man-made emissions of this gas, of which 50% is from the chemical process and 40% from burning fuel. These ratios are called carbon footprint.

Regrettably, cement, the main ingredient in concrete, has a significant carbon footprint. This business is responsible for around 7% of worldwide CO_2 emissions, making it the second largest contributor after water (IEA 2018; Korczak et al. 2022), and annually, it generates a minimum of 2.1 billion tons of carbon dioxide. CO_2 is emitted from various sources throughout the cement-making process:

- Half of the carbon dioxide (CO_2) released is a result of the calcination process of limestone ($CaCO_3$) into calcium oxide (CaO) and carbon dioxide (CO_2) (Mahasenan et al. 2003; Worrell et al. 2001).
- 40% of thermal energy production is derived from the combustion of fossil fuels (Schneider 2019).
- The remaining 10% of CO_2 is released through transportation and electrical consumption.

The cement industry in Egypt, which was founded in 1927, has traditionally played a crucial role in promoting economic growth and urbanization. The first construction of cement facilities along the Nile River was strategically done to facilitate efficient shipping (Askar et al. 2010). The industry makes up over 3.7% of the gross domestic product (GDP), directly providing employment to 50,000 people and indirectly supporting 200,000 jobs (Abdou 2017). The construction and real estate sectors are paramount, as they collectively account for over 40% of Egypt's national GDP. Nevertheless, the business has a significant ecological footprint, accounting for 5.3% of Egypt's overall energy consumption and responsible for 14% of the nation's CO_2 emissions, double the global average (EBRD 2016).

The industry received criticism for advocating the use of imported coal after a shortage of natural gas in 2016, which critics claim worsens its carbon emissions. According to the International Energy Agency (IEA), in 2018, Egypt's cement manufacturing released an average of 750 kg of carbon dioxide (CO_2) per ton, which is more than the global average of 620 kg. Additionally, reports are suggesting that the emissions from Egypt's cement production could be much higher, as indicated by the Global Network for Rights and Development (GNR) in 2020. The disparities in the production capacity and the maturity of sustainable practices among regions are the main factors that cause these inequalities (Rissman et al. 2020). Hence, it is imperative to customize carbon emission reduction plans according to the unique circumstances of each location.

Given its geographical and socioeconomic settings, Egypt is highly susceptible to the effects of climate change. The MENA area, where Egypt is located, is already facing significant climate-related challenges, including increasing temperatures and limited water resources. The coastal regions of Egypt face the threat of rising sea levels, which could result in flooding and erosion (UNFCCC 2023). Economic limitations add an additional layer of complexity to the efforts made to adapt.

In the aftermath of Egypt's hosting of COP27, the nation revised its National Determined Contributions (NDCs) to incorporate measures for addressing climate change, specifically targeting the reduction of carbon emissions in the cement industry (UNFCCC 2023). The June 2023 update priorities were enhancing efficiency and decreasing reliance on fossil fuels by integrating alternative energy. In

March 2021, a directive was issued requiring cement mills to replace 10% of their energy consumption with alternative fuels (EEAA 2022). However, despite efforts, the cement industry still faces a big problem due to its high-carbon footprint. Egypt's cement production emissions are considerably higher than the global average, as reported by the European Bank for Reconstruction and Development (EBRD) in 2016.

Considering the industry's significant impact on the economy, policy measures, aimed at reducing its carbon emissions, should consider its economic contributions. Cement facilities can utilize different types of fuel, which creates a chance to convert waste into valuable resources and promote a sustainable, circular economy (Busch et al. 2022). However, the industry is confronted with obstacles like overproduction capacity, elevated energy expenses, restricted access to financing, and insufficient legislative backing for a move toward sustainability (EBRD 2022).

This study aims to assess Egypt's recent initiatives to decrease the carbon emissions of the cement sector and offer valuable perspectives on moving toward a more environmentally friendly and productive future.

3 Decarbonizing the Global Cement Industry

The United Nations introduced the 2030 Agenda for Sustainable Development in 2015, emphasizing the pressing requirement to address climate change as its 13th objective. In 2016, the world leaders signed the Paris Agreement during the 21st session of the Conference of the Parties to the United Nations Framework Convention on Climate Change. This agreement aims to restrict global temperature rises to less than 2 °C above preindustrial levels (IEA 2018).

Several countries have since implemented Nationally Determined Contributions (NDCs), including voluntary goals for mitigating greenhouse gas emissions. For example, the United States has set a goal to reduce emissions by 50% compared to 2005 levels by 2030. The European Union wants a 40% reduction below 1990 levels by 2030. According to the International Energy Agency, China has committed to reaching the highest point of emissions by 2030 and achieving carbon neutrality by 2060 (IEA 2018).

These pledges have been shaped by the escalating frequency and intensity of extreme weather phenomena worldwide, such as the wildfires in Australia (2019–2020), the heatwave in the Pacific Northwest (June 2021), and the floods in Pakistan (2022) (Clark et al. 2022). Furthermore, there has been increasing pressure from investors on high-carbon corporations to improve their environmental reporting and develop plans for reducing emissions. However, the impact of this pressure has been somewhat limited thus far (Baines and Hager 2023).

The practice of sustainable and responsible investing has experienced substantial expansion, integrating environmental, social, and governance (ESG) considerations into investment choices. The sustainable investment market reached an anticipated value of $35.3 trillion in 2020, reflecting a 15% growth compared to the previous year. Investors are progressively pressuring corporations to confront climate change and reveal their climate-related risks and possibilities, resulting in a more significant number of companies establishing and reporting on climate targets (Biswas and Rahman 2023).

After the Paris Agreement, the Intergovernmental Panel on Climate Change (IPCC) published multiple assessments that monitored the advancement toward Nationally Determined Contributions (NDCs). The IPCC, in 2021, highlighted the worldwide delay in attaining these objectives and emphasized the urgency of intensifying efforts to achieve carbon reduction targets (IPCC 2021).

Cement production, which accounts for around 7–8% of worldwide carbon emissions, is a crucial sector that requires decarbonization. The production is significantly centralized, with China and other Asian nations representing over 80% of the total production (IEA 2018). The industry's decentralized structure poses a challenge due to its reliance on locally sourced raw materials such as limestone, clay, and sand, which exhibit variations in quality and composition. China's per capita cement production is 1818 kg, significantly higher than the global average of around 600 kg and the Middle East's average of 827 kg (IEA 2018). Therefore, it is necessary to customize decarbonization policies according to specific regional and national circumstances.

The decarbonization of cement is particularly challenging due to the intricate chemical reactions it undergoes. The primary cause of CO_2 emissions is the calcination process, in which limestone ($CaCO_3$) is broken down into lime (CaO) and CO_2, constituting around 90% of the total emissions. According to Monteiro et al. (2017), the remaining 10% comes from indirect sources such as power and transportation. This procedure necessitates substantial heat, often achieved through the combustion of fossil fuels at temperatures ranging from 1400 to 1500 °C (Monteiro et al. 2017). Local conditions, including the accessibility of raw materials and alternative fuels, influence the carbon footprint reduction capacity of cement factories. Hence, to accomplish decarbonization in the cement industry, it is imperative to adopt a location-specific strategy (Busch et al. 2022).

4 Sustained Cement Demand Through 2050

Cement, which is the main constituent of concrete, is the second most widely consumed substance worldwide, following water (Czigler et al. 2020). The demand for cement is expected to increase by 12–20% by 2050 due to continuous urbanization and growing infrastructure requirements (IEA 2018). The cement sector substantially contributes to greenhouse gas emissions, representing around 7% of world carbon dioxide emissions. Each ton of cement manufacturing releases around 600 kg of CO_2 (IEA 2018). The cement industry urgently needs transformation as countries aim to decrease their carbon emissions and mitigate the environmental consequences of heavy industries.

Leaders in the global heavy industry sectors are under increasing pressure from investors and shareholders to implement more sustainable practices due to the rise of carbon pricing legislation (Rissman et al. 2020). The Cement Sustainability Initiative (CSI), formed in 1999 and consisting of 24 prominent cement manufacturers worldwide, has collaborated with the International Energy Agency (IEA) to release a low-carbon roadmap. This roadmap seeks to synchronize carbon emissions reductions with the 2 degrees scenario (2DS), which delineates a trajectory for the energy system to restrict global warming to less than 2 °C over preindustrial levels, following the Paris Agreement. The plan prioritizes the ambitious implementation of carbon reduction technology in the cement industry and advocates for broad international collaboration and policy assistance to advance the use of low-carbon cement (IEA 2018).

Additional industries with significant environmental impacts, such as iron and steel, chemicals, and plastics, are also under pressure to reduce carbon emissions. However, these businesses have different obstacles when it comes to finding alternative options. The ongoing need for cement and other goods from heavy industries highlights the crucial significance of implementing rules and making efforts to decrease emissions (Rissman et al. 2020).

5 Reducing Emissions in the Cement Industry

There is an expected increase in the worldwide demand for cement, especially in emerging nations, due to urbanization, the affordable price of cement, the availability of required raw materials, and the absence of viable alternatives (Monteiro et al. 2017). While several technical solutions for decarbonization have been suggested (Fennell et al. 2017; Habert et al. 2022; Miller and Moore 2020), the successful implementation of these technologies necessitates enacting complementary legislative measures.

A comprehensive analysis of the decarbonization of the cement sector reveals that over 35 recent studies mostly focus on strategies to reduce emissions. In contrast, only a few studies discuss policies or barriers (Busch et al. 2022). Multiple industry associations, governments, and international organizations have developed quantitative models for reducing emissions, as evidenced by the roadmaps and paths provided by the IEA (2018), and Czigler et al. (2020). Although developing nations have higher cement emissions than developed countries (Rissman et al. 2020), most decarbonization literature is based on research conducted in developed countries. This emphasizes the need for more studies specifically addressing middle- and low-income countries (Busch et al. 2022).

Typical emission reduction strategies in the cement sector are enhancing energy efficiency, utilizing alternative fuels, and implementing carbon capture and storage. Enhancing energy efficiency, while not commonly mentioned, is economically appealing and extensively adopted, potentially decreasing worldwide cement production CO_2 emissions by 12% (Miller and Moore 2020). However, without legislative mandates, capital-intensive improvements are only feasible in economically viable areas, such as the Middle East and China, where there is an excess of cement supply.

Adopting alternative fuels, such as biomass, refuse-derived fuels (RDF), and industrial by-products like fly ash and slag, can reduce emissions from the cement industry by 9% by the year 2050 (ETC 2020). This strategy is economically efficient and promotes a circular economy. At the same time, it may necessitate additional energy for processing and does not modify the primary production process that is accountable for the majority of emissions. Challenges encompass the requirement for pretreatment and modifications in cement plants and the development of local infrastructure and regulatory frameworks (ETC 2020; EBRD 2016).

The clinker-to-cement ratio can be reduced by integrating alternative materials such as fly ash, slag, or calcined clay, resulting in a potential reduction of CO_2 emissions by up to 35% by the year 2050, according to the International Energy Agency (2018). Nevertheless, the availability of materials in specific regions and variations in regulatory standards present obstacles to the extensive implementation of this technology.

Carbon capture, storage, and utilization (CCSU) technologies, such as calcium looping and chemical absorption, can reduce CO_2 emissions by up to 35% by the year 2050, according to the International Energy Agency (2018). Although there is considerable promise, the high costs and the requirement for policy and research and development (R&D) assistance restrict the current practicality. Excess heat recovery (EHR) has the potential to produce extra energy or heat, but it currently lacks economic viability without further financial support or regulatory incentives (ETC 2020).

Alternative cement binding materials, such as the carbonation of calcium silicates and belite cement, are now being assessed for their technical feasibility. However, before they can be commercially viable, joint efforts and policy actions are needed (Liu et al. 2017).

Obstacles to implementing these solutions include expensive initial investments, limited technology development, and stringent regulatory requirements. According to Rissman et al. (2020), it typically takes 15–30 years for modern technologies to develop and become widely adopted in the market fully. Policy measures such as implementing carbon pricing, enhancing public procurement, and providing research and development support are crucial for mitigating emissions in the cement industry (Busch et al. 2022).

CarbonCure has been identified by the Global CO_2 Initiative (GCI) as the most scalable option among the examined CO_2 utilization technology providers. It is particularly suitable for the building materials industry, as it can effectively transform CO_2 into final products. CarbonCure's technology has been adopted by over 62 concrete manufacturers, resulting in the production of over 1.16 million tons of concrete utilizing the CarbonCure process Fig. 1. Consequently, a reduction of almost 15,000 tons of CO_2 emissions has been achieved in comparison (Table 1) to the amount released during the production of an equivalent quantity of regular concrete.

Fig. 1 Chemistry of CarbonCure process

Essentially, existing research discusses technological solutions and regulatory interventions to decrease emissions in the cement industry. However, limited attention is given to obstacles faced by stakeholders and practical strategies, which hinders the identification of the most appropriate techniques for specific markets (Busch et al. 2022). In addition, the focus on creating groundbreaking technologies may disregard the possibility of implementing demand-side interventions and exploring alternatives to cement (Ahman 2020).

6 Decarbonization of the Cement Industry in Egypt

Historically, the issue of reducing carbon emissions in cement production has not been given much importance in the discussions within Egypt's cement sector. Egypt exhibited a delayed response in implementing regulatory standards to restrict the release of dust emissions and dangerous pollutants from cement factories. The Egyptian Environmental Affairs Agency (EEAA) introduced more stringent emissions restrictions with online monitoring in 2010 (Askar et al. 2010). These guidelines revised the 1994 environmental protection law to conform with global norms for air pollutants (Fig. 2), although they did not specifically tackle CO_2 emissions.

At this time, there were suggestions to utilize alternative fuels and adopt cement production methods to reduce dangerous waste rather than resorting to landfilling. These ideas were influenced by practices observed in other developing nations (Askar et al. 2010). Regulatory amendments in 2015 permitted the utilization of coal and petcoke in cement manufacturing, resulting in a projected rise in CO_2 emissions of up to 15% (EBRD 2016). The 2016 environmental impact assessment

Table 1 Different commercialized technologies in the production of low-carbon cement and concrete

Technology location	Technology location	Method	CO_2 reduction potential	Reference
CarbonCure UK, Canada	CarbonCure UK, Canada	Injection of CO_2 in the concrete mix	231,319 MT CO_2 saved till now	Niven et al. (2014) and Monkman et al. (2020)
Solidia USA, Canada	Solidia USA, Canada	CO_2 curing cement concrete	1.5 gigatons of CO_2 sequestrated per year	Sahu et al. (2018, 2020a, b), Tas et al. (2018), and Atakan (2017)
CarbiCrete Canada	CarbiCrete Canada	Carbonation of steel slag to replace cement (cement-free concrete)	200 Kg per day	Shao et al. (2018, 2020)
TecEco Tasmania (Australia)	TecEco Tasmania (Australia)	Developed magnesium cements using magnesium carbonate	0.8 Kg CO_2/t cement	Harrison (2004, 2005)
Novacem London, UK	Novacem London, UK	Carbon negative cement using magnesium silicates as raw materials	0.57 Kg CO_2/t cement sequestrated	Vlasopoulos et al. (2009) and Vlasopoulos (2012a, b)
Heidelberg Germany	Heidelberg Germany	Ternocem clinker at lower burning temperature (1150–1280 °C)	Cut 20–30% CO_2 emission	Bullerjahn et al. (2012) and Heidelberg Cement (2016a, b)
Carbstone Belgium	Carbstone Belgium	Developed high-quality materials by adding CO_2 from flue gas to steel slag	180–200 g CO_2 per Kg of steel slag	Quaghebeur et al. (2014) and Kazemi et al. (2021)

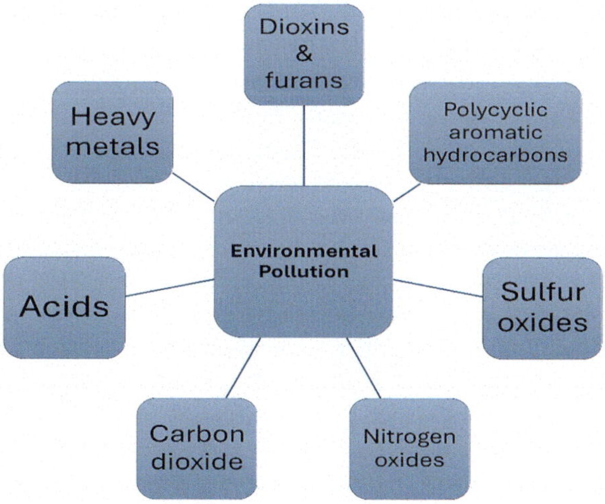

Fig. 2 Environmental pollution from the cement industry

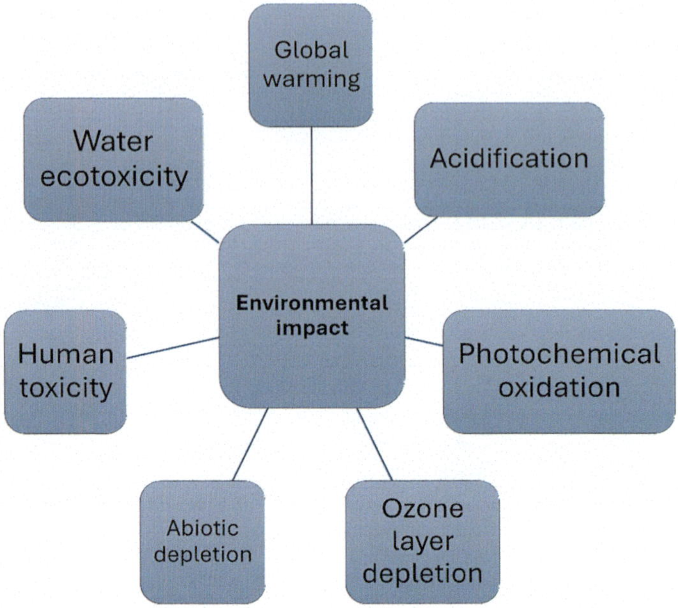

Fig. 3 Environmental impacts from the cement industry

identified insufficient surveillance of greenhouse gas emissions (Fig. 3), specifically pointing out that the EEAA (Abdou 2017) only conducted automated monitoring for carbon monoxide.

In late 2016, a notable paper called the "Low-Carbon Roadmap for the Egyptian Cement Industry" was published. The project, financed by the European Bank for Reconstruction and Development (EBRD) in collaboration with Egyptian authorities, offered a strategic plan for reducing CO_2 emissions caused by using coal and high-carbon fuels. If no measures are taken to reduce emissions, Egypt's cement sector has the potential to rank among the top 2% of cement companies worldwide in terms of carbon emissions. In this scenario, emissions would reach 820 kg of CO_2 per ton of cement (EBRD 2016). The roadmap also included ambitious strategies to decrease CO_2 emissions to levels lower than those observed in the past, before the transition to coal and other high-carbon fuels.

One of the main suggestions was to create a monitoring, reporting, and verification (MRV) system that would ensure the provision of precise CO_2 emissions data following international standards. Egypt's 2023 NDC report acknowledged the need to build an MRV platform but mentioned that finance for this initiative was still lacking (UNFCCC 2023). Additional suggestions included reducing the clinker ratio in cement, which poses a difficulty for Egypt due to the limited availability of appropriate materials such as fly ash and pozzolana. The (EBRD) European Bank for Reconstruction and Development analysis recommended the importation of these commodities, even though construction authorities and cement industries opposed it (EBRD 2016).

Table 2 Different Types of Cements and their Equivalent CO_2 Reduction Ratios

Type of cement	Raw material	Process temperature	CO_2 reduction
Belite-rich Portland cement (BPC)	Limestone, marl, sand, (Ca/Si) ratio 1–2	1350 °C	~10%
Calcium sulfoaluminate cement (CSA)	Bauxite, Limestone, Sand/Clay, Gypsum	1200–1300 °C	30–40%
BYF (Aether cement)	Limestone, clay, and bauxite, calcium sulfate	1250–1350 °C	~6% lower than ordinary Portland cement (OPC)
Geopolymer	Fly ash, wastes, Alkaline solutions	Ambient	~70%
Celitement	Limestone, marl, sand, slags, glasses, and fly ash (Ca/Si ratio 1–2)	150–200 °C (Hydrothermal)	~50%
Novacem	Magnesium silicates	700 °C, 180 bar	>100%
TecEco	Magnesium carbonate + OPC + fly ash	<450 °C (Tec-kilo)	>100%
Solidia cement	Limestone, chalk, clay, sand (Ca/Si ratio 1–1.5)	1200 °C	~70%

The report also recommended the compulsory use of alternative fuels and raw materials (AFR) and proposed the implementation of a new regulatory framework for solid waste management, which was established in 2022. Furthermore, it is advised to carefully consider licensing new cement plants in relation to the long-term demands of the market. This is particularly important as the current market is functioning at only 70% capacity, which may impede enterprises' capacity to allocate resources towards implementing emission reduction strategies. The potential of novel cements in CO_2 emission reduction (Naqi et al. 2019; Lehne et al. 2018; Ojan et al. 2016) are shown in Table 2.

Additional research on green transition pathways for Egypt's cement sector is urgently required. This chapter should employ a thorough strategy to involve a broader audience and pertinent stakeholders, going beyond technical conversations to examine possibilities for a circular economy and sustainable development. This study seeks to evaluate the effects of the low-carbon roadmap after 6 years of implementation, specifically within its defined boundaries.

7 Recommendations to Support Green Transition in Cement

Using the findings and analysis from this chapter, the next part will provide a series of policy recommendations that align with the low-carbon roadmap for the cement industry in Egypt. These ideas are formulated after talking with many stakeholders, such as senior executives in the cement sector, regulatory consultants, and technical specialists.

The EBRD's 2015 low-carbon plan for the Egyptian cement sector is outdated due to the introduction of new production lines and Egypt's endorsement of the Paris Accord. The cement industry is now targeted for emission reduction, necessitating revised research to align with the country's climate change policy for 2050. Egypt should set a specific CO_2 emission reduction target for the cement industry, incorporated into its Nationally Determined Contributions by 2030. This would signal the government's commitment to emission reduction and encourage cleaner technologies. The policy would also streamline funding for green transition solutions and capacity-building programs, demonstrating Egypt's commitment to a low-carbon economy.

Offer more practical protocols to facilitate and enhance the utilization of refuse-derived fuels (RDFs). The newest regulations by the EEAA require cement factories to employ a minimum of 10% RDFs in their energy consumption and to implement solid waste management practices. Regulations have established the basic structure, and the following suggestions can offer immediate enhancements:

1. Eliminate road charges for waste transportation to cement plants, as it contradicts Egypt's Nationally Determined Contributions (NDCs) sustainability goals. This policy encourages landfill disposal, reduces greenhouse gas emissions, and promotes a circular economy. Eliminating these fees would also reduce costs for waste management firms and end users.
2. A collaborative framework between cement enterprises, the Waste Management Regulatory Authority (WMRA) in Egypt, and private sector entities is needed to overcome regulatory and logistical challenges in waste handling. This will promote a circular economy, encourage reuse and recycling, and enforce the "polluter pays principle." Implementing gate fees could also promote the disposal of waste as an alternative fuel.
3. Accelerate the necessary measures to reduce the clinker ratio in cement. This will involve leveraging partnerships and adopting best practices from countries such as India and other similar nations that have already implemented such measures. The aim is to modify local construction codes and standards that enforce higher clinker ratios than global requirements, resulting in elevated CO_2 emissions, increased energy consumption, and consequently, higher costs.
4. Foreign money for industrial use. Although the most recent Nationally Determined Contributions (NDCs) explicitly mentioned a decrease in clinker ratio by 2030, there is no indication of a specific schedule or tangible improvement. This initiative will involve the cooperation of several governing bodies and sectors and cement manufacturers. Moreover, it will encompass end users such as construction and real estate firms, architects, and official regulatory agencies.
5. Implement a framework for monitoring and validating carbon emissions measurement in cement plants to improve accessibility and accuracy. This would ensure adherence to global norms, fair competition, and the investigation of

incentive programs for businesses exceeding their goals, thereby enhancing the efficiency of cement plants.
6. Introduce a reward system to encourage emission reduction in cement businesses, incorporating existing rules that penalize those not meeting 10% refuse-derived fuel usage. Incentives could include increased export subsidies, green public procurement agreements, and reduced import or quarry fees, aiming to achieve a 25% global average.
7. The maximum threshold for substituting renewable energy in cement plants should be examined or eliminated, as it contradicts Egypt's NDC's goal of 42% renewable electricity generation by 2030. Additionally, the EEAA laws should be reconsidered once a thermal substitution rate of 40% is achieved.
8. The Egyptian Competition Authority (ECA) is considering extending the production quota in Egypt's cement industry for 2 or 3 years, possibly with periodic revisions. This move would address the 30–35% gap between production capacity and demand, allowing cement producers to make more strategic investment decisions, protect the market from disruptions, and ensure sustainable practices. The policy would also allow for periodic updates to market conditions and long-term green financing for capital investments (Miller and Moore 2020).

Table 3 summarizes some advantages and disadvantages of different types of low-carbon cement.

8 Implications for Future Research

This chapter focused on evaluating the decarbonization strategies implemented in the cement sector in Egypt. However, it has also indicated other potential areas for further research. Based on the findings mentioned above, it is evident that there is a requirement to carry out technical studies to develop more sophisticated techniques that may effectively decrease the amount of CO_2 emissions produced by the cement. Emerging technologies like carbon capture and utilization hold potential for application in industries such as oil and gas. However, it is important to note that these technologies are in their early phases and would require substantial research and development.

Furthermore, this thesis specifically focused on investigating the feasibility of implementing environmentally friendly practices in the cement sector, primarily in cement manufacturing. It did not, however, assess the many possibilities for reducing the environmental impact of cement through modifications made by end users, such as construction or concrete firms. Additional investigation is necessary to comprehend the elements that influence the demand for cement, such as urbanization and infrastructure development, to pinpoint the chances of decreasing the overall demand for cement.

Furthermore, the results of this study highlight the pressing necessity to comprehend the waste management situation in Egypt and its potential to support

Table 3 Advantages and disadvantages of different types of low-carbon cement (Qiao et al. 2014; Adarlo 2017)

Type	Advantages	Disadvantages
Belite rich cement (BPC)	Low limestone demand leads to lower CO_2 emission than (OPC) ordinary Portland cement Less sintering temperature (~1350 °C) than (OPC) ordinary Portland cement Higher later age strength Lower shrinkage, resistance to sulfate attack, permeability, and carbonation	Low early age strength More grinding energy
Calcium sulfoaluminate cement (CSA)	Low sintering temperature of 1300 °C Higher early age strength 5000 psi within 24 h Less alkaline nature so minimizes the scope of alkali-silica reaction Reduces 0.61 kg CO_2 per kg of cement	Rapid setting High water demand Expensive
Geopolymer	Higher compressive and tensile strength than (OPC) ordinary Portland cement Less permeable, very low creep, lower shrinkage Fire and chemical resistant Reduces CO_2 footprint by ~70%	High cost of alkaline materials Geo-polymerization is sensitive
Celitement	Requires less energy and CO_2 emission during production No reactive aluminate phase, i.e., C3A is present so low heat of hydration (120–150 J/g) Low water demand Highly durable, sulfate resistant because of complete pure C–S–H phases	High curing temperature Economically not feasible
Magnesia	Excellent rheology, durability, acid resistance, good resistance to abrasion Quick setting and high early strength	More expensive Poor water resistance Limited large-scale production
Solidia	Low lime calcium silicate minerals instead of high lime calcium silicate phases in (OPC) ordinary Portland cement Cured in CO_2 rich environment Sequester about 300 kg of CO_2 per ton of cement in concrete Reduces total 70% emission in both the calcination and fuel combustion process	Low pH may corrode steel bars in reinforced concrete Applicable only in the precast system

sustainable growth in various sectors, as well as the implementation of circular economy concepts. In light of the recent changes in regulations for waste management in Egypt and the anticipated increase in demand from industries such as cement, it is crucial to conduct additional research to investigate the obstacles and feasible solutions for achieving sustainable waste management.

9 Conclusion

To summarize, this chapter highlights the long-term objective of decreasing the carbon emissions produced by the cement industry in Egypt. This goal can only be accomplished through a step-by-step approach that involves the active participation of various stakeholders, such as cement producers, regulatory agencies, construction companies, and waste management companies. This collaboration necessitates additional interdisciplinary study to provide legislators with more informed laws for business.

References

Abdou D (2017) An assessment of the oligopoly cement industry in Egypt: is it a curse or a blessing? Int J Green Econ. https://doi.org/10.1504/IJGE.2017.082713

Adarlo S (2017) The impact of large companies getting serious about green concrete research. Princeton University. https://acee.princeton.edu/acee-news/the-andlinger-center-speaks-the-impact-of-large-companies-getting-serious-a bout-green-concrete-research/. Accessed 21 Dec 2022

Ahman M (2020) Unlocking the "Hard to Abate" sectors: WRI expert perspective. WRI World Appl Energy. https://doi.org/10.1016/j.apenergy.2016.12.057

Albeanu G, Madsen H, Popentiu F, Thyregod P (2004) Computer-aided statistical modeling and optimization for pollution control in cement plants

Askar Y, Jago P, Mourad M, Huisingh D (2010) The cement industry in Egypt. In: ERSCP-EMSU conference, Delft, The Netherlands, October 25–29, 2010

Atakan V (2017) Steam-assisted production of metal silicate cements, compositions and methods thereof. Solidia Technologies, Inc. WO2017075126A1. https://patents.google.com/patent/WO2017075126A1/en?oq=PatentPatent+WO2017075126A1

Baines J, Hager SB (2023) From passive owners to planet savers? Asset managers, carbon majors and the limits of sustainable finance. Compet Change 27(3–4):449–471. https://doi.org/10.1177/10245294221130432

Biswas R, Rahman A (2023) Adaptation to climate change: a study on regional climate change adaptation policy and practice framework. J Environ Manag 336:117666. https://doi.org/10.1016/j.jenvman.2023.117666

Bullerjahn F et al (2012) Magnesia based binder composition. Heidelberg Cement. EP2508496A1. https://patents.google.com/patent/EP2508496A1/en?oq=PatentPatent+EP2508496A1

Busch P, Kendall A, Murphy CW, Miller SA (2022) Literature review on policies to mitigate GHG emissions for cement and concrete. Resour Conserv Recycl 182:106278. https://doi.org/10.1016/j.resconrec.2022.106278

Clark B, Otto F, Stuart-Smith R, Harrington L (2022) Extreme weather impacts of climate change: an attribution perspective. IOP Sci. https://doi.org/10.1088/2752-5295/ac6e7d

Czigler T, Reiter S, Schulze P, Somers K (2020) Laying the foundation for a zero- carbon cement industry. McKinsey & Company. https://www.mckinsey.com/industries/chemicals/our-insights/laying-the-foundation-for-zerocarbon-cement/

EBRD, European Bank for Reconstruction and Development (2016) Annual report 2016. European Bank for Reconstruction and Development, London

EBRD, European Bank for Reconstruction and Development (2022) Annual report 2022. European Bank for Reconstruction and Development, London

EEAA, Egyptian Environmental Affairs Agency (2022) Annual report 2022. Egyptian Environmental Affairs Agency, Cairo

ETC, Energy Transitions Commission (2020) Making mission possible: delivering a net-zero economy. https://www.energy-transitions.org/publications/making-mission-possible/

Fennell P, Mac Dowell N, Shah N et al (2017) The role of CO2 capture and utilization in mitigating climate change. Nat Clim Change 7:243–249. https://doi.org/10.1038/nclimate3231

Habert G, Miller SA, John VM et al (2022) Environmental impacts and decarbonization strategies in the cement and concrete industries. Nat Rev Earth Environ 1:559–573. https://doi.org/10.1038/s43017-020-0093-3

Harrison AJW (2004) A hydraulic Cement Comprising A Reactive Magnesium Oxide. Tececo Pty Ltd. NZ520426A. https://patents.google.com/patent/NZ520426A/en?oq=PatentPatent+NZ520426A

Harrison AJW (2005) TecEco eco-cement masonry product update–carbonation = sequestration. In: 10th Canadian Masonry Conference, Banff, Alberta. http://www.tececo.com.au/files/conference%20papers/Eco-CementBricksBlocksPaversandMortars301005.pdf

HeidelbergCement Report (2016a) Leading the way to carbon neutrality. Germany. https://www.heidelbergcement.com/sites/default/files/assets/document/7e/8c/co2strategie_factsheets_en.pdf

Heidelberg Cement Report (2016b) Sustainability report 2016. Germany. https://www.lehighhanson.com/docs/default-source/site-assets/heidelbergcement-sustainability-report2016.pdf?sfvrsn=71d74a3d_0

International Energy Agency (2018) Technology roadmap – low-carbon transition in the cement industry. https://www.iea.org/reports/technology-roadmap-low-carbon-transition-in-the-cement-industry

IPCC, Intergovernmental Panel on Climate Change (2021) Climate change 2021: the physical science basis. In: Contribution of Working Group I to the sixth assessment report of the intergovernmental panel on climate change. Cambridge University Press

Kazemi KH et al (2021) Carbstone pavers: a sustainable solution for the urban environment. Appl Sci 11(14):6418. https://doi.org/10.3390/app11146418

Korczak K et al (2022) Mitigation options for decarbonization of the non-metallic minerals industry and their impacts on costs, energy consumption and GHG emissions in the EU-Systematic literature review. J Clean Prod:132006. https://doi.org/10.1016/j.jclepro.2022.132006

Laschober C, Limbeck A, Rendl J, Puxbaum H (2004) Particulate emissions from on-road vehicles in the Kaisermühlen-tunnel (Vienna, Austria). Atmos Environ 38(14):2187–2195

Lehne J et al (2018) Making concrete change: innovation in low-carbon cement and concrete. Chatham House Report. https://www.chathamhouse.org/2018/06/making-concrete-change-innovation-low-carbon-cement-and-concrete-0/3-overcoming-barriers

Liu X, Yuan Z, Xu Y, Jiang S (2017) Greening cement in China: a cost-effective roadmap. Appl Energy. https://doi.org/10.1016/j.apenergy.2016.12.057

Mahasenan N et al (2003) The cement industry and global climate change: current and potential future cement industry CO2 emissions. In: Greenhouse gas control technologies-6th international conference, pp 995–1000. https://doi.org/10.1016/B978-008044276-1/50157-4

Marland G, Boden TA, Griffin RC, Huang SF, Kanciruk P, Nelson TR (1989) Estimates of CO2 emissions from fossil fuel burning and cement manufacturing, based on the United Nations energy statistics and the US Bureau of Mines cement manufacturing data (No. ORNL/CDIAC-25; NDP-030). Oak Ridge National Lab (ORNL), Oak Ridge, TN

Miller S, Moore F (2020) Climate and health damages from global concrete production. Nat Clim Change 10:439. https://doi.org/10.1038/s41558-020-0733-0

MonkmanSetal(2020)Integratedcarbondioxidecapture.CarboncureTechnologiesInc.US10570064B2. https://patents.google.com/patent/US10570064B2/en?oq=Patent+US10570064B2

Monteiro P, Miller S, Horvath A (2017) Towards sustainable concrete. Nat Mater 16:698. https://doi.org/10.1038/nmat4930

Naqi A et al (2019) Recent progress in green cement technology utilizing low- carbon emission fuels and raw materials: a review. Sustainability 11(2):537. https://doi.org/10.3390/su11020537

Niven R et al (2014) Carbon dioxide treatment of concrete upstream from product mold. Carboncure Technologies Inc. US8845940B2. https://patents.google.com/patent/US8845940B2/en?oq=Patent+US8845940B2
Ojan M et al (2016) Development of new types of low carbon cement. https://www.wbcsd.org/Sector-Projects/Cement-Sustainability-Initiative/News/CSI-climate-and-energy-workshop. Accessed 12 July 2022
Qiao H et al (2014) The application review of magnesium oxychloride cement. J Chem Pharm Res 6:180–185. https://www.jocpr.com/articles/the-application-review-of-magnesium-oxychloride-cement.pdf
Quaghebeur M et al (2014) Production of a mainly carbonate bonded article by carbonation of alkaline materials. Carbstone Innovation. US12/989,932. https://patentimages.storage.googleapis.com/51/1b/62/7dfac2cc4a5d6b/US8709151.pdf
Rissman J et al (2020) Technologies and policies to decarbonize global industry: review and assessment of mitigation drivers through 2070. Appl Energy 266:114848. https://doi.org/10.1016/j.apenergy.2020.114848
Sahu S et al (2018) Microstructured calcium silicate clinker capable of carbonation and method for producing the same. Lafarge SA, Solidia Technologies Inc. JP2018508459A. https://patents.google.com/patent/JP2018508459A/en?oq=PatentPatent+JP2018508459A
Sahu S et al (2020a) Sustainability and durability of solidia cement concrete. Concr Int 42(8):2934. https://www.concrete.org/publications/internationalconcreteabstractsportal.aspx?m=details&id=51728105
Sahu S et al (2020b) White carbonatble calcium silicate-based cements & methods of preparation. Solidia Technologies Inc. EP3423421B1. https://patents.google.com/patent/EP3423421B1/en?oq=PatentPatent+EP3423421B1
Schneider M (2019) The cement industry on the way to a low-carbon future. Cem Concr Res 124:105792. https://doi.org/10.1016/j.cemconres.2019.105792
Shao Y et al (2018) Carbonate-bonded construction products from steel-making residues and method for making the same. Carbicrete Inc. US10112871B2. https://patents.google.com/patent/US10112871B2/en?oq=PatentPatent+US10112871B2
Shao Y et al (2020) Carbonate-bonded construction products from steel-making residues and method for making the same. Carbicrete Inc. US10633288B2. https://patents.google.com/patent/US10633288B2/en?oq=PatentPatent+US10633288B2
Tas AC et al (2018) Carbonatable calcium silicate-based cements and concretes having mineral additives, and methods thereof. Solidia Technologies Inc. US20180273430A1. https://patents.google.com/patent/US20180273430A1/en?oq=PatentPatent+US20180273430A1
UNFCCC, United Nations Framework Convention on Climate Change (2023) Annual report 2023. United Nations Framework Convention on Climate Change, Bonn
Vlasopoulos N (2012a) Integrated process for producing compositions containing magnesium. Novacem Limited. WO 2012/028418 A1. https://patents.google.com/patent/WO2012028418A1/en?oq=PatentPatent+WO+2012%2f028418+A1
Vlasopoulos N (2012b) Process for producing cement binder compositions containing magnesium. Novacem Limited. WO 2012028471A1. https://patents.google.com/patent/WO2012028471A1/en?oq=PatentPatent+WO+2012028471A1
Vlasopoulos N et al (2009) Binder composition. Novacem Limited. WO 2009/156740A1. https://patents.google.com/patent/WO2009156740A1/en?oq=PatentPatent+WO+2009%2f156740A1
Worrell E et al (2001) Carbon dioxide emissions from the global cement industry. Annu Rev Energy Environ 26(1):303–329. https://doi.org/10.1146/annurev.energy.26.1.303. https://www.annualreviews.org/doi/abs/10.1146/annurev.energy.26.1.303

The Role of Woody Trees in Climate Changes Through Treatments

Mam Chand Singh, Eram Arzoo, and Lokendra Singh

Abstract Climate change (CC) is a highly regressive and devastating phenomenon, especially in developing nations. Climatological mitigation has to be taken up as a necessary strategy for better survival in the future. Naturally, woody trees are the best option for this process because they are the largest and cheapest factory of carbon mitigation on earth. Among all climate change factors, carbon dioxide (CO_2) is the major factor in trapping heat and global warming. Its proper mitigation and sequestration will not only lead to the absorption of CO_2 but will also help in releasing O_2 to the environment free of cost and shall maintain humidity, which leads to a decrease in temperature and cool down the earth. Many species of trees have been found best for the same process. These trees have drought and heat stress tolerance and can sequester and store carbon and perform hydraulic regulation. Healthy forest ecosystems contribute to sequestering and storing more than 80% of all terrestrial aboveground Carbon and more than 70% of all soil organic carbon. Therefore, to make the earth liveable, planting more trees is required to stop deforestation and should rather follow burlapping.

Keywords Woody trees · Climate mitigation · Photosynthesis · Burlapping · Photosynthesis · Air pollution tolerance index (APTI)

M. C. Singh (✉) · L. Singh
Division of Floriculture and Landscape Architecture & Centre for Protected Cultivation Technology (CPCT), ICAR-Indian Agricultural Research Institute, New Delhi, India
e-mail: mam.singh@icar.gov.in

E. Arzoo
Division of Floriculture and Landscape Architecture, ICAR-Indian Agricultural Research Institute, New Delhi, India

A. A. H. Abdel Latef et al. (eds.), *Sustainable Remediation for Pollution and Climate Resilience*, https://doi.org/10.1007/978-981-96-5674-5_22

1 Introduction

Climate largely refers to *variations in climatic patterns occurring over the last century, caused by anthropogenic activities that release greenhouse gases* (e.g., CO_2, CH_3, NO_2, etc.). Effects of rising greenhouse gases (GHG) viz. carbon dioxide (CO_2), warming, drought, and flooding, ozone (O_3) depletion, nitrogen (N), and phosphorus (P) fluxes are the major causes of climate change (CC) and irregularity in seasons. The increase in summer temperatures is projected to be very high (exceeding 6 °C in some parts of the world) by the end of the twenty-first century. Similarly, annual rainfall is expected to decrease by 10% (Yaduvanshi et al. 2021). The last three decades have witnessed a net increase in average temperatures and a significant variation in precipitation patterns. Here, the interaction of climate change (CC) and associated air pollution (AP) is the intrinsic component (CC–AP) that causes global warming (Matyssek et al. 2017). Woody trees include native and cultivated species found in various ecosystems worldwide, including forests, woodlands, savannahs, and even urban areas. These include evergreen and deciduous broad-leaved trees, conifers trees, tropical hardwood trees, fruit and nut trees, etc. Woody trees and climate change do intrinsically have a complicated and multidimensional relationship. Woody trees assist in slowing down climate change by storing carbon and lowering the amount of greenhouse gases in the atmosphere. At the same time, the ecosystems of woody trees are seriously threatened by climate change, which includes temperature modifications, precipitation patterns, and the frequency and severity of extreme weather events. These trees are the largest terrestrial carbon sink, present as forests and other woody vegetation in urban areas, which absorb substantial amounts of carbon dioxide (CO_2) from the atmosphere through photosynthesis. These carbons are then stored in various forms within the trees, including in their biomass, roots, and soil (Dewar 1991). Because of their possible effects on ecosystem dynamics and carbon cycling, the problem of woody plant encroachment has been receiving more attention to climate mitigation. Studies indicate that the encroachment of woody plants could be caused by an increase in atmospheric CO_2 accumulation in biomass, this phenomenon is referred to as CO_2 fertilization. Through improved intrinsic water use efficiency (iWUE) and increased CO_2 assimilation rates through stomata, this mechanism improves the physiological processes of CO_2 uptake in woody plants. Research has revealed that woody plants' iWUE has increased in response to rising atmospheric CO_2 levels, although there have been differences between functional and taxonomic categories of plants (Walker et al. 2021). Trees can transpire hundreds of litres of water every day using the energy from the sun. The amounts of 100 L of transpired water can provide a cooling power of 70 kWh, sufficient to replace two standard residential central air conditioners for a day (Omer 2010). For this reason, the ability of trees to block heat and provide cooling by evapotranspiration contributes to energy conservation. Climate change and CO_2 fertilization may encourage plant growth and photosynthesis, increasing plant biomass stocks and land carbon storage. But at the same time, nutrient absorption shortages may decrease the efficiency of CO_2 fertilization and eventually reduce its effect on the productivity of woody plants and CO_2 mitigation (Lukac et al. 2010). Although the concern about rising CO_2 levels in the earth's atmosphere reaches a record up to 420 ppm (parts per million), nowadays it is crucial to recognize the vital role of woody trees in mitigating climate change. A single woody tree can absorb 10 kg of carbon dioxide per year for the first 20 years, but these numbers can increase as the tree gets older, reaching up to 40 kg (Gowtham et al. 2022). These woody trees cover around 4 billion hectares of land (forest) that serve as a large carbon sink, taking in billions of metric tons of CO_2, every year. Many researchers have adopted an approach to identify tree species that can perform best in polluted areas (with the help of the air pollution tolerance index (APTI)) to mitigate climate change. Forests exhibit resilience and the ability to sequester carbon, which makes them crucial for offsetting carbon emissions even in the face of disturbances and a slower rate of degradation. But rising temperatures also correspond with rising carbon dioxide concentrations, intensifying the deleterious effects of climate change. There is an obvious need to cut carbon dioxide emissions immediately, and using natural remedies offers potential ways to mitigate the problem caused by woody trees. Furthermore, they provide shade and evapotranspiration, lowering air temperatures and aiding in cooling the environment. The provision of vital ecosystem services by both urban trees and wild forests, such as air purification, temperature regulation, and climate moderation, highlights the significant role that tree conservation and sustainable management techniques play in combating climate change. This chapter deals with the theme of woody trees for mitigating climate change in detail.

2 Role of Woody Trees in Climate Change

A famous quote "*The world is not going to protect the trees but trees are helping to protect the planet from climate change*" relates to the importance of the climate and humanity relationship which has a prime significance. People breathe in oxygen (O_2) and exhale CO_2, while trees breathe in CO_2 and release a "white gold" oxygen (O_2) out to the environment, the importance of which we have felt during COVID-19 (Gowtham et al. 2022). Besides some indirect negative roles of trees, they have a huge direct positive role for the environment and fauna, including humans on earth. All of the roles of trees contributing to climate change (Dar et al. 2020; Suliman et al. 2024) have been described below:

1. *Photosynthesis and Carbon Sequestration*:
   - These woody trees absorb CO_2 from the atmosphere through photosynthesis.
   - During photosynthesis, trees use CO_2, water (H_2O), minerals, and light energy to produce carbohydrates (CH_2O) as the food of trees and release O_2 into the atmosphere.

- This process allows trees to sequester carbon, storing it in their biomass, roots, and soil.

2. *Carbon Storage and Respiration*:

    - Trees store significant carbon in their bodies, acting as carbon sinks.
    - While they use some of the carbohydrates that will be produced through respiration, they release oxygen and emit some carbon dioxide. Despite emitting CO_2 through respiration, trees store most of the carbon they absorb, finally contributing to overall carbon sequestration.

3. *Carbon Decomposition and Release*:

    - When trees die, eventually decompose and release some of the stored carbon back into the atmosphere.
    - In the above concern, fungi and other decomposers break down lignin (the tough material in wood), contributing to the decomposition process.
    - While burning of trees in wildfires or deliberate clearing for agriculture harvest also releases stored carbon back into the atmosphere.

4. *Net Carbon Absorption by Trees and Forests*:

    - Despite disturbances such as decomposition and burning, forests remain a net sink for carbon dioxide.
    - Earth's forests collectively absorb billions of metric tons of carbon dioxide each year, outweighing the carbon they emit through respiration and decomposition.

5. *Regional Differences in Carbon Sink Efficiency*:

    - Not all forests are equally effective at carbon sequestration. For instance, tropical rainforests are highly biodiverse places that can become net sources of carbon emissions due to disturbances such as fires and deforestation.
    - While the Congo River basin's rainforest stands out as a significant carbon sink among the world's largest tropical rainforests.

6. *Importance of Trees in Carbon Storage and Soil Health*:

    - Wooden trees play a crucial role in capturing and storing CO_2, contributing to climate change mitigation efforts.
    - A significant portion of carbon bound in soil comes from tree roots and the activities of associated fungi. Therefore, tree planting and forest conservation efforts are essential for enhancing carbon storage and soil health.

7. *Impact of Deforestation on Climate Change*:

    - Increase in deforestation significantly reduces the capacity of forests to sequester carbon and mitigate climate change. These trees are vital for reducing the effects of climate change, making deforestation a major environmental concern.

After a proper understanding of the role of woody trees in carbon sequestration, storage, and ecosystem health, anyone can develop effective treatments and management strategies to enhance their contribution to climate change adaptation and mitigation.

3 Carbon Sequestration for Climate Mitigation

Woody trees can utilize the carbon they absorb through photosynthesis for various metabolic processes and structural growth (as shown below in the equation).

$$6\,CO_2 + 6\,H_2O + \text{light energy} \rightarrow C_6H_{12}O_6\,(\text{glucose}) + 6\,O_2\,(\text{process of photosynthesis}).$$

The process of carbon sequestration and mitigation involves the distribution of assimilated carbon to different parts of the tree, including stems, branches, leaves, roots, and belowground structures such as mycorrhizal fungi (Bhattacharyya et al. 2023). In 1991, Rowntree and Nowak estimated that urban forests in the USA sequester about 712 million tonnes of C in biomass. In addition to storing C, these trees cool down the ambient air and provide shade, which reduces energy costs spent on air conditioning, which will lead to savings of $24 per year for each mature tree. In addition to planting on ground green roof systems can also help in the same. The scientific mechanisms of carbon sequestration and mitigation involve:

1. *Assimilation of Carbon Compounds*: During photosynthesis, CO_2 is captured from the atmosphere and converted into organic carbon compounds (primarily glucose), through a series of biochemical reactions. This process takes place in specialized cell organelles called chloroplasts, located within the leaf cells of woody trees. Enzymes such as ribulose-1,5-bisphosphate carboxylase/oxygenase (RuBisCO) catalyze the fixation of CO_2 and its conversion into sugars, including glucose, through the Calvin cycle.
2. *Translocation of Assimilates*: Once synthesized, glucose and other assimilated carbon compounds are translocated within the tree through the phloem (a specialized vascular tissue responsible for transporting organic nutrients). The phloem carries these assimilates from the leaves, where they are produced, to other parts of the tree for growth, maintenance, and storage. These assimilates are primarily composed of sucrose (a disaccaride formed from glucose and fructose) for transport to the different part of trees.
3. *Growth and Maintenance*: The above-allocated carbon is used in the growth and maintenance processes after supporting cellular metabolism. This helps produce new tissues and maintain existing structures. In actively growing woody trees, carbon will be allocated to the synthesis of structural components such as cellulose, hemicellulose, and lignin, which form the cell walls of stems, branches, and roots. These carbons will also be utilized to produce proteins, lipids, and

other essential compounds required for growth, enzyme function, and defense against biotic and abiotic stress.

4. *Storage in Biomass*: Excess assimilated carbon beyond immediate metabolic needs will be allocated for long-term storage in various tree tissues, including stems, branches, roots, and storage organs such as tubers or bulbs. These carbons stored in woody biomass, contribute to the accumulation of structural carbohydrates, lignin, and other organic compounds that make up the tree biomass. Some woody trees, especially those adapted to seasonal or intermittent growth patterns, store excess carbon reserves in specialized storage tissues such as parenchyma cells, starch granules, lipid droplets, etc.

After understanding the allocation of carbon in woody trees, it is essential to assess the role of woody trees in carbon sequestration and storage and to design effective strategies to enhance the capacity of carbon sinks. By elucidating the scientific mechanisms underlying carbon allocation, researchers can develop models to predict carbon dynamics in forest and urban ecosystems and optimize management practices for sustainable carbon sequestration and climate change mitigation.

4 The Mechanisms Through Which Trees Are Treating Climate Change

Besides a different role in the understanding of woody trees, it's also important to understand various natural mechanisms to cope with the effects of climate change, as naturally done in the initial twenties, but becoming worse in the twenty-first century as average indicated in Fig. 1. Therefore, woody trees are working to act as a clean mechanism for CO_2 sequestration (Adiaha et al. 2019). But all the preventive measures and treatments done by trees can reverse the process again. Here are some key treatments or mechanisms exhibited by woody trees (Table 1 & 2):

1. *Drought Tolerance*: Woody trees can easily adapt to drought conditions through physiological mechanisms such as stomatal regulation, which reduces water loss through transpiration. Other than this, some species can develop deep root systems to access water stored deeper in the soil during dry periods.
2. *Heat Stress Adaptations*: Few woody trees can exhibit heat tolerance by adjusting their metabolic processes and leaf morphology to minimize heat damage. These can also undergo phenological changes, such as adjusting flowering and leaf emerging time to avoid extreme heat stress.
3. *Carbon Sequestration and Storage*: These woody trees play a crucial role in carbon sequestration i.e. absorbing atmospheric CO_2 through photosynthesis and storing carbon in their biomass and soils. Some species have higher rates of carbon sequestration than others, making them valuable assets in mitigating climate change. This is the major treatment done by wooden trees which is going to be

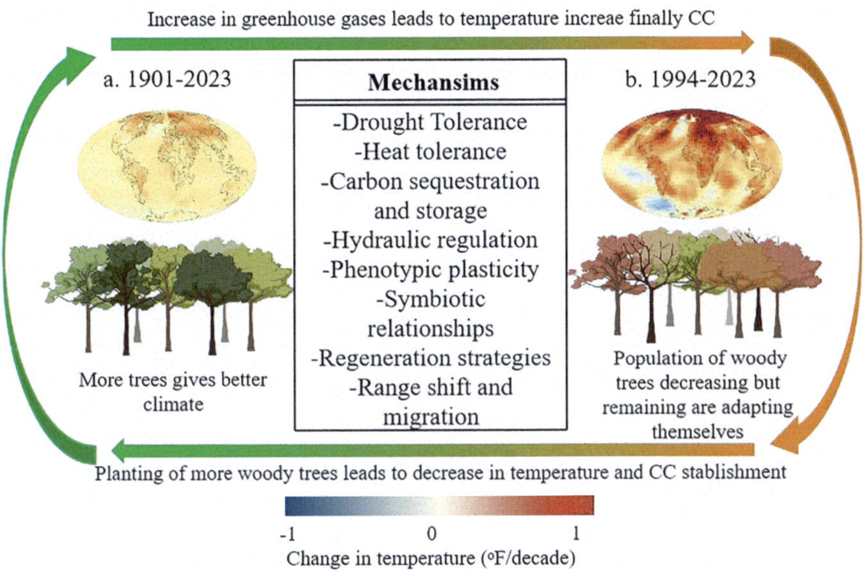

Fig. 1 The illustration shows how woody trees can cope with climate change in relevance to (**a**) 13 decades, and (**b**) 3 decades and how more planting of trees will be beneficial for the future. Most of these tree species are beneficial against the climate change

Table 1 List of tree species and mechanisms to cope with climate change

Trees species	Mechanism to cope with CC
Acer truncatum, *Cercis canadensis mexicana*, *C. canadensis texensis*, *C. canadensis texensis* "Oklahoma", *Quercus muehlenbergii*, *Q. robur*.	*Drought tolerance by*: improvement in root system, leaf structure, osmotic adjustment, relative water content, and stomata regulation. In addition, calcium and phytohormone (abscisic acid, jasmonic acid, salicylic acid, auxins, gibberellins, ethylene, etc.) signaling pathways and scavenging of reactive oxygen species
Alstonia scholaris, *Castanospermum austral*, *Heritiera trifoliolata*, *Sloanea woollsii*, *Populus nigra* (L.), *Eucalyptus* sp., *Picea* sp., *Platanus* sp.,	*Heat stress tolerance by*: accumulation of osmolytes including sugars and sugar alcohols (polyols), proline, tertiary and quaternary ammonium compounds (glycine betaine, GB), and tertiary sulphonium co.; expression and synthesis of Heat Shock Proteins (HSPs); synthesis of volatile organic compounds (VOC) (e.g. Isoprene) in response to high-temperature exposure

explained in detail ahead. Wang et al. (2015) have formulated an equation for the calculation of stored carbon and calculation for some species using this equation is shown below:

$$C_{\text{single\_tree}} = V \times \text{BD} \times \text{BEF} \times (1 + R) \times \text{CF}$$

where C_{single_tree} = individual tree carbon storage (tons), V = individual tree volume (m^3), BD = basic wood density (tons/m^3), BEF = biomass expansion factor, R = root:shoot ratio, CF = forest carbon fraction (Table 2).

4. *Hydraulic Regulation*: These huge woody trees regulate water flow within their vascular systems to maintain hydration and prevent damage during drought or excessive heat. This regulation helps trees withstand water availability fluctuations and maintain their physiological functions.
5. *Adaptive Phenotypic Plasticity*: Trees can exhibit phenotypic plasticity, meaning they can adjust their physical traits in response to environmental cues. For example, trees may produce smaller leaves or thicker cuticles in response to drought stress, helping to conserve water and mitigate the effects of water scarcity.
6. *Symbiotic Relationships*: Some of the woody trees form symbiotic relationships with mycorrhizal fungi, which enhance nutrient uptake and improve the resilience of trees to environmental stress. Additionally, nitrogen-fixing bacteria associated with certain tree species can increase soil fertility and promote tree growth in nutrient-poor environments.
7. *Regeneration Strategies*: Wooden trees can employ various strategies for regeneration and reproduction, such as producing large quantities of seeds, developing seed banks in the soil, or sprouting from dormant buds after disturbances like wildfires.
8. *Range Shifts and Migration*: With changes in climatic conditions, woody trees may undergo range shifts or migrate to more suitable habitats. This migration allows trees to navigate favourable environmental conditions and maintain viable environmental populations to face climate change.

After understanding these natural treatments and adaptive responses of woody trees, it is essential to develop effective strategies to conserve and manage forest ecosystems in a changing climate. Human interventions viz. afforestation, reforestation, and sustainable forest management can complement these natural treatments and enhance the resilience of woody tree populations to climate change impacts.

There are hundreds of genera and thousands of species of woody trees found on earth that have adaptive mechanisms towards climate change, but a few species were also reported that got extinct due to these phenomena. These trees can be classified into different categories according to their morphology, habitat, utility, and

Table 2 List of tree species and storage of carbon (ton) in individual trees after sequestration

Tree species	C_{tree} (ton)
Chinese banyan (*Ficus microcarpa*)	3.95
Madagascar almond (*Terminalia boivinii*)	0.37
Golden raintree (*Koelreuteria henryi*)	0.59
Camphor tree (*Cinnamomum camphora*)	0.62
Golden shower tree (*Cassia fistula*)	0.62
Blackboard tree (*Alstonia scholaris*)	3.47

Table 3 Tree species having crucial role in climate change and mitigation

Types of trees	Species responsible for CC
Broadleaf trees	Deciduous trees that shed their leaves annually, such as Oak (*Quercus* sp.), Maple (*Acer* sp.), Beech (*Fagus* sp.), and Birch (*Betula* sp.)
	Evergreen trees that retain their leaves year-round, such as Pine (*Pinus* sp.), Spruce (*Picea* sp.), Fir (*Abies* sp.), and Cedar (*Cedrus deodar*).
Coniferous trees	Douglas fir (*Pseudotsuga menziesii*), Norway spruce (*Plicae abies*), and Lodgepole pine (*Pinus contorta*)
Tropical hardwood trees	Mahogany (*Swietenia* spp.), Teak (*Tectona grandis*), Ebony (*Diospyros kaki*), and Rosewood (*Dalbergia sissoo*)
Fruit trees	Apple (*Malus domestica*), Mango (*Mangifera indica*), Litchi (*Litchi chinensis*), Avocado (*Persea americana*), Citrus (*Citrus* sp.), etc.
Nut trees	Walnut (*Juglans regia*), Chestnut (*Castanea* sp.), Almond (*Prunus dulcis*), Pecan (*Carya illinoinensis*), etc.
Urban trees	London plane (*Platanus* × *acerifolia*), Japanese maple (*Acer palmatum*), and Flowering cherry (*Prunus serrulata*)

place of growing (Thurman et al. 2020). Some of the trees that are responsible for climate mitigation are listed below in Table 3:

Nowadays, urban society needs more attention to mitigate climate change. Many societies/organizations like the Center for Science and Environment (CSE), National Green Tribunal (NGT), etc., are working towards mitigation of climate change after limiting the use of pollution emitter sources, controlling the deforestation or cutting of trees, and promoting afforestation or planting more and more trees. Center for Science and Environment (CSE) is a public interest research organization based in New Delhi, India, they are mostly working on reducing the air quality index (AQI) to revert climate change. The National Green Tribunal (NGT) was established on 18.10.2010 under the NGT Act 2010 for effective and expeditious disposal of cases relating to environmental protection and conservation of forests and other natural resources. They are working on restoring degraded land, enhancing soil fertility, and fostering biodiversity through sustainable land use practices, reforestation, and effective water management to stop desertification and climate change. Other than organizations, different researchers have studied the tree species and found some of the species more effective for air pollution mitigation in urban heat islands of the most polluted cities in the world. Same as Singh et al. (2024) evaluated tree species from New Delhi (one of the top polluted cities in the world) using (APTI). They concluded that higher APTI value tree species can be recommended to replace the low APTI value tree species. Because trees have a higher APTI values, they can help reduce the severity of air pollution in Delhi. They estimated APTI using Singh and Rao's (1983) equation.

$$\text{APTI\%} = \frac{\left[A\left(T+P\right)\right]+R}{10}$$

Table 4 List of trees and air pollution tolerance index (APTI) done at Delhi for polluted (Pl) and non-polluted (NP) places to compare the efficiency of pollution control

		Relative water content (%)		Total chlorophyll content (mg/g)		APTI	
Tree species	Location	Winter	Summer	Winter	Summer	Winter	Summer
Azadirachta indica	NP	85.95	82.25	2.45	2.65	16.21	15.02
	Pl	83.25	80.25	1.76	2.25	14.01	16.57
Alstonia scholaris	NP	92.25	86.25	1.95	3.02	14.38	15.11
	Pl	91.25	82.23	1.91	2.58	19.43	16.85
Ficus religiosa	NP	94.55	88.23	1.75	2.95	19.21	20.33
	Pl	88.56	86.47	1.65	2.78	23.23	20.28
Pongamia pinnata	NP	82.43	76.25	2.93	2.98	20.85	19.12
	Pl	80.25	72.46	2.50	2.57	15.58	19.66
Cassia fistula	NP	81.25	78.63	2.15	2.75	15.51	14.12
	Pl	80.15	71.52	2.10	2.55	14.80	16.53
Polyalthia longifolia	NP	83.52	72.36	2.18	2.91	14.63	11.14
	Pl	82.44	70.52	2.10	1.95	18.94	12.58
Ficus infectoria	NP	94.95	84.52	1.98	2.15	17.95	16.90
	Pl	92.56	75.86	1.95	1.98	19.05	17.26

Source: Singh et al. (2024)

where, A = ascorbic acid (mg/g); T = total chlorophyll (mg/g); P = pH of leaf extract, and R = relative water content (%).

They found that *Ficus religiosa*, *F. infectoria*, *Alstonia scholaris*, and *Polyalthia longifolia* have higher APTI values, which can vary according to seasons, as shown in Table 4 than others. Similarly, Gowtham et al. (2022) have identified ten net carbon sequestration and net oxygen release species among 58 tree species as enlisted here: *Albizia lebbeck* followed by *Tamarindus indica, Parkia biglandulosa, Delonix regia, Kigelia africana, Peltophorum pterocarpum, Ficus religiosa, Leucaena leucocephala, Pterospermum acerifolium*, and *Azadirachta indica*. These trees have been found to improve air quality and have many economic benefits in urban areas.

5 Conclusion

Increasing population leads to environmental pollution, which is detrimental to flora, fauna, and especially humans. This increase in pollution changes climate day by day, leading to temperature increase, ultimately affecting the climate pattern. Therefore, the mitigation of climate change by natural means is the only option. Among natural ways, a woody tree is the largest found body on earth which can convert 60% contributing GHG, i.e., CO_2 into O_2. Other than CO_2, other GHGs can also be mitigated by woody trees. Woody trees have the ability to perform photosynthesis through which they can do carbon sequestration followed by its storage and use in different parts of trees. Other than C sequestration and storage, trees can cool down the ambient air and provide shade, which reduces energy costs spend on air conditioning (approx. $24 per year by each mature tree). Due to an incrased awareness, green roof systems are being followed by urban population on a larger scale. Many species of trees have good potential to cope with climate change after modifying their morphology, anatomy, and physiology. These trees should have a high APTI value, which will help to mitigate pollution efficiently. But, nowadays, an increase in tree cutting and deforestation reduces carbon sequestration and climate mitigation. An increase in tree planting, burlapping (transfer of tree from one place to another), and their proper maintenance will be good to increase the efficiency of pollution mitigation to address climate change effectively.

References

Adiaha MS, Buba AH, Tangban EE, Okpoho AN (2019) Mitigating global greenhouse gas emission: the role of trees as a clean mechanism for CO2 sequestration. J Agric Sci Sri Lanka 15:101

Bhattacharyya SS, Mondaca P, Shushupti O, Ashfaq S (2023) Interplay between plant functional traits and soil carbon sequestration under ambient and elevated CO2 levels. Sustainability 15(9):7584

Dar JA, Subashree K, Bhat NA, Sundarapandian S, Xu M, Saikia P, Kumar A, Kumar A, Khare PK, Khan ML (2020) Role of major forest biomes in climate change mitigation: an eco-biological perspective. In: Socio-economic and eco-biological dimensions in resource use and conservation: strategies for sustainability. Springer, Cham, pp 483–526

Dewar RC (1991) Analytical model of carbon storage in the trees, soils, and wood products of managed forests. Tree Physiol 8(3):239–258

Gowtham S, Prasanthrajan M, Maheswari M, Sritharan N, Rajkishore SK, Keerthana S (2022) Oxygen production and carbon capturing capacity of various tree species in Coimbatore city, India. Int J Environ Clim Chang 12(11):601–615

Lukac M, Calfapietra C, Lagomarsino A, Loreto F (2010) Global climate change and tree nutrition: effects of elevated CO2 and temperature. Tree Physiol 30(9):1209–1220

Matyssek R, Kozovits AR, Wieser G, King J, Rennenberg H (2017) Woody-plant ecosystems under climate change and air pollution—response consistencies across zonobiomes ? Tree Physiol 37(6):706–732

Omer AM (2010) Environmental and socio-economic aspects of possible development in renewable energy use. J Agric Ext Rural Dev 2(1):001–021

Rowntree RA, Nowak DJ (1991) Quantifying the role of urban forests in removing atmospheric carbon dioxide. J Arboric 17:269

Singh SK, Rao DN (1983) Evaluation of plants for their tolerance to air pollution. In: Proceedings of the symposium on air pollution control. IIT, Delhi, pp 218–224

Singh L, Sindhu SS, Singh MC, Jaiswal S, Kumar S, Sharma DK, Singh B (2024) Evaluation of tree species using Air Pollution Tolerance Index for urban landscaping in Delhi. Indian J Agric Sci 94(1):086–091

Suliman M, Khan SM, Ali S, Abdullah A, Rahman AU, Ullah H, Muhammad D (2024) Role of forest's woody vegetation in the climate change mitigation through carbon sequestration in the northern Pakistan. In: Agroforestry for carbon and ecosystem management. Academic Press, pp 191–202

Thurman LL, Stein BA, Beever EA, Foden W, Geange SR, Green N, Gross JE, Lawrence DJ, LeDee O, Olden JD, Thompson LM (2020) Persist in place or shift in space? Evaluating the adaptive capacity of species to climate change. Front Ecol Environ 18(9):520–528

Walker AP, De Kauwe MG, Bastos A, Belmecheri S, Georgiou K, Keeling RF, McMahon SM, Medlyn BE, Moore DJ, Norby RJ, Zaehle S (2021) Integrating the evidence for a terrestrial carbon sink caused by increasing atmospheric CO2. New Phytol 229(5):2413–2445

Wang XL, Feng Y, Swail VR (2015) Climate change signal anduncertainty in CMIP5-based projections of global ocean surface-wave heights. J Geophys Res Oceans 120:3859–3871. https://doi.org/10.1002/2015JC010699

Yaduvanshi A, Nkemelang T, Bendapudi R, New M (2021) Temperature and rainfall extremes change under current and future global warming levels across Indian climate zones. Weather Clim Extrem 31:100291

The Role of Ornamental Plants and Remediation

Mam Chand Singh, Eram Arzoo, and Lokendra Singh

Abstract Contamination of soil, water, and air has resulted in serious health issues in humans and the extinction of flora and fauna. Remediation of the toxic limits at a certain stage becomes either too costly or even a threat to other lives, posing a great challenge to survival. The answer, to how to remediate these, lies in the existence of plants, which can extract the hazardous toxic pollutants and sustain mother earth. Certain plants, by nature, are hyperaccumulators that bioaccumulate these chemicals effectively with varying degrees of remediation effects in different species. However, these toxic and heavy metals (HMs) cannot be depraved while all organic pollutants can, making them easy targets for phytoremediation. An enormous number of ornamental plant species have the ability to ingest pollutants, toxic metals, and contaminants due to anthropogenic responses to our environment. Remediation with ornamental plants (OPs) provides wildlife habitat and has significant public recognition and acceptance, despite being so cheap, sustainable, and environmentally sound solution for treating commercial trash in polluted areas.

Keywords Phytoremediation · Pollutant · Ornamental plants · Phytoextraction · Climate

M. C. Singh (✉) · L. Singh
Division of Floriculture and Landscape Architecture & Centre for Protected Cultivation Technology (CPCT), ICAR-Indian Agricultural Research Institute, New Delhi, India
e-mail: mam.singh@icar.gov.in

E. Arzoo
Division of Floriculture and Landscape Architecture, ICAR-Indian Agricultural Research Institute, New Delhi, India

A. A. H. Abdel Latef et al. (eds.), *Sustainable Remediation for Pollution and Climate Resilience*, https://doi.org/10.1007/978-981-96-5674-5_23

1 Introduction

The global population has exploded dramatically and is exposed to harmful toxic chemicals and pollutants directly released to the environment (soil, water, and air) causing serious health issues in humans and the extinction of different flora and fauna. Traditional remediation of these chemicals and heavy metals (HMs) contaminate soil, water, or air because of excavation or management that takes place on the site, followed by disposal in landfills. In such situations, the task is so risky to handle and overcome bioaccumulating factors. The use of microbes or chemical decomposition of pollutants was started at a particular site but these methods can be utilized to cover smaller areas and turn out to be very costly with all possibilities of leakage that might pose serious health risks to both flora and fauna. These methods become costly and even harmful to other lives and generate curiosity as to how and what remediation options are required for remediation. The answer lies in the existence of plants, which can extract hazardous toxic pollutants/contaminants from soil, water, as well as air can be used to do the same. The process that aims to "*serve floras and concurrent microorganisms in addition to appropriate agronomic practices and remedies to either ingest, abolish or delineate toxic environmental pollutants mild and harmless*" is known as "phytoremediation" (Das 2018). This term came from the Greek word *"phyto,"* which means plant, and the Latin word "*remedium*," which means restoring balance, and both together are known as restoring balance by plants.

Ornamental plants (OPs) are garden plants mainly grown for aesthetic/beauty purposes (flowers, foliage, fruits, barks, their textures, and sometimes for beautiful thorns and spines) but their scent just melts anyone's soul/heart. Other than aesthetic value, these plants are also utilized in many processing industries such as essential oils (Rose, Champak, Lavender, etc.), dyes (Marigold, Bougainvillea, Celosia, Magnolia, etc.), potpourris, dried products, poultry feed (Marigold petals), and cosmetics. Botanically these include woody terrestrial, herbaceous, and other aquatic plants of annual, biennial, or perennial nature. While non-botanically OPs include bedding plants, houseplants, and hedges plants to use as cut flowers, loose flowers, and foliage. The study of the growth of OPs is known as floriculture (growing of all OPs) and arboriculture (growing of trees). OP production in gardens started from ancient civilizations (~2000 BC), while the use of OPs in remediation was scientifically proven by NASA in 1984. It is a worthwhile plant-based proposition to remediation the pollutants and chemical compounds from the proximate environment and remove them. The congregating effects are found in numerous plants called hyperaccumulators and bioaccumulate these toxic wastes effectively, but the amending effects in various diverse species are different. All these toxic and heavy metals (HMs) cannot be dissolute while all organic pollutants can, which makes them easy targets for phytoremediation. OPs have the potential to absorb and make a bad deed as pollutants or contaminants given by humans (anthropogenic) to the environment. Other than OPs, food crops are eaten at a larger (primary) level by consumers of different trophic levels, and biomagnification of these harmful contaminants reaches humans at last, which may lead to many diseases and disorders and then it permeates even to their progenies. In the following sections, we will discuss in detail about the mechanism of phytoremediation, the most appropriate species, and their features that hasten remediation.

Along with these, here we will also discuss the hastening techniques for phytoremediation, like adding amendments, microbes, and nanoparticles (NPs) in the media of pollution, as well as the genetic engineering technique and genes responsible for hastening the phytoremediation. Here we will also discuss one of the famous case studies of OPs in treating contaminants. Along with remediation, post-remediation management legislations for the harvested biomass and advantages, disadvantages, and limitations will be discussed here in this chapter.

2 How Ornamental Are Beneficial for Remediation and Phytoremediation

2.1 The Pollutants and Ornamental Species for Ecologically Sound and Sustainable Remediation

The plants can provide an eco-friendly solution to wipe up harmful emissions at a much lesser energy as well as their own cost than conventional means of detoxification for environmental tidiness because they use solar energy as well as their own physiology. Ornamental plants are more beneficial for remediation than other food and forage plants (*lower risk*) because the chances of bio-fortification of harmful organic contaminants, heavy metals, or gaseous pollutants are reduced as the remediated ornamental plants are not eaten by consumers of any tropic level of the food chain so accumulated pollutant inside of plant do not transfer from producer level to others consumer levels. OPs have a high capacity to lower the contaminants from different habitats, which includes water, air, and soil, and as an indirect perk, carbon sequestration will be embellished and ascribed to remediation practices. The reason is that the excavation of toxic contaminants from the environment will lower the array of atmospheric CO_2. Remediation, when constituted with homely site landscaping with beautiful ornamentals, generates broader advantages, such as social, ecologic, and fiscal well-being, especially in place of industries.

Identifying potential remedial aspects in ornamental species begins with observing their characteristics listed as follows:

1. *Physiology*: The capacity of plants to grow and flourish in bumped or depraved environments showing the presence of some level of tolerance and the pace of growth and production of biomass, as the usage of certain species may become impractical due to diminished growth.
2. *Morphology*: Some morphological characteristics, like edaphoclimatic acclimation, biomass production, root depth, growth rate, and maintenance feasibility, including the potential to develop micro-symbionts, have an immediate effect on

the ornamental plant species chosen for cleanup. These OP entities must be computed for

- Root size, density, and surface area affect foreign substances. Intake, as well as microorganism proliferation and reproduction in the rhizosphere through exudates from the root.
- Plant height, stem diameter, and leaf area index have a positive correlation with contamination endurance and building up, as they affect photosynthesis and production of biomass, which affects chemical volatilization and excretion during treatment.
- For instance: The optimum plants for created wetlands are those with large leaves and an extensive root system; these qualities improve the plant transpiration surface and effective distribution of the roots in the water column, ensuring considerable intake and flushing of toxins from water.

3. *Microbial interactions*: It is an important factor as microbes contribute to the degradation and affect plant uptake and tolerance to polluting substances.
4. *Stress tolerance*: OPs are resistant to both abiotic and biotic stresses, including temperature, drought, salinity, pollutants, diseases, and pests. Plant species should be chosen for phytoremediation based on their potential to activate antioxidant systems, as well as their stress tolerance. Stress activates physiological defense systems, leading to increased formation of reactive oxygen species (ROS) and oxidative damage. Antioxidants play a role in this process.

   - *Nonenzymatic antioxidants*: The metabolism of plants produces chemicals such as ascorbate, glutathione, phytochelatins, tocopherols, and phenolic substances that can reduce or remove ROS.
   - *Enzymatic antioxidants*: These include peroxidases and ROS (e.g., ascorbate peroxidase, and peroxiredoxin), catalases (CAT), superoxide dismutase (SOD), thioredoxin, and glutathione reductase and glutathione-S-transferases (GST).

Nonenzymatic antioxidant levels and enzyme activity can indicate plant contamination tolerance and cleanup capabilities.

Contaminant distribution and deposition Aside from the features mentioned earlier, the decontamination ability of OPs is intimately tied to the antioxidant system's reactions as well as the distribution and accumulation of pollutants in cell walls, vacuoles, and metabolically dormant tissues.

After considering the parameters listed above, numerous ornamental plants exhibit significant properties that render them viable restorative species due to their short growth cycles and quick expulsion of toxins. Any can detect and store physiological reactions to stress after flowering (<1 year). All of these factors combine to give an ornamental species a high aesthetic value as well as decontaminating properties (Rocha et al. 2022).

3 Mechanism/Methods of Remediation and Species Suitable for These Techniques

3.1 Phytoextraction

It is additionally referred to as phytoaccumulation or phytosequestration, and it primarily involves the building up of maximal levels of contaminants (at or above 0.1% of dry weight) through root systems to the different tissues of ornamental plants derived from water and the soil. It is the most widely recognized and used phytoremediation approach for eliminating hazardous metals from polluted areas. Plant selection parameters are based on the degree of pollutant transfer from roots to shoots. Ornamental plants having strong growth rates (huge amounts of plant biomass) and deeper root systems are thought to be best for this purpose (for example, *Eucalyptus* sp.). This strategy is mostly applied in the case of HMs extraction, so in that concern, it can be divided into two methods: (i) *continuous phytoextraction*, which uses endemic plants with inherent abilities to accumulate high contents of HMs (hyperaccumulators), (ii) *induced phytoextraction* enhance plant metal colligate by addition of chemical substrates like chelate. The idea of using floras to extract metals from contaminated soil was developed and introduced by Utsunamyia (1980) and Chaney (1983), for instance, *Alternanthera bettzickiana* found that it accumulates twice as much Cu in shoots than control when added chelates. The bioaccumulation factor (BAF) or biological accumulation coefficient (BAC) and phytoextraction efficiency (PE) can be evaluated by calculating the metal concentration or nutrients in the harvestable biomass vs. the metal concentration in the soil where the species grown, as illustrated in the equation (Zhou et al. 2016):

$$BAF \text{ or } BAC = \frac{\text{Value of metal in the shoot of plant}}{\text{Value of metal in the soil}}$$

while,

$$\text{Phytoextraction efficiency}\left(PE\%\right) = \frac{\left(\text{C} - \text{plant} \times \text{M} - \text{plant}\right)}{\left(\text{C} - \text{soil / sludge} \times \text{M} - \text{rooted zone}\right)} \times 100$$

Here, M-plant is the biomass of the aboveground plant collected in one harvest, C-plant is the value/concentration of metal in the plant biomass, M-rooted zone is the soil volume weight under the rooted portion of the species, and C-soil is the value/concentration of metal in the same soil volume.

3.2 Phytostabilization/Phytoimmobilization

It is also known as in situ or in place deactivation; in this strategy, contaminants are retained by plant roots or deposited by root secretions (phytostabilization), reducing the flow of pollutants and preventing their migration to groundwater (phytoimmobilization). The main qualities of plants ideal for this technique are their morphology and root depth, as well as maximum interaction with the soil matrix by an exceptional root (fibrous) system across the soil. For example, *Salix alba* is an important tree for Cd and Cu stabilization in contaminated soils and rivers.

3.3 Phytofiltration

It is an inexpensive method for removing or adsorbing contaminants, usually metals, from water and aquatic wasteland. After following the filtration, the below parameters can be calculated by using the equations:

$$\text{Tolerance index } (TI\%) = \frac{\text{Dry weight of } OPs \text{ planted in remediation place}}{\text{Dry weight of } OPs \text{ planted in control place}} \times 100$$

$$\text{Relative growth } (RG\%) = \frac{\text{Final fresh weight of } OPs \text{ (after remediation)}}{\text{Initial fresh weight of } OPs \text{ (before remediation)}} \times 100$$

$$\text{Bioaccumulation factor } (BAF) = \frac{\text{Metal conc. in } OPs \text{ tissue}}{\text{Metal conc. in solution}}$$

$$\text{Translocation factor } (TF) = \frac{\text{Conc.of metal in leaves and stem}}{\text{Concentration of metal in roots}}$$

$$\text{Enrichment coefficient } (EC) = \frac{\text{Conc.of metal in experimental } OPs}{\text{Conc.of metal in control } OPs}$$

$$\text{Root tolerance index } (TI)\% = \frac{\text{Roots length of experimental } OPs}{\text{Roots length of control } OPs} \times 100$$

It can be done by three ways: rhizofiltration, blastofiltration, or caulofiltration, respectively, described the following sections:

3.3.1 Rhizofiltration

It is a procedure for removing contaminants from water and liquid waste that involves precipitating pollutants onto surface roots or absorbing soluble chemicals into the roots. This approach relies heavily on fibrous root systems with huge surface areas. OPs that can take away a high amount of water from the soil are ideal for this use. It acts through a variety of physical and biochemical procedures, including adsorption, precipitation, rhizodegradation, and bioaccumulation. The remediation with ornamental wetland plants is an eco-friendly, visually striking, economic, reliable, and robust wastewater treatment technology that is helpful in sweeping up nutrient overloads, toxic metals, and organic pollutants present in municipal solid waste, industrial wastewater, wastewaters, aquaculture, agricultural, and dairy farm effluent, storm runoff, sludge, pharmaceutical waste, mine drainage, as well as polluted river and lake waters. For example, *Typha latifolia* is efficacious in removing methyl parathion (MeP) from hydromorphic soils like the above-contaminated areas.

3.3.2 Caulofiltration

This word came from the Latin word *caulis,* which means shoot and filtration of solvent from the solute. This process existence was first proposed by Mesjasz-Przybyłowicz et al. in 2004. It is the process when excised plant shoots were used for the excavation of heavy metals from polluted water; by this method, metals could easily preoccupied, and thus, their leaching into underground water finally be reduced.

3.3.3 Blastofiltration

This concept is more similar to that of rhizofiltration, but within blastofiltration, ornamental plant saplings can be utilized for the removal of heavy metals from aqueous matter. It may prove to be a substitute ornamental plant-based wastewater treatment technology that takes advantage of the rapid increase in surface area-to-volume ratio following germination and the ability of the seedlings used to absorb large amounts of heavy metals, ensuring this method is fit for treating wastewater. This process involves chemisorption, complexation, ion exchange, micro-precipitation, hydroxide condensation on a bio-surface, and adhesion to the surface.

3.4 Phytovolatilization

It is the process of uptake of the contaminants by the roots and translocated to the shoot to convert them into less harmful volatile forms and further volatilized into the atmosphere through the leaves stomata via transpiration. It is mostly used for eminently volatile, especially organic pollutants and sometimes inorganic and heavy metals. For example, *Arabidopsis thaliana* can convert organic Hg salts to a volatile form. The main leverage of this technique is that it helps in excavating the pollutants from the site beyond the need for plant reaping and disposition, as with the other approach. According to Limmer and Buken (2016), phytovolatilization is classified into two types: Direct and indirect. Direct phytovolatilization occurs when plants absorb volatile toxins from polluted soil or water and emit them into the surrounding environment by transpiration of the trunk, stem, and leaves. In the case of indirect phytovolatilization, plant root activities cause a flux of volatile contaminants from the subsoil. In this way, flora transfers massive amounts of water (globally, ~62,000 km^3/yr) while simultaneously exploring enormous quantities of soil.

The below equation calculates the rate of direct phytovolatilization:

$$\text{Direct phytovolatilization}\left(DP\right)\% = \frac{\text{Mass of shoot phytovolatilized}}{\left(\text{Mass of shoot phytovolatilized} + \text{Mass of harvestable part in plant}\right)} \times 100$$

3.5 Phytodegradation

It encompasses the breakdown of contaminants ingested by plants via metabolic pathways or the dissolution of pollutants outside the plant by root-produced enzymes (such as dehalogenases, nitro reductases, and peroxidases). Many ornamental plants can synthesize enzymes that catalyze and improve the degradation of organic pollutants. That is why organic contaminants are split down into simpler molecular forms by different enzymes, so the process is termed an *explanta* metabolic process, and these molecules are ingested into plant cells to aid their growth (McGrath and Zhao 2003). Therefore, green plants are considered "green livers of the biosphere." Through this process, plants can transform highly hazardous pollutants into nonhazardous components. For example, genetically modified *Liriodendron tulipifera* (yellow poplar) have higher efficiency in transforming immensely deleterious Hg^{2+} to a less harmful Hg^{o}. Similarly, *Populus deltoid* plants could transform (cyclonite or hexogen) : hexahydro-1,3,5-trinitro-1,3,5-triazine (RDX) into metabolic components, while rhizomes of *Typha* found effective against contaminated wetlands with Terbuthylazine (TER).

3.6 Rhizodegradation

This process is also referred to as microbe-assisted phytoremediation, phytostimulation, plant-assisted bioremediation, plant-aided in situ biodegradation, rhizoremediation, and improved rhizosphere biological degradation. In this process, plant plant-supplied substrates promote the establishment of microbial communities in the root zone, therefore breaking out organic contaminants in the soil. The rhizosphere of plants offers an exclusive environment for microorganisms, which are able to convert dangerous contaminants into benign and nontoxic substances via their metabolic processes. Compared to phytodegradation, this method is slow. Biodegradation occurs as an outcome of nutrients supplied by the plant in the form of exudes from the root. The mere existence of such fluid can boost the number as well as productivity of microorganisms within the root zone, resulting in a faster breakdown of organic contaminants in the environment.

3.7 *Phytodesalination*

It involves the process of removing salts from saline soils using halophytic ornamental plants, allowing the soils to help with normal plant development. Compared to glycophytic plants, halophytic ornamental plants may be more suited to handle heavy metals. Through processes like osmotic regulation and ion separation within their vacuoles, these plants have evolved to survive in saline environments like seawater, saltwater marshes, or salt deserts (Fig. 1).

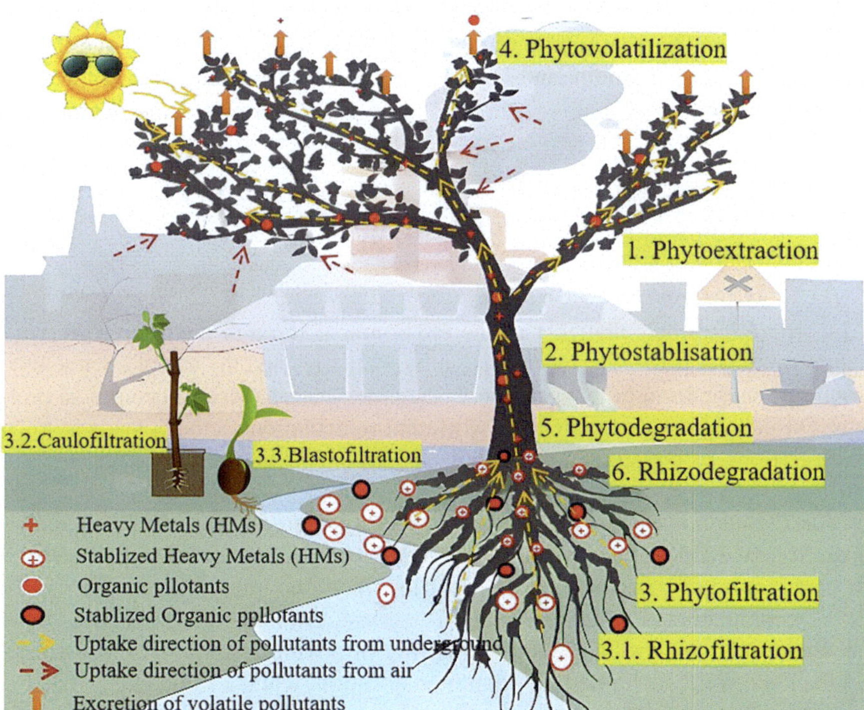

Fig. 1 Schematic illustration showing the mechanism of remediation of different types of pollutants

The success of implementing a mechanism for remediation depends on the selection of plants. Here, the ornamental plants should be selected so that they are tolerant or resistant towards the particular pollutant and adapt themselves to the same environment (Table 1). These OPs should be effective in proper remediation and provide economic, social, and environmental benefits along with the remediation of the pollutants. We have a wide range of ornamentals, so we can try to use each OPs for remediation. Many OPs all found effective for different remediation techniques, as enlisted in Tables 1 and 3. The different OPs responsible for the remediation of various types of pollutants depend on their physiology, morphology, microbial interactions, stress tolerance by other pollutants, and different contaminant distribution and deposition in their biomass as discussed above in potential features of OPs species to be used in remediation programs. As ornamental plants are nonedible and are used only for decoration purposes, the pollutant accumulation will be not very hazardous for other faunas.

4 Types of Contaminants or Pollutants

Our surroundings contain a variety of toxins and pollutants. It extends from soil, water, and up to air, nowhere is free from pollution and pollutants. There is a massive collection of OPs that help in remediating heavy metals, organic contaminants, and volatile organic compounds or air pollutants from the environment as described below:

4.1 Water and Soil Pollutants

A. *Heavy Metals or Trace Elements*

Incorporating trace elements in the environment through anthropogenic ways such as mining and wastewater expulsion by industries and the use of chemicals in agriculture in the form of fertilizers and pesticides leads to serious ecological problems. Despite this, the use of OPs by different remediation methods leads to lower concentrations of contaminants from the environment and also protects the soil fertility.

Bioaccumulation factor (BCF) is a vital factor in determining phytoremediation capability; BCF >1 implies considerable metal transfer from soil to plants. Within the same period of time readings equal or less than one imply that the plants may not be accumulators or metal excluders. The translocation factor (TF) is the second accessible index for evaluating the possibility of phytoextraction by OPs; a value less than one suggests limited pollutant translocation, while a value larger than one implies higher phytoextraction potential. The enrichment coefficient (EC) is the third major component determining the movement of pollutants from the environment to the plants; values ranging from 2 to 5 indicate moderate, 5 to 40 are

Table 1 Various ornamental plants responsible for the remediation of different pollutants by different mechanisms

The phytoremediation category and ornamental plant species associated	Pollutants
1. Phytoextraction	
Alternanthera bettzickiana	Cu
Althaea rosea	Pb, Ni, Cd, Cu
Mirabilis jalapa	PAHs with Cd
Pelargonium hortorum	Pb
Petunia hydrida	Cu, Cd, Ni, Cr, Pd
Quercus robur	Cd
Salix alba	Cd/Cu
Iris lactea and *Iris tectorum*	Cd
2. Phytostabilization/ Phytoimmobilization	
Atriplex halimus	Cd
Acanthus ilicifolius	Cd
Euphorbia sp.	Cd, Cu, Mn, Pb, Zn
Celosia cristata, Helianthus annus, Tagetes patula	Fe, Ca, Cr, Pb, Cu, Zn
3. Rhizofiltration	
Salvinia molesta	Cr/Zn/As
Phragmites australis	U
Genipa americana	Cr
Iris pseudacorus	Cr & Zn
Nymphaea spontanea	Cr
Calendula alata	Cs, Pb
Calendula officinalis	Cd
4. Phytovolatilization	
Arundo donax	As
Stanleya pinnata	Se
Zantedeschia aethiopica	Fe
Helianthus annus	Cd, Zn & Hg
5. Phytodegradation	
Rhizophora mangle	PAHs
Liriope muscari	MCP
Salix viminalis	TCE
Myriophyllum spicatum	TNT
Liriodendron tulipifera	Hg
6. Phytodesalination	
Suaeda maritima	NaCl
Sesuvium portulacastrum	NaCl

[a]*PAHs* polycyclic aromatic hydrocarbons, *MCP* 1-Methylcyclopropene, *TCE* trichloroethylene, *TNT* 2,4,6-trinitrotoluene

Table 2 Categories of ornamental plants to remediate metals from soil/sludge

Metal excluders	Metal indicators	Metal accumulators
Thoroughly restrict overall concentrations of heavy metal translocation throughout tissues as well as keep very moderate concentrations of heavy metal in its biomass.	Control the quantities of heavy metal translocation inside them as well as collect heavy metals within their aboveground OP parts.	Enhance the rate of heavy metal absorption throughout them in order to concentrate metals in their aboveground OP parts.
Safeguarding leaf tissues, particularly metabolically active photosynthetic cells, avoiding heavy metal injury.	These plant species perish when exposed to heavy metals for an extended period of time.	Whenever plants continue to absorb metals, they will ultimately perish.
The proportion of the metals present in plant: soil is $\ll 1$.	The proportion of the metals present in plant: soil ~1.	The proportion of the metals present in plant: soil >1.
Examples: *Morus alba*, *Populus deltoides*, *Salix viminalis* L., *Quercus laevis*, *Leucaena leucocephala*.	Examples: *Tagetes patula* L., *Tagetes erecta* L., *Euphorbia milii*, *Elsholtzia splendens*.	Examples: *Prosopis glandulosa*, *Azadirachta indica*, *Dendrocalamus strictus*.

substantial, and values greater than 40 are exceptionally high EC. Last but not least, the tolerance index (TI) is derived by comparing the total length of roots in plants that undergo mental stress to the baseline circumstance. According to their capacity to eliminate metals, OPs are categorized as metal excluders, indicators, or accumulators, as shown in Table 2.

Among all the metal accumulators, especially hyperaccumulators, are found best for remediation. Metal hyperaccumulators are defined as plants that accumulate higher than 10 mg Hg kg^{-1}, 100 mg Cd kg^{-1}, 1000 mg Pb/Cr/Co kg^{-1}, or 10,000 mg Ni/Zn kg^{-1}. Worldwide, cadmium (Cd) is the largest found trace material that causes contamination, followed by lead (Pb) and arsenic (As) are also a worldwide concern for humans and are contaminated through air, food, and drinking water. Many OPs are found to be tolerant of these trace elements and can easily be overtaken by these plants. This tolerance is linked to enhanced antioxidant enzyme activity, including ascorbate peroxidase (APX), glutathione reductase (GR), and CAT. These enzymes remove surplus H_2O_2 while retaining intracellular thiol-redox balance. Following the triggering of enzymes like SOD, peroxidase (POD), and CAT, OPs subjected to metals are benefited due to the activation of various additional protective systems, such as oxide reducers, lignin peroxidase, laccase, veratryl alcohol oxidase, tyrosinase, and azo-reductase, which aid in plant survival under stressful circumstances.

Remediation not only limits the utilization of garden and flowering plants for aesthetic elegance but it also assists in annuals and perennials to enhance both soil and air health (through carbon sequestration and dendro-remediation). This protects the native wildlife and maintains temperature balance by expelling water vapor. Additionally, ornamental trees can be used to produce timber, pulp, and paper (Khan et al. 2021). Consequently, OPs have acquired popularity in detoxifying heavy metals coming from polluted locations, particularly in metropolitan industries wherein OPs enhance visual appeal, cleanliness, and landscape infrastructure.

B. *Organic Contaminants (OCs)*

Organic contaminants are manufactured or natural substances that are not yet being observed in the surroundings or are still being analyzed but possess a high potential to penetrate the ecosystem and harm individuals or biodiversity. Organic pollutants are among the two most toxic to biodiversity. OCs comprise medicines, pesticides, solvents, fertilizers, household cleaners, PAHs, and hormones (Rathi et al. 2021). Delgado et al. (2020b) found that effluent can release OCs in amounts that vary from ng L^{-1} to mg L^{-1} if dumped inappropriately onto urban soil or aquatic bodies. Tables 3 and 4 show that OPs can successfully remove developing pollutants.

Table 3 List of trace element pollutants and ornamental plants used for remediation

Pollutants	Ornamental plants are used for remediation	Exposition roots mg kg^{-1}	Concentration in biomass	References
1. Heavy metals				
Cd	*Alcea rosea* L	100	236	Liu et al. (2009)
	Chrysanthemum maximum	10 mg L^{-1}	384	González-Chávez and Carrillo-González (2013)
	Chlorophytum comosum	200	2387.5	Wang et al. (2012)
	Tagetes erecta L.	100	185.22	Liu et al. (2018)
	Zinnia elegans Jacq			
	Tagetes patula L.			
	Gerbera jamesonii			
	Mirabilis jalapa L	30	40.18	Li et al. (2020)
	Echinodorus cordifolius	8 mg L^{-1}	6141.60	Woraharn et al. (2021)
	Heliconia psittacorum L.f.	–	2750.00	
Pb	*Antirrhinum majus* L	1000	709.81	Cui et al. (2013)
	Viola × *wittrockiana* Gams	1000	732.27	Sychta et al. (2020)
	Helianthus annuus L	1000	618.60	Chauhan et al. (2020)
	Calendula officinalis L	1000	15,213.2	Lu et al. (2020)
	Miscanthus sinensis Andersson	500	180.62	Song et al. (2020)
	Hemerocallis fulva L	1000	1089.74	
	Chrysanthemum maximum var. Shasta	100 mg L^{-1}	115	González-Chávez and Carrillo-González (2013)
Sn	*Tagetes erecta* L.	2000	106.10	Liu et al. (2021)
Zn	*Echinodorus cordifolius* (L.) Griseb	40 mg L^{-1}	8766.60	Woraharn et al. (2021)
	Zinnia elegans L	75	29.64	Panda et al. (2020)
	Tagetes erecta L.	12.5 mg L^{-1}	15900.00	Coelho et al. (2017)
Cu and Ni	*Chrysanthemum maximum* var. Shasta	10 mg L^{-1}	398.00	González-Chávez and Carrillo-González (2013)

Table 4 List of organic contaminants as pollutants, ornamental plants, and their efficiency used for remediation

Ornamental plants	Pollutants	Exposition	Removal (%)	References
2. Organic contaminants (OCs)				
2.1. Wastewater				
N	*Canna indica* L	14.56 mg g^{-1}	>57.4	Li et al. (2013)
P	*Iris pseudacorus* L.	1.93 mg g^{-1}	>37.06	
COD	*Canna flaccida, Agapanthus africanus, Zantedeschia aethiopica, Chrysopogon zizanioides, Watsonia borbonica, Spathiphyllum wallisii, Zantedeschia aethiopica, Canna indica.*	481 mg L^{-1}	>99	Calheiros et al. (2015)
BOD		182 mg L^{-1}	>98	
PO_4^{3-}		19.1 mg L^{-1}	>92	
NH^{4+}		21.6 mg L^{-1}	>84	
TSS		283 mg L^{-1}	87	
NO_3		11.95 mg L^{-1}	>59	Sandoval-Herazo et al. (2018)
Fecal coliforms		3319.31NMP 100 mL^{-1}	>65	
NO_3_N	*Chrysopogon zizanioides*	29 mg g^{-1}	89	Panja et al. (2020)
2.2. Pharmaceuticals				
Acetaminophen	*Helianthus annuus* L	15 mg L^{-1}	50	Ryšlavá et al. (2015)
Carbamazepine	*Canna flaccida* Salisb	5905.5 mgL^{-1}	100	Hwang et al. (2020)
		234.5 mg L^{-1}	>36.7	
	Iris sibirica L	15 mg L^{-1}	31.1	Tejeda and Zurita (2020)
	Cyperus haspan L.	0.2 mg L^{-1}	10	Delgado et al. (2020a)
Caffeine and Ibuprofen	*Heliconia rostrata* Ruiz & Pav	27 mg L^{-1}	97	de Oliveira et al. (2019)
Sildenafil	*Heliconia* spp.	0.2 mg L^{-1}	>97	Delgado et al. (2020a)
Ciprofloxacin	*Chrysopogon zizanioides* (L.) Roberty	10 mg L^{-1}	94	Panja et al. (2020)
Tetracycline	*Mirabilis jalapa* L, *Tagetes patula* L.	5 mg kg^{-1}	99	Li et al. (2020)
	Chrysopogon zizanioides (L.) Roberty.	10 mg L^{-1}	100	Panja et al. (2020)
2.3. Pesticides				
Organochlorine	*Eichhornia crassipes* (Mart.) Solms	1 mg L^{-1}	>62	Riaz et al. (2017)
Pyrethroid	*Pistia stratiotes* L.	1 mg L^{-1}	>72	
Atrazine	*Canna flaccida* Salisb	226.5 mg L^{-1}	>91.7	Hwang et al. (2020)
2.4. Other compounds				
Petroleum hydrocarbons	*Iris dichotoma* Pall and *Iris lactea* Pall.	40,000 mg kg^{-1}	30.79	Cheng et al. (2017)

(continued)

Table 4 (continued)

Ornamental plants	Pollutants	Exposition	Removal (%)	References
Methylparaben	*Cyperus haspan* L., *Heliconia* spp.	0.2 mg L^{-1}	>97	Delgado et al. (2020a)
3. Gaseous/air pollutant				
Formaldehyde	*Hedera helix* L.	1.0 mg kg^{-1}	70	Lin et al. (2017)
	Chamaedorea elegans Mart.	16.4 mg m^{-3}	100	Teiri et al. (2018)
Benzo [a] pyrene	*Tagetes patula* L	5 mg kg^{-1}	79.2–92.4	Sun et al. (2011)
Benzene	*Dracaena sanderiana* Sander ex Mast.	18 mg kg^{-1}	46	Treesubsuntorn and Thiravetyan (2012)
	Schefflera arboricola (Hayata) Merr.	29.5 mg m^{-3}	94	Parseh et al. (2018)
	Spathiphyllum wallisii Regel		93	

[a]*BOD* Biochemical oxygen demand, *PAH* polycyclic aromatic hydrocarbon, *COD* chemical oxygen demand, *TSS* total suspended solids

4.2 Air Pollutants

Rapid development and industrialization lead to increased pollution, especially air pollution from industries, transport, and indoor concrete jungles. OPs could indicate, monitor, and mitigate environmental gaseous containment (Parseh et al. 2018). Other than major air pollutants in the environment (including nitrogen dioxide [NO_2], particulate matter [PM], carbon monoxide [CO], sulfur dioxide [SO_2], and ozone [O_3]), many carcinogenic volatile organic compounds (VOCs) are also present in indoor air, which affect different flora and fauna health. These VOCs include acrylamide, acrolein, vinyl chloride, acrylonitrile, ethylene oxide (EtO), formaldehyde, benzene, 1-bromopropane, carbon disulfide, 1,3-butadiene (BD), tetrachloroethylene (TTE), propylene oxide, toluene, styrene, xylene, and trichloroethylene (by USPA-United States Environmental Protection and Agency). These pollutants and VOCs, which can be mitigated by different OPs, thus provide a viable green alternative to enhance the quality of the air (Table 3). Through various remediation techniques, flora can directly (by absorption) or indirectly (by biotransformation by rhizosphere-associated bacteria) promote the reduction or elimination of such chemicals. The rate of absorption is directly influenced by the length of time that stomata open or close as well as the many pathways that allow for the absorption of volatile organic compounds. As a result, species with diverse photosynthetic processes (C_3, C_4, and Crassulacean Acid Metabolism (CAM)) should have varying levels of VOC elimination effectiveness, with facultative CAM species likely to have higher levels of efficiency. In the dark, CAM species absorb CO_2, which is linked to a greater absorption of numerous volatile organic compounds during this time. Low concentrations of these VOCs may result in minor alterations within the chloroplasts, intercellular space, vacuole, mitochondrion, etc., whereas higher levels enable plants to show visible symptoms, including hydrolysis, holonecrosis, yellow leaf tips, and permanent anatomical changes.

Formaldehyde represents the most prevalent VOC produced from common household items and is linked to health hazards. In order to choose the most suitable OPs for controlling pollution in the indoor environment, one should consider the traits of morphological resistance (leaf damage rate, leaf damage time, and survival rate) when subjected to contaminants (Liu et al. 2022). In a similar manner, Wolverton et al. (1984) at NASA's Life Science Researchers Centre examined the application of decorative potted plants for minimizing VOCs in the indoor environment. The potential of potted plants for phytoremediation of significant indoor levels of formaldehyde, trichloroethylene, and benzene was investigated in the present investigation. Although the epiphytic bacterial communities found in host plant foliage can contain genes implicated in hydrocarbon breakdown, the characterization of microbes linked to the philosopher is an additional essential characteristic (Franzetti et al. 2020). Nevertheless, with decreasing phytoextraction ability, heavy metal pollution frequently coexists alongside organic pollutants in ecosystems, and higher levels of pollutants have a detrimental effect on plant development, biomass, germination, and survival (Reddy et al. 2020). Ornamental plants are often utilized on green roofs, vertical gardens, interior landscaping, etc., that utilize rainwater to cool down cities and are additionally employed to give businesses the opportunity to trade carbon credits. This helps decrease air pollution in urban areas.

5 Hasten the Mechanism of Remediation by Ornamental Plants

OPs have been found effective and more appropriate for the remediation of pollutants than other food crops. Nonetheless, the effectiveness of OPs for remediation can be improved by using various aids or amendments such as synthetic and natural additives, microorganisms, and genetic engineering. These aids promote OP resistance and tolerance to various pollutants, the creation of pollutant-chelate complexes, and the transport and deposition of pollutants within organs, as mentioned hereafter (Kafle et al. 2022). These all together increase overall remediation efficiency by OPs.

5.1 Amendments

These are some aids provided to OPs via soil, water, or air to enhance remediation rate mainly by improving translocation efficiency, plant surge, water and nutrient retention, and vitalization efficiency. These amendments can be natural or synthetic. The natural ones are developed by processing natural materials such as sugar beet baggage, paper, sugarcane byproducts, crop straw, sewage sludge, bio-char, compost, peat moss, bio-solids, vermicomposting, and fertilizers, while chemical amendments include synthetic chemicals such as ethylenediaminetetraacetic acid (EDTA), sodium dodecyl sulfate (SDS) and ethylene glycol tetraacetic acid (EGTA), and some organic amendments. For example, when sunflowers were grown for remediation, wastewater sludge raised the metal bioavailability of Cd and Ni, and chelating agents such as EDTA and EGTA were added to *Alcea rosea*, *Mirabilis jalapa*, and *Calendula officinalis* to increase metal uptake and remediation.

5.2 *Microbial Enhanced/Stimulated Remediation by Ornamentals*

Microbes can influence the phytoremediation process by degrading organic containment and uptake by microbial-mediated HMs. This could be activated by biostimulation (i.e., adding nutrients to damaged soil stimulates existing microbes to break down hazardous contaminants) and bioaugmentation (adding natural or genetically modified microbes to polluted soil can eliminate toxic pollutants). An example of bioaugmentation is 2,4 dinitrotoluene degradation with *Pseudomonas fluorescens* to reduce its toxicity for good growth of *Arabidopsis thaliana.*

5.3 *Nanoparticle-Assisted Phytoremediation*

This unique technology improves pollution removal efficiency through several strategies such as adsorption/redox reactions, stimulating plant development, and aiding remediation. NPs promote plant growth by improving rhizospheric microbes and fungi; these have a beneficial impact on increasing phytoremediation. For example, the use of zero-valent iron (nZVI) NP augments the antioxidative system and phytoextraction efficiency of Cd and Pb, in *Boehmeria nivea* and *Lolium perenne*, subsequently.

5.4 *Role of Genetic Engineering in Phytoremediation*

It has been demonstrated as an intriguing option for enhancing the ability of OPs to remediate pollutants. To develop genetically modified OPs, a foreign gene source from another living creature (such as plant species, bacteria, or animal) needs to be introduced and inserted within the genetic material of the intended decorative species. Here, the modified plants with desirable traits (pollutants/heavy metal tolerance and accumulation mechanisms in plants) can be remedied in a much shorter time with more efficiency, such as hyperaccumulators (Table 5).

Table 5 Genetically modified ornamental plants (GMOPs) responsible for remediating the pollutant

Targeted GMOPs	Genes responsible	Enzymes expressed	Remediation benefits
Arabidopsis thaliana	gsh1, AsPCS1	Glutathione synthase and phytochelatin synthase	Enhanced Cd and As tolerance
Festuca arundinacea	PaGCS, PaPCS	–	Cd
Sedum alfredii	SaNramp1	Iron-regulated transporter	Enhanced Zn uptake
Sedum plumbizincicola	HMA4, HMA2	P1B-type ATPase transporters	Cd hyperaccumulation
Populus sp.	ECS	g-glutamylcysteine synthetase	Enhanced Cd flux and detoxification
Petunia	CAXcd	–	Cd accumulation enhanced
Petunia hybrida	RsMYB1	–	Cu, Cr, Mn, and Zn
Pelargonium sp.	HPS, PHI	3-hexulose6-phosphate synthase and 6-phosphate-3-hexuloisomerase	Increased formaldehyde (CH_2O) tolerance
Dianthus superbus	NtGST	–	Improved copper (Cu) tolerance and accumulation
Populus alba	ScYCF1	–	As, Cd, Cr, Cu, Na, Ni, Pb, and Zn

6 Case Study: The Wastewater Remediation Project by Using Ornamental *Typha latifolia* Grass

In India, one of our centers, the Water Technology Centre (WTC) constructed a permanent wetland for long-term sludge/wastewater treatment at ICAR- Indian Agricultural Research Institute, New Delhi (a premier agricultural research institute of India) in 2012, having capacity of 2.2 MLD (mega liter day- (Chaney 1983)) on an area of 1.42 ha with a capital cost of ₹1.2 crores and operations and maintenance costs of ₹1335/day. The idea behind this project (Fig. 2a) is to provide an alternate water source for providing water to the agricultural fields institute. For this purpose, wastewater from nearby Loha-Mandi and Krishi-Kunj colony drains is collected in a tank (Fig. 2b) and then transported to a wastewater treatment plant comprised three cells for treatment (each of 80 × 40 m) with *Typha latifolia* grass grown on the floor with a layer of sand, gravel, and stone (Fig. 2c, d), here consecutively primary (cell 1), secondary (cell 2), and tertiary (cell 3) treatments take place one by one. The movement of wastewater from another cell is regulated by complete sub-surface flow and collected in a tank (Fig. 2e). This treatment can remediate turbidity by 99%, BOD by 87%, nitrate by 95%, phosphate by 90%, and heavy metal by 81–99%. All of these processes require less than 1% energy, no chemical use, and no sludge regeneration.

Fig. 2 Pictorial representation of wastewater treatment by *Typha latifolia* at ICAR-IARI, New Delhi, under a project that seeks to provide an alternative source of water for irrigation after remediating drainage wastewater. Here, (**a**) Project details, (**b**) grit chamber, (**c**) primary treatment cell with Typha grass, (**d**) closer view of *T. latifolia* root extracting pollutants from sewage, and (**e**) treated/remediated water collection tank

7 Post-remediation Disposal Legislation for the Harvestable OPs

OPs are found to be best for remediating pollutants other than food crops. Here the major concern required is the last and efficient disposal or utilization of mass developed as a result of bioremediation, especially in the case of HMs and toxic pharmaceutical chemicals. Therefore, following the plant biomass utilization/disposal legislation after remediation is necessary. For this purpose, there are many legal codes, acts, and rules formulated all around the world. Nevertheless, no acts, regulations, or laws are really concerned with post-remediation of developed biomass decomposition or use. Among the most important codes of acts for the mobility of toxic waste are:

- Management and Transboundary Movement of Hazardous and Other Wastes Rule, 2016 in India
- Hazardous Substance Rule, 2013 in Pakistan
- Waste Control Act in South Korea
- "Urban Waste Water Treatment Directive" and "Sewage Sludge Directive" in Europe
- "Solid Waste Disposal Act" in USA

8 Advantages, Limitations, and Disadvantages of Remediation by OPs

Remediation by OPs has gained tremendous popularity in recent years because of its less impact on other fauna. Other than that, its prior advantage is that it is approximately ten times cheaper than other methods. Some of the advantages are listed below:

- OPs along with aesthetic and mental pleasing effects, help in the remediation of the broad range of pollutants found in the environment, therefore acting as *solar energy-driven cleanup technology.*
- There is very little chance to use OPs as food material, so the transfer of toxic pollutants from the biomass of OP will not transfer to other tropic levels of the food chain.
- These are the easiest ways to do remediation because OPs are easy to grow without much effort (no harm from other animals) and easy to monitor the site.
- OPs could readily utilized for much larger-scale cleanup operations than other chemical or microbial methods.
- Remediation with OPs is the cheapest way and has no negative impact on other flora, fauna, and the environment.
- Remediation with OPs causes less environmental disturbance, and in situ treatment secures topsoil that is why acts as a soil stabilizer.
- Remediation with OPs secures wildlife ecology and has very more communal recognition.

Despite the fact that remediation is a low-cost, ecologically benign, and widely accepted technology for removing industrial waste from contaminated areas, it has various drawbacks and limits that may make it unsuitable. These are listed below:

- Remediation with OPs is species, pollutant, and climate dependent, and slower than other methods.
- Some OPs are edible by some fauna or are used in compost making, as these contain toxic chemicals and HMs after remediation, so they may pose a risk to biomagnification. They can cause harm at different tropic levels.
- A larger area is required for remediation by OPs.
- Toxic pollutants can be discharged back into the ecosystem through litter fall.
- Remediation using OPs is a time-consuming process that may require multiple growing seasons to complete.
- Some contaminants are difficult to remediate due to their large molecular weight, such as metals that are strongly linked to organic materials or those located far from the root zone.

9 Conclusion

In conclusion, we can say that OPs are more excellent for remediating pollutants/contaminants from environments than other food crops because there are very few chances to enter the toxic contaminant in the food chain of different tropic levels (biomagnification). These kinds of plants are liable for the beautification (from flowers as well as foliage and trunk) and provide revenue in rural areas (ecotourism) and metropolitan societies (vertical, roof, and indoor gardening), and along with can also be used to remediate pollution from industrial and populated areas. In the current era and future programs related to mitigating the pollutants/contaminants are going to increase, and as the population is going toward clean environment creation for the next generation OPs must come at first position for this purpose because of their advantages mentioned above. NASA has also proved the effectiveness of OPs in phytoremediation and many efforts have been made in India also, as given in the IARI case study. Some techniques like adding amendments, microbes, nanoparticles, and genetic engineering can accelerate the remediation speed. Whether the use of OPs for remediation is safe but following different policies, rules, acts, and codes for its biomass decomposition needs to be followed, and some more particular rules need to be enacted to avoid biomagnification.

References

Calheiros CSC, Bessa VS, Mesquita RBR et al (2015) Constructed wetland with a polyculture of ornamental plants for wastewater treatment at a rural tourism facility. Ecol Eng 79:1–7

Chaney RL (1983) Plant uptake of inorganic waste. In: Land treatment of hazardous wastes. Noyes Data Corp, Park Ridge, pp 50–76

Chauhan P, Rajguru AB, Dudhe MY, Mathur J (2020) Efficacy of lead (Pb) phytoextraction of five varieties of *Helianthus annuus* L. from contaminated soil. Environ Technol Innov 18:100718

Cheng L, Wang Y, Cai Z et al (2017) Phytoremediation of petroleum hydrocarbon-contaminated saline-alkali soil by wild ornamental *Iridaceae* species. Int J Phytoremediation 19:300–308

Coelho LC, Bastos ARR, Pinho PJ et al (2017) Marigold (*Tagetes erecta*): the potential value in the phytoremediation of chromium. Pedosphere 27:559–568

Cui S, Zhang T, Zhao S et al (2013) Evaluation of three ornamental plants for phytoremediation of Pb-contamined soil. Int J Phytoremediation 15:299–306

Das PK (2018) Phytoremediation and nan remediation: emerging techniques for treatment of acid mine drainage water. Def Life Sci J 3(2):190–196

de Oliveira M, Atalla AA, Frihling BEF et al (2019) Ibuprofen and cafeine removal in vertical fow and free-foating macrophyte constructed wetlands with Heliconia rostrata and Eichornia crassipes. Chem Eng J 373:458–467

Delgado N, Bermeo L, Hoyos DA et al (2020a) Occurrence and removal of pharmaceutical and personal care products using subsurface horizontal flow constructed wetlands. Water Res 187:116448

Delgado C, Feio MJ, Pardo I, Almeida SFP (2020b) Effects of water temperature over benthic diatom communities: insights from thermal springs. Plant Ecol Divers 13(3–4):325–337

Franzetti A, Gandolfi I, Bestetti G, Schioppa EP, Canedoli C, Brambilla D, Cappelletti D, Sebastiani B, Federici E, Papacchini M, Ambrosini R (2020) Plant-microorganisms interaction promotes the removal of air pollutants in Milan (Italy) urban area. J Hazard Mater 384:121021

González-Chávez MCA, Carrillo-González R (2013) Tolerance of Chrysantemum maximum to heavy metals: the potential for its use in the revegetation of tailings heaps. J Environ Sci 25:367–375

Hwang JI, Li Z, Andreacchio N et al (2020) Potential use of foating treatment wetlands established with Canna faccida for removing organic contaminants from surface water. Int J Phytoremediation 22:1304–1312

Kafle A, Timilsina A, Gautam A, Adhikari K, Bhattarai A, Aryal N (2022) Phytoremediation: mechanisms, plant selection and enhancement by natural and synthetic agents. Environ Adv 8:100203

Khan AHA, Kiyani A, Mirza CR, Butt TA, Barros R, Ali B, Iqbal M, Yousaf S (2021) Ornamental plants for the phytoremediation of heavy metals: present knowledge and future perspectives. Environ Res 195:110780

Li L, Yang Y, Tam NFY et al (2013) Growth characteristics of six wetland plants and their infuences on domestic wastewater treatment efciency. Ecol Eng 60:382–392

Li X, Zhu W, Meng G et al (2020) Phytoremediation of alkaline soils co-contaminated with cadmium and tetracycline antibiotics using the ornamental hyperaccumulators *Mirabilis jalapa* L. and *Tagetes patula* L. Environ Sci Pollut Res 27:14175–14183

Lin M, Chen L, Chuah Y (2017) Investigation of a potted plant (*Hedera helix*) with photoregulation to remove volatile formaldehyde for improving indoor air quality. Aerosol Air Qual Res 17:2543–2554

Limmer M, Burken J (2016) Phytovolatilization of organic contaminants. Environmental Science & Technology 50(13):6632–6643

Liu JN, Zhou QX, Wang S, Sun T (2009) Cadmium tolerance and accumulation of *Althaea rosea* Cav. and its potential as a hyperaccumulator under chemical enhancement. Environ Monit Assess 149:419–427

Liu Z, Chen W, He X (2018) Evaluation of hyperaccumulation potentials to cadmium (Cd) in six ornamental species (compositae). Int J Phytoremediation 20:1464–1469

Liu Y, Xu W, Wang Y et al (2021) Growth responses and accumulation characteristics of three ornamental plants to Sn contamination in soil. Agriculture 11:205

Liu Z, Chen Q, Lin M, Chen M, Zhao C, Lu Q, Meng X (2022) Electric field enhanced cadmium accumulation and photosynthesis in a woody ornamental hyperaccumulator—*Lonicera japonica* Thunb. Plan Theory 11(8):1040

Lu W, Li Z, Shao Z et al (2020) Lead tolerance and enrichment characteristics of several ornamentals under hydroponic culture. Bull Environ Contam Toxicol 105:166–172

McGrath SP, Zhao FJ (2003) Phytoextraction of metals and metalloids from contaminated soils. Curr Opin Biotechnol 14:277–282

Mesjasz-Przybylowicz J, Nakonieczny M, Ligula P, Augustyniak M, Tarnawska M, Reimold WU, Koeberl C, Przybylowicz W, Glowacka E (2004) Uptake of cadmium, lead, nickel and zinc from soil and water solutions by the nickel hyper-accumulator *Berkheya coddii*. Acta Biol Cracov Bot 46:75–85

Panda A, Patra DK, Acharya S et al (2020) Assessment of the phytoremediation potential of *Zinnia elegans* L. plant species for hexavalent chromium through pot experiment. Environ Technol Innov 20:101042

Panja S, Sarkar D, Datta R (2020) Removal of antibiotics and nutrients by Vetiver grass (*Chrysopogon zizanioides*) from secondary wastewater effluent. Int J Phytoremediation 22:764–773

Parseh I, Teiri H, Hajizadeh Y, Ebrahimpour K (2018) Phytoremediation of benzene vapors from indoor air by *Schefflera arboricola* and *Spathiphyllum wallisii* plants. Atmos Pollut Res 9:1083–1087

Rathi BS, Kumar PS, Vo DVN (2021) Critical review on hazardous pollutants in water environment: occurrence, monitoring, fate, removal technologies and risk assessment. Sci Total Environ 797:149134

Reddy KR, Chirakkara RA, Martins Ribeiro LF (2020) Effects of elevated concentrations of co-existing heavy metals and PAHs in soil on phytoremediation. J Hazard Toxic Radioact Waste 24(4):04020035

Riaz G, Tabinda AB, Iqbal S et al (2017) Phytoremediation of organochlorine and pyrethroid pesticides by aquatic macrophytes and algae in freshwater systems. Int J Phytoremediation 19:894–898

Rocha CS, Rocha DC, Kochi LY, Carneiro DNM, Dos Reis MV, Gomes MP (2022) Phytoremediation by ornamental plants: a beautiful and ecological alternative. Environ Sci Pollut Res 29:3336–3354

Ryšlavá H, Pomeislová A, Pšondrová Š et al (2015) Phytoremediation of carbamazepine and its metabolite 10,11-epoxycarbamazepine by C3 and C4 plants. Environ Sci Pollut Res 22:20271–20282

Sandoval-Herazo LC, Alvarado-Lassman A, Marín-Muñiz JL et al (2018) Efects of the use of ornamental plants and diferent substrates in the removal of wastewater pollutants through microcosms of constructed wetlands. Sustain For 10:1594

Song X, Zhang C, Chen W et al (2020) Growth responses and physiological and biochemical changes in five ornamental plants grown in urban lead-contaminated soils. Plant-Environ Interact 1:29–47

Sun Y, Zhou Q, Xu Y et al (2011) Phytoremediation for co-contaminated soils of benzo [a] pyrene (B[a]P) and heavy metals using ornamental plant *Tagetes patula*. J Hazard Mater 186:2075–2082

Sychta K, Słomka A, Kuta E (2020) Garden pansy (*Viola* × *wittrockiana* Gams.) — a good candidate for the revitalisation of polluted areas. Plant Soil Environ 66:272–280

Teiri H, Pourzamani H, Hajizadeh Y (2018) Phytoremediation of VOCs from indoor air by ornamental potted plants: a pilot study using a palm species under the controlled environment. Chemosphere 197:375–381

Tejeda A, Zurita F (2020) Capacity of two ornamental species (*Iris sibirica* and *Zantedeschia aethiopica*) to take up, translocate, and accumulate carbamazepine under hydroponic conditions. Water (Switzerland) 12:8–17

Treesubsuntorn C, Thiravetyan P (2012) Removal of benzene from indoor air by *Dracaena sanderiana*: effect of wax and stomata. Atmos Environ 57:317–321

Utsunamyia T (1980) Japanese patent application no. 55-72959

Wang Y, Yan A, Dai J et al (2012) Accumulation and tolerance characteristics of cadmium in Chlorophytum comosum: a popular ornamental plant and potential Cd hyperaccumulator. Environ Monit Assess 184:929–937

Wolverton BC, McDonald RC, Watkins EA Jr (1984) Foliage plants for removing indoor air pollutants from energy-efficient homes. Econ Bot 38:224–228

Woraharn S, Meeinkuirt W, Phusantisampan T, Chayapan P (2021) Rhizofiltration of cadmium and zinc in hydroponic systems. Water Air Soil Pollut 232:204

Zhou H, Yang WT, Zhou X, Liu L, Gu JF, Wang WL, Zou JL, Tian T, Peng PQ, Liao BH (2016) Accumulation of heavy metals in vegetable species planted in contaminated soils and the health risk assessment. Int J Environ Res Public Health 13(3):289

Groundwater Management and Treatment: A Resilient Source for Environmental Protection

Mohamed A. Dawoud and Waleed Abo Al Hassan

Abstract Groundwater serves as a critical resource for drinking water, irrigation, and industrial activities, particularly in regions with limited surface water availability. Effective management and treatment of groundwater are essential for ensuring its sustainability and quality, safeguarding both environmental and economic systems. Groundwater treatment methods, including reverse osmosis, ion exchange, electrodialysis, and advanced oxidation, play a pivotal role in mitigating contamination from nitrates, heavy metals, and industrial pollutants, enhancing its suitability for various applications. These technologies not only improve water quality but also help address public health concerns and support agricultural productivity in water-scarce regions. From an environmental perspective, treating groundwater reduces risks to ecosystems by preventing the spread of contaminants to soils and surface water bodies. Additionally, adopting sustainable groundwater management practices, such as artificial recharge and aquifer remediation, protects biodiversity, and promotes ecosystem resilience. Economically, treated groundwater contributes to long-term cost savings by reducing dependence on alternative and more expensive water sources, while also fostering water security critical for socio-economic development. Furthermore, advanced treatment technologies can create economic opportunities through the development of water treatment infrastructure and innovation. However, achieving these outcomes requires integrated approaches that balance technological, economic, and environmental considerations, emphasizing the need for collaborative efforts among stakeholders to ensure groundwater remains a resilient source for environmental protection and sustainable development.

M. A. Dawoud (✉)
National Water Research Center, El Qanatir El Khayriyah, Egypt

Environment Agency-Abu Dhabi, Abu Dhabi, United Arab Emirates
e-mail: mdawoud@ead.gov.ae

W. A. Al Hassan
Food and Agriculture Organization of the United Nations, Sub Regional Office for the GCC States and Yemen, Abu Dhabi, United Arab Emirates
e-mail: waleed.abouelhassan@fao.org

A. A. H. Abdel Latef et al. (eds.), *Sustainable Remediation for Pollution and Climate Resilience*, https://doi.org/10.1007/978-981-96-5674-5_24

Keywords Groundwater · Natural contaminations · Risk assessment · Remediation · Desalination

1 Introduction

Groundwater constitutes approximately 30% of the world's freshwater resources, yet it remains one of the most overlooked components of the water cycle. As the largest reservoir of liquid freshwater on Earth, groundwater holds between 8 to 10 million km^3, accounting for 96% of the planet's nonfrozen freshwater. It supplies nearly 51% of the world's potable water supply and 42% of agricultural irrigation needs, as shown in Fig. 1. Upon infiltration into the soil, water embarks on a complex subterranean journey, interacting chemically with soil and rock as it moves through the unsaturated and saturated zones. These processes, influenced by natural factors and human activities such as pumping and agricultural contamination, continuously alter groundwater quality. Groundwater is critical for sustaining surface water systems, especially in arid and hyper-arid regions, where it supports rivers during dry seasons. Unlike surface water, groundwater flow is significantly slower, with some water taking months or even years to re-emerge, thereby regulating

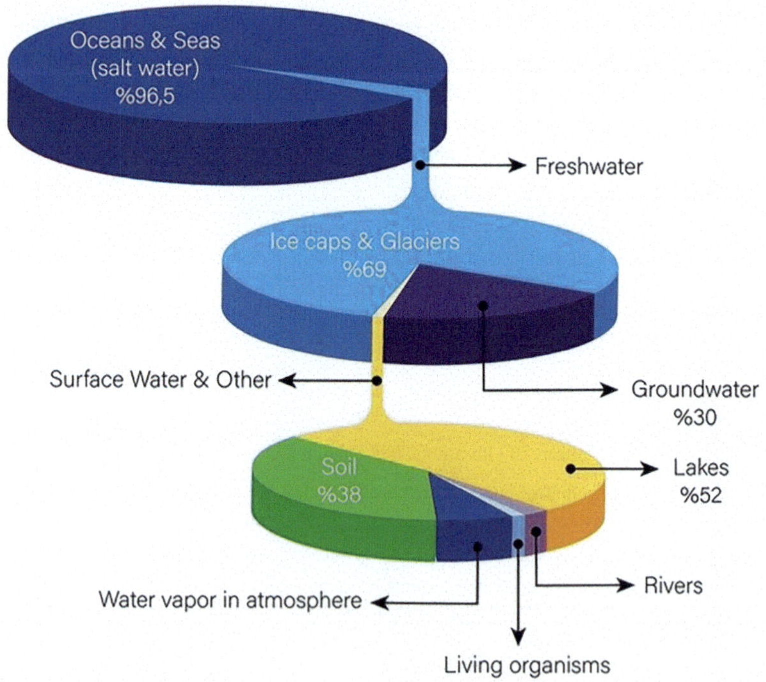

Fig. 1 Global water distribution

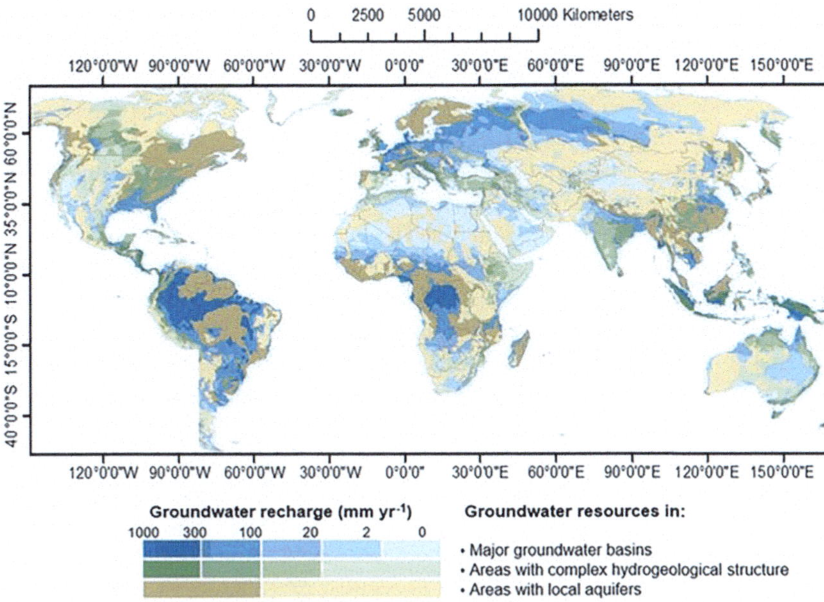

Fig. 2 Overview of global groundwater resources and recharge

surface water availability during severe droughts. Figure 2 highlights the global distribution, quality, and vulnerability of groundwater resources, serving as a vital tool for hydrologists and decision-makers. In regions like the central Mediterranean, substantial recharge occurs in complex local aquifers, though data scarcity underscores the need for further research, planning, and management. This delicate balance faces increasing stress in areas with pronounced disparities between rainy and dry seasons, high evapotranspiration, or rapid development. While some regions maintain sustainable abstraction rates, overextraction elsewhere has led to problems such as land subsidence. In desert areas, minimal recharge results in fossil water, classified as a nonrenewable resource due to its ancient origin. Moreover, groundwater quality varies significantly, affected by geological factors, rock weathering, and water-rock contact time, often leading to brackish conditions. Although groundwater is naturally more resistant to contamination than surface water, remediation is far more challenging and costly once contamination occurs (Machiwal et al. 2018).

The availability and quality of groundwater are fundamental to human health, livelihoods, food security, and economic development, while also sustaining aquatic ecosystems and their biodiversity. However, the mismanagement of groundwater threatens these vital benefits. Over the past 50 years, global groundwater abstraction has tripled, resulting in significant declines in major aquifers, reduced stream flows, and the degradation of riparian and wetland ecosystems (Lapham et al., 1997). This unsustainable exploitation jeopardizes irrigation-based economies, undermining global food security and the drinking water supply for hundreds of millions. Furthermore, groundwater degradation reduces the resilience of communities and economies to climate change, exacerbating vulnerabilities. Effective groundwater management is indispensable for biodiversity conservation, as alterations in groundwater levels, timing, quantity, or quality can lead to biodiversity loss in ecosystems reliant on groundwater, such as springs, rivers, lakes, and specific wetlands and forests. Sustainable management of these ecosystems is essential for maintaining groundwater recharge, fostering a reciprocal relationship between groundwater conservation and ecosystem health. To ensure sustainable and climate-resilient development, groundwater management must be integrated with surface water management, requiring the coordinated involvement of local users, technical experts, and policymakers. Collaborative efforts at local, basin, national, and international levels are vital for implementing participatory strategies that address the diverse needs of stakeholders. These strategies are crucial for safeguarding groundwater resources to support food security, energy development, access to safe drinking water, and the sustainability of urban and industrial systems (Clemens et al. 2020).

Enhancing groundwater management requires a comprehensive understanding of groundwater systems, which remain poorly understood and inadequately managed in many regions. Mismanagement can result in irreversible damage, as groundwater systems are highly variable, and their sustainable management hinges on understanding their responses to different pressures. Key factors include the geological structure of aquifers, recharge mechanisms, pollution risks, and sustainable extraction limits (Dawoud, 2020) The two primary threats to groundwater are overuse and pollution. Groundwater is stored within porous and fractured rocks, with recharge occurring primarily through precipitation or leakage from rivers, lakes, and canals. This recharge infiltrates the ground slowly and moves toward natural discharge points such as springs, rivers, wetlands, or the sea. Intensive abstraction, often through pumping, reduces natural discharge and depletes groundwater reserves. Pollution arises when recharge water carries contaminants from agricultural runoff, industrial activities, or waste disposal into the aquifer. Effective groundwater management requires a nuanced understanding of these pressures and their interactions with surface water systems. Addressing current and future challenges necessitates detailed information about aquifer conditions, regular monitoring of groundwater levels, and water quality assessments. National monitoring systems are vital for timely and informed interventions. Mapping soil and rock properties and assessing water balance—comparing recharge, natural discharge, and abstraction—are essential for estimating an aquifer's capacity and vulnerability. Key parameters such as storage capacity, transmissivity, hydraulic conductivity, and porosity must be analyzed to predict aquifer responses to pumping or contamination and to develop reliable groundwater models. Sound planning and preventative measures are crucial for groundwater protection and remediation, as depletion and pollution are often costly or only partially reversible. Land-use zoning ensures compatibility with the protection of vulnerable aquifers, while technical interventions like regulating pumping, controlling discharges, and managing aquifer recharge are critical. Although remediation of contaminated groundwater is possible, it is often expensive and only partially effective (Witkowski et al. 2017).

2 Groundwater Characteristics

Groundwater bearing layers can be divided into two main zones based on the water table: The saturated and unsaturated zones. Aquifer systems are further classified into confined and unconfined aquifers, characterized by physical hydraulic parameters such as effective porosity, permeability, transmissivity, specific yield, specific storage, and vertical and horizontal hydraulic conductivities. The unsaturated zone, or vadose zone, consists of the soil moisture zone, intermediate vadose zone, and capillary zone. This zone contains void spaces partially filled with air and water. Below the water table lies the saturated zone, where all void spaces are fully filled with water under hydrostatic pressure. Aquifers are categorized primarily as confined or unconfined, with permeable or fractured geological formations. An unconfined aquifer, shown in Fig. 3, lacks a confining layer at its upper boundary, resulting in a water table at atmospheric pressure. Groundwater levels in unconfined aquifers fluctuate due to factors such as pumping, permeability, and the dynamics of recharge and discharge areas, affecting well water levels. The water table, also called the phreatic or free surface, rises and falls freely. Contour graphs and water table profiles created from well elevations help quantify available groundwater and analyze its distribution and movement. Perched water sources, a type of unconfined aquifer, are highly vulnerable to contamination, as surface pollutants can directly penetrate and affect groundwater reserves. One distinctive characteristic of groundwater systems is their immense storage capacity, which far exceeds that of the largest man-made reservoirs (Driscoll 1986). This vast storage enables groundwater to move slowly and continuously from natural recharge areas, where rainfall surpasses plant requirements, to discharge points such as springs, watercourses, wetlands, and coastal zones, as illustrated in Fig. 4. This gradual movement underscores groundwater's critical role in sustaining ecosystems and water availability over extended periods.

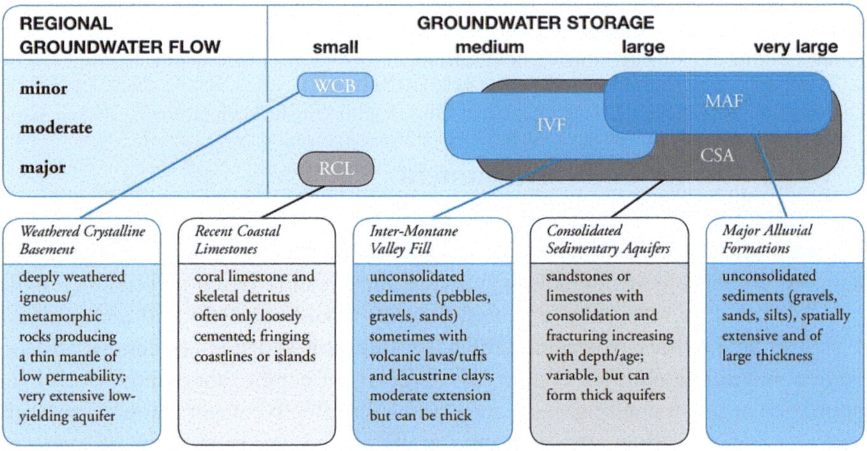

Fig. 3 Key properties of the most global aquifer systems

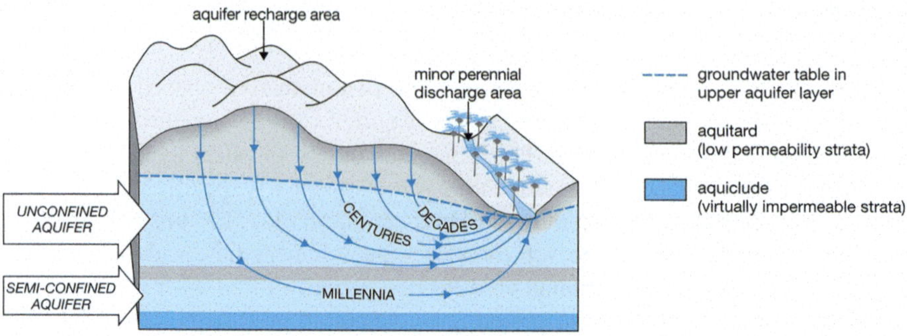

Fig. 4 Typical groundwater flow patterns of major aquifers

Groundwater naturally contains dissolved inorganic components such as sodium, calcium, magnesium, potassium, and chloride, as well as trace metals such as manganese, iron, zinc, arsenic, and fluoride, which are often present in microgram-per-liter concentrations (ppb). The Earth's strata play a critical role in regulating these elements, significantly impacting water quality. However, when these substances naturally exceed permissible levels for specific uses, they become contaminants, posing risks to human health. This natural process, termed geogenic contamination, results from the leaching or weathering of rock materials. Over the past few decades, contaminations and pollution loads from manmade activities have become a persistent concern in groundwater studies. Anthropogenic sources, such as agricultural runoff, industrial discharges, and improper waste disposal, contribute to the degradation of groundwater quality. According to the World Health Organization (WHO), alterations in groundwater's physical, chemical, or biological characteristics can have detrimental effects on human health. Research on groundwater contamination is crucial for understanding the evolution of subsurface aquifers and for providing decision-makers with essential information to protect these vital resources. This review explores the major sources of groundwater contamination, emphasizing the need for sustainable management practices to address both natural and human-induced challenges to groundwater quality.

3 Groundwater and Environment

Both groundwater and surface water are interconnected components of the hydrologic system, with their interactions varying significantly based on climatic conditions. In humid regions, they are often directly linked, while in arid and semiarid areas, they are commonly separated by a thick unsaturated zone. Excessive abstraction or contamination of one system can adversely affect the other, underscoring the importance of understanding their interactions for effective water resource management. Water flow follows the hydraulic gradient, moving from areas of higher to lower water tables. Determining the flow direction requires understanding the hydraulic head at both ends of the flow path, helping to identify whether water flows from surface to groundwater or vice versa. Groundwater, recharged by rainfall in upper catchments, can take months or years to discharge into streams, lakes, or wetlands downstream. Conversely, surface water from rainfall or runoff can percolate through streambeds and soil layers, merging with groundwater (Siegrist et al., 2001). These interactions occur at varying physical scales and over extended periods, manifesting as groundwater discharging into streams (baseflow), supplying springs and seeps, or contributing to streamflow recharge of groundwater. Large-scale hydrologic exchanges are influenced by factors like hydraulic properties, the relationship between stream stage and groundwater levels, stream channel geometry, and climatic conditions such as rainfall patterns (Xanke et al. 2017). Groundwater contamination differs from surface water contamination in complexity and remediation. Unlike surface water, groundwater contamination is often invisible, with contaminants being colorless and odorless, complicating detection and mitigation. Once contaminated, groundwater is difficult and expensive to remediate due to its subsurface location, long residence times, and slow natural purification processes, which may take decades or centuries. Contaminants originate from both point sources, such as leaking underground pipelines, septic systems, or chemical spills, and nonpoint sources, including agricultural runoff, roadways, and urban landscapes (Postel and Richter 2003). Certain contaminants, such as nonaqueous-phase liquids (NAPLs) like gasoline or solvents, persist as distinct phases from water, gradually dissolving and acting as long-term sources of pollution. The significant role of organic contaminants and NAPLs in groundwater pollution necessitates focused study and targeted management strategies.

3.1 Natural Contaminations Sources

Natural sources, or geogenic contamination, occur without human activity and depend on the geological materials through which groundwater moves. These natural contaminants can include minerals, salts, seawater intrusion, brackish water, and radionuclides, all of which can degrade water quality. Groundwater flowing through sedimentary rocks and soils can absorb various inorganic compounds like magnesium, calcium, and chlorides, while some aquifers naturally contain higher levels of dissolved elements such as iron, manganese, arsenic, fluoride, sulfate, and boron. Additionally, bacteria, viruses, and other microorganisms can be present in groundwater, particularly in shallow aquifers and sand wells. Wildlife and soil pollution, especially after flooding, can further contaminate groundwater, and radioactive elements like uranium and radon pose serious health risks through decay and erosion (NGWA 2003).

3.2 Anthropogenic Contaminations Sources

Anthropogenic groundwater contamination refers to introducing chemical pollutants into the subsurface due to human activities. Several human-induced sources contribute to the degradation of groundwater quality worldwide. Economic growth, industrialization, population increases, and changing land-use patterns are key factors that significantly impact groundwater resources. These anthropogenic changes alter various aspects of the hydrological cycle, exacerbating existing conditions and introducing new challenges to groundwater quality. Groundwater pollution is predominantly caused by anthropogenic activities, which encompass a wide range of practices contributing to environmental degradation and resource contamination. The disposal of wastewater, solid waste, and industrial effluents are primary sources of pollution, introducing harmful substances directly into the soil and water systems. Agricultural activities, including the extensive use of fertilizers, pesticides, and insecticides, further exacerbate contamination levels, as these chemicals leach into groundwater during irrigation or rainfall. Mining operations generate substantial by-products and waste, which, if improperly managed, release heavy metals and toxic substances into aquifers. Additionally, the disposal of nuclear energy waste presents a unique and long-term risk due to its persistent radioactive nature. Human activities such as over-pumping groundwater lead to reduced aquifer pressure and increased vulnerability to contamination. The uncontrolled application of fertilizers and other agricultural chemicals contributes to nutrient pollution, while waste dumps and improper waste management practices often result in leachate infiltration. Rapid urban development disrupts natural recharge areas and increases pollution from runoff and waste. Changes in microclimatic conditions caused by urbanization and deforestation can alter recharge patterns and groundwater flow, further complicating management efforts. Poorly maintained water basins, misuse of chemicals, and improper disposal of organic and inorganic substances add to the contamination burden. Additionally, leakage from sewage storage systems and waste heaps infiltrates aquifers, introducing pathogens and toxic compounds. The interruption of river networks and mineral processing activities, particularly involving radioactive materials, also pose significant risks to groundwater systems. Even cemeteries contribute to pollution, with organic matter and embalming chemicals potentially leaching into subsurface water. The cumulative impact of these diverse and widespread activities underscores the urgent need for comprehensive management, effective regulations, and sustainable practices to mitigate contamination and safeguard groundwater resources for future generations.

Groundwater pollution arises from both anthropogenic and natural sources, significantly impacting water quality and posing risks to ecosystems and human health. Elevated nutrient concentrations, particularly nitrogen and phosphorus ions or organic components, are closely associated with agricultural activities, such as the excessive use of fertilizers, livestock farming, and sewage leaks. These activities introduce nutrients that lead to eutrophication and degrade groundwater quality. Other contaminants linked to manmade activities include halogens, biphenyls, hydrocarbons, and biological pollutants such as bacteria, and viruses. Additionally, mining activities, sewage infiltration, and waste disposal contribute to groundwater contamination through the leaching of toxic inorganic compounds, often elevating salinity levels due to high concentrations of calcium, magnesium, sodium, chloride, and fluoride. Groundwater pollution also occurs naturally due to geogenic processes and environmental phenomena. Easily dissolved rocks, such as gypsum and mineral salts, release salts into the water, while intense evaporation, especially in shallow aquifers, raises groundwater levels and causes salt deposition. Geothermal and volcanic fields degrade water quality by introducing thermal and chemical pollutants, while rock oxidation releases harmful substances into aquifers. Seawater intrusion is another significant natural pollutant in coastal regions, where over-abstraction of groundwater allows saltwater to infiltrate freshwater aquifers. Decay of radioisotopes from uranium-rich bedrock and natural chemical reactions between elements in air or water also contribute to contamination, with the latter often being exacerbated by anthropogenic influences. The interplay between these natural and human-induced factors underscores the complexity of groundwater pollution and the need for integrated management approaches. Protecting groundwater requires understanding the sources and pathways of contaminants, mitigating anthropogenic activities that exacerbate natural processes, and adopting sustainable water resource management strategies. This holistic approach is essential to preserve groundwater quality for future generations while supporting current ecological and human needs.

3.2.1 Industrial Contaminations Sources

Industrial wastewater disposal from factories and mining activities is a major source of groundwater pollution, posing significant threats to both surface water and groundwater resources. These wastes are often released onto the Earth's surface and can drain into surface water, eventually directly or indirectly infiltrating groundwater. Both developed and developing countries have made efforts to identify, assess, and address sources of pollution through monitoring, treatment, and cleanup activities. Statistics indicate that industrial pollutants occupy 37% of land surfaces in urban areas, including mineral waste, spills, and solid and liquid wastes. The manufacturing sector, particularly mining, generates more waste than urban and agricultural activities. Rapid recent industrialization, especially in the energy sector and production of raw materials, has led to the release of a diverse range of waste products into the environment. Globally, over 83% of industrial raw wastewater is discharged into the marine or terrestrial environment, exacerbating the issue of groundwater contamination.

3.2.2 Agriculture Contaminations Sources

Agriculture, a widespread human activity, significantly impacts both surface and subsurface water. Major agricultural practices, including crop cultivation, livestock management, fish farming, pesticides, fertilizers, and animal waste, are key sources of groundwater contamination. These contaminations often occur when permeable soil formations allow chemicals from these substances to infiltrate into aquifer systems and water-bearing layers. Contaminants can also enter groundwater through pesticide or fertilizer spills during handling, runoff from pesticide sprayers, or improper storage of chemicals near wells. Pesticides, used to control pests and promote crop yields, play a crucial role in agriculture and pose environmental risks. These chemicals, including herbicides, nematicides, insecticides, rodents, and fungicides, can persist in the subsurface due to varying breakdown and absorption rates, making groundwater contamination a complex issue. Groundwater moves slowly, so it can take decades to wash and dilute contaminated water from polluted areas. Additionally, extensive fertilizer use, particularly nitrogen and phosphorus compounds, contributes to water pollution. Nitrogen contaminants, such as nitrate, nitrite, and ammonia, are prevalent in agricultural runoff and can cause severe health issues, including methemoglobinemia, or "blue baby syndrome" in children. Phosphorus, another critical element in fertilizers, can also pollute water when improperly handled. In developing countries, agriculture is a major contributor to both surface and groundwater pollution from nitrogen compounds.

3.2.3 Urban Contaminations Sources

Substantial population load growth, land-use changes, and land cover patterns have driven urbanization worldwide, converting natural surfaces like wetlands, forests, and water bodies into impervious commercial, industrial, and residential surfaces. This transformation is closely linked to increased polluted runoff, which degrades water quality. Urban activities contribute to groundwater contamination through municipal waste, improper land use, poor sanitation practices, and inadequate landfill management. Municipal waste, mainly from household activities, releases pollutants into groundwater through surface runoff. In densely populated urban areas, septic tanks can also contaminate local groundwater supplies, leading to waterborne diseases such as diarrhea, typhoid, and cholera. Groundwater contamination includes both inorganic and organic pollutants. Inorganic pollutants, such as nitrogen, fluoride, and heavy metals, naturally occur but can reach harmful concentrations, particularly in developing countries. High levels of nitrogen are particularly damaging to the environment and living organisms. Groundwater may also contain elevated total dissolved solids (TDS), indicating a mix of organic and inorganic substances. Organic pollutants, including, fertilizers, hydrocarbons, residue of pesticides, and pharmaceuticals, are widely used in agriculture, urban areas, and industries, posing significant risks to the environment and human health. These contaminants can appear in groundwater as pure substances or as mixtures, contributing to environmental degradation.

3.2.4 Mining Contaminations Sources

Oil and minerals exploration and development in Africa have profound implications for groundwater quantity and quality. Mining activities are often accompanied by chemical pollution, most notably acid mine drainage (AMD), a process in which sulfide minerals like pyrite oxidize upon exposure to air and water. This reaction produces highly acidic water capable of dissolving heavy metals, which then contaminate surrounding water bodies, including groundwater. Additionally, the practice of dewatering underground mines to facilitate operations can significantly lower the water table, disrupting surface water flows and depleting shallow aquifers. This often results in water shortages for nearby communities, as springs and surface water bodies dry up. These hydrological disturbances can persist for decades, emphasizing the need for comprehensive environmental impact assessments, robust monitoring systems, and detailed contingency plans. The geochemical conditions of ore bodies, waste rock, and tailings evolve over time, necessitating continual adjustments in water control and treatment strategies, even long after mining operations cease. Effective post-mining site management is often required for a long time, depending on the type of mine, the extent of the disturbance, and the methods used in ore processing. Permanent mine closure typically involves a range of activities, including the removal of residual chemicals, demolition of infrastructure, detoxification of waste materials, capping of tailings, backfilling of pits, and active water management to meet established water quality standards (Table 1).

Table 1 Real case studies for groundwater pollution

No	Pollution source	Contaminants	Case study/area	Reference
1	Agriculture	Pesticide residues Nitrate	Australia	Salvestrina and Hagare (2009)
2	Agriculture	Pesticide residues Nitrate NH_4 levels	Malaysia	Ahmad et al. (1994)
3	Industrial (aluminum refining, ceramic production, and coal combustion)	Fluoride Trace Metals	Pakistan	Masood et al. (2021)
4	Waste dumping sites Wastewater discharge	Nitrate Chloride Trace Metals	India	Gupta et al. (2018)
5	Oil refinery activities	Benzene, toluene, ethylbenzene, and xylenes compounds (BTEX)	USA	Li et al. (2024)

4 Groundwater Remediation

4.1 *Risk Assessment*

Since the late 1970s, the approach to risk assessment and remediation of contaminated sites has undergone significant evolution, largely driven by the recognition of groundwater contamination as a critical threat to water resources. Two primary factors have shaped this shift: The sheer number of contaminated sites combined with the high costs of remediation, and the intricate nature of subsurface contamination zones, which often extend the duration of cleanup efforts. These challenges have emphasized the need for risk-based approaches and societal prioritization in managing contaminated sites. It is now widely accepted that not all contaminated sites require complete remediation. This has led to the adoption of holistic strategies that assess and prioritize risks based on factors such as land use, potential exposure pathways, and the severity of contamination. Additionally, there is a growing demand for the development of faster, more efficient, and environmentally sustainable remediation technologies, particularly for cleaning soil and groundwater. Risks from contaminated sites are typically associated with site-specific factors, including indoor air quality, agricultural productivity, direct soil contact, or the migration of contaminants into groundwater or surface water systems. Although addressing these risks often requires diverse remedial actions, some methods can mitigate multiple risks simultaneously. This evolving framework underscores the importance of remediation technologies aimed at safeguarding water resources and addressing contaminant sources and plumes. Particular attention is given to managing contaminants such as chlorinated solvents and hydrocarbons, which pose significant environmental and health risks. By focusing on innovative and adaptable remediation strategies, the field continues to advance toward protecting water resources and ensuring the sustainable management of contaminated sites (Diodato and Bellocchi, 2024).

4.2 *On-site, In situ, and Off-site Remediation Technologies*

Remediation technologies are classified into three main categories: Off-site, on-site, and in situ, as outlined in Table 2 (Bjerg et al. 2013). Off-site technologies involve the excavation of contaminated soil, which is then transported to a treatment facility for physical, chemical, or biological processing. This method is effective but often costly due to transportation and the need for clean backfill soil, along with potential environmental impacts. On-site technologies treat contaminated soil, water, or vapors directly at the site, eliminating the need for transport and reducing some environmental costs. However, these methods can cause localized disturbances and are often less cost-efficient than centralized treatment facilities (ITRC 2008). In situ remediation addresses contaminants within the subsurface, minimizing surface-level disturbances. This approach typically involves minimal above-ground infrastructure, such as injection or monitoring wells, as employed in techniques like

Table 2 Commonly used in situ remediation technologies for chlorinated aliphatic (CAH) and petroleum hydrocarbons (PHC)

Technology			Remediation type		Contaminant phase	Target compound
			Mass transfer	Mass destruction		
Physical	Excavation	On-site Off-site	✓		All	CAH PHC
	Pump-and-treat	In situ On-site	✓		Primarily dissolved	CAH PHC
	Soil-vapor Extraction	In situ On-site	✓		All	CAH PHC
	Multiphase Extraction	In situ On-site	✓		All	CAH PHC
	Air sparging	In situ On-site	✓		Dissolved and vapor	CAH PHC
	Thermal treatments	In situ On-site	✓		Primarily DNAPLs	CAH PHC
	In situ flushing	In situ On-site	✓		All but vapor	CAH PHC
Chemical	ISCO	In situ		✓	DNAPL and dissolved	CAH PHC
	Chemical reduction, permeable reactive barriers or direct injection of reactant	In situ		✓	DNAPL and dissolved	CAH
Bioremediation	Enhanced aerobic or anaerobic degradation	In situ		✓	Dissolved	PHC
	Enhanced reductive dechlorination	In situ		✓	All, but primarily dissolved	CAH
	Monitored natural Attenuation	In situ		✓	Dissolved	CAH PHC

bioremediation or in situ chemical oxidation (ISCO). In situ methods are advantageous for their reduced environmental disruption and lower handling requirements. Some remediation technologies are hybrids, incorporating elements from multiple categories. For example, thermal treatment technologies, although primarily in situ, may require the above-ground capture and treatment of vapors or water, blending the characteristics of in situ and on-site approaches. Each remediation method is selected based on-site–specific conditions, contamination types, and cost-effectiveness, ensuring an optimal balance between efficiency and environmental impact.

In the 1970s and 1980s, groundwater remediation efforts primarily focused on addressing contaminant plumes, with the preferred method being the abstraction of contaminated water followed by above-ground treatment. However, this approach often proved ineffective due to the continuous release of contaminants from the source zone, particularly in cases involving dense non-aqueous-phase liquids (DNAPLs) like chlorinated solvents, and from low-permeability zones that trap contaminants for centuries, known as secondary sources. While pump-and-treat systems have been effective in containing contaminant plumes, source zone remediation is now considered the ultimate solution for groundwater contamination. Plume management technologies such as pump-and-treat, permeable reactive barriers, and monitored natural attenuation (MNA) continue to play an essential role, especially in specific contamination scenarios or when combined with source zone remediation. Widely used and successful remediation technologies for in situ source and plume treatment of groundwater contamination by PHC and CAH are examined in the next sections. The final section discusses considerations for selecting remedial actions, including the use of life cycle assessment methods. Table 3 summarizes the principles of these technologies, including target compounds, zones (source/plume), and their advantages and disadvantages.

4.3 Life Cycle Assessment and Monitoring

When a decision is made for the remediation of contaminated sites, a detailed screening process is conducted to address and evaluate various applicable technologies and select the most technically and economically viable ones. This process involves technical, environmental, and economic considerations, with key decision criteria including cost, technical applicability, and performance metrics such as cleanup levels and time frames. Other factors, such as the impact on nearby residents (e.g., noise, dust, and vibrations from excavation work) or the less disruptive, yet prolonged, effects of in situ remediation, may also be important. Recently, there has been growing interest in adopting a holistic decision-making approach and using life cycle assessment (LCA) for decision support in selecting remediation methods (Siegrist et al. 2011). While remediation technologies aim to eliminate local contamination, they can also contribute to environmental impacts on local, regional, and global scales due to the energy, chemicals, raw materials used, and the emissions and waste generated. These secondary environmental impacts, which are distinct from the primary impacts directly caused by on-site contamination, are also a significant consideration. LCA, a widely recognized environmental assessment tool, is a quantitative method that compares environmental impacts associated with fulfilling a specific function or service, encompassing all life cycle stages from raw material extraction to production, use, and final disposal. As remediation efforts can lead to problem shifting, LCA is particularly useful for assessing environmental impacts due to its comprehensive scope and systems perspective. Numerous studies have applied LCA to evaluate the environmental impacts of remediation, primarily

Table 3 Principles for remediation technologies selection

Principles		Target compound	Target zone	Advantages	Disadvantages
Thermal treatment	Enhanced mass transfer from soil and water to the vapor phase is achieved through heating, followed by above-ground treatment of the water and vapor	CAH PHC	Source	Fast and highly efficient, requiring large above-ground installations during cleanup, capable of targeting various compounds with moderate to high vapor pressure across different geological settings	High energy consumption, expensive
ISCO	In situ destruction of contaminants using strong oxidants, with a potential activation step needed for oxidation	CAH PHC	Source	Fast, efficient, and safety issues	High environmental costs due to the large quantity of oxidant used, with limitations in low-permeability settings
Bioremediation	Enhancement of naturally occurring in situ biodegradation processes by adding electron acceptors, electron donors, or nutrients, with the introduction of specific bacteria to stimulate these processes known as bioaugmentation	CAH PHC	Source Plume	Requires few above-ground installations, generally low environmental costs, and is suitable for easily degradable compounds.	May have long time frames, generate toxic by-products, and face limitations in low-permeability settings
Monitored natural attenuation	Leveraging naturally occurring attenuation processes, particularly microbial degradation, where part of the aquifer acts as a natural reactor	CAH PHC	Plume	Requires very few above-ground installations, generally low environmental costs, and is suitable for easily degradable compounds	May have long time frames and limited experience in low-permeability settings

focusing on off-site remediation and soil pollutants like polycyclic aromatic hydrocarbons (PAHs), hydrocarbons, and other trace metals. More recent studies have expanded to include in situ soil and groundwater remediation, with LCA being used to compare permeable reactive barriers with pump-and-treat solutions for groundwater plumes, as well as assessing chlorinated solvent remediation via enhanced bioremediation and off-site treatment (Siegrist et al. 2011).

5 Groundwater Treatment

5.1 Oxidation

In groundwater treatment, oxidation is typically initiated at the plant headworks. The choice of oxidant, dosage, and reaction time depends on the specific contaminants being targeted and the overall quality of the raw water. The simplest oxidation method is aeration, commonly used for oxidizing iron, manganese, and arsenic, as well as removing hydrogen sulfide and volatile organic compounds (VOCs). Aeration is a physical process that transfers oxygen from the air to the water, reduces dissolved carbon dioxide, and strips volatile compounds. The efficiency of this exchange process is generally enhanced by maximizing the surface area. Aeration systems are typically categorized into four types: waterfall aerators, diffusion or bubble aerators, mechanical aerators, and pressure aerators (AWWA and ASCE 1998). Waterfall aerators, which are widely used for groundwater oxidation, include spray aerators, multiple-tray aerators, cascade aerators, cone aerators, and packed columns. The specific aerator system selection depends on many factors including the required transfer efficiency, available space, and costs. Chemical oxidants used in groundwater treatment include chlorine, permanganate, chlorine dioxide, and ozone. Table 4 outlines the oxidation potential of common groundwater treatment oxidants. Many of these oxidants serve as disinfectants and can be added at the plant headworks for both oxidation and disinfection in the treatment process for disinfection (Singer and Reckhow 1999).

Table 4 Oxidation potential of most common groundwater treatment oxidants

Oxidizing compound	Standard half-cell potential (V)
Ozone	1.99
Hydrogen peroxide	1.67
Hypochlorite	1.64
Permanganate	1.59
Chlorine dioxide	0.95
Hypochlorous acid	1.48
Monochloramine	1.39
Chlorine gas	1.36
Dichloramine	1.32

5.2 *Clarification*

Following oxidation, the clarification step is typically used to aggregate and settle dissolved and suspended materials, such as clay, silt, organic matter, soluble compounds, algae, and microbes. Conventional clarification involves three stages: First, a coagulant or pH-adjusting chemical is added in the flash mixer to destabilize particles and promote their agglomeration; second, flocculation occurs, where gentle mixing forms larger, heavier flocs; and third, settling, where these flocs settle by gravity at the bottom of the tank. Common coagulants include alum, iron salts, and polyaluminum chloride (PAC), with dosages varying based on water pH, alkalinity, turbidity, suspended solids quality, total organic carbon (TOC), and temperature. Higher doses are required for groundwaters with elevated TOC and lower temperatures. Polymers may also be added to enhance particle removal and filterability. If hardness removal is needed, chemical precipitation using lime or sodium hydroxide can be employed. Clarification performance depends on several factors, including facility design, particulate concentration and type, coagulant type and dosage, pH, ionic strength, temperature, and natural organic matter (NOM) concentration. Turbidity, which indicates water clarity, is often used to assess clarification efficiency, with a common target being less than 2 NTU. For high-quality groundwater with low turbidity, flocculation, and sedimentation may not be necessary; in such cases, in-line coagulation before direct filtration can be used to enhance filter performance by improving particle filterability (LeChevallier and Au 2004).

5.3 *Filtration*

After clarification, filtration is used as the final step in water treatment to remove any remaining particulates. In this process, water passes through a bed of granular filter media, such as sand, anthracite, or other materials, which retain most of the suspended matter. The filtered water should ideally exhibit low turbidity, preferably below 0.2 NTU. The filter's performance is influenced by factors such as media size, filtration rates, the effectiveness of the preceding clarification step, and the addition of filter-aid polymers. In the United States, the EPA mandates that filtered water in drinking water plants must have a turbidity level not exceeding 1 NTU, with 95% of daily samples remaining below 0.3 NTU. These filters consist of concrete boxes filled with one or more porous media layers, often including granular activated carbon to adsorb dissolved chemicals. Water flows by gravity through the filter, and solids are removed via a backwash process, where water flow is reversed to clean the filter media. The backwash process requires skilled operators to restore the filter to service without compromising water quality. Slow sand filters, which have been in use for nearly two centuries, can treat water with turbidity up to 50 NTU. They operate at lower hydraulic loading rates and do not require coagulation or flocculation. The filter media, consisting of fine-grained sand supported by

Table 5 Case studies for groundwater treatment with filtration

No	Objectives	Case study	Reference
1	using anthracite-sand filter methods for treating groundwater for drinking water supply	Azogues city, Ecuador	García-Ávila et al. (2020)
2	The treatment of groundwater using simple ceramic filters involves a combination of dilution with rainwater and filtration through ceramic filters. This method, which integrates both dilution and filtration, offers a practical and simple alternative to more complex modern or chemical treatment methods	Malaysia	Musa et al. (2016)

gravel, removes suspended solids through physical straining and biological activity in the "schmutzdecke," a biologically active layer on the sand surface. Although slow sand filters are effective, they are less capable of removing clay particles and color. Their simplicity, low cost, and lack of power requirements make them ideal for low-income urban and rural communities. Prefabricated slow sand filter units are also available for small communities and emergencies (Cleasby and Logsdon 1999). Pressure filters use the same media as rapid sand filters but operate under pressure within a steel vessel, eliminating the need for repumping and reducing the risk of air binding. However, they cannot visually monitor the filtration process and may experience turbidity breakthroughs due to the high pressure (USEPA 1999). Table 5 shows two case studies for groundwater treatment with filtration methods including sand filters and ceramic filtration (Fig. 5).

Ceramic filters are effective at removing bacteria and parasites from water by passing them through a porous ceramic cartridge, but they have some limitations. Typically used in kitchen sink units to produce small amounts of drinking and cooking water, these filters cannot remove all viruses, requiring additional disinfection. They also tend to clog quickly if the water contains high levels of particulate matter. One notable "low-tech" ceramic filtration system, known as Filtron, was developed by Dr. Mazariegos for using in low-income areas and emergencies (Potters for Peace 2007). This system consists of a porous clay filter, a storage container, and a spigot for dispensing water. Made locally from terracotta clay mixed with a combustible material like sawdust, the filter creates fine pores when the combustible material burns out during firing. It has been found that the filter could reduce diarrhea by 51%, as it removes most bacteria and protozoa through mechanical filtration, while colloidal silver inactivates bacteria. However, its effectiveness against viruses remains uncertain (Table 6).

5.4 Membranes-Based Filtration System

Membranes represent a new generation of water filters, increasingly replacing conventional filtration methods due to their ease of operation and reliable high performance. They are especially effective at removing pathogens larger than the membrane's pore size, effectively acting as a form of disinfection as shown in

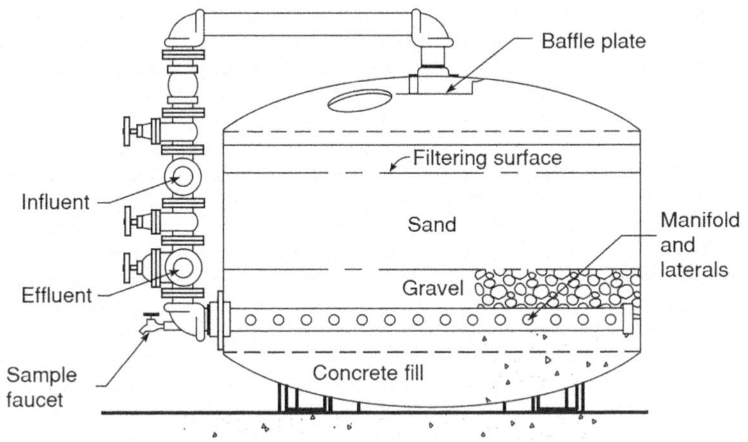

Fig. 5 Pressure filter cross section

Table 6 Comparison of different groundwater filtration methods

Methods	Cost	Filtration rate	Treatment level	Power
Slow sand filter	High	Low	Moderate	No
Three stages integrated filter	Low	High	High	No
Granular media filter	High	Moderate	Low	No
Activated carbon	Low	Moderate	High	No
Ceramic filter	High	High	High	Yes
Membrane filter	High	High	High	Yes

Fig. 6. Over time, the costs associated with membrane filtration, including capital and operational expenses, have decreased, making it an attractive treatment option. Membrane filtration, driven by pressure or vacuum, forces water through a semipermeable membrane, retaining impurities. Commonly used membrane types in domestic water treatment include ultrafiltration (UF), microfiltration (MF), nanofiltration (NF), and reverse osmosis (RO). These membranes, typically made from synthetic polymers, are characterized by their ability to exclude solutes based on pore size or molecular weight cutoff (MWCO). UF membranes, with pore sizes ranging from 0.01 to 0.05 μm, effectively remove viruses and larger organic macromolecules. With larger pores, MF membranes primarily remove turbidity and larger pathogens such as Giardia and Cryptosporidium. RO and NF membranes, made from semipermeable cellulose acetate or polyamide, are used to remove dissolved contaminants, including salts and synthetic organic compounds. RO is particularly effective for desalination, as shown in Fig. 7. However, NF and RO are not suitable for treating water with high suspended solids, as they are prone to fouling. Additional types of membrane filtration, such as electrodialysis (ED) and electrodialysis reversal (EDR), remove dissolved ionic species but do not provide a physical barrier to pathogens (Cui et al. 2010).

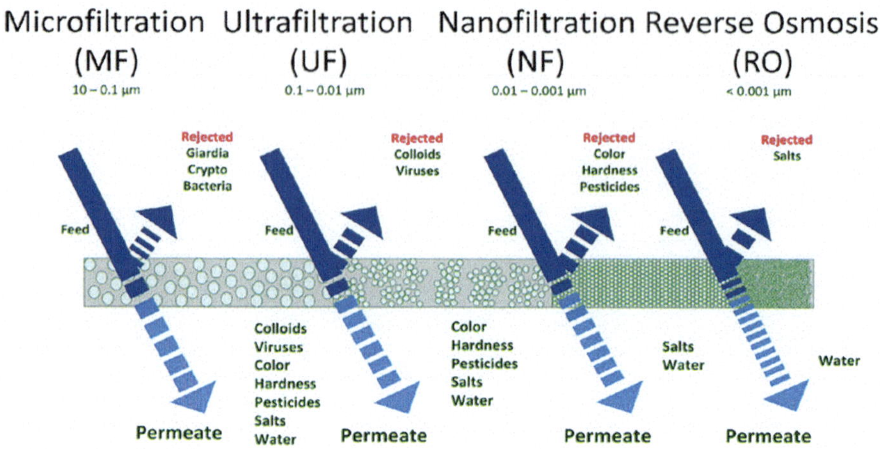

Fig. 6 Pressure-driven membranes classification

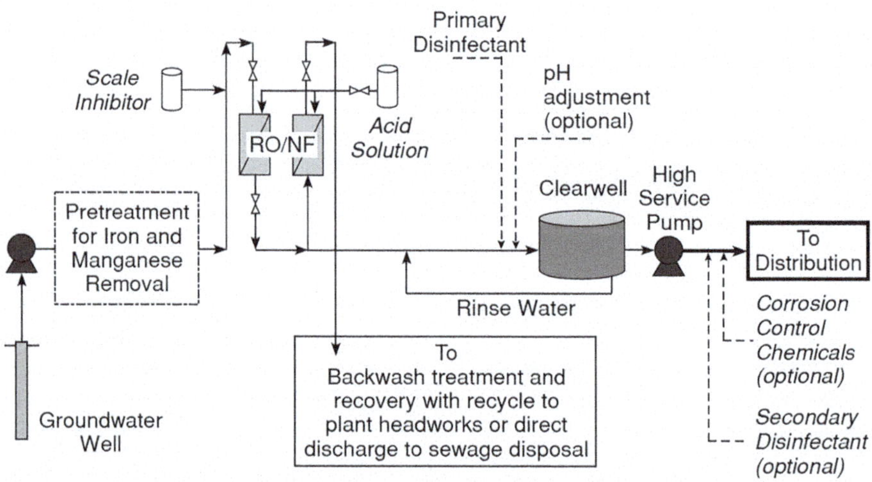

Fig. 7 Reverse osmosis/nanofiltration treatment system schematic diagram

5.5 *Inorganic Adsorption and Ion-Exchange Method*

Ion exchange is a reversible chemical process in which a charged molecule is removed from a solution by exchanging it for a similarly charged ion attached to a solid matrix, as illustrated in Fig. 8. This matrix is typically composed of synthetic or natural high-molecular-weight polymeric or inorganic resins with high porosity, offering a large surface area for the adsorption of pollutants. These resins, usually in the form of small beads, are packed into filter columns and can effectively remove charged constituents such as calcium (hardness), nitrates, heavy metals, arsenic,

Fig. 8 Ion-exchange treatment system

fluoride, and radionuclides. Ion exchange is particularly effective for water sources with stable quality, such as groundwater, and low levels of suspended solids and organic matter. However, the disposal of water used to regenerate ion-exchange resins, which often contain high salt concentrations, can present environmental challenges, and impact wastewater plant performance. Despite this, ion exchange remains a relatively simple and cost-effective process, with low capital and labor requirements. Specialized ion-exchange media have been developed to target specific compounds, enhancing the technology's effectiveness when multiple ions compete for adsorption sites. A similar process, activated alumina, removes ions through adsorption. In this method, water flows through an alumina-packed bed, where contaminants such as arsenic, uranium, and fluoride are adsorbed onto the alumina's surface. Once its adsorptive capacity is exhausted, the activated alumina media must be replaced or regenerated (Clifford 1999).

5.6 Biological Treatment

Biological water treatment processes, commonly used in European municipal water treatment plants, are gaining attention in North America. These processes optimize conditions to promote a stable, active biofilm that facilitates the biodegradation or conversion of unwanted constituents, such as biodegradable organic matter, taste- and odor-causing substances, some synthetic organic compounds (SOCs), as well as iron, manganese, and arsenic. Granulated activated carbon (GAC) is particularly effective in supporting bacterial growth, especially when preceded by ozonation, and can enhance the biological removal of certain compounds. However, managing biological growth is critical to prevent operational issues, such as the sloughing off of microbial clumps from the filter or the transport of microbes and fine-activated carbon particles into the distribution system. These problems can result in elevated bacterial counts and taste and odor issues in the distribution system (Elyanow and Persechino, 2005). To mitigate these concerns, disinfection of the GAC effluent is recommended to ensure its microbiological quality before it is released into receiving surface water bodies. Biological treatment for groundwater, while effective for removing contaminants such as organic matter, iron, manganese, and certain synthetic organic compounds, does come with a set of biosecurity issues. These concerns typically revolve around microbial control, potential pathogen proliferation, and the management of biofilms. Key biosecurity issues include:

1. *Microbial Contamination and Pathogen Growth*: Biological treatment processes rely on the use of microorganisms, such as bacteria, to degrade contaminants. If not properly controlled, these microorganisms can include opportunistic pathogens, potentially leading to contamination of the treated water or the distribution system. For example, pathogenic bacteria or viruses could proliferate in the biofilm or in the treated effluent, posing a risk to human health.
2. *Biofilm Formation*: The development of biofilms (layers of microbial growth) on filters or other treatment media is essential for the biological treatment process. However, if these biofilms are not regularly monitored and maintained, they can harbor harmful microorganisms, including pathogens. Additionally, biofilms can detach from surfaces and be carried into the water distribution system, leading to microbial contamination.
3. *Release of Harmful By-products*: While biological treatment processes can degrade contaminants, they can also produce by-products that may be harmful. For instance, the biodegradation of certain organic compounds can result in the formation of intermediate by-products, which might be more toxic or recalcitrant than the original contaminants. Some biological processes may also release unwanted gases (like hydrogen sulfide), which can contribute to water quality issues or present safety risks.
4. *Antibiotic Resistance*: There is growing concern that the widespread use of biological treatment systems could promote the development of antibiotic-resistant bacteria. In some cases, biological filters can serve as environments where bacteria are exposed to antibiotics or other antimicrobial agents, potentially leading to the development of resistance. If resistant bacteria are released into the groundwater or water distribution system, they could pose a serious public health risk.
5. *Disinfection Challenges*: While biological treatment is effective at reducing certain contaminants, it may not always be sufficient to eliminate pathogens entirely. As such, post-treatment disinfection (e.g., chlorination and UV treatment) is often required. However, this may not fully eliminate the risk of microbial contamination if biofilms or resistant microorganisms are not adequately treated.

6. *Nutrient Imbalance*: Biological treatment often requires the addition of nutrients (e.g., carbon, nitrogen, or phosphorus) to support microbial growth. If not properly balanced, excessive nutrients could encourage the growth of harmful microorganisms, including pathogens, leading to water quality issues or an imbalance in the microbial community that impedes effective treatment.
7. *Cross-Contamination Between Systems*: If not properly isolated or contained, microorganisms used in biological treatment could be transferred between systems or different stages of treatment, leading to cross-contamination. This is especially critical when treating multiple water sources or when handling contaminated groundwater that may contain pathogens.

5.7 Desalination

Desalination of brackish and saline groundwater is essential for enhancing water quality and supply in many regions worldwide. Among the various desalination methods (thermal and membrane), membrane-based technologies, particularly reverse osmosis (RO) and electrodialysis (ED), are predominantly used for treating brackish groundwater as shown in Fig. 9. RO operates through pressure-driven processes, while ED relies on the electric potential to achieve desalination (Dawoud et al. 2024). The choice between these technologies depends on factors such as the composition of the feedwater, the desired quality of the product water, environmental impact, and cost considerations. Energy consumption and the economics of desalination plants are critical aspects, with brine disposal being a significant cost factor, especially for inland facilities. The need for environmentally friendly alternatives to current brine disposal practices and conventional energy sources is increasingly important. Since geography heavily influences desalination decisions, regional and national trends in desalination capacity, investment, technology adoption, and source water types are also closely examined to guide future developments in this vital field. Table 7 shows the comparison between the thermal and RO desalination technologies used for groundwater treatment for potable water supply. Desalination processes, particularly reverse osmosis (RO), are associated with carbon emissions. The carbon footprint for seawater RO desalination typically ranges between 0.4 to 6.7 kg CO_2 equivalent per cubic meter (kg CO_2eq/m^3), while brackish water RO desalination is lower, at 0.4 to 2.5 kg CO_2eq/m^3. This variability is influenced by factors such as location, technology, and the stages of the desalination process. Although RO technologies produce fewer emissions than thermal desalination methods, the overall carbon footprint remains a concern, especially regarding energy consumption and the disposal of brine.

Emerging desalination technologies are increasingly being explored for their potential to provide efficient, cost-effective, and sustainable solutions for groundwater treatment (USEPA 2005). Forward osmosis (FO) is one of the promising technologies that use a natural osmotic pressure gradient to draw water through a semipermeable membrane, requiring less energy compared with RO. FO is

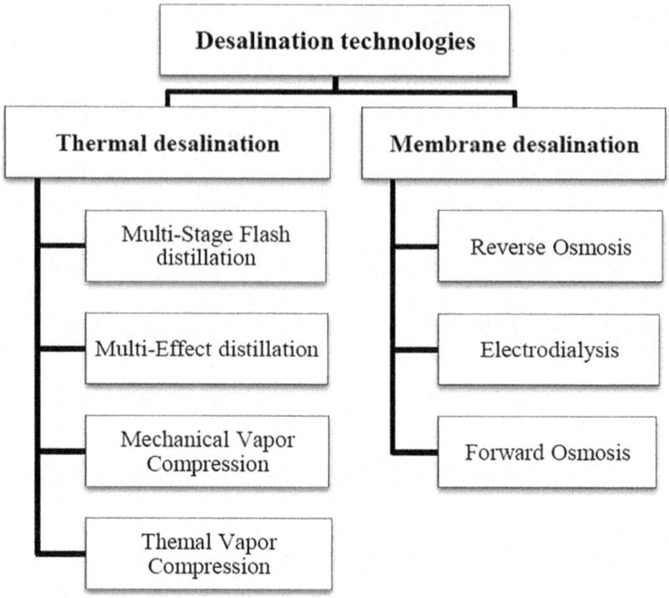

Fig. 9 Mature desalination technologies used for groundwater treatment

Table 7 Comparison of thermal and RO desalination systems

		Thermal desalination		Membrane desalination	
Paramter	Unit	MED	MSF	ED	RO
Feed water salinity	ppm	25,000–50,000	25,000–50,000	25,000–50,000	5000–40,000
Unit capacity	m^3/day	50,000–75,000	5000–25,000	100,000–150,000	25,000–50,000
Electrical energy consumption	kWh/m^3	5–7	2–4	2.5–5	4–6
Thermal energy consumption	KJ/kg	200–410	225–400	None	None
Electrical equivalent of thermal energy	kWh/m^3	10–20	6–9	None	None
Carone footprint (CO_2 emissions)	Kg CO_2/m^3	20–24	16–19	5–7	6–8
Permeate water salinity	ppm	Less than 15	Less than 15	150–500	150–500
Capital cost	US$/$m^3$	1200–1500	1400–1600	1700–2200	900–1200
Fresh water cost	US$/$m^3$	1.1–1.75	0.75–1.2	0.65–1.10	0.45–0.65

particularly useful for low salinity and brackish groundwater and has gained attention for its potential in remote and off-grid applications. Capacitive deionization (CDI) is another emerging technology, which uses an electric field to remove ions from water, offering advantages in terms of energy efficiency and the ability to selectively treat groundwater with varying salinity levels. CDI is particularly suitable for small-scale or decentralized desalination systems. Solar desalination, including solar stills and solar-powered reverse osmosis, utilizes solar energy to distill or pressurize water, reducing reliance on conventional energy sources and making it an environmentally friendly option. Membrane distillation (MD) is a thermal desalination process that uses hydrophobic membranes to separate water from saline solutions, requiring lower energy compared to traditional thermal processes. These emerging technologies offer the promise of reducing the energy consumption, operational costs, and environmental impact of groundwater desalination, making them viable options for addressing water scarcity in regions with limited freshwater resources.

5.8 Disinfection

The final step in the water treatment process is disinfection, ensuring that the water quality is safe before consumption. However, since water quality can deteriorate once it enters the distribution system, many utilities apply a secondary disinfectant to preserve its microbiological integrity. Contamination in the distribution system can occur through backflow, pipe leaks, intrusion, and bacterial regrowth within pipes. In the USA, EPA mandates that treated water must contain sufficient disinfectant to maintain a residual level in the distribution system, preventing microbial regrowth and recontamination as the water is delivered. However, the requirement for maintaining a residual disinfectant is not universally accepted. The advantages and disadvantages of different disinfection methods, along with their effectiveness against different pathogens, are key considerations in the disinfection process, as illustrated in Table 8.

6 Groundwater Contamination Parameters

6.1 Hardness

Excessive water hardness, caused primarily by high concentrations of dissolved calcium and magnesium, can lead to issues like soap scum, scale buildup in plumbing and appliances, and reduced cleaning efficiency of soaps and detergents. Total hardness, measured as mg $CaCO_3$/L, is classified into carbonate hardness, associated with bicarbonate compounds often found in limestone or chalk aquifers, and

Table 8 Disinfection techniques advantages and disadvantages

Parameter	Disinfection techniques Chlorine	Chloramines	Ozone	Chlorine dioxide	UV
Equipment reliability	Good	Good	Good	Good	Medium
Relative complexity of technology	Less	Less	More	Medium	Medium
Safety concerns	Low to high[a]	Medium	Medium	High	Low
Bactericidal	Good	Good	Good	Good	Good
Virucidal	Good	Medium	Good	Good	Medium
Efficacy against Protozoa	Medium	poor	Good	Medium	Good
By-products of possible health concerns	High	Medium	Medium	Medium	None
Persistent residual	High	High	None	Medium	None
pH dependency	High	Medium	Low	Low	None
Process control	Well developed	Well developed	Developing	Developing	Developing
Intensiveness of operations and maintenance	Low	Moderate	High	Moderate	Moderate

[a]Safety concern is high for gaseous chlorine, but it is low for hypochlorites

noncarbonate hardness, linked to chloride and sulfate anions. To address these issues, groundwater is often treated, though some systems avoid softening due to concerns over potential links between softened water and heart disease. For large potable systems, precipitative softening using lime, lime–soda ash, or caustic soda is common. The choice of method depends on factors like total TDS, sludge production, and chemical storage requirements. Lime softening generally reduces TDS and produces more sludge than caustic soda, which, in contrast, can increase TDS but is more accessible to store and handle. Maintaining proper pH is crucial in precipitative softening to avoid lime deposition in filter beds, which can damage the media. After softening, recarbonation with carbon dioxide is often used to lower pH and control scaling (Kawamura 2000). For smaller systems, ion exchange, where bivalent ions are swapped with sodium ions, is a more practical alternative due to its lower costs and operational simplicity. In some cases, electrodialysis reversal (EDR) and reverse osmosis/nanofiltration (RO/NF) are also used to reduce hardness, though they are less common.

6.2 Nitrates

Nitrate contamination of groundwater is a widespread concern, primarily due to agricultural practices, such as the use of nitrogen-based fertilizers, livestock operations, and sewage disposal, although natural sources can also contribute. High nitrate levels in drinking water pose significant health risks, including methemoglobinemia, or "blue baby syndrome" in infants under 6 months, and the potential for carcinogenic nitrosamine formation. EPA has set the maximum contaminant level (MCL) for nitrate at 10 mg/L as nitrate-nitrogen (NO_3–N), while the WHO guideline is 50 mg/L as nitrate (NO_3). To mitigate nitrate contamination, treatment methods such as ion exchange, RO, and EDR are commonly employed. In ion exchange, water passes through a resin that replaces nitrate anions with chloride ions. Specialized resins are used to prevent "nitrate dumping" when the ion-exchange column becomes saturated. In North Carolina, ion exchange reduced nitrate levels from over 10 mg/L to an average of 3 mg/L. RO systems typically achieve a 90% nitrate rejection rate, sufficient for most applications. In Milan, Italy, GE Italba used RO to reduce nitrate levels from 50 to 60 mg/L to below 40 mg/L, blending treated water with raw water to minimize operational costs. Electrodialysis reversal (EDR) leverages electrostatic properties for high recovery rates, often exceeding 90%, and is used in places like Bermuda, Delaware, and Italy. While RO is the most common method at the point-of-use scale due to its cost-effectiveness for small volumes, ion exchange, and distillation are less frequently used due to their respective limitations. Ion-exchange resins specifically designed for nitrate removal help mitigate the risks associated with nitrate dumping, ensuring effective treatment and protecting water quality. Table 9 shows examples of electrodialysis reversal for nitrate removal with high levels of Nitrate in groundwater.

6.3 *Total Dissolved Solids (TDS)*

TDS represent the combined concentration of dissolved substances in water, including inorganic salts including magnesium, calcium, potassium, sodium, bicarbonates, chloride, and sulfate, as well as trace metals and organic materials such as iron, manganese, arsenic, aluminum, copper, and lead. TDS originates from natural sources like mineral dissolution in geological formations, as well as from anthropogenic sources including sewage, industrial wastewater, and urban runoff. TDS is typically measured by filtering a water sample, evaporating the filtrate, and weighing the residue. While high TDS levels do not directly pose significant health risks, they can lead to aesthetic issues such as reduced detergent effectiveness, corrosion

Table 9 Examples of electrodialysis reversal for nitrate removal from groundwater

Parameter	Unit	Bermuda	Delaware	Italy
Production	m^3/hour	94.6	62.1	47.3
Recovery ratio	%	90	90.1	90.08
Desalting stages		3	3.2	3.1
Feed nitrate concentration	mg/L NO_3	66	61.1	120.4
Product nitrate concentration	mg/L NO_3	8.7	4.55	37.1
Feed TDS	mg/L	277	11.12	473.4
NO_3 removal ratio	%	86.7	91.98	68.2
TDS removal ratio	%	81.5	88.1	53.12

Table 10 Groundwater quality classifications based on TDS

Water source	Total dissolved solids (mg/L)
Potable water	<1000
Low brackish water	1000–3000
Moderately brackish water	3000–15,000
High brackish water	15,000–36,000
Sea water	More than 36,000

of plumbing, scale formation, and undesirable taste. EPA has set a secondary maximum contaminant level (SMCL) for TDS at 500 mg/L, while the WHO does not have a health-based guideline but considers water with TDS concentrations above 1200 mg/L undesirable. For comparison, seawater has a TDS concentration of about 35,000 mg/L. The presence of different dissolved species determines the treatment approach for elevated TDS. For example, high levels of calcium, magnesium, or iron can be addressed using softening or ion-exchange processes, whereas high concentrations of sodium or potassium salts require techniques like RO, ED, or distillation. Advances in RO and ED technologies have made them increasingly effective for desalination in water-scarce regions, such as arid inland areas and coastal zones, with desalination plants in operation globally, including in the Western United States, the Middle East, and North Africa. These technologies are used to treat water for various applications, from potable water production to irrigation and industrial uses. Though elevated TDS levels generally do not pose direct public health threats, they affect the water's taste and quality, and the USEPA's secondary limit of 500 mg/L for TDS aims to prevent such aesthetic problems. Table 10 shows the classification of groundwater quality based on the total dissolved solids (TDS) in mg/l.

6.4 *Arsenic*

Arsenic, a naturally occurring element in rock and soil, can dissolve into groundwater due to its solubility, causing significant contamination. This issue has become increasingly important globally. The WHO, EPA, and the European Union have set a maximum standard of 10 parts per billion (ppb) for arsenic in drinking water. However, the National Academy of Sciences (NAS) has found that even at this concentration, the health risks remain substantial, especially regarding cancer risk. Consuming 2 liters of water per day with 10 ppb of arsenic can result in a lifetime risk of fatal bladder or lung cancer exceeding 1 in 300, which is much higher than the EPA's acceptable cancer risk of 1 in 10,000. In response, the EPA identified seven Best Available Technologies (BATs) for arsenic removal in municipal systems, based on their removal efficiency, cost-effectiveness, operational history, geographic applicability, and compatibility with other water treatment processes. These technologies' effectiveness depends on the oxidation state of arsenic, with Arsenic

V being easier to remove than Arsenic III, which is more challenging and may require peroxidation before treatment. The alluvial aquifer beneath Bangladesh and India contains high arsenic levels, which pose serious environmental and public health risks. Al-Muyeed and Afrin (2006) report that groundwater arsenic concentrations exceed 0.05 mg/L in most areas of Bangladesh, and iron and arsenic are often found together, with approximately 65% of groundwater samples containing iron levels above 2 mg/L. Centralized water treatment systems are scarce in these regions, and most of the population relies on individual or community tube wells for water. As a result, addressing arsenic contamination in the short term requires the development of low-cost, community-based water treatment technologies. Al-Muyeed and Afrin (2006) conducted a study on conventional iron removal plants for arsenic removal in small Bangladeshi communities. These plants, which generally use aeration, sedimentation, and filtration, were found to remove arsenic at rates of 60–80% in 60% of the systems surveyed, while the remaining 40% showed lower removal rates. Water pH was a significant factor, with higher pH levels leading to better arsenic removal, particularly when the pH was above 7.0. The correlation between iron and arsenic removal was also evident, with higher arsenic removal rates observed when iron concentrations ranged between 6 and 8 mg/L. Additionally, point-of-use (POU) technologies have been implemented to treat arsenic-contaminated water in rural communities, offering cost-effective solutions despite certain technological limitations. Among the most effective POU technologies are the 3-Kolshi and SONO filters, which have become widely used in Bangladesh as they help mitigate arsenic exposure, especially in less affluent areas. These simple and affordable technologies, while not without limitations, play a vital role in safeguarding public health in rural regions affected by arsenic contamination.

7 Economics of Groundwater Treatment

The cost of groundwater treatment projects varies significantly due to several site-specific factors, including plant capacity, treatment, raw water quality, site conditions, land costs, and local economic conditions. In the USA the construction cost for a basic treatment plant, which includes well construction, chemical addition, filtration, pumping, storage, and disinfection, typically ranges from \$250 to \$500 per cubic meter (\$1.10 to \$2.20 per gallon), depending on factors such as the plant size and location are presented in Table 11. The costs of specific treatment technologies, such as conventional clarification and lime softening, are influenced by construction and land requirements, while others depend more on equipment costs. Average capital and operational costs for various treatment technologies, based on U.S. installations, are provided in industry guidelines. However, land requirements, pretreatment, and sludge disposal costs are site-specific and need individual evaluation (EPA 2005).

Table 11 Capital and operational costs of groundwater treatment process

Process unit	Capital cost (including construction costs) ($/m³) ($/1000 gal)			O&M ($/m³) ($/1000 gal)		
	Small plant <4000 m³/d (<1 mgd)	Medium plant 4000–40,000 m³/d (1–10 mgd)	Large plant >40,000 m³/d (>10 mgd)	Small plant <4000 m³/d (<1 mgd)	Medium plant 4000–40,000 m³/d (1–10 mgd)	Large plant >40,000 m³/d (>10 mgd)
Oxidation						
Chlorine	14 (53)	5 (19)	1.3 (4.9)	0.021 (0.080)	0.006 (0.021)	0.001 (0.004)
Permanganate	15 (55)	4.3 (16.4)	1.3 (4.9)	0.024 (0.090)	0.013 (0.050)	0.001 (0.004)
Chlorine dioxide	95 (360)	15 (56)	6.3 (24)	0.021 (0.080)	0.005 (0.020)	0.003 (0.010)
Ozone (5 mg/L dose)	33 (125)	21 (78)	10 (36)	0.026 (0.100)	0.007 (0.026)	0.002 (0.007)
Aeration (packed tower)	84 (318)	40 (150)	12 (45)	0.011 (0.040)	0.005 (0.020)	0.004 (0.015)
Clarification						
Conventional	284 (1075)	73 (275)	30 (114)	0.001 (0.004)	0.013 (0.050)	0.003 (0.010)
Upflow	140 (528)	63 (237)	27 (101)	0.023 (0.090)	0.008 (0.030)	0.003 (0.010)
Media filtration						
Rapid sand	268 (1014)	146 (551)	50 (190)	0.050 (0.190)	0.018 (0.070)	0.018 (0.070)
Greensand	269 (1018)	147 (556)	52 (198)	0.050 (0.190)	0.018 (0.070)	0.018 (0.070)
Pressure	455 (1725)	89 (340)	83 (314)	0.016 (0.060)	0.034 (0.013)	0.018 (0.070)
Softening						
Lime softening	205 (777)	64 (241)	41 (155)	0.026 (0.100)	0.011 (0.040)	0.001 (0.005)
Ion exchanger softener	201 (761)	90 (339)	55 (208)	0.021 (0.080)	0.006 (0.022)	0.004 (0.015)
Membrane filtration						
MF/UF	302 (1145)	183 (691)	167 (633)	0.40 (1.5)	0.066 (0.250)	0.028 (0.110)
RO/NF	957 (2623)	507 (1920)	320 (1210)	4 (15)	1.3 (4.8)	0.276 (0.015)
Electrodialysis reversal	172 (650)	121 (456)	102 (385)	0.150 (0.580)	0.120 (0.470)	0.250 (0.940)
GAC	819 (3101)	599 (2266)	553 (2093)	0.558 (2.113)	0.390 (1.490)	0.231 (0.873)
Ion exchange	14 (54)	9.8 (37)	6.7 (25)	0.05 (0.19)	0.036 (0.14)	0.025 (0.009)

(continued)

Table 11 (continued)

Process unit	Capital cost (including construction costs) ($/m³) ($/1000 gal)			O&M ($/m³) ($/1000 gal)		
	Small plant <4000 m³/d (<1 mgd)	Medium plant 4000–40,000 m³/d (1–10 mgd)	Large plant >40,000 m³/d (>10 mgd)	Small plant <4000 m³/d (<1 mgd)	Medium plant 4000–40,000 m³/d (1–10 mgd)	Large plant >40,000 m³/d (>10 mgd)
Disinfection						
Chlorine	19 (71)	6.7 (25)	2.6 (9.9)	0.025 (0.090)	0.007 (0.026)	0.001 (0.004)
Ozone (5 mg/L)	44 (166)	27 (103)	17 (64)	0.026 (0.100)	0.009 (0.034)	0.002 (0.008)
Chloramines	26 (98)	9.2 (35)	3.2 (12)	0.0004 (0.0016)	0.0004 (0.0016)	0.0004 (0.0016)
Ultraviolet radiation	44 (165)	29 (111)	23 (87)	0.090 (0.034)	0.03 (0.007)	0.003 (0.012)
Corrosion control (lime or soda ash)	30 (113)	6.1 (23.2)	2.8 (11)	0.004 (0.015)	0.0024 (0.009)	0.001 (0.004)

8 Environmental Aspects of Groundwater Management

Groundwater management is crucial for sustaining rural livelihoods, urban growth, irrigation, and industrial activities. Key considerations for environmentally sound groundwater management include:

- *Sustainability of Groundwater Development*: Abstraction rates must not exceed recharge rates to prevent groundwater mining. Excessive abstraction leads to declining groundwater levels, increased pumping costs, and eventual economic unfeasibility for uses such as agriculture.
- *Protection of Groundwater Quality*: Measures must be taken to control activities that could impair groundwater quality. This includes managing the leaching of chemicals from agriculture, preventing contamination from landfills and waste disposal, ensuring proper sewage treatment, and addressing issues such as salinity from inefficient irrigation or over-pumping in coastal areas.
- *Mitigation of Environmental Impacts*: Poor groundwater management can lead to adverse environmental effects, such as the desiccation of wetlands and reduced baseflow. Environmental impact assessments are valuable tools for planners to anticipate and address the potential impacts of groundwater development activities, helping to select options that maximize benefits while mitigating negative effects.

Groundwater treatment and integrated management can better support sustainable development and environmental health by addressing these considerations.

References

Ahmad AR, Zulke DP, Ahmed M, Aminuddinl BY, Sharma ML, Mohd ZM (1994) Environmental Impact of agricultural inorganic pollution on groundwater resources of the Kelantan Plain, Malaysia. In: Agricultural impacts on groundwater quality, the proceedings of an international workshop held in Kota Bharu, Kelantan, Malaysia. Australian Centre for International Agricultural Research

Al-Muyeed A, Afrin R (2006) Investigation of the efficiency of existing iron and arsenic removal plants in Bangladesh. J Water Supply Res Technol – AQUA 55(4):293–299

AWWA and ASCE (American Society of Civil Engineers) (1998) Water Treatment Plant Design, 3rd ed. AWWA and ASCE, McGraw-Hill, New York, 806 p.

Bjerg P, Broholm M, Lemming G, Heron G (2013) Groundwater remediation. In: Encyclopedia of environmetrics, 2nd edn. John Wiley & Sons, Ltd

Cleasby JL, Logsdon GS (1999) Granular bed and precoat filtration. In: Letterman RD (ed) Water quality and treatment, 4th edn. American Water Works Association, McGraw-Hill, New York, pp 8.1–8.92

Clifford DA (1999) Ion exchange and inorganic adsorption. In: Letterman, R.D. (ed.), Water Quality and Treatment, 4th ed. American Water Works Association, McGrawHill, New York, pp. 9.1–9.87

Clemens M, Khurelbaatar G, Merz R, Siebert C, van Afferden M, Rödiger T (2020) Groundwater protection under water scarcity; from regional risk assessment to local wastewater treatment solutions in Jordan. Sci Total Environ 706:1–12

Cui ZF, Jiang Y, Field RW (2010) Fundamentals of pressure-driven membrane separation processes. In: Membrane technology. Elsevier

Dawoud M (2020) Groundwater economics in arid regions: Abu Dhabi Emirate case study. Desalin Water Treat 176:84. https://doi.org/10.5004/dwt.2020.25501

Dawoud MA, Sallam GR, Abdelrahman M, Emam M (2024) The performance and feasibility of solar-powered desalination for brackish groundwater in Egypt. Sustain For 16(4):1630

Diodato N, Bellocchi G (2024) Millennium-scale changes in the Atlantic Multidecadal Oscillation influenced groundwater recharge rates in Italy. Commun Earth Environ 5(1):56

Driscoll FG (1986) Groundwater and wells, 2nd edn. Johnson Filtration Systems, St. Paul

Elyanow D, Persechino J (2005) Advances in nitrate removal. Available at: http://www.gewater.com/pdf/Technical%20Papers Cust/Americas/English/TP1033EN.pdf. GE technical papers. Accessed 1 Aug 2007

EAP (Environmental Protection Agency) (2005) Source Water Assessment and Protection Programs Show Initial Promise, But Obstacles Remain, Evaluation Report No. 2005-P-00013

García-Ávila F, Zhindón-Arévalo C, Donoso-Moscoso S, Tonon-Ordoñez M, Flores del Pino L (2020) Optimization of water use in a rapid filtration system: a case study. Water-Energy Nexus 3(2020):1–10

Gupta R, Srivastava P, Sardar KA, Kanaujia A (2018) Ground water pollution in india- a review. IJTAS 10(1):79–82

ITRC (2008) In situ bioremediation of chlorinated ethene: dnapl source zones. The Interstate Technology & Regulatory Council

Kawamura S (2000) Integrated design and operation of water treatment facilities, 2nd edn. John Wiley & Sons, New York, p 691

Lapham WW, Franceska WD, Koterba MT (1997) Guidelines and standard procedures for studies of ground-water quality: selection and installation of wells and supporting documentation. U.S. Geological Survey Water-Resources Investigations Report 96-4233, Reston

Li H, Gu X, Song J, Hui K, Chen G, Tan W, Wang H, Jiang Y, Yuan Y (2024) Effects of soil-groundwater environmental factors on BTEX transport and transformation: a review. J Environ Chem Eng 12(5):113697

LeChevallier MW, Au K (2004) Water Treatment and Pathogen Control: Process Efficiency in Achieving Safe Drinking Water. WHO Drinking Water Quality Series. IWA Publishing

Machiwal D, Jha MK, Singh VP, Mohan C (2018) Assessment and mapping of groundwater vulnerability to pollution: current status and challenges. Earth-Sci Rev 185:901–927. https://doi.org/10.1016/j.earscirev.2018.08.009

Masood N, Batool S, Farooqi A (2021) Chapter 22: Global groundwater: source, scarcity, sustainability, security, and solutions. In: Groundwater pollution in Pakistan. Elsevier, pp 309–322

Musa S, Ariff NA, Kadir MNA, Denan F (2016) Simplified method for groundwater treatment using dilution and ceramic filter. Mater Sci Eng 136(2016):012074. https://doi.org/10.1088/1757-899X/136/1/012074

National Ground Water Association (NGWA) (2003) The ground water supply and its use. Available at: http://www.wellowner.org/agroundwater/gwsupplyanduse.shtml/. Accessed 4 Aug 2003

Postel S, Richter BD (2003) Rivers for life—managing water for people and nature. Island Press, Washington, p 253

Potters for Peace (2007) Filters https://www.pottersfor-peace.org/?page_id=9

Salvestrina H, Hagare P (2009) Removal of nitrates from groundwater in remote indigenous settings in arid Central Australia. Desalin Water Treat 11(2009):151–156

Siegrist RL, Urynowicz MA, West OR, Crimi ML, Lowe KS (2001) Principles and practices of in situ chemical oxidation using permanganate. Battelle Press, Columbus

Siegrist RL, Crimi M, Simpkin TJ (2011) In situ chemical oxidation for groundwater remediation. Springer, New York, p 678

Singer PC, Reckhow DA (1999) Chemical oxidation. In: Letterman RD (ed) Water quality and treatment, 4th edn. American Water Works Association, McGraw-Hill, New York, pp 12.1–12.46

USEPA (1999) Guidance manual for compliance with the interim enhanced surface water treatment rule: turbidity provisions. USEPA Office of Water. EPA 815-R-99-010

USEPA (2005) Technologies and costs document for the final long term 2 enhanced surface water treatment rule and final stage 2 disinfectants and disinfection byproducts rule. Office of Water. EPA 815-R-05-013. Available at: www.epa.gov/safewater/disinfection/lt2/regulations.html – 21k. Accessed 10 Oct 2007

Witkowski AJ, Kowalczyk A, Jaroslav V (2017) Groundwater vulnerability assessment and mapping, 1st edn IAH-Selected Papers 11. CRC Press

Xanke J, Liesch T, Goeppert N, Klinger J, Gassen N, Goldscheider N (2017) Contamination risk and drinking water protection for a large-scale managed aquifer recharge site in a semi-arid karst region, Jordan. Hydrogeol J 25:1795

Decoration Plants and Their Role in Remediation

Jaspreet Kaur, Gaurav Mudgal, and Sanjoy Kumar Dhar

Abstract This chapter investigates the innovative integration of decorative plants in environmental remediation, examining their dual functionality in ecological restoration and aesthetic enhancement. The study covers various dimensions, encompassing the use of decorative plants in phytoremediation, their impact on soil cleanup, contributions to improving air quality, and their role in water filtration by aquatic plants. This analysis emphasizes the economic and ecological advantages of employing nonedible decorative plants for phytoremediation, including their potential to convert contaminated areas into ecotourism sites and the substantial market demand for these plants. This explores specific ornamental species acknowledged for their phytoremediation capabilities, scrutinizing their capacity to absorb, accumulate, and translocate heavy metals. A scientific examination extends to the multifaceted influences that govern remediation through ornamental plants, taking into account environmental variables like substrate composition and exposure durations. The investigation into the intricate relationship between aesthetic greenery and air quality improvement underscores the diverse mechanisms by which green spaces contribute to minimizing emissions, boosting oxygen production, and enhancing overall urban well-being. Moreover, this delves into the distinct role of decorative aquatic plants in water filtration, emphasizing their potential to address water pollu-

J. Kaur (✉)
Department of Biotechnology, Chandigarh University, Mohali, Punjab, India

G. Mudgal
Department of Biotechnology, Chandigarh University, Mohali, Punjab, India

Center for Waste Management and Renewable Energy, Saveetha Dental College and Hospitals, Saveetha Institute of Medical and Technical Sciences, Saveetha University, Chennai, India

S. K. Dhar
Center for Waste Management and Renewable Energy, Saveetha Dental College and Hospitals, Saveetha Institute of Medical and Technical Sciences, Saveetha University, Chennai, India

Banophool Adibashi Green Heart College, Mirpur, Dhaka, Bangladesh

A. A. H. Abdel Latef et al. (eds.), *Sustainable Remediation for Pollution and Climate Resilience*, https://doi.org/10.1007/978-981-96-5674-5_25

tion. The analysis surpasses individual plant-centric examinations, considering the interplay of environmental factors affecting water filtration. This reveals the interconnected eco-physiological dynamics that determine the efficacy of ornamental aquatic plants in contributing to cleaner and healthier aquatic environments. This interdisciplinary study bridges the gap between traditional remediation methods and visually appealing green solutions, providing a thorough understanding of the active role played by decorative plants in establishing cleaner, healthier, and more aesthetically pleasing environments. The findings strongly advocate continuously exploring and integrating decorative plants in sustainable environmental management practices.

Keywords Phytoremediation · Biophilic · Aquatic · Indoor plants · Heavy metals · Rhizofiltration · Contamination

1 Introduction of Decorative Plants in Environmental Remediation

Decorative plants utilized in environmental remediation encompass the deployment of ornamental or aesthetically pleasing flora to rehabilitate and enhance environmental conditions. In contrast to conventional practices where plants are primarily employed for ecological functionality, such as phytoremediation to address soil contamination, decorative plants fulfill a dual role. They not only aid in environmental rehabilitation but also elevate the visual aesthetics of a given area. This innovative approach is gaining traction as it harmoniously blends the effectiveness of traditional remediation methods with the visual allure of ornamental vegetation (Sadasivam 2022).

Decorative plants are meticulously chosen for their visual appeal, rendering them suitable for various landscaping and beautification undertakings. Their vivid hues, distinctive shapes, and seasonal blooms can metamorphose polluted or deteriorated spaces into alluring and aesthetically pleasing environments. Beyond their ornamental value, these plants are strategically selected to contribute to the restoration and rehabilitation of compromised ecosystems. They play a role in enhancing soil quality, offering wildlife refuge, and augmenting a region's overall ecological equilibrium. Specific decorative plants can act as natural filters, purifying the air and water. They can aid in ameliorating air pollution, mitigating the urban heat island effect, and absorbing contaminants from stormwater runoff (Lajayer et al. 2019). By incorporating a variety of decorative plant species, it is possible to foster biodiversity within a particular area. This, in turn, supports the presence of pollinators and other wildlife, ultimately bolstering the ecosystem's overall health (Chowdhuri and Deka 2019).

Integrating decorative plants into environmental remediation efforts has added the benefit of engaging and educating the public about the significance of environmental conservation. People are more likely to connect with and appreciate green spaces that are functional and visually captivating. Landscape architects and designers play a pivotal role in seamlessly integrating decorative plants into environmental remediation projects. They can craft attractive and functional designs incorporating diverse ornamental species (Cantor et al. 2008). Decorative plants are particularly valuable in urban and suburban regions where environmental challenges like pollution, soil degradation, and limited green spaces are prevalent. The use of these plants can effectively address these challenges and enhance the overall quality of life for residents. Thoughtful planning is essential for maintaining decorative plants in environmental remediation projects. This encompasses considerations for irrigation, fertilization, pruning, and pest management to ensure plants' long-term vibrancy and vitality (Rocha et al. 2022; Dela Cruz et al. 2014).

Decorative plants in environmental remediation exemplify a comprehensive approach to ecological and aesthetic enhancement. It exemplifies the potential to amalgamate the practical advantages of environmental restoration with the visual allure of ornamental flora, creating spaces that are cleaner, healthier, and more enjoyable for people to experience.

2 Use of Decorative Plants in Phytoremediation

Phytoremediation represents an eco-friendly approach employing plants to remediate pollutants from metal-contaminated soils and waters. However, using edible plants for this purpose can pose a risk of contaminating the food chain. To address this concern, cultivating nonedible plants, such as decorative plants, presents a viable and economically rewarding alternative (Lal et al. 2008; Ramana et al. 2008, 2009). Decorative plants, which are grown for their aesthetic appeal, come in various shapes, sizes, and colors suitable for diverse climates and landscaping preferences. They encompass various plant types, (herbaceous to woody, lower level to higher), including both terrestrial and aquatic varieties (Liu et al. 2007b; Sharma and Messar 2017). Notably, some decorative plants enhance the environment and follow the cleanup approach for contaminated soils, especially in urban settings (Sadasivam 2022; Han et al. 2007).

The application of decorative plants for phytoremediation offers the prospect of both ecological and economic benefits. Contaminated areas can be transformed into ecotourism sites by cultivating decorative plants, and preserving the land's aesthetic beauty. Additionally, through artificial cultivation, decorative plants can be sold as houseplants, leading to economic gains and reducing disposal costs (Liu et al. 2006). The global plant tourism industry, valued at billions of dollars annually, underscores the substantial market for decorative plants (Tao et al. 2015). As the demand for decorative plants grows, the industry is actively pursuing new plant varieties that exhibit superior characteristics such as improved anatomy, stress tolerance, and disease resistance. This quest aims to expand the options available for phytoremediation purposes (Noman et al. 2017).

Decorative plants present a diverse reservoir of germplasm resources on a global scale, boasting around 30,000 species. This wide array enables the selection of remediation plants from diverse decorative species, including those thriving in different climate zones. Although decorative plants might not always meet the stringent hyperaccumulator standards concerning heavy metal concentration, their substantial biomass suggests they can accumulate heavy metals at levels comparable to or exceeding hyperaccumulators. The scientific literature underscores a preference for herbaceous plants possessing aesthetic appeal in studies focused on phytoremediation (Liu et al. 2013; Marques et al. 2013). The quick life cycle of herbaceous ornamental plants permits ongoing observation of plant responses to stress throughout an entire growing season in a greenhouse environment, making them well-suited for evaluating potential heavy metal accumulation or hyperaccumulation. Furthermore, the aboveground biomass of herbaceous ornamental plants utilized for phytoextraction can typically be harvested shortly after flowering, often within 1 year of initial planting.

3 Ornamental Plants and Soil Cleanup

The exploration of ornamental plants delves into the intricate scientific interplay between specific plants and their ability to remediate contaminated soils with various pollutants. In a world struggling with an escalating issue of soil pollution, this critically examines the scientific intricacies of harnessing ornamental plants as active agents in the ecological restoration of soil ecosystems. Initiating with a robust recognition of the growing global concern regarding soil contamination, the need to develop scientifically sound, sustainable, and environmentally conscious solutions. Ornamental plants, traditionally celebrated for their aesthetic appeal, emerge as potential contributors to soil cleanup, prompting a shift in the traditional perception of their ecological role.

The scientific categorization of ornamental species is renowned for their phytoremediation capabilities. Species such as *Alyssum maritima*, *Aptenia cordifolia*, *Chrysanthemum*, *Brassica juncea*, *Mirabilis jalapa*, *Brassica oleracea*, *Calendula*, *Aster*, *Dahlia*, *Impatiens balsamina*, *Tagetes erecta*, and *Vinca rosea* are meticulously examined (Erdogan et al. 2011; Rathore et al. 2019; Kusznierewicz et al. 2012; Liu et al. 2021; Biswal et al. 2022; Wei et al. 2018; Ehsan et al. 2016b). Each species undergoes scrutiny for its metal-specific absorption, accumulation, and translocation capacities, shedding light on the biochemical and physiological mechanisms underpinning phytoremediation.

This employs a scientific lens to analyze experimental findings related to heavy metal remediation, encompassing hexavalent chromium (Cr), lead (Pb), copper (Cu), and arsenic (As). Complex dynamics, including metal uptake kinetics, translocation patterns, and the resulting impact on plant morphology, are thoroughly investigated. This scientific rigor provides a comprehensive understanding of how ornamental plants actively engage in the intricate phytoremediation processes.

Moving beyond individual plant-centric examination, this broadens its scientific scope to encompass the multifaceted influences governing ornamental plant-mediated remediation. Environmental variables, such as substrate composition, metal concentrations, and exposure durations, undergo scientific scrutiny, unveiling the intricate eco-physiological interactions that dictate the efficacy of ornamental plants in soil cleanup endeavors (Awad et al. 2021).

Scientific scrutiny and appreciation extend beyond their ornamental allure, positioning these plants as dynamic entities actively contributing to the intricate fabric of ecological restoration. As the scientific community struggles with the complexities of soil contamination, ornamental plants emerge not merely as passive greenery but as dynamic entities, exhibiting scientific promise in pursuing a resilient and ecologically balanced environment.

4 Aesthetic Greenery and Air Quality Improvement

The enduring value of aesthetic greenery, encompassing lush gardens, meticulously planned parks, and thoughtfully designed landscapes, has long been acknowledged for its capacity to elevate the visual appeal of urban settings. Nevertheless, the advantages conferred by these green spaces transcend the realm of aesthetics. In this section, we explore the intricate relationship between aesthetic greenery and its pivotal role in enhancing air quality, underscoring its profound impact on urban life and the environment.

Aesthetic greenery, characterized by its vibrant color and serene beauty, contrasts the concrete expanses of urban areas. It fosters an inviting ambiance that resonates with urban residents and visitors, providing a source of tranquility and rejuvenation amid the chaotic pace of city life. The contribution of aesthetic greenery to the amelioration of air quality is multifaceted and indispensable for the sustainability of urban environments. Several fundamental mechanisms support this enhancement: (a) Carbon dioxide (CO_2) absorption: Through the process of photosynthesis, plants and trees assimilate CO_2, reducing greenhouse gas emissions and aiding in efforts to mitigate climate change. This not only bolsters the global carbon equilibrium but also aids in reducing air pollution within local urban areas (Suhaimi et al. 2017; Lee and Kang 2015). (b) Filtration of Particulate Matter (PM): The leaves and surfaces of plants serve as natural filters, capturing minute airborne particles (PM 2.5 and PM 10). This filtration capability helps diminish the amount of these detrimental pollutants in the atmosphere, rendering the air safer for inhalation (Pettit et al. 2017). (c) Oxygen production: Photosynthesis engenders the production of oxygen, an essential element for human and animal life. The presence of greenery leads to elevated oxygen levels, thereby enriching overall air quality (Bao et al. 2022). (d) Temperature reduction: Trees and vegetation furnish shade and lower ground temperatures, thereby mitigating the formation of ground-level ozone, a noxious air pollutant. Moreover, this cooling effect reduces the energy demand for cooling, reducing emissions (Nievola et al. 2017). (e) Encouragement of eco-friendly transportation: The aesthetic appeal of green spaces can serve as an incentive for adopting alternative modes of transportation, such as walking and cycling. This decreased reliance on motor vehicles contributes to reduced emissions and positively impacts air quality (Namoun et al. 2021; Duque et al. 2014).

Urban green spaces can provide a conducive environment for diverse flora and fauna. Biodiversity within urban areas plays a pivotal role in preserving ecological balance, including natural pest control, pollination, and overall environmental health. Green spaces offer a platform for community engagement, social interaction, and physical activity. The air quality and the aesthetic charm of these spaces make significant contributions to the overall well-being and mental health of residents, fostering a sense of connection with the environment (Sinou et al. 2023; Brilli et al. 2018).

In view of the multifaceted benefits associated with aesthetic greenery in enhancing air quality, numerous cities are integrating green infrastructure into their urban planning and policy endeavors. These initiatives encompass tree-planting programs, establishing urban forests, and promoting rooftop gardens, thereby nurturing a greener and healthier urban milieu. The dual role of aesthetic greenery in elevating urban aesthetics and augmenting air quality is undeniable. This section underscores the symbiotic association between green spaces and cleaner air, elucidating the profound impact of their coexistence on urban life, environmental sustainability, and the welfare of residents. As the process of urbanization unfolds, it becomes imperative to acknowledge and prioritize the preservation and expansion of aesthetic greenery in the ongoing pursuit of healthier, more visually captivating, and sustainable urban environments.

5 Water Filtration by Decorative Aquatic Plants

Ornamental aquatic plants explore the multifaceted contributions to the purification of specific water bodies. The intricate mechanisms through which these decorative plants enhance water quality address concerns related to water pollution and environmental sustainability. Beginning with acknowledging the escalating global issues concerning water contamination, this emphasizes the importance of exploring nature-based solutions for water filtration. Decorative aquatic plants, renowned for their aesthetic appeal, emerge as pivotal elements in the ecological restoration of aquatic ecosystems, signaling a paradigm shift in their perceived role from mere adornments to active contributors to water purification. Diverse species such as *Echinodorus*, *Cryptocoryne*, *Nymphaea*, and *Pistia stratiotes* take center stage in this analysis, undergoing meticulous scrutiny for their nutrient absorption, oxygen release, and overall impact on water quality. The underlying physiological and biochemical processes that govern the efficacy of these plants in water filtration. Parameters such as nutrient removal kinetics, bioaccumulation of contaminants, and the influence of plant density on water clarity are rigorously examined. This scientific approach provides a comprehensive understanding of the active role played by decorative aquatic plants in mitigating water pollution (Cule et al. 2022; Belmont and Metcalfe 2003).

Beyond individual plant-centric analyses, the part broadens its scientific perspective to encompass the intricate interplay of environmental factors influencing water filtration. Variables such as water temperature, pH levels, nutrient concentrations, and the presence of other aquatic organisms undergo scientific scrutiny, revealing the interconnected eco-physiological dynamics that dictate the efficacy of ornamental aquatic plants in water purification endeavors. A diverse array of fast-growing ornamental macrophytes, well-suited to various aquatic conditions, holds the potential for wastewater treatment, offering improved water quality and economic benefits while enhancing ecological aesthetics (Madera-Parra et al. 2015; Sandoval et al. 2019).

Certain tropical plant species, such as *Colocasia esculenta* (L.) and *Heliconia psittacorum* Schott, demonstrate remarkable phytoremediation potential for heavy metals in landfill leachate when cultivated in constructed wetlands (Ali et al. 2020). In the specific case of *Episcia cordifolius*, higher biomass production was observed under elevated Cd and Zn treatments, accompanied by an extensive fibrous root system. However, this increased biomass was associated with lower percentage relative growth rates, possibly linked to factors, such as low plant height, root length, carbon-to-nitrogen ratio (CN), and carbon-to-biomass ratio (CB).

Research findings indicate that cadmium (Cd) can impede root growth and elongation, resulting in reduced carbon (C) and nitrogen (N) levels (Sricoth et al. 2018). Cd inhibitory effect on nitric oxide synthase activity can reduce endogenous nitric oxide, crucial for crown root primordia initiation in plants (Xiong et al. 2009). Likewise, abnormal zinc levels, whether excessive or deficient, can affect crown root development, carbon-to-nitrogen ratio (CN), and biomass in different plant species (Sricoth et al. 2018; Nanda and Wissuwa 2016). In contrast, *H. psittacorum* and *H. spathocircinata*, while exhibiting modest biomass yields, displayed percentage relative growth rates surpassing those of other plants in the study. This underscores its resilience and suitability for phytoremediation in contaminated environments. On the other hand, *P. cordata* demonstrated low dry biomass and relative growth rates, possibly due to its lower heavy metal tolerance in harsh conditions. The combined Cd-Zn treatments, while affecting the growth performance of the study plants, showed lower toxicity compared to individual Cd or Zn treatments. As corroborated by prior research, the phenomenon of dilution effects in metal mixtures played a role in the reduced percentage of relative growth rates recorded in *E. cordifolius* cultivated in zinc (Zn) solutions. Elevated levels of Zn, which are recognized for their capacity to elicit toxic responses in plants, could result in impaired growth performance (Chayapan et al. 2015; Mwamba et al. 2016).

Decorative aquatic plants emphasize the pivotal role in the scientific discourse surrounding sustainable water management. Beyond their visual appeal, these plants are recognized as dynamic entities contributing actively to the intricate water filtration processes. As societies struggle with the challenges of water pollution, ornamental aquatic plants emerge as indispensable allies, providing a scientifically grounded approach to achieving cleaner and healthier aquatic environments.

6 Indoor Plants: Mechanism and Role in Purification

Plants play a crucial role in environmental cleanup through a process referred to as "phytoremediation," which encompasses various techniques for the elimination of contaminants from the surroundings (Fig. 1) (Pilon-Smits 2005). Regarding indoor environments, plants are regarded as natural purifiers that employ multiple methods to cleanse the environment. These methods include absorption, dilution, precipitation, and filtration, as supported by several studies (Kim et al. 2018; Lee et al. 2020).

One of the most well-known processes orchestrated by plants is photosynthesis during which they enhance air quality by absorbing CO_2 and releasing O_2. Conversely, respiration is another process in which plants take in O_2 and release CO_2. These intricate processes involve the exchange of gases through stomata, which serve as plants' primary apparatus for their absorption and filtration mechanisms, as pointed out by Jones (1998).

Fundamentally, plants exhibit an extraordinary capability to restore ecological equilibrium and capture airborne molecules in the surrounding atmosphere, as

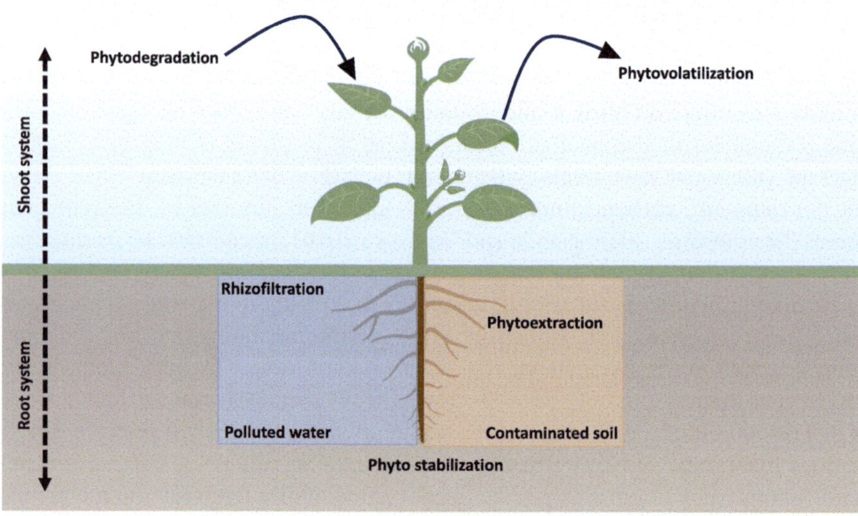

Fig. 1 Various phytoremediation techniques utilized by plants across diverse environments, classified into five primary categories: Phyto-stabilization: Entails trapping contaminants and pollutants in the soil to reduce their availability for biological processes, thereby preventing their leaching into water or dispersion in the air (Morikawa and Erkin 2003). Phytoextraction: Accumulation of pollutants within the tissues of plants (Lee 2013). Phytodegradation: Transforming toxic pollutants into substances that are less harmful or nontoxic (Cunningham et al. 1995; Newman et al. 1997; Lee 2013). Rhizofiltration (Phyto-filtration): Approach utilizes various plant components such as roots (rhizofiltration), seedlings (blasto-filtration), or shoots (caulofiltration), to eliminate pollutants from water or wastewater (Mesjasz-Przybyłowicz et al. 2004; Yan et al. 2020). Phytovolatilization: Involves the transformation of toxic volatile elements, absorbed from the soil, into less harmful forms (Mahar et al. 2016)

evidenced by Agarwal and group (Agarwal et al. 2019). Furthermore, indoor plants demonstrate the capacity to cleanse indoor air by efficiently eliminating pollutants, i.e., volatile organic compounds (VOCs), particulate matter, organic compounds, carbon dioxide, sulfates, carbonyls, nitrates, calcium, ammonia, carbonates, and ozone. Consequently, indoor plants offer a cost-efficient remedy that notably decreases indoor pollutant levels, thereby reducing human exposure to numerous harmful compounds (Pegas et al. 2011; Soreanu et al. 2013; Abbass et al. 2017; Parseh et al. 2018; Wei et al. 2021).

Indoor relative humidity (RH) serves as a crucial element that impacts the intricate dynamics of humans, viruses, and plants. Generally, the suggested comfort range of indoor RH levels for humans is 30–60%, as indicated by Fan and coworkers (Fan et al. 2017). Furthermore, maintaining indoor humidity levels at 40–60% has been shown to hinder viral transmission, according to Audi and the group (Audi et al. 2020). Research demonstrates that plants possess the capacity to absorb minute airborne water vapor particles, contributing to the enhancement of indoor humidity, as outlined in studies (Jeong et al. 2008; Pérez-Urrestarazu et al. 2016).

A relevant concern arises in the transpiration process of indoor plants via stomata, which emits water droplets into the nearby air, thereby increasing RH levels. Nevertheless, as humidity levels rise, the transpiration rate of plants decreases, aiding in sustaining indoor humidity at levels favorable for human comfort. Pegas and colleagues underscore the significant correlation between humidity and the proliferation of molds, mildew, bacteria, and biological pathogens, all of which can substantially affect human health (Pegas et al. 2011).

Notably, plants are recognized for their ability to modulate air humidity, particularly through foliar water uptake, and this action helps reduce bioaerosols, as documented in studies (Lohr and Pearson-Mims 1996; Limm et al. 2009; Gawrońska and Bakera 2015; Brilli et al. 2018). Consequently, plants can regulate air humidity, keeping it at intermediate levels that can diminish the viability of certain viruses. Plants emit minor amounts of secondary metabolites and their derivatives, including alkaloids and polyphenols, into the atmosphere. The process is referred to as allelochemicals, and these compounds have reported antimicrobial properties, interacting with airborne microbes in the vicinity of the plant. Research conducted by Wolverton and coworkers illustrated that indoor houseplants can potentially decrease airborne microbes by around 50% compared to indoor environments devoid of plants. The volatile compounds emitted by houseplants are considered a significant contributor to regulating the population of airborne microbes in environments where plants are present, as emphasized in the study conducted by Wolverton and Wolverton (1996).

Recent studies indicate that plants play a remarkable role in reducing air microbiomes, thereby fostering a healthier environment and lowering the risk of exposure to airborne diseases (Li et al. 2021). Additionally, an investigation examined 11 distinct household plant species, uncovering a considerable decrease in ultrafine particles within indoor settings (Stapleton and Ruiz-Rudolph 2018).

Although there is substantial evidence suggesting that plants can influence airborne microorganisms such as bacteria and fungi, there is a lack of direct experimental proof supporting the notion that plants can effectively reduce virus transmission in indoor settings (Wolverton and Wolverton 1996; Pegas et al. 2011; Li et al. 2021). While various research studies have shown the effects of indoor plants on bacteria and fungi, extra research is needed to identify how various plants can mitigate airborne microbes and determine which plant species are most efficient in reducing both bacterial and viral transmission indoors.

7 Biophilic Design and Its Influence on Human Well-Being

Biophilic design, an ideology embedded in present-day architectural and interior design approaches, signifies a deep perception of the interconnectedness between humans and the natural world. In an era characterized by swift urbanization and technological progress, this design philosophy emerges as a transformative influence, aiming to bridge the widening gap between individuals and the natural world (Huntsman and Bulaj 2022). Fundamentally, biophilic design embodies a comprehensive method in crafting built environments, acknowledging the intrinsic human inclination toward nature. It surpasses mere considerations of aesthetics, delving into the realms of human psychology, biology, and overall well-being. The fundamental premise is evident: By incorporating natural elements into the spaces we inhabit, there exists the potential to enhance not only the visual appeal of our surroundings but also our mental, emotional, and physical states (Kellert et al. 2011; Barbiero and Berto 2021).

Natural light, a foundational aspect of biophilic design, transcends its practical purpose. The deliberate integration of windows, skylights, and other openings aims to optimize exposure to natural light, aligning with the circadian rhythms governing our daily lives. Beyond its energy-efficient attributes, this focus on natural light positively influences mood, sleep patterns, and overall cognitive function, highlighting the intricate interplay between our environment and well-being (McGee and Park 2022; Noser et al. 2022; Garcia et al. 2021).

The integration of greenery emerges as another pivotal dimension of biophilic design. Indoor plants, vertical gardens, and green walls are not merely decorative elements but active contributors to improved air quality, stress reduction, and a reconnection to the natural world. The use of nature-inspired materials, such as wood and stone, further deepens this connection, infusing spaces with warmth, authenticity, and a tangible reminder of the organic world (Yin et al. 2020). Water features, ranging from indoor fountains to reflective ponds, introduce a sensory layer to biophilic design. Beyond their aesthetic appeal, these elements bring in water's calming sounds and visual dynamism, creating spaces that transcend the purely visual realm. The therapeutic effects of water on stress reduction and mental well-being align seamlessly with the overarching goals of biophilic design (McGee and Park 2022; Yin et al. 2020).

The emphasis on views and access to the outdoors becomes a cornerstone of this design philosophy. The provision of expansive windows, outdoor terraces, or interconnected spaces blurs the boundaries between interior and exterior, providing occupants with vistas of natural landscapes. This deliberate connection to the outdoors offers moments of respite, opportunities for reflection, and a constant reminder of the broader natural context. Dynamic and interactive elements constitute yet another layer of biophilic design, ensuring that spaces evolve and engage their occupants. Environments that respond to seasonal changes, installations influenced by weather patterns, and designs encouraging movement and exploration contribute to a more immersive and vital living experience. The dynamic nature of these spaces resonates with the inherently dynamic and ever-changing aspects of the natural world (Hahn et al. 2021).

Growing research supports the positive impacts of biophilic design on human well-being. Studies reveal that exposure to nature within the built environment reduces stress levels and enhances cognitive function, creativity, and overall psychological resilience. In healthcare settings, incorporating biophilic elements has been linked to expedited recovery times and improved patient outcomes. This empirical evidence underscores the potential of biophilic design to transcend its conceptual underpinnings and emerge as a tangible catalyst for positive change in the lives of individuals (Yao et al. 2021; McMahan and Estes 2015).

Biophilic design serves as a guiding principle in our endeavor to create environments that foster a harmonious coexistence between humanity and the natural world (Fig. 2). Beyond its visual allure, it represents a conscientious effort to address the multifaceted well-being of individuals in a rapidly evolving world. As urban landscapes continue to expand, biophilic design principles offer a blueprint for a sustainable, regenerative, and deeply enriching approach to the spaces we inhabit. The influence of biophilic design on human well-being extends far beyond the visual and physical realms, permeating the very essence of living in harmony with nature.

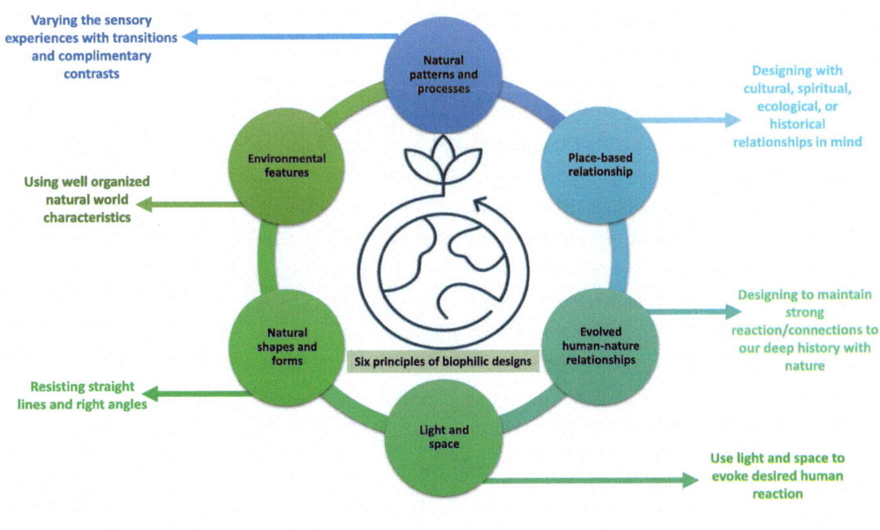

Fig. 2 Principles of biophilic design

8 Integrating Decorative Plants into Urban Landscapes

Urban environments display varying degrees of human influence, particularly in large cities with high population density, increased pollution levels, and limited green space. The rapid urbanization trend, predicted to encompass nearly 70% of all-round populations residing in urban areas by 2050, presents significant challenges such as water scarcity, climate change, pollution, and decreased human well-being. To tackle these challenges, cities must actively seek solutions (Revi et al. 2017; Organization 2016). One approach involves enhancing urban ecosystems to amplify their environmental benefits by adopting green urban infrastructure. Green infrastructure (GI), as delineated by the European Biodiversity and GI Strategies, consists of a strategically planned network of natural areas designed to provide various ecosystem services, including air quality improvement, water purification, climate mitigation, recreational spaces, and adaptation.

Although ornamental plants are commonly acknowledged as crucial living elements within urban environments, their potential for providing essential ecosystem services is frequently underestimated. When chosen thoughtfully, ornamental plants can substantially contribute to human well-being. Ecosystem services, classified according to the Common International Classification of Ecosystem Services (CICES), include the provision of material and energy requirements. Maintenance and regulation of the environment, and non-material aspects that influence both physical and mental well-being (Assessment 2005; Elmqvist et al. 2015; Nawrath et al. 2022).

In densely populated industrialized nations, major urban centers can face challenges with air and soil pollution caused by human activities such as indoor combustion, industrial emissions, and vehicular traffic. Particulate matter (PM) pollution, which contains heavy metals, can negatively impact environmental quality and public health. Moreover, this pollution hinders the selection of suitable ornamental plant species and constrains the use of contaminated soil for cultivation (Schwela 2000; Li et al. 2022).

Maintaining and enhancing green infrastructure becomes crucial for sustainable development, offering a range of ecosystem services. Despite the scientific evidence supporting the relationship between green infrastructure, plants, and life quality, the role of ornamental plants is often underestimated or negatively perceived, particularly when they are nonnative. Ornamental plants, cultivated for aesthetic purposes, contribute significantly to the visual quality of both indoor and outdoor environments. Beyond their ornamental value, these plants can serve practical purposes, such as providing edible flowers or supporting phytoremediation efforts (Kendal et al. 2012). Recognizing the potential of ornamental plants is essential for addressing environmental challenges, improving well-being, and restoring degraded landscapes. Ornamental plants have played essential roles in various civilizations, connecting with myths and cultural practices. In contemporary urbanized environments, ornamental plants are not merely luxuries but necessities, fostering a connection with nature. Their cultivation and relevant horticulture practices offer numerous benefits, including economic, environmental, social, and aesthetic advantages for human well-being (Savé 2007; Wilson et al. 2016; Gopal and Nagendra 2014).

Integrating ornamental plants into urban environments presents an excellent opportunity to enhance the visual allure of the surroundings, foster biodiversity, and elevate the overall quality of life for urban residents. Some concepts and considerations for integrating ornamental plants into urban settings: (a) Plant selection: Select plants that match the local climate, soil conditions, and sunlight availability. Indigenous plants, well-suited to the regional environment, are frequently favored because they generally demand less upkeep. (b) Vertical gardening: Make efficient use of vertical spaces for planting. Vertical gardens, or green walls, can be installed on building facades, fences, or other structures, introducing greenery without occupying valuable ground space. (c) Street-side trees: Enhance the beauty of streets by planting trees, providing not just aesthetic appeal but also shade, improved air quality, and support for biodiversity. Choose tree species resistant to urban stressors with minimal impact on sidewalks and infrastructure. (d) Community gardening: Establish community gardens where residents can collectively plant and maintain ornamental plants. This fosters community spirit, encourages sustainable practices, and offers access to fresh produce. (e) Public parks and plazas: Infuse well-designed landscaping into public spaces like parks and plazas. Consider a diverse selection of plant species, seating areas, and walking paths to craft inviting and serene environments. (f) Green rooftops: Implement green roofs on buildings to maximize urban green space. Beyond insulation benefits, green roofs support biodiversity and mitigate stormwater runoff. (g) Seasonal plant arrangements: Plan for seasonal variations by selecting plants that bloom or change colors at different times of the year. This approach ensures a dynamic and ever-changing urban landscape. (h) Public art and plant synergy: Integrate public art installations with plantings to create visually captivating landscapes. Sculptures, murals, or other artistic elements can complement and amplify the beauty of ornamental plants. (i) Educational signage: Include informative signage about the plants used in urban landscaping. This educational aspect informs residents and visitors about the advantages of greenery and promotes environmental awareness. (j) Maintenance and irrigation: Implement a sustainable maintenance plan to ensure the longevity of the greenery. Consider water-efficient irrigation systems and involve the community in plant care activities. (k) Biodiversity considerations: Strive for a diverse array of plants to support local ecosystems. This approach attracts pollinators, birds, and other wildlife, contributing to a healthier urban environment. (l) Collaboration with local entities: Collaborate with local environmental organizations, horticultural societies, or community groups to garner support for urban landscaping initiatives. Collaboration brings in expertise, resources, and community engagement. The successful integration of ornamental plants into urban landscapes necessitates meticulous planning, community participation, and consistent maintenance. By infusing greenery into urban spaces, cities can fashion more vibrant, sustainable, and enjoyable environments for both residents and visitors.

9 Case Studies: Successful Implementation of Decorative Plant Remediation

Based delve into the multifaceted advantages of decorative plant remediation, spanning from air quality enhancement to psychological well-being, through our meticulously curated selection of compelling case studies.

Based on the findings of various experiments, formaldehyde has been observed to have detrimental effects on living organisms, impacting their respiratory systems, and blood health, and increasing the risk of conditions such as leukemia and brain-related ailments (Duan et al. 2018; Ge et al. 2019; Zhang et al. 2009). Furthermore, it is linked to the phenomenon known as "sick building syndrome," which manifests in symptoms such as nausea, headaches, eye irritation, dizziness, mucous membrane irritation, fatigue, drowsiness, and overall discomfort (Ghaffarianhoseini et al. 2018; Nakayama et al. 2019). Recent advancements have enforced a formaldehyde threshold below 1% to address these health issues to mitigate the sick building syndrome (Suzuki et al. 2019).

Several researchers have conducted experiments using indoor ornamental plants to eliminate formaldehyde from indoor environments (Meng and Ji 2012; Wang et al. 2020). The study aimed to assess the potted Areca palm's success in diminishing total organic pollutants indoors. This investigation involved nine potted plants deployed over 4 months in a real indoor setting, achieving an elimination efficiency of nearly 90% (Bhargava et al. 2021). Furthermore, the *Areca* palm has been reported for its capability to remove xylene and toluene (Wolverton et al. 1989; Wolverton and Wolverton 1993; Samudro and Mangkoedihardjo 2021). Similarly, research involving three species of tropical plants, *Nephrolepis exaltata*, *Rhapis excelsa,* and *Dracaena fragrans*, demonstrated their capacity to purify indoor air by eliminating various gases (Aini Jasmin et al. 2012). A specific palm species, *Chamaedorea elegans*, effectively reduced formaldehyde levels indoors when exposed to different concentrations of up to 16 mg/m^3. Results indicated a maximum purification rate of approximately 1.5 $mg/m^2/h$ achieved for 2 days at the highest concentration. It was noted that maintaining a plant area-to-floor area ratio of 70%:30% yielded optimal results, with well-lit areas showing superior formaldehyde elimination compared to darker spaces (Teiri et al. 2018).

Research comparing formaldehyde removal in illuminated and dark environments using *Ficus japonica* and *Ficus benjamina L.* plants indicated that aboveground parts of these species narrowed concentration of formaldehyde during daylight hours; their impact was minimal at night. In contrast, the roots of these plants efficiently removed significant formaldehyde amounts during the day and night. The efficacy of the roots in formaldehyde removal was primarily attributed to microorganisms and root activity, highlighting enhanced rhizo-degradation as the primary process for formaldehyde elimination (Kim et al. 2008; Nakayama et al. 2019; Aydogan and Montoya 2011). Additionally, experiments involving plants from the *Araceae*, *Agavaceae*, and *Liliaceae* families, conducted in a space measuring 0.8 m^3 initially containing formaldehyde at a concentration of 15 mg/m^3, indicated that most species from these plant families effectively reduced formaldehyde levels within 3 days. However, *Philodendron selloum* exhibited less resistance to formaldehyde exposure (Zhou et al. 2011).

A study focusing on *Nephrolepis obliterata*, a commonly found indoor fern species, aimed to eradicate formaldehyde from the environment, with a supreme formaldehyde vapor concentration of 11 mg/m^3 over 48 h. These results highlight the significant potential of these plant species in improving indoor environments by reducing formaldehyde exposure (Teiri et al. 2018). Effective removal of toluene and xylene was achieved using plants such as *Schefflera actinophylla* and *Ficus benghalensis*. *S. actinophylla* exhibited eliminating efficiencies of approximately 13 μg/m^3/m^2 of leaf area for the whole day, while *F. benghalensis* achieved about half of this efficiency. It is noteworthy that, approximately half of the toluene and xylene concentrations were transferred from the air to the root via the stem, indicating the significance of the root zone (Kim et al. 2016).

Intensive research has been conducted to identify numerous plant species capable of effectively eliminating various VOC gases. When exposed to gas concentrations as high as 10 ppm over 6 h, *Hedera helix*, *Hemigraphis alternata*, *Asparagus densiflorus*, and *Hoya carnosa* exhibited the greatest effectiveness in removing pollutants, as compared to other plant types (Yang et al. 2009). Various plant species displayed differing and generally lower removal efficiencies, highlighting the crucial role of biodiversity in the potential of plants to process pollutants (Samudro and Mangkoedihardjo 2020; Ren et al. 2017). Furthermore, four decorative plants—*Chlorophytum comosum*, *Epipremnum aureum*, *Echinopsis tubiflora*, and *Hedera helix*—attained a 70% average benzene removal efficiency. *Echinopsis tubiflora*, specifically, showcased outstanding removal efficiency attributed to its elevated transpiration rate and chlorophyll content, distinguishing it from the other three plant species (Gong et al. 2019). The increased transpiration of these plants implies the need for more frequent watering when employed for indoor remediation. Similarly, heightened chlorophyll content underscores the necessity for adequate sunlight exposure to optimize its utilization. Therefore, plant care and maintenance considerations are imperative (Mangkoedihardjo and Samudro 2014; Ni et al. 2019) to support the plants' maximum potential to eliminate polluting gases.

In a study conducted by Liu and colleagues, significant benzene removal efficiency was observed in *Dracaena anderiana*, *Chlorophytum comosum*, *Hedera helix*, *Hydrangea macrophylla*, *Clitori aternatea*, *Sansevieria trifasciata*, *Cymbidium* "Golden elf," *Syngonium podophyllum*, *Euphorbia milii*, *Crassula portulacea* (Liu et al. 2007a). These plants were also observed to remove various other pollutants in the water (Kaur et al. 2023). Toluene phytoremediation can be accomplished through the utilization of *Philodendron spp.* "Sunlight," *Schefflera elegantissima*, and *Hedera helix* (Kim et al. 2011). *Zamioculcas zamiifolia* has demonstrated efficacy in remediating xylene (Sriprapat et al. 2014), and to some extent, ethylbenzene (Toabaita et al. 2016). *Fittonia argyroneura*, *Hoya carnosa*, *Asparagus densiflorus*, and *Hemigraphis alternata*, have shown efficient removal of VOCs, including toluene, benzene, octane, trichloroethylene (TCE), and α-pinene, whereas *Ficus benjamina* has been acknowledged for its capability to remove octane and α-pinene.

According to findings by Wood and collaborators, the Dracaena deremenis "Janet Craig" has been identified as an excellent species for VOC removal (Wood et al. 2006). Similarly, *Ficus spp.* have exhibited significant efficacy in eliminating VOCs such as formaldehyde, ethylbenzene, styrene, benzene, acetaldehyde, toluene, and xylene (Hong et al. 2017). Importantly, decorative indoor plants have demonstrated effectiveness in reducing VOC levels and decreasing concentrations of carbon dioxide and carbon monoxide, as well as mitigating particulate matter (Panyametheekul et al. 2018).

Most ornamental plants tend to amass heavy metals primarily in their roots. Examples include *Nerium oleander* (Trigueros et al. 2012), *Quamoclit pinnata*, *Antirrhinum majus* (Cui et al. 2013), *Erica andevalensis*, and *Erica australis* (Pérez-López et al. 2014), *Calendula officinalis* (Liu et al. 2008b), and *Tagetes patula* (Chaturvedi et al. 2014; Chintakovid et al. 2008). This natural characteristic enables these plants not only to remediate contaminated soils through phytostabilization but also to enhance the visual appeal of the environment, providing a potential source of income.

Three decorative plants, that is, *Tagetes erecta*, *Salvia splendens*, and *Abelmoschus manihot*, showed remediation potential against the most prevalent contaminant named cadmium. The investigation showed minimal effects on all three seed germination and shoot elongation for *Salvia splendens*. However, a noticeable inhibitory effect on root elongation was observed in tested plants, along with a significant impact on shoot elongation in *Tagetes erecta*. Cd-tolerance indices indicated that *A. manihot* exhibited the highest tolerance to Cd, while *S. splendens* was found to be the most sensitive (Wang and Zhou 2005). Bosiacki documented substantial Cd accumulation in *S. splendens* and *Helianthus annuus* shoots and leaves—an ornamental species extensively studied for remediation. *Tagetes erecta* demonstrated an average potential to accumulate Cd, with the highest concentration found in roots, followed by shoots and leaves, and the lowest in inflorescences (Bosiacki 2008).

Three varieties of *Tuberose* were exposed to five different concentrations of cadmium (Cd) (0–100 mg kg^{-1} soil), and none of the varieties displayed any toxic macroscopic symptoms. *Tuberose* showed Cd accumulation in shoots exceeding 100 μg g^{-1} dry weight, with a Cd ratio >1 in shoots to bulbs, indicating its potential effectiveness as a Cd accumulator (Ramana et al. 2012). *Chlorophytum comosum* exhibited potential as a Cd accumulator, with a tolerance index exceeding 100 at a soil Cd concentration of 100 mg kg^{-1} in a pot experiment. In-depth research into the interaction between *Canna indica* and Cd highlighted its significant potential for cadmium accumulation, although the root concentration factor exceeded the bioconcentration factor, suggesting limited translocation (Wang et al. 2012). *Calendula officinalis* demonstrated normal growth in soils containing 100 mg kg^{-1} Cd, accumulating high metal levels in both shoots and roots. In a hydroponic culture, *Althaea rosea* and *C. officinalis* exhibited good accumulation capacity and tolerance to cadmium and lead (Pb) (Liu et al. 2008a).

Brassica juncea, Alyssum maritima, Aptenia cordifolia, and *Brassica oleracea* underwent investigation to assess their ability to translocate and uptake chromium (VI) through irrigation. Elevated concentrations of Cr noteworthy increased translocation and accumulation of the metal in the shoots and roots of species, with the maximal values observed in *Alyssum maritima* shoots and *Brassica juncea* roots (Budak et al. 2011). Study involving exposure to Cr for four ornamental plants—*aster*, *chrysanthemum*, *calendula*, and *dahlia*—the metal led to a substantial reduction in plant growth at 10 mg kg^{-1}, causing widespread mortality in *Chrysanthemum* at 25 mg kg^{-1}. Overall, only *calendula* emerged as a potential for phytoremediation of soils contaminated with low levels of Cr (Ramana et al. 2013). A study involving three ornamental species cultured in pots with substrate containing Cr concentrations revealed a decline in biomass for *Impatiens balsamina* as the Cr dose increased. *Vinca rosea*, when grown in pots with soil containing chromium levels ranging from 10 to 60 mg kg^{-1}, exhibited decreased plant height, and dry and fresh weight at higher chromium contamination levels (Ehsan et al. 2016b).

In the presence of lead (Pb), *V. rosea* showed remediation potential exceeding 1 at Pb concentrations ranging from 20 to 40 mg kg^{-1}, but this potential decreased at concentrations between 50 to 90 mg kg^{-1} (Ehsan et al. 2016b). The same trends were observed with *Zinnia elegans* grown in pots with soils containing levels of cadmium and lead. Plants cultivated in lead-contaminated soils exhibited better health compared to those in soils contaminated with chromium (Ehsan et al. 2016a). *C. officinalis* displayed robust tolerance to copper contamination, withstanding levels up to 400 mg kg^{-1}, and showing maximum copper accumulation in soil treated with 300 mg kg^{-1} (Goswami and Das 2016). *Syngonium sp.* and *T. patula* were evaluated in pots containing soil with arsenic (As) concentrations of up to 10 mg kg^{-1}, demonstrating significant As accumulation, particularly in the roots (Huq et al. 2005). In hydroponic system, various ornamental plant species were tested for their potential in arsenic (As) remediation, including switchgrass (*Panicum virgatum*), iris (*Iris savannarum*), *Coreopsis lanceolata*, *Tithonia rotundiflora*, marigold (*Tagetes erecta*), sunflower (*H. annuus*), *Tithonia*, and *Coreopsis* exhibited reductions of 85% and 65% in dry weight, respectively, at an As concentration of 0.75 mg l^{-1}. At the highest As concentration, sunflower and marigold demonstrated uptake ratios of 16.6 and 7.4 (Reed et al. 2013).

Hydroponic conditions were utilized to examine the impact of arsenic (As) on *Zinnia elegans* and *Gomphrena globosa*. Arsenic was predominantly found in the leaves, roots, flowers, and stems, suggesting that these species were tolerant of arsenic but lacked significant remediate ability (Signes-Pastor et al. 2015). Regarding the remediation of mixed heavy metals, a study involving *Alternanthera bettzickiana*, commonly used as an ornamental edging plant, indicated substantial accumulation of cadmium (Cd) and lead (Pb) at concentrations up to 1.0 mM, with greater total uptake of metals in shoots compared to roots (Tauqeer et al. 2016). *Canna indica* was identified as a proficient accumulator of Cd, Pb, nickel (Ni), zinc (Zn), and chromium (Cr), displaying high translocation factors (TF) for Ni and Cr (Subhashini and Swamy 2014). The macrophyte *Iris pseudacorus* showed promise for chromium (Cr) rhizofiltration and zinc (Zn) phytoextraction, exhibiting strong tolerance and accumulation capacity for these metals (Caldelas et al. 2012). *Helianthus annuus* and *Hydrangea paniculata* accumulated noteworthy levels of copper (Cu) and lead (Pb) (Forte and Mutiti 2017). *Tanacetum vulgare* demonstrated environmental adaptability to high industrial pollution and exhibited a remarkable capacity for mercury and lead uptake (Stevovi and Calic-Dragosavac 2010).

10 Aquatic Decorative Plants for Remediation

Both freshwater and marine resources face contamination from diverse toxic elements due to human activities and natural sources. Addressing polluted aquatic environments is just as crucial as tackling terrestrial pollution. Heavy metals, known for their significant toxicity even in trace amounts, endure in the environment, presenting hazards to organisms and readily infiltrating the food web. The surveillance and elimination of these pollutants are expensive endeavors. Deterioration in water quality arises from various sources of pollution, including both localized and diffuse origins, with agricultural and urban/rural practices serving as primary culprits. As a substantial user of chemical fertilizers, pesticides, and manure-derived amendments, agriculture exacerbates the issue.

In wetland ecosystems, plants serve as crucial components, aiding in prolonging water retention, decreasing flow speed, and improving the settling of particles and contaminants. Aquatic vegetation plays a vital role in managing eutrophication by absorbing nitrogen (N) and phosphorus (P) through photosynthesis while releasing oxygen simultaneously. The constructed wetlands approach is extensively utilized for eliminating heavy metals from wastewater, utilizing mechanisms like settling, absorption, coprecipitation, ion exchange, phytoaccumulation, and microbial processes (Dhir 2013).

Excessive iron levels in waterlogged soils present a significant hurdle, as iron, vital for plant growth, can become toxic in high concentrations. Under anaerobic conditions with low pH, a substantial portion of insoluble Fe^{3+} converts into more soluble Fe^{2+}. Wetland plants evolve aerenchyma cells to facilitate the transfer of O_2 from aboveground parts to the roots, triggering the precipitation of FeO_2 on root surfaces. This formation, known as iron plaque, is observed in many wetland plants and possesses a notable capacity for functional groups on iron hydroxides, aiding in the sequestration of metals through adsorption and/or coprecipitation (Zhong et al. 2010). Constructed wetlands for phytoremediation of heavy metals offer an environmentally friendly and cost-effective solution compared to traditional cleanup methods. Since metals primarily accumulate in roots, the biomass harvested from these wetlands holds promise for diverse nonfood industries, including biogas, compost, and fiber production.

Aquatic plants (ornamental) are known for their robust growth and resilience in polluted environments. Certain species exhibit remarkable effectiveness in purging various pollutants, such as total nitrogen (N), phosphorus (P), biological oxygen demand, chemical oxygen demand, and heavy metals like chromium (Cr), iron (Fe), lead (Pb), cadmium (Cd), manganese (Mn), and copper (Cu), from sewage. Addressing the crucial challenge of eutrophication control, research conducted by

Li and colleagues demonstrated that *Oenanthe javanica* contributed to a reduction of 28–34% in N and 25.2–33.4% in P, while *Potamogeton crispus* exhibited 61.5% N removal and 67.5% P removal (Li et al. 2015). *Zantedeschia aethiopica* also significantly impacted N removal (Belmont and Metcalfe 2003).

Aquatic ornamental plants serve a twofold purpose, aiding in water purification and enhancing flower yield, providing economic benefits. *Iris* species, recognized for their phytoremediation capabilities, demonstrate potential for cleansing water contaminated with Cd, Cr, and Zn (Han et al. 2007; Caldelas et al. 2012). Their significant biomass production and ability to extract metals make *Iris* species suitable for Cr filtration through roots and Zn extraction through plants. Limiting the transfer of Cr to leaves could have positive effects on flower production, aboveground yield, and human safety (Caldelas et al. 2012). Another promising candidate for phytoremediation is *Zantedeschia aethiopica*, commonly referred to as white *Calla lilies*, which shows potential for remediation of iron contamination (Casierra-Posada et al. 2014).

The lotus plant (*Nelumbo nucifera*) showcases the capacity to gather contaminants like heavy metals and pesticides from water and sediment. The bioaccumulation of polychlorinated-biphenyls predominantly takes place in the leaves, followed by the stems and roots of lotus plants (Dai et al. 2014). The Hindu lotus (*N. nucifera*) effectively treats total phosphorus and total nitrogen from zoo wastewater, achieving removal rates of up to 58% and 38%, respectively (Jiang and Xinyuan 1998).

Water lilies (*Nymphaea*), which grow in the mud with their leaves floating on the water's surface, demonstrate tolerance to the highest levels of toxic metals and can accumulate significantly. According to Twining (1993), the uptake of metals tends to increase with the age of the plant tissue. These water lilies, known to accumulate heavy metals such as chromium (Cr), Lead (Pb), and cadmium (Cd) exhibit unique characteristics, including epidermal glands located on the submerged surfaces of their leaf blades, petioles, and rhizomes. The co-crystallization of cadmium (Cd) and calcium (Ca) in these epidermal glands leads to the depletion of manganese (Mn) and occasionally other elements like iron (Fe), lead (Pb), and mercury (Hg) in the epidermal glands of the leaves (Lavid et al. 2001). Cadmium (Cd) deposition and storage in water lilies occur without apparent translocation, indicating immobilization in the epidermal glands during the initial stages of heavy metal accumulation. This immobilization contributes to the water lily's high tolerance to heavy metals. Photosynthetic energy or products likely play a role in the uptake of heavy metals by the epidermal glands of water lilies, as metal uptake is lower in the absence of light, and the green tissues in the leaves accumulate higher levels of metals compared to the non-green rhizome tissues.

Lotus and water lily are commonly found in wetlands or ponds. The widespread use of lotus, often employed for decorative purposes in various festivals, necessitates a significant supply of essential nutrients such as phosphorus and nitrogen for successful growth. These nutrients are typically obtained from fertilizers or compost. Maintaining a balance in nutrient levels in lotus ponds is essential for effectively treating nonpoint sources, especially agricultural wastewater discharged from farmlands (Seo et al. 2010). The lotus, also known as the sacred lotus, or arrowroot, holds considerable cultural and culinary value as an aquatic plant cultivated and consumed across Asia. Practically every part of the lotus plant, including stamens, pollen, flowers, stems, rhizomes, and leaves, finds utility in culinary practices and various therapeutic applications in countries like India and China. Apart from being a food source, the lotus plant serves multiple purposes, with its stems used as fresh vegetables and fuel, rhizomes as canned food, fresh vegetables, starch, dessert, and seeds as both dessert and medicine and flowers as religious ornaments. Moreover, various parts of the plant serve as raw materials for cosmetic production. Scientifically, many components of the lotus plant, such as seeds, rhizomes, stamens, and leaves, have demonstrated notable pharmacological activities (Leong et al. 2012).

However, it is important to highlight that lotus plants exhibit an accumulation strategy for heavy metals. Research by Leong and colleagues emphasized that the maximal concentrations of phenolic content, lead (Pb), and chromium (Cr) were found in the seeds of lotus plants cultivated in a former tin-mining pond, surpassing permissible limits set by food safety regulations for Pb (Leong et al. 2012). Although reports suggest that heavy metal contents in lotus seeds from mining areas were comparable to those in reference areas, and heavy metal accumulation in lotus roots was primarily independent or partially negatively correlated with soil, assessing metal concentrations remains a crucial priority before utilizing contaminated areas for cultivating edible plants (Dingjian 2012; Xiong et al. 2013).

11 Challenges and Future Perspectives for Decorative Plant-Based Remediation

Decorative plant-based remediation offers a profound exploration of the innovative and eco-conscious utilization of decorative plants in environmental remediation. In the face of mounting environmental challenges, this section provides a compelling vision of the potential of decorative plants in tackling issues such as soil contamination, air pollution, and water quality.

This underscores the multifaceted advantages of integrating decorative plants into environmental remediation endeavors. These plants, far from being purely ornamental, possess inherent qualities that make them natural agents for remediation. Notably, their root systems exhibit an extraordinary capacity to absorb, sequester, and even detoxify contaminants, rendering them invaluable in purifying soil and water. This intrinsic ability to capture pollutants, encompassing heavy metals, hydrocarbons, and organic compounds, opens the door to sustainable environmental cleanup solutions. Additionally, the significant role of decorative plants is in ameliorating air quality by filtering harmful pollutants, thus contributing to enhanced human health (Bartucca et al. 2023; Khan et al. 2022).

Preference of relevant plant species and their adaptability to diverse environmental conditions represent prominent hurdles, given the complexity of ensuring the survival and efficacy of these plants in varying contexts. This part also underscores the need for a nuanced understanding of the distinct requirements and limitations of different decorative plant species, while also considering the potential risks of introducing nonnative species to new environments. Another key concern that, the enduring effectiveness of decorative plant-based remediation. While these plants serve as nature's environmental custodians, sustaining their performance over time proves to be a significant challenge. Factors such as plant health, soil conditions, and ongoing maintenance require meticulous monitoring and management to achieve successful, long-term remediation (Fig. 3) (Shinomol and Bhanu 2021).

In spite of these challenges, spotlights innovative techniques, such as phytoremediation, which leverages the natural abilities of plants to remove contamination in various areas. Furthermore, integrating green infrastructure into urban planning is highlighted as a progressive approach that harmonizes functionality with environmental benefits, paving the way for a greener and more sustainable future. It underscores the significance of merging aesthetics with environmental responsibility. It serves as an invaluable resource for environmental scientists, policymakers, and researchers committed to creating cleaner, healthier, and more sustainable ecosystems. As we look forward to an era marked by heightened environmental awareness, this encourages us to embrace decorative plants as influential allies in our pursuit of a cleaner and more sustainable planet.

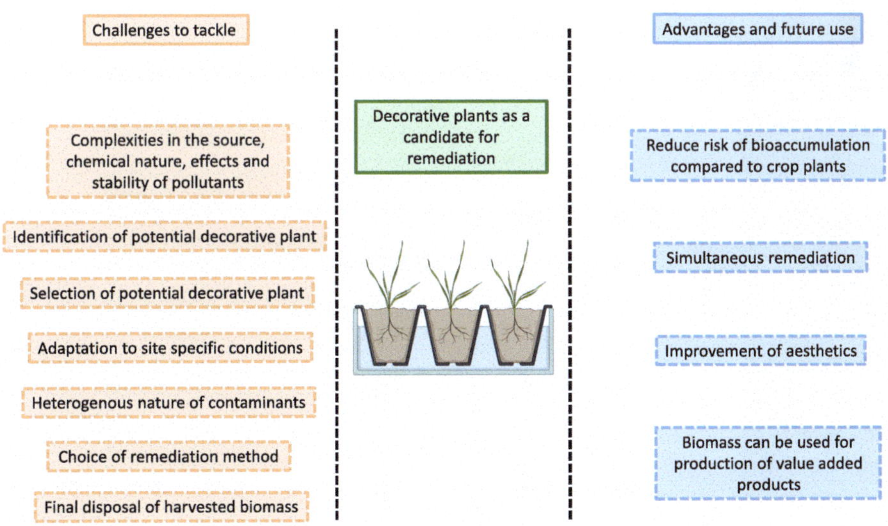

Fig. 3 Challenges and future perspective of remediation with decorative plants

12 Conclusion: The Beauty and Benefits of Decorative Plants in Remediation

Integrating decorative plants into environmental remediation enhances the visual appeal of surroundings and yields a myriad of advantages (Fig. 4). Carefully selected ornamental plants play a remarkable role in restoring and strengthening environmental quality. Attractiveness of decorative plants is intrinsically linked to their functional contributions to remediation processes. Whether by purifying the air or aiding in soil restoration, these plants are pivotal in mitigating the impacts of human-induced activities. Recognizing that their ornamental aspects should not overshadow their potential to contribute to broader ecological objectives is crucial.

Acknowledging the dual nature of decorative plants becomes essential as we confront the challenges posed by urbanization and environmental decline. Beyond their aesthetic charm, these plants function as allies in the pursuit of sustainable and resilient ecosystems. Leveraging their remediation capabilities allows us to craft visually pleasing landscapes and actively engage in restoring our natural environments. In the convergence of beauty and functionality, decorative plants emerge as indispensable components for fostering environmental health and well-being. Their integration into remediation strategies symbolizes a harmonious coexistence between human aesthetics and ecological resilience, charting a course toward a more sustainable and visually gratifying future.

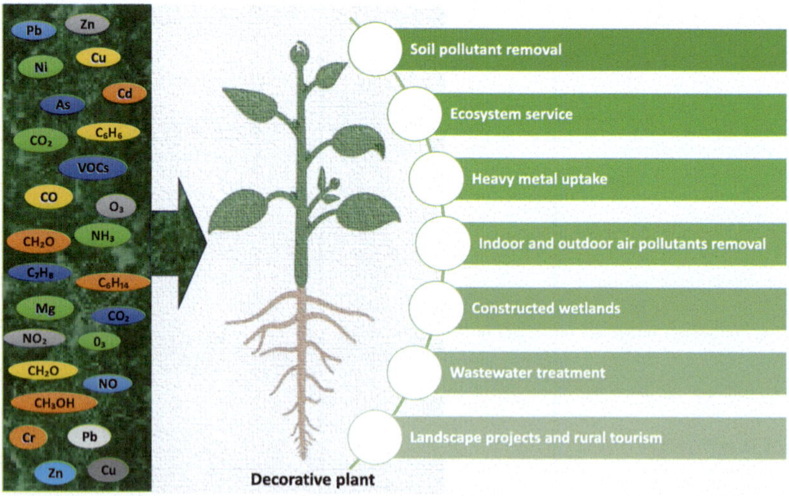

Fig. 4 Phytoremediation potentials of decorative plants

References

Abbass OA, Sailor DJ, Gall ET (2017) Effectiveness of indoor plants for passive removal of indoor ozone. Build Environ 119:62–70

Agarwal P, Sarkar M, Chakraborty B, Banerjee T (2019) Phytoremediation of air pollutants: prospects and challenges. In: Phytomanagement of polluted sites. Elsevier, pp 221–241

Aini Jasmin G, Noorizan M, Suhardi M, Murad AG, Ina K (2012) The use of plants to improve indoor air quality in small office space. Pertanika J Soc Sci Humanit 20:493–503

Ali S, Abbas Z, Rizwan M, Zaheer IE, Yavaş İ, Ünay A, Abdel-Daim MM, Bin-Jumah M, Hasanuzzaman M, Kalderis D (2020) Application of floating aquatic plants in phytoremediation of heavy metals polluted water: a review. Sustainability 12:1927

Assessment, Millennium Ecosystem (2005) Ecosystems and human well-being. Island Press, Washington

Audi A, AlIbrahim M, Kaddoura M, Hijazi G, Yassine HM, Zaraket H (2020) Seasonality of respiratory viral infections: will COVID-19 follow suit? Front Public Health 8:567184

Awad M, El-Desoky MA, Ghallab A, Jan Kubes SE, Abdel-Mawly SD, Ratnasekera D, Islam MS, Skalicky M, Brestic M (2021) Ornamental plant efficiency for heavy metals phytoextraction from contaminated soils amended with organic materials. Molecules 26:3360

Aydogan A, Montoya LD (2011) Formaldehyde removal by common indoor plant species and various growing media. Atmos Environ 45:2675–2682

Bao S, Li B, Zhao M (2022) Study on oxygen production capacity of green window system based on subsistence building. Int Core J Eng 8:88–94

Barbiero G, Berto R (2021) Biophilia as evolutionary adaptation: an onto-and phylogenetic framework for biophilic design. Front Psychol 12:700709

Bartucca ML, Cerri M, Forni C (2023) Phytoremediation of pollutants: applicability and future perspective. Plan Theory 12:2462

Belmont MA, Metcalfe CD (2003) Feasibility of using ornamental plants (Zantedeschia aethiopica) in subsurface flow treatment wetlands to remove nitrogen, chemical oxygen demand and nonylphenol ethoxylate surfactants—a laboratory-scale study. Ecol Eng 21:233–247

Bhargava B, Malhotra S, Chandel A, Rakwal A, Kashwap RR, Kumar S (2021) Mitigation of indoor air pollutants using Areca palm potted plants in real-life settings. Environ Sci Pollut Res 28:8898–8906

Biswal B, Singh SK, Patra A, Mohapatra KK (2022) Evaluation of phytoremediation capability of French marigold (Tagetes patula) and African marigold (Tagetes erecta) under heavy metals contaminated soils. Int J Phytoremediation 24:945–954

Bosiacki M (2008) Accumulation of cadmium in selected species of ornamental plants. Acta Sci Pol Hortorum Cultus 7:21–31

Brilli F, Fares S, Ghirardo A, de Visser P, Calatayud V, Muñoz A, Annesi-Maesano I, Sebastiani F, Alivernini A, Varriale V (2018) Plants for sustainable improvement of indoor air quality. Trends Plant Sci 23:507–512

Budak F, Zaimoglu Z, Basci N (2011) Uptake and translocation of hexavalent chromium by selected species of ornamental plants. Pol J Environ Stud 20:857–862

Caldelas C, Araus JL, Febrero A, Bort J (2012) Accumulation and toxic effects of chromium and zinc in Iris pseudacorus L. Acta Physiol Plant 34:1217–1228

Cantor M, Buta E, Chiş L, Hărşan E (2008) Improving the perennial ornamental plants collection for summer ladscaping design, Scientific papers of the Research Institute for Fruit Growing Pitesti

Casierra-Posada F, Blanke MM, Guerrero-Guío JC (2014) Eisentoleranz von Calla (Zantedeschia aethiopica). Gesunde Pflanzen 66:63–68

Chaturvedi N, Ahmed MJ, Dhal NK (2014) Effects of iron ore tailings on growth and physiological activities of Tagetes patula L. J Soils Sediments 14:721–730

Chayapan P, Kruatrachue M, Meetam M, Pokethitiyook P (2015) Phytoremediation potential of Cd and Zn by wetland plants, Colocasia esculenta L. Schott., Cyperus malaccensis Lam., and Typha angustifolia L. grown in hydroponics. J Environ Biol 36:1179

Chintakovid W, Visoottiviseth P, Khokiattiwong S, Lauengsuchonkul S (2008) Potential of the hybrid marigolds for arsenic phytoremediation and income generation of remediators in Ron Phibun District, Thailand. Chemosphere 70:1532–1537

Chowdhuri TK, Deka K (2019) Biodiversity and conservation of ornamental crops. In: Conservation utilization of horticultural genetic resources. Springer, Singapore, pp 139–216

Cui S, Zhang T, Zhao S, Li P, Zhou Q, Zhang Q, Han Q (2013) Evaluation of three ornamental plants for phytoremediation of Pb-contamined soil. Int J Phytoremediation 15:299–306

Cule N, Lucic A, Nesic M, Veselinovic M, Mitrovic S, Bosanac LB (2022) The efficiency of sodium removal by decorative plant species and algae in the floating treatment wetland. Fresenius Environ Bull 31:5243–5254

Cunningham SD, Berti WR, Huang JW (1995) Phytoremediation of contaminated soils. Trends Biotechnol 13:393–397

Dai S, Wong CS, Qiu J, Wang M, Chai T, Fan L, Yang S (2014) Enantioselective accumulation of chiral polychlorinated biphenyls in lotus plant (Nelumbonucifera spp.). J Hazard Mater 280:612–618

Dela Cruz M, Christensen JH, Thomsen JD, Müller R (2014) Can ornamental potted plants remove volatile organic compounds from indoor air?—a review. Environ Sci Pollut Res 21:13909–13928

Dhir B (2013) Phytoremediation: role of aquatic plants in environmental clean-up. Springer

Dingjian C (2012) Contents of heavy metals in lotus seed from REEs mining area. J Saudi Chem Soc 16:175–176

Duan J, Kang J, Qin W, Deng T, Liu H, Li B, Wei Y, Gong S, Yang X, Chen M (2018) Exposure to formaldehyde and diisononyl phthalate exacerbate neuroinflammation through NF-κB activation in a mouse asthma model. Ecotoxicol Environ Saf 163:356–364

Duque RB, Gray D, Harrison M, Davey E (2014) Invisible commuters: assessing a university's eco-friendly transportation policies and commuting behaviours. J Transp Geogr 38:122–136

Ehsan N, Nawaz R, Ahmad S, Arshad M, Umair M, Sarmad M (2016a) Remediation of heavy metal-contaminated soil by ornamental plant Zinnia (Zinnia elegance L.). Asian J Chem 28:1338

Ehsan N, Nawaz R, Ahmad S, Khan MM, Hayat J (2016b) Phytoremediation of chromium contaminated soil by an ornamental plant Vinca (Vinca rosea L.). J Environ Agric Sci 7:29–34

Elmqvist T, Heikki Setälä SN, Handel S, van der Ploeg J, Aronson JN, Blignaut EG-B, Nowak DJ, Kronenberg J, de Groot R (2015) Benefits of restoring ecosystem services in urban areas. Curr Opin Environ Sustain 14:101–108

Erdogan R, Zaimoglu Z, Budak F, Koseoglu C (2011) Use of sewage sludge in growth media for ornamental plants and its effects on growth and heavy metal accumulation. J Food Agric Environ 9:632–635

Fan G, Xie J, Yang H, Liu J (2017) Improving indoor air humidity in heated rooms based on natural evaporation of water from wetted textile fabric-A pilot study. Procedia Eng 205:1302–1309

Forte J, Mutiti S (2017) Phytoremediation potential of Helianthus annuus and Hydrangea paniculata in copper and lead-contaminated soil. Water Air Soil Pollut 228:1–11

Garcia LM, Birckhead BJ, Krishnamurthy P, Sackman J, Mackey IG, Louis RG, Salmasi V, Maddox T, Darnall BD (2021) An 8-week self-administered at-home behavioral skills-based virtual reality program for chronic low back pain: double-blind, randomized, placebo-controlled trial conducted during COVID-19. J Med Internet Res 23:e26292

Gawrońska H, Bakera B (2015) Phytoremediation of particulate matter from indoor air by Chlorophytum comosum L. plants. Air Qual Atmos Health 8:265–272

Ge S, Yan B, Huang J, Chen Y, Chen M, Yang X, Wu Y, Shen D, Ma P (2019) Diisodecyl phthalate aggravates the formaldehyde-exposure-induced learning and memory impairment in mice. Food Chem Toxicol 126:152–161

Ghaffarianhoseini A, AlWaer H, Omrany H, Ghaffarianhoseini A, Alalouch C, Clements-Croome D, Tookey J (2018) Sick building syndrome: are we doing enough? Archit Sci Rev 61:99–121
Gong Y, Zhou T, Wang P, Lin Y, Zheng R, Zhao Y, Bin X (2019) Fundamentals of ornamental plants in removing benzene in indoor air. Atmos 10:221
Gopal D, Nagendra H (2014) Vegetation in Bangalore's slums: boosting livelihoods, well-being and social capital. Sustain For 6:2459–2473
Goswami S, Das S (2016) Copper phytoremediation potential of Calandula officinalis L. and the role of antioxidant enzymes in metal tolerance. Ecotoxicol Environ Saf 126:211–218
Hahn N, Essah E, Blanusa T (2021) Biophilic design and office planting: a case study of effects on perceived health, well-being and performance metrics in the workplace. Intell Build Int 13:241–260
Han Y-L, Yuan H-Y, Huang S-Z, Guo Z, Xia B, Jiguang G (2007) Cadmium tolerance and accumulation by two species of Iris. Ecotoxicology 16:557–563
Hong S-H, Hong J, Jihyeon Y, Lim Y (2017) Study of the removal difference in indoor particulate matter and volatile organic compounds through the application of plants. Environ Health Toxicol 32:e2017006
Huntsman DD, Bulaj G (2022) Healthy dwelling: design of biophilic interior environments fostering self-care practices for people living with migraines, chronic pain, and depression. Int J Environ Res Public Health 19:2248
Huq SMI, Joardar JC, Parvin S (2005) Marigold (Tagetes patula) and ornamental arum (Syngonia sp.) as phytoremediators for arsenic in pot soil. Bangladesh J Bot 34:65–70
Jeong SJ, Song JS, Kim WS, Lee DW, Kim HD, Kim KJ, Yoo EH, Cho JG (2008) Evaluation of selected foliage plants for improvement of indoor humidity. Hortic Environ Biote 49:439–446
Jiang Z, Xinyuan Z (1998) Treatment and utilization of wastewater in the Beijing zoo by an aquatic macrophyte system. Ecol Eng 11:101–110
Jones HG (1998) Stomatal control of photosynthesis and transpiration. J Exp Bot 49:387–398
Kaur J, Mudgal G, Negi A, Tamang J, Singh S, Singh GB, Jagadeesh Chandra Bose K, Debnath S, Wadaan MA, Khan MF, Ruokolainen J, Kesari KK (2023) Reactive Black-5, Congo red and methyl orange: chemical degradation of azo-dyes by agrobacterium. Water 15:1664
Kellert SR, Heerwagen J, Mador M (2011) Biophilic design: the theory, science and practice of bringing buildings to life. John Wiley & Sons
Kendal D, Williams NSG, Williams KJH (2012) A cultivated environment: exploring the global distribution of plants in gardens, parks and streetscapes. Urban Ecosyst 15:637–652
Khan AU, Khan AN, Waris A, Ilyas M, Zamel D (2022) Phytoremediation of pollutants from wastewater: a concise review. Open Life Sci 17:488–496
Kim KJ, Kil MJ, Song JS, Yoo EH, Son K-C, Kays SJ (2008) Efficiency of volatile formaldehyde removal by indoor plants: contribution of aerial plant parts versus the root zone. J Am Soc Hortic Sci 133:521–526
Kim KJ, Yoo EH, Jeong MI, Song JS, Lee SY, Kays SJ (2011) Changes in the phytoremediation potential of indoor plants with exposure to toluene. HortScience 46:1646–1649
Kim KJ, Kim HJ, Khalekuzzaman M, Yoo EH, Jung HH, Jang HS (2016) Removal ratio of gaseous toluene and xylene transported from air to root zone via the stem by indoor plants. Environ Sci Pollut Res 23:6149–6158
Kim KJ, Khalekuzzaman M, Suh JN, Kim HJ, Shagol C, Kim H-H, Kim HJ (2018) Phytoremediation of volatile organic compounds by indoor plants: a review. Hortic Environ Biote 59:143–157
Kusznierewicz B, Bączek-Kwinta R, Bartoszek A, Piekarska A, Huk A, Manikowska A, Antonkiewicz J, Namieśnik J, Konieczka P (2012) The dose-dependent influence of zinc and cadmium contamination of soil on their uptake and glucosinolate content in white cabbage (Brassica oleracea var. capitata f. alba). Environ Toxicol Chem 31:2482–2489
Lajayer A, Behnam NK, Moghadam MR, Maghsoodi MG, Kariman K (2019) Phytoextraction of heavy metals from contaminated soil, water and atmosphere using ornamental plants: mechanisms and efficiency improvement strategies. Environ Sci Pollut Res 26:8468–8484

Lal K, Minhas PS, Chaturvedi RK, Yadav RK (2008) Cadmium uptake and tolerance of three aromatic grasses on the Cd-rich soil. J Indian Soc Soil Sci 56:290–294
Lavid N, Barkay Z, Tel-Or E (2001) Accumulation of heavy metals in epidermal glands of the waterlily (Nymphaeaceae). Planta 212:313–322
Lee JH (2013) An overview of phytoremediation as a potentially promising technology for environmental pollution control. Biotechnol Bioprocess Eng 18:431–439
Lee J, Kang H (2015) The effect of improving indoor air quality using some C. Indian J Sci Technol 8:26
Lee BX, Yi TH, Yuniarto A (2020) Phytoremediation mechanisms in air pollution control: a review. Water Air Soil Pollut 231:437
Leong ES, Tan S, Chang YP (2012) Antioxidant properties and heavy metal content of lotus plant (Nelumbo nucifera Gaertn) grown in ex-tin mining pond near Kampar, Malaysia. Food Sci Technol Res 18:461–465
Li J, Yang X, Wang Z, Shan Y, Zheng Z (2015) Comparison of four aquatic plant treatment systems for nutrient removal from eutrophied water. Bioresour Technol 179:1–7
Li H, Zhi-Feng W, Yang X-R, An X-L, Ren Y, Jian-Qiang S (2021) Urban greenness and plant species are key factors in shaping air microbiomes and reducing airborne pathogens. Environ Int 153:106539
Li C, Daolin D, Gan Y, Ji S, Wang L, Chang M, Liu J (2022) Foliar dust as a reliable environmental monitor of heavy metal pollution in comparison to plant leaves and soil in urban areas. Chemosphere 287:132341
Limm EB, Simonin KA, Bothman AG, Dawson TE (2009) Foliar water uptake: a common water acquisition strategy for plants of the redwood forest. Oecologia 161:449–459
Liu JN, Zhou QX, Wang XF, Zhang QR, Sun T (2006) Potential analysis of ornamental plant resources applied to contaminated soil remediation. Floriculture 3:245–252
Liu Y-J, Yu-Jing M, Zhu Y-G, Ding H, Arens NC (2007a) Which ornamental plant species effectively remove benzene from indoor air? Atmos Environ 41:650–654
Liu J-N, Zhou Q-X, Sun T, Wang X-F (2007b) Feasibility of applying ornamental plants in contaminated soil remediation. J Appl Ecol 18:1617–1623
Liu J-n, Zhou Q-x, Sun T, Ma LQ, Wang S (2008a) Growth responses of three ornamental plants to Cd and Cd–Pb stress and their metal accumulation characteristics. J Hazard Mater 151:261–267
Liu J-n, Zhou Q-x, Sun T, Ma LQ, Wang S (2008b) Identification and chemical enhancement of two ornamental plants for phytoremediation. Bull Environ Contam Toxicol 80:260–265
Liu Z, He X, Chen W, Zhao M (2013) Ecotoxicological responses of three ornamental herb species to cadmium. Environ Toxicol Chem 32:1746–1751
Liu Y, Weili X, Wang Y, Hao W, Zhou Q, Liu J (2021) Growth responses and accumulation characteristics of three ornamental plants to Sn contamination in soil. Agriculture 11:205
Lohr VI, Pearson-Mims CH (1996) Particulate matter accumulation on horizontal surfaces in interiors: influence of foliage plants. Atmos Environ 30:2565–2568
Madera-Parra CA, Peña-Salamanca EJ, Peña MR, Rousseau DPL, Lens PNL (2015) Phytoremediation of landfill leachate with Colocasia esculenta, Gynerum sagittatum and Heliconia psittacorum in constructed wetlands. Int J Phytoremediation 17:16–24
Mahar A, Wang P, Ali A, Awasthi MK, Lahori AH, Wang Q, Li R, Zhang Z (2016) Challenges and opportunities in the phytoremediation of heavy metals contaminated soils: a review. Ecotoxicol Environ Saf 126:111–121
Mangkoedihardjo S, Samudro G (2014) Research strategy on kenaf for phytoremediation of organic matter and metals polluted soil. Adv Environ Biol 8:64–67
Marques APGC, Moreira H, Franco AR, Rangel AOSS, Castro PML (2013) Inoculating Helianthus annuus (sunflower) grown in zinc and cadmium contaminated soils with plant growth promoting bacteria–effects on phytoremediation strategies. Chemosphere 92:74–83
McGee B, Park N-K (2022) Colour, light, and materiality: biophilic interior design presence in research and practice. Interiority 5:27–52

McMahan EA, Estes D (2015) The effect of contact with natural environments on positive and negative affect: a meta-analysis. J Posit Psychol 10:507–519
Meng G-Z, Ji K-S (2012) Comparisons of the capacity of four ornamental plants on purifying indoor formaldehyde pollution. In: 2012 Fifth international conference on information and computing science. IEEE, pp 225–227
Mesjasz-Przybyłowicz N, Migula A, Monika Tarnawska W, Reimold CK, Przybyłowicz W, Głowacka E (2004) Uptake of cadmium, lead nickel and zinc from soil and water solutions by the nickel hyperaccumulator Berkheya coddii. Acta Biol Crac Ser Bot 46:75–85
Morikawa H, Erkin ÖC (2003) Basic processes in phytoremediation and some applications to air pollution control. Chemosphere 52:1553–1558
Mwamba TM, Ali S, Ali B, Lwalaba JL, Liu H, Farooq MA, Shou J, Zhou W (2016) Interactive effects of cadmium and copper on metal accumulation, oxidative stress, and mineral composition in Brassica napus. Int J Environ Sci Technol 13:2163–2174
Nakayama Y, Nakaoka H, Suzuki N, Tsumura K, Hanazato M, Todaka E, Mori C (2019) Prevalence and risk factors of pre-sick building syndrome: characteristics of indoor environmental and individual factors. Environ Health Prev Med 24:1–10
Namoun A, Tufail A, Mehandjiev N, Alrehaili A, Akhlaghinia J, Peytchev E (2021) An eco-friendly multimodal route guidance system for urban areas using multi-agent technology. Appl Sci 11:2057
Nanda AK, Wissuwa M (2016) Rapid crown root development confers tolerance to zinc deficiency in rice. Front Plant Sci 7:428
Nawrath M, Elsey H, Dallimer M (2022) Why cultural ecosystem services matter most: exploring the pathways linking greenspaces and mental health in a low-income country. Sci Total Environ 806:150551
Newman LA, Strand SE, Choe N, Duffy J, Ekuan G, Martin Ruszaj B, Shurtleff B, Wilmoth J, Heilman P, Gordon MP (1997) Uptake and biotransformation of trichloroethylene by hybrid poplars. Environ Sci Technol 31:1062–1067
Ni J, Leung AK, Ng CWW (2019) Unsaturated hydraulic properties of vegetated soil under single and mixed planting conditions. Géotechnique 69:554–559
Nievola CC, Carvalho CP, Carvalho V, Rodrigues E (2017) Rapid responses of plants to temperature changes. Temperature 4:371–405
Noman A, Aqeel M, Deng J, Khalid N, Sanaullah T, Shuilin H (2017) Biotechnological advancements for improving floral attributes in ornamental plants. Front Plant Sci 8:530
Noser AE, Klages KL, Gamwell KL, Brammer CN, Hommel KA, Ramsey RR (2022) A systematic evaluation of primary headache management apps leveraging behavior change techniques. Cephalalgia 42:510–523
Organization, World Health (2016) Global report on urban health: equitable healthier cities for sustainable development
Panyametheekul S, Rattanapun T, Ongwandee M (2018) Ability of artificial and live houseplants to capture indoor particulate matter. Indoor Built Environ 27:121–128
Parseh I, Teiri H, Hajizadeh Y, Ebrahimpour K (2018) Phytoremediation of benzene vapors from indoor air by Schefflera arboricola and Spathiphyllum wallisii plants. Atmos Pollut Res 9:1083–1087
Pegas PN, Alves CA, Evtyugina MG, Nunes T, Cerqueira M, Franchi M, Pio CA, Almeida SM, Cabo Verde S, Freitas MC (2011) Seasonal evaluation of outdoor/indoor air quality in primary schools in Lisbon. J Environ Monit 13:657–667
Pérez-López R, Márquez-García B, Abreu MM, Nieto JM, Córdoba F (2014) Erica andevalensis and Erica australis growing in the same extreme environments: phytostabilization potential of mining areas. Geoderma 230:194–203
Pérez-Urrestarazu L, Fernández-Cañero R, Franco A, Egea G (2016) Influence of an active living wall on indoor temperature and humidity conditions. Ecol Eng 90:120–124
Pettit T, Irga PJ, Abdo P, Torpy FR (2017) Do the plants in functional green walls contribute to their ability to filter particulate matter? Build Environ 125:299–307

Pilon-Smits E (2005) Phytoremediation. Annu Rev Plant Biol 56:15–39
Ramana S, Biswas AK, Ajay SRA, Rao AS (2008) Phytoextraction of lead by marigold and chrysanthemum. Indian J Plant Physiol 13:297–299
Ramana S, Biswas AK, Ajay SAB, Rao AS (2009) Phytoextraction of Cadmium by African marigold and chrysanthemum. Natl Acad Sci Lett 32:333–336
Ramana S, Biswas AK, Singh AB, Naveen Kumar P, Ahirwar NK, Behera SK, Rao AS (2012) Phytoremediation of cadmium contaminated soils by tuberose. Indian J Plant Physiol 17:61–64
Ramana S, Biswas AK, Singh AB, Ajay NKA, Rao AS (2013) Phytoremediation ability of some floricultural plant species. Indian J Plant Physiol 18:187–190
Rathore SS, Shekhawat K, Dass A, Kandpal BK, Singh VK (2019) Phytoremediation mechanism in Indian mustard (Brassica juncea) and its enhancement through agronomic interventions. Proc Natl Acad Sci India Sect B Biol Sci 89:419–427
Reed ST, Ayala-Silva T, Dunn CB, Gordon GG, Meerow A (2013) Screening ornamentals for their potential as as accumulator plants. J Agric Sci 5:20
Ren Y, Ge Y, Ma D, Song X, Shi Y, Pan K, Zelong Q, Guo P, Han W, Chang J (2017) Enhancing plant diversity and mitigating BVOC emissions of urban green spaces through the introduction of ornamental tree species. Urban For Urban Green 27:305–313
Revi A, Satterthwaite DE, Aragón-Durand F, Corfee-Morlot J, Kiunsi RBR, Pelling M, Roberts DC, Solecki W, Field CB, Barros VR (2017) Urban areas. Cambridge University Press
Rocha CS, Rocha DC, Kochi LY, Carneiro DNM, Reis MVD, Gomes MP (2022) Phytoremediation by ornamental plants: a beautiful and ecological alternative. Environ Sci Pollut Res 29:3336–3354
Sadasivam S (2022) Role of ornamental plants in phytoremediation–an overview. Chem Sci Rev Lett 11(44):478–488
Samudro G, Mangkoedihardjo S (2020) Mixed plant operations for phytoremediation in polluted environments–a critical review. J Phytology 12:99–103
Samudro H, Mangkoedihardjo S (2021) Indoor phytoremediation using decorative plants: an overview of application principles. J Phytol 13(6):28–32
Sandoval L, Zamora-Castro SA, Vidal-Álvarez M, Marín-Muñiz JL (2019) Role of wetland plants and use of ornamental flowering plants in constructed wetlands for wastewater treatment: a review. Appl Sci 9:685
Savé R (2007) What is stress and how to deal with it in ornamental plants? In: VI International symposium on new floricultural crops, vol 813, pp 241–254
Schwela D (2000) Air pollution and health in urban areas. Rev Environ Health 15:13–42
Seo DC, DeLaune RD, Han MJ, Lee YC, Bang SB, Eun Jeong O, Chae JH, Kim KS, Park JH, Cho JS (2010) Nutrient uptake and release in ponds under long-term and short-term lotus (Nelumbo nucifera) cultivation: influence of compost application. Ecol Eng 36:1373–1382
Sharma R, Messar Y (2017) Transgenics in ornamental crops: creating novelties in economically important cut flowers. Curr Sci 113:43–52
Shinomol GK, Bhanu RK (2021) Phytoremediation and the way forward: challenges and opportunities. Appl Water Sci 2:405–436
Signes-Pastor AJ, Munera-Picazo S, Burlo F, Cano-Lamadrid M, Carbonell-Barrachina AA (2015) Phytoremediation assessment of Gomphrena globosa and Zinnia elegans grown in arsenic-contaminated hydroponic conditions as a safe and feasible alternative to be applied in arsenic-contaminated soils of the Bengal Delta. Environ Monit Assess 187:1–9
Sinou M, Skalkou K, Nourentini I, Roubaki M, Lampropoulos S (2023) Sustainable and aesthetic improvement of highways through multisensory landscape design: the case of the urban part of PA TH. E. in the prefecture of Attica. In: IOP conference series: earth and environmental science, vol 1196. IOP Publishing, p 012031
Soreanu G, Dixon M, Darlington A (2013) Botanical biofiltration of indoor gaseous pollutants–a mini-review. Chem Eng J 229:585–594

Sricoth T, Meeinkuirt W, Saengwilai P, Pichtel J, Taeprayoon P (2018) Aquatic plants for phytostabilization of cadmium and zinc in hydroponic experiments. Environ Sci Pollut Res 25:14964–14976

Sriprapat W, Boraphech P, Thiravetyan P (2014) Factors affecting xylene-contaminated air removal by the ornamental plant Zamioculcas zamiifolia. Environ Sci Pollut Res 21:2603–2610

Stapleton E, Ruiz-Rudolph P (2018) The potential for indoor ultrafine particle reduction using vegetation under laboratory conditions. Indoor Built Environ 27:70–83

Stevovi S, Calic-Dragosavac D (2010) Environmental study of heavy metals influence on soil and Tansy (Tanacetum vulgare L.). Afr J Biotechnol 9:2392–2400

Subhashini V, Swamy AVVS (2014) Phytoremediation of metal (Pb, Ni, Zn, Cd and Cr) contaminated soils using Canna indica. Curr World Environ 9:780–784

Suhaimi MM, Leman AM, Afandi A, Hariri A, Idris AF, Dzulkifli SNM, Gani P (2017) Effectiveness of indoor plant to reduce CO2 in indoor environment. In: MATEC web of conferences, vol 103. EDP Sciences, p 05004

Suzuki N, Nakaoka H, Hanazato M, Nakayama Y, Takaya K, Mori C (2019) Emission rates of substances from low-volatile-organic-compound paints. Int J Environ Sci Technol 16:4543–4550

Tao Z, Ge Q, Wang H, Dai J (2015) Phenological basis of determining tourism seasons for ornamental plants in central and eastern China. J Geogr Sci 25:1343–1356

Tauqeer HM, Ali S, Rizwan M, Ali Q, Saeed R, Iftikhar U, Ahmad R, Farid M, Abbasi GH (2016) Phytoremediation of heavy metals by Alternanthera bettzickiana: growth and physiological response. Ecotoxicol Environ Saf 126:138–146

Teiri H, Pourzamzni H, Hajizadeh Y (2018) Phytoremediation of formaldehyde from indoor environment by ornamental plants: an approach to promote occupants health. Int J Prev Med 9:70

Toabaita M, Vangnai AS, Thiravetyan P (2016) Removal of ethylbenzene from contaminated air by Zamioculcas Zamiifolia and microorganisms associated on Z. Zamiifolia leaves. Water Air Soil Pollut 227:1–11

Trigueros D, Mingorance MD, Oliva SR (2012) Evaluation of the ability of Nerium oleander L. to remediate Pb-contaminated soils. J Geochem Explor 114:126–133

Twining JR (1993) A study of radium uptake by the water-lily Nymphaea violacea (Lehm) from contaminated sediment. J Environ Radioact 20:169–189

Wang X-F, Zhou Q-X (2005) Ecotoxicological effects of cadmium on three ornamental plants. Chemosphere 60:16–21

Wang Y, Yan A, Dai J, Wang NN, Dan W (2012) Accumulation and tolerance characteristics of cadmium in Chlorophytum comosum: a popular ornamental plant and potential Cd hyperaccumulator. Environ Monit Assess 184:929–937

Wang L, Sheng Q, Zhang Y, Jingyuan X, Zhang H, Zhu Z (2020) Tolerance of fifteen hydroponic ornamental plant species to formaldehyde stress. Environ Pollut 265:115003

Wei S, Lei X, Dai H, Yahu H (2018) Ornamental hyperaccumulator Mirabilis Jalapa L. phytoremediating combine contaminated soil enhanced by some chelators and surfactants. Environ Sci Pollut Res 25:29699–29704

Wei Z, Van Le Q, Peng W, Yang Y, Yang H, Haiping G, Lam SS, Sonne C (2021) A review on phytoremediation of contaminants in air, water and soil. J Hazard Mater 403:123658

Wilson A, Kendal D, Moore JL (2016) Humans and ornamental plants: a mutualism? Ecopsychology 8:257–263

Wolverton BC, Johnson A, Bounds K (1989) Interior landscape plants for indoor air pollution abatement (No. NASA-TM-101766)

Wolverton BC, Wolverton JD (1993) Plants and soil microorganisms: removal of formaldehyde, xylene, and ammonia from the indoor environment

Wolverton BC, Wolverton JD (1996) Interior plants: their influence on airborne microbes inside energy-efficient buildings. J Miss Acad Sci 41:99–105

Wood RA, Burchett MD, Alquezar R, Orwell RL, Tarran J, Torpy F (2006) The potted-plant microcosm substantially reduces indoor air VOC pollution: I. Office field-study. Water Air Soil Pollut 175:163–180

Xiong J, Han L, Kaixing L, Duan Y, An L, Zhu C (2009) Cadmium decreases crown root number by decreasing endogenous nitric oxide, which is indispensable for crown root primordia initiation in rice seedlings. Planta 230:599–610

Xiong C, Zhang Y, Xiaoguang X, Yongen L, Ouyang B, Ye Z, Li H (2013) Lotus roots accumulate heavy metals independently from soil in main production regions of China. Sci Hortic 164:295–302

Yan A, Wang Y, Tan SN, Yusof MLM, Ghosh S, Chen Z (2020) Phytoremediation: a promising approach for revegetation of heavy metal-polluted land. Front Plant Sci 11:359

Yang DS, Pennisi SV, Son K-C, Kays SJ (2009) Screening indoor plants for volatile organic pollutant removal efficiency. HortScience 44:1377–1381

Yao W, Chen F, Wang S, Zhang X (2021) Impact of exposure to natural and built environments on positive and negative affect: a systematic review and meta-analysis. Front Public Health 9:758457

Yin J, Yuan J, Arfaei N, Catalano PJ, Allen JG, Spengler JD (2020) Effects of biophilic indoor environment on stress and anxiety recovery: a between-subjects experiment in virtual reality. Environ Int 136:105427

Zhang L, Steinmaus C, Eastmond DA, Xin XK, Smith MT (2009) Formaldehyde exposure and leukemia: a new meta-analysis and potential mechanisms. Mutat Res Rev Mutat Res 681:150–168

Zhong S, Shi J, Jianming X (2010) Influence of iron plaque on accumulation of lead by yellow flag (Iris pseudacorus L.) grown in artificial Pb-contaminated soil. J Soils Sediments 10:964–970

Zhou J, Qin F, Jie S, Liao J-w, Hui-lian X (2011) Purification of formaldehyde-polluted air by indoor plants of Araceae, Agavaceae and Liliaceae. Food Agric Environ 9:1012–1018

The manufacturer's authorised representative in the EU is Springer Nature Customer Service Centre GmbH, Europaplatz 3, 69115 Heidelberg, Germany. If you have any concerns regarding our products, please contact ProductSafety@springernature.com

Printed and bound by CPI Group (UK) Ltd, Croydon, CR0 4YY
30/09/2026
02225240-0017